Fluid Mechanics & Machinery

Fluid Mechanics and Machinery

Second Edition

Chanamala Ratnam
Professor
Department of Mechanical Engineering
Andhra University
Visakhapatnam (Andhra Pradesh)

Arun Vikram Kothapalli
Faculty
Department of Industrial Production Engineering
GITAM Institute of Technology
GITAM University
Visakhapatnam (Andhra Pradesh)

I.K. International Publishing House Pvt. Ltd.

NEW DELHI • BANGALORE

Published by
I.K. International Publishing House Pvt. Ltd.
S-25, Green Park Extension
Uphaar Cinema Market
New Delhi – 110 016 (India)
E-mail: info@ikinternational.com
Website: www.ikbooks.com

ISBN 978-93-81141-41-0

Reprint 2013

Published by Krishan Makhijani for I.K. International Publishing House Pvt. Ltd., S-25, Green Park Extension, Uphaar Cinema Market, New Delhi – 110 016 and Printed by Rekha Printers Pvt. Ltd., Okhla Industrial Area, Phase II, New Delhi – 110 020.

“Troubles give a chance to self analysis”

— Dr A.P.J. Abdul Kalam

“Scientists discover the world that exists; engineers create the world that never was.”

—Theodore Von Kármán

“I came to realize that exaggerated concern about what others are doing can be foolish. It can paralyze effort, and stifle a good idea. One finds that in the history of science, almost every problem has been worked out by someone else. This should not discourage anyone from pursuing his own path.”

— Theodore Von Kármán

Preface to the Second Edition

The aim of this revised and enlarged edition is same as that of the earlier edition, as to intend primarily for easiness in understanding of the subject of **Fluid Mechanics and Machinery** and to raise the standard of the reader, step by step in understanding the concepts of Fluid Mechanics and its applications in fluid machinery, with clear pictorial explanations.

In this edition, all the content has been revised thoroughly and modified. Apart from it the errors in previous edition have also been rectified. Additional information wherever necessary has been added. Some additional problems are also been included.

In this edition, a new topic "**Buoyancy and Floatation**" has been added as an extension of Chapter 4.

Suggestions and comments from the readers for further editions of the text will be greatly appreciated.

Chanamala Ratnam
(chratnam_au@yahoo.co.in)
Arun Vikram Kothapalli
(arunvikram@gmail.com)

Preface to the First Edition

This book is intended primarily for easiness in understanding of the subject of **Fluid Mechanics and Machinery**. It is designed to raise the standard of the reader, step by step in understanding the concepts of fluid mechanics and its application in fluid machinery, with clear pictorial explanation. The contents are been developed in transition from basics of simple chapters to complexity of the remaining chapters, including the fundamental formulae used for deriving equations, without leaving the connection between chapters and their applications.

An attempt of more number of diagrams are been introduced for enhancing easiness in understanding of the changes and assumptions taken in derivations. Important formulae including their units, important points and values to be remembered, are been given in a tabular format at the end of each chapter, so as to provide quick reference of these, by the reader.

This book has been organized in two parts, namely, **1. Fluid Mechanics** and **2. Fluid Machinery**, so that the reader can have the flexibility to switch over to the subject of Fluid mechanics or Fluid mechanics application in machinery, as per his necessity and need.

Part 1 of this book, structured on the basics and fundamental theories behind fluids (Liquids and Gasses) and consists of nine chapters, **Chapter 1** Discusses concepts of Fluid properties like their density, specific weight, specific volume, etc. and also explained fluid behavioral aspects like viscosity, surface tension, cohesion and adhesion phenomenon, capillarity effects. The fluid classification is provided basing on their behavior like Newtonian and non-Newtonian fluids.

The important property of fluid, like pressure, which is of main concern in case of fluids and its measurements, when in hydrostatic condition, based on Pascal's law (**manometers**) is dealt in **Chapters 2** and **Chapter 3** describes the hydrostatic forces arise on the surface immersed in fluids and hydrostatic condition.

Chapters 4 and 5 deals with the **Kinematics** and **Dynamics** fluid flows respectively, considering Circulation and Vorticity in flow and assuming the flow to be continuous and incompressible. Equation of Continuity and Bernoulli's equation has been derived and application of Bernoulli's equation in devices like Venturimeter, Orifice meter and Pitot tube are discusses. Moreover, vortex motion is also explained.

The viscous flow of fluid in closed conduit in one dimensional basing on Couette flow, Hagen Poiseulle law and friction factors and two-dimensional (**Navier Stokes equation**) are explained. Flow potentiality and its resistance are also explained including minor frictional losses like due to sudden expansion, sudden contraction, losses due to pipe bends and obstructions are also taken up in **Chapter 6**. **Chapter 7** explains the Boundary layers over flat plates and their equations based on Displacement, energy, Momentum thickness in Laminar and Turbulent flows and their transition from laminar to Turbulent, is explained while discussing about the fully developed boundary layers and **laminar sub-layers**. Moreover, boundary layer separation and its control are been included in this chapter.

Chapter 8 details the analysis of dimensional homogeneity of various physical variables using **Releigh and Buckingham method**. Model analyzing using correlation of prototype with modal using non-dimensional parameters and different types of similarities are discussed, which the reader gets the capability to be familiar with the **Model similitude** of the actual prototype, which is to be manufactured. **Chapter 9** Elaborates the gas dynamics (**Compressible flow**) based on Mach number significance and stagnation properties of the compressible fluid. And also discusses the application of the compressible flow in Venturimeter, Diffuser nozzles and Pitot tube for Maximum efficiency.

Part 2 of this book focuses on the application (**Fluid Machinery**), based on the theory of Fluid mechanics and consists of three chapters, namely, **Impact of jets**, which gives clear explanation of the jet motions on Pelton wheel, its impact on various surface and the velocity diagrams and the efficiency of jet stories. **Hydraulic Turbines** categorizes differences in different types of hydraulic turbines and also mentions draft tube theory used behind the reaction turbines. **Specific speed** for transition of the reader to study the performances, the characteristic curves and governing of the hydraulic turbines are been discussed. Finally, last chapter discusses hydraulic pumps like for **Centrifugal** and **Reciprocating pump** significances, functioning, efficiency and requirement of the above pumps are taken up in detail, in these chapters, so as to achieve the reader's attention, for easy understanding.

Chanamala Ratnam
Arun Vikram Kothapalli

Acknowledgements

We sincerely submit our due respect to our beloved parents for their unfailing love and affection without which we would not be what we are.

We express our heartfelt gratitude to our colleagues and respective administrations of A.U. College of Engineering and GITAM Institute of Technology, GITAM University, Visakhapatnam.

Lastly, Ch. Ratnam thanks his wife (Suhasini), children (Vijay and Ajay) and K. Arun Vikram thanks his wife (Vijaya Laxmi) and his two daughters (Reshma Priya and Geetha Madhuri) for extending warmful support.

Suggestions and comments from the readers for further improvement of the text will be greatly appreciated.

Chanamala Ratnam
(chratnam_au@yahoo.co.in)
Arun Vikram Kothapalli
(arunvikram@gmail.com)

Contents

A Brief History

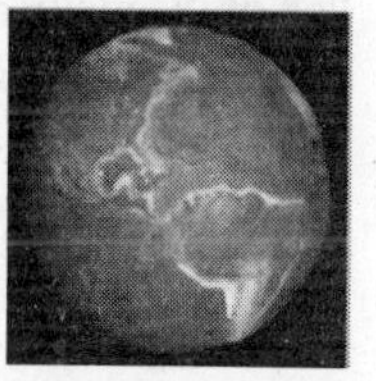

Earth comprises many systems that are often unpredictable like air, water, land and life—including humans. **Fluids** (means liquids or gases, which conform to the shape of the container containing them) cover at least three-fourths of Earth's surface. Fresh water exists in the liquid phase only within a narrow temperature span (0° C to 100° C). And earth is the only planet where it exists in its liquid form on the surface. This is probably part of the reason that the earth is the only planet known to contain life.

Human beings have been searching from the past till date for understanding and analyzing of these fluids and their behavioural aspects, for their convenient application in simple to complex machinery, for making human life a comfort and endeavor over nature.

The term **fluid** in real time refers other than solids, like liquids and gasses. In reality, fluid describes any liquids or gases that conform to the shape of the container containing it.

DIFFERENCES BETWEEN SOLIDS AND FLUIDS

Solids are rigid bodies, possessing a definite volume and shape, whereas liquids; though they possess a definite volume, have no definite shape and gases have neither a definite shape nor a volume. Due to the rigid shape and volume, the solids are treated as incompressible under sufficient amount of external pressures. Example: a glass tube when pressed with hand will resist small amount hand pressure and once the amount of pressure exceeds some limit, the glass tube will break. Whereas fluids can be compressed with a little amount of external pressure. When considered a liquid, it can have small amount of external pressure resistance whereas gases possess no resistance to external pressure and hence comparatively liquids are treated as incompressible over gases.

Fluid is **a substance that can flow** and the behavioral aspects of fluids can be explained with their molecular movements. Solids do have a strong molecular resisting force towards displacement from one place to another, as the solid molecules are placed close to each other whereas fluids deform with no little forces acting on them, as their molecules are placed at a considerable amount of distance and are capable of flowing. And hence solids remain a rigid body once it hits the ground after it is released from a height whereas a liquid splits into droplets once it hits the ground and, on the other hand, as gas molecules are lighter, they mix up with the atmosphere.

FLUIDS AND CONTINUUM

Fluids can be viewed at microscopic level as well as macroscopic level. In this context throughout, the fluids are treated as macroscopic and called **Continuum**, i.e. the fluids are considered continuous mass substance irrespective to their atomic arrangement and hence the mean free flow paths and average parameters of pressure, velocity, etc. are being treated.

Fluids viewed at microscopic level are studied under kinetic theory approach or statistical mechanics; where in the mean free path is larger than the size of the body around which the flow takes place.

On the other hand, fluids can be categorized based on their implication of acceptance of stress-strain relations, due to the shear effect between molecule layers, and due to the external force application; namely, **Newtonian** and **non-Newtonian** fluids. All the fluids which obey Newton's law of viscosity are termed Newtonian fluids and those which do not obey are termed non-Newtonian fluids. And further the non-Newtonian fluids can be subcategorized into Newtonian fluids with initial yield stresses and complete non-Newtonian.

Fluid Mechanics is a subject concerned with understanding and analyzing of the fluids and their behavior at rest as well as in motion. Fluid mechanics is basic to such varied fields as aeronautics; chemical, civil, and mechanical engineering; meteorology; naval architecture; and oceanography. Applications of fluid mechanics include jet propulsion, hydraulics, turbines, compressors, and pumps.

Fluid Mechanics is divided into three groups of application like Fluid Statics, Kinematics and Dynamics, wherein Dynamics is subdivided into two groups like **hydrodynamics** (study of liquid flow) and **aerodynamics** (study of compressible flow). And apart from this, **Fluid Machinery** is the concept of application of Fluid Mechanics, in real time like in hydraulic machinery, air plane shape profiles, flood defense constructions, dams, pipeline networks, pumps and turbines, sub-merged body constructions, groundwater flows, etc.

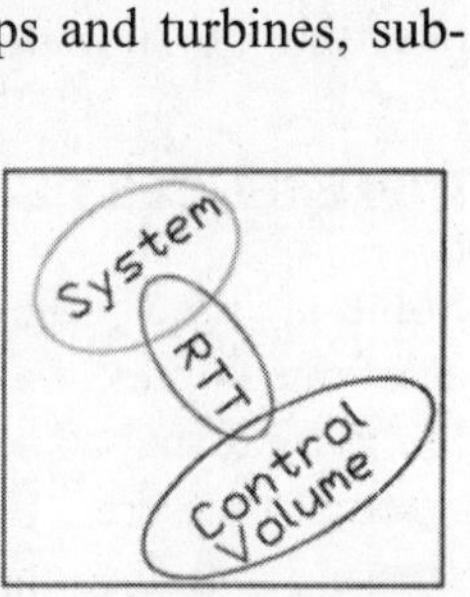

Fluid Mechanics mainly relies on three fundamental theories like Conservation of Mass, Conservation of Energy and Conservation of Momentum of a system. But **Eulerian description methods** are added over **Lagrangian description methods** hence a control volume analysis is required. So to convert a system analysis using the conservation laws into control volume analysis, where **System** is a fixed quantity of matter which can change its shape and position of its matter but still the volume of mass does not change, but on the other hand, **Control volume** is a fixed quantity of matter, which has fixed arbitrary region. We require a theorem called **Reynolds Transport Theorem** (RTT). The Reynolds transport theorem uses Partial Differential Equations to transform Lagrangian description to Eulerian description and use **integral equation analysis** for transformation of a System analysis to Control volume analysis.

HYDROSTATIC PRESSURE AND BUOYANCY

Greek mathematician, physicist, and inventor Archimedes (c. 287–212 B.C.) discovered a concept of fluid statics called hydrostatic pressure and buoyancy of an immersed object in a fluid, which had greater importance as that of Pascal's principle in analyzing the point of application of hydrostatic pressure and its direction of action. According to Archimedes' principle, **"the Hydrostatic force**

exerted on the object immersed in a fluid is equal to the weight of the fluid it displaced". This phenomenon of hydrostatics and buoyancy has applications on how a ship floats on water, how a balloon floats in the air. If an object is immersed in fluid, like a balloon in the air or a ship on the ocean—weighs less than that of the fluid it displaces, it will float. If it weighs more, it will sink or fall. The balloon itself may be 'heavier than air,' but it is not as heavy as the air it has displaced.

PRESSURE OF A FLUID

Pressure plays an important role within both the statics and dynamics behaviors of fluids. A number of interesting principles describe the response to pressure, on the part of fluids at rest and also on fluids in motion.

Blaise Pascal (1623–1662), a French mathematician and physicist who discovered that the external pressure applied on a fluid is transmitted uniformly throughout its entire body. The understanding offered by Pascal's principle later became the basis for one of the most important machines ever developed—the hydraulic press.

Henri Pitot (May 3, 1695, Aramon, France – December 27, 1771) a French hydraulic engineer, disproved the idea that the velocity of flowing water increases with depth and invented Pitot tube to measure flow velocity either in liquids or gases in 1732. Nowadays, Pitot tube has a system of pressure-sensitive instruments that is most often used in aviation to determine an aircraft's airspeed, Mach number, altitude, and altitude trend. Pitot derived that the height of the fluid column is proportional to the square of the fluid's velocity. Later, Pitot's model was improved and modified by aeronautic engineer **Henry Darcy**.

Conservation of Energy given by Daniel Bernoulli (8 Feb 1700 in Groningen, Netherlands—17 March 1782 in Basel, Switzerland) was the son of Johann Bernoulli (held the chair of mathematics there). Bernoulli formulated that "**as the speed of a moving fluid (liquid or gas) increases, the pressure within (static) the fluid decreases**" and this phenomenon described by taking a fluid moving from a larger cross-section to smaller cross-section, a definite volume of fluid must move through a greater distance forward in the smaller cross-section and thus have a greater speed, making the kinetic energy of that definite volume increased. But by the law of conservation of energy, this increase in kinetic energy must be balanced by a decrease in the pressure-volume (work done) product but as the volumes are equal, it can be balanced by decrease in static pressure. And vice versa for a fluid flow from smaller cross-section to larger cross-section, there will be increase in static pressure and decrease in kinetic energy, for the same amount of volume. And hence "**the velocity is inversely proportional to the Static pressure**".

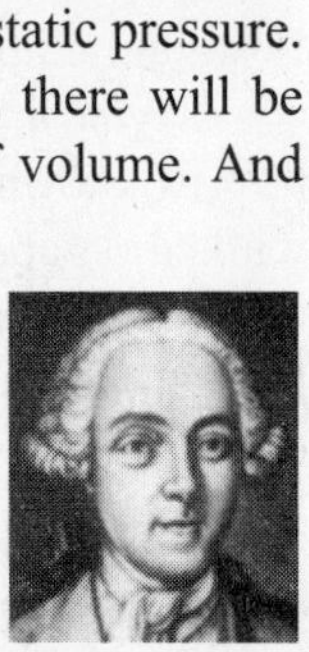

When a fluid at rest having "Static pressure" begins to move in a pipe, the static pressure decreases as per Bernoulli's principle, but however, a portion of the static pressure—proportional to the speed of the fluid—is converted to the pressure of movement called "Dynamic pressure". And in a cylindrical pipe, static pressure is exerted perpendicular to the surface of the pipe, whereas dynamic pressure is parallel to the flow and hence, he formulated that "**for all changes in movement**

of a fluid, the sum of static and dynamic pressure in a fluid remains the same and greater the velocity of flow of a fluid, the greater the dynamic pressure and the less the static pressure". The phenomenon of Bernoulli's principle has many practical applications; like in the carburetor for atomization of fuel, design of an aerofoil airplane's wing, airplanes, Wind tunnel (wind tunnel was built in England in 1871, and years later, aircraft pioneers Orville (1871–1948) and Wilbur Wright (1867–1912) used a wind tunnel to improve their planes and Researchers today use helium to create wind blasts at speeds up to Mach 50).

Efforts to obtain a mathematical formulation of a fluid flow took shape during the century following the publication of the Principles, with the contributions of Daniel Bernoulli (1700–82), Jean le Rond d'Alembert (1717–83) and Leonhard Euler—all well-known heavy hitters in classical physics. Among the three, **Leonhard Euler** (1707–1783) a Swiss mathematician who made enormous contributions to a wide range of mathematics and physics including analytic geometry, trigonometry, geometry, Calculus and number theory was the most instrumental in conceptualizing the mathematical description of a fluid flow. He described flow in terms of spatially varying three-dimensional pressure and velocity fields and modeled the flow as a continuous collection of infinitesimally small fluid elements. By applying the basic principles of mass conservation and Newton's second law. Euler obtained two coupled, non-linear partial differential equations involving the flow fields of pressure and velocity. Although those Euler equations were an intellectual breakthrough in theoretical fluid dynamics, obtaining general solutions of them was quite another matter. Moreover, Euler did not account for the effect of friction acting on the motion of the fluid elements that is, he ignored viscosity.

Major advance in the development of fluid mechanics had to await the formulation of the laws of motion by English mathematician and physicist Sir Isaac Newton. These laws were first applied to fluids by Swiss mathematician Leonhard Euler. Euler first recognized that dynamical laws for fluids can only be expressed in a relatively simple form if the effects of friction or viscosity can be neglected. Because this is never the case for real fluids in motion, the results of analysis can serve only as an estimate, and only for those flows where viscous effects are small.

It was another hundred years before the Euler equations were modified to account for the effect of internal friction within a flow field. The resulting equations, a system of even more elaborate nonlinear partial differential equations now called the **Navier-Stokes equations**, were first derived by **Claude-Louis Navier** in 1822, and then independently derived by George Stokes in 1845. To this day, those equations are the gold standard in the mathematical description of a fluid flow, and no one has yet obtained a general analytical solution of them.

Navier is remembered today, not as the famous builder of bridges for which he was known in his own days, but rather for the Navier-Stokes equations of fluid dynamics. He worked on applied mathematics topics such as engineering, elasticity and fluid mechanics and, in addition, he made contributions to Fourier series and their application to physical problems. He gave the well-known Navier-Stokes equations for an incompressible fluid in 1821 while in 1822 he gave equations for viscous fluids.

The first carefully documented friction experiments in low-speed pipe flow were carried out independently in 1839 by the French physiologist **Jean Leonard Marie Poiseuille** (1797–1869),

who was interested in the characteristics of blood flow. An attempt to include the effects of viscosity into the mathematical equations was made first in 1827 by the French engineer Claude Louis Marie Navier, and independently by the British mathematician Sir George Gabriel Stokes, who in 1845 perfected the basic equations for viscous incompressible fluids. These are now known as the Navier-Stokes equations, and they are so complex that they can be applied only to simple flows. One such flow is that of a real fluid through a straight pipe. Here Bernoulli's principle is not applicable because part of the total mechanical energy is dissipated as a result of viscous friction, resulting in a pressure drop along the pipe. The equations suggest that this pressure drop for a given pipe and a given fluid should be linear with the flow velocity. Experiments first conducted near the middle of the 19th century showed that this was only true for low velocities; at higher velocities, the pressure drop was more nearly proportional to the square of the velocity. This problem was not resolved until 1883 when the British engineer Osborne Reynolds showed the existence of two types of viscous flows in pipes. At low velocities the fluid particles follow the streamlines (laminar flow) and results match the analytical prediction. At higher velocities the flow breaks up into a fluctuating velocity pattern or eddies (turbulent flow) in a form that cannot be fully predicted even today.

Hagen Poiseuille law was developed independently by **Gotthilf Heinrich Ludwig Hagen** (3 March 1797—3 Feb 1884, German hydraulic engineer) and published in 1846. Hagen did his experiments in 1839. The Hagen-Poiseuille equation is a physical law that describes slow viscous incompressible flow through a constant circular cross-section. It is also known as the **Hagen-Poiseuille law, Poiseuille law** and **Poiseuille equation**.

Reynolds also established that the transition from laminar to turbulent flow was a function of a single parameter that has since become known as the Reynolds number. If the Reynolds number, which is the product of velocity, fluid density, and pipe diameter, divided by the fluid viscosity, is less than 2100, the pipe flow will always be laminar; at higher values it will normally be turbulent. The concept of a Reynolds number is basic too much of modern fluid mechanics.

The **Darcy-Weisbach equation** is an important and widely used phenomenological equation in hydraulics. It relates the head loss or pressure loss due to friction along a given length of pipe to the average velocity of the fluid flow. This variant was developed by **Henry Philibert Gaspard Darcy** (June 10, 1803 – January 3, 1858) of France, who made several important contributions to hydraulics and further refined into the form used today by **Julius Ludwig Weisbach** of Saxony (10 August 1806, Mittelschmiedeberg (now Mildenau), Erzgebirge—24 February 1871, Freiberg, was a German mathematician) in 1845. Initially, data on the variation of f with velocity was lacking, so the Darcy-Weisbach equation was outperformed by the empirical Prony equation in many cases.

The Darcy-Weisbach friction factor is 4 times larger than the Fanning friction factor, so attention must be paid to note which one of these is meant in any "friction factor" chart or equation being used. Of the two, the Darcy-Weisbach factor is more commonly used by civil engineers, and the

Fanning factor by chemical and mechanical engineers, but care should be taken to identify the correct factor regardless of the source of the chart or formula. Most charts or tables indicate the type of friction factor, or at least provide the formula for the friction factor with laminar flow. If the formula is $f = 16/\text{Re}$, it is the Fanning factor, and if the formula is $f = 64/\text{Re}$, it the Darcy-Weisbach factor.

Maurice Marie Alfred Couette (9 January 1858, Tours—18 August 1943, Angers) a French physicist known for his studies of fluidity, whose name is associated with theory of flow and rheometer and with a correction method for end effects in capillary flows. Couette is best known for his contributions to rheology and the theory of fluid flow. He designed a concentric cylinder viscometer that he used to accurately measure the viscosity of fluids. He studied the boundary conditions of a fluid and showed that the "no slip" condition was satisfied for the fluids and wall materials tested. He designed a continuously rotating concentric cylinder apparatus and measured the viscosity of water and air with remarkable accuracy.

Couette flow refers to the laminar flow of a viscous fluid in the space between two parallel plates, one of which is moving relative to the other. The flow is driven by virtue of viscous drag force acting on the fluid and the applied pressure gradient parallel to the plates.

Turbulent flows cannot be evaluated solely from computed predictions and depend on a mixture of experimental data and mathematical models for their analysis, with much of modern fluid-mechanics research still being devoted to better formulations of turbulence. The transitional nature from laminar to turbulent flows and the complexity of the turbulent flow can be observed as cigarette smoke rises into very still air. At first it rises in a laminar streamline motion but after some distance it becomes unstable and breaks up into an intertwining eddy pattern.

The complexity of viscous flows, especially turbulent flows, severely restricted progress in fluid dynamics until the German engineer **Ludwig Prandtl** recognized in 1904 that many flows could be divided into two principal regions. The region close to the surface consists of a thin boundary layer where the viscous effects are concentrated and where the mathematical model can be greatly simplified. Outside the boundary layer viscous effects can be disregarded and the simpler mathematical equations for inviscid flows can be used. The boundary-layer theory has made possible much of the development of modern aircraft wings and the design of gas turbines and compressors. The boundary-layer model not only permitted a much simplified formulation of the Navier-Stokes equations in the region close to the body surface but also led to further developments of the flow of inviscid fluids that can be applied outside the boundary layer. Much of the modern development of fluid mechanics was made possible by the boundary-layer concept and it has been carried out by such key contributors as the Hungarian-born American aeronautical engineer Theodore von Kármán, and the German mathematician Richard von Mises, by the British physicist and meteorologist Sir Geoffrey Ingram Taylor.

THE PRANDTL BOUNDARY LAYER

Ludwig Prandtl (4 Feb 1875, Freising, Germany–15 Aug 1953, Göttingen, West Germany) German physicist is considered the father of aerodynamics. He taught at the University of Göttingen (1904–

53). During the week of 8 August 1904, Heidelberg was a natural venue for the 3rd International Mathematics Congress. One of the presenters at the Congress was Ludwig Prandtl, a 29-year-old professor at the Technische Hochschule (equivalent to a US technical university) in Hanover, presented the concept that would revolutionize the understanding and analysis of fluid dynamics at the surface of a body moving in air or water led to an understanding of skin friction drag and of the way streamlining (streamline) reduces the drag of airplane wings and other moving bodies. Things only started to change after Ludwig Prandtl, in 1904, suggested a way to bring the **hydrodynamics** (neglecting viscous-friction and density change) and **Hydraulic** (experimental data) together by introducing a new concept, called the **Boundary Layer**. He theorized that an effect of friction was to cause the fluid immediately adjacent to the surface to stick to the surface —in other words, he assumed the no-slip condition at the surface —and that frictional effects were experienced only in a boundary layer, a thin region near the surface.

Paul Richard Heinrich Blasius (1883–1970) was a German fluid dynamics engineer. He was one of the first students of Prandtl who provided a mathematical basis for boundary-layer drag but also showed as early as 1911 that the resistance to flow through smooth pipes could be expressed in terms of the Reynolds number for both laminar and turbulent flow.

One of his most notable contributions involves a description of the steady two-dimensional boundary-layer that forms on a semi-infinite plate which is held parallel to a constant unidirectional flow *U*.

Correlations

First law of Blasius for turbulent fanning friction factor: $\frac{f}{2} = 0.039 * R_e^{-0.26}$

Second law of Blasius for turbulent fanning friction factor: $\frac{f}{2} = 0.023 * R_e^{-0.26}$

Theodore von Kármán (11 May 1881, in Budapest, Hungary—7 May 1963 Aachen, Germany) an engineer and physicist, who was active primarily in the fields of aeronautics during the seminal era in the 1940s and 1950s. He is personally responsible for many key advances in aerodynamics, notably his work on supersonic and hypersonic airflow characterization.

Theodore von Karman is considered one of the great aeronautical scientists of the 20th century. In 1906, von Kármán received a scholarship to the University of Göttingen in Germany, earning a Ph.D. in engineering there in 1908. In March of that year, he made a trip to Paris, where he watched an airplane flight by French aviation pioneer Henri Farman. As a result, von Kármán began a lifelong interest in the application of mathematics to aeronautics. In 1911, after conducting wind tunnel experiments, he made an important analysis of the alternating double row of vortices behind a flat body in a fluid flow, now known as Kármán's Vortex Street.

When World War I broke out in August 1914, von Karman was recalled to active duty. In February 1915, he designed gun emplacements for the defense of Budapest. In August 1915, he was made director of research for the Austro–Hungarian Aviation Corps and given extensive facilities and staff. It was an ideal assignment, and he converted a Zeppelin hangar into a laboratory that investigated a host of ideas, including self-sealing fuel tanks and machine gun synchronizers. He emphasized development of the PKZ 1 helicopter, called the Schraubenfesselflieger ("rotor-driven tethered aircraft"). It was intended to substitute for the dangerously flammable observation balloons used in artillery spotting. Powered by an Austro–Daimler electric motor, it had a rotor speed of 700 rpm and made at least four flights.

CHAPTER 1

Fluids and their Properties

INTRODUCTION

Humans as such like engineers, are analyzing as well as experimenting with fluids in all the engineering branches like mechanical, civil, aerodynamics etc. for hydraulic machinery applications and channel flows, irrigation, dams and aerospace design applications respectively. So there is an intense necessity of having the knowledge of the properties and the response of the fluids, on which they are studying and experimenting.

In nature we find three substances of matter like solids, liquids and gases. The difference in properties and their behavior can be studied by their molecular properties. If substances of the three fluid states like solids, liquids and gases are kept out without any container surrounding them, at a place on ground and seen after sometime, it can be observed that the solid remains as it is placed, but while the liquid floods all over the ground and the gases mix up with the atmosphere. By this we can clearly understand that the solids do not move/displace without some external force acting on it and can resist an applied shear stress by deforming but liquids and gases do flow without much of external source acting on them. And hence due to this reason, liquids and gases can be termed **fluids**, where a fluid can be defined as **'any object which has the capability to flow is called a fluid'**. And the continuous deformation with shear distinguish a fluid from a solid and also when the fluid are at rest the shear stress do completely disappears.

In addition to this, in solids **stress is proportional to strain**, but in fluids **stress is proportional to strain rate**.

When a constant shear force is applied, a solid eventually stops deforming, at some fixed strain angle, whereas a fluid never stops deforming and approaches a certain rate of strain.

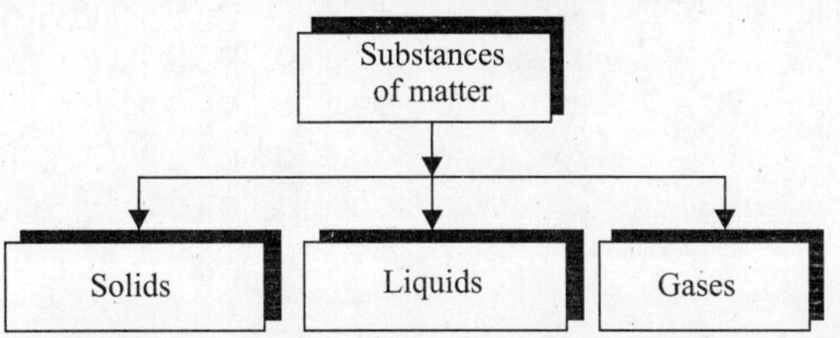

Figure 1.1: Different types of substances of matter

Solid is a rigid body (as shown in Figure 1.2) which has molecules very closely packed, due to which the intermolecular attraction force (i.e. force of attraction between molecules is high) which makes solids compact, rigid shape, mass possessive and also gives compressive strength, tensile strength and shear stresses bearing capacity.

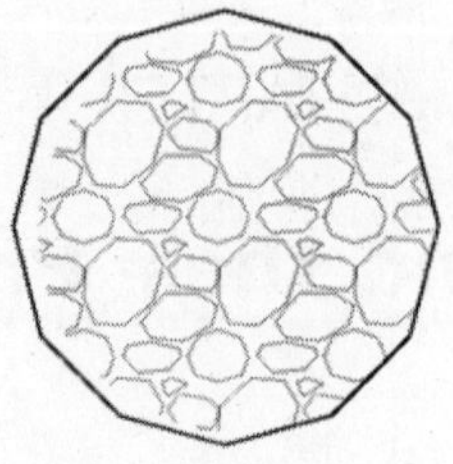

Figure 1.2: Solid body

Solids will deform when sheared and this shear force may cause some displacement but the solid does not continue to move indefinitely and position of equilibrium is soon reached and further deformation ceases, due to the reason that the elastic forces induced by the deformation of the solid exactly counter balance the applied shear force.

In **liquids**, as shown in Figure 1.3, the molecules are loosely packed and are free to move, due to which the intermolecular cohesive (attraction) force is comparatively low, due to which liquids have only compressive stress bearing capacity. A liquid has no shape of its own but confirms to the shape of the container containing it and even small amount of shear stress can cause the deformation. Liquids even though are compressible up to some extent but in comparison to gases the

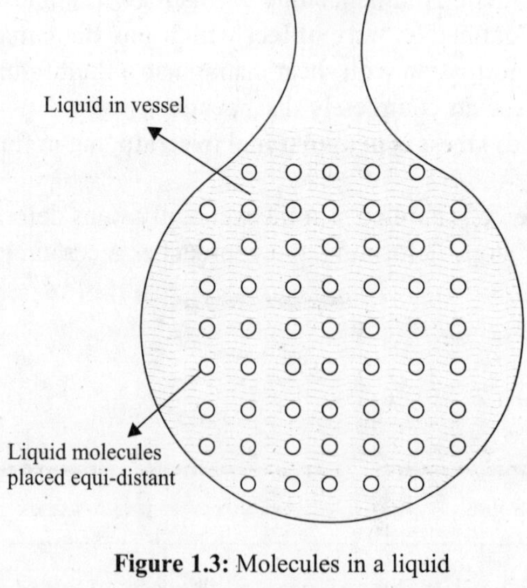

Figure 1.3: Molecules in a liquid

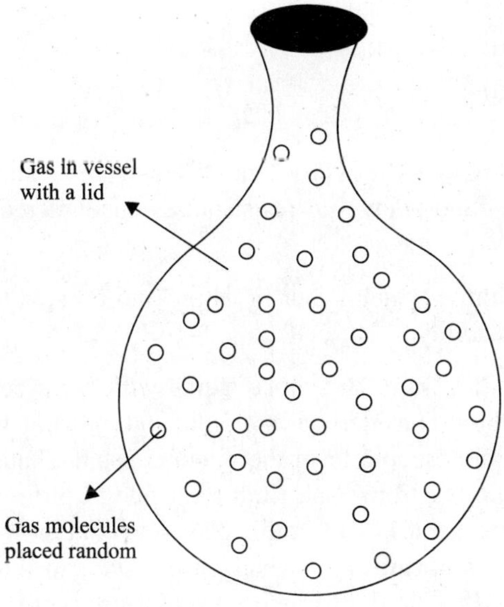

Figure 1.4: Molecules in a gas

compressibility is very low and hence they are treated as not compressible and are known **incompressible fluids** and gases as **compressible fluids**. As the liquid molecules are placed at equidistant from each other, they form a **free surface layer** as on of its boundary, whereas gasses don't do so.

In **gases**, as shown in Figure 1.4, the molecules are very loosely packed and are free to fly, due to which the intermolecular cohesive (attraction) force is comparatively lower than the liquids. As gases have wide gap in between their molecules and hence are compressible due to which gases are treated as compressible fluid.

	Solids	**Fluids**
Short time application of weak stress	Behaves like elastics	Flow is viscous
Long time application of weak stress	Initially deform and then resists	Flows continuously
Application of High stress	Deforms irreversible after yield stress	Uncontrollable flow

Fluid: Fluid is a substance, which has the capability to flow. It has no definite shape of its own but conforms to the shape of the vessel containing it. Fluids undergo deformation for small amount of shear force.

One may notice the similarities between the solids and fluids is that these two substances follow similar kind of laws like Hook's law and Newton's law respectively. The solids obey Hook's law in which the stress is directly proportional to strain and the constant of proportionality is called modulus of elasticity, whereas fluids obey Newton's law of viscosity, wherein the shear stress is directly proportional to the velocity gradient/strain rate and the constant of proportionality is called as absolute viscosity.

Types of Fluids

a) Ideal fluid
b) Real fluid
c) Continuum fluid

Ideal fluids: Ideal fluids are the fluids with no viscosity and surface tension and they are incompressible, i.e. as these fluids move no resistance is encountered. These fluids are however imaginary.

Real fluids: Real fluids are those which occur in nature and possess the properties like viscosity, surface tension and compressibility.

Continuum fluids: Continuum fluids are those fluids which are continuous and homogenous having steady properties without considering its atomic and molecular structure.

When fluids viewed at microscopic level there is a constant change or disturbance of matter and the properties will be varying from each fluid molecule to molecule, which is studied under kinetic theory or statistical mechanics. But in this complete context we assume the fluids to be at macroscopic level (called as "**Continuum**") where in the average fluid properties are taken into account and as such each and every fluid is presumed to have constant fluid properties and the fluids are considered as continuous mass substance irrespective to their atomic arrangement.

FLUID MECHANICS

The physical science that deals with both stationary and moving bodies under the influence of external forces is called **Mechanics** (i.e. the branch of mechanics that deals with bodies at rest is called **static's,** while that deals with bodies in motion is called **dynamics).**

Fluid Mechanics is a science concerned with the response of fluids to the forces exerted upon them. It is a branch of classical physics with applications of great importance in hydraulics, aeronautical engineering, chemical engineering, meteorology and Zoology.

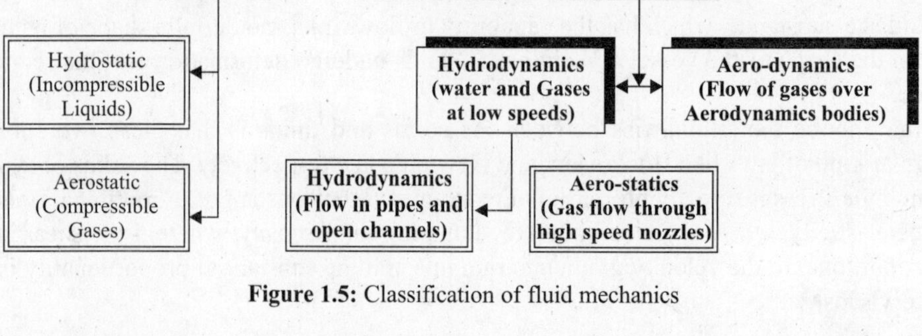

Figure 1.5: Classification of fluid mechanics

Fluid Mechanics, as its name suggests, is the branch of Applied Mechanics concerned with the Statics and Dynamics of fluids – both liquids and gases. The analysis of the fluid behaviors are based on the fundamental laws of mechanics which relate continuity of mass and energy with force and momentum together along with the familiar solid mechanics properties. Fluid mechanics which deals with the behavior of fluids under the conditions of rest or motion and it can be classified in three categories like Fluid Statics, Fluid Kinematics and Fluid Dynamics.

Fluid Statics: Fluid statics deals with fluids under rest, without any motion and their properties related to the fluid in rest.

Fluid Kinematics: Fluid kinematics deals with velocities, acceleration, and pattern of flow but does not consider the forces causing them.

Fluid Dynamics/Kinetics: Fluid dynamics deals with relation between velocities, acceleration of fluids and the forces/energies causing them like shear forces or compressive forces.

IMPORTANCE AND APPLICATIONS OF FLUID MECHANICS

Fluid mechanics play a vital role in all the aspects starting from oldest technology to new technology like in predicting the seasons change in the past to tornado, hurricane prediction in the recent times etc. below listed are some of the applications of fluid mechanics in the field of technology like,

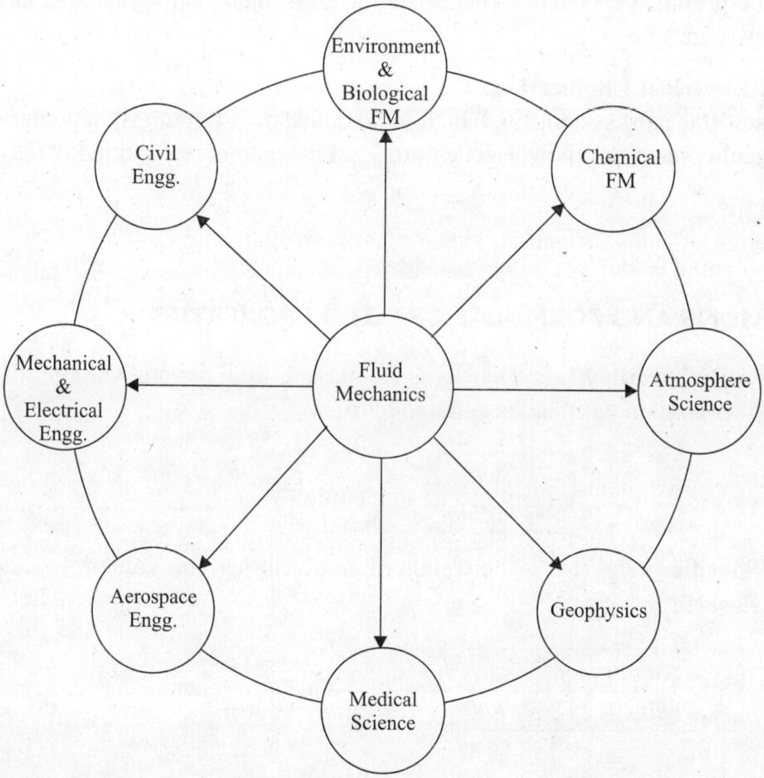

Environmental and Biological Fluid Mechanics

- Motion and mixing of fluids in the environment and interests in substances and properties transported by the fluid like wastewater discharge into stream, ocean etc.
- junction of two rivers, smokestack discharge into atmosphere
- contaminant spill into ocean or river causing biological creatures effected like, fish eggs, protozoa, bacteria, and viruses etc.
- mixing of salt and fresh water in estuaries, mixing of warm and cold water in lakes
- cellular processes.

Chemical fluid mechanics: like

- Salinity, dissolved oxygen, dissolved solids, metals, pesticides, BOD, pH

Atmospheric sciences:

- Analysis of climatic changes, Tornados, Hurricane etc.

Geophysics:

- Earthquakes, volcanoes, earth's molten core analysis etc.

Medical Science:

- Artificial heart, kidney dialysis machines, insulin pumps etc.

Aerospace Engineering:

- Aerospace propulsive systems, rocket engines, turbojets, aerodynamics, aircraft design and analysis etc.

Mechanical and Electrical Engineering:

- All types of transport systems in vehicles and domestic use (pumps), automatic pneumatic and hydraulic machines, power generation, turbines, compressors, industrial etc.

Civil Engineering:

- Construction of dams, irrigation, pipeline network etc.

BASIC DEFINITIONS AND FORMULAE OF FLUID PROPERTIES

Density/Mass Density/Specific Mass: Density is the mass of fluid per unit volume space occupied by it, in the container containing it and is denoted by ρ.

$$\rho = \frac{mass}{volume} \quad \text{units are } \frac{\text{kg}}{\text{m}^3}$$

Weight Density/Specific Weight: It is the weight of an object per unit volume space occupied by the object and is denoted by γ .

$$\gamma = \frac{weight}{volume} = \frac{mass * gravity}{volume} = \rho g \quad \text{units are } \frac{\text{KN}}{\text{m}^3} \text{ or } \frac{\text{kg}-\text{f}}{\text{m}^3}$$

Specific Gravity (*S*): Specific gravity is defined as the ratio of specific weight of the liquid with respect to the specific weight of water at 4°C.

$$S = \frac{\gamma_{liquid}}{\gamma_{Water}} = \frac{\rho_{liquid}}{\rho_{water}}$$ units are dimensionless

Specific Volume (*v*): Specific volume is defined as the volume of fluid per unit mass of the fluid and is reciprocal to density of the fluid.

$$v = \frac{volume}{mass} = \frac{1}{\rho}$$ units are $\frac{m^3}{kg}$

Vapour Pressure: Vapour is often termed as Gas, when it is above the critical temperature. A liquid boils at a certain pressure called "**Vapour pressure**". If considered normal cooking process, we can observe that the vapor formed due to cooking flows up to the atmosphere. But when the cooking vessel is covered with a lid, the velocities of the Vapour molecules are stopped and thus the internal pressure of the vapor rises. The rise in vapor pressure adds some additional rise in temperature, which cause the cooking process to be faster. Hence, due to this reason the pressure cooker does operate at less time for cooking rather with normal process cooking.

Vapour pressure will be of main intention when studying about **Cavitations** (or in other words "**bubble of vapor**"), which is a hazardous activity and is caused due to the increase in velocities and drop in vapor pressure, like in turbo-machines. Cavitation is not desirable for several reasons like Cavitation bubbles collapse when they migrate into regions of higher pressure causing noise and then leading to inefficiencies and reduction of heat transfer in turbo-machines by causes pitting and corrosion of blades etc.

Example 1.1: Find mass density ρ and specific gravity *S*, if 5.6 m^3 of an oil which weighs 48,000 N.

Solution: Given data: Weight = 48,000 N

Mass of oil, $$m = \frac{weight}{gravity} = 48{,}000/9.81 = 4892.96 \text{ kg}$$

Density of oil, $$\rho = \frac{mass}{volume} = 4892.96/5.6 = 873.74 \text{ kg/m}^3$$

Specific gravity of oil, $$S = \frac{\rho_{oil}}{\rho_{water}} = {}^{873}\!/_{1000} = 0.873$$

Example 1.2: A liquid having specific gravity of 0.82 is filled in a vessel. Calculatets density, specific volume and specific weight.

Solution: Given data: specific gravity of liquid, *S* = 0.82

Specific gravity of oil, $S = \dfrac{\rho_{Liquid}}{\rho_{water}} \Rightarrow \rho_{Liquid} = \rho_{water} * S$

Density of liquid is $\rho_{Liquid} = 1000 * 0.82 = 820 \text{ kg/m}^3$

Specific volume $\nu = \dfrac{1}{density} = 1/820 = 0.0012 \text{ m}^3\text{/kg}$

Specific weight $\gamma = \rho_{liquid} * g = 820 * 9.81 = 8044.2 \text{ N/m}^3$

Example 1.3: Determine the mass density, specific volume and specific weight of CO_2 at 35°C and pressure of 2×10^{-6} atmospheres absolute. Take molecular weight of CO_2 as 44 and universal gas constant $R_o = 14 \text{ Nm/mole} - \text{K}$.

Solution: Given data: pressure = 2 Pascal (absolute), temperature =35+273=308 K

Characteristic gas constant $R = {}^{R_0}\!/_{M} = {}^{14}\!/_{44} = 0.318 \text{ N/mole} - \text{K}$

As per Ideal gas equation, mass density $\rho = \dfrac{P}{RT} = \dfrac{2}{0.318 * 308} = 0.204 \text{ kg/m}^3$

Specific volume $\nu = \dfrac{1}{density} = 1/00.0204 = 49 \text{ m}^3\text{/kg}$

Specific weight $\gamma = \rho * g = 49 * 9.81 = 480.8 \text{ N/m}^3$

Compressibility (Z) and **Bulk modulus** (K)**:** Under application of external pressure, fluids undergo change in volume and this property by virtue of which fluids undergo change in volume, under the action of external pressure is called as compressibility and is the reciprocal of bulk modulus, i.e. $Z = \dfrac{1}{\text{K}}$ and Bulk Modulus is the measuring of elasticity and is given by

$$\text{K} = \frac{-dp}{\left(\dfrac{\delta v}{Volume}\right)} = \frac{Compressive\ stress}{Volumetric\ strain}.$$

Consider a frictionless plunger moving in a rigid cylinder having a fluid filled in it, as shown in figure 1.6. As the plunger moves forward, the fluid gets compressed and the volume decreases, giving rise to a volumetric strain in the fluid and also this negative change in volume does exert increase the fluids internal pressure. So the ratio of negative change in volumetric strain to increase in pressure is known as **compressibility**. The graph between compressible stress and volumetric strain must

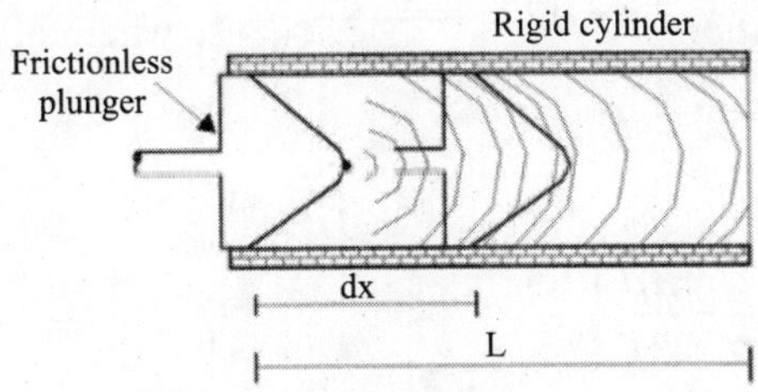

Figure 1.6: Compressibility

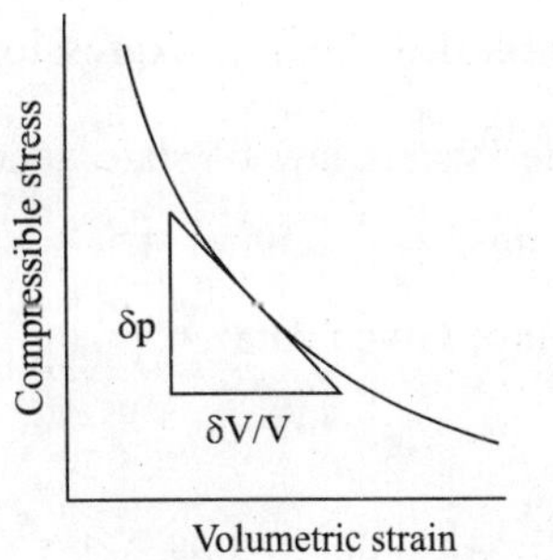

Figure 1.6a: Graph between stress and volumetric strain

be directly proportional but as there is a negative change in volume, the graph tends to be indirectly proportional as seen in the Figure 1.6a.

Example 1.4: If bulk modulus of water is 2.1×10^9 N/m^2, then find the change in pressure required to reduce the volume of water by 0.7%.

Solution: Given data:

$$K = 2.1\times10^9 \text{ N/m}^2 \text{ and } \left(\frac{\delta v}{Volume}\right) = 0.7\%$$

We know $$\text{K} = \frac{-dp}{\left(\frac{\delta v}{Volume}\right)} \Rightarrow dp = -\left(\frac{\delta v}{Volume}\right) * K$$

Change in pressure required is $dp = -\left(\frac{0.7}{100}\right) * 2.1\times10^9 = -14.7\times10^6$ pascal

Example 1.5: If an oil in a container is compressed by a change in pressure of 155 pascal, which gives a decrease in volume of the oil by 0.75%, determine the bulk modulus and compressibility.

Solution: Given data:

$$dp = 155 \text{ pascal and } \left(\frac{\delta v}{Volume}\right) = 0.75\%$$

We know bulk modulus $$\text{K} = \frac{-dp}{\left(\frac{\delta v}{Volume}\right)} = \frac{155 * 100}{0.75} = 20.67 \text{ kPa}$$

Compressibility $$Z = \frac{1}{\text{K}} = \frac{1}{20670} = 4.84\times10^{-5} \text{ m}^2/\text{N}$$

Example 1.6: Find an expression for isothermal bulk modulus of elasticity for a gas which obeys van der Waals law of state according to the equation: $p = \rho RT\left[\frac{1}{1-b\rho} - \frac{a\rho}{RT}\right]$, where a, b, are constants.

Solution: Given data:

$$p = \rho RT\left[\frac{1}{1-b\rho} - \frac{a\rho}{RT}\right]$$

We know bulk modulus $$K = \left[-\frac{dp}{dV/V}\right] = -V\left(\frac{dp}{dV}\right)$$

But $$v = \frac{1}{\rho} \Rightarrow \frac{dv}{v} = \frac{-d\rho}{\rho}$$

So $$K = \rho\left(\frac{dp}{d\rho}\right)$$

From the given data, we have $$p = \rho RT\left[\frac{1}{1\text{-}b\rho} - \frac{a\rho}{RT}\right]$$

$$\Rightarrow \frac{dp}{d\rho} = RT\left[\frac{1}{1\text{-}b\rho} - \frac{a\rho}{RT}\right] + \rho RT\left[\frac{-b}{(1\text{-}b\rho)^2} - \frac{a}{RT}\right]$$

$$\Rightarrow \frac{dp}{d\rho} = \left[\frac{RT(1\text{-}b\rho)}{(1\text{-}b\rho)^2} - 2a\rho\right]$$

$$K = \rho\left(\frac{dp}{d\rho}\right) = \rho\left[\frac{RT(1\text{-}b\rho)}{(1\text{-}b\rho)^2} - 2a\rho\right]$$

Hence, we get bulk modulus as

$$K = \frac{\rho RT(1\text{-}b\rho)}{(1\text{-}b\rho)^2} - 2a\rho^2$$

Which is the required expression.

COHESION AND ADHESION

Some liquid molecules have the tendency to attract the intermolecules, while other liquid molecules have the tendency to stick to the boundaries of the vessels, in which they are contained. The tendency of liquid molecules to attract the intermolecules is called '**cohesion**' and the tendency of liquid molecules to stick to the boundaries of the vessel, in which they are contained, is called '**adhesion**'. Due to cohesion, the molecules tend to come closer towards each other, and this phenomenon makes the liquids (like mercury) not to wet the manometer tube, which is dipped in the liquid, as shown in Figure 1.7.

Cohesion is the attraction of intermolecules of a liquid, which gives them resistance towards tensile stress and which gives surface tension to a liquid.

Adhesion is the attraction of molecules of a liquid to its boundary, which gives a sticking nature to the liquids, with the other bodies with which they are in contact. Due to adhesion, the molecules tend to come closer towards the boundary wall of the container in which they are contained, and this phenomenon makes the liquid (like water) to wet the container vessel or manometer tube, which is dipped in the liquid.

Due to adhesion phenomenon, water in a manometer tube leads to an upward concave shape at the walls of the tube and contributes to capillary action. This concave shape at the surface is called **meniscus**, as shown in Figure 1.8.

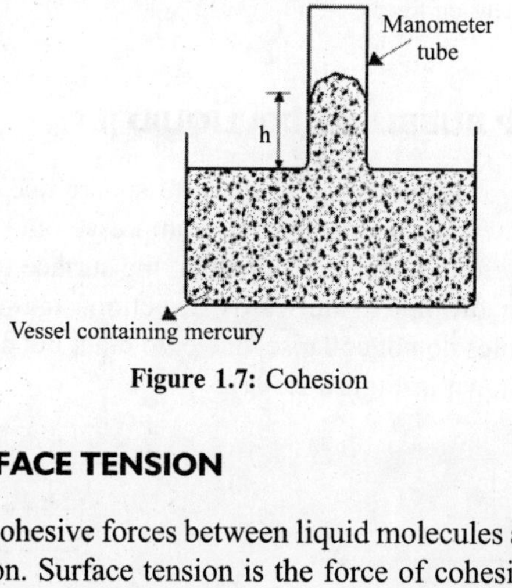

Figure 1.7: Cohesion

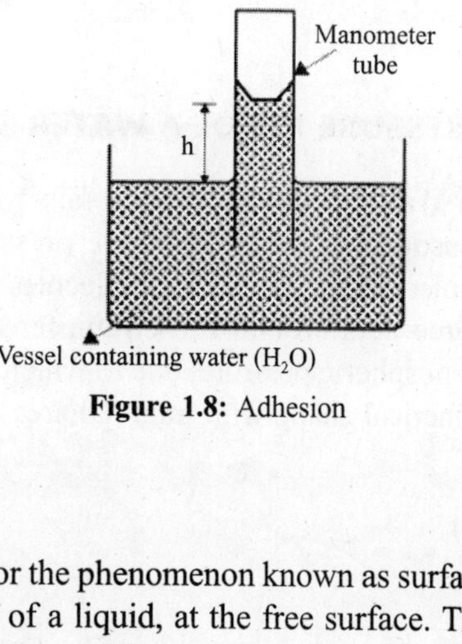

Figure 1.8: Adhesion

SURFACE TENSION

The cohesive forces between liquid molecules are responsible for the phenomenon known as surface tension. Surface tension is the force of cohesion of molecules of a liquid, at the free surface. The molecules at the surface do not have other like molecules on all sides of them and consequently they

are attracted more strongly to those directly associated with them on the surface. This forms a surface 'film in tension' and is difficult to move small particles through the surface. Hence, surface tension is the force required to break the surface force film by a unit length and has units N/m, e.g., small insects such as the water strider (see below) can walk on water because their weight is not enough to penetrate the surface.

Every molecule of a liquid exerts equal pressure force on all directions according to the Pascal's law. As the molecules on the frees surface of the liquid are pulled by the inward forces by molecules in the below layers, these surface layer molecules tend to give a layer of pressure forces in tension on the surface of liquid, due to which any small particle falling on the free surface of a liquid, floats on the surface without sinking.

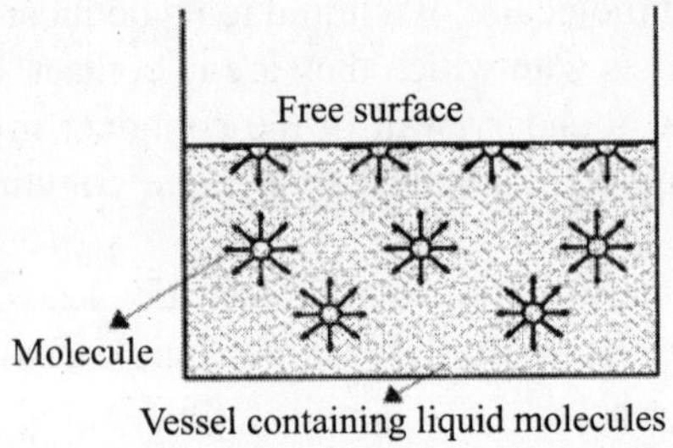

Figure 1.9: Surface tension forces

PRESSURE INSIDE A WATER DROPLET, SOAP BUBBLE AND A LIQUID JET

a) Water droplet: When a rain water droplet falls, it acquires the shape of a sphere due to the reason that the atmospheric pressure all around the rain water droplet, compresses the water molecules, but once the molecules of water droplet are too near to each other, the surface tension force acts around the circumference of the water droplet in outwards direction, against the atmospheric pressure, due to which the water molecules do not collapse, one onto other but acquire spherical shape with minimal area occupation (as shown in Figure 1.10a).

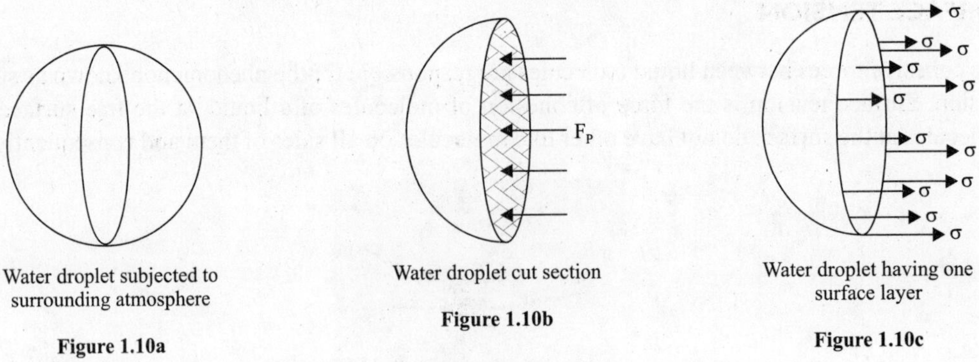

Water droplet subjected to surrounding atmosphere

Figure 1.10a

Water droplet cut section

Figure 1.10b

Water droplet having one surface layer

Figure 1.10c

Figure 1.10

Under equilibrium conditions, the difference in outside atmosphere pressure and inside pressure of the water droplet must be equal to the surface tension force, so as to keep the water droplet intact with spherical shape.

i.e. $$Net\ force(F_P) = Surface\ tension\ force(\sigma_F) \tag{1.0}$$

Let the outside pressure and inside pressure around water droplet be F_{P_0}, F_{P_i} respectively.

Then net pressure force

$$F_P = \left(F_{P_i} - F_{P_o}\right)$$

$$\Rightarrow F_P = P * \left(\frac{\Pi}{4} d^2\right) \tag{1.1}$$

Surface tension forces around the water droplet

$$\sigma_F = \sigma * \Pi d \tag{1.2}$$

σ is surface tension.

Under equilibrium Eq. (1.1) = Eq. (1.2) i.e. $\sigma_F = F_P \ldots \Rightarrow \sigma * \Pi d = P * \left(\frac{\Pi}{4} d^2\right)$

$$\Rightarrow P = \frac{4\sigma}{d} \tag{1.3}$$

$$\Delta P = P_i - P_o = \frac{4\sigma}{d} \tag{1.3a}$$

Equation for pressure acting on a water droplets

Note:

1. If the curvature of the spherical droplet is not uniform through out i.e. if the surface is a general surface with differing radius of curvature as shown in the below figure, then Eq. (1.3a) differs as

$$\Delta P = P_i - P_o = 4\sigma\left(\frac{1}{d_1} + \frac{1}{d_2}\right) \tag{1.3b}$$

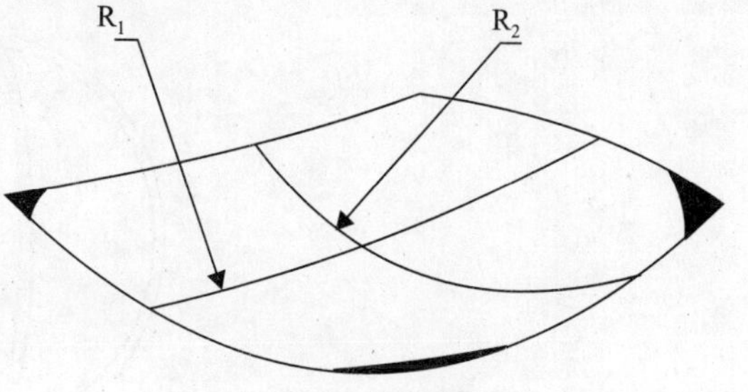

Irregular surface having differing radius of curvature

Figure 1.10d

2. Considering Eq. (1.3), we can observe that as the size of the droplet decreases, the effect of pressure difference increases, i.e.

$$\Delta P \alpha \frac{1}{d} \text{ and } \Delta P \alpha \sigma$$

Example 1.7: A glass tube of 3.5 mm internal diameter is immersed in a crude oil having 99.3 pascals of surrounding pressure. If a pressure of 185 pascals is needed to form an oil bubble, determine the surface tension of the oil.

Solution: Given data: diameter $d = 0.0035$ m, $P_o = 99.3$ Pa, $P_i = 185$ Pa

We know as per the equation 1.3a,

$$\Delta P = \frac{4\sigma}{d} \qquad \Rightarrow \sigma = \Delta P * \frac{d}{4}$$

$$\sigma = \Delta P * \frac{d}{4} = (185 - 99.3)\frac{0.0035}{4} = 0.0749 \text{ N/m}$$

Example 1.8: If pressure difference between the inside and outside of the air bubble of diameter 0.01 mm is 29.2 kPa, what will be the surface tension at the air water interface? (AMIE, Summer 2000)

Solution: Given data:

$$d = 0.01 \times 10^{-3} \text{ m}, \Delta P = 29.2 \times 10^3 \text{ Pa}$$

We know as per the Eq. 1.3a,

$$\Delta P = \frac{4\sigma}{d} \Rightarrow \sigma = \Delta P * \frac{d}{4}$$

$$\Rightarrow \sigma = 29.2 \times 10^3 * \frac{0.01 \times 10^{-3}}{4} = 0.073 \text{ N/m}$$

b) Soap bubble: Just as water droplet, soap bubble also experiences the pressure difference. But the soap bubble has a thin film cover, which can be treated as two surfaces separating inside and outside pressures.

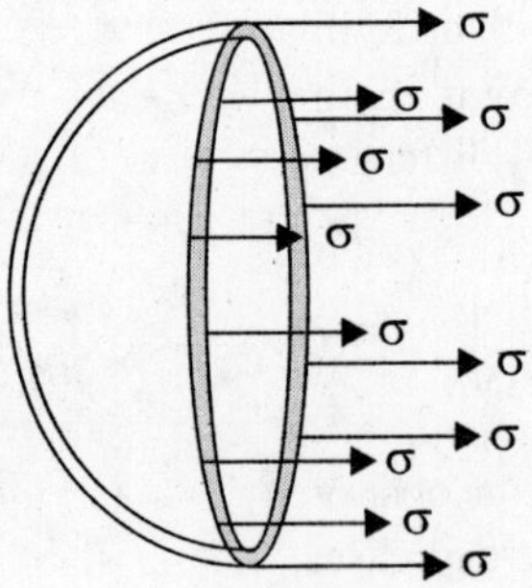

Figure 1.11: Soap bubble having two surface layers

Under equilibrium conditions, the difference in outside and inside pressures of the soap bubble must be equal to twice the surface tension force, so as to keep the soap bubble intact without burst.

i.e. $$\text{Net force}(F_P) = \text{Surface tension force}(\sigma_F) \tag{1.4}$$

Let the outside pressure and inside pressure around water droplet be F_{P_0}, F_{P_i} respectively.

Then net pressure force be

$$F_P = \left(F_{P_i} - F_{P_o}\right)$$

$$\Rightarrow F_P = P * \left(\frac{\Pi}{4} d^2\right) \tag{1.5}$$

Surface tension forces around the water droplet,

$$\sigma_F = 2(\sigma * \Pi d) \tag{1.6}$$

where σ is surface tension.

Under equilibrium Eq. (1.5) = Eq. (1.6) i.e. $\sigma_F = F_P$.

$$\Rightarrow 2(\sigma * \Pi d) = P * \left(\frac{\Pi}{4} d^2\right)$$

$$P = \frac{8\sigma}{d} \tag{1.7}$$

$$\Delta P = P_i - P_o = \frac{8\sigma}{d} \tag{1.7a}$$

Equation for pressure acting on a soap bubble

Example 1.9: Calculate the work done in blowing a soap bubble of diameter 100 mm, assuming the surface tension to be 0.35 N/m

Solution: Given data:

$$d = 0.01 \text{ m}, \ \sigma = 0.35 \text{ N/m}$$

Work done = surface tension × Total surface area

But the soap bubble has two surface interfaces and hence the work done becomes

Work done = surface tension × 2 × Total surface area

$$\Rightarrow \textit{Work done} = 0.35 * 2 * \left(4 * \pi * \left(\frac{0.01}{2}\right)^2\right)$$

$$\Rightarrow \textit{Work done} = 0.0002198 \text{ Nm}$$

c) Liquid jet: A liquid jet experiences the pressure difference along its surface though its length 'L'. And so the surface tension force will be acting along its length.

Under equilibrium conditions, the difference in outside and inside pressures of the liquid jet will be equal to twice the surface tension over the length, so as keep the liquid jet intact without splitting.

i.e. $$Net\ force(F_P) = Surface\ tension\ force(\sigma_F) \quad (1.8)$$

Let the net pressure force, outside pressure force and inside pressure force around water droplet be F_P, F_{P_o}, F_{P_i} respectively.

Then net pressure force

$$F_P = \left(F_{P_i} - F_{P_o}\right)$$

$$\Rightarrow F_P = P*(L*d) \quad (1.9)$$

Surface tension force around the liquid jet

$$\sigma_F = \sigma(2L) \quad (1.10)$$

where σ is surface tension. Eq. (1.9) = Eq. (1.10) and hence

$$\sigma_F = F_P \Rightarrow \sigma(2L) = P*(L*d)$$

$$\Rightarrow P = \frac{2\sigma}{d} \quad (1.11)$$

$$\Delta P = P_i - P_o = \frac{2\sigma}{d} \quad (1.11a)$$

Equation for pressure acting on a liquid jet

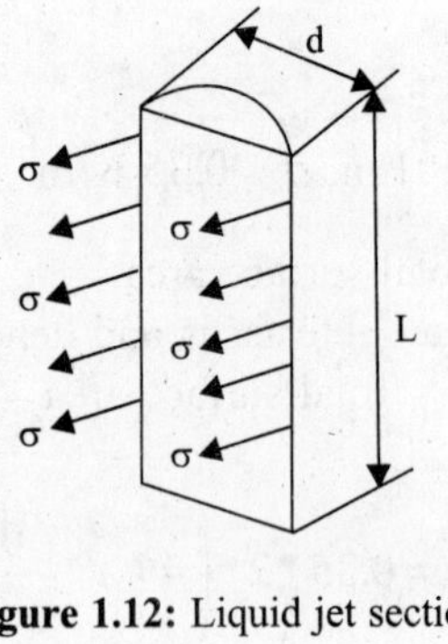

Figure 1.12: Liquid jet section

Capillarity

Capillarity is the phenomenon of a liquid to raise or fall of its surface level to its general surface level, in a thin glass tube, as shown in Figure. 1.13a and 1.13b. It is the combined effect of cohesion and adhesion properties of fluids.

For example, water has high adhesion capacity, due to which its surface level raises up to its general surface level and hence wets the boundary of the contact object, as shown in Figure 1.13a, while mercury (Hg) has cohesion capacity due to which its surface level falls down to its general surface level and hence do not wet the boundary of the contact object, as shown in Figure 1.13b.

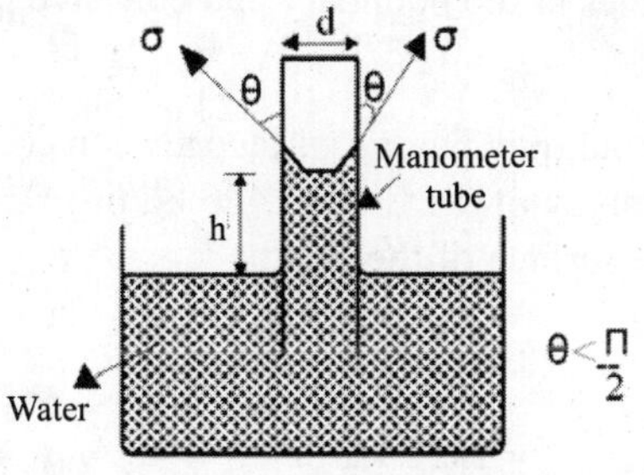

Figure 1.13a: Adhesion > Cohesion

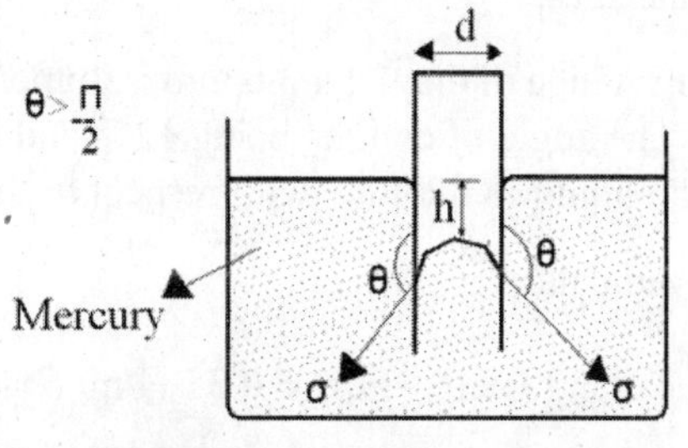

Figure 1.13b: Adhesion < Cohesion

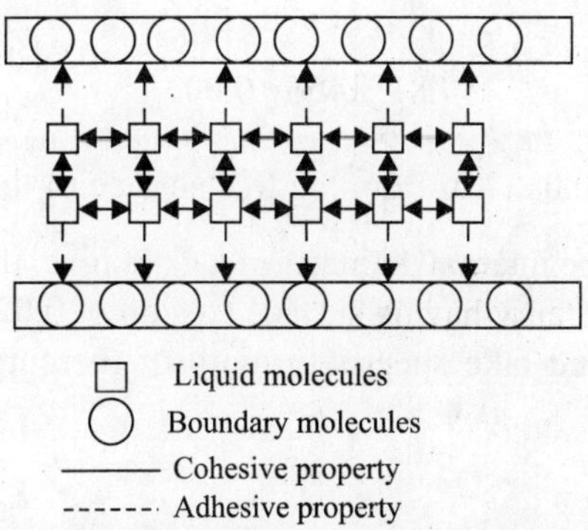

Figure 1.13c

Consider a manometer tube dipped in water. The water raises up to some height in the manometer tube and is called **capillary head/Height "h"**, as shown in Figure 1.13a.

This is due to the action of surface tension force (lifting force) but the lifting of liquid column is opposed after some capillary height, by the weight of the water raised in the tube column (gravity force) due to which the possibility of overflow of water from the tube is stopped. Hence, at equilibrium conditions, the lifting force is equal to the gravity force of the liquid column raised.

i.e. surface tension upward component = Weight of the water column (gravity force)

$$(2\Pi r * \sigma)\cos\theta = \left(\frac{\Pi}{4}d^2\right) * h * \gamma \tag{1.12}$$

Where θ is the angle of contact of liquid with the tube, d is the diameter of tube, σ is the surface tension, h is the capillary head/height, γ is the specific weight of liquid.

By rearranging

$$h = \frac{4\sigma\cos\theta}{\gamma d} \tag{1.13}$$

Equation for Capillary head of raise or fall

For better understanding of the adhesiveness and cohesiveness of the intermolecules of a liquid, Figure 1.13c depicts the adhering property of the molecules at the boundary and cohesive property within the molecules.

Example 1.10: A tube of diameter 3.0 mm is immersed in a liquid having surface tension coefficient of 0.35 N/m. The angle of contact being 125° and the density of the liquid 1456 kg/m³. Determine the level of the liquid in the tube with respect to the free surface of the liquid.

Solution: Given data:

$$d = 3.0\times10^{-3}\,\text{m}, \sigma = 0.35\ \text{N/m},\ \theta = 125^0, \rho_{liquid} = 1456\ \text{kg/m}^3$$

We have capillary head as

$$h = \frac{4\sigma\cos\theta}{\rho_{liquid} * gd} \Rightarrow h = \frac{4*0.35*\cos125}{9.81*1456*0.003} = -0.0187\ \text{m}$$

Hence, there is a fall in level of 0.0187 m from the free surface of the liquid.

Example 1.11: What should be the internal diameter of a glass tube, if immersed in (i) Water having surface tension 0.074N/m (ii) Mercury having surface tension of 0.02 N/m, if the capillary raise or fall must not exceed 2.0 mm and take specific gravity of mercury and water as 13.6 and 1.0 respectively.

Solution: Given data:

a) $\sigma = 0.074\ \text{N/m}, \theta = 0^0\ for\ water,\ h = 0.002\ \text{m},\ \gamma = 9810\ \text{N/m}^3$

We have capillary head as

$$h = \frac{4\sigma\cos\theta}{\gamma d} \Rightarrow d = \frac{4\sigma\cos\theta}{\gamma h} = \frac{4*0.074*1}{9810*0.002} = 0.015\ \text{m}$$

So the internal diameter of the glass tube is 0.015 m.

b) $\sigma = 0.02\ \text{N/m},\ \theta = 140^0\ for\ Mercury,\ h = 0.002\ \text{m},\ \gamma = 13.6*9810 = 133.4\ \text{kN/m}^3$

We have capillary head as

$$h = \frac{4\sigma\cos\theta}{\gamma d} \Rightarrow d = \frac{4\sigma\cos\theta}{\gamma h} = \frac{4*0.02*\cos140}{13.6*9810*0.002} = -0.000229\ \text{m}$$

So the internal diameter of the glass tube is 0.022 mm and negative sign indicates that there is fall of level from the free surface.

Example 1.12: If a glass tube having two bores of 1.3 mm and 1.9 mm respectively, is immersed vertically in a an oil having surface tension of 0.578 N/m and has zero contact angle with difference in two meniscus levels as 22 mm. Determine the mass density of the oil.

Solution: Given data:

$$\sigma = 0.578 \text{ N/m}, d_1 = 0.0013 \text{ m}, d_2 = 0.0019 \text{ m}, h_1 - h_2 = 0.022 \text{ m}$$

Refer Eq. (1.13), we have

$$h = \frac{4\sigma\cos\theta}{\rho g d}$$

$$h_1 - h_2 = \frac{4\sigma}{\rho g}\left(\frac{1}{d_1} - \frac{1}{d_2}\right) \Rightarrow 0.022 = \frac{4*0.578}{\rho*9.81}\left(\frac{1}{0.0013} - \frac{1}{0.0019}\right) = \frac{57.24}{\rho}$$

$$\Rightarrow \rho = \frac{57.24}{0.022} = 2602 \text{ kg/m}^3$$

So the mass density of the oil is 2602 kg/m^3.

Example 1.13: Find the smallest diameter of a manometer tube such that the error due to capillary action in the measured gauge pressure 120 Pa is less than 5%. The manometric liquid is water and has surface tension $\sigma = 0.073$ N/m

Solution: Given data:

$$\sigma = 0.073 \text{ N/m}, Error = 5\%, \theta = 0^\circ, \rho = 1000 \text{ kg/m}^3, P_{gauge} = 120 \text{ Pa}$$

We have

$$h_1 = \frac{4\sigma\cos\theta}{\rho g d}$$

$$\Rightarrow h_1 = \frac{4*0.073*1}{1000*9.81*d} = \frac{0.002976}{d} \text{ cm} \qquad (1)$$

According to hydrostatic law, we have

$$h_2 = \frac{P_{gauge}}{\gamma} = \frac{120}{9810} = 1.22 \text{ cm} \qquad (2)$$

So % error is equal to the ratio of Eq. (1) and Eq. (2),

$$\%Error = 0.05 = \frac{h_1}{h_2}$$

$$\Rightarrow 0.05 = \frac{0.002976}{1.22 * d}$$

$$\Rightarrow d = 0.0487 \text{ cm}$$

The smallest diameter of the tube 0.0487 cm.

Viscosity (μ)

The rate at which a fluid deforms continuously depends not only on the magnitude of the applied force but also on a property of the fluid called its viscosity or resistance to deformation and flow.

Viscosity is the property of a fluid by virtue of which fluid offers resistance to the flow of one layer of molecules over the other adjacent layer of molecules of the fluid. And this is basically due to cohesion and molecular momentum exchange between the fluid layers of the fluid. And hence viscosity – a fluid property that relates the shear stress in a fluid to the angular rate of deformation.

To find viscosity, we use Newton's law of viscosity.

Newton's law of viscosity: Consider a plate of negligible thickness and weight, pulled with a force "F" and a velocity (u), over a fluid (say oil) which is on a fixed plate and let "h" be the depth of the fluid as shown in Figure 1.14.

As the molecule layers of the fluid at the surface are in contact with the moving plate, consider a molecule at point "B" on the surface layer, as shown in figure, slides to new point "B^1" with the velocity "u", due to the pull given by the moving plate. But the bottom layer molecules of the fluid below the surface, resists the change and due to which they offer resistance towards the velocity (**u**) and have less velocity compared to the surface layer and so the velocity "**u**" is a function of the depth of the fluid. i.e.

$$u = u(y) \tag{1.14}$$

and so the molecule layers at the fixed plate has zero velocity.

Consider the triangle ABB^1, wherein

$$\tan(\delta\phi) = \frac{BB^1}{h}$$

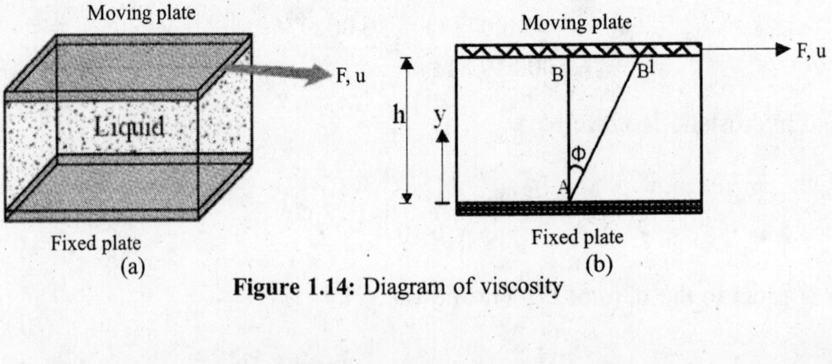

Figure 1.14: Diagram of viscosity

$$\Rightarrow \tan(\delta\phi) = \frac{u * \delta t}{h}$$ since Displacement (BB[1]) = velocity(u) * time (dt)

But $\tan(\delta\phi) \approx \delta\phi$ and so $\delta\phi = \dfrac{u * \delta t}{h}$; where $\delta\phi$ is the shear strain.

Rate of shear strain/angular deformation is $\dfrac{\delta\phi}{\delta t} = \dfrac{u}{h}$ where $\dfrac{u}{h}$ is the velocity gradient

According to **Hooke's Law**, shear stress is directly proportional to shear strain and hence, we have

$$\tau \propto \lim_{\delta t \to 0} \frac{\delta\phi}{\delta t}$$

As the velocity $u = u(y)$, the velocity gradient can be written in form of differential, as $\dfrac{u}{h} = \dfrac{du}{dy}$

and so

$$\tau \propto \lim_{\delta t \to 0} \frac{\delta\phi}{\delta t} = \frac{du}{dh} \Rightarrow \tau \propto \frac{du}{dy}$$

So from the above relation, we can state that the angle of deformation $\delta\phi$ is directly proportional to the change in time but however the shear strain rate $\dfrac{\delta\phi}{\delta t}$ remains unchanged until and unless thc shear stress varies.

$$\Rightarrow \tau = \mu\left(\frac{du}{dy}\right) \tag{1.15}$$

μ Absolute or Dynamic Viscosity, units Ns/m² or Poise.

Increase in temperature decreases the cohesive bond between the liquid molecule layers which mades the shearing of the fluids heated easy and so increase in temperature for liquids decreases viscosity $\mu \; \alpha \; \dfrac{1}{\Delta T}$. In case of gases as a the molecules are placed randomy, the cohesive effect of molecules is negligible. As the gas temperature is raised, molecular movement increases, thereby causing continuous exchange of molecular momentum exchange between layers and drag. And hence the increase in temperature increases viscosity, in gasses. i.e. $\mu\Delta T$.

Example 1.14: The density of an oil is 775 kg/m³. Find its specific gravity and kinematic viscosity of the oil, if the dynamic viscosity is 3 × 10⁻³ kg/ms.

Solution: Given data: $\rho_{oil} = 775$ kg/m³ and $\rho_{water} = 1000$ kg/m³

Specific gravity $S = \rho_{oil}/\rho_{water} = 775/1000 = 0.775$

Kinematic viscosity $\nu = \dfrac{\mu}{\rho} = \dfrac{3\times10^{-3}}{1000} = 3\times10^{-6}\ \text{m}^2/\text{s}$

Example 1.15: The velocity distribution of a viscous liquid (dynamic viscosity $\mu = 0.8$ Ns/m^2) flowing over a fixed plate is given by $u = 1.5y^2 + y$ (u is velocity in m/s and y is the distance from the plate in m). What are the shear stresses at the plate surface and at $y = 0.33$ m?

Solution: Given data $\quad u = 1.5y^2 + y$ and $\mu = 0.8$ Ns/m^2

$$\frac{du}{dy} = 3y + 1$$

At the plate face $y = 0$ m,

$$\frac{du}{dy} = 1.0$$

Shear stress at the plate face

$$\tau_{y=0} = \mu\left(\frac{du}{dy}\right)_{y=0} = 0.8\times1.0 = 0.8\ \text{N/m}^2$$

At

$$y = \frac{1}{3},\ \frac{du}{dy} = 3\left(\frac{1}{3}\right) + 1 = 2$$

Shear stress at $y = \dfrac{1}{3}$ $\quad \tau_{y=0.33} = \mu\left(\dfrac{du}{dy}\right)_{y=0.33} = 0.8 * 2.0 = 1.6\ \text{N/m}^2$

Example 1.16: Given a shaft in a bush bearing having oil (viscosity is 45 poise) filled in between, is rotated at a speed of 190 rpm. If the oil film thickness is 0.08 m and diameter and length of shaft are 12 cm and 25 cm respectively, determine shear stress, shear force, torque, and power.

Solution: Given data:

$$\mu = 4.5\ \text{Ns/m}^2,\ N = 190\ \text{rpm},\ t = 0.08\ \text{m},\ d = 0.12\ \text{m},\ L = 0.25\ \text{m}$$

Velocity of the shaft rotation is

$$u = \frac{\pi dN}{60} = \frac{3.14 * 0.12 * 190}{60} = 1.19\ \text{m/s}$$

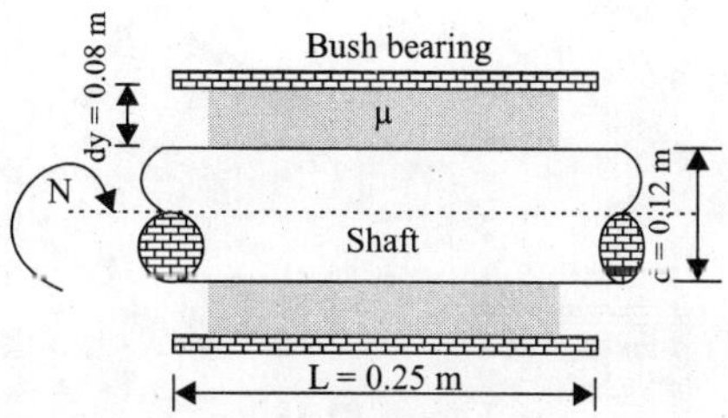

Figure P1.16

Shear stress $$\tau = \mu \frac{du}{dy} = 4.5 * \left(\frac{1.19}{0.08}\right) = 66.9 \text{ Pa}$$

Shear force $F = \tau * A = \tau * (\pi d L) = 66.9 * (3.14 * 0.12 * 0.25) = 6.3$ N

Torque $$T = F * \frac{d}{2} = 6.3 * 0.06 = 0.378 \text{ Nm}$$

Power $$P = \frac{2\pi N}{60} T = \frac{2 * 3.14 * 190}{60} * 0.378 = 7.52 \text{ watt}$$

Example 1.17: In a fluid having linear velocity distribution flowing over a flat plate has absolute viscosity 0.064 poise and of specific gravity 0.85, has a velocity of 1.78 m/s at a height of 78 mm. What is the shear stress at the contact of fluid and surface?

Solution: Given data $\mu = 0.0064$ Ns/m^2

Specific gravity, $S = 0.85$

Velocity, $du = 1.78$ m/s

Height, $dy = 0.078$ m

Shear stress at the contact of fluid and surface is

$$\tau_{y=0} = \mu \left(\frac{du}{dy}\right)_{y=0} = 0.0064 \times \frac{1.78}{0.078} = 0.146 \text{ N/m}^2$$

Example 1.18: A plate of area 0.3 m × 0.3 m slides on an inclined plane inclined at 30° to horizontal, moves with a velocity 0.41 m/s and has a weight of 250 N. An oil film of 0.20 m thickness is filled in between plate and inclined plane. Find the viscosity of the oil.

Solution: Given data:

$$A = 0.09\,\text{m}^2,\ \theta = 30^0,\ du = 0.41\,\text{m/s}$$
$$W = 250\,\text{N},\ dy = 0.2\,\text{m}$$

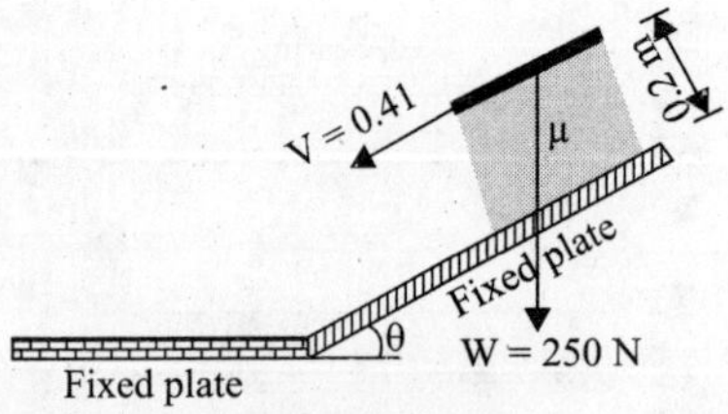

Figure P1.18

The weight of the plate can be resolved into two components, one component along the motion (*W* sin θ) and the other component along the normal to the fixed plate (*W* cos θ) and so the driving force F = (W sin θ)

$$\tau = F/A = \frac{250 * \sin 30}{0.09} = 1388.9 \text{ Pa}$$

As per Newton's law

$$\tau = \mu\left(\frac{du}{dy}\right) \Rightarrow \mu = \tau * \left(\frac{dy}{du}\right)$$

$$\Rightarrow \mu = 1388.9 * (0.2/0.41) = 677.5 \text{ Ns/m}^2$$

Example 1.19: A moving plate having area 0.3 m × 0.3 m is being pulled by a 140 N force. The moving plate is in between two fixed plates which are 15 m apart, having filled by two different oils having viscosities in the ratio of 1:3, on the either sides of the moving plate. Find the velocity ratio with which the moving plate moves.

Solution: Given data:

$$\frac{\mu_1}{\mu_2} = \frac{1}{3}, \text{ area of plate} = 0.09 \text{ m}^2$$

Shear stress ratio

$$\frac{\tau_1}{\tau_2} = \frac{F_1/A}{F_2/A} = \frac{F_1}{F_2} = \frac{\mu_1}{\mu_2} * \left(\frac{du_1}{dy_1}\right) * \left(\frac{dy_2}{du_2}\right)$$

$$\Rightarrow \frac{140}{140} = \frac{1}{3} * \left(\frac{du_1}{7.5}\right) * \left(\frac{7.5}{du_2}\right)$$

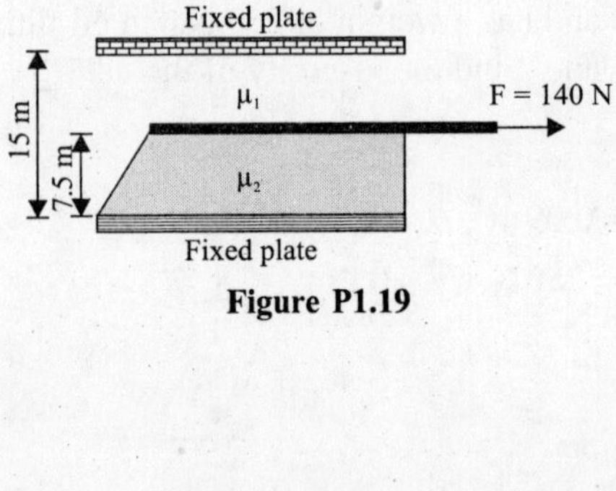

Figure P1.19

$$\Rightarrow \frac{u_1}{u_2} = \frac{3}{1}$$

Hence, the velocity ratio is 3:1.

Example 1.20: A horizontal cylinder of 0.075 m diameter is inserted concentrically in a drum of 0.085 m internal diameter. Oil is filled in the gap between cylinder and drum to a length of 0.2 m. The torque required to rotate the cylinder in the drum is 4 Nm, when the speed of rotation is 7.5 rev/s, assuming the end effects to be negligible. Calculate the coefficient of viscosity of the oil filled, shear force and power utilized in overcoming resistance.

Solution: Given data:

$$d_1 = 0.085\,\text{m},\ d_2 = 0.075\,\text{m},\ T = 4\,\text{Nm},\ N = 7.5\,\text{rev/s},\ L = 0.2\,\text{m}$$

Velocity of cylinder

$$u = \frac{2\pi r_2 N}{60} = \frac{2*3.14*0.0375*7.5}{60} = 0.029 \text{ m/s}$$

Torque needed to rotate the cylinder is

$$T = \tau * \text{ surface area } * \text{ perpendicular distance}$$

$$\Rightarrow T = \tau * (2\pi r_2 L) * (d_1 - d_2)$$

$$\Rightarrow \tau = 8.49 \text{ kPa}$$

Using Newton's law of viscosity, we have

$$\tau = \mu \frac{du}{dy}$$

$$\Rightarrow \mu = 1464 \text{ Ns/m}^2$$

Shear force $\Rightarrow F_\tau = \tau * (2\pi r_2 L) = 399 \text{ N}$

Power utilized $\Rightarrow P = T * (2\pi N)/60 = 3.14 \text{ watt}$

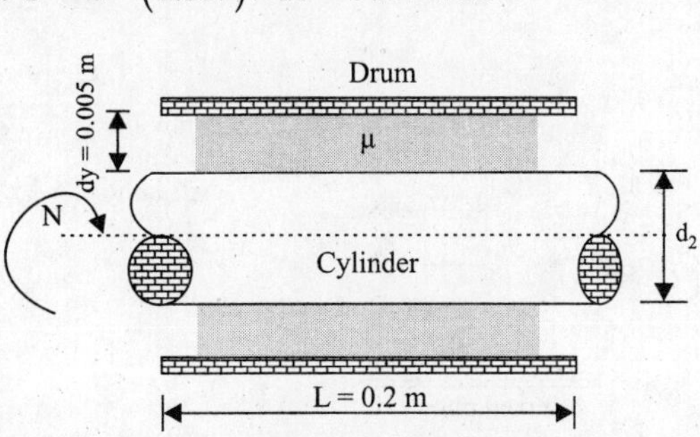

Figure P1.20

Example 1.21: The velocity profile distribution of flow over a plate is parabolic $\left(u = Ay^2 + By + C\right)$ with vertex 30 cm from the plate, where the velocity is 0.18 m/s. If the viscosity of the fluid is 0.9 Ns/m^2, find the velocity gradient and shear stress at a distance of 0 cm, 15 cm and 30 cm from the plate.

Solution: Given data:

$$u = 0.18 \text{ m/s}, \mu = 0.9 \text{ Ns/m}^2$$

$$u = Ay^2 + By + C$$

Applying the boundary conditions, we have

At $\quad y = 0$ m, $u = 0$ m/s and so $C = 0$

At $\quad y = 0.3$ m, u = 0.18 m/s and

So
$$0.18 = 0.09A + 0.3B \tag{1}$$

At $\quad$ y = 0.3 m, $\dfrac{du}{dy} = 0$ and

So
$$\frac{du}{dy} = 2Ay + B$$

$$0 = 0.6A + B \tag{2}$$

Equating Eq.(1) and Eq.(2), we get $A = -2.0$, $B = 1.2$ and so velocity distribution becomes

$$u = -2.0y^2 + 1.2y$$

Velocity gradient
$$\frac{du}{dy} = -4y + 1.2$$

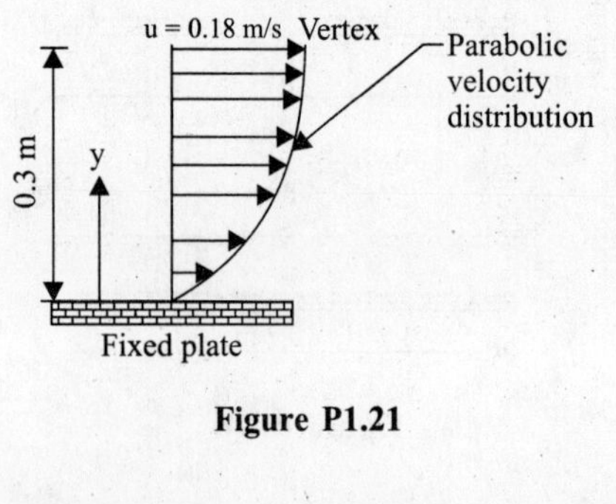

Figure P1.21

At $$y = 0, \left(\frac{du}{dy}\right)_{y=0} = -4*0+1.2 = 1.2$$

At $$y = 0.15, \left(\frac{du}{dy}\right)_{y=0.15} = -4*0.15+1.2 = 1.8$$

At $$y = 0.3, \left(\frac{du}{dy}\right)_{y=0.3} = -4*0.3+1.2 = 0$$

Shear stresses $$\tau = \mu\frac{du}{dy} = \mu(2Ay+B)$$

At $$y = 0, \ \tau = \mu\left(\frac{du}{dy}\right)_{y=0} = 0.18(-8*0+1.2) = 0.216 \text{ Pa}$$

At $$y = 0.15, \ \tau = \mu\left(\frac{du}{dy}\right)_{y=0.15} = 0.18*(-8*0.15+1.2) = 0.0 \text{ Pa}$$

At $$y = 0.3, \ \tau = \mu\left(\frac{du}{dy}\right)_{y=0.3} = 0.18*(-8*0.3+1.2) = -1.2 \text{ Pa}$$

Example 1.22: A solid cone of maximum radius *R* and vertex angle of 2θ is to rotate at angular velocity of ω. An oil of viscosity μ and thickness *t* fills the gap between the cone and the housing, as shown in the figure. Derive an expression for the torque required and the rate of heat dissipation in the bearing.

Solution: Given data:

$$R, 2\theta, \omega, \mu, t$$

Consider an elementary area dA at radius of the cone, then

$$dA = 2\pi r ds = 2\pi r * \frac{dr}{\sin\theta}$$

Shear stress $\tau = \mu\frac{du}{dy} = \mu\frac{u}{t}$ and so torque on the elementary strip will be product of shear force and perpendicular distance, i.e. $dT = \tau * dA * r = \mu\frac{u}{t} * 2\pi r * \frac{dr}{\sin\theta} * r$

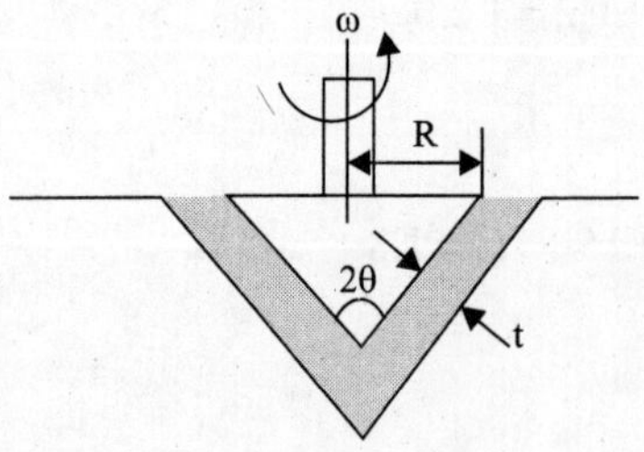

Figure P1.22a

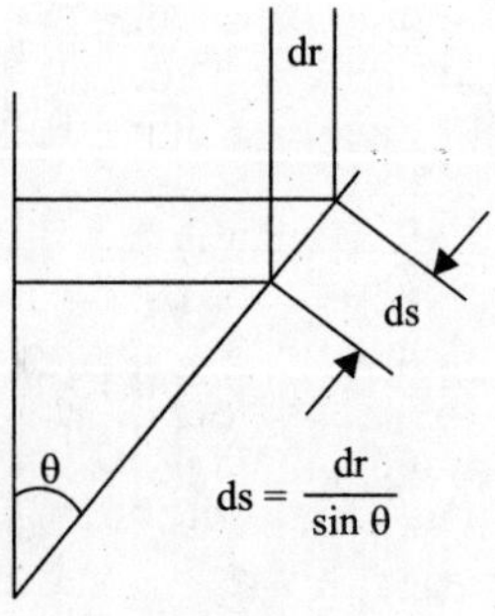

Figure P1.22b

Total torque

$$T = \int_0^R \mu \frac{r\omega}{t} * 2\pi r * \frac{dr}{\sin\theta} * r \text{ since } u = r\omega$$

$$\Rightarrow T = \frac{\pi\mu\omega^2}{2t\sin\theta} R^4$$

Power utilized in overcoming the resistance (rate of heat dissipation in the bearing)

$$P = T * \omega = \frac{\pi\mu\omega^2}{2t\sin\theta} R^4$$

Example 1.23: A square block of 2.0 m × 0.18 m side and 0.18 m thick, weighing 70 N is to be lifted through a vertical gap of 0.35 m between two fixed plates of infinite length, filled with oil in between. The oil has viscosity 3.2 Ns/m^2 and specific gravity of 0.85. Find the force and power required to lift the block with uniform velocity of 0.13 m/s.

Solution: Given data:

$$A = 2.0 * 0.18 = 0.36\,\text{m}^2,\ dy = t = (0.35 - 0.18)/2 = 0.085\text{ m}$$
$$du = 0.13\,\text{m/s},\ \mu = 3.2\,\text{Ns/m}^2,\ S = 0.85$$

We know as per Newton's law of viscosity

$$\tau = \mu \frac{du}{dy}$$

$$\Rightarrow \tau = 3.2 * \frac{0.13}{0.085} = 4.89\,\text{Pa}$$

Force required

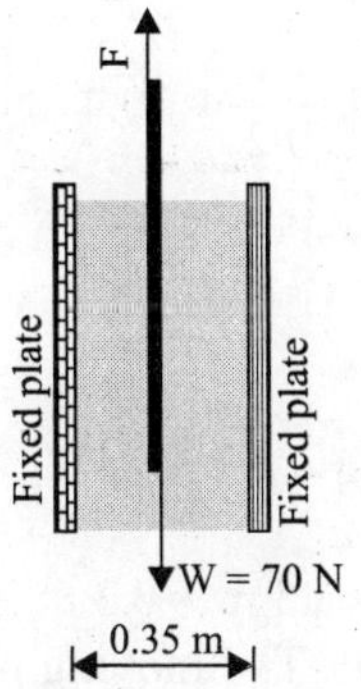

Figure P1.23

$$F = W + 2(\tau * A)$$

$$\Rightarrow F = 70 + 2*(4.89*0.36) = 73.56 \text{ N}$$

Power required

$$P = F * u$$

$$\Rightarrow P = 73.56*0.13 = 9.56 \text{ watt}$$

Example 1.24: A plate moves with a velocity 0.41m/s and requires 250 N of force per unit area, to be pulled on oil film of 0.20 m thickness, filled in between the plate and fixed plate. Find the viscosity of the oil.

Solution: Given data:

$$\tau = 250 \text{ N/m}^2,\ du = 0.41 \text{ m/s},\ dy = 0.2 \text{ m}$$

As per Newton's law

$$\tau = \mu\left(\frac{du}{dy}\right) \Rightarrow \mu = \tau*\left(\frac{dy}{du}\right)$$

$$\Rightarrow \mu = 250*(0.2/0.41) = 121.95 \text{ Ns/m}^2$$

Example 1.25: A plate of area 5 m × 3 m slides with a velocity 0.41m/s relative to a fixed plate at the bottom and an oil film filled in between plate and fixed plate of 0.20 m thickness. Find the force required to drag the moving plate if viscosity of the oil is 2 Poise.

Solution: Given data:

$$A = 15 \text{ m}^2,\ du = 0.41 \text{ m/s},\ \mu = 0.2 \text{ Ns/m}^2,\ dy = 0.2 \text{ m}$$

As per Newton's law $\tau = \mu\left(\frac{du}{dy}\right)$

$$\Rightarrow \tau = 0.2\left(\frac{0.41}{0.2}\right) = 0.41 \text{ N/m}^2$$

And the shear stress

$$\tau = F / A$$
$$\Rightarrow F = \tau * A = 0.41 * 15 = 6.15 \text{ N}$$

Example 1.26: A moving plate having area 3 m × 0.3 m is being pulled with a velocity of 0.55 m/s. And the viscosity of the oil is 0.85 Ns/m^2. The moving plate is in between two fixed plates which are 15 m apart. Calculate the drag force

a) If the moving plate is placed in between the two fixed plates.
b) If the moving plate is at a distance of 5 m from the bottom fixed plate.

Solution: Given data:

$$A = 0.9 \text{ m}^2,\ \mu = 0.85 \text{ Ns/m}^2,\ dy = 15 \text{ m},\ du = 0.55 \text{ m/s}$$

a) If the moving plate is placed in between the two fixed plates (i.e. dy =7.5 m):
Shear stress on one side of the moving plate

$$\tau = \mu\left(\frac{du}{dy}\right) \Rightarrow \tau_1 = 0.85\left(\frac{0.55}{7.5}\right) = 0.062 \text{ N/m}^2$$

Shear force on one side of the moving plate $F_1 = \tau_1 * A = 0.062 * 0.9 = 0.0558 \text{ N}$. The total drag force will be twice of the shear force acting on one side of the moving plate and so

Total drag force is $F = 2 * F_1 = 0.0558 * 2 = 0.11 \text{ N}$

b) If the moving plate is at a distance of 5 m from the bottom fixed plate:
Shear stress on one side of the moving plate

$$\tau = \mu\left(\frac{du}{dy}\right) \Rightarrow \tau_1 = 0.85\left(\frac{0.55}{5}\right) = 0.0935 \text{ N/m}^2$$

Shear force on one side of the moving plate

$$F_1 = \tau_1 * A = 0.0935 * 0.9 = 0.08415 \text{ N}$$

Shear stress on another side of the moving plate

$$\tau = \mu\left(\frac{du}{dy}\right) \Rightarrow \tau_2 = 0.85\left(\frac{0.55}{10}\right) = 0.0497 \text{ N/m}^2$$

Shear force on another side of the moving plate

$$F_2 = \tau_2 * A = 0.0497 * 0.9 = 0.042\,\text{N}$$

So total drag force is

$$F = F_1 + F_2 = 0.08415 + 0.042 = 0.126\,\text{N}$$

Classification of Fluid Behaviors

1) Newtonian fluids.
2) Non-Newtonians fluids.

The classification of fluid behaviors can be shown by the graphical representation shown below by taking velocity gradient (*du*/*dy*) on the X-axis and the shear stress on the Y-axis.

All the fluids that follow the Newton's law of viscosity are termed Newtonian fluids and fluids that follow Newton's law of viscosity after some yield are termed Newtonian fluids with initial yield stress (like Bingham plastics) and remaining fluids are termed Non-Newtonian fluids (i.e. which do not obey Newton's law of viscosity like Thixotropic substances, Pseudo plastics, Dilatent fluids).

The generalized equation for distinguishing between Newtonian and non-Newtonian fluids can be given as

$$\tau = c + \mu\left(\frac{du}{dy}\right)^n \tag{1.16}$$

where *c* is a constant and *n* is the power index.

Ideal Fluids: Ideal fluids are the fluids which do not possess viscosity and hence with almost no shear stress they tend to have high velocity gradient and so they behave like fluids with no shear stress and follow the velocity gradient line, on the X-axis.

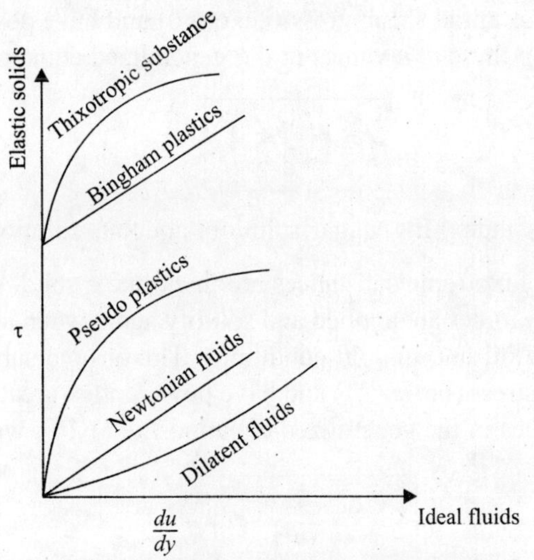

Figure 1.15: Classification of fluid behaviors

Newtonian fluids: The fluids which obey Newton's law of viscosity are known as Newtonian fluids and are generalized by the equation by $c = 0$, $n = 1$ when substituted in Eq. (1.16) gives

$$\tau = \mu\left({du}/{dy}\right) \tag{1.16a}$$

i.e. all the Newtonian fluids follow a straight line from the origin of the graph between velocity gradient (du/dy) on X-axis and the shear stress τ on the Y-axis.

E.g.: air, water, kerosene, etc.

Non-Newtonian Fluids: All the fluids which do not obey Newton's law of viscosity are known as non-Newtonian fluids and can be generalized by the Eq. (1.16) by giving different values to c, n and when substituted in Eq. (1.16) gives different types of fluid flow behaviors like Pseudo Plastics, Dilatants fluids, Thixotropic substances and Bingham plastics.

Pseudo Plastics: Pseudo plastics are non-Newtonian fluids having no initial shear stress (i.e. $c = 0$) and have power index less than unity (i.e. $n < 1$). By substituting these c, n values in the generalized equation Eq. (1.16), we get

$$\tau = k\left({\partial u}/{\partial y}\right) \tag{1.16b}$$

k is the consistency index.

E.g. blood, milk, liquid, cement, paper pulp.

Dilatants fluid: A dilatant substance turns to a thick paste, when stirred two different substances. Like mixing water with starch powder forms a paste. Hence, for these substances, when external disturbing forces are applied, the bonding action between the two substances activates and becomes more viscous, which conclude to a single substance called Dilatant fluid. Dilatants fluids are non-Newtonian fluids having no initial shear stress (i.e. c = 0) and have power index greater than unity (i.e. $n > 1$). By substituting these c, n values in the generalized equation Eq. (1.16), we get

$$\tau = k\left({\partial u}/{\partial y}\right)^n \tag{1.16c}$$

where k is the consistency index. E.g. sugar solution, aqueous solutions, starch solution sauce.

Thixotropic substance: Thixotropic substances are thick like a solid, but flows like a liquid when external sideway (rubbing) forces are applied and solidify again when left standing. E.g.: Rubbing a frozen oil bottle to get the oil out, in cold conditions. Thixotropic substances are non-Newtonian fluids having initial shear stress (i.e. $c = \tau_0$) and have power index greater than unity (i.e. $n < 1$). By substituting these c, n values in the generalized equation Eq. (1.16), we get

$$\tau = \tau_0 + \mu\left(\frac{\partial u}{\partial y}\right)^n \tag{1.16d}$$

E.g. Toothpaste and gel.

Bingham/Ideal plastics: Bingham plastics are non-Newtonian fluids having initial shear stress (i.e. $c = t_0$) and have power index of unity (i.e. $n = 1$). By substituting these *c, n* values in the generalized Eq. (1.16), we get

$$\tau = \tau_0 + \mu\left(\frac{\partial u}{\partial y}\right) \tag{1.16e}$$

E.g. drilling mud, sewage sludge, water suspension of clay and flash.

Example.1.27: Determine the height to which water at 20°C will rise in a capillary glass tube exposed to atmosphere and has 4 mm internal diameter. Take surface tension as 0.074 N/m.

Solution: Given data: $\sigma = 0.074$ N/m, $d = 0.004$ m, $\rho = 1000$ kg/m^3, $\theta = 0°$

We have capillary head as $$h = \frac{4\sigma\cos\theta}{\rho g d}$$

So $$h = \frac{4*0.074*\cos 0}{1000*9.81*0.004} = 0.754 \text{ cm}$$

Example.1.28: Determine the viscosity and shear strain rate for oil filled in between parallel plate separated by a distance of 15 mm. If the upper plate is pulled with a velocity of 3.5 m/s imposing an amount of 130 Pa of shear on the moving plate.

$$\mu = \tau\left(\frac{du}{dy}\right) = 130*\frac{3.5}{0.015} = 30.3 \text{ kNs/m}^2$$

$$\Rightarrow \frac{d\theta}{dt} = \frac{du}{dy} = \frac{3.5}{0.015} = 233.3 \text{ s}^{-1}$$

Example.1.29: Oil fills the 1.5 mm gap between two concentric cups having flat bottoms as shown in Figure P1.29. The inner cup rotates at a speed of 50 rpm, while the outer is stationary. The torque required to rotate the inner cup is 6 Nm. Assuming the end effects to be negligible and the velocity variation to be linear, calculate the viscosity of the oil filled, shear force and power utilized in overcoming resistance.

Solution:

Given data: $d_1 = 0.1$ m, $d_2 = 0.103$ m, $T = 6$ Nm, $N = 50$ rev/s, $L = 0.15$ m

Velocity of inner cup $$u = \frac{\pi d_2 N}{60} = 0.269 \text{ m/s}$$

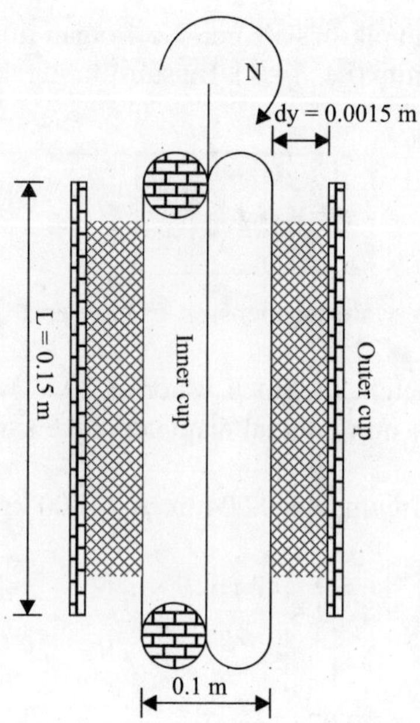

Figure P1.29

Torque needed to rotate the inner cup is

$$T = \tau * \text{surface area} * \text{perpendicular distance}$$

$$\Rightarrow \tau = 41.226 \text{ kPa}$$

Using Newton's law of viscosity, we have

$$\tau = \mu \frac{du}{dy} \Rightarrow \mu = 229.8 \text{ Ns/m}^2$$

Shear force $\Rightarrow F_\tau = \tau * (2\pi r_2 L) = 1999 \text{ N}$

Power utilized $\Rightarrow P = T * (2\pi N)/60 = 31.4 \text{ watt}$

$$\Rightarrow P = 4 * (2 * 3.14 * 7.5)/60 \Rightarrow P = 3.14 \text{ watt}$$

EXERCISE

Objective

1. Fluid is a substance that
   a) Cannot be subjected to shear forces

b) Always expands until it fills any container
c) Has the same shear stress at point regardless of its motion
d) Cannot remain at rest under action of any shear force
e) Flows and all the above

2. Fluid is a substance which offers no resistance to change of
a) Pressure
b) Flow
c) Shape
d) Volume
e) all the above

3. Real fluids
a) Are viscous
b) Possess surface tension
c) Are compressible
d) Possess all the above.

4. A fluid is said to be ideal, if it is
a) Incompressible
b) Inviscous
c) Viscous and incompressible
d) Inviscous and incompressible
e) Inviscous and compressible

5. If no resistance is encountered by displacement, such a substance is known as
a) Fluid
b) Water
c) Gas
d) Perfect solid
e) Ideal fluid

6. The volumetric change of the fluid caused by a resistance is known as
a) Volumetric strain
b) Compressibility
c) Adhesion
d) Cohesion

7. Liquids
a) Cannot be compressed
b) Occupy definite volume
c) Are not affected by change in pressure and temperature
d) Are not viscous
e) None

8. Mass density of liquid (ρ) is given by

(a) $\rho = \dfrac{mass}{volume}$

(b) $\rho = \frac{metric\ Slug}{m^2}$

(c) $\rho = \frac{kg - \sec}{m^2}$

(d) None

9. Property of a fluid by which its own molecules are attracted is called
   a) Adhesion
   b) Cohesion
   c) Viscosity
   d) Compressibility
   e) Surface tension
10. Mercury does not wet glass. This is due to property of liquid known as
   a) Adhesion
   b) Cohesion
   c) Surface tension
   d) Viscosity
   e) Compressibility
11. Property of a fluid which enables it to resist tensile stress is known as
   a) Compressibility
   b) Surface tension
   c) Cohesion
   d) Adhesion
   e) Viscosity
12. Specific weight of water in S.I. units is equal to
   a) 1000 N/m^3
   b) 10000 N/m^3
   c) 9.81×10^3 N/m^3
   d) 9.81×10^6 N/m^3
   e) 9810 N/m^3
13. Which of the following is dimensionless?
   a) Specific weight
   b) Specific volume
   c) Specific speed
   d) Specific gravity
14. The tendency of a liquid surface to contract is due to the following property
   a) Cohesion
   b) Adhesion
   c) Viscosity
   d) Surface tension
   e) Elasticity
15. The bulk modulus of elasticity with increase in pressure
   a) Increases

b) Decreases
c) Remains constant
d) Unpredictable

16. A balloon lifting on air follows the following the principle
a) Law of gravitation
b) Archimedes principle
c) Principle of buoyancy
d) All of the above

17. The increase of temperature results in
a) Increase in viscosity of gas
b) Increase in viscosity of liquid
c) Decrease in viscosity of gas
d) Decrease in viscosity of liquid
e) a and d above

18. Surface tension has the units of
a) Newton-s/m^2
b) Newton/m^2
c) Newton/m
d) Newton
e) Newton-m

19. Barometer is used to measure
a) Pressure in pipes
b) Atmospheric pressure
c) Very low pressure
e) Rain level

20. The property of fluid by virtue of which it offers resistance to shear is called
a) Surface tension
b) Adhesion
c) Cohesion
d) Viscosity
e) All of the above

21. The magnitude of rise of pressure due to water hammer in a rigid and non-elastic pipe carrying water of density ρ and bulk modulus K will be equal to

(a) $\sqrt{\frac{k}{\rho}}$

(b) $\sqrt{k\rho}$

(c) $\sqrt{\frac{\rho}{k}}$

(d) $\frac{k}{\rho}$

(e) $k*\rho$

22. The shear stress in a fluid flowing in a round pipe
    a) is constant over the cross-section
    b) is zero at the wall and increases linearly to the centre
    c) is zero at center and varies linearly with radius
    d) varies parabolically across the section
    e) unpredictable
23. If 5.6 m^3 of oil weighs 46,800 N, what is the mass density in kg/m^3?
    a) 1.2×10^{-4}
    b) 852.0
    c) 8357.1
    d) 8.52
    e) 8.36
24. A fluid has absolute viscosity, μ, of 0.048 Pa. If at point A, 75 mm from the wall the velocity is measured as 1.125 m/s, calculate the intensity of shear stress at point B 50 mm from the wall in N/m^2. Assume a linear (straight line) velocity distribution from the wall.
    a) 15
    b) 0.048
    c) 0.72
    d) 0.0032
    e) 0.032

Answers

1. e	2. e	3. d	4. d	5. e	6. b
7. b	8. a	9. b	10. b	11.a	12. e
13. d	14. d	15.a	16. b	17. e	18. c
19. b	20. d	21.a	22. c	23. b	24. c

True/False

a) Fluids are capable of flowing. ☐ Yes ☐ No
b) Fluids conform to the shape of the containing vessels. ☐ Yes ☐ No
c) When in equilibrium, fluids cannot sustain tangential forces. ☐ Yes ☐ No
d) When in equilibrium, fluids can sustain shear forces. ☐ Yes ☐ No
e) Fluids have some degree of compressibility and offer little resistance to form. ☐ Yes ☐ No
f) Viscosity is the property of a fluid which determines the amount of its resistance to a shearing force. ☐ Yes ☐ No
g) Viscosity is primarily due to interaction between fluid molecules. ☐ Yes ☐ No
h) Viscosity of liquids decreases with increase in temperature. ☐ Yes ☐ No
i) Viscosity of liquids is appreciably affected by change in pressure. ☐ Yes ☐ No
j) Viscosity is expressed as poise, stoke, or say bolt seconds. ☐ Yes ☐ No
k) Kinematic viscosity is equal to density/dynamic viscosity. ☐ Yes ☐ No

l) Free surface of a liquid tends to contract to the smallest possible area due to force of surface tension. ☐ Yes ☐ No
m) A liquid would wet the solid, if adhesion as compared to cohesion is more. ☐ Yes ☐ No
n) Capillary action is due to the surface tension. ☐ Yes ☐ No
o) Newton's law of viscosity is a relationship between shear stress and the rate of angular distortion. ☐ Yes ☐ No
p) When a fluid is subjected to a steady shear stress, it will reach a state of equilibrium in which no further motion occurs. ☐ Yes ☐ No
q) Pressure and shear stress are two examples of a force per unit area. ☐ Yes ☐ No
r) If ρ is density in g/cm^3 and μ is viscosity in g/cm s, then the kinematic viscosity $\nu = \mu/\rho$ is in stokes. ☐ Yes ☐ No
s) The density of water is 1,000 kg/m^3 and 9810 N/m^3. ☐ Yes ☐ No
t) The density of an ideal gas depends only on its absolute temperature and its molecular weight. ☐ Yes ☐ No
u) A fluid at rest does have no shear stress ? ☐ Yes ☐ No
v) The shear strain rate does remains constant until and unless the shear stress varies. ☐ Yes ☐ No

Theory

1. Define a fluid and distinguish between ideal and real fluids.
2. Define mass density, specific volume and specific gravity.
3. Explain theory behind cohesion and adhesion.
4. Explain the phenomenon of surface tension.
5. Explain the phenomenon of capillarity.
6. Differentiate and explain dynamic and kinematic viscosity.
7. Enunciate Newton's law of viscosity. Explain the importance of viscosity in fluid motion.
8. Distinguish between Newtonian and non-Newtonian fluids.
9. Write a note on different types of fluid behaviors.
10. Define compressibility. How is it related to bulk modulus of elasticity? Obtain an expression for the bulk modulus of elasticity K of a fluid in terms of the pressure p and the mass density.
11. Explain from microscopic point of view, the concept of viscosity. Explain how viscosity of liquids and gases behave with temperature.

Problems

1. If oil having 4.0 KN weight occupies a volume 4.5 m^3 of the container in which it is contained, calculate specific weight, specific gravity, specific volume.
2. Determine mass density, specific volume, and specific weight if the specific gravity of a liquid is 0.71.
3. Determine the pressure increase required to reduce the volume of water by 1.5%, if its bulk modulus of elasticity is 2.2×10^9 Pa.
4. The surface tension of water in contact with air at 20°C is 0.07 N/m. The pressure inside a water droplet is 0.15 kPa greater than the outside pressure. Calculate the diameter of the water droplet.

5. A thin plate of large area is placed midway in a gap of height *h* filled with oil of viscosity μ_0 and the plate is pulled at a constant velocity *V*. If a lighter oil of viscosity μ_1 is then substituted in the gap, it is found that for the same velocity *V*, the drag force will be the same as before if the plate is located unsymmetrical in the gap but parallel to the walls. Find μ_1 in terms of μ_0 and the distance from the nearer wall to the plate.
6. A circular disc of diameter *d* is slowly rotated in a liquid of large viscosity μ at a small distance *h* from a fixed surface. Draw the sketch and derive an expression for the torque "T" necessary to maintain an angular velocity μ.
7. Given a shaft in a bush bearing having oil filled in between, is rotated at a speed of 210 rpm. Determine the shear stress, torque and power if the velocity is 4 centipoises and film thickness is 0.5 mm and diameter of shaft is 110 mm.
8. A 30 cm diameter shaft revolves in a guide bearing of 60 cm long and at 500 rpm. If the oil film bearing is 0.13 mm and viscosity of oil is 0.05 Ns/m^2. Find the power absorbed.
9. A 120 mm disc rotates on a table separated by an oil film of 1.8 mm thickness. Find the viscosity of oil if the torque required to rotate the disc at 60 rpm is 3.6×10^{-4} Nm.
10. A metal plate 1.25 m × 1.25 m × 6 mm thick and weighing 90 N is placed midway in the 24 mm gap between the two vertical plane surfaces. The gap is filled with an oil of specific gravity 0.85 and dynamic viscosity 3.0 Ns/m^2. Determine the force required to lift the plate with a constant velocity of 0.15 m/s.
11. A moving plate having area 0.25 m^2 is being pulled with a velocity of 0.3 m/s. And the viscosity of the oil is 9.72 Poise. The moving plate is in between two fixed plates which are 12 m apart. Calculate the drag force
    a) If the moving plate is placed in between the two fixed plates.
    b) If the moving plate is at a distance of 4mm from the bottom fixed plate.
12. A moving plate having area 0.3m × 0.3m is sliding down an inclined plane at 45° to the horizontal with a velocity of 0.55m/s. And there is a cushion of an oil film with 2 mm thickness. Find the viscosity of the oil if the plate weighs 250 N.

QUICK REFERENCE OF FORMULAE

Property	Formulae	Units	Remarks, if any
Pressure (p)	$p=\frac{F}{a}$	N/m^2 or Pascal	
Density (ρ)	$\rho=\frac{m}{V}$	Kg/m^3	Also called specific mass or mass density
Specific weight (γ)	$\gamma=\frac{W}{V}=\rho g$	N/m^2	Also called weight density
Specific gravity (S)	$S=\frac{\gamma_{liquid}}{\gamma_{water}}=\frac{\rho_{liquid}}{\rho_{water}}$	Nil	Nil
Specific volume (ν)	$\nu=\frac{Volume}{mass}$	m^3/kg	$=1/\rho$
Capillarity head (h)	$h=\frac{4\sigma\cos\theta}{\gamma d}$	m	
Bulk modulus (K)	$K=-\frac{dp}{\left(\frac{\delta v}{V}\right)}$	N/m^2 or pascal	
Compressibility (Z)	$Z=\frac{1}{K}$	m^2/N	
Viscosity or Dynamic viscosity (μ)	$\tau=\mu\left(\frac{\partial u}{\partial y}\right)=\mu\left(\frac{\partial\theta}{\partial t}\right)$	Ns/m^2 or poise	Newton's law of viscosity) (Ns/m^2 = 10 poise)
Kinematic viscosity (υ)	$\upsilon=\frac{\mu}{\rho}$	cm^2/sec or stokes	

Table 1.1: Values of some properties to remember

Property	Material	Value	Remarks
Specific weight (γ)	Water	$9810\ N/m^3$	$1000\ kgf/m^3$
Specific gravity (S)	Water	1	
	Mercury	13.6	
Angle of contact of liquid with container (θ)	Water	0°	
	Mercury	140°	
$P_{atmosphere}$		101 KPa or 760 mm of Hg or 10.3 m of water	

CHAPTER 2

Pressure and Pressure Measuring Devices

INTRODUCTION

The main properties of fluids, which are considered in real-time applications are pressure, velocity and temperature. Pressure and temperatures play a major role in the fields of pipeline constructions, water supply, hydraulics for irrigation, thermal power, hydraulic power, aviation of flights and even in human blood, etc. Fluids under pressure can exert forces that can be used by us to do work. Fluids form the basis for pneumatic and hydraulic systems which are used in industries such as automobile manufacturing (brakes), hoists found in service stations, dentist's chair, aeronautics, submarines, the shipping industry and the list can go on and on.

A small pressure, spread over a very large area, can add up to be very large force. When a fluid moves over or through an object; it gives small pushes on the surface of the object. These pushes, over the entire surface, are defined as pressure and are measured as force per unit area.

To understand the fluid flow, we need to study the fluid properties like pressures, velocity and temperature. And we deal with study of pressure and its ways of measuring, in this section.

PRESSURE OF A FLUID (*P*)

All fluid molecules will be in constant and random motion called "**Brownian motion**", due to which fluid at rest in a vessel, does exerts force on all the walls of the vessel, with which it is in contact,

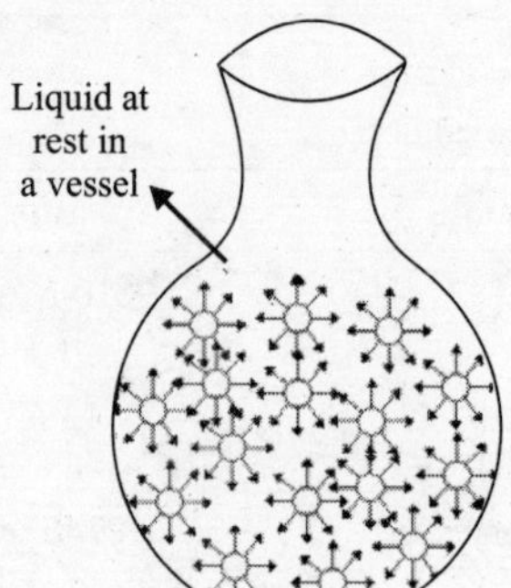

Figure 2.0

as shown in Figure 2.0, due to the reason that it (fluid) is composed of several energetic molecules which when kept in a vessel, the molecules collide with the walls of the vessel and interchange the molecular momentum between molecules, due to which force is exerted on the walls of the vessel at normal directions by the molecules. This force exerted by the fluid molecules on the vessel walls on unit area is called **pressure** and has units N/m^2 *or* pascal

i.e.
$$Pressure = \frac{Normal\ force}{area}$$

$$\Rightarrow P = \frac{F_N}{a} \qquad (2.0)$$

In a fluid at **rest**, the walls of the container in which the fluid exists eliminate shear stress, and thus a fluid at rest is at a state of zero shear stress. But when the walls are removed or a liquid container is tilted, a shear develops and the liquid splashes or moves to attain a horizontal free surface.

TYPES OF PRESSURE EXISTING

a) Atmospheric pressure.
b) Gage pressure.
c) Negative/vacuum gage pressure.
d) Absolute zero pressure.

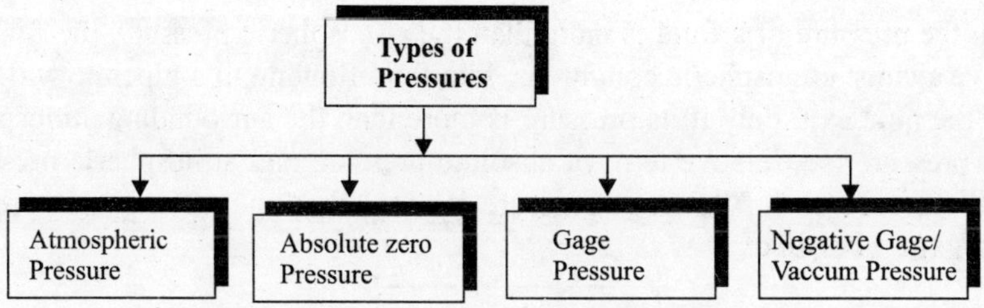

Figure 2.1: Classification of pressures

Atmospheric Pressure (P_{atm})

Atmospheric pressure is the pressure of atmospheric air, which it exerts on all the surfaces with which it is in contact and is denoted by P_{atm} and has units N/m^2 or pascal.

Atmospheric pressure usually varies with the altitude and surrounding conditions, at which the atmospheric pressure is measured by a device called **barometer**, invented by **Evangelista Torricelli** and hence atmospheric pressure can also be called **barometric pressure**.

$$1P_{atm} = 101.104\,\text{kN/m}^2 = 10.3 \text{ m of } H_2O = 76 \text{ cm of Hg} = 760 \text{ torr}$$

If observed hydrostatic pressure equation, we get the pressure exerted to be depending on the height of the column and the density of the fluid, due to which the atmospheric pressure decrease as the altitude increase from the earth surface, and hence breathing becomes difficult at high altitudes because of reduced atmospheric pressrue.

Absolute pressure: All fluids have **Brownian** motion, but if their molecular motion is ceased by freezing the fluid to "absolute zero at –459.4°F", then pure vacuum exist and the pressure at this point is termed "Absolute zero pressure".

Gage Pressure

Any fluid will be traveling with high or low pressures, compared to the atmospheric pressure existing in the surroundings and the pressure with which a fluid travels is called **Gage pressure**.

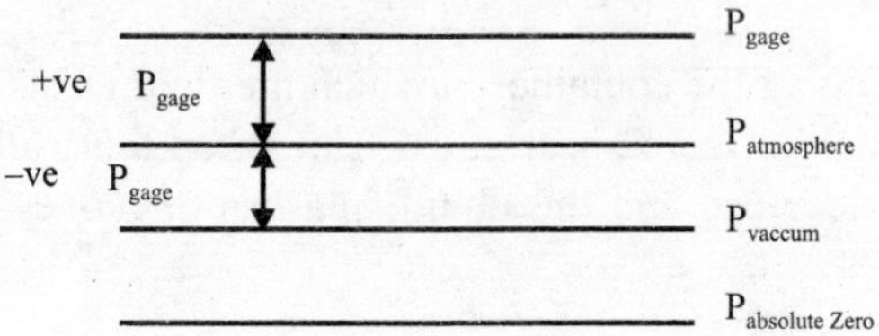

Figure 2.2: Relation between pressures

Fluid molecules do not collide with each other to exhibit molecular colliding force and hence absolute pressure can be defined in reference to perfect vacuum conditions. But in nature, atmospheric pressure acts against the absolute pressure of the fluid and hence there is a need to measure the relative pressure difference of absolute and atmospheric pressures, to obtain the actual pressure of a fluid, which is the gage pressure.

Only when the pressure of a fluid is more than the atmospheric pressure, the fluid will be in a position to move against atmospheric conditions, like fluid flowing in a pipeline will be capable of flowing out of the pipe exit, only if its pressure is more than the surrounding atmosphere pressure and hence this pressure is a relative term of absolute pressure and atmospheric pressure which is known as gage pressure and if the gage pressure is more than the atmospheric pressure then is called **positive gage pressure**.

i.e.
$$P_{abs} = P_{atm} + P_{gage} \qquad (2.1)$$

If a fluid is traveling with pressure less than the atmospheric pressure, then the fluid will not be in a position to move out into the open atmospheric conditions due to the obstruction created by the atmosphere pressure, like fluid flowing in a vacuum pipeline. And so there exists vacuum conditions around the fluid flowing and hence this pressure is a negative relative term of absolute pressure and atmospheric pressure and hence this negative gage pressure is called **negative gage/vacuum pressure**.

i.e.
$$P_{abs} = P_{atm} - P_{vaccum} \qquad (2.2)$$

PASCAL'S LAW

Pascal stated that "At any point in a liquid at rest, the intensity of pressure is same in all directions". i.e. $P_{xy} = P_{yz} = P_z$.

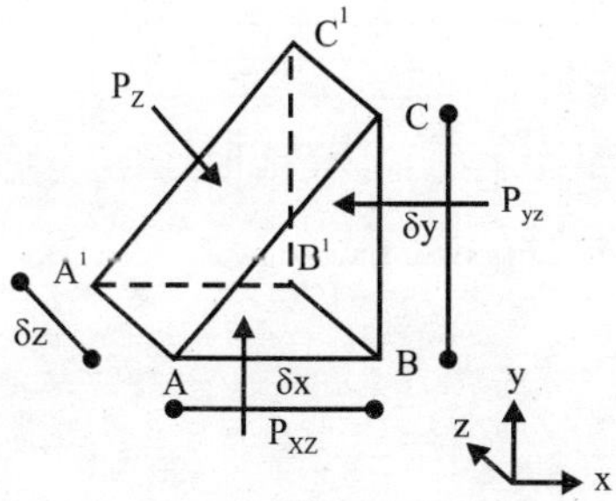

Figure 2.3a

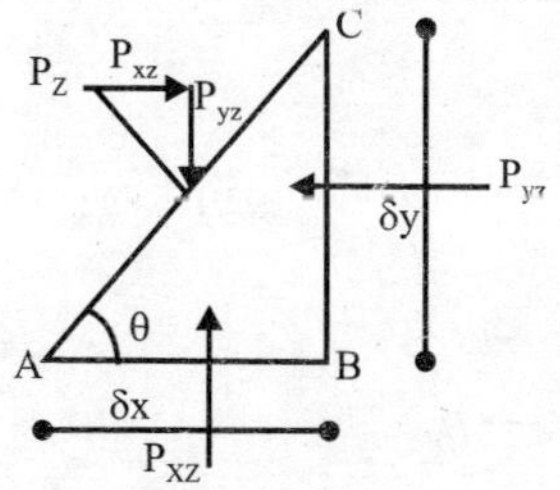

Figure 2.3b

Consider a liquid drop element in shape of a wedge with unit thickness (i.e. $\delta z = 1$), having two triangular faces $ABC, A'B'C'$ and three rectangular faces $AA'CC', AA'BB', BB'CC'$ as shown in the Figures 2.3a and 2.3b.

Let pressure intensities p_{xz}, p_{yz}, p_z be acting along the three rectangular faces.

Let total pressures acting along faces $AA'BB', BB'CC', AA'CC'$ be P_{xz}, P_{yz}, P_z respectively.

Then total pressure $\quad P_{xz} = p_{xz} * (AB * BB')$ since $P_{xz} = p_{xz} * (area)$

$$\Rightarrow P_{xz} = p_{xz} * (\delta x * 1) \tag{2.3a}$$

Similarly, total pressure $P_{yz} = p_{yz} * (BC * BB')$ since $P_{yz} = p_{yz} * (area)$

$$\Rightarrow P_{yz} = p_{yz} * (\delta y * 1) \tag{2.3b}$$

Similarly, total pressure $P_z = p_z * (AC * AA')$ since $P_z = p_z * (area)$

$$\Rightarrow P_z = p_z * (AC * 1) \tag{2.3c}$$

Under equilibrium condition, sum of all forces is equal to zero in horizontal and vertical directions. i.e.

$$\sum F_x = \sum F_y = 0$$

Hence, for sum of forces along horizontal direction is $\sum F_x = 0$

$$\text{i.e. } P_{yz} = P_z * (\sin\theta)$$

$$\Rightarrow p_{yz} * (\delta y * 1) = (p_z * AC * 1) * (\sin\theta)$$

$$\Rightarrow p_{yz} = p_z \tag{2.4}$$

since $\delta y = AC * \sin\theta$

Similarly, for sum of forces along vertical direction is $\sum F_y = 0$

$$\text{I.e. } P_{xz} = P_z * (\cos\theta)$$

$$\Rightarrow p_{xz} * (\delta x * 1) = (p_z * AB * 1) * (\cos\theta)$$

$$\Rightarrow p_{xz} = p_z \tag{2.5}$$

since $\delta x = AB * \cos\theta$.

So. from Eq. (2.4) and Eq. (2.5), we get

$$p_{xz} = p_{yz} = p_z = p \tag{2.6}$$

Hence, **Pascal's law is proved** and is independent of the wedge angle θ or independent of surface angular orientation and is a scalar quantity.

Example 2.1: Find the weight that can be lifted by a hydraulic press when the force applied at the plunger is 350 N and has diameters of 250 mm and 40 mm of ram and plunger respectively.

Solution: Given data:

$$F = 350\text{N}, d_{ram} = 0.25\text{m}, d_{plunger} = 0.04\,\text{m}$$

Area of the ram

$$A_{ram} = \frac{\pi d_{ram}{}^2}{4} = \frac{3.14 * 0.25^2}{4} = 0.049\,\text{m}^2$$

Area of the plunger

$$A_{plunger} = \frac{\pi d_{plunger}{}^2}{4} = \frac{3.14 * 0.04^2}{4} = 0.001256\,\text{m}^2$$

Normal pressure intensity due to the force applied

$$p = \frac{F}{A_{plunger}} = \frac{350}{0.001256} = 278.6\,\text{kPa}$$

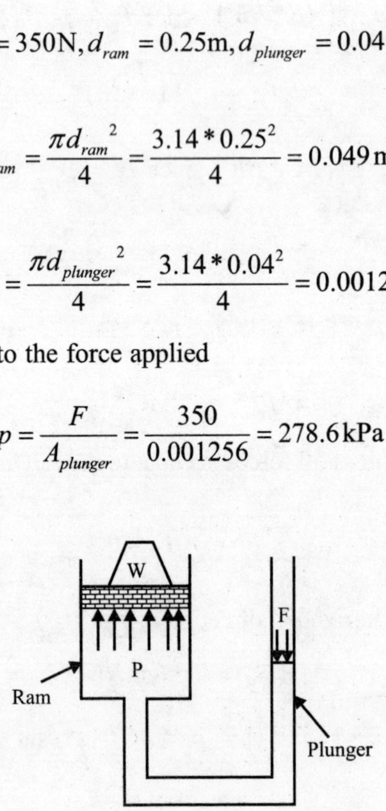

Figure P2.1

As the pressure intensity due to the force in the plunger lifts the weight, so the total pressure acting on the weight in the ram must be equal or greater to the weight, and hence

$$p = \frac{W}{A_{ram}}$$

$$\Rightarrow W = p * A_{ram} = 278.6 * 0.049 = 13.65\text{kN}$$

HYDROSTATICS LAW : VARIATION OF PRESSURE IN A STATIC FLUID

Hydrostatic law defines the variation of pressure with respect to the elevation of fluid particles under static condition and states that "as the distance from a datum surface level increases the pressure drops". When an object pushed in a fluid to its bottom, it compresses the fluid. so there arise an increase in pressure force around it and is termed as "**Hydrostatic force.**"

A projectile soaring higher altitudes from earth surface experiences decrease in pressure and similarly underwater missiles experience increase in hydrostatic force as it penetrates into the depths of sea waters, as illustrated in the below pictures.

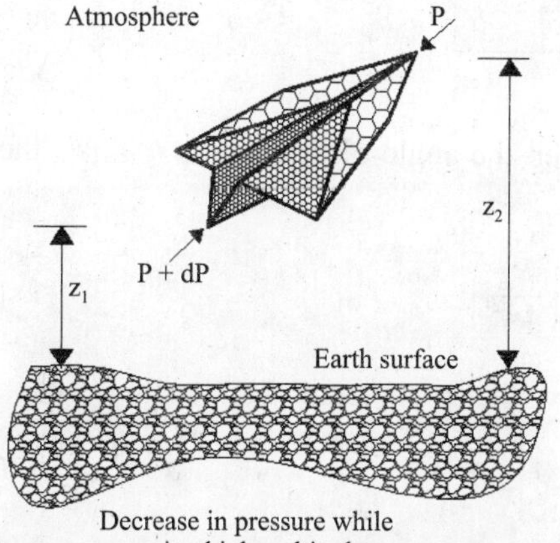

Decrease in pressure while soaring higher altitudes

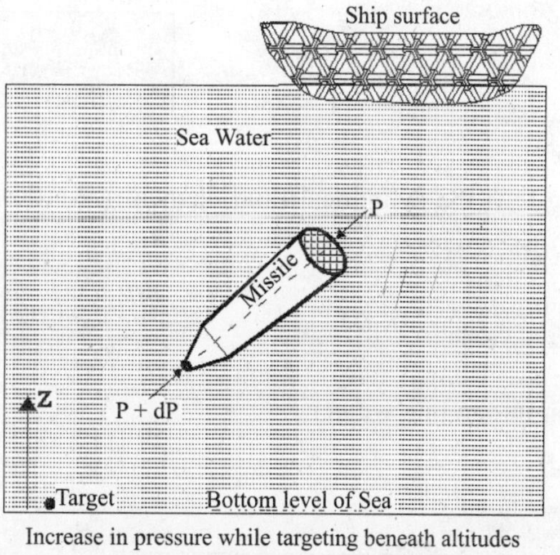

Increase in pressure while targeting beneath altitudes

Similarly, a diver experiences more pressure as he goes down to the depths of the sea levels and similarly a tumbled mug when pushed to bottom of a water filled bucket, experience more upward force and this can be proved by taking a cylindrical fluid element having an area and length as dA, ds respectively, at some height from the datum, as shown in Figure 2.4.

The cylindrical element will be subjected to pressure force at one end with height z_1 is P and at the other end of height z_2 is $P + dP$ and also the weight of the fluid element dw which acts vertically downwards.

Under equilibrium of the element, the resultant of forces in any direction is zero and resolving the forces in the direction along the central axis gives

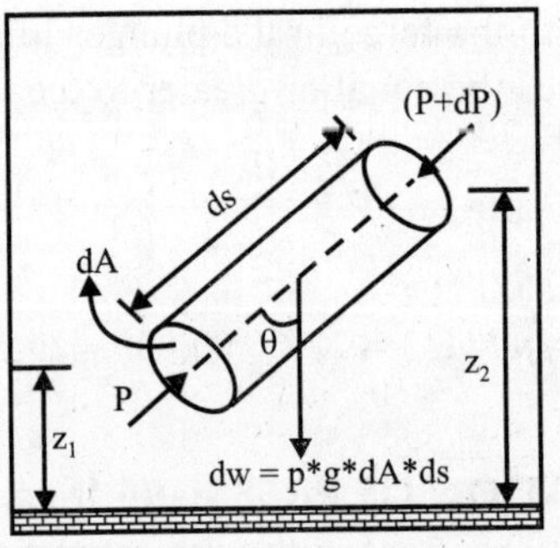

Figure 2.4: Cylindrical fluid element with an arbitrary orientation in a container

$$(P * dA) - \left[(P + dP)dA\right] - (\rho g * dA * ds * \cos\theta) = 0$$

By rearranging

$$(dP) = -(\rho g * ds * \cos\theta)$$

$$\Rightarrow \frac{dP}{ds} = -(\rho g * \cos\theta) \tag{2.7}$$

Note:

Case 1: If the cylindrical fluid element is vertical, then the angle of orientation $\theta = 0^o$, then Eq. (2.7) becomes.

$$\frac{dP}{dz} = -(\rho g) = -\left(\frac{Pg}{RT}\right) \tag{2.7a}$$

so on integrating we get

$$P = P_0 * e^{-gZ/RT} \tag{2.7b}$$

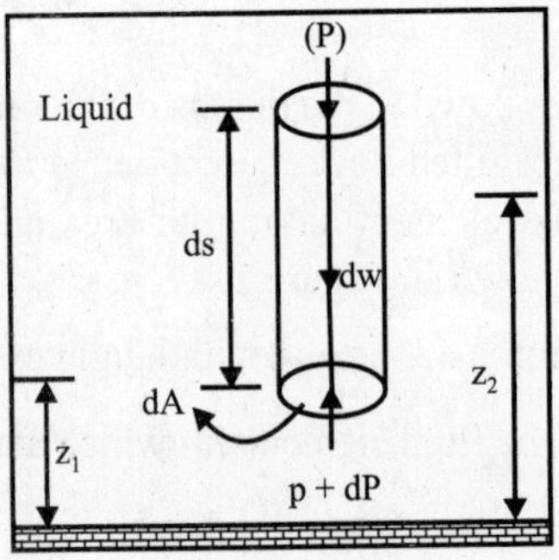

Figure 2.4a: Cylindrical fluid element subjected to variation of pressure through out its length

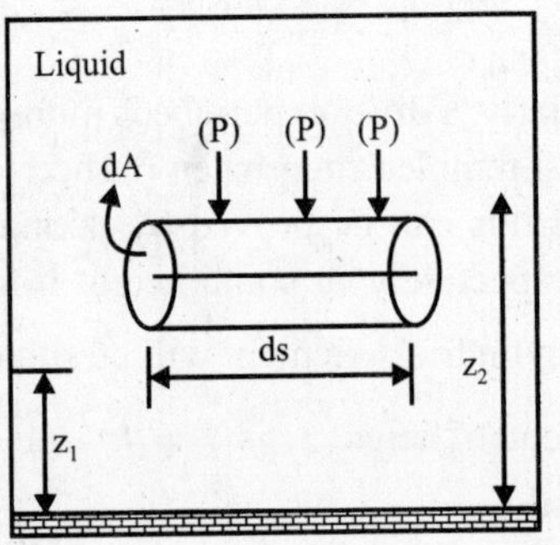

Figure 2.4b: Cylindrical fluid element subjected to uniform pressure through out its length

From Eq. (2.7) it is evident that when a diver dives into sea, the datum elevation will be reducing, as he goes into the depths of the sea and hence the pressure increases at the depths.

Case 2: If the cylindrical fluid element is horizontal, then the angle of orientation θ = 90° then Eq. (2.7) becomes

$$\frac{dP}{dz} = 0 \tag{2.7c}$$

and so pressure is constant (i.e. P = constant) throughout in a static liquid, at a given height.

PRESSURE HEAD (h)

When pressure expressed in terms of the height of fluid raised/fallen is called fluid "Pressure head". A diver experiences more pressure as he goes below the sea level, due to reason of hydrostatic forces which increases the level of fluid by an amount equal to pressure raise and is expressed as "mm of Hg" on "meters of water".

If considered a cylindrical bar of length L and cross-section area A immersed in a water/liquid with an inclination θ, as shown in figure.

Under equilibrium conditions sum of all the forces acting on the cylindrical bar is equal to zero and hence there will be three forces acting like pressure forces $\left(P_1A, \left(P_1 + \Delta P\right)A\right)$ at both the ends and weight force $\gamma lA\cos\theta$ due to the weight of the cylindrical bar.

Then $$\left(P_1 + \Delta P\right)A - P_1A - \gamma lA\cos\theta = 0$$

But $$h = L\cos\theta \text{ so } P_1A - P_2A - \gamma Ah = 0$$

$$\Rightarrow h = \frac{\Delta P}{\gamma} \tag{2.7d}$$

where γ is specific weight of liquid, h is depth or else called **pressure head.**

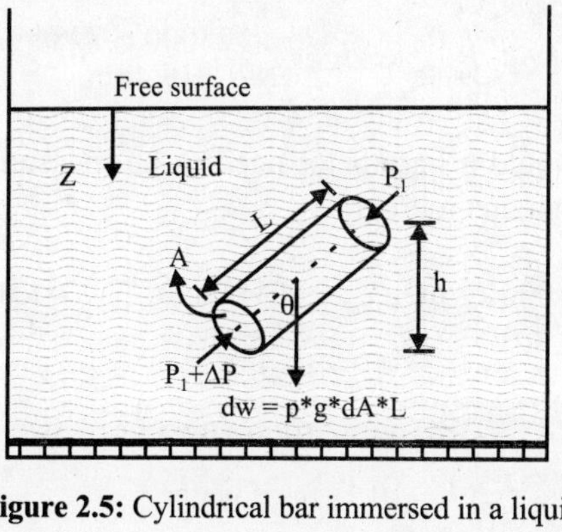

Figure 2.5: Cylindrical bar immersed in a liquid

Example 2.2: Determine increase in pressure (ΔP) in Pascal at the bottom tip of a torpedo, moving into the depths of sea waters ($\rho_{sea\ water}$ = 1.2 kg/m^3) and is at a depth of 6 m below the free surface of a sea water surface.

Solution: Given data: Gage pressure $\rho_{sea\ water}$ = 1.2 kg/m^3, h = 6 m

Increase in pressure $\Delta P = \rho gh$ = 1.2*9.81*6 = 70.63 Pa

Example 2.3: Determine the absolute pressure in pascal at a depth of 6 m below the free surface of a tank of water, when a barometer reads 760 mm mercury (assume relative density 13.6).

Solution: Gage pressure $P_{gauge} = \rho gh = 1000 * 9.81 * 6 = 58.86\,\text{kPa}$

Atmospheric pressure, pressure due to 760 mm of mercury

$$P_{atm} = \rho gh = 13600 * 9.81 * 0.76 = 101.396\,\text{kPa}$$

Absolute pressure = Gage pressure +Atmospheric pressure

$$P_{abs} = P_{gauge} + P_{atm} = 58860 + 101396 = 160256\,\text{Pa}$$

Example 2.4: What will be the gage pressure and absolute pressure of crude oil at a depth of 20 m below the earth's surface? At what depth below the surface of the oil, will produce a pressure of 125 kPa. Assume the density of crude oil is 775 kg/m^3 and atmospheric pressure 101 kPa.

Solution: Given data: $\rho_{oil} = 775\,\text{kg/m}^3,\ P_{atm} = 101\,\text{kPa},\ h = 20\,\text{m}$

a) Gage pressure $P_{gauge} = \rho gh = 775 * 9.81 * 20 = 152.055\,\text{kPa}$

Absolute pressure $P_{abs} = P_{gauge} + P_{atm} = 152.055 + 101 = 253\ \text{KPa}$

b) Given data: $\rho_{oil} = 775 kg/m^3, P_{atm} = 101\,\text{kPa}, P_{gauge} = 125\,\text{kPa}$

Depth below the surface of oil $P_{gauge} = \rho gh \Rightarrow 125000 = 775 * 9.81 * h \Rightarrow h = 16.44\text{m}$

Example 2.5: What heights would a barometer need to be measured atmospheric pressure, if it contains water and also if it contains mercury?

Solution: Given data: $\rho_{water} = 1000\,\text{kg/m}^3, P_{atm} = 101\,\text{kPa}, \rho_{Mercury} = 13600\,\text{kg/m}^3$

a) Height of Water Barometer $P_{atm} = \rho gh$

$$\Rightarrow 101000 = 1000 * 9.81 * h$$

$$\Rightarrow h = 10.19\,\text{m of water}$$

b) Height of Mercury Barometer $P_{atm} = \rho gh$

$$\Rightarrow 101000 = 13600 * 9.81 * h$$
$$\Rightarrow h = 0.75\,\text{m of mercury}$$

Example 2.6: An oil of specific gravity 0.8 is under pressure of 137.2 kPa, then determine pressure head expressed in terms of meters of oil.

Solution: Given data: $\rho_{water} = 1000\,\text{kg/m}^3, P_{gauge} = 137.2\,\text{kPa}, S = 0.8$

Pressure head $P_{gauge} = \rho_{oil} gh$

$$\Rightarrow P_{gauge} = S * \rho_{water} gh$$

$$\Rightarrow 137.2 * 1000 = 0.8 * 1000 * 9.81 * h$$
$$\Rightarrow h = 17.48\,\text{m of oil}$$

So pressure head in terms of oil is 17.48 m.

Example 2.7: Find the weight that can be lifted by a hydraulic press when the force applied at the plunger is 350 N and has diameters of 250 mm and 40 mm of ram and plunger respectively, take the specific weight of the liquid in the press as 1000 N/m³.

Solution: Given data:

$$F = 350\,\text{N}, \text{d}_{ram} = 0.25\,\text{m}, d_{plunger} = 0.04\,\text{m}$$
$$\gamma = 1000\,\text{N/m}^3, h = 0.3$$

Area of the ram

$$A_{ram} = \frac{\pi d_{ram}^{\ 2}}{4} = \frac{3.14 * 0.25^2}{4} = 0.049\,\text{m}^2$$

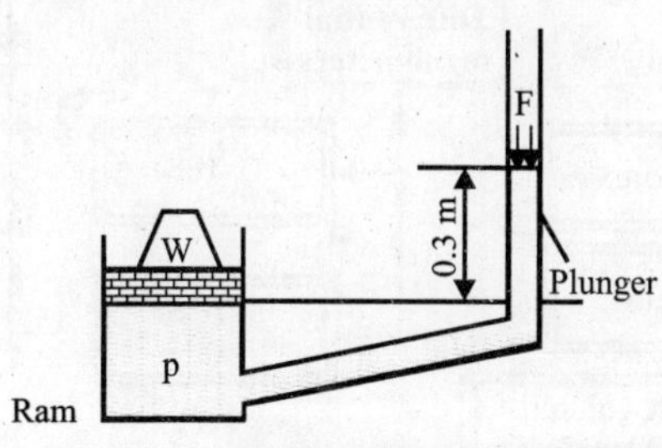

Figure P2.7

Area of the plunger

$$A_{plunger} = \frac{\pi d_{plunger}^{\ 2}}{4} = \frac{3.14 * 0.04^2}{4} = 0.001256\,\text{m}^2$$

Normal pressure intensity due to the force applied in the plunger

$$p = \frac{F}{A_{plunger}} = \frac{350}{0.001256} = 278.6\,\text{kPa}$$

As the level of force and weight are not of the same and so there exists a pressure intensity due to head of 0.3 m height in the plunger then the total pressure intensity on the weight will be

$$p = \frac{F}{A_{plunger}} + \gamma h = \frac{350}{0.001256} + 1000 * 0.3 = 278.96\,\text{kPa}$$

As the total pressure intensity due to the force in the plunger lifts the weight, so the total pressure acting on the weight in the ram must be equal or greater to the weight, and hence

$$p = \frac{W}{A_{ram}}$$
$$\Rightarrow W = p * A_{ram} = 278.96 * 0.049 = 13.669\,\text{kN}$$

Types of Pressure Measuring Devices

a) Manometers
b) Mechanical gages.

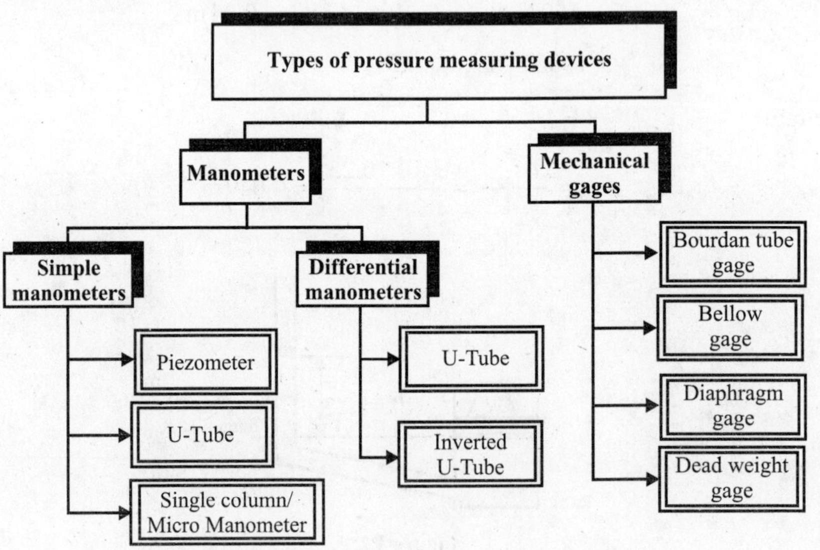

Figure 2.6: Different types of pressure mesuring devices

Manometers

Manometer is a device used to measure pressure at a single or multiple points in a single or multiple pipelines, by balancing the fluid column by the same or another column of fluid. Manometers can be categorized into two types, namely simple manometer and differential manometer. Simple manometric devices measure pressure at a single point in a fluid, whereas differential manometric devices measure pressure at two or more number of points, in a single or multiple flow lines.

The simplest type of manometer is simple manometers, wherein we have three types of simple manometers like **Piezometer**, **U-tube manometer** and **Single column/Micro manometer**. Simple manometers consist of a glass tube straight or bent, whose one end is connected to the point at which the measurement of pressure is required and the other end is left open to the atmosphere.

A. Piezometer

Piezometer is a simple manometric device which measures pressure at a point in a fluid, without balancing any other fluid column.

Consider a cylindrical vessel having a fluid filled in it and is open to atmosphere. Connect a piezometer glass tube as shown in Figure 2.7, at the point where pressure is to be measured, such that the other end of the piezometer tube is open to the atmosphere. As soon as the piezometer glass tube is connected to the cylindrical tank, some amount of the fluid rushes into the tube, due to atmospheric pressure outside the cylindrical tank. But as the other end of the piezometer glass tube is exposed to atmosphere, atmospheric pressure acts at that open end also and hence fluid raises up to some level in the glass tube and does not overflow from the tube.

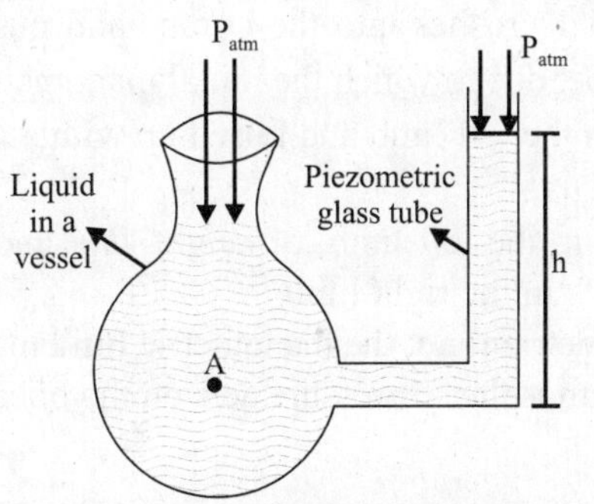

Figure 2.7: Piezometer

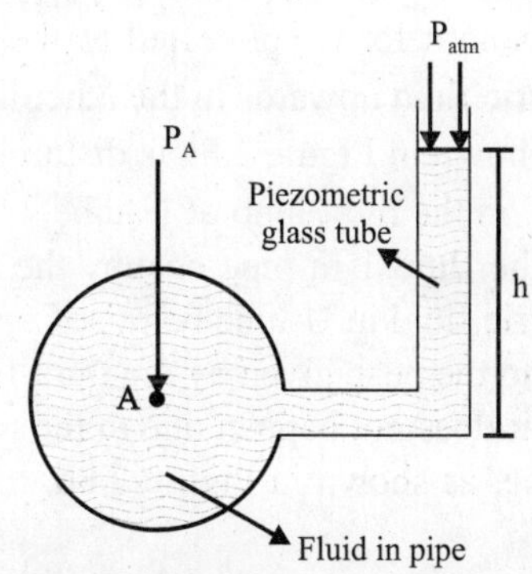

Figure 2.7a: Piezometer connected to pipeline

If pressures considered in the cylindrical vessel $(P_{atm} + P_A)$ and piezometer tube $(P_{atm} + \gamma h)$ then they will be equal, due to which the overflow of fluid does not take place,

i.e. $$P_{atm} + P_A = P_{atm} + \gamma h$$

$$\Rightarrow P_A = \gamma h \tag{2.8}$$

where P_A is the pressure at the point "A" where pressure is required to measure in an open container and (γh) is the specific weight of the fluid raised in the piezometer column.

If the container is not an open one, for instance like a pipeline and if a piezometer is connected to it, as shown in the Figure 2.7a, to find the pressure inside the pipeline, the hydrostatic pressure equation varies. i.e.

$$P_A = P_{atm} + \gamma h \tag{2.8a}$$

And if needed in terms of pressure heads, dividing the Eq. (2.8a) with the specific weight of water (γ_{water}), we get

$$\Rightarrow h_A = h_{atm} + Sh = 10.3 + Sh \tag{2.8b}$$

Hence, Eqs. (2.8), (2.8a) and Eq. (2.8b) are the equations of **hydrostatic pressure** and **hydrostatic pressure head**.

B. U-tube Simple Manometer

U-tube manometer is a simple manometric device used to measure pressure at a point in a fluid, by balancing the fluid column by the same or another column of fluid. It has a glass tube bent in "U" shape with some amount of same or other type of fluid, called manometric fluid like mercury, as shown in Figures 2.8a and 2.8b.

Consider a pipe carrying a fluid called "fluid 1" with specific weight γ_1 and let pressure P_A to be measured at a point "A" in the pipe. Let the manometric fluid have specific weight γ_2 - As the U-tube manometer is connected to the pipe, the "fluid 1" rushes into the U-tube and pushes the manometric fluid upwards in the other limb of the U-tube, due to which the initial manometric fluid level as shown in Figure 2.8a is disturbed and lowers in the left limb and raised up with a pressure head "h_2" in the right limb of U-tube.

Let the fluid 1 in pipe occupy the U-tube column in the left limb, of height "h_1" and let the manometric fluid in U-tube be raised up to a height "h_2", in the right limb.

Due to the push given by the fluid 1 onto the manometric fluid, the manometric fluid may overflow but it does not happen due to the reason that the atmospheric pressure acts on the open end of the U-tube, as shown in Figure 2.8b.

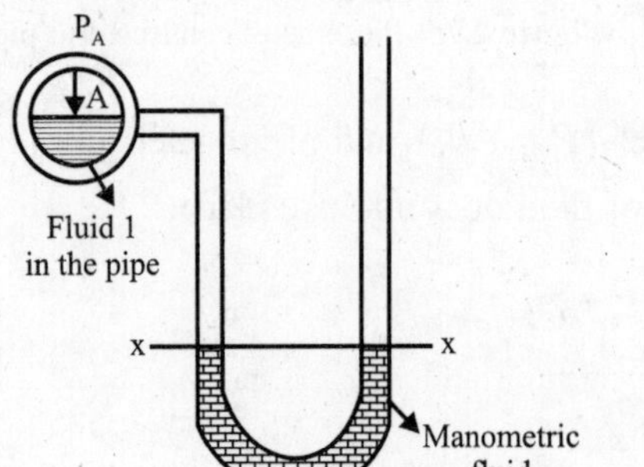

Figure 2.8a: Just before connecting the U-Tube

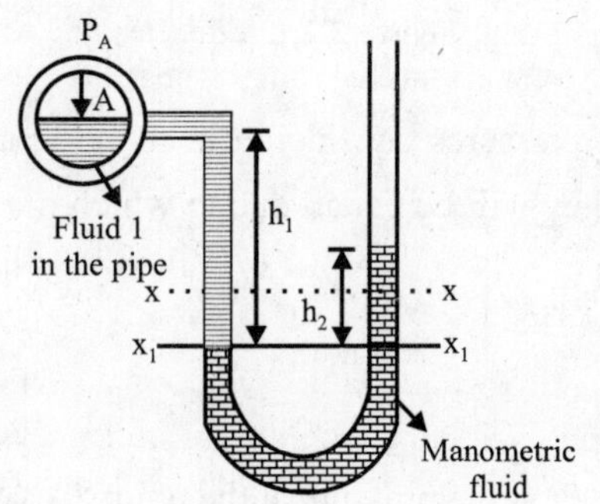

Figure 2.8b: After connecting the U-Tube

Under equilibrium the pressures on the left limb of U-tube must be equal to the pressures in the right limb, so

$$P_A + \gamma_1 h_1 = \gamma_2 h_2 + P_{atm} \tag{2.9}$$

$$\Rightarrow P_A + \gamma_w S_1 h_1 = \gamma_w S_2 h_2 + P_{atm} \tag{2.9a}$$

$$\Rightarrow \frac{P_A}{\gamma_w} = h_2 S_2 - h_1 S_1 \quad \text{meters of water} \tag{2.9b}$$

$$\therefore h = S_2 h_2 - h_1 S_1 \tag{2.10}$$

or

If hydrostatic pressure is considered, then

$$h = S_2 h_2 - h_1 S_1 + 10.3 \tag{2.10a}$$

Where h is the pressure head at the point "A" where pressure is required.

Example 2.8: Pipeline carrying water has a connection of a mercury manometer connected to it and the readings are measured as shown in Figure P2.8. Find the pressure of the water flowing in the pipeline.

Solution: Given data:

$$h_1 = 0.1\,\text{m},\ h_2 = 0.8\,\text{m},\ S_{water} = 1,\ S_{Mercury} = 13.6$$

Equating the pressure heads on both sides of the limbs of the manometer, we have

$$h + h_1 S_{water} = h_2 S_{Mercury}$$

$$\Rightarrow h = 0.8 * 13.6 - 0.1 * 1 = 10.78\,\text{m}$$

So pressure in the pipeline will be $P_A = \rho_{water} gh = 1000 * 9.81 * 10.78 = 105.75\,\text{kPa}$

If hydrostatic pressure is considered, then we have $P_A = \rho_{water} g(h + 10.3) = 206.75$ kPa

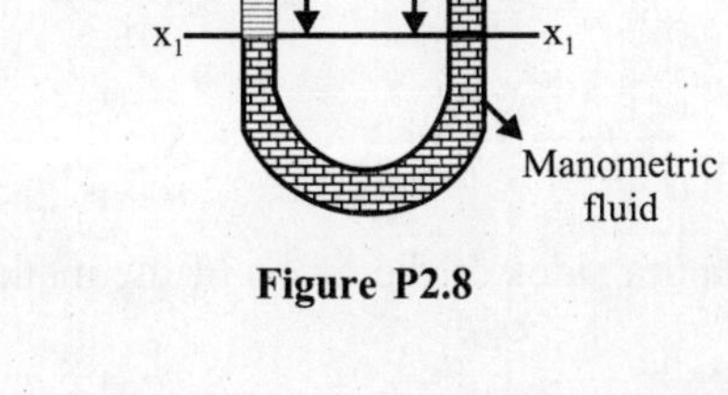

Figure P2.8

Example 2.9: Pipeline carrying an oil of specific gravity 0.8 has pressure head of 3 m and has a connection of a mercury manometer connected to it and the readings are measured as shown in the Figure P2.9. Find the pressure of the oil flowing in the pipeline and also the pressure head in the left limb.

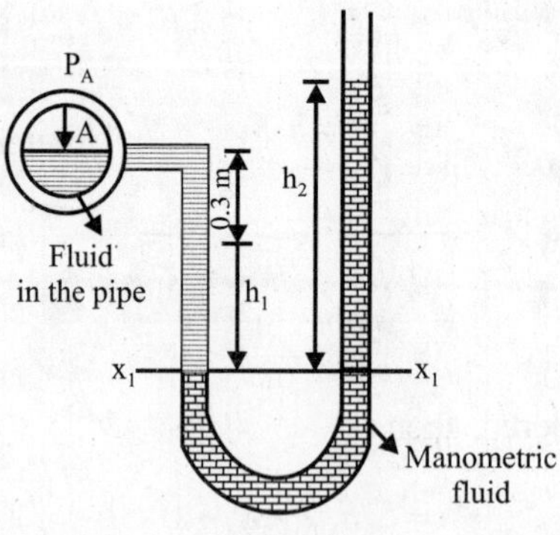

Figure P2.9

Solution: Given data:

$$h = 3\,\text{m},\ h_2 = 0.5\,\text{m},\ S_{oil} = 0.8,\ S_{Mercury} = 13.6$$

Pressure of oil in the pipeline will be

$$P_A = \rho_{oil} g h$$
$$P_A = \rho_{water} S_{oil} g h = 2400 \text{ Pa}$$

Hence, the pressure of oil in the pipeline is 2400 Pa.
Equating the pressure heads on both sides of the limbs of the manometer, we have

$$h + (0.3 + h_{oil}) S_{oil} = h_2 S_{Mercury}$$
$$\Rightarrow 3 + (0.3 + h_1) * 0.8 = 5 * 13.6$$
$$\Rightarrow h_1 = 80.95 \text{ m}$$

Hence, the pressure head in the left limb will be 0.3 m + 80.25 m = 81.25 m.

Example 2.10: Pipeline carrying water has a connection of a U-tube mercury manometer connected to it and the other end is exposed to atmosphere. The readings are measured to be vacuum, as shown in the Figure P2.10. Find the pressure in the water flowing in the pipeline.

Solution: Given data:

$$h_1 = 0.3\,\text{m},\ h_2 = 0.5\,\text{m},\ S_{water} = 1,\ S_{Mercury} = 13.6$$

Equating the pressure heads on both sides of the limbs of the manometer, we have

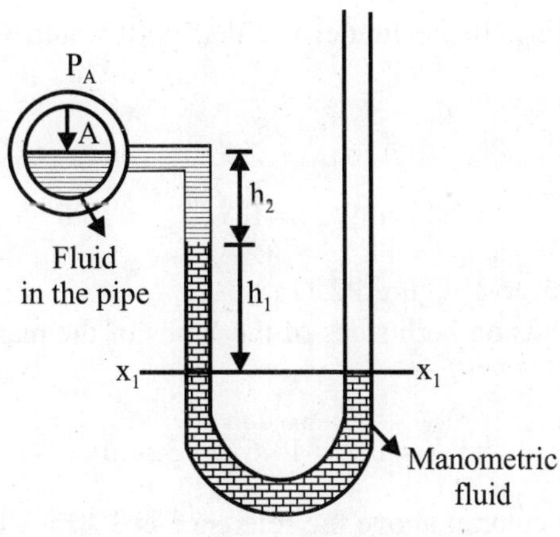

Figure P2.10

$$h + h_1 S_{water} + h_2 S_{Mercury} = 0$$

$$\Rightarrow h = -0.3 * 1 - 0.5 * 13.6 = -7.1\text{m}$$

So pressure in the pipeline will be

$$P_A = \rho_{water} gh = 1000 * 9.81 * 7.1 = 69.651\text{kPa (vaccum)}$$

If hydrostatic pressure considered, then we have

$$P_A = \rho_{water} g(h + 10.3) = 1000 * 9.81 * 3.2 = 31.4 \text{ kPa}$$

Example 2.11: A funnel connected to a U-tube mercury manometer having water in it, as shown in Figure P2.11a. The other end of the manometer is exposed to atmosphere. The level of mercury is

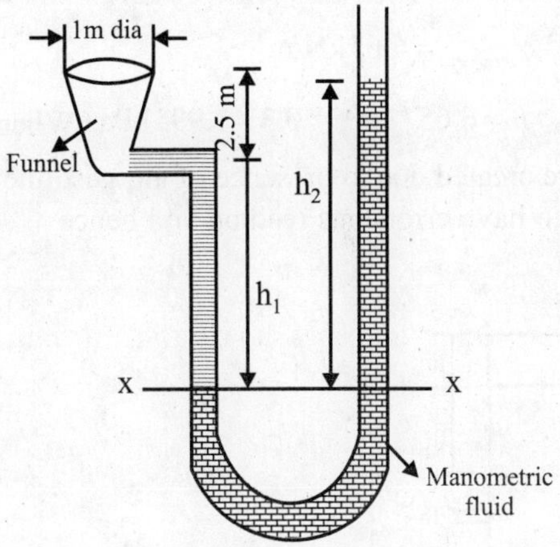

Figure P2.11a

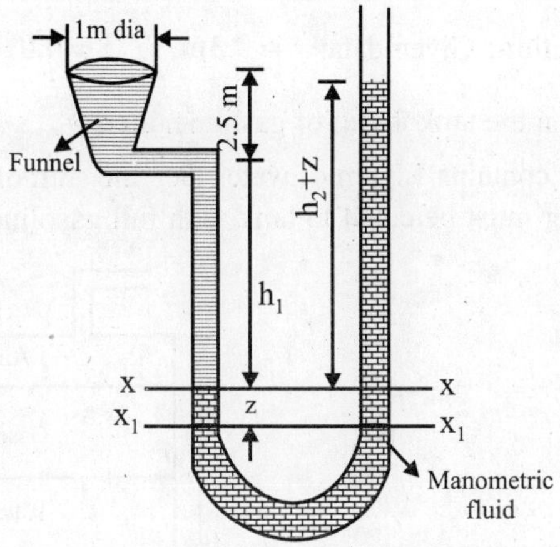

Figure P2.11b

0.25 m from the reference line. If the funnel is filled with water, find the level of manometer readings.

Solution: Given data:

$$h_2 = 0.25\,\text{m}, S_{water} = 1, S_{Mercury} = 13.6$$

When the funnel is empty: Observe Figure P2.11a.

Equating the pressure heads on both sides of the limbs of the manometer, we have

$$h_1 S_{water} = h_2 S_{Mercury}$$
$$\Rightarrow h_1 = 0.25 * 13.6/1 = 3.4\ \text{m}$$

Hence, the level of the water column above the reference is 3.4 m, when the funnel is empty.

When the funnel is filled completely: Observe Figure P2.11b. The reference line moves to new position to X_1-X_1 and so by equating the pressure heads on both sides of the limbs of the manometer, we have

$$h_1 S_{water} + z + 2500 = (h_2 + 2z) S_{Mercury}$$
$$\Rightarrow 3.4 * 1 + z + 2500 = 0.25 * 13.6 + 2 * z * 13.6$$
$$\Rightarrow z = 95.4\,\text{mm}$$

Hence, the reading of manometer is 95.4 mm.

Example 2.12: Figure below shows a gasoline tank in a car, which reads proportional to the bottom gage. The tank is 30 cm deep and accidentally contains 1.8 cm of water in addition to the gasoline. Determine the height of air remaining at the top when the gauge erroneously reads full. Assume

$$\gamma_{water} = 0.0118\,\text{kN/m}^3, \gamma_{mercury} = 6.65\,\text{kN/m}^3$$

Solution: Given data: $h = 0.3\,\text{m}, \gamma_{water} = 0.0118\,\text{kN/m}^3, \gamma_{mercury} = 6.65\,\text{kN/m}^3$

When the tank is full of gasoline, then $P_{gauge} = \gamma_{gasoline} * h = 6.65 * 1000 * 0.3 = 1.995\,\text{kPa}$. When the tank contains 1.8 cm of water then the sum of pressure created due to presence of air, gasoline and water must be equal to tank with full gasoline, so as to have erroneous reading and hence

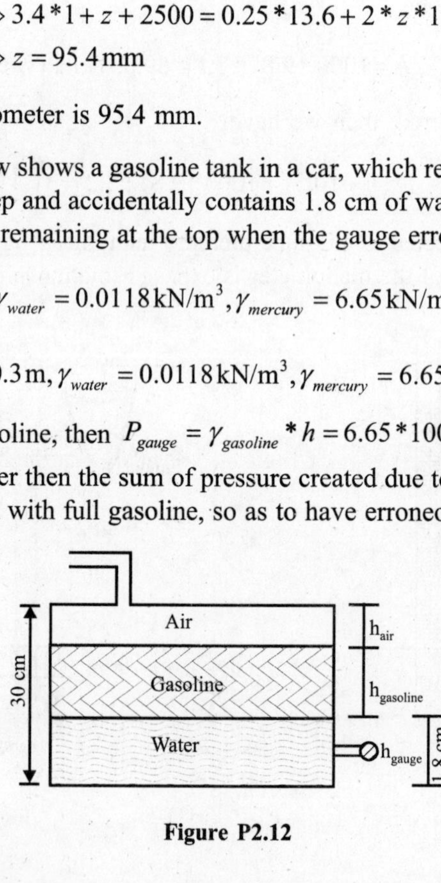

Figure P2.12

$$P_{Air} + P_{gasoline} + P_{water} = P_{Gage}$$

$$\Rightarrow \gamma_{Air} * h_{Air} + \gamma_{gasoline} * h_{gasoline} + \gamma_{water} * h_{water} = 1.995\,\text{kPa}$$

$$\Rightarrow 0.0118 * h_{Air} + 6.65 * (0.3 - 0.018 - h_{air}) + 9.81 * 0.018 = 1.995\,\text{kPa}$$

$$\Rightarrow h_{Air} = 0.857\,\text{cm}$$

Hence, the height of air remaining at the top is 0.857 cm.

c) Single Column/Micro-manometer

Single column/micro-manometers are just similar to the U-tube manometer, only with a difference of having a small reservoir in the tube, as shown in Figures 2.9a and 2.9b.

Micro-manometer has a glass tube bent in "U" shape or inclined and has a small reservoir in the tube, having some amount of same or other type of fluid called manometric fluid. Consider a pipe

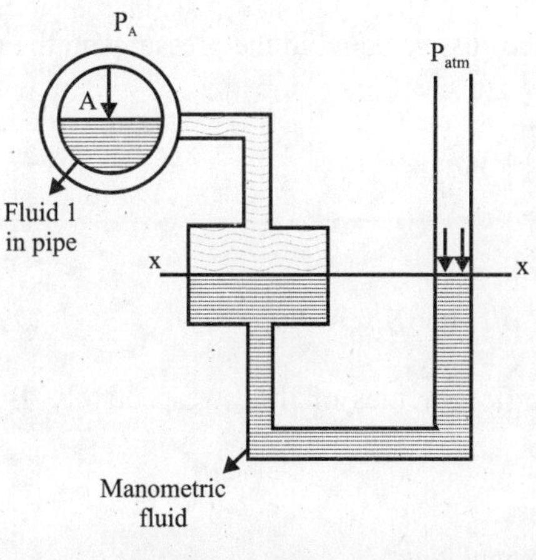

Figure 2.9a: just before connecting micro-manometer

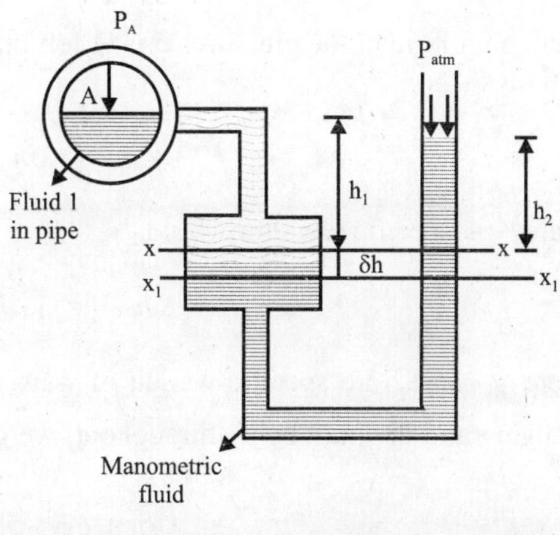

Figure 2.9b: After connecting micro-manometer

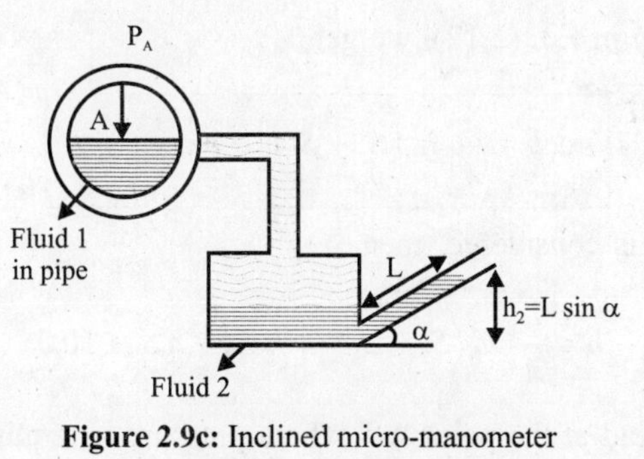

Figure 2.9c: Inclined micro-manometer

carrying a fluid called "fluid 1" with specific weight γ_1 and let pressure P_A be measured at a point "*A*" in the pipe.

Let the micro-manometer containing a manometric fluid having specific weight γ be filled in the pipe, as shown in Figure 2.9a. And as the micro-manometer is connected to the pipe, the "fluid 1" rushes into the micro-manometer and pushes the manometric fluid upwards into the other limb of the U-tube, due to which the initial manometric fluid level as shown in Figure 2.9a, is disturbed and lowers in the left limb and raised up, in the right limb of micro-manometer.

Let the cross-section area of the manometer tube and small reservoir be "*a*" and "*A*" respectively.

Let the fluid in pipe occupy manometer tube column in the left limb, of height "h_1".

Let the fluid 1 displace some volume of manometric fluid in the reservoir by a height of δh.

Let the manometric fluid in tube be raised up to a height "h_2", in the right limb. Due to the push given by the fluid 1 onto the manometric fluid, the volume of manometric fluid in the reservoir displaced by the fluid 1 must be equal to the volume occupied by the manometric fluid in the right limb, i.e. $A * \delta h = a * h_2$ $(\text{since } volume = area * length)$

$$\Rightarrow \delta h = \left(\frac{a}{A} * h_2\right) \tag{2.11}$$

Under equilibrium, the pressures on the left limb of tube must be equal to the pressures in the right limb, so

$$P_A + \gamma_1 h_1 + (\gamma_1 \delta h) = (\gamma_2 \delta h) + \gamma_2 h_2 + P_{atm}$$

where γ is specific weight of fluid.

$$\Rightarrow P_A + \gamma_w S_1 h_1 + (\gamma_w S_1 \delta h) = (\gamma_w S_2 \delta h) + \gamma_w S_2 h_2 + P_{atm} \tag{2.11a}$$

where γ_w, S_1, S_2 are specific weight of water and specific gravities of fluids respectively. By re-arranging and dividing by γ_w throughout, we get

$$\frac{P_A}{\gamma_w} = \left[\delta h(S_2 - S_1)\right] + h_2 S_2 - h_1 S_1 \tag{2.12}$$

Substituting Eq. (2.11) in Eq. (2.12), we get

$$\therefore h = \frac{a}{A}\left[h_2 (S_2 - S_1)\right] + h_2 S_2 - h_1 S_1 \tag{2.13}$$

Or if hydrostatic pressure is considered, then

$$h = \frac{a}{A}\left[h_2 (S_2 - S_1)\right] + h_2 S_2 - h_1 S_1 + 10.3 \tag{2.13a}$$

Where *h* is the pressure head at the point "*A*" where pressure is required to be measured.

If a<<<<A, then $\frac{a}{A} \approx 0$ which give

$$h = h_2 S_2 - h_1 S_1 \tag{2.13b}$$

If the tube is **inclined** to the reservoir at angle α as shown in Figure 29c, then $h_2 = l \sin\alpha$ and so

$$h = \frac{a}{A}\left[l \sin\alpha\left(S_2 - S_1\right)\right] + l \sin\alpha S_2 - h_1 S_1 \tag{2.14}$$

And if $\frac{a}{A} \approx 0$ then

$$h = l \sin\alpha S_2 - h_1 S_1 \tag{2.14a}$$

Example 2.13: A micro-manometer having a ratio of reservoir to limb areas as 40 was used to determine the pressure in a pipe containing water. Determine the pressure in the pipe for manometer readings, as shown in Figure P2.13.

Solution:
Given data:

$$S_{water} = 1, S_{mercury} = 13.6, P_{atm} = 101 \text{kPa}, \frac{A}{a} = 40$$
$$h_1 = 50 \text{ mm}, h_2 = 80 \text{ mm}$$

Equating the pressure heads on both sides of the manometer limbs, we have

$$h = \frac{a}{A}\left[h_2\left(S_{mercury} - S_{water}\right)\right] + h_2 S_{mercury} - h_1 S_{water}$$

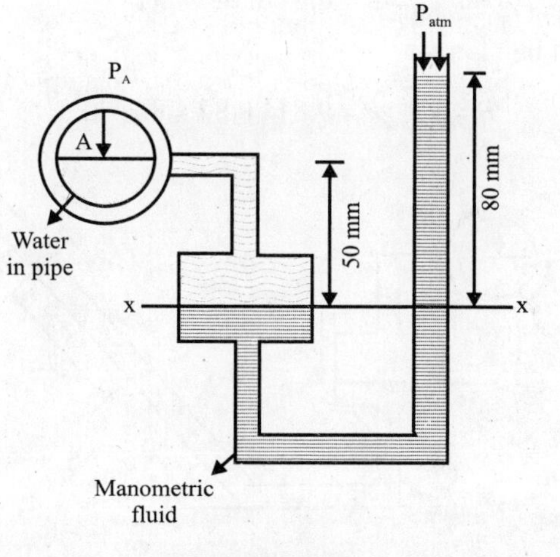

Figure P2.13

$$\Rightarrow h = \frac{1}{40}\left[0.08(13.6-1)\right] + 0.08*13.6 - 0.05*1$$

$$\Rightarrow h = 1.06\,\text{m of water}$$

So pressure in the pipe will be

$$P = \gamma_{water} h = 9.81*1.06 = 10.4\,\text{kPa}$$

Example 2.14: An inclined micro-manometer having a ratio of reservoir to limb areas as 10 was used to determine the pressure in a pipe containing water. Determine the pressure in the pipe for manometer readings, if the inclination of the manometer limb has a slope of 4:1, as shown in Figure P2.14.

Solution:
Given data:

$$S_{water} = 1, S_{mercury} = 13.6, \frac{A}{a} = 10, l = 0.25\,\text{m}$$

$$h_1 = 50\,\text{mm}$$

$$h_2 = l\sin\alpha = 0.25 * \sin 14° = 0.06\,\text{m}$$

Equating the pressure heads on both sides of the manometer limbs, we have

$$h = \left(\frac{a}{A}\right)\left[l\sin\alpha(S_2 - S_1)\right] + l\sin\alpha S_2 - h_1 S_1$$

$$\Rightarrow h = \frac{1}{10}\left[0.06(13.6-1)\right] + 0.06*13.6 - 0.05*1$$

$$\Rightarrow h = 0.84\,\text{m of water}$$

So pressure in the pipe will be

$$P = \gamma_{water} h = 9.81*0.84 = 8.25\,\text{kPa}$$

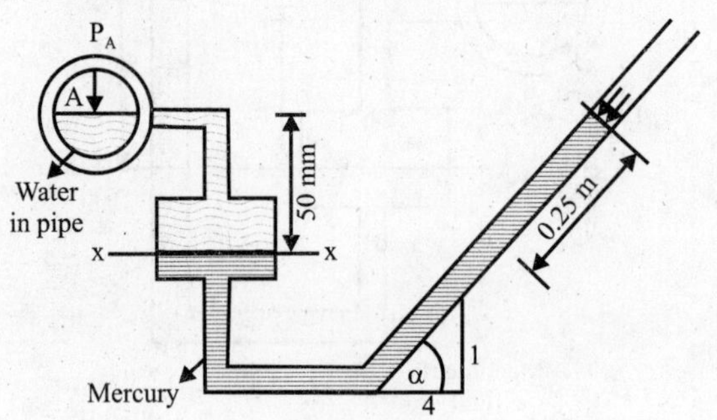

Figure P2.14:

Differential Manometers

Differential manometer is a device used to measure pressure difference at two different points in a single or different pipe, carrying two fluids of same or different kinds, by balancing the manometric fluid column.

The differential manometers are of two types like **Differential U-tube manometer** and **Inverted Differential U-tube manometer**. Differential manometer consists of a glass tube bent in "U" shape, whose one end is connected to a point "A" (at which the measurement of pressure required in first pipe) and the other end is connected to point "B" (at which the measurement of pressure required, in first or second pipe). Hence, differential manometer gives the pressure difference between two points "A" and "B".

a) Differential U-tube Manometer: Differential U-tube manometer is a device used to measure pressure difference ΔP at two different points "A" and "B" in a single pipe or different pipes, carrying two fluids of same or different kinds, as shown in Figures 2.10a and 2.10b.

Consider a differential U-tube having a manometric fluid in it and connected to two different pipes wherein the pressure difference is to be measured. (i.e. at points "A" and "B").

As the differential U-tube is connected to the two pipes, the fluids "fluid 1" and "fluid 2", rushes into the differential U-tube, touching the manometric fluid at the either ends, as shown in Figure 2.10a. If let the pressure at point "A" in the pipe 1 be more than the pressure at point "B" in pipe 2 (i.e. $P_A > P_B$ then the fluid pressure at point "A" pushes the manometric fluid downwards due to which the fluid 1 column level increases and in turn the manometric fluid pushes the fluid 2 column level upwards, as shown in Figure 2.10b.

Let the fluid 1 column level, manometric column level, fluid 2 column level be h_1, h_2, h_3 after displacement, respectively.

Under equilibrium, the pressures in the left limb must be equal to the pressures in the right limb and hence

$$P_A + \gamma_1 h_1 = \gamma_2 h_2 + \gamma_3 h_3 + P_B$$

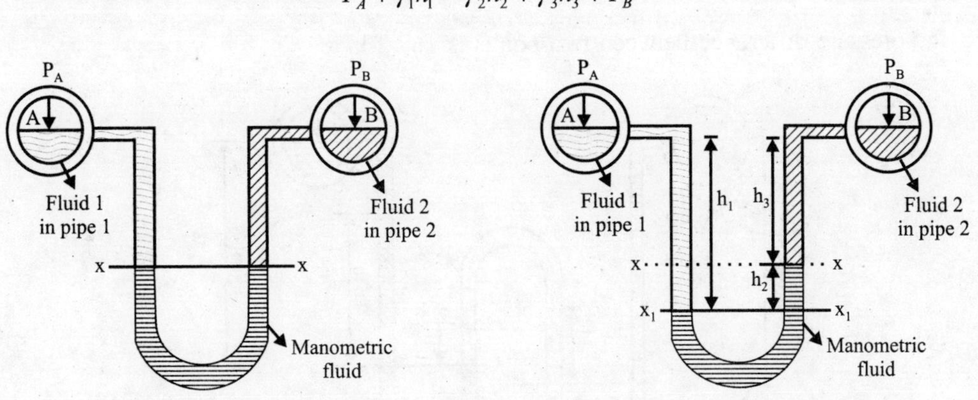

Figure 2.10a **Figure 2.10b**

Figure 2.10: Differential U-tube

where γ is the specific weight of fluids.

$$\Rightarrow P_A + \gamma_w S_1 h_1 = \gamma_w S_2 h_2 + \gamma_w S_3 h_3 + P_B \tag{2.15}$$

$\left(\text{since } S = \gamma/\gamma_w\right)$ where γ_w, S_1, S_2 are specific weight of water and specific gravities of fluids respectively. By rearranging and dividing by γ_w throughout, we get

$$\frac{(P_A - P_B)}{\gamma_w} = h_3 S_3 + h_2 S_2 - h_1 S_1 \tag{2.16}$$

meters of water. When written in terms of pressure heads, we get

$$(h_A - h_B) = h_2 S_2 + h_3 S_3 - h_1 S_1 \tag{2.17}$$

If the "fluid 1" and "fluid 2" are of same kind, then $S_1 = S_3$.

$$(h_A - h_B) = h_2 (S_2 - S_1) \tag{2.17a}$$

Example 2.15: Find the pressure difference at two points A and B in the pipe carrying water and connected to a mercury differential manometer and also has oil filled in between it, as shown in figure P2.15. Take $\gamma_{water} = 10\,\text{kN/m}^3, \gamma_{mercury} = 136\,\text{kN/m}^3, \gamma_{oil} = 8.5\,\text{kN/m}^3$

Solution: Given data: $\gamma_{water} = 10\,\text{kN/m}^3, \gamma_{mercury} = 136\,\text{kN/m}^3, \gamma_{oil} = 8.5\,\text{kN/m}^3$

Equating using differential manometer equations, we have

$$P_A + \gamma_{water} * 1.08 - \gamma_{mercury} * 0.72 + \gamma_{oil} * 0.48 - \gamma_{mercury} * 0.6 - \gamma_{water} * 0.36 = P_B$$

$$\Rightarrow P_A - P_B = 10 * 1.08 - 136 * 0.72 + 8.5 * 0.48 - 136 * 0.6 - 10 * 0.36 = 168.2\,\text{kPa}$$

Hence, the pressure difference between two points is 168.2 kPa.

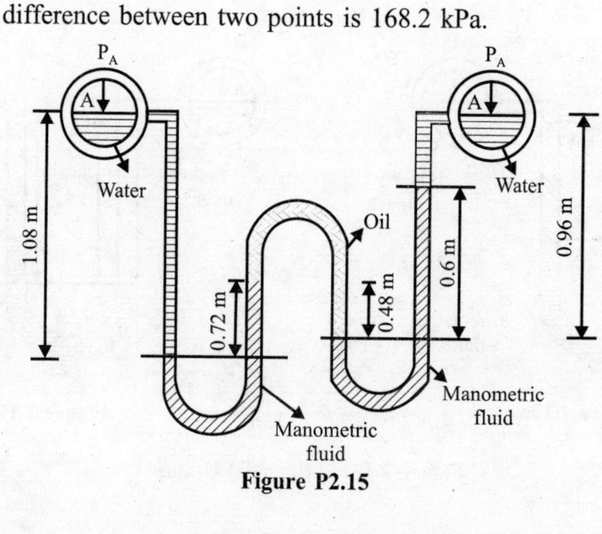

Figure P2.15

Example 2.16: Two pipes A and B contain carbon tetrachloride of specific gravity 1.59 at 103 kPa pressure and oil of specific gravity 0.8 at 171.6 kPa respectively. The manometric liquid is mercury. Find the difference "h" between the levels of mercury.

Take $S_{ccl2} = 1.594, S_{mercury} = 13.6, S_{oil} = 0.8$

Solution: Given data:

$$S_{ccl2} = 1.594, S_{mercury} = 13.6, S_{oil} = 0.8$$

$$P_A = 103\,\text{kPa}, P_B = 171.6\,\text{kPa}$$

Equating the pressure heads on both sides of the manometer limbs, we have

$$h_A + 4 * S_{ccl2} + h * S_{mercury} = h * S_{oil} + S_{oil} * 1.5 + h_B$$

$$\frac{P_A}{\gamma_{water}} + 4 * S_{ccl2} + h * S_{mercury} = h * S_{oil} + S_{oil} * 1.5 + \frac{P_B}{\gamma_{water}}$$

$$\frac{103}{9810} + 4 * 1.59 + h * 13.6 = h * 0.8 + 0.8 * 1.5 + \frac{171.6}{9810}$$

$$\frac{103}{9.81} + 4 * 1.59 + h * 13.6 = h * 0.8 + 0.8 * 1.5 + \frac{171.6}{9.81}$$

$$\Rightarrow h = 402 \text{ mm}$$

Hence, the difference in mercury levels is 402 mm.

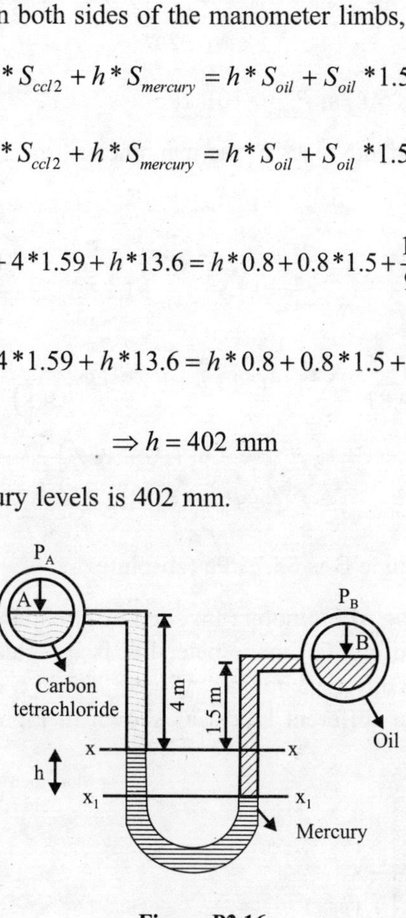

Figure P2.16

Example 2.17: Two pipes A and B contain water and air above in pipe A with 78.5 kPa pressure of air and oil of specific gravity 0.8 in pipe B respectively. The manometeric liquid is mercury. Find the pressure in pipe B (absolute). Take $P_{atm} = 101\,\text{kPa}$

Solution: Given data: $S_{water} = 1$, $S_{mercury} = 13.6$, $S_{oil} = 0.8$

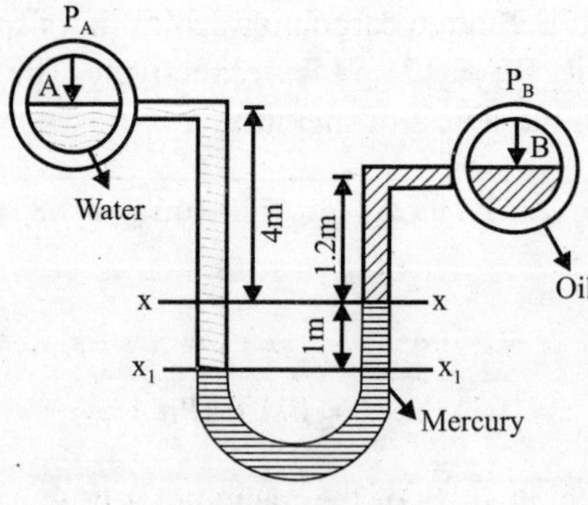

Figure P2.17

$$P_A = 78.5 \text{ kPa}, P_{atm} = 101 \text{ kPa}$$

Equating the pressure heads on both sides of the manometer limbs, we have

$$h_A + 5 * S_{water} = 1 * S_{mercury} + 1.2 * S_{oil} + h_B$$

$$\frac{P_A}{\gamma_{water}} + 5 * S_{water} = 1 * S_{mercury} + 1.2 * S_{oil} + \frac{P_B}{\gamma_{water}}$$

$$\frac{78.5}{9.81} + 5 * 1 = 1 * 13.6 + 1.2 * 0.8 + \frac{P_B}{9.81}$$

$$\Rightarrow P_B = -15.3 \text{ kPa } (vacuum)$$

$$\Rightarrow P_{B(absolute)} = P_{atm} + P_B = 101 - 15.3 = 85.7 \text{ kPa}$$

Hence, the pressure inside the pipe B is 85.7 kPa (absolute).

b) Inverted Differential U-tube Manometer: Inverted differential U-tube manometer is a device just as same that of U-tube differential manometer but is inverted in construction. It measures pressure difference ΔP at two different points "A" and "B" in a single pipe or different pipes, carrying two fluids of same or different kinds, as shown in Figure 2.11a. As small change in

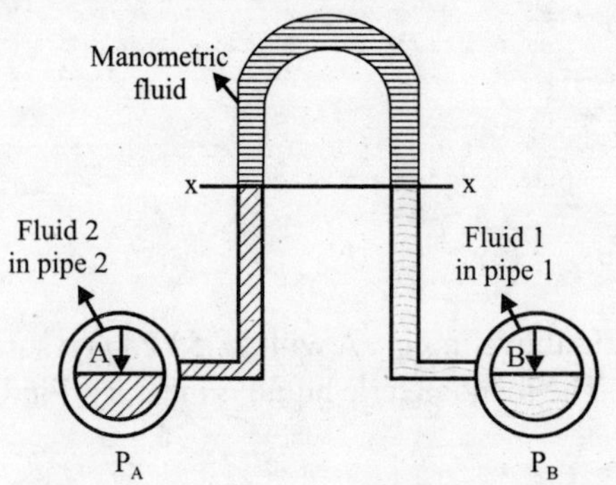

Figure 2.11a: Just before connecting inverted differential U-tube

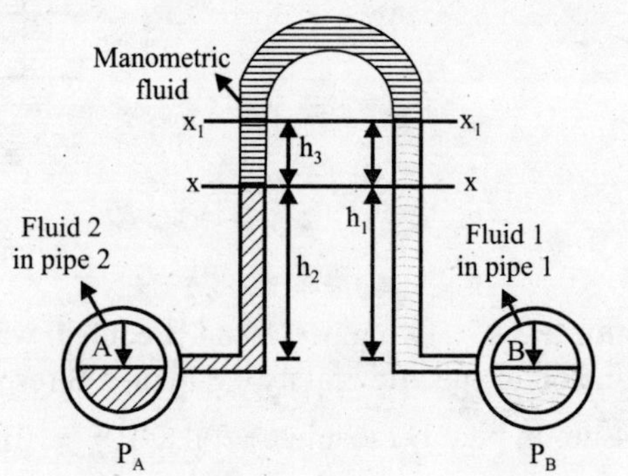

Figure 2.11b: After connecting inverted differential U-tube

pressure level vary manometric fluid level column and so this type of inverted differential U-tube manometer is very much sensitive to pressure changes and hence gives accurate measurements, ΔP.

Consider an inverted differential U-tube connected to two different pipes wherein the pressure difference is to be measured. (i.e. at points "B" and "A").

As the inverted differential U tube is connected to the two pipes, the fluids "fluid 2" and "fluid 1", rush into the differential U-tube, touching the manometric fluid at the either ends, as shown in Figure 2.11a.

Let the pressure at point "B" in the pipe 1 be more than the pressure at point "A" in pipe 2 (i.e. $P_B > P_A$, then the fluid pressure in fluid 2 pushes the manometric fluid upwards due to which the fluid 2 column level increases and in turn the manometric fluid pushes the fluid 1 column level downwards, as shown in Figure 2.11b.

Let the fluid 2 column level, fluid 1 column level, manometric column level be h_1, h_2, h_3 after displacement, respectively.

Under equilibrium, the pressures in the left limb must be equal to the pressures in the right limb and hence

$$P_B + \gamma_1 h_1 = \gamma_2 h_2 + \gamma_3 h_3 + P_A \Rightarrow P_B + \gamma_w S_1 h_1 = \gamma_w S_2 h_2 + \gamma_w S_3 h_3 + P_A \left(\text{since } S = \gamma / \gamma_w\right)$$

Where γ_w, S are specific weight of water and specific gravities of fluids.

By rearranging and dividing by γ_w throughout, we get

$$\frac{(P_B - P_A)}{\gamma_w} = h_3 S_3 + h_2 S_2 - h_1 S_1 \text{ meters of water} \quad (2.18)$$

When written in terms of pressure heads, we get

$$(h_B - h_A) = h_2 S_2 + h_3 S_3 - h_1 S_1 \quad (2.19)$$

If the "fluid 1" and "fluid 2" are of same kind, then $S_1 = S_2$ substituting in Eq. (2.19), we get.

$$(h_B - h_A) = h_3 (S_3 - S_1) \quad (2.19a)$$

Advantages and Disadvantages of Manometers

Advantages	**Disadvantages**
Manometers are easy to fabricate and are inexpensive.	As these are made up of glass tubes, they are fragile and easily broken.
Manometers give accurate reading.	They are bulky.
Manometers are highly sensitive to pressure changes.	Readings are affected by change in temperature and locations.
Manometers are suitable for low pressures.	For better accuracy, meniscus has to be measured by accurate means.
Manometers are not affected by vibrations.	Capillary effect disturbs readings.
Manometers require low maintenance.	

Example 2.18: Find the absolute pressure in the pipe A carrying an oil having specific gravity of 0.8 and has a mercury manometric fluid, as shown in figure P2.18. Take

$$h_1 = 0.66\,\text{m}, h_2 = 0.3\,\text{m}, h_3 = 0.165\,\text{m}$$

$$h_4 = 0.11\,\text{m}, P_{atm} = 105\,\text{kPa}$$

Solution: Given data:

$$S_{oil} = 0.8, S_{Mercury} = 13.6, S_{Water} = 1$$

$$h_1 = 0.66\,\text{m}, h_2 = 0.3\,\text{m}, h_3 = 0.165\,\text{m}$$

$$h_4 = 0.11\,\text{m}, P_{atm} = 105\,\text{kPa}$$

Equating using differential manometer equations, we have

$$P_A = \gamma_{oil}\left(h_4 + h_3\right) - \gamma_{mercury} h_3 - \gamma_{mercury} h_2 + \gamma_{water} h_1 + P_{atm}$$

$$P_A = 0.88 * 9.81 * (0.11 + 0.165) - 13.6 * 9.81 * 0.165 - 1.6 * 9.81 * 0.33 + 9.841 * 0.66 + 105 = 88.6 kPa$$

Hence, pressure in the pipe is 88.6 kPa.

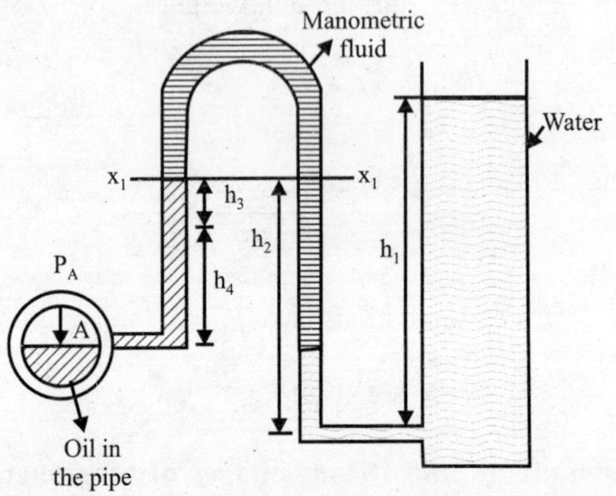

Figure P2.18

Example 2.19: A pipe has two points A and B and carries water and is connected to an inverted U-tube differential manometer having manometeric liquid as mercury. Find the difference in pressures at the two points A and B in absolute. Take atmospheric pressure as 101 kPa, as shown in Figure P2.19

Solution: Given data: $S_{water} = 1, S_{mercury} = 13.6$

Equating the pressure heads on both sides of the manometer limbs, we have

$$h_A + 1.7 * S_{water} + 0.8 * S_{water} = 0.8 + S_{mercury} + 0.7 * S_{water} + h_B$$

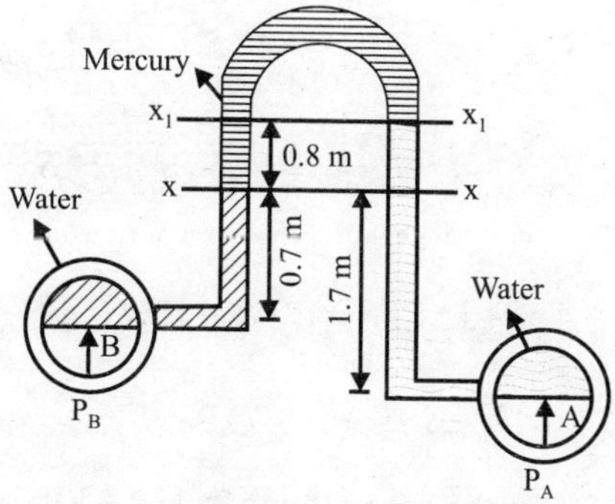

Figure P2.19

$$\Rightarrow h_A + 1.7*1 + 0.8*1 = 0.8*13.6 + 0.7*1 + h_B$$

$$\Rightarrow h_A - h_B = -9.08\,\text{m of water}$$

$$P_A - P_B = -9.88*9810 = -89.07\text{ kPa}$$

$$\Rightarrow (P_A - P_B)_{(absolute)} = P_{atm} = (P_A - P_B) = 101 - 89.7 = 11.91\text{ kPa}$$

Hence, the pressure difference inside the two points A and B in a pipe is 4.1 kPa (absolute).

Example 2.20: Two pipelines carrying oil and water respectively are connected by a inverted u-tube differential manometer at points A and B in the pipes 1 and 2 respectively. Thc inverted U-tube differential manometer has manometeric liquid as mercury. Find the pressure in second pipe at the point B, if the pressure at point A is 78.5 kPa. Take $S_{oil} = 0.8, S_{mercury} = 13.6$

Solution: Given data:

$$S_{oil} = 0.8, S_{mercury} = 13.6, P_A = 78.5\,\text{kPa}$$

Equating the pressure heads on both sides of the manometer limbs, we have

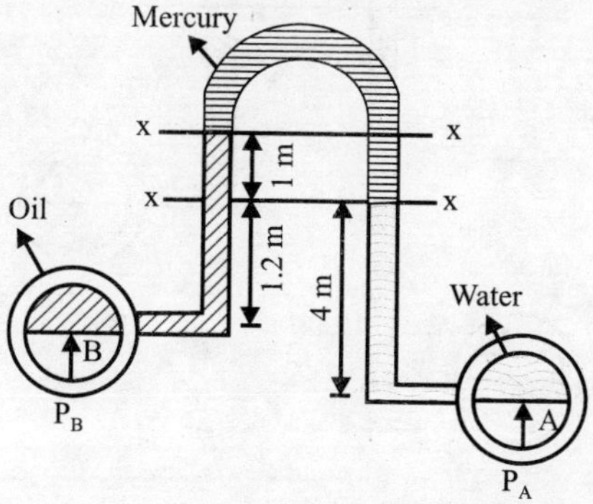

Figure P2.20

$$h_A + 4 * S_{water} + 1 * S_{mercury} = 1 * S_{oil} + 1.2 * S_{oil} + h_B$$

$$\frac{P_A}{\gamma_{water}} + 4 * S_{water} + 1 * S_{mercury} = 1 * S_{oil} + 1.2 * S_{oil} + \frac{P_B}{\gamma_{water}}$$

$$\frac{78.5}{9.81} + 4 * 1 + 1 * 13.6 = 1 * 0.8 + 1.2 * 0.8 + \frac{P_B}{9.81}$$

$$\Rightarrow P_B = 139.7\,\text{kPa}$$

Hence, the pressure at the point B in the second pipe is 139.7 kPa.

Example 2.21: An inverted differential U-tube manometer having an oil of specific gravity 0.8 as manometeric liquid is connected to two different pipes carrying water under pressure. Determine the pressure in pipe B, if the pressure head in the pipe A is 2.0 m of water as shown in Figure P2.21.

Solution: Given data: $S_{oil} = 0.8$, $H_A = 2.0$ m

Equating the pressure heads on both sides of the manometer limbs, we have

$$(h_B - h_A) = 0.15 * S_{oil} + 0.1 * S_{water} - 0.3 * S_{water}$$

$$\Rightarrow (h_B - 2.0) = 0.15 * 0.8 + 0.1 * 1 - 0.3 * 1$$

$$\Rightarrow h_B = 1.92\,\text{m}$$

$$\Rightarrow P_B = h_B * \gamma_{water} = 1.92 * 9.81 = 18.8\,\text{kPa}$$

Hence, the pressure at the point B in the second pipe is 18.8 kPa.

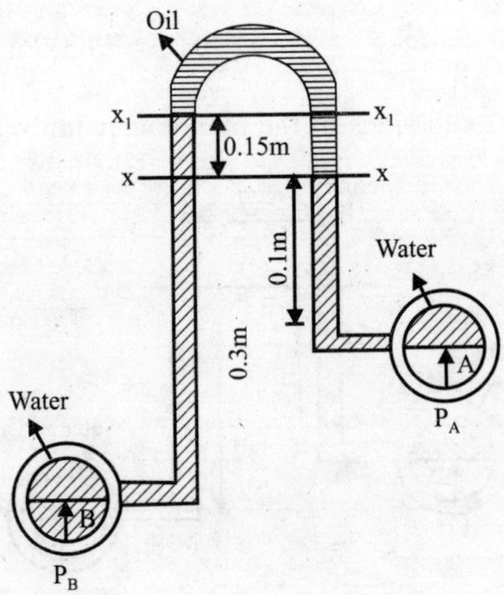

Figure P2.21

MECHANICAL GAGES

Manometers are suitable only for small pressure measurements. In real life we are accustomed to work with high pressure working devices like pumps, compressors, etc. and so manometers do not facilitate our need. Hence, additional pressure measuring devices like **mechanical gages** came into existence, and these serve the purpose of measuring high pressures by taking the pressure of the fluid to be measured as input and show the reading/measurement in terms of deflection of one of its components as output. These gages incorporate elastic systems like tubes, diaphragms, bellows, etc. through which the pressurized fluid is sent there by deflecting the dial indicators or pointers attached to them. They are of four types like:

a) Bourdon tube gage
b) Bellow tube gage
c) Diaphragm gage
d) Dead weight gage.

One has to choose the most appropriate type of mechanical gage basing on their specific requirements.

Example 2.22: Show that the pressure variation with respect to the height can be given as $P = P_0\left[\frac{T_0}{T_0 + Az}\right]^{g/AR}$ if the temp. varies is taken as a function of height and is given by $T = T_0 + A_z$, where T_0, P_0 are the temperature and pressure at the datum, "A" is a constant and "R" is gas constant. (**Hint:** use hydrostatic equation).

Solution: Given data: $T = T_0 + A_z$

According to hydrostatic law, we have $\frac{dP}{dz} = -\left(\frac{Pg}{RT}\right)$

$$\Rightarrow \int_{P_0}^{P}\frac{dP}{P} = -\int_{0}^{Z}\left(\frac{g}{R(T_0 + Az)}\right)dz$$

Assuming $x = (T_0 + Az) \Rightarrow dx = Adz$ and as $z \to 0; x \to T_0$ & $z \to Z; x \to T$

$$\Rightarrow \int_{P_0}^{P}\frac{dP}{P} = -\frac{g}{RA}\int_{T_0}^{T}\left(\frac{1}{(x)}\right)dx$$

$$\Rightarrow \log(P) - \log(P_0) = -\frac{g}{RA}\left[\log(T_0 + Az) - \log(T_0)\right]$$

$$\Rightarrow P = P_0\left[\frac{T_0}{T_0 + Az}\right]^{g/AR}$$

and is proved.

EXERCISE

Objective

1. In a static fluid
   a) Resistance to shear stress is small
   b) Fluid pressure is zero
   c) Linear deformation is small
   d) Only normal stresses can exist
   e) Viscosity is nil
2. Liquids transmit pressure equally in all the directions. This is according to the
   a) Boyle's law
   b) Archimedes' principle
   c) Pascal's law
   d) Newton's formula
3. Manometer is used to measure
   a) Pressure in pipes, channels, etc.
   b) Atmospheric pressure
   c) Very low pressures
   d) Difference of pressure between two points
   e) Velocity
4. Gage pressure is equal to
   a) Absolute pressure + atmospheric pressure
   b) Absolute pressure – atmospheric pressure
   c) Atmospheric pressure – absolute pressure
   d) Atmospheric pressure – vacuum
   e) Atmospheric pressure + vacuum
5. Piezometer is used to measure
   a) Pressure in pipes, channels
   b) Atmospheric pressure
   c) Very low pressures
   d) Difference of pressure between two points
   e) Flow
6. Differential manometer is used to measure
   a) Pressure in pipes channels, etc.
   b) Atmospheric pressure
   c) Very low pressure
   d) Difference of pressure between two points
   e) Velocity in pipes
7. 10 m of water column is equal to
   a) 10 kN/m^2
   b) 1 kN/m^2
   c) 100 kN/m^2
   d) 0.1 kN/m^2
   e) None of the above

8. Determine the pressure in bar at a depth of 10 m in oil of relative density 0.750.
   a) 735575
   b) 0.075
   c) 735575 × 10^5
   d) 73.575
   e) 98100
9. What depth of oil (in m), relative density 0.75, will give a gage pressure of 275000 Pa?
   a) 37.38
   b) 367
   c) 0.027
   d) 20.2 × 10^4
   e) 28.03
10. Express the pressure head of 15 m of water in meters of oil of relative density 0.75.
    a) 110.36
    b) 11.25
    c) 11250
    d) 15.0
    e) 20.0

Answers

1. d	2. c	3. c	4. b	5. a	6. d
7. c	8. b	9. b	10. c		

Theory

1. Explain different types of pressures.
2. Give the relation between different types of pressures.
3. State and prove the Pascal's law.
4. Explain theory behind hydrostatic law.
5. Write a detailed note on pressure head.
6. Write a detailed note on manometers.
7. Explain the differences between Simple and Differential manometers.
8. Write the merits and demerits of manometers.

Problems

1. A manometer connected to a pipe indicates a negative gage pressure of 60 mm of mercury. What is the absolute pressure in pipe, if the atmosphere pressure is 101 kPa?
2. Calculate the pressure due to column of 0.4 m of (a) Water (b) Oil of specific gravity 0.9 (c) Mercury.
3. A simple manometer (U-Tube) containing mercury is connected to a pipe in which an oil of specific gravity 0.8 is flowing. The pressure in the pipe is vacuum. The other end of the manometer is open to atmosphere. Find the vacuum pressure in the pipe, if the difference

of mercury levels in the two limbs is 20 cm and the height of the oil in the left limb from the center of the pipe is 15 cm below.

4. A U-tube differential manometer connects two pressure pipes A and B containing carbon tetrachloride having specific gravity 1.594 under pressure of 11.77 N/cm^2 and pipe B containing oil of specific gravity of 0.8 under pressure of 11.77 N/cm^2 respectively. The pipe A lies 2.5 cm above pipe B. Find the difference of pressure measured by Mercury as fluid filling U-Tube.
5. A single column vertical manometer is connected to a pipe containing oil of specific gravity of 0.9. The area of the reservoir is 80 times the area of the manometer tube. The reservoir contains mercury. The level of mercury in the reservoir is at a height of 30 cm below the center of the pipe and difference of mercury levels in the reservoir and right limb is 50 cm. Find the pressure in the pipe.
6. Find the gage readings at L and M in Figure E2.6, if the local atmosphere pressure is 755 mm of mercury.

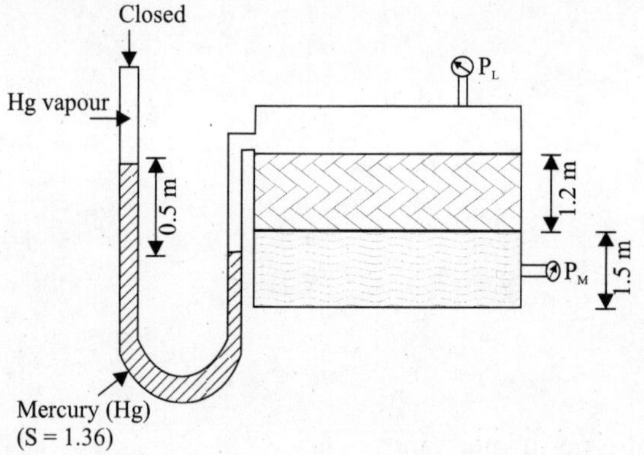

Figure E2.6

Table 2.1: Values of some important properties to remember

Property	Material	Value	Remarks, if any
Atmospheric pressure (P_{atm})	Atmosphere	$1P_{atm}$ = 101.104 kN/m^2	At one atmosphere the mercury raises
		= 10.3 m of H_2O	up to 76 cm of height and water up to
		= 76 cm of Hg	10.0 m height.
		= 760 torr	
Specific gravity (S)	Mercury	13.6	
	Water	1.0	
Specific weight (g)	Mercury	13.6*9.81 = 13000 kg/m^3	
	Water	9810 N/m^3.	

QUICK REFERENCE OF FORMULAE

Property	Formulae	Remarks, if any
Absolute pressure (P_{abs})	$P_{abs} = P_{atm} - P_{vacuum}$ $P_{abs} = P_{atm} + P_{gauge}$	
Specific weight (γ)	$\gamma = \frac{W}{V} = \rho g$	
Pressure head in piezometer (h)	$h = \frac{P}{\gamma}$	
Pressure head (h) in U-Tube simple manometer	$P_A + \gamma_1 h_1 = \gamma_2 h_2 + P_{atm}$ $h = (h_2 S_2 - h_1 S_1)$ If hydrostatic head considered, then $h = (h_2 S_2 - h_1 S_1) + 10.3$	(for +ve gage pressure) (h_1, S_1), (h_2, S_2) are the pressure heads and specific gravities of "fluid 1", manometric fluid respectively.
	$h = -(h_2 S_2 + h_1 S_1)$	(for –ve gage pressure) (h_1, S_1), (h_2, S_2) are the pressure heads and specific gravities of "fluid 1", manometetric fluid respectively.
Pressure head in single column/micro manometer (h)	$\frac{P_A}{\gamma_w} = [\delta h(S_2 - S_1)] + h_2 S_2 - h_1 S_1$ $h = \frac{a}{A}[h_2(S_2 - S_1)] + h_2 S_2 - h_1 S_1$ If hydrostatic head considered, then $h = \frac{a}{A}[h_2(S_2 - S_1)] + h_2 S_2 - h_1 S_1 + 10.3$	(h_1, S_1), (h_2, S_2) are the pressure heads and specific gravities of "fluid 1", manometric fluid respectively.
	$h = \left(\frac{a}{A}\right)[l \sin\alpha(S_2 - S_1)] + l \sin\alpha S_2 - h_1 S_1$ $h = l \sin\alpha S_2 - h_1 S_1$	(for inclined) (h_1, S_1), (h_2, S_2) are the pressure heads and specific gravities of "fluid 1", manometric fluid respectively.
Pressure head in differential U-Tube ($h_A - h_B$)	$\frac{(P_A - P_B)}{\gamma_w} = h_3 S_3 + h_2 S_2 - h_1 S_1$ $(h_A - h_B) = h_2 S_2 + h_3 S_3 - h_1 S_1$	(h_1, S_1), (h_2, S_2), (h_3, S_3) are the pressure heads and specific gravities of "fluid 1", manometric fluid "fluid 2" respectively.
Pressure head in inverted differential U-tube ($h_B - h_A$)	$\frac{(P_B - P_A)}{\gamma_w} = h_3 S_3 + h_2 S_2 - h_1 S_1$ $(h_B - h_A) = h_2 S_2 + h_3 S_3 - h_1 S_1$	(h_1, S_1), (h_2, S_2), (h_3, S_3) are the pressure heads and specific gravities of "fluid 1", "fluid 2" manometric fluid respectively.

CHAPTER 3

Hydrostatic Forces and Buoyancy

INTRODUCTION

When a solid object like a plane or curved surface is immersed in a static fluid (fluid at rest), the surface experiences an upward pushing pressure exerted by the fluid. As the surface penetrates into interior depths of the fluid, the upward pushing pressure increases due to increase in pressure head, as seen in hydrostatic law.

As per the Pascal's law, the pressure exerted by the fluid is uniform in all the directions throughout the surface immersed in fluid and hence we can treat all the pressure intensities acting on the immersed surface as a single pressure (P) acting at a single point (**center of pressure** *C*) on the surface. In addition to the center of pressure, the surface does possess **center of gravity** *G* where the whole weight of the surface concentrates. Hence, in this section we will be focusing on these two center points, namely, **center of pressure** and **center of gravity** for carrying out derivations on pressures exerted on horizontal plane surface, vertical plane surface, inclined plane surface and curved surface.

PRESSURE EXERTED ON HORIZONTAL PLANE SURFACE

When a plane surface immersed in a liquid horizontally the **center of pressure** *C* where the total pressure acts and **center of gravity G** where the center of gravity of the body concentrated, coincides with each other.

Consider a horizontal plane surface having area "A" immersed in a liquid having specific weight γ as shown in Figure 3.1.

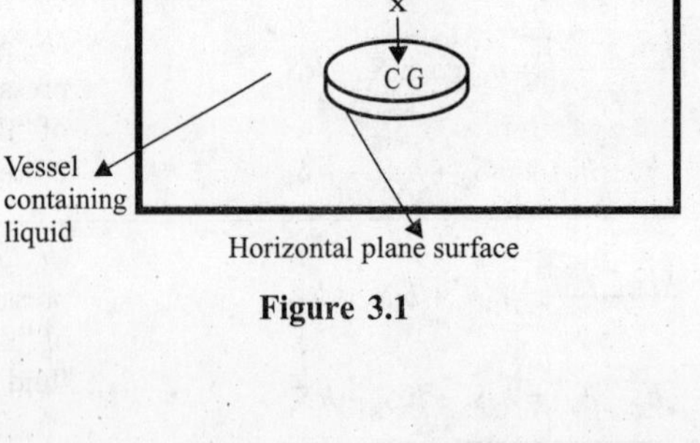

Figure 3.1

Let the distance of center of pressure and center of gravity on the plane surface from the liquid surface be $\bar{h}$ and $\bar{x}$ respectively and as the two center points coincide, their distances from the liquid surface also coincide, i.e.

$$\bar{h} = \bar{x} \cdot$$

The plane surface immersed in liquid displaces some volume of the liquid, which is equal to the weight of the plane surface and so the total pressure that acts on the plane surface will be equal to the weight of the liquid displaced above the immersed plane surface, i.e. **total pressure** P is equal to the weight of the liquid above the immersed surface.

$$\Rightarrow P = W_{liquid\ above\ immersed\ surface}$$

$$\Rightarrow P = \left(\gamma * (A\bar{x})\right) \tag{3.0}$$

since $\dfrac{Weight}{Volume} = \dfrac{W}{A\bar{x}} = Specfic\ Weight(\gamma)$

Eq. (3.0) gives the **total pressure acting on a object immersed horizontally**, in a liquid.
Note: According Hydrostatic law, in a fluid at rest the pressure varies linearly in the vertical direction so long as the fluid density is constant. So the pressure distribution over a plane surface immersed horizontally has equally distributed over the surface, as shown in Figure 3.1a.

PRESSURE EXERTED ON VERTICAL PLANE SURFACE

Consider a vertical plane surface having area "A" and width "b" immersed in a liquid having specific weight γ. As the plane surface is immersed vertically, the **center of pressure** C and **center of gravity** G may or may not coincide with each other (it depends on the shape of the plane surface), as shown in Figure 3.2.

Let the distance of center of pressure and center of gravity on the plane surface from the liquid surface be $\bar{h}$ and $\bar{x}$ respectively.

Total Pressure

Considering a strip of the immersed object of dimension (b, dx) the **total pressure** p_s **acting on the strip** due to pressure intensity of the liquid on the strip is

$$p_s = p(b * dx)$$

$$\Rightarrow p_s = \gamma x(b * dx) \quad \text{since } p = \frac{weight}{area} = \frac{\gamma(b * dx)x}{(b * dx)} = \gamma x$$

Total pressure acting on the plane surface will be

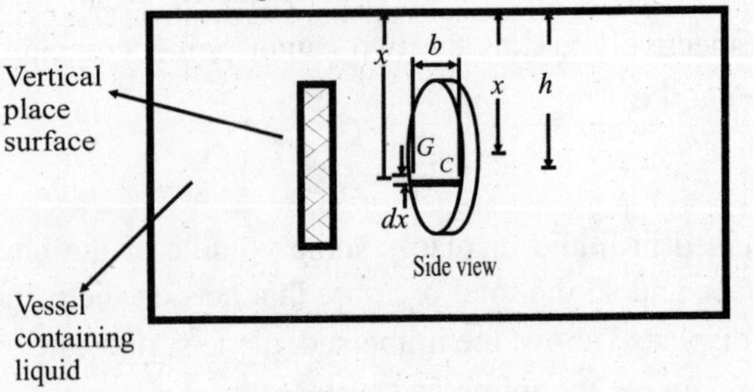

Figure 3.2: A vertical plane surface immerse in a liquid

$$P = \int p_s = \int(\gamma x)(b * dx) \tag{3.1}$$

But $\int(b * dx)x$ is the **moment of inertia** of the immersed plane surface about liquid surface level (**first moment of inertia**).

i.e.
$$I_o = \int(b * dx)x = A\overline{x} \tag{3.1a}$$

Substituting Eq. (3.1a) in Eq. (3.1), we have
Total pressure acting on the plane surface immersed vertically as

$$\Rightarrow P = \gamma * (A\overline{x}) \tag{3.2}$$

Total Moment
Moment of the strip due to total pressure p_s is $M_s = p_s * x$. So the total moment on the object immersed will be $M = P * \overline{h}$ $\because$ Moment = Pressure* perpendicular distance.
Substituting Eq. (3.1a), Eq. (3.2) in Eq. (3.3), we have

$$M = P * \overline{h} = \gamma I_o$$

$$\Rightarrow \gamma * (A\overline{x}) * \overline{h} = \gamma I_o$$

$$\overline{h} = \frac{I_o}{A\overline{x}} \tag{3.3}$$

Parallel Axis Theorem: Consider a body having a centroid "G" through which an axis named G-G passes. Then consider an infinitely small area "dA" in the body at a distance of "y" from the centroid, where the second moment about another axis named "O-O" which is parallel to the centroid axis by $(\overline{x})$ is to be determined.

The second moment of the area "dA" about the axis "O-O" will be the product of the their distance square and the area, i.e. $(I_o)_{dA} = y_0^2 * dA$.

Hence, the total second moment of the area will be

$$I_o = \int_A (I_o) = \int_A y_0^2 * dA$$

$$\Rightarrow I_o = \int_A (\bar{x} - y)^2 * dA = \int_A y^2 * A + A * (\bar{x})^2 - 2 * \bar{x} \int_A y * dA$$

$$\Rightarrow I_o = I_G + A * (\bar{x})^2 \qquad (3.3a)$$

If converted in terms of **moment of inertia of gravity** I_G about the surface level, then according to parallel axis theorem.
Substituting Eq. (3.3a) in Eq. (3.3), we have

$$\bar{h} = \frac{I_G}{A\bar{x}} + \bar{x} \qquad (3.4)$$

Example 3.1: An open tank of 5 m × 3 m × 6 m is completely filled with oil (shown in Figure P3.1) of specific gravity 0.8. Determine the total pressure force and center of pressures on all the faces including the base.

Solution: Given data:

$$L = 5\,\text{m},\ B = 3\,\text{m},\ H = 6\,\text{m},\ S = 0.8$$

Base EFGH is horizontal and has area $A = 5\,\text{m} * 3\,\text{m} = 15\,\text{m}^2$ and center of gravity $\bar{x} = 6\,\text{m}$ and hence total pressure force

$$P = \gamma A\bar{x} = 0.8 * 9810 * 15 * 6 = 706.32\ \text{kN}$$

Faces AEHD and BFGC are vertical and has area of

$$A = 6\,\text{m} * 3\,\text{m} = 18\,\text{m}^2 \text{ and } \bar{x} = 3\,\text{m}.$$

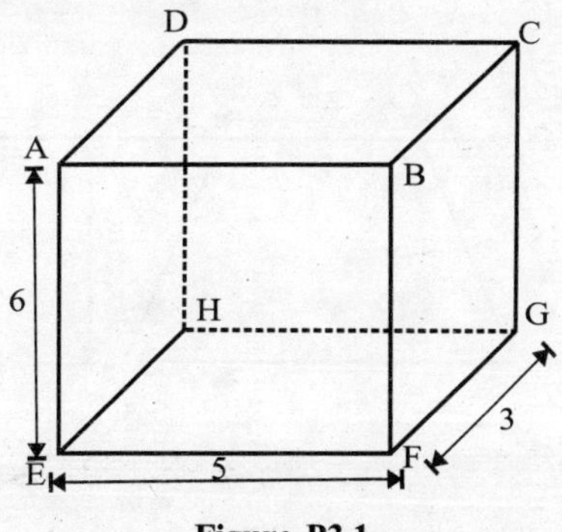

Figure P3.1

And hence total pressure force

$$P = \gamma A\bar{x} = 0.8 * 9810 * 18 * 3 = 423.792 \text{ kN}$$

And center of pressure

$$\bar{h} = \frac{I_G}{A\bar{x}} + \bar{x} = \frac{\left(\frac{bd^3}{12}\right)}{A\bar{x}} + \bar{x} = 4 \text{ m}$$

Faces AEFB and DHGC are vertical and has area of

$$A = 5\,\text{m} * 6\,\text{m} = 30\,\text{m}^2 \text{ and } \bar{x} = 3\,\text{m}.$$

And hence total pressure force

$$P = \gamma A\bar{x} = 0.8 * 9810 * 30 * 3 = 706.32\,\text{kN}$$

And center of pressure

$$\bar{h} = \frac{I_G}{A\bar{x}} + \bar{x} = \frac{\left(\frac{bd^3}{12}\right)}{A\bar{x}} + \bar{x} = 4 \text{ m}$$

Hence the center of gravity is 3.0 m and center of pressure is 15 m.

Example 3.2: A plane surface in the shape of a trapezium is immersed in a liquid having specific gravity of 0.8 such that the smaller side is at the surface level of the liquid free surface and is of dimensions as shown in the figure. Determine the total pressure force acting on the trapezium plane surface and also the center of pressure.

Solution: Given data:

$$a = 2\,\text{m},\ b = 5\,\text{m},\ h = 3\,\text{m},\ S = 0.8$$

The area of the trapezium

$$\text{A=h}\left(\frac{\text{a+b}}{2}\right) \quad \text{(refer Table 3.1)}$$

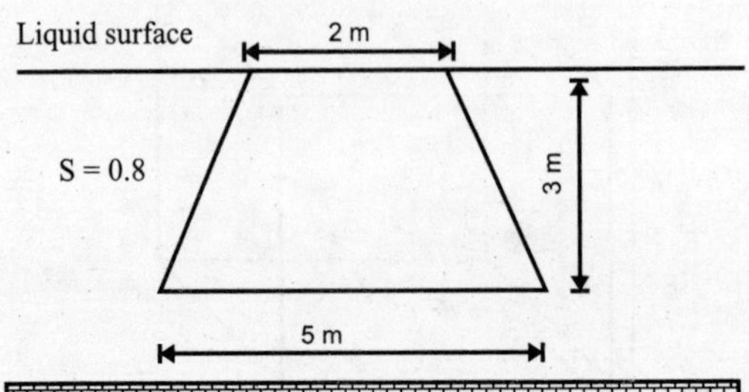

Figure P3.2

Table 3.1: Center of gravity, area and moment of inertia of different types of surfaces.

Shape	Center of Gravity (G)	Area (A)	Movement of Inertia (I_G)
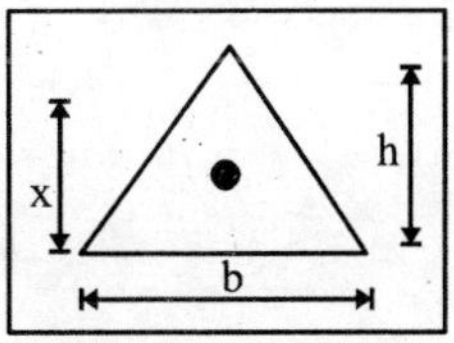	$\frac{h}{3}$	$\frac{bh}{2}$	$\frac{bh^3}{36}$
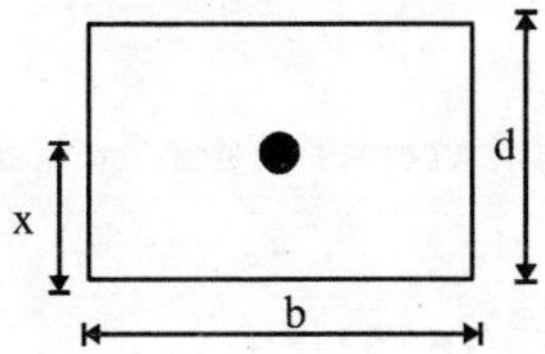	$\frac{d}{2}$	b*d	$\frac{bd^3}{12}$
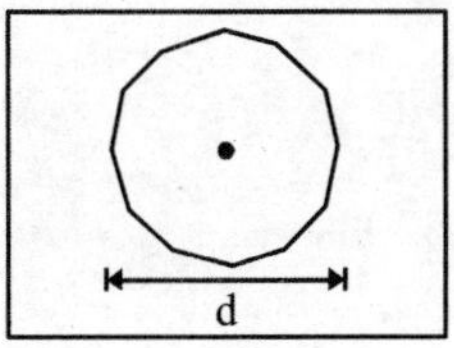	$\frac{d}{2}$	$\frac{\Pi d^2}{4}$	$\frac{\Pi d^4}{64}$
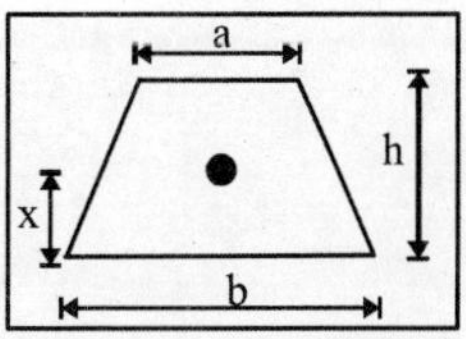	$\frac{h}{3}\left(\frac{2a+b}{a+b}\right)$	$h\left(\frac{a+b}{2}\right)$	$h^3\left(\frac{a^2+b^2+4ab}{36(a+b)}\right)$
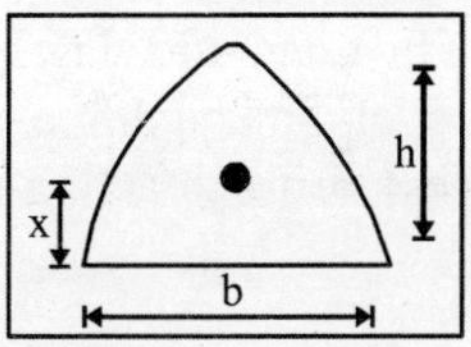	$\frac{2}{2}h$	$\frac{2}{3}bh$	$\frac{8}{175}bh^3$
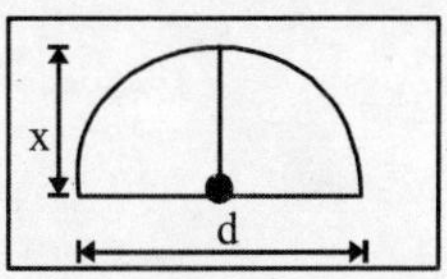	$\frac{4r}{3\pi}$	$\frac{\pi d^2}{8}$	$\frac{\pi d^4}{128}$
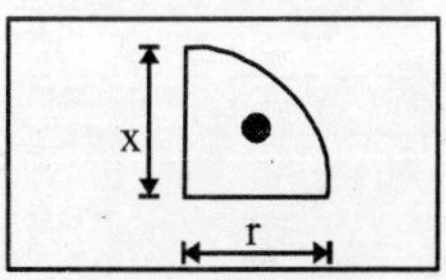	$\frac{4r}{3\pi}$	$\frac{\pi d^2}{8}$	$\frac{\pi d^4}{128}$

$$\Rightarrow A = 3*\left(\frac{2+5}{2}\right) = 10.5\,\text{m}^2$$

Center of gravity from the bigger side as the base is

$$\bar{x}_t = \frac{h}{3}\left(\frac{2a+b}{a+b}\right) = 1.285 \text{ m}$$

Hence, the **center of gravity** from the smaller side is

$$\bar{x} = 3 - \bar{x}_t = 3 - 1.285 = 1.715\,\text{m}$$

Or else if the trapezium is divided into a rectangle and two triangles, then center of gravity in terms of area will be

$$\bar{x} = \frac{\{A_1 * \bar{x}_1 + A_2 * x_2 + A_3 * x_3\}}{\{A_1 + A_2 + A_3\}} = \frac{\{A_{rect} * \bar{x}_{rect} + 2 * A_{tri} * \bar{x}_{tri}\}}{\{A_{rect} + 2 * A_{tri}\}} = \frac{\{2*2*1.5 + 2*0.5*1*3*2\}}{\{2*2 + 2*0.5*1*3\}} = 1.715 \text{ m}$$

And hence total pressure force

$$P = \gamma A\bar{x} = S * \gamma_{water} * A * \bar{x} = 0.8 * 10.5 * 1.715 = 141.32 \text{ kN}$$

Center of pressure

$$\bar{h} = \frac{I_G}{A\bar{x}} + \bar{x} = \frac{h^3\left(\dfrac{a^2+b^2+4ab}{36(a+b)}\right)}{A\bar{x}} + \bar{x} = 2.125 \text{ m}$$

Hence, the center of gravity is 1.715 m and center of pressure is 2.125 m.

Example 3.3: A plane surface in the shape of a trapezium is immersed in a liquid having specific gravity of 0.8 such that the smaller side is at 2 m below the surface level of the liquid free surface and is of dimensions as shown in the figure. Determine the total pressure force acting on the trapezium plane surface and also the center of pressure.

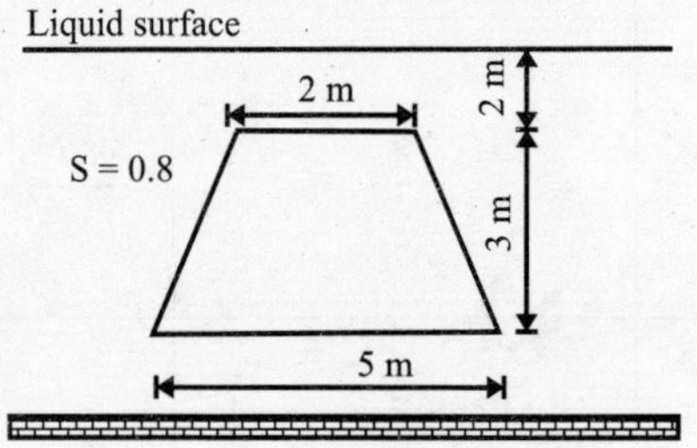

Figure P3.3

Solution: Given data:

$$a = 2\,\text{m}, b = 5\,\text{m}, h = 3\,\text{m}, S = 0.8$$

The area of the trapezium

$$\text{A}=\text{h}\left(\frac{\text{a+b}}{2}\right) \quad \text{(refer Table 3.1)}$$

$$\Rightarrow \text{A}= 3\left(\frac{2+5}{2}\right) = 10.5\,\text{m}^2$$

Center of gravity from the bigger side as the base is

$$\overline{x}_t = \frac{\text{h}}{3}\left(\frac{\text{2a+b}}{\text{a+b}}\right) = 1.285\ \text{m}$$

Hence the **center of gravity** from the free surface of the liquid is

$$\overline{x} = 5 - \overline{x}_t = 5 - 1.285 = 3.715\,\text{m}$$

Or else if the trapezium is divided into a rectangle and two triangles, then center of gravity in terms of area will be

$$\overline{x} = \frac{\{A_1 * \overline{x}_1 + A_2 * x_2 + A_3 * x_3\}}{\{A_1 + A_2 + A_3\}} = \frac{\{A_{rect} * \overline{x}_{rect} + 2 * A_{tri} * \overline{x}_{tri}}{\{A_{rect} + 2 * A_{tri}\}} = \frac{\{2*2*3.5 + 2*0.5*1*3*4\}}{\{2*2 + 2*0.5*1*3\}} = 3.715\ \text{m}$$

And hence total pressure force

$$P = \gamma A\overline{x} = S * \gamma_{water} * A * \overline{x} = 0.8 * 9810 * 10.5 * 3.715 = 306.12\ \text{kN}$$

Center of pressure

$$\overline{h} = \frac{I_G}{A\overline{x}} + \overline{x} = \frac{h^3\left(\dfrac{a^2 + b^2 + 4ab}{36(a+b)}\right)}{A\overline{x}} + \overline{x} = 3.778\ \text{m}$$

Hence, the center of gravity is 3.715 m and center of pressure is 3.778 m.

Example 3.4: A plane surface in the shape of a trapezium having a circular hole in it is immersed in a liquid having specific gravity of 0.8 such that the smaller side is at 2 m below the surface level of the liquid free surface and is of dimensions as shown in the figure. Determine the total pressure force acting on the trapezium plane surface with circular hole in it and also the center of pressure.

Solution: Given data:

$$a = 2\,\text{m},\ b = 5\,\text{m},\ h = 3\,\text{m},\ S = 0.8,\ d = 1\,\text{m}$$

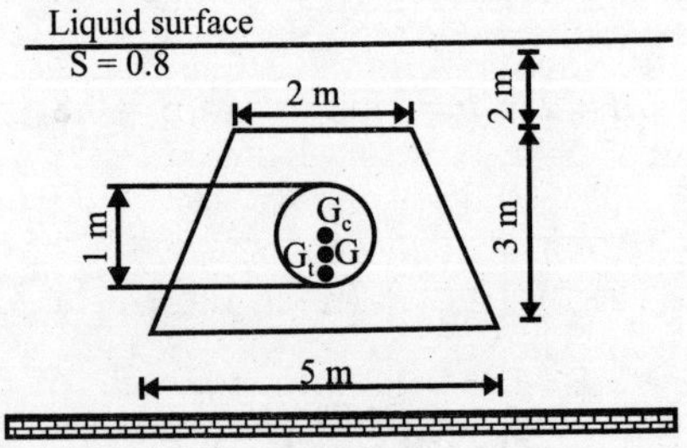

Figure P3.4

The area of the trapezium

$$A_t = h\left(\frac{a+b}{2}\right) \quad \text{(refer Table 3.1)}$$

$$\Rightarrow A_t = 3*\left(\frac{2+5}{2}\right) = 10.5\,\text{m}^2$$

The area of the circle

$$\frac{\pi \text{d}^2}{4} \quad \text{(refer Table 3.1)}$$

$$\Rightarrow A_c = \frac{3.14*1^2}{4} = 0.785\ \text{m}^2$$

Total area of the plane surface is

$$\Rightarrow A = A_t - A_c = 10.5 - 0.785 = 9.715\ \text{m}^2$$

Center of gravity of trapezium from the bigger side at the base is

$$\bar{x}_t = \frac{h}{3}\left(\frac{2a+b}{a+b}\right) = 1.285\ \text{m}$$

Center of gravity of circle from the bottom circumference of circle is

$$\bar{x}_c = \frac{\text{d}}{2}$$

$$\Rightarrow \bar{x}_c = \frac{\text{d}}{2} = \frac{1}{2} = 0.5\ \text{m}$$

Hence, the center of gravity of the plane surface from the free surface of the liquid is

$$\bar{x} = 2 + \left[\left(1.5 - \bar{x}_t + \bar{x}_c\right)\right] = 2 + (1.5 - 1.285 + 1.5) = 3.715\ \text{m}$$

Or else if the trapezium is divided into a rectangle and two triangles, then Center of gravity in terms of area will be

$$\bar{x} = 2 + \frac{\{A_1 * x_1 - A_2 * \bar{x}_2\}}{\{A_1 - A_2\}} = 2 + \frac{\{A_{trap} * \bar{x}_{trap} - A_{cir} * \bar{x}_{cir}\}}{\{A_{trap} - A_{cir}\}} = 2 + \frac{\{10.5 * 1.715 - 0.785 * 1.5\}}{\{10.5 - 0.785\}} = 3.75 \text{ m}$$

And hence total pressure force

$$P = \gamma A\bar{x} = 0.8 * 9810 * 9.715 * 3.715 = 283.5 \text{ kN}$$

Center of pressure

$$\bar{h} = \left(\frac{(I_G)_t - (I_G)_c}{A\bar{x}} + \bar{x}\right) = \left(\frac{h^3\left(\dfrac{a^2 + b^2 + 4ab}{36(a+b)}\right) - \dfrac{\pi * d^4}{64}}{A\bar{x}} + \bar{x}\right) = 3.918 \text{ m}$$

Hence, the center of gravity is 3.715 m and the center of pressure is 3.918 m.

Example 3.5: A plane surface in the shape of a rectangle having a square hole in it is immersed in a liquid having specific gravity of 0.8 such that the one of its longer side is at 2 m below the surface level of the liquid free surface and is of dimensions as shown in the figure. Determine the total pressure force acting on the rectangular plane surface with square hole in it and also the center of pressure.

Solution: Given data:

$$\text{Rectangle}: L = 8\,\text{m},\ b = 6\,\text{m}\ \&\ S = 0.8$$

$$Square: a = 2\,\text{m},\ a = 2\,\text{m}$$

The area of the rectangle

$$A_r = L * b \Rightarrow A_r = 8 * 6 = 48\,\text{m}^2$$

The area of the square

$$A_s = a * a$$

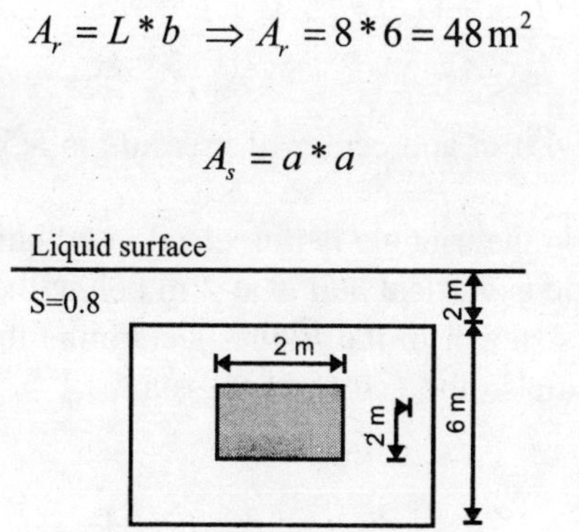

Figure P3.5:

$$\Rightarrow A_s = 2*2 = 4\,\text{m}^2$$

Total area of the plane surface is

$$\Rightarrow A = A_r - A_s = 48 - 4 = 44\,\text{m}^2$$

Center of gravity of rectangle from the bigger side at the base is

$$\bar{x}_r = \frac{b}{2}$$

$$\Rightarrow \bar{x}_r = \frac{6}{2} = 3\,\text{m}$$

Center of gravity of square from bigger side as base is

$$\bar{x}_s = \frac{a}{2}$$

$$\Rightarrow \bar{x}_s = \frac{2}{2} = 1\,\text{m}$$

Hence, the center of gravity of the plane surface from the free surface of the liquid is

$$\bar{x} = 2 + \bar{x}_{rect} = 2 + \bar{x}_{sq} = 2 + 3 = 5\,\text{m}$$

Or else center of gravity in terms of area can be

$$\bar{x} = 2 + \frac{\{A_1 * \bar{x}_1 - A_2 * \bar{x}_2\}}{\{A_1 - A_2\}} = 2 + \frac{\{A_{rect} * x_{rect} - A_{sq} * \bar{x}_{sq}\}}{\{A_{rect} - A_{sq}\}} = 2 + \frac{\{48*3 - 4*3\}}{\{48 - 4\}} = 5\text{ m}$$

And hence total pressure force $\quad P = \gamma A\bar{x} = 0.8*9810*44*5 = 1726.5\text{ kN}$

Center of pressure $\quad \bar{h} = \left(\frac{(I_G)_r - (I_G)_s}{A\bar{x}} + \bar{x}\right) - \left(\frac{Lb^3 - a^4}{12*Ax} + \bar{x}\right) = 5.594\text{ m}$

Hence, the center of gravity is 4.0 m and center of pressure is 5.594 m.

Example 3.6: A plane surface in the triangle is immersed in a liquid having specific gravity of 1.1 such that the base of the triangle is vertical and is at 2 m below the surface level of the liquid free surface and is of dimensions as shown in the figure. Determine the total pressure force acting on the triangular plane surface and also the center of pressure.

Solution: Given data:

$$b = 5\,\text{m},\ h = 5\,\text{m},\ S = 1.1$$

The area of the triangle

$$A = \left(\frac{bh}{2}\right) \quad \text{(refer Table 3.1)}$$

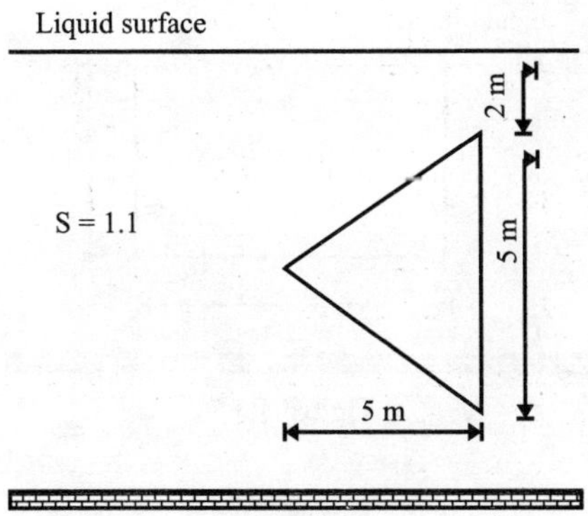

Figure P3.6

$$\Rightarrow A = \left(\frac{5*5}{2}\right) = 12.5\,\text{m}^2$$

Center of gravity of the triangle lie on the line passing through the mid point of the height "*h*" and so

$$\bar{x}_t = \frac{h}{2} = \left(\frac{5}{2}\right) = 2.5\,\text{m}$$

Hence, the **center of gravity** from the free surface of the liquid is

$$\bar{x} = 2 + \bar{x}_t = 2 + 2.5 = 4.5\,\text{m}$$

And hence total pressure force

$$P = \gamma A\bar{x} = 1.1 * 12.5 * 4.5 = 606.95\ \text{kN}$$

Center of pressure

$$\bar{h} = \frac{I_G}{A\bar{x}} + \bar{x} = \frac{\left(\frac{bh^3}{36}\right)}{A\bar{x}} + \bar{x} = 4.808\ \text{m}$$

Hence, the center of gravity is 4.5 m and center of pressure is 4.808 m.

Example 3.7: A plane surface in the shape of a rectangle is immersed in a liquid having specific gravity of 0.8 such that the one of its longer side is at the surface level of the liquid free surface and is of dimensions as shown in the figure. Determine the total pressure force acting on the rectangular plane surface and also the center of pressure.

Solution: Given data:

$$L = 8\,\text{m},\ b = 6\,\text{m},\ S = 0.8$$

The area of the rectangle

A=L*b

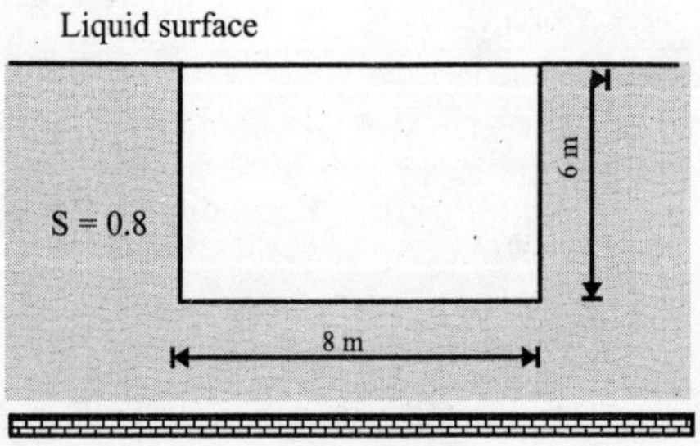

Figure P3.7

$$\Rightarrow A=8*6=48\,\text{m}^2$$

Center of gravity of rectangle from the bigger side as the base is and from the free surface of the liquid is

$$\bar{x} = \frac{b}{2}$$

$$\Rightarrow \bar{x} = \frac{6}{2} = 3\,\text{m}$$

And hence total pressure force

$$P = \gamma A\bar{x} = 0.8*9810*48*3 = 1130.1\text{ kN}$$

Center of pressure

$$\bar{h} = \left(\frac{I_G}{A\bar{x}} + \bar{x}\right) = \left(\frac{\text{Lb}^3}{12A\bar{x}} + \bar{x}\right)$$

$$\bar{h} = \left(\frac{8*6^3}{12*48*3} + 3\right) = 4.0\,\text{m}$$

Hence the center of gravity is 3.0 m and center of pressure is 4.0 m.

Example 3.8: A plane surface in shape of a triangle is immersed in a liquid having specific gravity of 0.8 such that the base of the triangle horizontal and is at the surface level of the liquid free surface and is of dimensions as shown in the figure. Determine the total pressure force acting on the triangular plane surface and also the center of pressure.

Solution: Given data:

$$b = 8\,\text{m},\ h = 6\,\text{m},\ S = 0.8$$

The area of the triangle

$$A = \left(\frac{bh}{2}\right) \quad \text{(refer Table 3.1)}$$

$$\Rightarrow A = \left(\frac{8*6}{2}\right) = 24\text{ m}^2$$

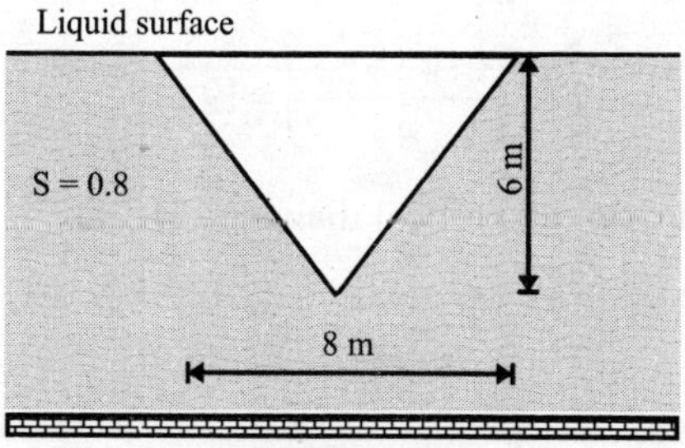

Figure P3.8

Centers of gravity of the triangle lie on the vertical line passing through the mid point of the base "*b*" and so

$$\bar{x} = \frac{h}{3} = \left(\frac{6}{3}\right) = 2\,\text{m}$$

And hence total pressure force

$$P = \gamma A\bar{x} = 0.8 * 9810 * 12 * 2 = 188.35\ \text{kN}$$

Center of pressure

$$\bar{h} = \frac{I_G}{A\bar{x}} + \bar{x} = \frac{\left(\frac{bh^3}{36}\right)}{A\bar{x}} + \bar{x} = 4\ \text{m}$$

Example 3.9: A plane surface in shape of a circle is immersed vertically in a liquid having specific gravity of 1.1 such that its circumference touches the surface level of the liquid free surface and is of dimensions as shown in the figure. Determine the total pressure force acting on the circular plane surface and also the center of pressure.

Solution: Given data:

$$d = 5\,\text{m},\ S = 1.1$$

The area of the circle

$$\text{A} = \left(\frac{\pi \text{d}^2}{4}\right) \quad \text{(refer Table 3.1)}$$

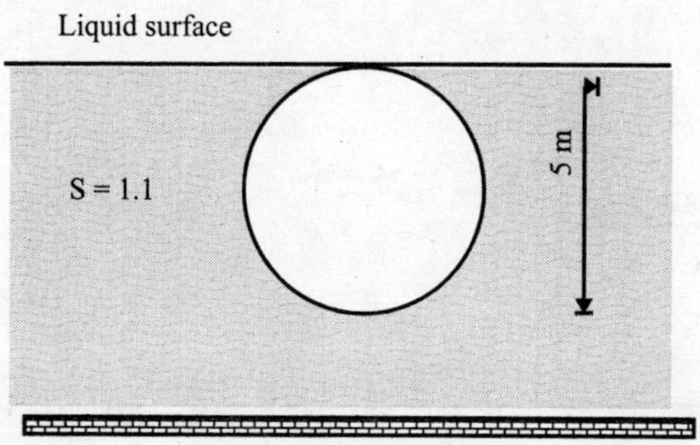

Figure P3.9

$$\Rightarrow A = \left(\frac{3.14 * 5 * 5}{4}\right) = 19.625\,\text{m}^2$$

Center of gravity of the circle lie on the vertical diameter line, so

$$\overline{x} = \frac{d}{2} = \left(\frac{5}{2}\right) = 2.5\,\text{m}$$

And hence total pressure force

$$P = \gamma A\overline{x} = 1.1 * 9810 * 19.625 * 2.5 = 529.4\ \text{kN}$$

Center of pressure

$$\overline{h} = \frac{I_G}{A\overline{x}} + \overline{x} = \frac{\left(\frac{\pi d^4}{64}\right)}{A\overline{x}} + \overline{x} = 3.125\ \text{m}$$

Hence, the center of gravity is 2.5 m and center of pressure is 3.125 m.

Example 3.10: A plane surface in the shape of a semicircle is immersed vertically in a liquid having specific gravity of 1.1 such that its diameter touches the surface level of the liquid free surface and is of dimensions as shown in the figure. Determine the total pressure force acting on the semicircular plane surface and also the center of pressure.

Solution: Given data:

$$d = 10\,\text{m},\ S = 1.1$$

The area of the semicircle

$$\text{A} = \left(\frac{\pi \text{d}^2}{8}\right) \quad \text{(refer Table 3.1)}$$

$$\Rightarrow \text{A} = \left(\frac{3.14*10*10}{8}\right) = 39.25\,\text{m}^2$$

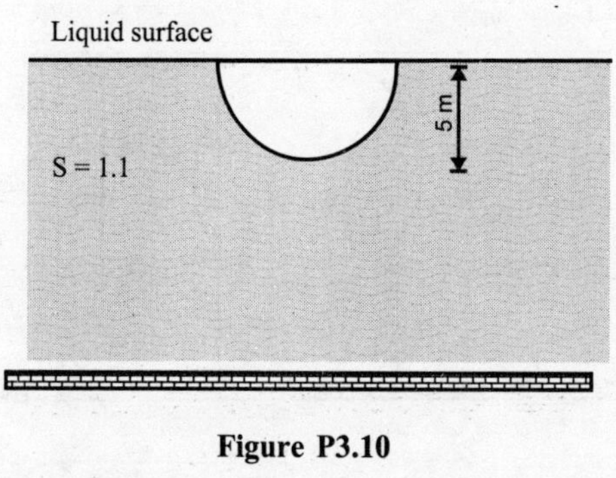

Figure P3.10

Center of gravity of the circle lie on the vertical diameter line, so

$$\bar{x} = \frac{4r}{3\pi} = \left(\frac{4*5}{3*3.14}\right) = 2.12\,\text{m}$$

And hence total pressure force

$$P = \gamma A\bar{x} = 1.1**9810*39.25*2.12 = 899.25\ \text{kN}$$

Center of pressure

$$\bar{h} = \frac{I_G}{A\bar{x}} + \bar{x} = \frac{\left(\frac{\pi d^4}{128}\right)}{A\bar{x}} + \bar{x} = 5.0\ \text{m}$$

Hence, the center of gravity is 2.12 m and center of pressure is 5.0 m.

Example 3.11: A plane surface in the shape of a rectangle having a circular hole in it and is immersed in a liquid having specific gravity of 0.8 such that the one of its longer sides is at 2 m below the surface level of the liquid free surface and is of dimensions as shown in the figure. Determine the total pressure force acting on the rectangular plane surface with circular hole and also the center of pressure.

Solution: Given data:

$$Rectangle: L = 8\,\text{m},\ b = 6\,\text{m}\ \&\ S = 0.8$$

$$Circle: d = 3\,\text{m}$$

The area of the rectangle

$$A_r = L*b$$

$$\Rightarrow A_r = 8*6 = 48\,\text{m}^2$$

The area of the circle hole

$$A_c = \frac{\pi}{4}*d^2$$

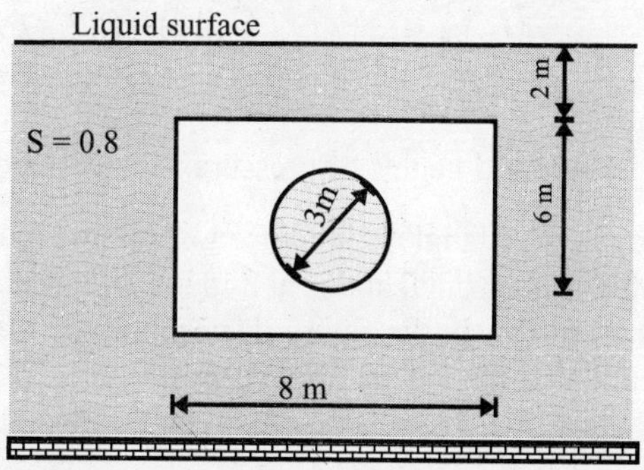

Figure P3.11

$$\Rightarrow A_c = \frac{3.14}{4} * 3^2 = 7.065\,\text{m}^2$$

Total area of the plane surface is

$$\Rightarrow A = A_r - A_C = 48 - 7.065 = 40.94\,\text{m}^2$$

Center of gravity of rectangle from the bigger side as the base is

$$\overline{x}_r = \frac{\text{b}}{2}$$

$$\Rightarrow \overline{x}_r = \frac{6}{2} = 3\,\text{m}$$

Center of gravity of circle is

$$\overline{x}_c = \frac{d}{2}$$

$$\Rightarrow \overline{x}_c = \frac{3}{2} = 1.5\,\text{m}$$

Hence, the center of gravity of the plane surface from the free surface of the liquid is

$$\overline{x} = 2 + \overline{x}_r = 2 + \overline{x}_c = 2 + 3 = 5\,\text{m}$$

Or else center of gravity in terms of area can be

$$\overline{x} = \frac{\{A_1 * \overline{x}_1 - A_2 * \overline{x}_2\}}{\{A_1 - A_2\}} = \frac{\{A_{rect} * \overline{x}_{rect} - A_{cir} * \overline{x}_{cir}\}}{\{A_{rect} - A_{cir}\}} = \frac{\{48 * 5 - 7.065 * 5\}}{\{48 - 7.065\}} = 5\,\text{m}$$

And hence total pressure force $P = \gamma A\overline{x} = 0.8 * 9810 * 40.94 * 5 = 1606.4$ kN

Center of pressure $\overline{h} = \left(\frac{(I_G)_{rect} - (I_G)_{cir}}{A\overline{x}} + \overline{x}\right) - \left(\frac{(Lb^3/12) - (\pi d^4/64)}{A\overline{x}} + \overline{x}\right) = 5.68$ m

Hence, the center of gravity is 5 m and center of pressure is 5.68 m.

Example 3.12: A plane surface in the triangle is immersed in a liquid having specific gravity of 0.98 such that the base of the triangle is horizontal and is at 3 m below the surface level of the liquid free surface and is of dimensions as shown in the figure. Determine the total pressure force acting on the triangular plane surface and also the center of pressure.

Solution: Given data:

$$b = 5\,\text{m},\ h = 5\,\text{m},\ S = 0.985$$

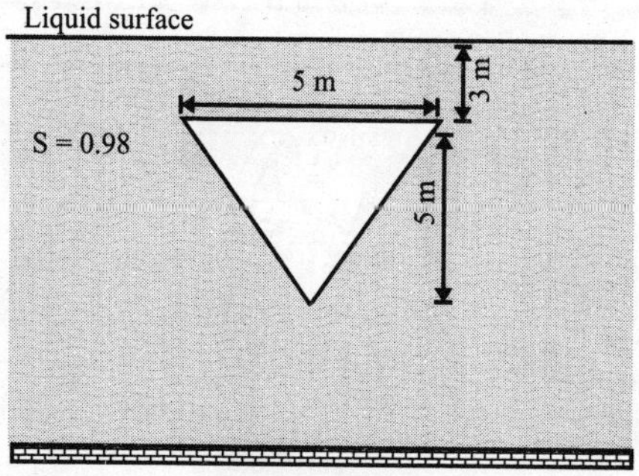

Figure P3.12

The area of the triangle

$$A = \left(\frac{bh}{2}\right)$$

$$\Rightarrow A = \left(\frac{5*5}{2}\right) = 12.5\,\text{m}^2$$

Center of gravity of the triangle lies on the line passing through the mid point of the base "*b*" and so

$$\bar{x}_t = \frac{h}{3} = \left(\frac{5}{3}\right) = 1.67\,\text{m}$$

Hence, the center of gravity from the free surface of the liquid is

$$\bar{x} = 3 + \bar{x}_t = 3 + 1.67 = 4.67\ \text{m}$$

And hence total pressure force

$$P = \gamma A\bar{x} = 0.98 * 12.5 * 4.67 = 561.2\ \text{kN}$$

Center of pressure

$$\bar{h} = \frac{I_G}{A\bar{x}} + \bar{x} = \frac{\left(\frac{bh^3}{36}\right)}{A\bar{x}} + \bar{x} = 4.729\ \text{m}$$

Hence, the center of gravity is 4.67 m and center of pressure is 4.729 m.

Pressure Exerted on Inclined Plane Surface

Consider an inclined plane surface having area "A" and width "b" immersed in a liquid having specific weight γ. As the plane surface is immersed inclined, the **center of pressure force (*C*)** and **center of gravity** G may or may not coincide with each other (it depends on the shape of the plane surface), as shown in Figure 3.3.

Let the distance of center of pressure force and center of gravity on the plane surface from the liquid surface be $\bar{h}$ and $\bar{x} = l\sin\theta$ respectively.

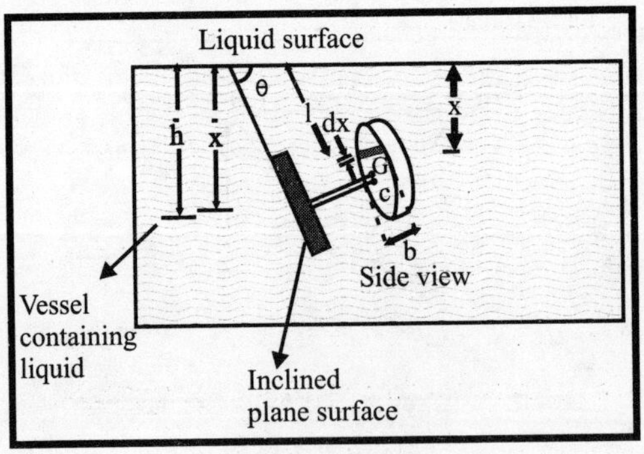

Figure 3.3: Inclined plane surface immersed in a liquid

Total Pressure

Considering a strip of the immersed object of dimension (b, dx), the **total pressure force p_s acting on the strip** due to pressure intensity of the liquid on the strip is $p_s = p\,(b*dx)$ and hence

$$p_s = \gamma l \sin\theta (b*dx)$$

$$\because p = \frac{weight}{Area} = \frac{\gamma(b*dx)l\sin\theta}{(b*dx)} = \gamma l \sin\theta$$

Total pressure force acting on the plane surface will be

$$P = \int p_s = \int(\gamma l \sin\theta)(b*dx) \tag{3.5}$$

But $\int(b*dx)l$ is the **moment of inertia** of the immersed plane surface about liquid surface level (**first moment of inertia**). i.e.

$$I_o = \int(b*dx)l = \frac{A\overline{x}}{\sin\theta} \tag{3.5a}$$

Substituting Eq. (3.5a) in Eq. (3.5), we have **total pressure force acting on the plane surface immersed vertically as**

$$\Rightarrow P = \gamma * (A\overline{x}) \tag{3.6}$$

Total Moment:

Moment of the strip due to total pressure p_s is $M_s = p_s * l\sin\theta$. So the total moment on the object immersed will be

$$M = \frac{P*\overline{h}}{\sin\theta} \tag{3.7}$$

$\because$ *Moment = Pressure * perpendicular distance.*

Substituting Eq. (3.5a), Eq. (3.6) in Eq. (3.7), we have

$$M = \frac{P * \bar{h}}{\sin\theta} = \gamma I_o \sin\theta$$

$$\Rightarrow \frac{\gamma * (A\bar{x}) * \bar{h}}{\sin\theta} = \gamma I_o \sin\theta$$

$$\bar{h} = \frac{I_o \sin^2\theta}{A\bar{x}} \tag{3.8}$$

If converted in terms of **moment of inertia of gravity** I_G about the surface level, then according to parallel axis theorem from engineering mechanics, we have

$$I_o = I_G + \frac{A(\bar{x})^2}{\sin^2\theta} \tag{3.8a}$$

Substituting Eq. (3.8a) in Eq. (3.8), we have

$$\bar{h} = \left(\frac{I_G \sin^2\theta}{A\bar{x}} + \bar{x}\right) \tag{3.9}$$

Example 3.13: A plane surface in rectangular shape is immersed inclined at angle of 30° to the free surface of a liquid having specific gravity of 0.8 such that its tip is 2 m below the surface level of the liquid free surface and is of dimensions as shown in the figure. Determine the total pressure force acting on the rectangular plane surface and also the center of pressure.

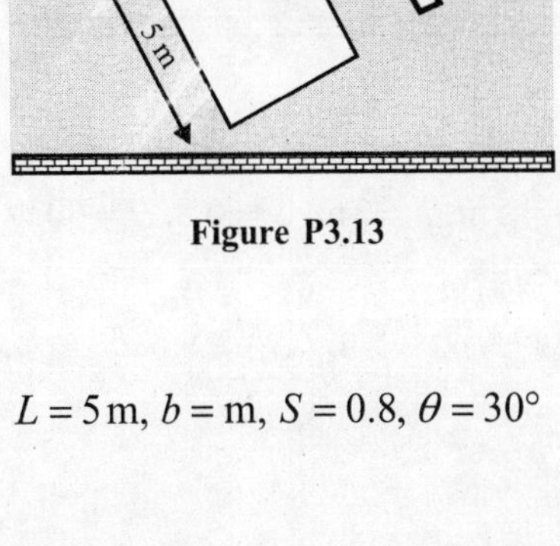

Figure P3.13

Solution: Given data:

$$L = 5\,\text{m},\ b = \text{m},\ S = 0.8,\ \theta = 30°$$

The area of the rectangle $A = L * b$

$$\Rightarrow A = 5 * 3 = 15\,\text{m}^2$$

Center of gravity of the rectangle

$$\bar{x} = 2 + \frac{L}{2}\sin\theta = 2 + \left(\frac{5}{2}\right)\sin 30° = 3.13\,\text{m}$$

And hence the total pressure force

$$P = \gamma A\bar{x} = 0.8 * 9810 * 15 * 3.13 = 369.0\ \text{kN}$$

Center of pressure

$$\bar{h} = \frac{I_G}{A\bar{x}} + \bar{x} = \frac{\left(\frac{Lb^3}{12}\right)}{A\bar{x}} + \bar{x} = 3.74\ \text{m}$$

Hence, the center of gravity is 3.13 m and center of pressure is 2.8 m.

Example 3.14: A plane surface in trapezium shape is immersed inclined at angle of 60° to the free surface of a liquid having specific gravity of 0.8 such that its tip is 2 m below the surface level of the liquid free surface and is of dimensions as shown in the figure. Determine the total pressure force acting on the trapezium plane surface and also the center of pressure.

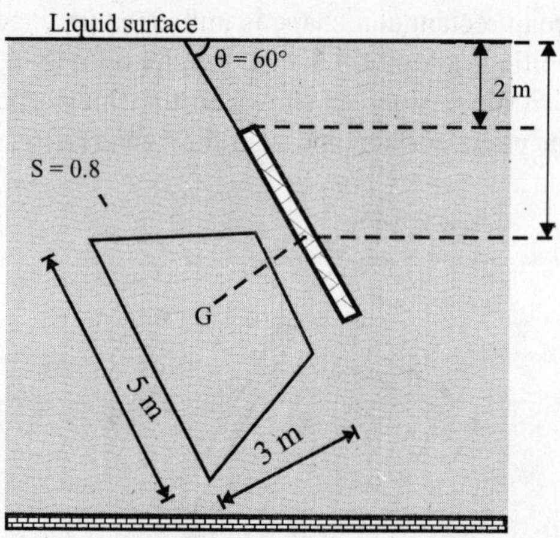

Figure P3.14

Solution: Given data: $a = 2\,\text{m}$, $b = 5\,\text{m}$, $h = 3\,\text{m}$, $S = 0.8$, $\theta = 60°$

The area of the trapezium

$$A = h\left(\frac{a+b}{2}\right)$$

$$A = 3*\left(\frac{2+5}{2}\right) = 10.5\,\text{m}^2$$

Center of gravity of the trapezium from the bigger side as the base

$$\bar{x}_t = \frac{h}{3}\left(\frac{2a+b}{a+b}\right) = \frac{3}{3}\left(\frac{2*2+5}{2+5}\right) = 1.285\,\text{m}$$

Hence, total center of gravity is

$$\bar{x} = 2 + (5/2)\sin 60° = 4.165\,\text{m}$$

And hence total pressure force $P = \gamma A\bar{x} = 0.8*9810*10.5*4.165 = 343.2\,\text{kN}$

Center of pressure

$$\bar{h} = \frac{I_G * \sin^2 60°}{A\bar{x}} + \bar{x} = \frac{h^3\left(\dfrac{a^2+b^2+4ab}{36(a+b)}\right)}{A\bar{x}} * \sin^2 60° + \bar{x}$$

$$\Rightarrow \bar{h} = \frac{3^3\left(\dfrac{2^2+5^2+4*2*5}{36(2+5)}\right)}{10.5*4.165} * \sin^2 60° + 4.165 = 0.169*0.75 + 4.165 = 4.29\,\text{m}$$

Hence, the center of gravity is 4.165 m and center of pressure is 4.29 m.

Example 3.15: A plane surface in circular shape is immersed inclined at angle of 60° to the free surface of a liquid having specific gravity of 1.2 such that its tip is 2 m below the surface level of the liquid free surface and is of dimensions as shown in the figure. Determine the total pressure force acting on the circular plane surface and also the center of pressure.

Solution: Given data: $d = 5\,\text{m},\ \theta = 60°,\ S = 1.2$
The area of the circle

$$A = \left(\frac{\pi d^2}{4}\right)$$

Center of gravity of the circle

$$\bar{x} = 2 + \left(\frac{d}{2}\right)\sin 60° = 4.165\,\text{m}$$

Total pressure force

$$P = \gamma A\bar{x} = 1.2*9810*19.625*4.165 = 962.2\ \text{kN}$$

Moment of inertia of circle about C.G. is

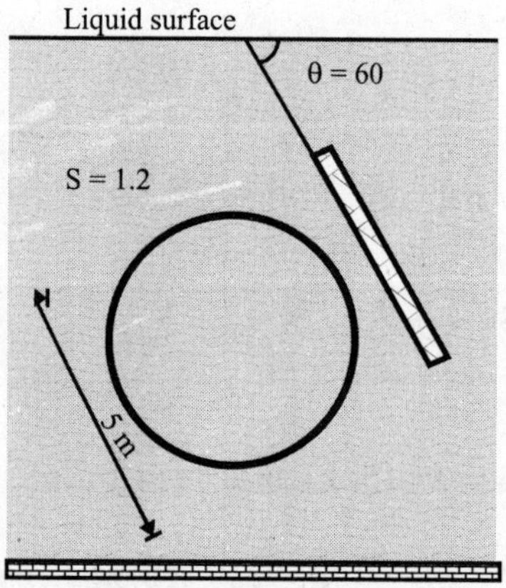

Figure P3.15

$$I_G = \left(\frac{\pi D^4}{64}\right) = \left(\frac{3.14 * 5^4}{64}\right) = 30.66\,\text{m}^4$$

Center of pressure of plane surface is

$$\bar{h} = \frac{I_G * \sin^2 60°}{A\bar{x}} + \bar{x} = \frac{30.66}{19.625 * 4.16} * \sin^2 60° + 4.16 = 4.415\,\text{m}$$

Hence, the center of gravity is 4.165 m and center of pressure is 4.415 m.

Example 3.16: A plane surface in trapezium shape having a circular hole is immersed inclined at angle of 60° to the free surface of a liquid having specific gravity of 0.8 such that its tip is 2 m below the surface level of the liquid free surface and is of dimensions as shown in the figure. Determine the total pressure force acting on the trapezium plane surface with circular hole in it and also the center of pressure.

Solution: Given data:

$$a = 2\,\text{m},\ b = 5\,\text{m},\ h = 3\,\text{m},\ d = 2.5\,\text{m},\ S = 0.8,\ \theta = 60°$$

The area of the trapezium

$$A_t = h\left(\frac{a+b}{2}\right)$$

$$\Rightarrow A_t = 3 * \left(\frac{2+5}{2}\right) = 10.5\,\text{m}^2$$

Area of the circle hole

$$A_c = \left(\frac{\pi d^2}{4}\right) \Rightarrow A_c = \left(\frac{\pi * 2.5^2}{4}\right) = 4.9\,\text{m}^2$$

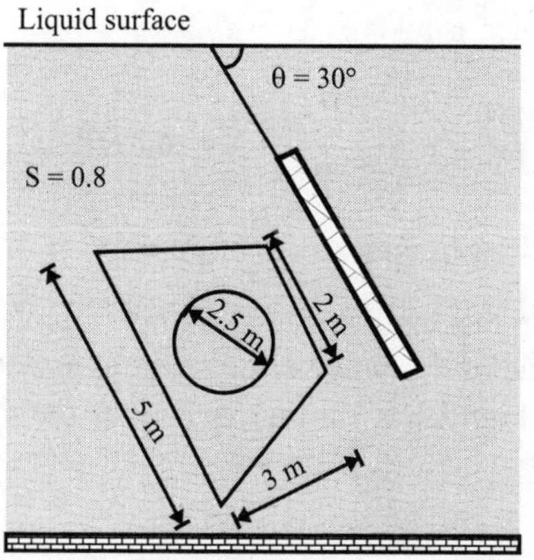

Figure P3.16

Total area of the plane surface

$$A = A_t - A_c = 10.5 - 4.9 = 5.59\,\text{m}^2$$

Center of gravity of the trapezium from the bigger side as the base

$$\bar{x}_t = \frac{h}{3}\left(\frac{2a+b}{a+b}\right) = \frac{3}{3}\left(\frac{2*2+5}{2+5}\right) = 1.285\,\text{m}$$

Center of gravity of the circle

$$\bar{x}_c = \frac{d}{2} = \frac{2.5}{2} = 1.25\,\text{m}$$

Hence, total center of gravity is

$$\bar{x} = 2 + (2/2)\sin 60° = 2 + (5/2)\sin 60 = 4.165\,\text{m}$$

And hence total pressure force

$$P = \gamma A\bar{x} = 0.8 * 9810 * 5.59 * 4.16 = 182.7\text{ kN}$$

Moment of inertia of trapezium about C.G. is

$$I_{G_t} = h^3\left(\frac{a^2+b^2+4ab}{36(a+b)}\right) = 3^3\left(\frac{2^2+5^2+4*2*5}{36(2+5)}\right) = 2.464\,\text{m}$$

Moment of inertia of circle about centre of gravity is

$$I_{GC} = \left(\frac{\pi d^4}{64}\right) = \left(\frac{3.14 * 2.5^4}{64}\right) = 1.916\,\text{m}$$

Total moment of inertia

$$I_G = I_{G_t} - I_{GC} = 2.464 - 1.916 = 0.547\,\text{m}$$

Center of pressure of plane surface is

$$\bar{h} = \frac{I_G * \sin^2 60°}{A\bar{x}} + \bar{x} = \frac{0.547}{5.59 * 4.16} * \sin^2 60° + 4.16 = 4.18\,\text{m}$$

Hence, the center of gravity is 4.16 m and center of pressure is 4.18 m.

Example 3.17: A plane surface in rectangular shape having a circular hole in it is immersed inclined at angle of 30° to the free surface of a liquid having specific gravity of 0.8 such that its tip is 2 m below the surface level of the liquid free surface and is of dimensions as shown in the figure. Determine the total pressure force acting on the rectangular plane surface with circular hole in it and also the center of pressure.

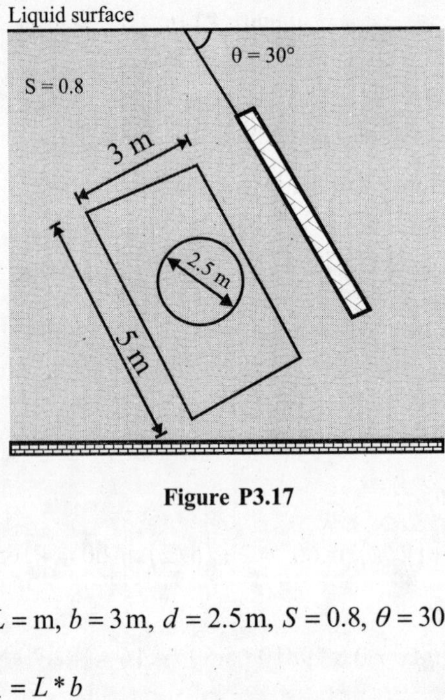

Figure P3.17

Solution: Given data:

$$L = \text{m},\ b = 3\,\text{m},\ d = 2.5\,\text{m},\ S = 0.8,\ \theta = 30°$$

The area of the rectangle $A_r = L * b$

$$\Rightarrow A_r = L * b = 5 * 3 = 15\,\text{m}^2$$

Area of the circle hole $A_c = \dfrac{\pi d^2}{4}$

$$\Rightarrow A_c = \frac{3.14 * 2.5^2}{4} = 4.906\,\text{m}^2$$

Total area of the plane surface

$$\Rightarrow A = A_r - A_c = 15 - 4.906 = 10.09\,\text{m}^2$$

Total center of gravity

$$\bar{x} = 2 + \left(\frac{L}{2}\right)\sin 30 = 2 + \left(\frac{5}{2}\right)\sin 30 = 3.135\,\text{m}$$

And hence total pressure force

$$P = \gamma A\bar{x} = 0.8 * 10.09 * 3.13 = 248.2\,\text{kN}$$

Moment of inertia of rectangle about centre of gravity is

$$I_{G_r} = \left(\frac{Lb^3}{12}\right) = \left(\frac{5 * 3^3}{12}\right) = 11.25\,\text{m}$$

Moment of inertia of circle about centre of gravity is

$$I_{G_C} = \left(\frac{\pi d^4}{64}\right) = \left(\frac{3.14 * 2.5^4}{64}\right) = 1.916\,\text{m}$$

Total moment of inertia

$$I_G = I_{G_r} - I_{G_C} = 11.25 - 1.916 = 9.334\,\text{m}$$

Center of pressure of plane surface is

$$\bar{h} = \frac{I_G * \sin^2 30°}{A\bar{x}} + \bar{x} = \frac{9.334}{10.09 * 3.13} * \sin^2 30° + 3.13 = 3.195\,\text{m}$$

Hence, the center of gravity is 3.13 m and center of pressure is 3.195 m.

Example 3.18: A plane washer surface is immersed inclined at angle of 60° to the free surface of a liquid having specific gravity of 1.2 such that its tip is 2 m below the surface level of the liquid free surface and is of dimensions as shown in the figure. Determine the total pressure force acting on the circular washer plane surface and also the center of pressure.

Solution: Given data:

$$D = 5\,\text{m},\ d = 2\,\text{m},\ \theta = 60°,\ S = 1.2$$

The area of the bigger circle

$$A_C = \left(\frac{\pi D^2}{4}\right)$$

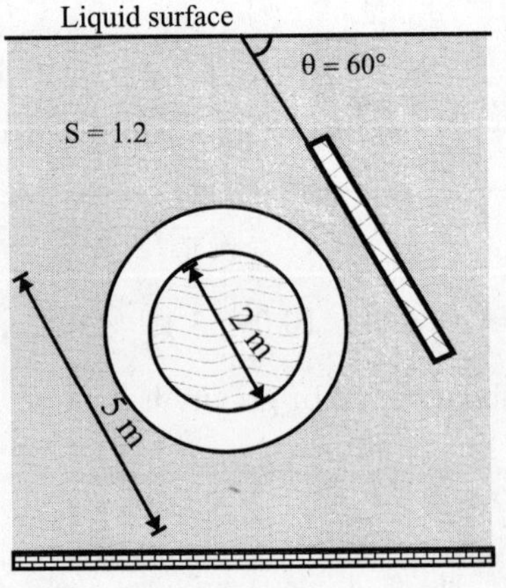

Figure P3.18

$$\Rightarrow A_C = \left(\frac{3.14 * 5 * 5}{4}\right) = 19.625\,\text{m}^2$$

The area of the smaller circle

$$A_c = \left(\frac{\pi d^2}{4}\right) \Rightarrow A_c = \left(\frac{3.14 * 2 * 2}{4}\right) = 3.14\,\text{m}^2$$

Total area of the plane surface is

$$A = A_C - A_c = 19.625 - 3.14 = 16.485\,\text{m}^2$$

Total center of gravity will be

$$\bar{x} = 2 + \left(\frac{d}{2}\right)\sin 60° = 4.165\text{ m}$$

And hence total pressure force

$$P = \gamma A\bar{x} = 1.2 * 16.485 * 4.16 = 808.26\,\text{kN}$$

Moment of inertia of the bigger circle about C.G. is

$$I_{G_C} = \left(\frac{\pi D^4}{64}\right) = \left(\frac{3.14 * 5^4}{64}\right) = 30.66\,\text{m}^4$$

Moment of inertia of smaller circle about C.G. is

$$I_{G_c} = \left(\frac{\pi d^4}{64}\right) = \left(\frac{3.14 * 2^4}{64}\right) = 0.785\,\text{m}^4$$

Total moment of inertia

$$I_G = I_{G_C} - I_{G_c} = 30.66 - 0.785 = 29.875\,\text{m}^4$$

Center of pressure of plane surface is

$$\bar{h} = \frac{I_G * \sin^2 60°}{A\bar{x}} + \bar{x} = \frac{29.875}{16.485 * 4.16} * \sin^2 60° + 4.16 = 4.45\,\text{m}$$

Hence, the center of gravity is 4.16 m and center of pressure is 4.45 m.

Example 3.19: A cover plate of 0.8 m × 0.8 m in dimension is hinged to a container at one of its ends to cover the oil contained in the container and oil has specific gravity of 0.85. The internal pressure due to the oil in the container is read as 20 kPa, by the pressure gage connected to the container. A force of "*F*" is applied to the cover plate at the other end so as to keep the cover plate tight, as shown in the figure. Determine the force "*F*" and the reaction "*R*" at the hinge.

Solution: Given data:

$$A = 0.8 \times 0.8 = 0.64\,\text{m}^2,\ \theta = 45°,\ S = 0.85,\ P = 20\ \text{kPa}$$

Center of gravity of the cover plate

$$\bar{x}_C = \frac{\text{L}}{2} * \sin 45° = \left(\frac{0.8}{2}\right) = 0.283\,\text{m}$$

As there is a pressure measured due to presence of oil in the container, we need to calculate the pressure head created due to the oil presence, and hence, the pressure head will be

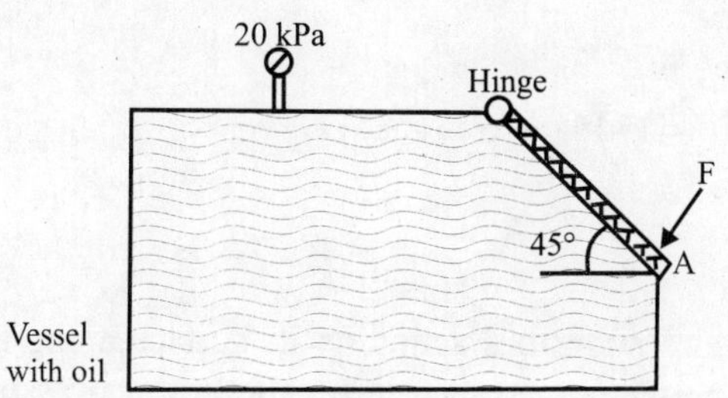

Figure P3.19a

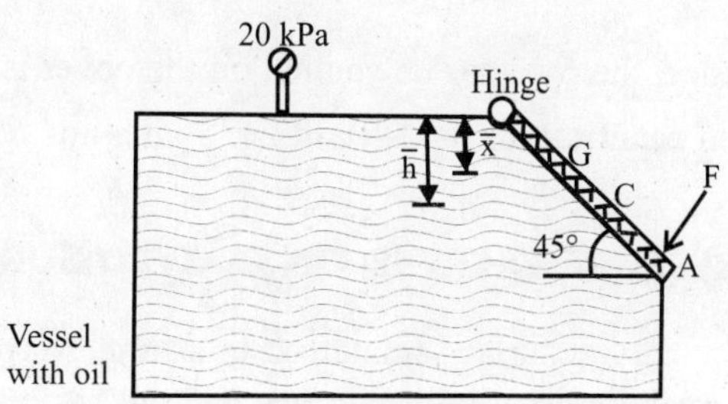

Figure P3.19b

$$h = \frac{P}{\gamma} = \left(\frac{20000}{0.85*1000*9.81} \right) = 2.399\,\text{m}$$

So the total center of gravity will be the sum of center of gravity and the pressure head, which gives total center of gravity as

$$\overline{x} = \overline{x}_C + h = 0.283 + 2.99 = 3.273\,\text{m}$$

Hence, total pressure force

$$P = \gamma A\overline{x} = 0.85 * 1000 * 9.81 * 0.64 * 3.27 = 17.465\,\text{kN}$$

Center of pressure of plane surface is

$$\overline{h} = \frac{I_G * \sin^2 60°}{A\overline{x}} + \overline{x}$$

$$\Rightarrow \overline{h} = \frac{\left(\frac{LB^3}{12}\right)}{0.64 * 3.273} * \sin^2 45° + 3.273 = \left(\frac{0.8*0.8^3}{12}\right) * 0.2386 + 3.273 = 3.28\,\text{m}$$

Hence, the center of gravity is 2.399 m and center of pressure is 3.28 m.
Taking moment about the hinge, we have two forces acting, like force applied "*F*" and the pressure force acting at the center of pressure due to the presence of oil in the container, and hence the moments about the hinge by these two forces must be equal
Hence,

$$F * L = P * \frac{\overline{h} - h}{\sin 45^0}$$

$$\Rightarrow F * 0.8 = 17465 * \frac{3.28 - 2.99}{\sin 45°}$$

$$\Rightarrow F = 8953\,\text{N}$$

Hence, the **force** to be applied on the cover is 101266 N.
And **reaction force** "R" will be "P–F", i.e. $R = p - F = 17465 - 8953 = 8511.5\,\text{N}$

PRESSURE EXERTED ON CURVED SURFACE

Any curved surface immersed in a liquid have pressure intensities acting on its surface along the normal direction to the surface and the force acting on the curved surface is called **total pressure force** and it can be resolved into components along the X-axis and Y-axis.

Consider a curved surface AB as shown in Figure 3.4a. If OA and OB are treated as rectangular faces, as shown, on which the horizontal component of pressure intensity dP_H and vertical component of pressure intensity acts. Then the total pressure "P" will be $P = pA = \int \gamma(dA)(x)$ acting on the curved surface AB and the **magnitude of total pressure force** will be

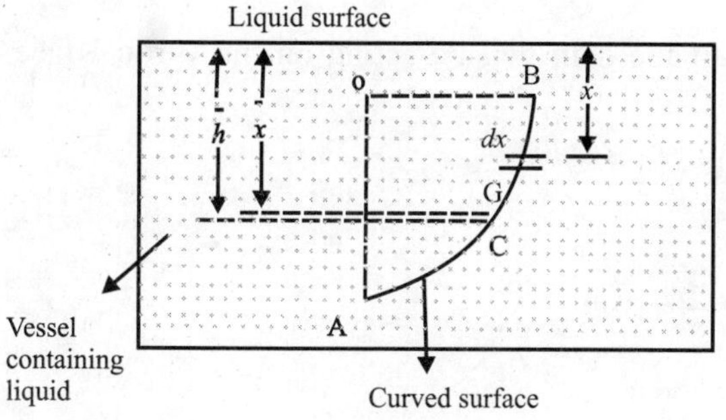

Figure 3.4a

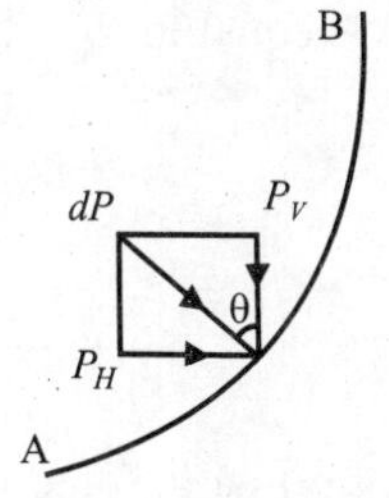

Figure 3.4b

$$P = \sqrt{P_H^2 + P_V^2} \tag{3.10}$$

And the total pressure force inclination is

$$\theta = \tan^{-1}\left(\frac{P_V}{P_H}\right) \tag{3.10a}$$

Note: If the liquid acts from below the curved surface, then the vertical pressure component dP_V will be acting in upward direction.

Example 3.20: A dam has the cross-sectional profile composed of a vertical face with a circular curved section at the base as shown in Figure P3.20. The dam is 2 m long. Calculate the resultant force and its direction of application on the curved surface of the dam design.

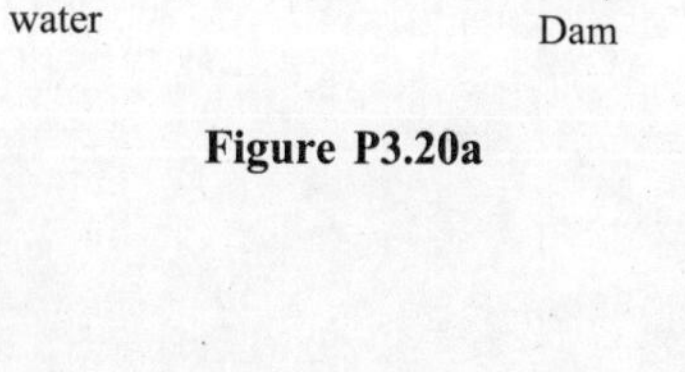

Figure P3.20a

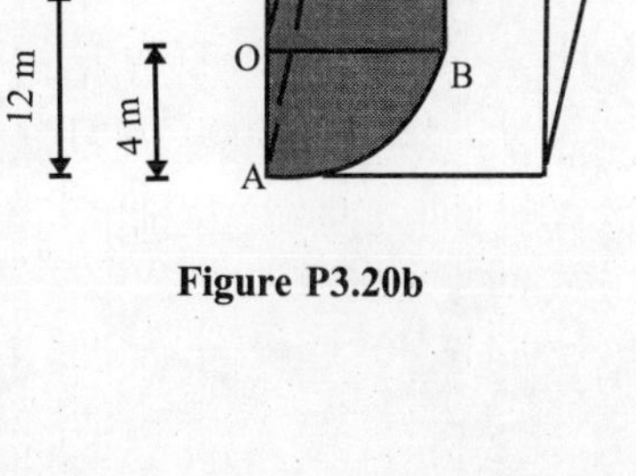

Figure P3.20b

Solution: Given data:

$$R = OA = OB = 4 \text{ m}, \; Width = AJ = DG = OF = 2\,\text{m}$$

$$Height = AD = 12\,\text{m}$$

Horizontal Pressure Force

Horizontal pressure force on the curved surface of the dam will be acting on the vertical face "OAJF" and is equal to

$$F_H = \gamma * Area * \bar{x} = \rho g * (OA * AJ) * \bar{x}$$

where

$$\bar{x} = (12 - 4) + \frac{4}{2} = 10\,\text{m}$$

So horizontal pressure force will be

$$F_H = 1000 * 9.81 * (4 * 2) * 8 = 627.8 \text{ kN}$$

Center of pressure force is

$$\bar{h} = \frac{I_g}{Area * \bar{x}} \sin^2\theta + \bar{x} = \frac{OF * OA^3/12}{Area * \bar{x}} * \sin^2\theta + \bar{x}$$

$$\Rightarrow \bar{h} = \frac{2 * 4^3}{12 * 8} * \sin^2 90° + 8 = 9.33\,\text{m}$$

Hence, the horizontal pressure force will be 627.8 kN acting at a point of $(12 - 9.33) = 2.67\,\text{m}$ from the base point "A".

Vertical Pressure Force

Vertical force on the dam will be equal to the weight of the water acting on the dams curved surface area and hence

$$F_V = W = \gamma * (volume) = \rho g * \left\{ (Area * Width)_{curved} + (Area * Width)_{rectangular} \right\}$$

$$\Rightarrow F_V = \rho g * \left\{ \left(\frac{\pi R^2}{4} * OF \right)_{curved} + (OB * OD * DG)_{rectangular} \right\}$$

$$\Rightarrow F_V = 1000 * 9.81 * \left\{ \left(\frac{3.14 * 4^2}{4} * 2 \right) + (4 * 8 * 2) \right\} = 874.27\,\text{kN}$$

Magnitude of the resultant force

$$F_R = \sqrt{F_h^2 + F_V^2} = \sqrt{627800^2 + 874267^2} = 1076.33\,\text{kN}$$

Direction of the resultant force $\theta = \tan^{-1}\left(\dfrac{F_V}{F_h}\right) = 55°$

Example 3.21: A concrete dam of width 10 m has the cross-sectional profile as shown in Figure P3.21a. Calculate the magnitude, direction and position of action of the resultant force exerted by the water on the dam.

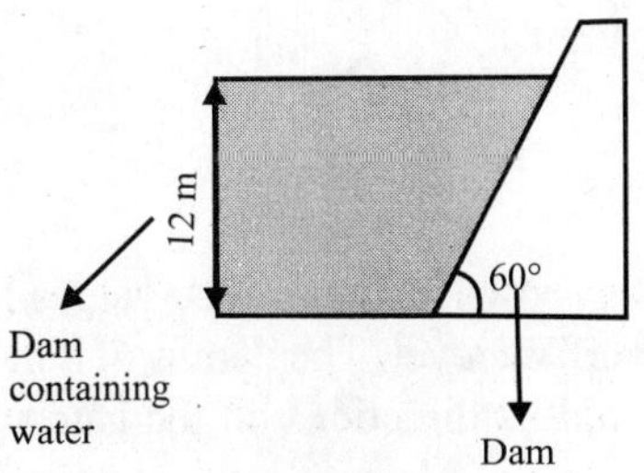

Figure P3.21a

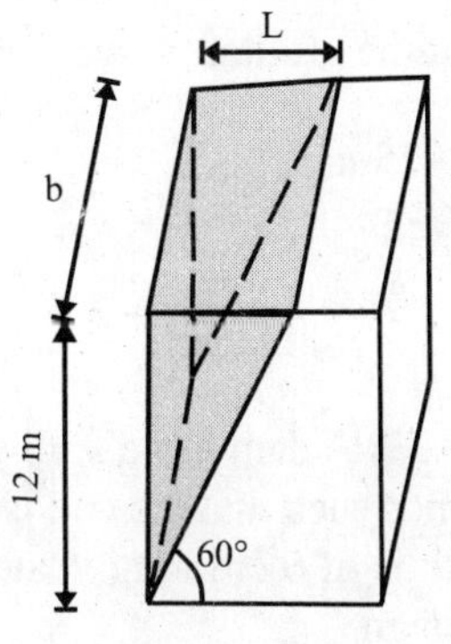

Figure P3.21b

Solution: Given data: Area of the dam wall = 10 m × 12 m = 120 m²
Vertical height of the water level (h) = 12 m
Width of dam (b) =10 m.

As per the Figure 3.21a, we have sloping wall length, $m = h/\sin 60 = \dfrac{12}{0.866} = 13.85\,\text{m}$ and $L = h\tan 30 = 6.928$ m. Vertical force on the dam will be equal to the weight of the water acting on the dam's slope area, and hence

$$F_V = W = \gamma * (volume) = \rho g * (Area * Width)$$

$$\Rightarrow F_V = \rho * g * \left(\frac{1}{2} h * L * b\right)$$

$$\Rightarrow F_V = 1000 * 9.81 * (0.5 * 12 * 6.928 * 10) = 4077820\,\text{N}$$

Horizontal force on the dam will be equal to the projection of water weight acting on the vertical area, and hence $F_H = W = \rho g * (Vertical\ \ Area * Projection\ \ height)$

$$\Rightarrow F_H = \rho * g * \left(h * b * \frac{h}{2} \right)$$

$$\Rightarrow F_H = 1000 * 9.81 * (12 * 10 * 6) = 7063200 \text{N}$$

Magnitude of the resultant force

$$F_R = \sqrt{F_h^2 + F_V^2} = \sqrt{7063200^2 + 4077820^2} = 8155820 \text{N}$$

Direction of the resultant force $\theta = \tan^{-1}\left(\frac{F_V}{F_h}\right) = 60^0$ $\bar{x}$ is the distance along the face to the centroid

= m/2 = 6.925 m

$$\bar{h} = \frac{I_g}{A\bar{x}} + \bar{x} = \frac{bm^3/12}{A\bar{x}} + \bar{x} = \frac{10 * 13.85^3}{12 * 120 * 6.925} + 6.925 = 9.58 \text{ m}$$

Example 3.22: A dam has a spillway gate at angle of 30°, as shown in figure. The gate is hinged at one of its ends such that the end is at the level of the water surface level. The dam is of 2 m in width and the gate is of 6 cm long. Calculate the resultant force and its directions on the gate surface of the dam design.

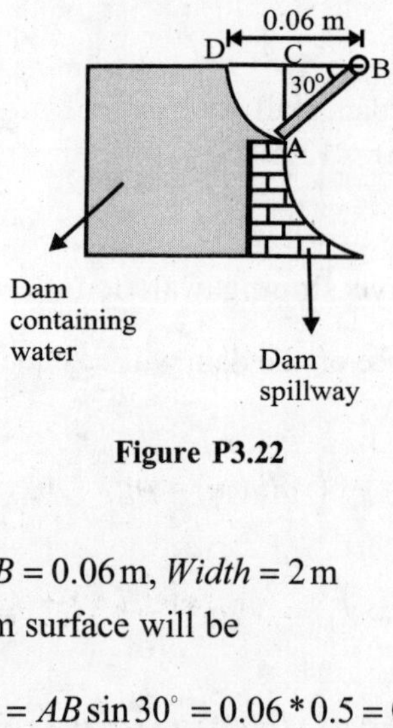

Figure P3.22

Solution: Given data: $R = AB = DB = 0.06\text{m}$, $Width = 2\text{m}$
Height of the spillway from the dam surface will be

$$h = AC = AB\sin 30^\circ = 0.06 * 0.5 = 0.03\text{m}$$

Horizontal Pressure Force

Horizontal pressure force on the curved surface of spillway will be acting on the vertical face "AC" and is equal to

$$F_H = \gamma * Area * \bar{x} = \rho g * (AC * Width) * \bar{x}$$

where $\bar{x} = \frac{h}{2} = 0.015m$

So horizontal pressure force will be $F_H = 1000 * 9.81 * (0.03 * 2) * 0.015 = 8.829 \text{ N}$. Hence, the horizontal pressure force will be 8.829 Pa acting at a point of 0.015 m from the surface level of the dam.

Vertical Pressure Force

Vertical force on the dam will be equal to the weight of the water acting on the dam's curved surface area and hence

$$F_V = W = \gamma * (volume) = \rho g * \{(Area * Width)_{ABD} - (Area * Width)_{ABC}\}$$

Since the sector "ABD" represents a 1/12th of a circle and so area of the sector will be $\frac{\pi R^2}{12}$.

i.e. $$\Rightarrow F_V = \rho g * \left\{\left(\frac{\pi R^2}{12} * 2\right) - \left(\frac{1}{2} * AC * BC * 2\right)\right\}$$

$$\Rightarrow F_V = 1000 * 9.81 * \left\{\left(\frac{3.14 * 0.06^2}{12} * 2\right) - \left(\frac{1}{2} * 0.03 * 0.06 \cos 30^\circ * 2\right)\right\} = 3.198 \text{N}$$

Magnitude of the resultant force $F_R = \sqrt{F_h^2 + F_V^2} = \sqrt{8.829^2 + 3.198^2} = 9.39 \text{N}$

Direction of the resultant force $\theta = \tan^{-1}\left(\frac{F_V}{F_h}\right) = 20^0$

BUOYANCY AND FLOTATION

Introduction to Buoyancy

When thought of fishes, sharks, sea creatures living in sea waters; why do they do not sink, when gravity acts on them and remains at the sea depth levels as per their requirement? The answer is **Buoyancy,** which is the "**upward or downward force of the fluids exerted on objects, tendening object immersed in the liquid to float or to rise**".

When fishes tend to go the depths, they give out the air in them through gas bladder make themselves denser than the sea waters. When required to go to surface levels, they inhale air through mouth into the gas bladder making their density lesser than the sea water. This concept of upward or downward lift of objects immersed in fluids was given by **Archimedes principle** (287

BC – 212 BC) is named after Archimedes of Syracuse who first discovered this law and was an ancient Greek mathematician, physicist, astronomer, engineer and defines as "**An immersed object is pushed up by a force equal to the weight of the fluid it displaces**" and the **principle of flotation** as "**A floating object displaces a quantity of fluid of weight equal to its own weight".** According to the principle of floatation, an object will float if the fluid displaced by it has a greater mass than the object's mass. For example, ships when they have their density lesser than the sea water, they float on the surface, but if density of the ship is greater than the water then it sinks.

Ships when designed, kept in mind that the hull (portion of ship in the sea waters) carries maximum weight with less surface of contact between hull and sea water. The increased weight carrying displaces maximum sea water to give maximum upward buoyancy force. At the same time minimal surface contact with sea water is maintained to reduce friction of drag.

To maintain at a particular level of the depth by the fishes or the ships, they should make their density of their bodies to be equal to the fluid that's been displaced by them, which is termed "**Neutral Buoyancy**". To float on surface, they need to be **positively buoyant** (i.e., object must not be denser than the fluid displaced by it) and to sink/go to depths they need to **be negatively Buoyant** (i.e., object must be denser than the fluid displaced by it). And thus among objects with equal masses, the one with greater volume has greater buoyancy.

The density of the immersed object with respect to the density of the fluid will be:

$$\frac{Density\ of\ object}{Density\ of\ fluid} = \frac{Weight}{Weight - Apparent\ immersed\ weight} \tag{3.11}$$

$$Apparent\ weight = weight\ of\ object - loss\ weight \tag{3.11a}$$

So the **Apparent weight is "the weight experienced by an object when immersed in a fluid**".

The buoyancy force acting through the center of gravity of the displaced fluid is termed **"Center of Buoyancy"** and it can be in three ways of equilibrium position, like:

a) Stable. The body returns to it's original position if given a small angular displacement.
b) Neutral. The body remains in a new position if given a small angular displacement.
c) Unstable. The body heals further over if given a small angular displacement.

$$F_{buoyancy} = \rho_{liquid} * \text{(Displaced liquid volume)} * g \tag{3.12}$$

$$F_{gravity} = \rho_{object} * \text{(Object volume)} * g \tag{3.13}$$

$$F_{total} = F_{buoyancy} - F_{gravity} \tag{3.14}$$

Example 3.23: What would be the apparent weight of a ball lifted up into the air. Assume the weight of ball as 75 grams and loss 10 grams of its weight in the air.

Solution: Given data: $W = 75$ g, $W_L = 10$ g
As the ball lifted up is losing its weight, the apparent weight will be

Apparent weight = weight of object – loss in weight or buoyancy force

Apparent weight = 75 – 10 = 65 grams
So the apparent weight of the ball in the air is 65 grams.

Example 3.24: Which weighs more underwater: a kilogram of lead or concrete?

Solution: Given data:

$$W_{lead} = W_{concrete} = 1 \text{ kg}$$

The concrete and lea weighs equally on earth surface.

But we know the lead is more denser than concrete and we know the mass of lead as

$$m = \rho_{lead} * (\text{Displaced liquid volume})$$

$$\Rightarrow \rho_{lead} = \frac{m}{(\text{Displaced liquid volume})}$$

As the volume displaced is inversely proportional to the density, less amount of water is displaced when lead is immersed in fluid than concrete and so less weight loss (or buoyancy) for lead.

Hence, as lead is denser than concrete and so weigh more in under water.

Stability of Fully Submerged Bodies

When an object (as shown in Figure 3.5a) when immersed fully immersed in a fluid, displaces an equal amount of fluid above it. Hence, the initial level of fluid (O-O) increases to new level (O′-O′).

Every object has center of gravity through which the whole weight of the body passes and if assumed the volume of object as "*v*", "γ" as specific weight of the fluid displaced by the object and "*W*" as weight of the object, then there arise a uniform pressure normal to the surface (according to

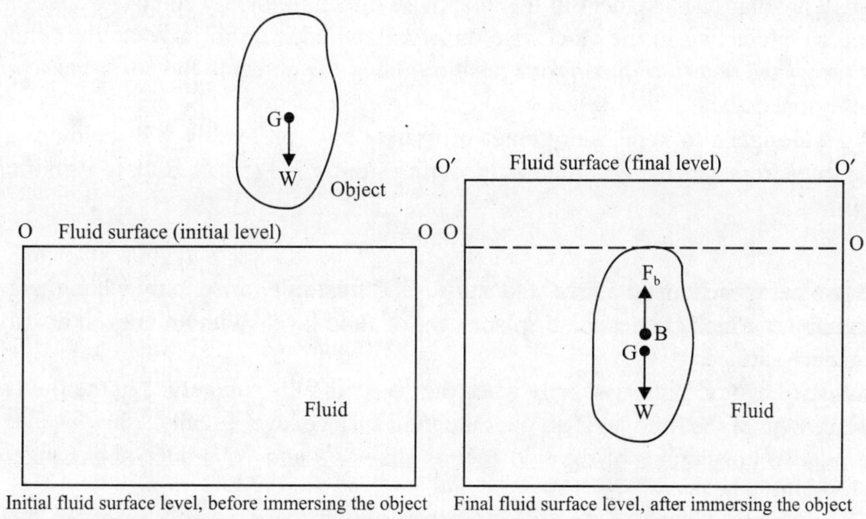

Fig. 3.5a: Object immersed showing buoyancy and center of gravity

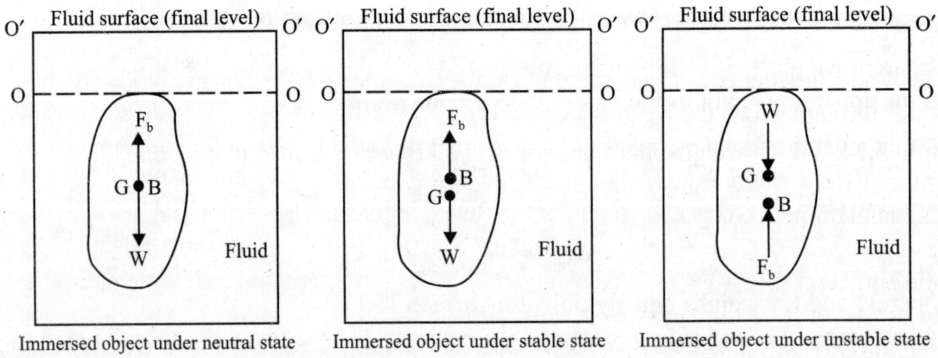

Fig. 3.5b: Object immersed showing nature of stability

Pascal law) tending the object upwards. Hence, the upward pressure when viewed in terms of single force results in upward force which will be in opposite direction of the weight of the object. This force is termed **Buoyancy force** (F_b) and passes through a point called **center of buoyancy** (*B*) which is the center of gravity of the displaced fluid. So there exist three different situations with respect of "center of gravity (*G*)" and "center of buoyancy (*B*)" positioning, like as shown in Figure 3.5b:

If "*B*" and "*G*" coincides, then the weight of the fluid displaced by the object and the weight of the object are equal and thereby the object remains neutral state.

If "*B*" is above "*G*", then the weight of the fluid displaced by the object is denser than weight of the object and there by the object remains floating and will be in a stable state. If the object is given a small angular displacement in the clockwise direction, then weight "*W*" and buoyancy F_b constitute a couple acting in the anticlockwise direction and brings the object to the original position. Thus, the object in the position shown is in stable equilibrium.

If "*B*" is below "*G*", then the weight of the fluid displaced by the object is less dense than weight of the object and thereby the object sinks and will be in an unstable state. If the object is given a small angular displacement in the clockwise direction, then weight "*W*" and buoyancy F_b constitute a couple acting in the clockwise direction and adds up to increase the tilting position, instead of brings the object to the original position. Thus, the object in the position shown is in Unstable equilibrium.

Ships are designed to keep their center of gravity low (preferably well below water line) to preventing ships from capsizing/titling easily, during storms and restore back to its normal position under equilibrium.

Note:

1. The two cases mentioned above (like **stable** and **unstable**) arise, only when there is a tilting of the object which creates the displacement of fluid levels with an uneven distribution over the object surface.
2. The case of neutral state arise only when the object is fully submerged in the fluid and creates displacement of the fluid level all over the fluid surface level equally.
3. The relative positioning of the two centers (like "*B*" and "*G*") does not change due to the rotation/tilting of the object.
4. The small tilting of object creates a rotating couple in between the buoyancy force and the weight of object.

METACENTER AND STABILITY OF PARTIALLY SUBMERGED BODIES

Observe a body floating over the horizontal fluid surface level ($O - O$). The body has a center of gravity "G" through which the weight due to gravity ("W") of the body passes vertically downwards and the fluid level displaced by its weight surrounds the partially immersed object uniformly and so the displaced fluid exerts equal amount weight force termed "Buoyancy force ("F_b") vertically upwards maintaining a constant normal line of action providing the neutral position equilibrium of floating.

If the body that is floating is tilted by an angle "θ", then the neutral state is disturbed and so the fluid surface level ($O - O$) displaces to new surface level ($O' - O'$), as shown in the Figure 3.6a. The center of gravity "G" and center of buoyancy "B" also get displaced along with the normal line of action to new positions "G_1 & B_1" and new normal line of action. At the same time the relative positions (line of action) of these two centers "G_1 & B_1" do not change even in the new normal line of action. Hence, if the line of action of the new centers, "G_1 & B_1" when extended, meet the old vertical line of action at a point termed "**Metacenter** (*M*)". So a **Metacenter** can be defined as "**the point where the normal line of actions of old and new positions meet is called Metacenter**" and the height between the center of gravity (*G*) and metacenter (*M*) is termed "**Metacentric Height** ("*MG*"). The new normal line of action develops a moment couple (*C*) about the original position due to the rotation of the body, as shown in Figure 3.6b. And the **rotating couple/righting moment** can be given as,

C = Weight * perpendicular distance about metacenter

$$C = W * GM \sin \theta \tag{3.15}$$

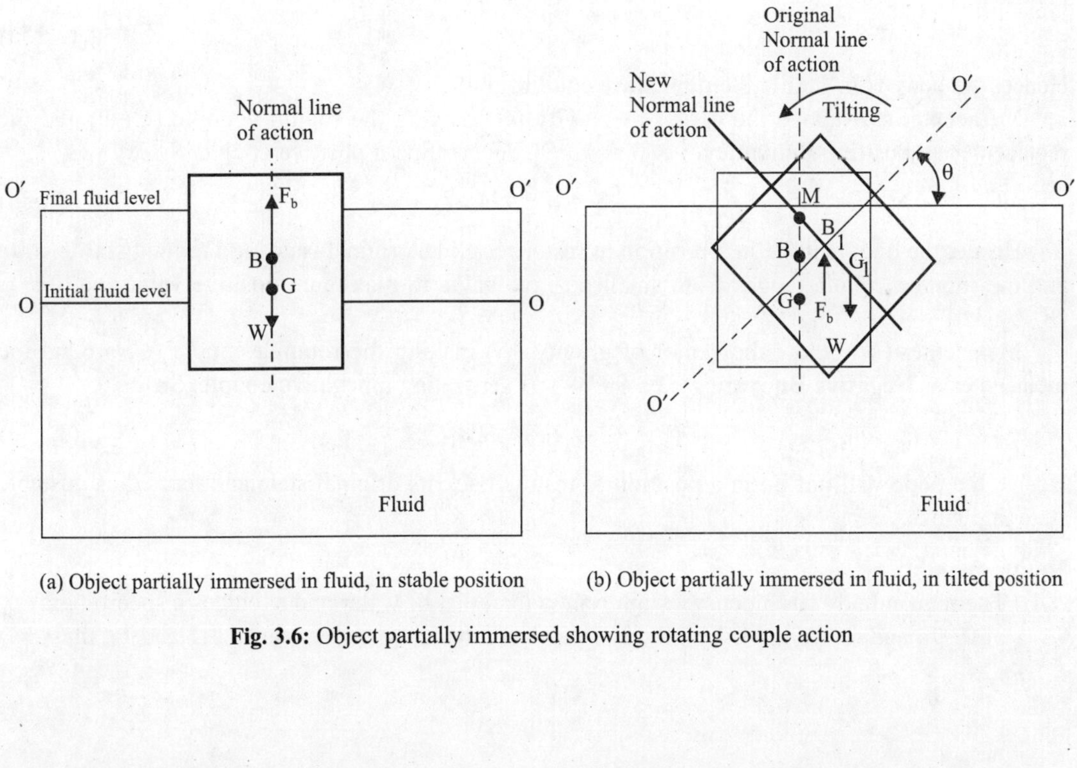

(a) Object partially immersed in fluid, in stable position (b) Object partially immersed in fluid, in tilted position

Fig. 3.6: Object partially immersed showing rotating couple action

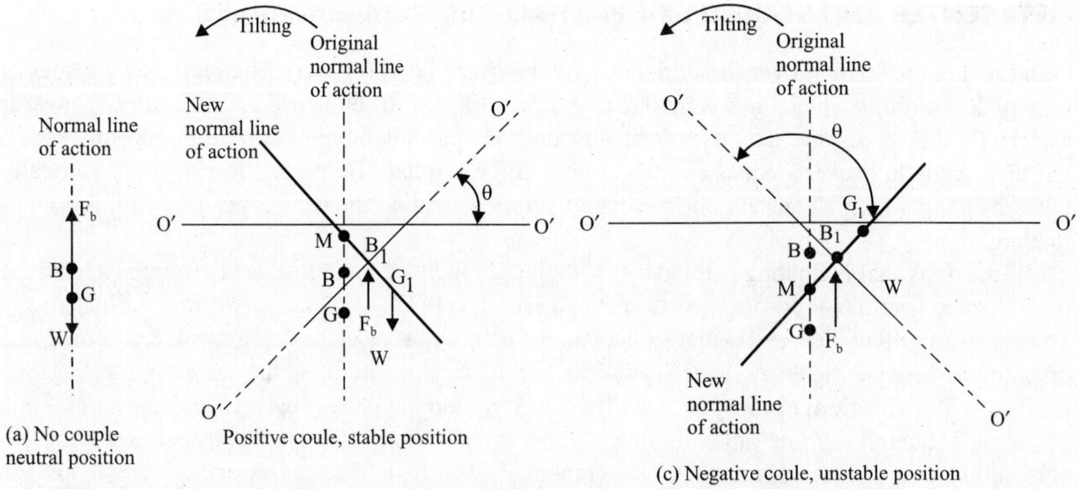

Fig. 3.7: Object partially immersed showing difference rotating couple action

As the rotating couple of new normal line of actions decides the stability state of a partially submerged body, the metacenter positioning with respect to the center of gravity of the body can be considered for analyzing the floating or sinking of the floating body. The positioning of the metacenter relative to the center of gravity "can be in three different positions "G_1 & B_1", as shown in Figure 3.7.

Metacenter (M) coincides with the center of gravity (G) making the rotating couple (C) around the metacenter will be **zero**, i.e. $\theta = 0°$ generating no couple. So

$$C = 0 \tag{3.15a}$$

Hence, the body is in **Stable Neutral state** equilibrium.

Metacenter (M) above the center of gravity (G) making the rotating couple (C) around the metacenter as **Positive Buoyancy**, i.e. $0 < \theta \leq 90°$ generating a positive couple. So

$$C = W * GM \tag{3.15b}$$

Hence, the body will be in a position to restore back its original state, and termed **stable state** and the rotating couple ranges from small positive value to maximum positive value (i.e. $0 < C$ ($W * GM$)).

Metacenter (M) below the center of gravity (G) making the rotating couple (C) around the metacenter as **Negative Buoyancy**, i.e., $90° - < \theta$ generating a negative couple. So

$$C = - W * GM \tag{3.15c}$$

Hence, the body will **not be** in a position to restore back its original state and termed as unstable state and sinks.

Note:

1. The human body has a density slightly less than that of water and averages a specific gravity of 0.974 and so we float. Lean body mass has a typical density near to 1.1 and fat mass has

a density of about 0.9. Hence, an obese individual floats more easily than a lean individual.

2. The buoyant force of a liquid depends on its density, its weight per unit of volume. The density of freshwater (28.3 kg/0.03 m^3) is less dense than the density of sea water (29 kg/0.03 m^3) and so people float better in sea water than in freshwater.

METACENTER HEIGHT DETERMINATION

The positioning of metacenter plays an important role in determining the stability of an object, as described and hence it is required to determine the **Metacentric height** (GM). To determine the metacentric height, we can use two methods like:

a) Analytical/Theoretical method.
b) Experimental method.

a. Analytical/Theoretical Method

Consider an object of dimensions with length "x" and height "h" submerged partially in fluid. Let the fluid level initially be O-O when the object is in neutral position. When given a small tilt of an angle ("θ"), the fluid gets displaced on either side of the object height in form of two wedges (EAA_1) and (EDD_1) respectively, as shown in Figure 3.8.

Due to the disturbance created by the tilting of the object, small portion of the object gets lifted up and other opposite small portion gets immersed fully into the fluid forming wedges on the either sides.

The wedge (EAA_1) formed due to lifting of small portion loses it fluid density and the same density is attained by the other wedge (EDD_1) formed due to the immersion in the fluid. So there is

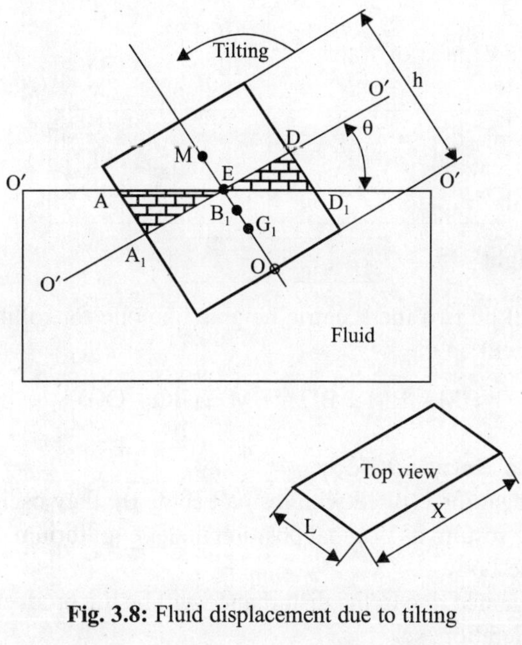

Fig. 3.8: Fluid displacement due to tilting

a shift in buoyancy center from point (B) to (B_1). Hence, the change in moment of buoyancy force will be

$$W * BM * \theta = (\gamma * \text{volume} * BM * \theta) \tag{3.16}$$

And the volume of the displaced wedge (EAA_1) is

$$\left(\frac{1}{2} * \frac{x}{2} * \frac{x}{2} * L\right) = \left(\frac{x^2 L}{8}\right) \tag{3.17}$$

The moment/rotating couple due to change in wedge displacement (C) will be

$$C = \gamma * \text{volume} * \text{perpendicular distance} = \gamma * \left(\frac{x^2 L}{8}\right)\theta * \frac{2x}{3} = \left(\frac{\gamma x^3 L}{12}\right)\theta \tag{3.18}$$

Under equilibrium, the moment of buoyancy point should be equal to the rotating couple and so equate Eq. (3.16) and Eq. (3.18)

i.e. $$(\gamma * \text{volume} * BM * \theta) = \left(\frac{\gamma x^3 L}{12}\right)\theta$$

$$\Rightarrow \left(\gamma * \left(\frac{x^2 L}{8}\right) * BM * \theta\right) = \left(\frac{\gamma x^3 L}{12}\right)\theta = \gamma I \theta$$

$\because$ Second moment inertia of area of water $I = \left(\dfrac{x^3 L}{12}\right)$

$$\Rightarrow BM = \left(\frac{x^3 L}{12}\right) * \left(\frac{8}{L * x^2}\right) = \left(\frac{I}{volume}\right) \tag{3.19}$$

Hence, the metacentric height $$GM = BM \pm BG \tag{3.20}$$

Note: For a ship, there will be two metacentric heights like one for rolling/rotating and other for pitching and so the height will be

$$\text{GM} = \text{BM} \pm \text{BG} = \text{BM} \pm (\text{BO} - \text{OG}) \tag{3.20a}$$

Period of Oscillation of a Rotating Body

When floating bodies are disturbed (tilted with a small angle θ), they oscillate for sometime like a simple pendulum and try to restore its original position under equilibrium. The period of oscillation depends on the mass of the body.

We know the torque attained due to the disturbance will be the product of moment of inertia of the body and angular acceleration, so

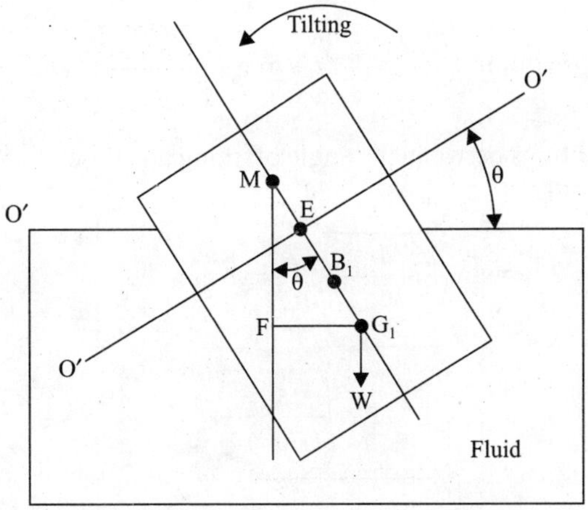

Fig. 3.9: Period of oscillation due to tilting

$$Torque = I * \alpha = \left(-\frac{W}{g} k^2 \right) * \alpha \tag{3.21}$$

Considering Figure 3.6, we have angular acceleration directed towards equilibrium position as

$$\alpha = \frac{d^2\theta}{dt^2} \tag{3.22}$$

Restoring couple will be

$$= W * GF = W * GM * \tan\theta = W * GM * \theta \tag{3.23}$$

So

$$Torque = \left(-\frac{W}{g} k^2 \right) * \left(-\frac{d^2\theta}{dt^2} \right) \tag{3.24}$$

In equilibrium, the restoring couple must be equal to the torque and so

$$W * G_1M * \theta + \left(\frac{W}{g} k^2 \right) * \left(\frac{d^2\theta}{dt^2} \right) = 0$$

By rearranging, we obtain $\left(\frac{d^2\theta}{dt^2} \right) + \frac{GM * g * \theta}{k^2} = 0$, which is an differential equation of second degree.

So the solution will be

$$\theta = A\sin\sqrt{\frac{GM*g}{k^2}}*t + b\cos\sqrt{\frac{GM*g}{k^2}}*t \tag{3.25}$$

Apply the boundary conditions between the angle of tilting and time of completion of an oscillation ("*T*"), we have at θ = 0°, t = T/2

So $A\sin\sqrt{\frac{GM*g}{k^2}}*\frac{T}{2} = 0 \Rightarrow \sin\sqrt{\frac{GM*g}{k^2}}*\frac{T}{2} = \pi$

$$T = 2\pi\sqrt{\frac{k^2}{GM*g}} \tag{3.26}$$

Hence, the time period of oscillation depends on the metacenter height and radius of gyration.

Example 3.25: Determine the metacentric height of a cylinder of 4 m diameter and height of 4 m, floating in water with its axis vertical, if period of oscillation is 6 seconds.

Solution: Given data: *GM* = *H*, *d* = *L* = 4 m, *T* = 6 sec.

The cylinder is immersed partially, so let "x" portion be immersed in the water out of the whole height "h".

We know, $BM = \left(\frac{I}{volume}\right) = \left(\frac{\pi*4^4}{64}\right)*\left(\frac{4}{\pi*x*L^2}\right) = \frac{1}{x}$

Radius of gyration for a cylinder is

$$k^2 = d^2\left(\frac{1}{16}+\frac{1}{12}\right) = 2.33 \text{ m}^2$$

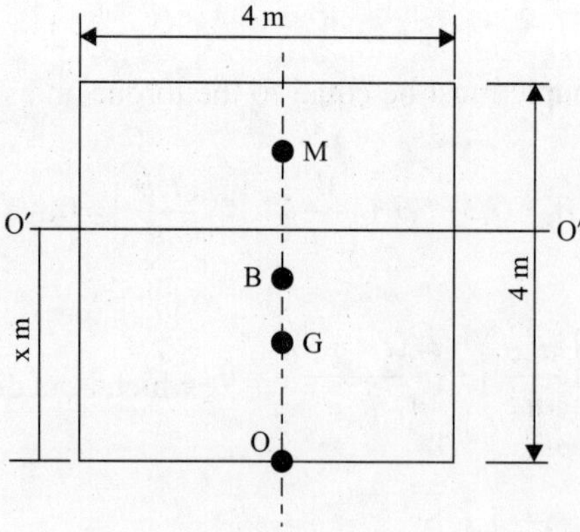

Fig. P3.25

We know, $T = 2\pi\sqrt{\dfrac{k^2}{GM * g}} \Rightarrow GM = H = \dfrac{4\pi^2 k^2}{gT^2} = 0.26$ m

Hence, metacentric height is 0.26 m.

Example 3.26: An iceberg in the ocean floats with 1/5th of its volume above the surface. What will be the specific gravity of the iceberg relative to the ocean water which weighs 50 N/m³. What portion of its volume would be above the surface in oil, if the iceberg floats with specific weight 13.6?

Solution: Given data: $\gamma_{ocean} = \rho_{ocean} * g = 50\ \text{N/m}^3$, volume = 1/5th

a) For equilibrium, the buoyancy force should be equal to the weight of the iceberg, so

$$F_b = W \Rightarrow \rho_{ocean} * \text{volume} * g = \rho_{iceberg} * \text{volume} * g$$

$$\Rightarrow \rho_{iceberg} * \text{volume} * g = \rho_{ocean} * \frac{1}{5} * \text{volume} * g$$

$$\Rightarrow \gamma_{iceberg} = \frac{1}{5} * \gamma_{ocean} = \frac{1}{5} * 50 = 10\ \text{N/m}^3$$

Relative specific gravity $S = \dfrac{\gamma_{iceberg}}{\gamma_{ocean}} = \dfrac{10}{50} = 0.2$

b) Specific weight of the oil will be $S = \dfrac{\gamma_{iceberg}}{\gamma_{oil}} = \dfrac{13.6}{\gamma_{oil}} = 0.2 \quad \Rightarrow \gamma_{oil} = 68\ \text{N/m}^3$

For equilibrium, the buoyancy force of oil should be equal to the weight of the iceberg, so

$$F_b = W \Rightarrow \gamma_{oil} * (\text{volume})_{oil} = \gamma_{iceberg} * (\text{volume})_{iceberg}$$

$$\Rightarrow \frac{(\text{volume})_{oil}}{(\text{volume})_{iceberg}} = \frac{\gamma_{iceberg}}{\gamma_{oil}} = 0.2$$

Volume above = total volume – (buoyancy volume in oil)

Divide through out by total volume, so

$$\frac{(\text{volume})_{above}}{(\text{volume})_{total}} = 1 - \frac{(\text{volume})_{oil}}{(\text{volume})_{iceberg}} = 1 - 0.2 = 0.8$$

So the portion of volume that will be 80% of the total volume.

EXERCISE

Objective

1. A depth of center of pressure in a rectangular lamina of height *h* with one side in the liquid surface is at
   a) *h* b) *h*/3 c) 2*h*/3 d) *h*/2 e) 3/4 *h*.
2. The horizontal component of force on a curved surface is equal to the
   a) Product of pressure at its centroid and area
   b) Weight of liquid retained by the curved surface
   c) Force on a vertical projection of the curved surface
   d) Weight of liquid vertically above the curved surface
   e) None of the above
3. The vertical component of pressure force on a submerged curved surface is equal to
   a) Weight of liquid vertically above the curved surface and extending upon the free surface
   b) The force on a vertical projection of the curved surface
   c) The product of pressure at centroid and surface area
   d) Horizontal component
   e) None of the above
4. When will the centre of pressure and the center of gravity of an immersed plane surface coincide?
   a) When surface immersed inclined
   b) The force on a vertical projection of the curved surface
   c) When surface immersed vertically
   d) When surface immersed horizontally
   e) None of the above
5. Upward or downward force of the fluids exerted on objects, tendening object immersed in the liquid to float or to rise
   a) Buoyancy
   b) Hydrostatic force
   c) Pressure force
   d) None of the above
6. Object immersed in a fluid is exerted to
   a) Gravity force downwards only
   b) Buoyancy force downwards only
   c) Gravity with downward and buoyancy with upward directions
   d) None of the above
7. To float on surface, an object needs to be **positively buoyant**, if immersed in a fluid.
   a) True
   b) False
   c) Not know
   d) None of the above

8. To stay at same level of depth, an object needs to be ____, if immersed in a fluid.
   a) Negatively buoyant
   b) Positively Buoyant
   c) Neutrally Buoyant
   d) None of the above
9. According to principle of flotation "A floating object displaces a quantity of fluid of weight ___ its own weight.
   a) Greater than
   b) Lesser than Positively Buoyant
   c) Equal to
   d) None of the above
10. The point in the immersed body through which the resultant pressure of the liquid may be taken to act is known as
    a) Metacenter
    b) Center of pressure
    c) Center of buoyancy
    d) Center of gravity
    e) None of the above
11. Center of pressure on an inclined plane is
    a) At the centroid
    b) Above the centroid
    c) At metacenter
    d) At center of pressure
12. The center of gravity of the volume of the liquid displaced by an immersed body is called
    a) Metacenter
    b) Center of pressure
    c) Center of buoyancy
    d) Center of gravity
13. Center of pressure compared to center of gravity is
    a) Above it
    b) Below it
    c) At same point
    d) Above or below depending on area of body
14. For a body floating in a liquid, the normal pressure exerted by the liquid acts at
    a) Bottom surface of the body
    b) C.G. of the body
    c) Metacenter
    d) All points on the surface of the body e) all of the above
15. In a stable immersed body, center of buoyancy should be ____
    a) At the center of gravity
    b) Above the center of gravity
    c) Below the center of gravity
    d) Could be above or below

16. When a block of wood is placed in a pail of water, the block displaces some of the water and the water level goes up. What would be the weight of the wood block if you could weigh the water that the wood displaces as 0.25 liters?
    a) 2.45 N
    b) 24.5 N
    c) 2.45 kg
    d) 24.5 kg

Answers

1. d	2. c	3. a	4. d	5. a	6. c	7. a	8. c
9. c	10. b	11. d	12. c	13. d	14. d	15. b	16. a

Theory

1. Write a note on immersion of plane surfaces.
2. Derive equation for pressure exerted on a horizontal plane surface.
3. Derive equation for pressure exerted on a vertical plane surface.
4. Derive equation for pressure exerted on an inclined plane surface.
5. Derive equation for pressure exerted on a curved plane surface.

Problems

1. A vertical isosceles triangle gate with its vertex up has a base width of 2 m and a height of 1.5 m. If the vertex of the gate is 1m below the free surface of water, find the total pressure force and the position of the center of the pressure on one side of the plane.
2. A cubical tank has sides of 1.5 m. It contains water in the lower 0.6 m depth. The upper remaining part is filled with oil of relative density 0.9. Calculate for one vertical side of the tank a.) The pressure force b.) Position of the center of pressure.
3. A vertical gate of width "W" controls the water level in the intake reservoir of a power plant. And the reservoir has a head level "H". What is the force of the water on the gate? What is the net hydrostatic force on the gate?
4. An inclined gate of 30° to the horizontal and has length of 3 m, width of 1.5 m, controls the water level in the irrigation channel. And the channel has a head level 1m. What is the force of the water on the gate? What is the net hydrostatic force on the gate?
5. A rectangular gate of length 1.5 m and width 0.5 m is fitted at the opening of the bottom of a cylindrical vessel containing kerosene. The gate is hinged on one of the larger edge corners and is fixed tight on the other larger side. The vessel contains kerosene to a depth of 5 m. Take the mass of the gate as 50 kg.
6. The caisson for closing the entrance to a dry dock is of trapezoidal form 15 m wide at the top and 12.2 m wide at the bottom and 8 m deep. If the water on the outside is just level with the top and dock is empty, find the total pressure on it and the depth of the center of pressure.

QUICK REFERENCE OF FORMULAE

Property	**Formulae**	**Units**	**Remarks, if any**
Pressure exerted on horizontal immersed plane surface (P)	$P = \gamma A\bar{x}$		$\bar{x}$ is distance of center of gravity of surface from free liquid surface.
Pressure exerted on vertical immersed plane surface (P)	$P = \gamma A\bar{x};\ \bar{h} = \frac{I_0}{A\bar{x}}$ or $\bar{h} = \frac{I_G}{A\bar{x}} + \bar{x}$ $I_0 = I_G + A\bar{x}$	N/m^3 or Pascal	$\bar{h}$ is distance of center of pressure on surface from free liquid surface.
Pressure exerted on inclined immersed plane surface (P)	$P = \gamma A\bar{x}$ $I_0 = I_G + \frac{A\bar{x}^2}{\sin^2\theta}$ $\bar{h} = \frac{I_0 \sin^2\theta}{A\bar{x}}$ *or* $\bar{h} = \left(\frac{I_G \sin^2\theta}{A\bar{x}} + \bar{x}\right)$		I_0 is Moment of inertia of surface about free liquid surface. I_G is Moment of inertia of surface gravity about free liquid surface.
Pressure exerted on curved immersed surface (P)	$P = \int \gamma A\bar{x} = \sqrt{P_H^2 + P_V^2},\ \theta = \tan^{-1}\left(\frac{P_V}{P_H}\right)$		θ is the pressure inclination P_H & P_V are the horizontal and vertical pressure components.
Density ratio of object and fluid in which it is immersed	$\frac{\textit{Density of object}}{\textit{Density of fluid}}$ $= \frac{\textit{Weight}}{\textit{Weight} - \textit{Apparent immersed weight}}$		
Apparent weight of object immersed in a fluid	*Apparent weight = Weight of object – Loss in weight*		
	$F_{buoyancy} = \rho_{liquid}$ * (Displaced liquid volume) * g $F_{gravity} = \rho_{object}$ * (Object volume) * g $F_{total} = F_{buoyancy} - F_{gravity}$		

CHAPTER 4

Fluid Flow Kinematics

INTRODUCTION

Fluid kinematics is primarily interested in the flow fields like velocity $v = f(x, y, z, t)$, acceleration, etc. and mass flow field $\overline{m} = f(x, y, z, t)$ while accompanying rotations effects.

Fluid kinematics is a branch of fluid mechanics, which deals with time and space relationship like displacement, velocity, accelerations, rotation of fluid particles, without the concern of forces/ energies causing them (like pressure energy, kinetic energy or potential energy) and describes the motion of fluids without necessarily considering the forces and moments that cause the motion, i.e. fluid kinematics tries to describe flow processes in field variables.

The motion of fluids can be described by two methods, namely, **Lagrangian** method and **Eulerian** method. Lagrangian method analyse any particular selected fluid particle, throughout its motion in space, i.e. pertaining to a flow particle, whereas Eulerian method analyse any fluid particle in space, at that point of location of fluid particle, i.e. pertaining to a flow field.

In this chapter, we adopt Eulerian method and introduce several kinematic concepts related to flowing fluids. We discuss the description of fluid flows and various ways to visualize flow fields—streamlines, streak lines, path lines, etc.

Fluids when in motion have several parameters coming into action, having time and space relationship like displacement, velocity, acceleration and rate of rotation of fluid particles, etc. which need to be analyzed for knowing the types of fluid and their behavior, which is done in fluid kinematics. To simplify the analysis, fluid kinematics does not consider the forces and energies that cause these parameters behavioral aspects and their impacts.

At last, the Reynolds transport theorem (RTT) is discussed emphasizing its role in transforming the equations of motion from Lagrangian method to Eulerian method. i.e. the conversion related with flow particle to a control volume field.

VELOCITY AND ACCELERATION OF A FLUID IN MOTION

Consider a fluid particle "P" moving in space with velocity $v = f(x, y, z, t)$. If the origin axis along X, Y, Z at time "t" is moved to the new particle location at time (t+dt), making the particle location as new location of the origin, then new axes becomes $i\ j\ k$ where $i\ j\ k$ are the unit vectors parallel

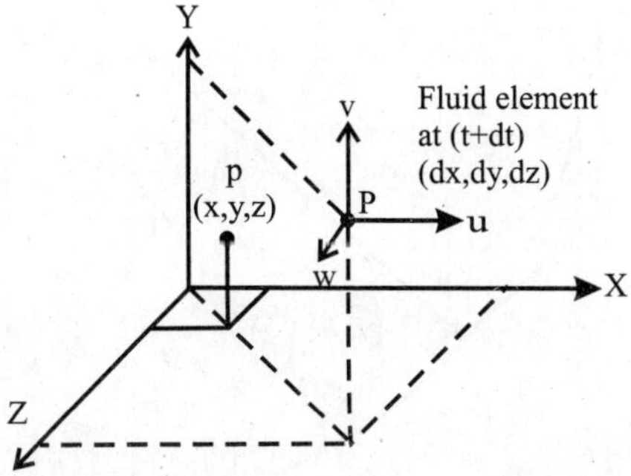

Figure 4.1: Motion of fluid element in space

to the X, Y, Z axes and with velocity components *u, v, w* along the original X, Y, Z axes, as shown in figure below. As the velocity is a function of space and time, the components u, v, w with respect to time, can be given as

$$\begin{Bmatrix} u = f(x, y, z, t) \\ v = f(x, y, z, t) \\ w = f(x, y, z, t) \end{Bmatrix} \tag{4.0}$$

The partial derivate of the velocity components can be written as

$$\begin{Bmatrix} du = \dfrac{\partial u}{\partial x}dx + \dfrac{\partial u}{\partial y}dy + \dfrac{\partial u}{\partial z}dz + \dfrac{\partial u}{\partial t}dt \\ dv = \dfrac{\partial v}{\partial x}dx + \dfrac{\partial v}{\partial y}dy + \dfrac{\partial v}{\partial z}dz + \dfrac{\partial v}{\partial t}dt \\ dw = \dfrac{\partial w}{\partial x}dx + \dfrac{\partial w}{\partial y}dy + \dfrac{\partial w}{\partial z}dz + \dfrac{\partial w}{\partial t}dt \end{Bmatrix} \tag{4.1}$$

If let the velocity components *u, v, w* be along *x, y, z* axis respectively, then

$$\Rightarrow u = \frac{dx}{dt}, v = \frac{dy}{dt}, w = \frac{dz}{dt} \tag{4.2}$$

So **velocity vector** *v* and **magnitude** *v* with respect to its components can be written as

$$\bar{V} = ui + vj + wk \tag{4.2a}$$

$$V = \sqrt{u^2 + v^2 + w^2} \tag{4.2b}$$

Acceleration is a measure of how the velocity of an object is changing over time. It can be found by computing the difference in velocities at initial and later times and dividing that by the difference in time.

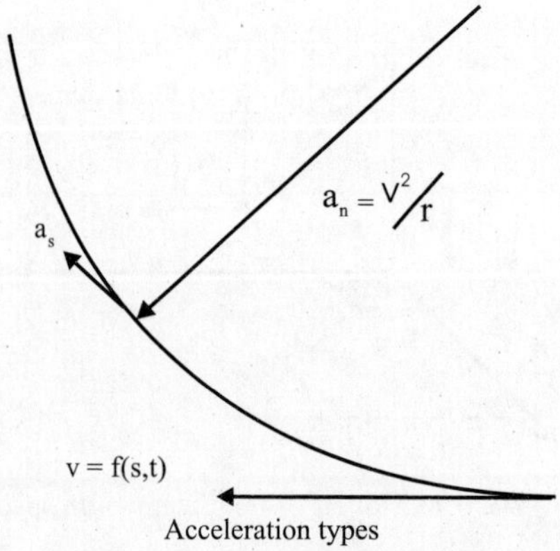

Figure 4.2: Acceleration types

i.e. $$a_s = \frac{V(x+dx, t+dt) - V(x,t)}{dt} = \frac{1}{dt}\left(V(x,t) + \frac{dV}{dx}dx + \frac{dV}{dt}dt - V(x,t)\right)$$

$$\Rightarrow a_s = \frac{dV}{dx} * \frac{dx}{dt} + \frac{dV}{dt} = V\frac{dV}{dx} + \frac{dV}{dt} \tag{4.3}$$

Or in other terms, we can write **Spatial Acceleration** components a_x, a_y, a_z with respect to time be given as

$$a_x = \frac{du}{dt}, a_y = \frac{dv}{dt}, a_z = \frac{dw}{dt} \tag{4.3a}$$

Since acceleration is the rate of change of velocity, substituting Eq. (4.2) in Eq. (4.3), we have

$$\left\{\begin{aligned} a_x &= \frac{\partial u}{\partial x}\frac{dx}{dt} + \frac{\partial u}{\partial y}\frac{dy}{dt} + \frac{\partial u}{\partial z}\frac{dz}{dt} + \frac{\partial u}{\partial t}\frac{dt}{dt} \\ a_y &= \frac{\partial v}{\partial x}\frac{dx}{dt} + \frac{\partial v}{\partial y}\frac{dy}{dt} + \frac{\partial v}{\partial z}\frac{dz}{dt} + \frac{\partial v}{\partial t}\frac{dt}{dt} \\ a_z &= \frac{\partial w}{\partial x}\frac{dx}{dt} + \frac{\partial w}{\partial y}\frac{dy}{dt} + \frac{\partial w}{\partial z}\frac{dz}{dt} + \frac{\partial w}{\partial t}\frac{dt}{dt} \end{aligned}\right\}$$

$$\Rightarrow \left\{\begin{aligned} a_x &= u\frac{\partial u}{\partial x} + v\frac{\partial u}{\partial y} + w\frac{\partial u}{\partial z} + \frac{\partial u}{\partial t} \\ a_y &= u\frac{\partial v}{\partial x} + v\frac{\partial v}{\partial y} + w\frac{\partial v}{\partial z} + \frac{\partial v}{\partial t} \\ a_z &= u\frac{\partial w}{\partial x} + v\frac{\partial w}{\partial y} + w\frac{\partial w}{\partial z} + \frac{\partial w}{\partial t} \end{aligned}\right\} \tag{4.4}$$

From Eq. (4.4) it can be generalized that the **spatial tangential acceleration vectors** and **magnitude** a_s as

$$\left\{\begin{aligned} a_s &= V\frac{\partial V}{\partial s} + \frac{\partial V}{\partial t} \\ a_s &= \sqrt{a_x^2 + a_y^2 + a_z^2} \end{aligned}\right\} \tag{4.5}$$

So the **total acceleration** a is a combination of spatial tangential a_s and normal acceleration a_n, as shown in figure above.

i.e. $$a = a_s + a_n \text{ where } a_n = \frac{V^2}{r} \tag{4.6}$$

i.e. Total acceleration $$a = V\frac{\partial V}{\partial s} + \frac{\partial V}{\partial t} + \frac{V^2}{r} \tag{4.7}$$

or $$a_s = u\frac{\partial V}{\partial x} + v\frac{\partial V}{\partial y} + w\frac{\partial V}{\partial z} + \frac{\partial V}{\partial t} \tag{4.7a}$$

where $V \cdot \frac{\partial V}{\partial s}$ is convective acceleration and

$V \cdot \frac{\partial V}{\partial t}$ is local acceleration.

Example 4.1: The velocity field in a fluid medium is given by $V = 3xy^2 i + 2xyj + (2zy - 3t)k$. Find magnitudes and directions of translation velocity and acceleration at point $P(1, 2, 1, 3)$.

Solution: Given data:

$$V - 3xy^2 i + 2xyj + (2zy - 3t)k \text{ and } P(1,2,1,3)$$

From the given equation, velocity components will be

$$u = 3xy^2, v = 2xy, w = (2zy - 3t)$$

Magnitude of translation velocity will be

$$V = \sqrt{(3xy^2)^2 + (2xy)^2 + ((2zy - 3t))^2}$$

$$\Rightarrow V_{1,2,1,3} = \sqrt{(3*1*4)^2 + (2*1*2)^2 + ((2*1*2 - 3*3))^2} = 13.6 \text{ m/s}$$

Direction of translation velocity will be

$$\bar{V} = 3xy^2 i + 2xyj + (2zy - 3t)k$$

$$\Rightarrow \overline{V}_{1,2,1,3} = 3*1*4i + 2*1*2j + (2*1*2 - 3*3)k$$

$$\Rightarrow \overline{V}_{1,2,1,3} = 12i + 4j - 5k$$

Spatial acceleration $$a_s = u\frac{\partial V}{\partial x} + v\frac{\partial V}{\partial y} + w\frac{\partial V}{\partial z} + \frac{\partial V}{\partial t}$$

$$\frac{\partial V}{\partial x} = \frac{\partial}{\partial x}\left(3xy^2 i + 2xyj + (2zy - 3t)k\right) = 3y^2 i + 2yj$$

$$\frac{\partial V}{\partial y} = \frac{\partial}{\partial y}\left(3xy^2 i + 2xyj + (2zy - 3t)k\right) = 6xyi + 2xj + 2zk$$

$$\frac{\partial V}{\partial z} = \frac{\partial}{\partial z}\left(3xy^2 i + 2xyj + (2zy - 3t)k\right) = 2yk$$

$$\frac{\partial V}{\partial t} = \frac{\partial}{\partial t}\left(3xy^2 i + 2xyj + (2zy - 3t)k\right) = -3k$$

So spatial acceleration will become

$$a_s = 3xy^2\left(3y^2 i + 2yj\right) + 2xy\left(6xyi + 2xj + 2zk\right) + (2zy - 3t)(2yk) - 3k$$

$$\Rightarrow a_s = 9xy^4 i + 6xy^3 j + 12x^2y^2 i + 4x^2yj + 4xyzk + 4zy^2k - 6tyk - 3k$$

$$\Rightarrow a_s = \left(9xy^4 + 12x^2y^2\right)i + \left(6xy^3 + 4x^2y\right)j + \left(4xyz + 4zy^2 - 6ty - 3\right)k$$

$$\Rightarrow (a_s)_{1,2,1,3} = \left(9*1*2^4 + 12*1^2*2^2\right)i + \left(6*1*2^3 + 4*1^2*2\right)j$$
$$+ \left(4*1*2*1 + 4*1*2^2 - 6*3*2 - 3\right)k$$

$$\Rightarrow (a_s)_{1,2,1,3} = 192i + 56j - 15k$$

$$\Rightarrow (a_s)_{1,2,1,3} = \sqrt{192^2 + 56^2 + 15^2} = 200.56 \text{ m/s}^2$$

Example 4.2: The velocity field in a fluid medium is given by $V = 10xyi + 5x^2 j + \left(z + xt^2\right)k$. Find magnitudes and directions of translation velocity and acceleration at point $P(1,2,3,1)$.

Solution: Given data: $V = 10xyi + 5x^2 j + \left(z + xt^2\right)k$ and $P(1,2,3,1)$

From the given equation, velocity components will be

$$u = 10xy, v = 5x^2, w = \left(z + xt^2\right)$$

Magnitude of translation velocity will be

$$V = \sqrt{(u)^2 + (v)^2 + (w)^2}$$

$$\Rightarrow V_{1,2,3,1} = \sqrt{(10*1*2)^2 + (5*1*1)^2 + (3+1*1*1)^2} = 21 \text{ m/s}$$

Direction of translation velocity will be $\bar{V} = 10xyi + 5x^2 j + (z + xt^2)k$

$$\Rightarrow \bar{V}_{1,2,1,3} = 10*1*2i + 5*1*1j + (3+1*1*1)k$$

$$\Rightarrow \bar{V}_{1,2,1,3} = 20i + 5j + 4k$$

Spatial acceleration $\quad a_s = u\dfrac{\partial V}{\partial x} + v\dfrac{\partial V}{\partial y} + w\dfrac{\partial V}{\partial z} + \dfrac{\partial V}{\partial t}$

$$\frac{\partial V}{\partial x} = \frac{\partial}{\partial x}\left(10xyi + 5x^2 j + (z + xt^2)k\right) = 10yi + 10xj + t^2 k$$

$$\frac{\partial V}{\partial y} = \frac{\partial}{\partial y}\left(10xyi + 5x^2 j + (z + xt^2)k\right) = 10xi$$

$$\frac{\partial V}{\partial z} = \frac{\partial}{\partial z}\left(10xyi + 5x^2 j + (z + xt^2)k\right) = k$$

$$\frac{\partial V}{\partial t} = \frac{\partial}{\partial t}\left(10xyi + 5x^2 j + (z + xt^2)k\right) = 2txk$$

So spatial acceleration will become

$$a_s = 10xy\left(10yi + 10xj + t^2 k\right) + 5x^2(10xi) + (z + xt^2)(k) + 2txk$$

$$\Rightarrow a_s = 100xy^2 i + 100yx^2 j + 10xyt^2 k + 50x^3 i + zk + xt^2 k + 2txk$$

$$\Rightarrow a_s = (100xy^2 + 50x^3)i + (100yx^2)j + (10xyt^2 + z + xt^2 + 2tx)k$$

$$\Rightarrow (a_s)_{1,2,,1} = (100*1*2^2 + 50*1^3)i + (100*2*1^2)j + (10*1*2*1^2 + 3 + 1*1^2 + 2*1*1)k$$

$$\Rightarrow (a_s)_{1,2,3,1} = 450i + 200j + 26k$$

$$\Rightarrow (a_s)_{1,2,3,1} = \sqrt{450^2 + 200^2 + 26^2} = 493.1 \text{ m/s}^2$$

Example 4.3: The velocity field in a fluid medium is given by $V = 4x^3yi - 10x^2y^2j + (2tz)k$. Find magnitudes and directions of translation velocity and acceleration at point $P(1,2,3,2)$.

Solution: Given data: $V = 4x^3yi - 10x^2y^2j + (2tz)k$ and $P(1,2,3,2)$. From the given equation, velocity components will be

$$u = 4x^3y, v = -10x^2y^2, w = 2tz$$

Magnitude of translation velocity will be

$$V = \sqrt{(u)^2 + (v)^2 + (w)^2}$$

$$\Rightarrow V_{1,2,3,2} = \sqrt{(4*1*2)^2 + (10*1*4)^2 + (2*2*3)^2} = 42.52 \text{ m/s}$$

Direction of translation velocity will be

$$\bar{V} = 4x^3yi - 10x^2y^2j + (2tz)k$$

$$\Rightarrow \bar{V}_{1,2,3,2} = 4*1*2i - 10*1*4j + 2*2*3k$$

$$\Rightarrow \bar{V}_{1,2,3,2} = 8i - 40j + 12k$$

Spatial acceleration

$$a_s = u\frac{\partial V}{\partial x} + v\frac{\partial V}{\partial y} + w\frac{\partial V}{\partial z} + \frac{\partial V}{\partial t}$$

$$\frac{\partial V}{\partial x} = \frac{\partial}{\partial x}\left(4x^3yi - 10x^2y^2j + (2tz)k\right) = 8x^2yi - 20xy^2j$$

$$\frac{\partial V}{\partial y} = \frac{\partial}{\partial y}\left(4x^3yi - 10x^2y^2j + (2tz)k\right) = 4x^3i - 20x^2yj$$

$$\frac{\partial V}{\partial z} = \frac{\partial}{\partial z}\left(4x^3yi - 10x^2y^2j + (2tz)k\right) = 2tk$$

$$\frac{\partial V}{\partial t} = \frac{\partial}{\partial t}\left(4x^3yi - 10x^2y^2j + (2tz)k\right) = 2zk$$

So spatial acceleration will become

$$a_s = 4x^3y\left(8x^2yi - 20xy^2j\right) - 10x^2y^2\left(4x^3i - 20x^2yj\right) + 2tz(2tk) + 2zk$$

$$\Rightarrow a_s = 32x^5y^2i - 80x^4y^3j - 40x^5y^2i + 200x^4y^3j + 4t^2zk + 2zk$$

$$\Rightarrow a_s = \left(32x^5y^2 - 40x^5y^2\right)i + \left(120x^4y^3\right)j + \left(4t^2z + 2z\right)k$$

$$\Rightarrow (a_s)_{1,2,3,4} = (32*1*4 - 40*1*4)i + (120*1*8)j + (4*4*3 + 2*3)k$$

$$\Rightarrow \left(a_s\right)_{1,2,3,1} = -32i + 960j + 54k$$

$$\Rightarrow \left(a_s\right)_{1,2,3,2} = \sqrt{32^2 + 960^2 + 54^2} = 962.05 \text{ m/s}^2$$

Example 4.4: The velocity field in a fluid medium is given by $V = (3x + 2y)i + (2z + 3x^2)j + (2t - 3z)k$. Find a) magnitudes and directions of translation velocity components at any point. b) Velocity at point $P(1,1,1,4)$. c) Magnitude and direction of acceleration at point $P(1,2,1,3)$. d) Classify the type of fluid flow.

Solution: Given data:

$$V = (3x + 2y)i + \left(2z + 3x^2\right)j + (2t - 3z)k \text{ and } P(1,1,1,4)$$

a) From the given equation, velocity components will be

$$u = (3x + 2y), v = \left(2z + 3x^2\right), w = (2t - 3z)$$

b) Magnitude of translation velocity will be

$$V = \sqrt{(3x + 2y)^2 + \left(2z + 3x^2\right)^2 + ((2t - 3z))^2}$$

$$\Rightarrow V_{1,1,1,4} = \sqrt{(3*1 + 2*1)^2 + (2*1 + 3*1)^2 + ((2*4 - 3*1))^2} = 8.66 \text{ m/s}$$

Direction of translation velocity will be

$$\bar{V} = 3xy^2i + 2xyj + (2zy - 3t)k$$

$$\Rightarrow \bar{V}_{1,1,1,4} = (3x + 2y)i + \left(2z + 3x^2\right)j + (2t - 3z)k = (3*1 + 2*1)i + (2*1 + 3*1)j + (2*4 - 3*1)k$$

$$\Rightarrow \bar{V}_{1,1,1,4} = 5i + 5j + 5k$$

c) Spatial acceleration $\quad a_s = u\frac{\partial V}{\partial x} + v\frac{\partial V}{\partial y} + w\frac{\partial V}{\partial z} + \frac{\partial V}{\partial t}$

$$\frac{\partial V}{\partial x} = \frac{\partial}{\partial x}\left((3x + 2y)i + \left(2z + 3x^2\right)j + (2t - 3z)k\right) = 3i + 6xj$$

$$\frac{\partial V}{\partial y}=\frac{\partial}{\partial y}\left((3x+2y)i+\left(2z+3x^2\right)j+(2t-3z)k\right)=2i$$

$$\frac{\partial V}{\partial z}=\frac{\partial}{\partial z}\left((3x+2y)i+\left(2z+3x^2\right)j+(2t-3z)k\right)=2j-3k$$

$$\frac{\partial V}{\partial t}=\frac{\partial}{\partial t}\left((3x+2y)i+\left(2z+3x^2\right)j+(2t-3z)k\right)=2k$$

So spatial acceleration will become

$$a_s=(3x+2y)(3i+6xj)+\left(2z+3x^2\right)(2i)+(2t-3z)(2j-3k)+2k$$

$$\Rightarrow a_s=9xi+18x^2j+6yi+12xyj+4zi+6x^2i+4tj-6tk-6zj+9zk+2k$$

$$\Rightarrow a_s=\left(9x+6y+4z+6x^2\right)i+\left(18x^2+12xy+4t-6z\right)j+(9z-6t+2)k$$

$$\Rightarrow (a_s)_{1,2,1,3}=(9*1+6*2+4*1+6*1)i+(18*1+12*1*2+4*3-6*1)j+(9*1-6*3+2)k$$

$$\Rightarrow (a_s)_{1,2,1,3}=31i+48j-7k$$

$$\Rightarrow (a_s)_{1,2,1,3}=\sqrt{31^2+48^2+7^2}=57.56\text{ m/s}^2$$

d) Classification of the flow:
As the flow is based on time "t", it is unsteady.
As the velocity has three velocity components, so it is three-dimensional flow.
As the velocity does not change its magnitude with respect to time, so it is uniform flow.

Example 4.5: The velocity distribution at an entry to a pump intake is inversely proportional to the square of the radial distance from the inlet to the suction pipe. If the velocity at a radial distance of 2.5 m is 0.99 m/s, determine the acceleration of flow at 1 m, 1.8 m from the inlet. Assume the streamlines to be radial and steady.

Solution: Given data: $V={C}/{r^2}$ and $V_{L=2.5m}=0.99\,\text{m/s}$.

From the given equation, $V={C}/{r^2}$ we have $C=V*r^2=0.99*2.5*2.5=6.18$

Acceleration $$a=a_s+a_n=V\frac{\partial V}{\partial s}+\frac{\partial V}{\partial t}+0 \text{ since } a_n=0$$

$$a=a_s+a_n=V\frac{\partial V}{\partial s}+0+0 \text{ since } \textit{steady \& Radial}$$

$$\Rightarrow a = V\frac{\partial V}{\partial r} = {C}/{r^2} * \frac{\partial}{\partial r}\left({C}/{r^2}\right)$$

$$\Rightarrow a = \frac{\partial}{\partial r}\left({C^2}/{r^4}\right) = \frac{-2C}{r^5} = \frac{-76.4}{r^5}$$

Hence, acceleration at r = 1 m, will be

$$\Rightarrow a_{r=1} = \frac{-76.4}{1^5} = -76.4 \text{ m/s}^2$$

$$r = 1.8 \text{ m, will be} \Rightarrow a_{r=1.8} = \frac{-76.4}{1.85} = -4.04 \text{ m/s}^2$$

Negative sign indicates the acceleration towards the intake.

TYPES OF FLUID FLOWS

When a fluid is flowing from one point to another, the flow behavior may be of any type like

a) Steady and unsteady flow.
b) Rotational or irrotational flow.
c) Uniform and non-uniform flow.
d) Laminar or turbulent flow.
e) Compressible or incompressible flow.
f) Internal and external flow.
g) Viscous and inviscid flow.

Steady and Unsteady Flow

When a fluid flow having its characteristics like velocity, pressure, density, temperature, etc. is unchanged with respect to the time, then the flow is said to be steady flow, i.e. when all the time derivatives of flow field vanish, is termed as steady flow

$$\frac{dV}{dt} = \frac{dp}{dt} = \frac{d\rho}{dt} = 0 \tag{4.8a}$$

and if any one of these characteristics or all the characteristics, changes with respect to time then it is said to be unsteady flow.

i.e.

$$\frac{dV}{dt} = \frac{dp}{dt} = \frac{d\rho}{dt} \neq 0 \tag{4.8b}$$

Rotational or Irrotational Flow

When a fluid flowing, having its fluid particles rotating, about their mass center, as shown in Figure 4.3, then the flow is said to be rotational flow. E.g. Motion of fluid in a rotating tank, flow near the solid boundaries is rotational.

If the fluid flow does not have rotation of fluid particles about their mass centers, then it is said to be having irrotational flow.

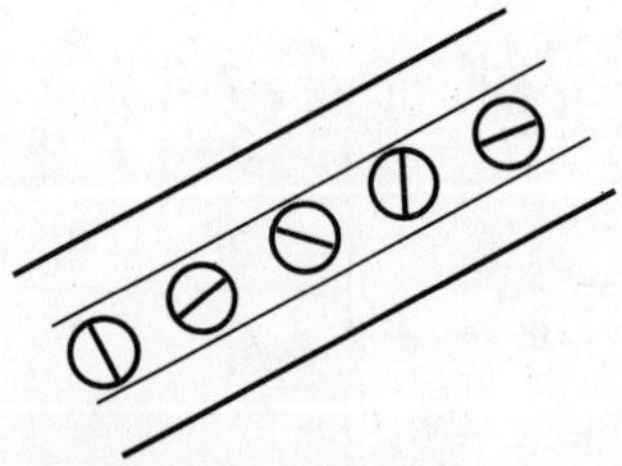

Figure 4.3: Rotational flow

For example, Flow above a drain hole of a wash basin.

Note: If the flow is irrotational and as well as steady, then the flow is said to be "**potential flow**".

Uniform and Non-uniform Flow

In a uniform flow, the velocity of fluid particles at any given time does not change in magnitude and direction, from point to point.

i.e. $\left(\frac{dV}{ds}\right) = \text{constant}$ where "s" represent any direction. For example, Flow through a straight pipe with constant diameter.

Whereas in non-uniform flow the velocity changes in magnitude and direction, at any given time, i.e. $\left(\frac{dV}{ds}\right) \neq \text{constant}$. For example, Flow through a curved pipe with constant diameter, flow through a non-prismatic conduit.

Laminar and Turbulent Flow

In laminar flow, each and every fluid particle moves in a consistent path without overlapping with other fluid particles due to which each fluid particle has well-defined paths, as shown in Figure 4.4. Flow of fluids at low velocities is smooth and has the fluid particles moving in straight lines along the direction of flow and its existence can be predicted by the dimensionless number like Reynolds number R_e given by **Osborne Reynold** where $R_e = {}^{\rho VD}\!/_{\mu}$ (if $R_e < 2000$ then the flow is said to be laminar, in a pipe).

Characteristics of laminar flow are as follows:

1) "No slip" at the boundary.

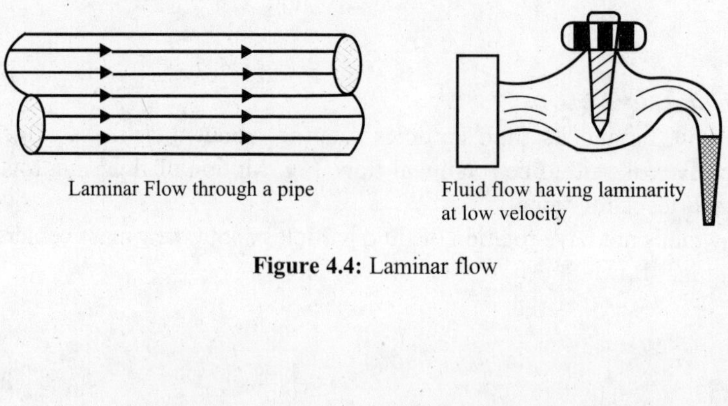

Figure 4.4: Laminar flow

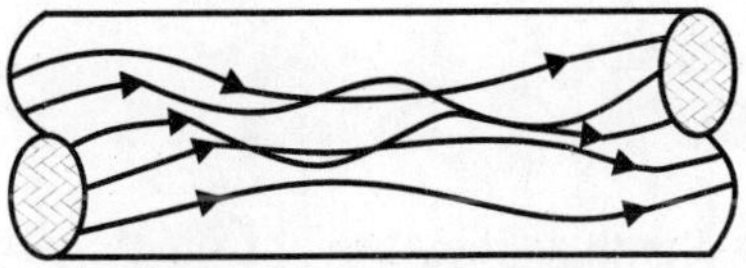

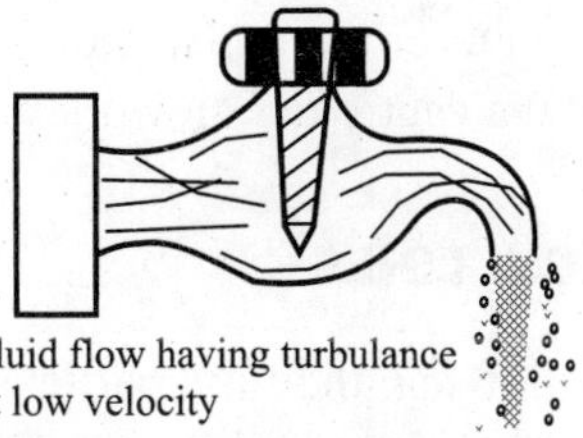

Figure 4.5: Turbulent flow

2) The flow is rotational.
3) No mixing between different layer molecules of the same liquid.
4) Continuous dissipation of energy due to viscous shear.

E.g.: Flow through capillary tube, flow of blood, underground flow, flow of oil in measuring instruments, rise of water in plates through their roots, flow past tiny bodies.

In turbulent flow, each particle does not move in a consistent path and they try to overlap with other fluid particles making the flow circulate with eddies formation and due to which the particles do not move in well-defined paths, as shown in Figure 4.5. And its existence can be predicted by the dimensionless number like Reynolds number R_e (if $R_e > 4000$ then the flow is said to be turbulent, in a pipe).

E.g.: Flow through water supply pipes, sewers.

Compressible or Incompressible Flow

When a fluid can be compressed by an external source due to which the density changes (i.e. $\rho \neq \text{constant}$) from point to point, then the fluid is said to be compressible fluid, for example, flow of gases through orifice, nozzles, turbines, etc.

When a fluid cannot be compressed, due to which the density changes (i.e. $\rho = \text{constant}$) from point to point, then the fluid is said to be incompressible fluid, for example, all liquids can be treated as incompressible.

Internal flow and external flow: Internal flow has flow within the boundary walls like pipe flow, channel flow, airflow in ducts and is controlled using valves, fans, pumps.

External flow has flow outside of a boundary or body like flow over aerodynamic profiles, flow over flat plates, flow over immersed bodies, air flow around bridges and buildings, etc. Internal flows are dominated by the influence of viscosity throughout the flow field. In external flows the viscous effects are limited to boundary layers near solid surfaces and to wake regions downstream of bodies.

Viscous and inviscid flows: In nature most of the fluid flows are in conjunction with viscosity and so the flows whose flow patterns are dominated by the viscous properties of the fluid are called viscous flows. For example, flow of liquid near to the walls of the pipes or channels.

When the viscous properties are not dominant, the flow is defined as inviscid flow. Eor example, Flow of fluid at the centre region while flowing through pipes.

TYPES OF FLOW LINES

When a fluid is flowing, the fluid particles follow a path as they travel from point to point and the flow paths may be of several differing types, like.

a) Path lines.
b) Stream lines.
c) Stream tube.
d) Streak line.

Path Lines

Path lines are the lines indicating the path followed by fluid particles or in other words path line is the curve traced out by a fluid particle from one point to other point, when in motion. Shown in Figure 4.6 are the path lines traced out by two fluid particles, P1 and P2.

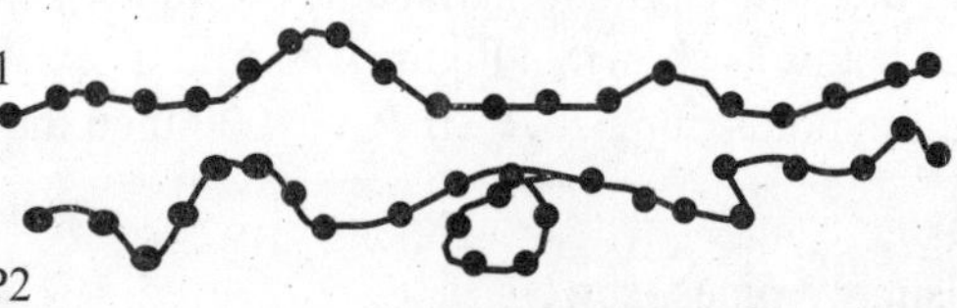

Figure 4.6: Path lines

Streamlines

Streamlines are the imaginary lines within the fluid flow along which individual particles of fluid is assumed to flow, such that the tangent at any point on the streamlines indicates the velocity at that point, as shown in Figure 4.7.

From Figure 4.7, we have

$$\frac{v}{u} = \tan\theta = \frac{dy}{dx}$$

i.e.
$$\frac{dx}{u} = \frac{dy}{v} = \frac{dz}{w} \tag{4.8c}$$

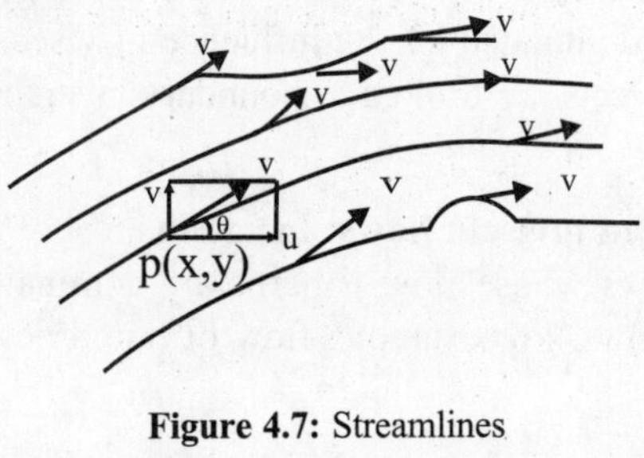

Figure 4.7: Streamlines

A streamline is continuous following of a bunch of fluid particles along a path line, for a steady non-fluctuating flows and where in the tangent at any point represent the velocity of the flow but if the flow is unsteady, the streamlines are in constant disturbance and hence the tangent gives the velocity of an element at an instant of time.

E.g.: Observe the picture of streamlines passing over a car surface.

Example 4.6: Determine the equation of streamline for a two-dimensional flow field, having the velocity components as

i) $u = 4xy + y^2, v = 6xy + 3x$

ii) $u = 2x^2 + y^2, v = -4x$ passing through point $P(3,0)$

Solution: Given data: Streamline passing through point $P(3,0)$.

i) for $u = 4xy + y^2, v = 6xy + 3x$. Equation of a stream line for 2-dimensional $\frac{dx}{u} = \frac{dy}{v}$

$$\Rightarrow \frac{dx}{4xy + y^2} = \frac{dy}{6xy + 3x}$$

$$\Rightarrow \int (6xy + 3x)\,dx = \int (4xy + y^2)\,dy$$

$$\Rightarrow 3x^2y + \frac{3x^2}{2} - \frac{y^3}{3} - 2xy^2 = c$$

Hence, **streamline equation** at point $P(3,0)$, will be $c = \frac{27}{2}$

$$\Rightarrow 3x^2y \frac{3x^2}{2} - \frac{y^3}{3} - 2xy^2 = \frac{27}{2}$$

$$\Rightarrow 6x^2y + 3x^2 - \frac{2y^3}{3} - 4xy^2 = 27$$

ii) for $u = 2x^2 + y^2, v = -4x$

Equation of a stream line for 2-dimensional $\frac{dx}{u} = \frac{dy}{v}$

$$\Rightarrow \frac{dx}{2x^2 + y^2} = \frac{dy}{-4x}$$

$$\Rightarrow -\int (4x)\,dx = \int (2x^2 + y^2)\,dy$$

$$\Rightarrow \frac{y^3}{3} + \frac{2x^3}{3} + 2x^2 = c$$

Hence, **streamline equation** at point $P(3,0)$ will be $c = 36$

$$\Rightarrow \frac{y^3}{3} + \frac{2x^3}{3} + 2x^2 = 36$$

Example 4.7: Determine the equation of streamline for a two-dimensional flow field, having the velocity components as $u = \frac{-y}{b^2}, v = \frac{x}{a^2}$ passing through point $P(a,0)$.

Solution: Given data: Streamline passing through point $P(a,0)$.

Equation of a streamline for 2-dimensional $\frac{dx}{u} = \frac{dy}{v}$

$$\Rightarrow \frac{dx}{\left(\frac{-y}{b^2}\right)} = \frac{dy}{\left(\frac{x}{a^2}\right)}$$

$$\Rightarrow \int\left(\frac{x}{a^2}\right)dx = \int\left(\frac{-y}{b^2}\right)dy$$

$$\Rightarrow \frac{x^2}{2a^2} + \frac{y^2}{2b^2} = c$$

$$\Rightarrow \frac{x^2}{a^2} + \frac{y^2}{b^2} = 2c$$

Hence, **streamline equation** at point $P(a,0)$, will be $c = \frac{1}{2}$

$$\Rightarrow \frac{x^2}{a^2} + \frac{y^2}{b^2} = 1$$

Example 4.8: The velocity field in a fluid medium is given by $V = axi + ayj + (-2az)k$. Find the equation of streamline at point $P(2,2,4)$.

Solution: Given data: $V = axi + ayj + (-2az)k$ and $P(2,2,4)$

From the given equation, velocity components will be $u = ax, v = ay, w = (-2az)$

Equation of a streamline for 3-dimensional $\frac{dx}{u} = \frac{dy}{v} = \frac{dz}{w}$

$$\Rightarrow \frac{dx}{ax} = \frac{dy}{ay} = \frac{dz}{-2az}$$

Consider expression of dx and dy only, we have $\frac{dx}{ax} = \frac{dy}{ay}$

$$\Rightarrow \int \frac{dx}{ax} = \int \frac{dy}{ay}$$

$$\Rightarrow \log_e x = \log_e y + \log_e c$$
$$\Rightarrow x = cy$$

Hence, at point $P(2,2,4)$, will be $c = 1 \Rightarrow x = y$

Similarly, consider expression of dx and dz only, we have $\frac{dx}{ax} = \frac{dz}{-2az}$

$$\Rightarrow \int \frac{dx}{ax} = \int \frac{dz}{-2az}$$

$$\Rightarrow \log_e x = \frac{-1}{2} \log_e z + \log_e c$$
$$\Rightarrow x = \frac{c}{\sqrt{z}}$$

Hence, at point $P(2,2,4)$, will be $c = 4$

$$x = \frac{4}{\sqrt{z}}$$

Therefore, we get the **streamline equation** as $x = y = \frac{4}{\sqrt{z}}$.

Example 4.9: Determine the equation of streamline for a two-dimensional flow field, having the velocity components as $u = 2x^3 - 3y^2, v = 3x^3 - 4y^2$ passing through point $P(2,0)$.

Solution: Given data: Streamline passing through point $P(2,0)$ and velocity components as $u = 2x^3 - 3y^2, v = 3x^3 - 4y^2$

Equation of a streamline for two-dimensional $\frac{dx}{u} = \frac{dy}{v}$

$$\Rightarrow \frac{dx}{2x^3 - 3y^2} = \frac{dy}{3x^3 - 4y^2}$$

$$\Rightarrow \int (3x^3 - 4y^2)\, dx = \int (2x^3 - 3y^2)\, dy$$

$$\Rightarrow \frac{3x^4}{4} - 4y^2 x = \frac{2y^4}{4} - 3\frac{y^3}{3} + c$$

$$\Rightarrow \frac{3x^3}{4} + y^3 - 4y^2 x + \frac{y^4}{2} = c$$

Hence, **streamline equation** at point *P*(2,0), will be $c = 12$

$$\Rightarrow \frac{3x^3}{4} + y^3 - 4y^2 x + \frac{y^4}{2} = 12$$

Example 4.10: Determine the equation of streamline for a two-dimensional flow field, having the velocity components as $u = x^2 - y^2, v = x^2 + y^2$ passing through point *P*(3, 3).

Solution: Given data: Streamline passing through point $P(3,3)$ and velocity components as $u = x^2 - y^2, v = x^2 + y^2$

Equation of a streamline for 2-dimensional $\frac{dx}{u} = \frac{dy}{v}$

$$\Rightarrow \frac{dx}{x^2 - y^2} = \frac{dy}{x^2 + y^2}$$

$$\Rightarrow \int (x^2 + y^2)\, dx = \int (x^2 - y^2)\, dy \Rightarrow \frac{x^3}{3} + \frac{y^3}{3} = \frac{x^3}{3} - \frac{y^3}{3} + c$$

$$\Rightarrow \frac{2y^3}{3} = c$$

Hence, **streamline equation** at point $P(3,3)$, will be $c = 18 \;\Rightarrow y = 3$.

Stream Tube

Steam tube is a group of streamlines bounded together on a fluid mass and so the shapes of steam tube vary with respect to the fluid mass flow. As the stream tube comprises of steam lines only, the fluids have only tangential velocity component and no chance of having normal velocity components and hence there will be no obstacle for a fluid to enter or exit from a pipe, i.e. fluid can only enter and leave a stream tube at its ends.

E.g.: Fluid entering or leaving a pipeline, fluid leaving a nozzle.

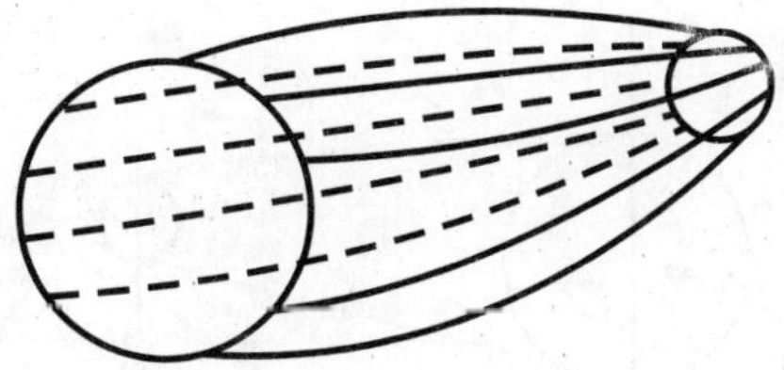

Figure 4.8: Stream tube

Streak Line

A **streak line** is the locus of fluid particles that have passed sequentially through a prescribed point in the flow and so streak lines are the curves which shows the instantaneous picture of the location of the fluid particles, passed through a given point.

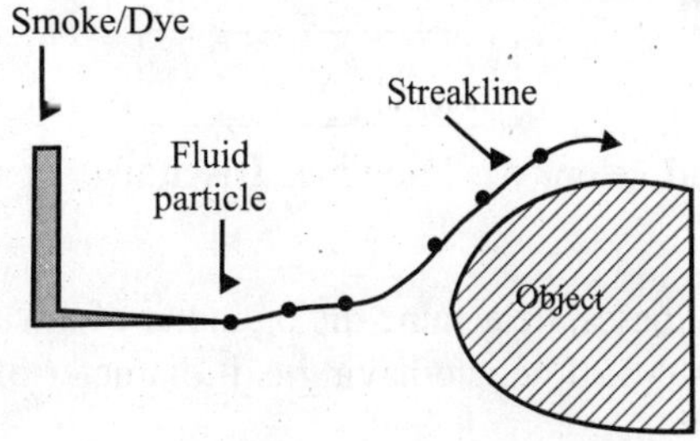

Figure 4.9: Streak line

E.g.: 1) The path taken by color dye, sprayed on to an automotive vehicle.
2) The path taken by smoke coming out of a chimney.

Note: For steady flow, streamlines, path lines and streak lines are identical.

CONTINUITY EQUATION

Continuity equation is based on the "conservation of mass (i.e. mass can neither be created nor destroyed). Continuity equation states that **"The mass of a fluid is same at any section, along any length of a pipe, unless some fluid is added or removed from the pipe"**.

Consider a pipe of non-uniform diameters D_1, D_2 at the inlet and outlet respectively and length of the pipe be L. Let ρ_1, A_1, V_1 and ρ_2, A_2, V_2 be density, cross-section area, velocity of fluid respectively at inlet section $1-1$ and outlet section 2-2 respectively, as shown in Figure 4.10.

As per the continuity equation, mass flow rate remains constant at any section in a pipe, along a length, unless some fluid is added or removed from the pipe. So in unit time the mass flow rates at inlet $\bar{m}_1$ and outlet section $\bar{m}_2$ are same and hence $\bar{m}_1 = \bar{m}_2$ where $\bar{m}_1 = \rho_1 A_1 V_1$ and $\bar{m}_2 = \rho_2 A_2 V_2$ and have units kg/s.

i.e. $$\Rightarrow \rho_1 A_1 V_1 = \rho_2 A_2 V_2 \quad (4.9)$$

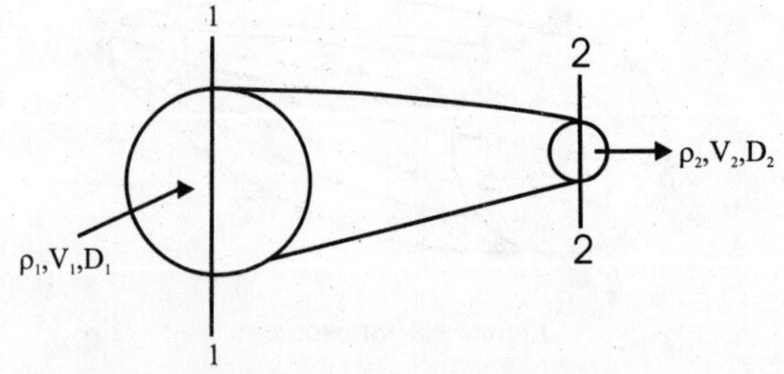

Figure 4.10: Continuity equation

Eq: (4.9) is applicable to compressible fluids like gases, due to the reason that the density is not constant for gases (i.e. $\rho_1 \neq \rho_2$).

If density is constant (i.e. $\rho_1 = \rho_2$) as in case of incompressible fluids like liquids, then eq. (4.9) reduces to $A_1V_1 = A_2V_2 \Rightarrow AV = \text{constant}$

i.e. $$\Rightarrow Q = AV \tag{4.10}$$

Where the product of area and velocity is known as **Discharge** or **volume rate of flow** Q and has units m^3/s.

Example 4.11: Assuming no losses, determine the diameter of a water jet traveling a height of 5 m vertically, from an exit of a vertical nozzle having exit diameter of 0.04 m and the velocity of jet leaving the nozzle is 15 m/s.

Solution: Given data: $d_1 = 0.04\,\text{m}, V_1 = 15\,\text{m}/s, h = 5\,\text{m}$. Using fundamental formulae of physics, we know that $V^2 - u^2 = 2gh$. Hence, taking the velocity of jet from the nozzle as the initial velocity, we have

$$V^2 - V_1^2 = -2gh$$

$$\Rightarrow V^2 - 15^2 = -2 * 9.81 * 5$$

$$\Rightarrow V = 11.27\,\text{m/s}$$

As per continuity equation, we have $Q = A_1V_1 = A_2V_2$

$$\Rightarrow A = {A_1V_1}/{V}$$

$$\Rightarrow \frac{\pi}{4} * d^2 = \frac{\pi * 0.04^2 * 15}{4 * 11.27}$$

$$\Rightarrow d = 0.046\ \text{m}$$

Hence, the **diameter of the jet at a height of 5 m** is 0.046 m.

Example 4.12: A pipeline 60 cm in diameter has water flow in it at a velocity of 4 m/s and it suddenly expands to a pipe diameter of 80 cm up to some length and then the flow bifurcates to a *p-Y*-junction into two branches 40 cm and 30 cm diameter (as shown in Figure 4.12) such that the

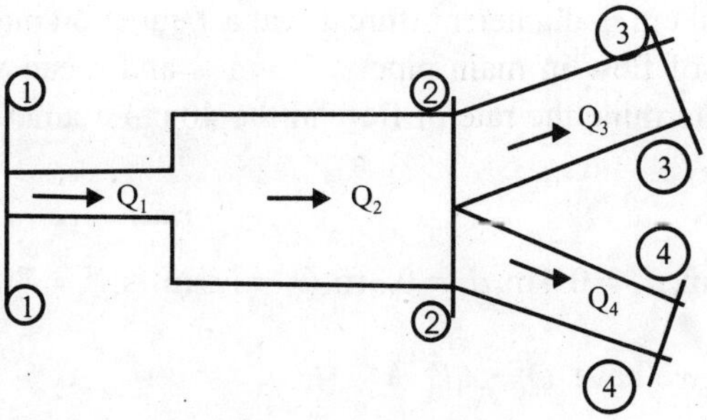

Figure P4.12: Stream tube

velocity of flow in branched pipe of 30 cm diameter has 7.5 m/s, determine the discharge through the 40 cm diameter pipe and also velocity of flow in the 40 cm diameter pipe.

Solution: Given data: $d_1 = 0.6\,\text{m}, d_2 = 0.8\,\text{m}, d_3 = 0.4\,\text{m}, d_4 = 0.3\,\text{m}, V_1 = 4.0\,\text{m/s}, V_4 = 7.5\,\text{m/s}$. As per continuity equation, we have $Q = A_1V_1$

$$\Rightarrow Q_1 = A_1V_1 = \frac{\pi * 0.6^2 * 4}{4} = 1.1\,\text{m}^3/\text{s}$$

As the discharge through the 60 cm diameter pipe is let to flow into the pipe diameter of 80 cm, the discharge $Q_2 = Q_1 = 1.1\,\text{m}^3/\text{s}$.

Similarly, as per the continuity equation, we have

$$Q_4 = A_4V_4 = \frac{\pi * 0.3^2 * 7.5}{4} = 0.529\,\text{m}^3/\text{s}$$

$$\Rightarrow Q_3 = A_3V_3 = \frac{\pi * 0.3^2}{4} * 7.5 = 0.529\,\text{m}^3/\text{s}$$

And also we know $Q_2 = Q_3 + Q_4$ $\quad \Rightarrow 1.1 = 0.529 + Q_4$

$$\Rightarrow Q_4 = 0.57\ \text{m}^3/\text{s}$$

As per continuity equation, we have $\quad Q_4 = A_4V_4$

$$\Rightarrow V_4 = \frac{Q_4}{A_4} = \frac{0.57 * 4}{\pi * 0.4^2} = 4.54\ \text{m/s}$$

Hence, the **discharge and velocity** through 40 cm diameter pipe are $0.57\,\text{m}^3/\text{s}, 4.54\,\text{m/s}$ respectively.

Example 4.13: A pipeline 60 cm in diameter bifurcates at a *Y*-junction into two branches 40 cm and 30 cm diameter. If the rate of flow in main pipe is 1.5 m³/s and mean velocity of flow in 30 cm diameter pipe is 7.5 m/s, determine the rate of flow in the 40 cm diameter pipe.

Solution: Given data:

$$d_1 = 0.6\,\text{m}, d_2 = 0.4\,\text{m}, d_3 = 0.3\,\text{m}, Q_1 = 1.5\,\text{m}^3/\text{s}, V_3 = 7.5\,\text{m/s}$$

As per continuity equation, we have $Q = A_1V_1$

$$\Rightarrow V_1 = {Q}/{A_1} = \frac{4*1.5}{\pi*0.6^2} = 5.3\,\text{m/s}$$

And also we know $Q_1 = Q_2 + Q_3$

$$\Rightarrow 1.5 = Q_2 + 0.529$$

$$\Rightarrow Q_2 = 0.97\,\text{m}^3/\text{s}$$

Hence, the **discharge through 40 cm diameter pipe** is 0.97 m³/s

Example 4.14: Water flows in a circular pipe which decreases in diameter from 500 mm at point 1 to 400 mm at point 2. The pipe then splits into two branches of diameters 0.25 m and 0.3 m discharging at 3 and 4 respectively) as shown in Figure 4.14). If the velocity at 1 is 1.5 m/s and at 2.0 m/s at 1 and 4 respectively, what are the discharges at 3 and 4 and the velocities at 2 and 3?

Solution:

$$d_1 = 0.5\,\text{m}, d_2 = 0.4\,\text{m}$$
$$d_3 = 0.25\,\text{m}, d_4 = 0.3\,\text{m}$$
$$V_1 = 1.5\,\text{m/s}, V_4 = 2.0\,\text{m/s}$$

Bottom of the vessel at the edge is $z_1 = 0, r_1 = 0.15\,\text{m}$

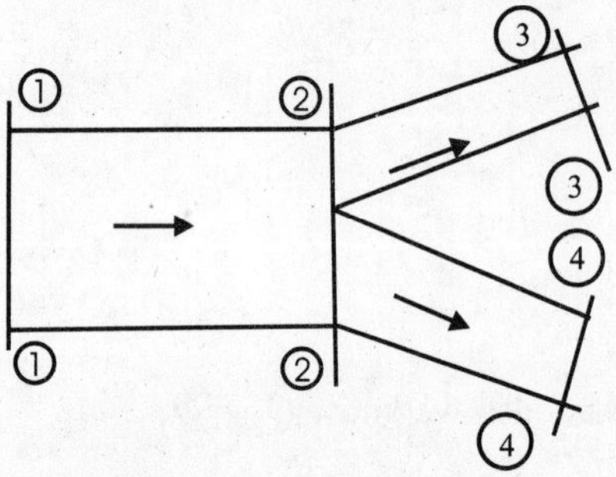

Figure P4.14: Stream tube

Point	*Diameter (d)* m	*Area (A)* m²	*Velocity (V)* m/s	*Discharge (Q = AV)* m³/s	*Remarks*
1	0.5	0.196	1.50	0.294	
2	0.4	0.125	2.35	0.294	$(A_1V_1 = A_2V_2)$
3	0.25	0.049	3.10	0.152	$(A_1V_1 = A_3V_3 + A_4V_4)$
4	0.3	0.071	2.0	0.142	

a) Continuity Equation in Cartesian Coordinates

Consider a small elemental parallelepiped of dimensions (dx, dy, dz), as shown in Figure 4.11 and let the velocity components along *x, y, z* axes be *u, v, w* respectively.

Let the mass rate entered into the parallelepiped from the face *ABCD* along x-axis be $\bar{m}_{x_I}$

i.e. $$\bar{m}_{x_I} = \rho_x A_x V_x$$

$$\Rightarrow \bar{m}_{x_I} = \rho(dy * dz)u \tag{4.11a}$$

Let the mass rate leaving the parallelepiped from the face *EFGH* along x-axis be $\bar{m}_{x_o}$.

i.e. $$\bar{m}_{xO} = (\rho_x A_x V_x) + \frac{\partial}{\partial x}((\rho_x A_x V_x)dx)$$

where $\frac{\partial}{\partial x}((\rho_x A_x V_x)dx)$ is the mass already existing in the parallelepiped.

$$\Rightarrow \bar{m}_{xO} = \rho(dy * dz)u + \left[\frac{\partial(\rho * dy * dz * u)}{\partial x}\right]dx \tag{4.11b}$$

Hence, **total mass** accumulated per unit time along X-axis be $\bar{m}_x = (\bar{m}_{x_i} - \bar{m}_{x_0})$.

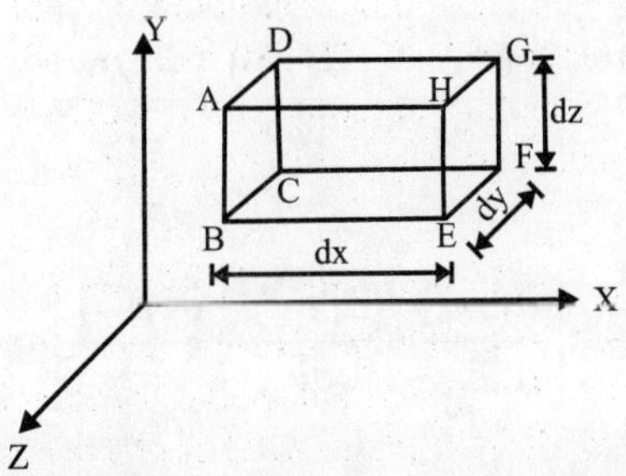

Figure 4.11: Continuity equation in Cartesian coordinates

$$\Rightarrow \bar{m}_x = -\left[\frac{\partial(\rho u)}{\partial x}\right]dx * dy * dz \tag{4.12a}$$

Similarly, the difference in mass rate left and entered into the parallelepiped from the faces *BCEF, ADGH* along

Y-axis respectively is $$\Rightarrow \bar{m}_y = -\left[\frac{\partial(\rho v)}{\partial y}\right]dx * dy * dz \tag{4.12b}$$

Similarly, the difference in mass rate left and entered into the parallelepiped from the faces *ABEH, DCFG* along Z-axis respectively is

$$\Rightarrow \bar{m}_z = -\left[\frac{\partial(\rho w)}{\partial z}\right] dx * dy * dz \tag{3.12c}$$

So, **total gain** in fluid mass per unit time along the three X, Y, Z axes be $\bar{m}$.

Where $\bar{m}$ is equal to sum of all the equations Eq. (4.12a) + Eq. (4.12b) + Eq. (4.12c)

i.e.

$$\bar{m} = -\left[\frac{\partial(\rho u)}{\partial x} + \frac{\partial(\rho v)}{\partial y} + \frac{\partial(\rho w)}{\partial z}\right] dx * dy * dz \tag{4.13}$$

Let the rate of change of mass of parallelepiped be $\frac{\partial m}{dt}$.

i.e.

$$\frac{\partial m}{dt} = \frac{\partial(\rho * dx * dy * dz)}{dt} \tag{4.14}$$

Since

$$\frac{\partial m}{dt} = \frac{\partial}{dt}(\rho * volume)$$

As the parallelepiped is same Eq. (4.14) must be equal to the Eq. (4.11)

i.e.

$$\frac{\partial(\rho * dx * dy * dz)}{dt} = -\left[\frac{\partial(\rho u)}{\partial x} + \frac{\partial(\rho v)}{\partial y} + \frac{\partial(\rho w)}{\partial z}\right] dx * dy * dz$$

$$\Rightarrow \frac{\partial(\rho * dx * dy * dz)}{dt} + \left[\frac{\partial(\rho u)}{\partial x} + \frac{\partial(\rho v)}{\partial y} + \frac{\partial(\rho w)}{\partial z}\right] dx * dy * dz = 0 \tag{4.15}$$

By taking the volume of parallelepiped $(dx * dy * dz)$ as common term, we get

$$\frac{\partial(\rho)}{dt} + \left[\frac{\partial(\rho u)}{\partial x} + \frac{\partial(\rho v)}{\partial y} + \frac{\partial(\rho w)}{\partial z}\right] \tag{4.16}$$

Equation (4.16) is the **general equation for Continuity in three-dimensional form** and is applicable for any type of flow. And hence any **possible flow** should satisfy the continuity equation.

If the flow is steady $\left(\frac{d\rho}{dt} = 0\right)$, 3-dimensional and incompressible flow $\rho = \text{constant}$, then we get

$$\frac{\partial(u)}{\partial x}+\frac{\partial(v)}{\partial y}+\frac{\partial(w)}{\partial z}=0 \tag{4.16a}$$

If the flow is steady $\frac{d\rho}{dt}=0$, 2-dimensional $\frac{\partial w}{dz}=0$ and incompressible flow ρ = constant, then we get

$$\frac{\partial(u)}{\partial x}+\frac{\partial(v)}{\partial y}=0 \tag{4.16b}$$

If the flow is steady $\frac{d\rho}{dt}=0$ 1-dimensional $\frac{\partial v}{dy}=\frac{\partial w}{dz}=0$ and incompressible flow ρ = constant, then we get

$$\frac{\partial(u)}{\partial x}=0 \Rightarrow u=\text{constant}=\frac{Q}{A},$$

$$\Rightarrow Q = Area * Velocity \tag{4.16c}$$

Example 4.15: A conical pipe diverges uniformly from 125 mm to 210 mm diameter of length 1.5 m. Determine acceleration at the mid section, when the rate of flow is $0.16\ \text{m}^3/\text{s}$ and is constant throughout.

Solution: Given data:

$$d_1 = 0.125\,\text{m}, d_2 = 0.21\,\text{m}, Q = 0.16\,\text{m}^3/\text{s}, L = 1.5\,\text{m}$$

Diameter at any section on the conical diverging pipe can be given by

$$d_x = d_1 + \left(\frac{d_2 - d_1}{L}\right)x$$

$$\Rightarrow d_x = 0.125 + 0.0567x$$

We know velocity

$$V_x = \frac{Q}{A_x} = \frac{4Q}{\pi(0.125+0.0567x)^2} = \frac{0.203}{(0.125+0.0567x)^2}$$

As this problem is of one-dimensional and steady state, we have spatial acceleration as

$$a_s = u\frac{\partial V}{\partial x}$$

$$\Rightarrow a_s = \frac{0.203}{(0.125+0.0567x)^2}\frac{\partial}{\partial x}\left(\frac{0.203}{(0.125+0.0567x)^2}\right)$$

$$\Rightarrow a_s = -0.0824*0.0567\left(\frac{1}{(0.125+0.0567x)^5}\right) = -0.0046\left(\frac{1}{(0.125+0.0567x)^5}\right)$$

Acceleration at mid section

$$\Rightarrow (a_s)_{x=0.75} = -0.0046\left(\frac{1}{(0.125+0.0567*0.75)^5}\right) = -35.4\,\text{m/s}^2$$

Example 4.16: For steady, incompressible flow, verify whether the following values of u and v are possible:

i. $u = 4xy + y^2, v = 6xy + 3x$

ii. $u = 2x^2 + y^2, v = -4xy$

Solution: Given data: Steady, incompressible flow.

For steady, incompressible flow to be possible, the continuity equation must be satisfied and hence

$$\frac{\partial u}{\partial x} + \frac{\partial v}{\partial y} = 0$$

a) for $u = 4xy + y^2, v = 6xy + 3x$

$$\frac{\partial u}{\partial x} = 4y, \frac{\partial v}{\partial y} = 6x \text{ and so } \frac{\partial u}{\partial x} + \frac{\partial v}{\partial y} = 4y + 6x \neq 0$$

And hence the flow is **not possible**.

b) for $u = 2x^2 + y^2, v = -4xy$

$$\frac{\partial u}{\partial x} = 4x, \frac{\partial v}{\partial y} = -4x \text{ and so } \frac{\partial u}{\partial x} + \frac{\partial v}{\partial x} = 4y - 4x = 0$$

And hence the flow is **possible**.

Example 4.17: The "u" and "v" components of velocity in a three-dimensional flow are given by $u = x^2 + z^2, v = y^2 + z^2$. Find the simplest "z" component of velocity that satisfies the continuity equation.

Solution: Given data: 3-dimensional continuity equation satisfies.

For satisfying 3-dimensional continuity equation $\frac{\partial u}{\partial x}+\frac{\partial v}{\partial y}+\frac{\partial w}{\partial z}=0$

And hence $\frac{\partial u}{\partial x}=2x, \frac{\partial v}{\partial y}=2y$ and so $\frac{\partial u}{\partial x}+\frac{\partial v}{\partial y}+\frac{\partial w}{\partial z}=2x+2y+\frac{\partial w}{\partial z}=0$

$$\Rightarrow \frac{\partial w}{\partial z}=-(2x+2y)$$
$$\Rightarrow \int \partial w=\int -(2x+2y)\partial z$$
$$\Rightarrow w=-(2xz+2yz)$$

Hence, the Z-component of velocity is $w=-(2xz+2yz)$.

Example 4.18: The "v" and "w" components of velocity in a three-dimensional flow are given by $v=ax^3-by^2+cz^2, w=bx^3-cy^2+az^2x$. . Find the simplest "$u$" component of velocity that satisfies the continuity equation.

Solution: Given data: 3-dimensional continuity equation satisfies, velocity components

$$v=ax^3-by^2+cz^2, w=bx^3-cy^2+az^2x$$

For satisfying 3-D continuity equation

$$\frac{\partial u}{\partial x}+\frac{\partial v}{\partial y}+\frac{\partial w}{\partial z}=0$$

And hence

$$\frac{\partial v}{\partial y}=-2by, \frac{\partial w}{\partial z}=2azx$$

So

$$\frac{\partial u}{\partial x}+\frac{\partial v}{\partial y}+\frac{\partial w}{\partial z}=\frac{\partial u}{\partial x}-2by+2azx=0$$

$$\Rightarrow \frac{\partial u}{\partial x}=(2by-2azx)$$

$$\Rightarrow \int \partial u=\int (2by-2azx)\partial x$$

$$\Rightarrow u=\left(2bxy-\frac{azx^2}{2}\right)$$

Hence, the U-component of velocity is

$$u = \left(2bxy - \frac{azx^2}{2}\right).$$

Example 4.19: Given $u = xy, v = 2yz$. Examine whether these velocity components represent 2-D or 3-dimensional incompressible continuity flow. If it is 3-dimensional, determine the third velocity component.

Solution: Given data: $u = xy, v = 2yz$.

For 2-dimensional flows, the continuity equation will be $\frac{\partial u}{\partial x} + \frac{\partial v}{\partial y} = 0$

And hence $\frac{\partial u}{\partial x} = y, \frac{\partial v}{\partial y} = 2z$ and so $\frac{\partial u}{\partial x} + \frac{\partial v}{\partial y} = y + 2z \neq 0$

And hence the flow is **not 2-dimensional.**

For satisfying 3-dimensional continuity equation $\frac{\partial u}{\partial x} + \frac{\partial v}{\partial y} + \frac{\partial w}{\partial z} = 0$

And hence $\frac{\partial u}{\partial x} = y, \frac{\partial v}{\partial y} = 2z$ and so $\frac{\partial u}{\partial x} + \frac{\partial v}{\partial y} + \frac{\partial w}{\partial z} = 0$

$$\Rightarrow y + 2z + \frac{\partial w}{\partial z} = 0$$

$$\Rightarrow -\int dw = \int (y + 2z)\, dz$$

$$\Rightarrow w = -yz - z^2$$

So the third component of velocity is $w = -yz - z^2$.

Example 4.20: Does it satisfy continuity incompressible equation having velocity components

$$u = (x^3 - y^3 - z^3), v = (y^3 - z^3), w = \left(-3x^2z - 3y^2z + \frac{z^3}{3}\right).$$

Solution: Given data: $u = (x^3 - y^3 - z^3), v = (y^3 - z^3), w = \left(-3x^2z - 3y^2z + \frac{z^3}{3}\right)$

For satisfying 3-dimensional continuity equation $\frac{\partial u}{\partial x} + \frac{\partial v}{\partial y} + \frac{\partial w}{\partial z} = 0$

And
$$\frac{\partial u}{\partial x}=3x^2,\frac{\partial v}{\partial y}=3y^2,\frac{\partial w}{\partial z}=-3x^2-3y^2+z^2$$

And
$$\frac{\partial u}{\partial x}+\frac{\partial v}{\partial y}+\frac{\partial w}{\partial z}=0$$
$$\Rightarrow 3x^2+3y^2-3x^2-3y^2+z^2\neq 0$$

Hence, the continuity equation **does not satisfy**.

CIRCULATION AND VORTICITY

Circulation (Γ)

Circulation is defined as the "**line integral of the tangential velocity along a closed path**" and is denoted by Γ, Consider flow of a fluid mass indicated by its streamlines and let a closed curve passing through them, cut these streamlines, as shown in Figure 4.12.

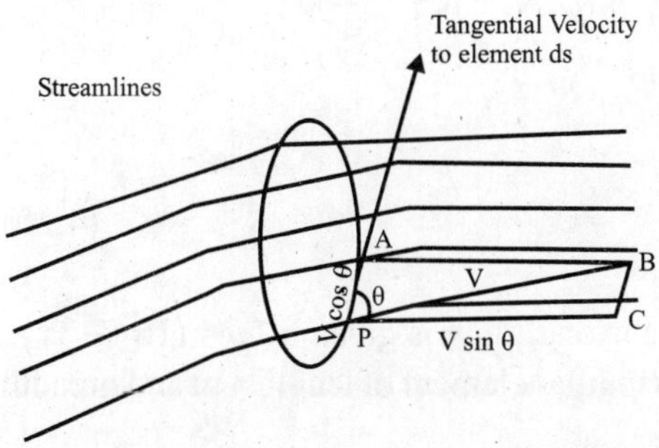

Figure 4.12: Streamlines cut by a curve

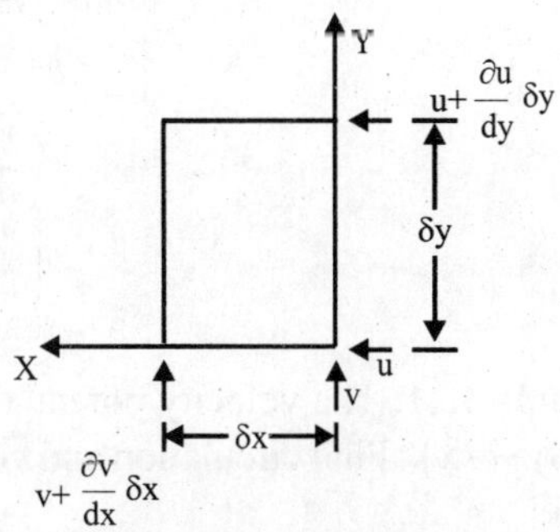

Figure 4.12a

Consider an intersection point of the closed curve and a fluid element in space "*ds*" on any one streamline and let this point be named as "P". The streamlines will have a tangential velocity "V", which can be resolved into two components, like one component along the velocity of streamline like $V\sin\theta$ and other component along tangential velocity to the streamline element "*ds*", like V sin θ, where θ is angle between tangential velocity of stream line and tangential velocity to element.

According to the definition

$$\Gamma=\oint(V\cos\theta)\,ds \tag{4.17}$$

If a rectangular fluid element having edges *CD, DA, AB, BC* and having velocity components *u v* along edges *CD, CB* and

$$DA=\left(u+\frac{\partial u}{dy}\delta y\right),AB=\left(v+\frac{\partial v}{dx}\delta x\right)$$

along edges *DA, AB* respectively, as shown in Figure 4.12a be considered, then circulation Γ of the rectangular element becomes

$$\Gamma = \Gamma_{AB} + \Gamma_{BC} + \Gamma_{CD} + \Gamma_{DA} \tag{4.18}$$

By substituting the velocity components $\left[\left(u + \frac{\partial u}{dy}\delta y\right), \left(v + \frac{\partial v}{dx}\delta x\right), u, v\right]$ along the rectangular fluid element edges *AB, BC, CD, DA* in Eq. (4.18), assuming clockwise as +ve, we get

$$\Gamma_{ABCD} = \left[-\left(u + \frac{\partial u}{dy}\delta y\right)\delta x - v\delta y + u\delta x + v\delta y + \left(\frac{\partial v}{dx}\delta x \delta y\right)\right]$$

$$\Gamma = \left(\frac{\partial v}{dx} - \frac{\partial u}{dy}\right)\delta x * \delta y \tag{4.19}$$

Vorticity (Ω)

Vorticity is defined as "**circulation per unit area enclosed**" and is denoted by Ω

where $$\Omega = \frac{Circulation}{Elemental\ Area} = \frac{\Gamma}{dA}$$

$$\Rightarrow \Omega = \left(\frac{\partial v}{dx} - \frac{\partial u}{dy}\right) \tag{4.20}$$

Example 4.21: The velocity potential function for 2-dimensional flow is given by $u = (16y - 8x)$, $v = (8y - 7x)$. Find circulation and vorticity around a rectangular element of length 4 m and breadth 6 m.

Solution: Given data: $u = (16y - 8x),\ v = (8y - 7x), \delta x = 4m, \delta y = 6m$

Circulation is given by $$\Gamma = \left(\frac{\partial v}{dx} - \frac{\partial u}{dy}\right) * \delta x * \delta y$$

$$\Rightarrow \Gamma = \left[\frac{\partial}{dx}(8y - 7x) - \frac{\partial}{dy}(16y - 8x)\right] * \delta x * \delta y$$

$$\Rightarrow \Gamma = [-7 - 16] * 4 * 6 = -552$$

Vorticity is given by $$\Omega = \frac{\Gamma}{dA} = \left(\frac{\partial v}{dx} - \frac{\partial u}{dy}\right)$$

$$\Rightarrow \Omega = \frac{-552}{4 * 6} = -23$$

Rotation of Fluid Elements

A fluid element in addition to its linear motion has rotation, linear strain and shear strain also due to the disturbance created by the adjacent fluid elemental flow (i.e. due to the angular momentum in the fluid flow).

To cause rotation for an elemental object of a solid which is moving, a single disturbance (angular displacement) perpendicular to its motion is sufficient enough but a fluid element will be having two perpendicular disturbances (angular displacements in y, z directions) along the perpendicular axis's to its flow.

If the angular displacement of the fluid element is in single perpendicular direction, then the fluid element will be having only rotation, as shown in Figure 4.13b. And if the angular displacement is in both the perpendicular directions to its flow then the fluid element is not only rotated but also shear strained, as shown in Figure 4.13c.

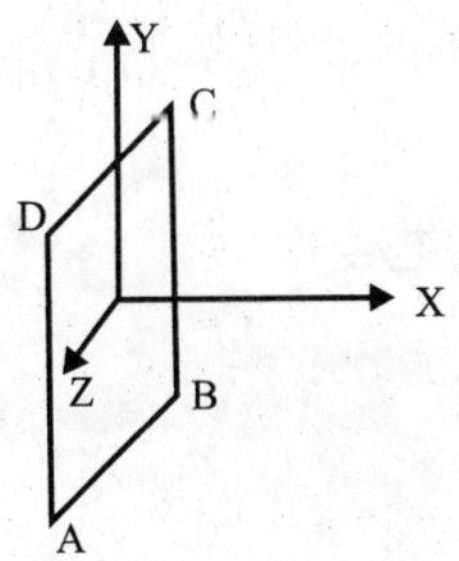

Figure 4.13a: Fluid element ABCD flowing in the direction X-axis

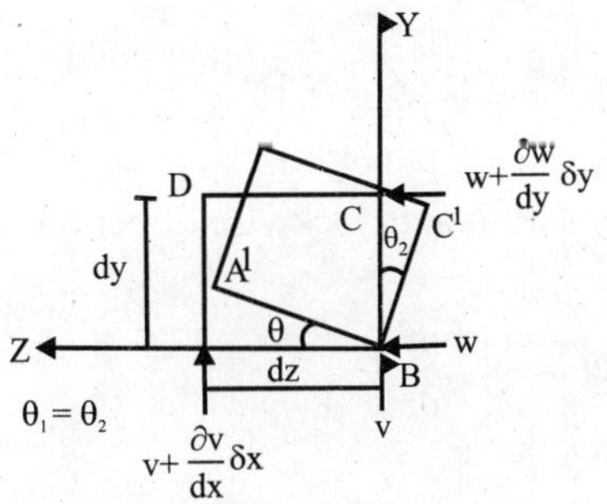

Figure 4.13b: Fluid element ABCD having rotation only along Z-axis

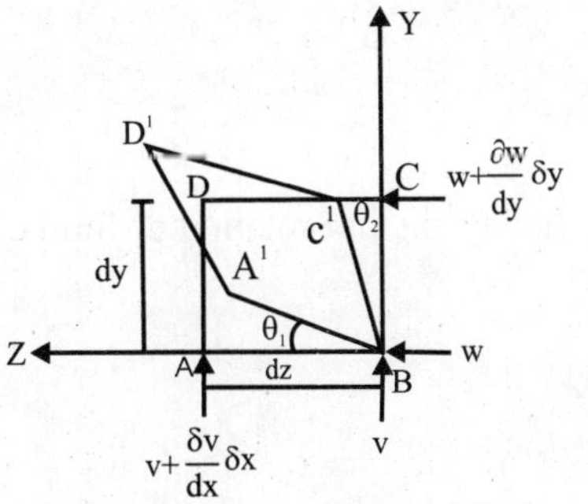

Figure 4.13c: Fluid element ABCD having rotation and shear strain

Considering the fluid element ABCD in Figures 4.13b and 4.13c, we observe that AAP^1 is the arc length of the fluid element rotation with an angular rotation of "θ". And hence AAP^1 is function of the radius (AB) and rotational angle θ.

i.e.
$$AA^1 = AB * \theta_1 \tag{4.21}$$

Since we know $V = r\omega$ & $\omega = \theta/_t$ and by replacing ω with θ, we get $\theta = \left(\frac{V}{r}\right)t$

$$\Rightarrow \theta_1 = \left(\frac{\partial v}{\partial z}\right)dt \tag{4.21a}$$

and
$$AB = dz \tag{4.21b}$$

Substituting Eq. (4.21a) and Eq. (4.21b) in Eq. (4.21), we get

$$AA^1 = \left(\frac{\partial v}{\partial z}\right) * dz * dt \tag{4.22}$$

Similarly, $$CC^1 = \left(\frac{\partial w}{\partial y}\right) * dy * dt \tag{4.23}$$

And hence angular rotations

$$\bar{\theta}_1 = \left(\frac{\partial v}{\partial z}\right) dt \tag{4.23a}$$

Clockwise direction

$$\bar{\theta}_2 = \left(\frac{\partial w}{\partial y}\right) dt \tag{4.23b}$$

As the total rotation of the fluid element comprises two angular displacements, the rotational velocity ω will be the average of these two angular displacements.

i.e. $$\omega = \frac{1}{2}\left(\bar{\theta}_1 + \bar{\theta}_2\right) \tag{4.24}$$

Taking angular rotation of fluid element in anticlockwise direction, we get

$$\omega_x = \frac{1}{2}\left(\left(\frac{\partial v}{\partial z}\right) - \left(\frac{\partial w}{\partial y}\right)\right) \tag{4.24a}$$

$$\omega_y = \frac{1}{2}\left(\left(\frac{\partial u}{\partial z}\right) - \left(\frac{\partial w}{\partial x}\right)\right) \tag{4.24b}$$

$$\omega_z = \frac{1}{2}\left(\left(\frac{\partial v}{\partial x}\right) - \left(\frac{\partial u}{\partial y}\right)\right) \tag{4.24c}$$

And so the rotational velocity vector $\bar{\omega}$ and magnitude ω can be written as

$$\bar{\omega} = \omega_x i + \omega_y j + \omega_z k \quad OR \quad \bar{\omega} = \frac{1}{2}\begin{vmatrix} i & j & k \\ \partial/\partial x & \partial/\partial y & \partial/\partial z \\ u & v & w \end{vmatrix} \tag{4.25}$$

$$\omega = \sqrt{\omega_x^2 + \omega_y^2 + \omega_z^2} \tag{4.26}$$

and $$\bar{\omega} = \frac{1}{2}\left(\nabla x \bar{V}\right) = \frac{\Omega}{2}$$ where Ω is **vorticity**

$$2\bar{\omega} = \frac{1}{2}\left(\nabla x \bar{V}\right) = \Omega \tag{4.27}$$

If a fluid element to be **irrotational**, then

$$\boxed{\omega_z = 0} \tag{4.27a}$$

i.e. from Eq. (4.27a), we have

$$\left(\frac{\partial v}{\partial x}\right) = \left(\frac{\partial u}{\partial y}\right)\left(\nabla x \bar{V}\right) = \frac{\Omega}{2} = 0$$

Hence, **vorticity will happen only when fluid elements rotate.**

Example 4.22: The velocity field in a fluid medium is given by $V = 3xy^2 i + 2xyj + (2zy - 3t)k$. Find magnitudes and directions of translation velocity and rotational velocity, vorticity at point $P(1, 2, 1, 3)$.

Solution: Given data: $V = 3xy^2 i + 2xyj + (2zy - 3t)k$ and $P(1, 2, 1, 3)$

From the given equation, velocity components will be

$$u = 3xy^2, v = 2xy, w = (2zy - 3t)$$

Magnitude of translation velocity will be

$$V = \sqrt{\left(3xy^2\right)^2 + (2xy)^2 + ((2zy - 3t))^2}$$

$$\Rightarrow V_{1,2,1,3} = \sqrt{(3*1*4)^2 + (2*1*2)^2 + ((2*1*2 - 3*3))^2} = 13.6 \text{m/s}$$

Direction of translation velocity will be

$$\bar{V} = 3xy^2 i + 2xyj + (2zy - 3t)k$$

$$\Rightarrow \bar{V}_{1,2,1,3} = 3*1*4i + 2*1*2j + (2*1*2 - 3*3)k$$

$$\Rightarrow \bar{V}_{1,2,1,3} = 12i + 4j + 5k$$

$$\omega = \frac{1}{2}\begin{vmatrix} i & j & k \\ \partial/\partial x & \partial/\partial y & \partial/\partial z \\ u & v & w \end{vmatrix}$$

Rotational velocity vector
$$\Rightarrow \omega = \frac{1}{2}\begin{vmatrix} i & j & k \\ \partial/\partial x & \partial/\partial y & \partial/\partial z \\ 3xy^2 & 2xy & (2zy-3t) \end{vmatrix}$$

$$\Rightarrow (\omega)_{1,2,1,3} = \frac{1}{2}(2z)i + \frac{1}{2}(0-0)j + \frac{1}{2}(2y-6xy)k = i - 4k$$

Rotational velocity magnitude

$$\omega = \sqrt{\omega_x^2 + \omega_y^2 + \omega_z^2} = \sqrt{1^2 + 4^2} = 4.1 \text{ rad/s}$$

Hence, flow is rotational.

Vorticity is given by
$$\Omega = \left(\frac{\partial v}{dx} - \frac{\partial u}{dy}\right)$$

$$\frac{\partial v}{\partial x} = (2y) = 2 * 2 = 4$$
$$\frac{\partial u}{\partial y} = (6xy) = 6 * 1 * 2 = 12$$

So vorticity
$$\Rightarrow \Omega = \left(\frac{\partial v}{dx} - \frac{\partial u}{dy}\right) = 4 - 12 = -8$$

VELOCITY POTENTIAL AND STREAM FUNCTION

Velocity Potential (ϕ)

Velocity potential is defined as the "**scalar function of space and time, such that in any direction, its negative derivative gives the fluid velocity in that direction**", i.e.

$$u = -\frac{\partial \phi}{\partial x}, v = -\frac{\partial \phi}{\partial y}, w = -\frac{\partial \phi}{\partial z} \tag{4.28}$$

where

$$\phi = f(x, y, t) \text{ for unsteady flow}$$
$$\phi = f(x, y) \text{ for steady flow}$$

Hence, total differential

$$d\phi = \frac{\partial \phi}{\partial x}dx + \frac{\partial \phi}{\partial y}dy$$
$$\Rightarrow d\phi = -udx - vdy \tag{4.28a}$$

As per the continuity equation,

$$\frac{\partial(u)}{\partial x} + \frac{\partial(v)}{\partial y} + \frac{\partial(w)}{\partial z} = 0 \tag{4.29}$$

for steady, incompressible fluid.

So, by substituting Eq. (4.28) in Eq. (4.29), we get

$$\frac{\partial}{dx}\left(-\frac{\partial \phi}{dx}\right) + \frac{\partial}{dy}\left(-\frac{\partial \phi}{dy}\right) + \frac{\partial}{dz}\left(-\frac{\partial \phi}{dz}\right) = 0$$

$$\Rightarrow \frac{d^2\phi}{dx^2} + \frac{d^2\phi}{dy^2} + \frac{d^2\phi}{dz^2} = 0$$
$$\Rightarrow \nabla^2(\phi) = 0 \tag{4.30}$$

where ∇ is the Laplace operator

$$\nabla = \frac{d^2}{dx^2} + \frac{d^2}{dy^2} + \frac{d^2}{dz^2}$$

This equation is called **Laplace equation in velocity potential** ϕ and it indicates that the flow is steady, incompressible and irrotational flow if satisfied.

Note: If velocity potential ϕ is constant throughout a stream line, then that line is called Equipotential line.

i.e. $$d\phi = 0$$

$$\Rightarrow \frac{dy}{dx} = -u/v$$

Stream Function (ψ)

Stream function is defined as: "**scalar function of space and time, such that its derivative with respect to any direction, gives the fluid velocity components at the right angles to that direction**".

i.e. $$u = \frac{\partial \psi}{\partial y}, v = -\frac{\partial \psi}{\partial x} \tag{4.31}$$

where

$$\psi = f(x, y, t) \;\; for\ unsteady\ flow$$
$$\psi = f(x, y) \;\; for\ \ steady\ \ flow$$

Hence, total differential

$$d\psi = \frac{\partial \psi}{\partial x}dx + \frac{\partial \psi}{\partial y}dy$$
$$\Rightarrow d\psi = -vdx + udy \tag{4.31a}$$

As per the continuity equation

$$\frac{\partial(u)}{\partial x} + \frac{\partial(v)}{\partial y} = 0 \tag{4.32}$$

for steady, two-dimensional and incompressible fluid.

So, by substituting Eq. (4.31) in Eq. (4.32), we get

$$\frac{\partial}{dx}\left(\frac{\partial \psi}{dy}\right) - \frac{\partial}{dy}\left(\frac{\partial \psi}{dx}\right) = 0$$

$$\Rightarrow \frac{\partial^2 \psi}{dxdy} - \frac{\partial^2 \psi}{dydx} = 0 \tag{4.33}$$

This equation decides whether the flow exists or not

i.e. for the flow to be rotational $\omega_z = \frac{1}{2}\left(\frac{\partial v}{\partial x} - \frac{\partial u}{\partial y}\right) \neq 0$

Substituting Eq. (4.31) in ω_z, we get

$$\omega_z = \frac{1}{2}\left(\frac{\partial^2 \psi}{\partial x^2} + \frac{\partial^2 \psi}{\partial y^2}\right) \tag{4.34}$$

Eq. (4.34) is called **Poisson equation**.

For irrotational flow $\omega_z = 0$, i.e.

$$\omega_z = -\frac{1}{2}\left(\frac{\partial^2 \psi}{\partial x^2} + \frac{\partial^2 \psi}{\partial y^2}\right) = 0$$

$$\Rightarrow \left(\frac{\partial^2 \psi}{\partial x^2} + \frac{\partial^2 \psi}{\partial y^2}\right) = 0$$

$$\Rightarrow \Delta \psi = 0$$

where Δ is the Laplace operator $\Delta = \frac{d^2}{dx^2} + \frac{d^2}{dy^2}$ and is called **Laplace equation in stream function ψ.**

Note: The stream function ψ is constant along a stream line, i.e. ψ = constant on any stream line.

i.e.

$$d\psi = 0$$

$$\Rightarrow \frac{dy}{dx} = \frac{v}{u}$$

Example 4.23: The velocity potential function for 2-dimensional flow is $\phi = x(2y-1)$. At point $P(4, 5)$, determine velocity vector and its magnitude and value of stream function.

Solution: Given data: $\phi = x(2y-1)$ and $P(4,5)$.

We know velocity potential

$$u = -\frac{d\phi}{dx}, \; v = -\frac{d\phi}{dy}$$

$$\Rightarrow u = -\frac{d}{dx}x(2y-1), \Rightarrow v = -\frac{d}{dy}x(2y-1)$$

$$\Rightarrow u = (1-2y), \Rightarrow v = -2x$$

Hence, **velocity vector** and **magnitude** $\bar{V} = (1-2y)i - 2xj$

$$V = \sqrt{(1-2y)^2 + (2x)^2}$$

$$\Rightarrow (V)_{4,5} = \sqrt{(1-2*5)^2 + (2*4)^2} = 12.04 \text{ m/s}$$

Stream function and **its value** $d\psi = udy - vdx$

$$\Rightarrow d\psi = (1-2y)dy + 2xdx$$

$$\Rightarrow \int d\psi = \int (1-2y)dy + \int 2xdx$$

$$\Rightarrow \psi = x^2 - y^2 + y$$

$$(\psi)_{4,5} = 16 - 25 + 5 = -4$$

Example 4.24: The stream function in a 2-D flow is $\psi = 6x - 4y + 7xy$. Verify whether the flow is irrotational or rotational.

Solution: Given data: $\psi = 6x - 4y + 7xy$ and $P(1,-1)$.

To have flow rotational, $$\omega_z = -\frac{1}{2}\left(\frac{\partial^2\psi}{\partial x^2} + \frac{\partial^2\psi}{\partial y^2}\right) = 0$$

i.e. $$\omega_z = -\frac{1}{2}\left(\frac{\partial^2\psi}{\partial x^2} + \frac{\partial^2\psi}{\partial y^2}\right) = -\frac{1}{2}\left(\frac{\partial^2}{\partial x^2}(6x - 4y + 7xy) + \frac{\partial^2}{\partial y^2}(6x - 4y + 7xy)\right)$$

$$\Rightarrow \omega_z = -\frac{1}{2}(0 + 0) = 0$$

Hence, the **flow is rotational.**

Example 4.25: The stream function in a 2-dimensional flow is $\psi = 6x - 4y + 7xy$. Estimate the vectors and magnitudes of velocity and acceleration at point $P(1, -1)$, if the flow is having a uniform discharge of 0.16 m³/s.

Solution: Given data: $\psi = 6x - 4y + 7xy$, $Q = 0.16\,\text{m}^3/\text{s}$, $P(1,-1)$

We know stream function

$$u = \frac{d\psi}{dy},\ v = -\frac{d\psi}{dx}$$

$$\Rightarrow u = \frac{d}{dy}(6x - 4y + 7xy),\ \Rightarrow v = -\frac{d}{dx}(6x - 4y + 7xy)$$

$$\Rightarrow u = (-4 + 7x),\ \Rightarrow v = -6 - 7y$$

Velocity vector and magnitude

$$\bar{V} = (-4 + 7x)i - (6 + 7y)j$$

$$(\bar{V})_{1,-1} = (-4 + 7*1)i - (6 - 7*1)j = 3i + j$$

$$V = \sqrt{3^2 + 1^2} = \sqrt{10}\text{ m/s}$$

Spatial acceleration

$$a_s = u\frac{\partial V}{\partial x} + v\frac{\partial V}{\partial y}$$

$$\frac{\partial V}{\partial x} = \frac{\partial}{\partial x}((-4 + 7x)i - (6 + 7y)j) = 7i$$

$$\frac{\partial V}{\partial y} = \frac{\partial}{\partial y}((-4 + 7x)i - (6 + 7y)j) = -7j$$

So spatial acceleration will become $a_s = (7x-4)7i + (7y+6)7j$

$$\Rightarrow a_s = (49x-21)i + (49y+42)j$$

$$\Rightarrow (a_s)_{1,-1} = (49*1-4)i + (7*(-1)+6)j$$

$$\Rightarrow (a_s)_{1,-1} = 45i - 1j$$

$$\Rightarrow (a_s)_{1,-1} = \sqrt{45^2+1^2} = 45.01\text{m/s}^2$$

Example 4.26: In a two-dimensional incompressible flow, the fluid velocity components are given by $u = x-4y, v = -y-4x$. Show that the velocity potential exists and determine its form for stream function as well.

Solution: Given data: $u = x-4y, v = -y-4x$

$$d\phi = udx + vdy$$

$$\Rightarrow \int d\phi = \int (x-4y)dx + \int (-y-4x)dy$$

$$\Rightarrow \phi = \frac{x^2}{2} - 4xy - \frac{y^2}{2} - 4xy$$

$$\Rightarrow \phi = \left(\frac{x^2}{2} - \frac{y^2}{2} - 8xy\right)$$

For velocity potential to exist $\qquad \dfrac{d^2\phi}{dx^2} + \dfrac{d^2\phi}{dy^2} = 0$

$$\Rightarrow \frac{d^2}{dx^2}\left(\frac{x^2}{2} - \frac{y^2}{2} - 8xy\right) + \frac{d^2}{dy^2}\left(\frac{x^2}{2} - \frac{y^2}{2} - 8xy\right) = 0$$

$$\Rightarrow 1 - 1 = 0$$

As L.H.S = R.H.S, hence **velocity potential exists**.

$$d\psi = udy - vdx$$

$$\Rightarrow \int d\psi = \int udy - \int vdx$$

$$\Rightarrow \psi = \int (x-4y)dy - \int (-y-4x)dx$$

$$\Rightarrow \psi = \left(2x^2 - 2y^2 + 2xy\right)$$

For stream function to exist

$$\frac{\partial^2 \psi}{dxdy} - \frac{\partial^2 \psi}{dydx} = 0$$

$$\Rightarrow \frac{\partial^2}{dxdy}\left(2x^2 - 2y^2 + 2xy\right) - \frac{\partial^2}{dydx}\left(2x^2 - 2y^2 + 2xy\right) = 0$$

$$\Rightarrow 2 - 2 = 0$$

As L.H.S = R.H.S, hence **stream function exists.**

Example 4.27: In a two-dimensional incompressible flow, the fluid velocity components are given by $u = x^2 - y^2, v = x^2 + y^2$. Show that velocity potential exists and determine its form for stream function as well.

Solution: Given data: $u = x^2 - y^2, v = x^2 + y^2$

$$d\phi = udx + vdy$$

$$\Rightarrow \int d\phi = \int\left(x^2 - y^2\right)dx + \int\left(x^2 + y^2\right)dy$$

$$\Rightarrow \phi = \frac{x^3}{3} - xy^2 + x^2 y + \frac{y^3}{3}$$

For velocity potential to exist $\frac{d^2\phi}{dx^2} + \frac{d^2\phi}{dy^2} = 0$

$$\Rightarrow \frac{d^2}{dx^2}\left(\frac{x^3}{3} - xy^2 + x^2 y + \frac{y^3}{3}\right) + \frac{d^2}{dy^2}\left(\frac{x^3}{3} - xy^2 + x^2 y + \frac{y^3}{3}\right) = 0$$

$$\Rightarrow \frac{d}{dx}\left(x^2 - y^2 + 2xy\right) + \frac{d}{dy}\left(-2xy + x^2 + y^2\right) = 0$$

$$\Rightarrow 2x + 2y - 2x + 2y = 4y \neq 0$$

As L.H.S = R.H.S, hence **velocity potential does not exist.**

$$d\psi = udy - vdx$$

$$\Rightarrow \int d\psi = \int udy - \int vdx$$

$$\Rightarrow \psi = \int (x^2 - y^2)\,dy - \int (x^2 + y^2)\,dx$$

$$\Rightarrow \psi = \left(x^2 y - \frac{y^3}{3} - \frac{x^3}{3} - xy^2 \right)$$

For stream function to exist

$$\frac{\partial^2 \psi}{dxdy} - \frac{\partial^2 \psi}{dydx} = 0$$

$$\Rightarrow \frac{\partial^2}{dxdy}\left(x^2 y - \frac{y^3}{3} - \frac{x^3}{3} - xy^2 \right) - \frac{\partial^2}{dydx}\left(x^2 y - \frac{y^3}{3} - \frac{x^3}{3} - xy^2 \right) = 0$$

$$\Rightarrow \frac{\partial}{dx}\left(x^2 - y^2 - 2xy \right) - \frac{\partial}{dy}\left(2xy - x^2 - y^2 \right) = 0$$

$$\Rightarrow 2x - 2y - 2x + 2y = 0$$

$$\Rightarrow 0 = 0$$

As L.H.S $\neq$ R.H.S, hence **stream function exists.**

Example 4.28: In a two-dimensional incompressible flow, the streamline is represented by $\psi = x^2 - y^2, \psi = x^2 + y^2$. Sketch streamlines.

Solution: Given data: $\psi = x^2 - y^2, \psi = x^2 + y^2$. Refer Table 4.1: we have on any streamline, the stream function is constant and hence for the function $\psi = x^2 - y^2$, we have

$$x = \sqrt{y^2 + \psi} \qquad (1)$$

Substituting different values of ψ, y we can plot the streamlines with constant stream functions, as given below in the table.

	y	$y = 0$	$y = 1$	$1 = 2$	$y = 3$
$\psi = 1$	$x = \sqrt{y^2 + \psi}$	± 1	$\pm\sqrt{2}$	$\pm\sqrt{5}$	$\pm\sqrt{10}$
$\psi = 2$	$x = \sqrt{y^2 + \psi}$	$\pm\sqrt{2}$	$\pm\sqrt{3}$	$\pm\sqrt{6}$	$\pm\sqrt{11}$
$\psi = 3$	$x = \sqrt{y^2 + \psi}$	$\pm\sqrt{3}$	± 2	$\pm\sqrt{7}$	$\pm\sqrt{11}$

Using the above tabulated data, we can plot the graph between x, y, ψ, as shown in the Figure P4.28.

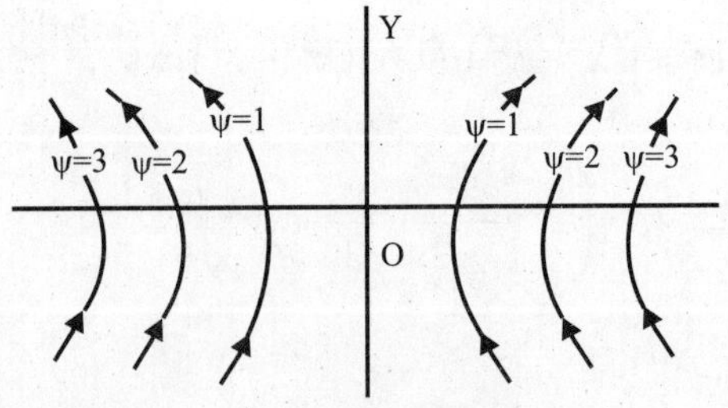

Figure P4.28

FLOW NETS

Flow nets can be defined as "**graphical representation of flow path of a fluid particle flowing from the upstream to the downstream denoting streamlines and series of equipotential lines (slope *dy/dx* and along any equipotential line, the energy available to cause flow is the same), in form of a grid**". A flow net may be drawn for a two-dimensional irrotational flow and it provides a simple, yet valuable indication of the flow pattern based on Laplace equation's and is a concept to illustrates graphically how the head or energy is lost as water flows from one point to other. Flow net provides a simple graphical technique for studying two-dimensional irrotational flows, especially in the cases where mathematical relations for stream function and velocity function are either not available or are rather difficult and cumbersome to solve.

There are four methods of representing flow nets and they are:

a) Analytical method
b) Graphical method
c) Hydraulic models
d) Electrical analogy.

a) Analytical method: Analytical method obtains the values of ϕ and ψ mathematically and are plotted to get the required flow net grid. This method is useful in drawing flow nets, when dealing with simple and ideal boundary conditions.

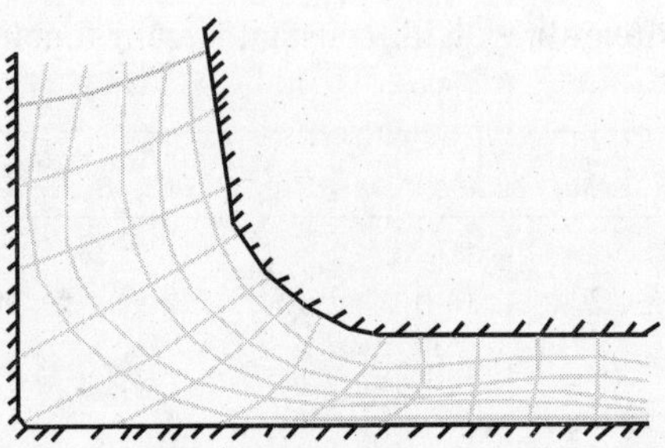

Figure 4.14a: Flow net

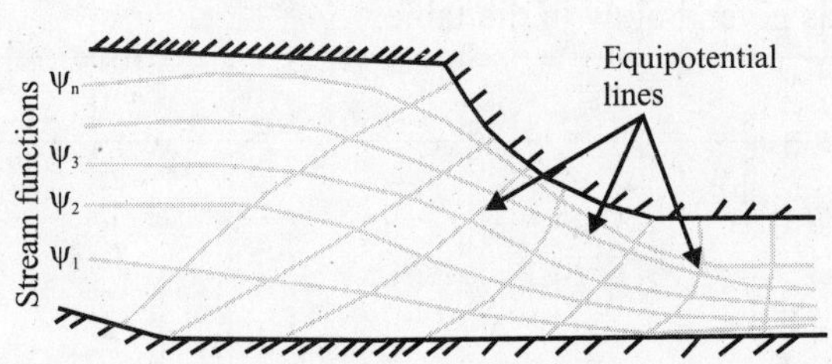

Figure 4.14b: Flow net

b) Graphical method: The product of stream flow lines slope $\left({}^{dy}/_{dx} = {}^{v}/_{u} \right)$ and the equipotential line $\left({}^{dy}/_{dx} = {}^{-u}/_{v} \right)$ is –1 and so, in graphical method the streamlines and equipotential lines are plotted on a graph, so that they cut each other orthogonally, so as to obtain curvilinear squares. And this method of drawing flow nets requires more time consumption for getting proper curvilinear shapes.

In constructing the graphical flow nets, all the equipotential lines are assumed to have equal head losses and same amount of seepage flows between two adjacent pairs of flow lines, i.e. discharge $Q = \mu L \Delta h \left({}^{n_f}/_{n_d} \right)$, where μ being coefficient of permeability or hydraulic conductivity, "L" length of aquifer perpendicular to the cross-section, "Δh" head loss through the flow net, n_f number of flow channels, n_d number of equipotential drops.

c) Hydraulic models: Hydraulic models like a seepage model is been used with a dye injected by it, so as to trace the streamlines and equipotential lines, and by drawing them the flow nets are completed.

d) Electrical analogy method: Electrical analogy method is based on assuming that the fluid flow is analogy to flow of electricity through a conductor. And so the electric current is taken as analogy to velocity of the flow and electric potential is taken as analogy to velocity potential. And this method is a practical method.

Applications and Limitations of Flow Nets

Uses of flow nets

1) Flow net can be used for determining streamlines and equipotential lines.
2) It is useful to determine velocity, pressure distributions for a given boundary of flow.
3) It is used to determine the design of the outlets of their streamlining.
4) To determine quantity of seepage and upward lift pressure below the hydraulic structures.

Limitations of flow nets:

1) Flow nets cannot be applied for close boundary, due to viscosity presence.
2) In case of flow of a fluid past over a solid body, it cannot give clear picture of streamlines and flow conditions at the boundaries of the grid, due to presence of eddies and separations. The flow should be steady and irrotational because streamline patterns will remain constant only for steady flow and the rotation of flow changes the flow pattern instantaneously and hence the flow need to be steady and irrotational.

REYNOLDS TRANSPORT THEOREM

Eulerian coordinates is fixed coordinate systems focusing on independent variables like $x, y, z, \ldots$ at particular time "t" and so Eulerian form can be written as $R = f(x_i, t)$. Whereas Lagrangian coordinates system focus on the fluid mass flow from one point to another and so can be written as $R = f(x_{oi}, t)$, where the coordinate x_{oi} indicates which fluid element is being considered.

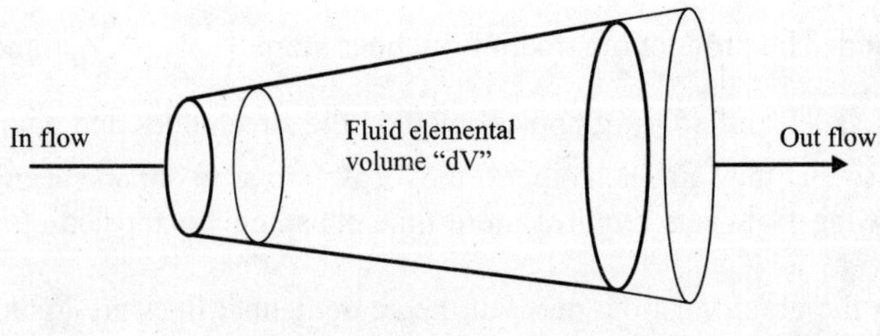

Figure 4.15: Fluid elemental volume

Consider an elemental control volume "*V*" having fluid flow in and out through its faces, as shown in Figure 4.15.

Once the elemental fluid control volume describes to new position "dV", all the variable properties "$\beta = f(\beta_i, t)$" of the fluid changes and to with respect to time "dt" and so the net change in elemental control volume properties (**Lagrangian property derivative**) can be expressed in mathematical as a form of total differentiation of time of each and every fluid particles (**Eulerian property derivatives**) which is given as

$$\frac{d}{dt}\left(\iiint_{dV} control\ \ volume\right)_{fixed\ \ net} = \left\{\frac{d}{dt}\left(\iiint_{dV} final\ \ control\ \ volume\right) - \frac{d}{dt}\left(\iiint_{V} \beta dV\right)\right\}$$

As the control volume "V" is fixed, inspite of its change to new position (i.e., from (t) to ($t + dt$), it remains constant and so

$$\frac{d}{dt}\left(\iiint_{V} \beta(t)\right) = \left\{\frac{d}{dt}\left(\iiint_{dV} \beta(t+dt)\right)\right\}$$

But still there will be **change in system** in spite of fixed control volume and hence,

$$\frac{d}{dt}\left(\iiint_{dV} \beta(t+dt)dV\right)_{NET} = \left\{\frac{d}{dt}\left(\iiint_{dV} \beta(t+dt)dV\right)_{final} - \frac{d}{dt}\left(\iiint_{dV} \beta(t)dV\right)_{initial}\right\}$$

$$\Rightarrow \frac{d}{dt}\left(\iiint_{dV} \beta dV\right)_{NET} = \lim_{dt\to 0}\left\{\frac{1}{dt}\left[\left(\iiint_{dV(t+dt)} \beta(t+dt)dV\right) - \left(\iiint_{dV(t)} \beta(t)dV\right)\right]\right\}$$

$$\Rightarrow \frac{d}{dt}\left(\iiint_{dV} \beta dV\right)_{NET} = \lim_{dt\to 0}\left\{\frac{1}{dt}\left[\left(\iiint_{dV(t+dt)} \beta(t+dt)dV\right) - \left(\iiint_{dV(t)} \beta(t+dt)dV\right)\right]\right.$$

$$+\left[\left(\iiint_{V(t)}\beta(t+dt)\,dV\right)-\left(\iiint_{V(t)}\beta(t)\,dV\right)\right]\Bigg\}$$

$$\Rightarrow \frac{d}{dt}\left(\iiint_{dV(t)}\beta(t)\,dV\right)_{NET} = \lim_{dt\to 0}\left\{\frac{1}{dt}\left[\left(\iiint_{dV(t+dt)-V(t)}\beta(t+dt)\,dV\right)\right]+\left(\iiint_{dV(t)}\frac{\partial\beta}{\partial t}dV\right)\right\} \quad (4.35)$$

The change in volume "dV" (m^3 units) of the control volume is due to inflow of mass flow rate and can be written as $dV = \overline{m}_i = \rho AV.n = V.n\delta tds$

where "ds" is the surface area, "V" is the velocity and "n" is the unit normal.

Hence, Eq. (4.35) becomes

$$\frac{d}{dt}\left(\iiint_{dV(t)}\beta(t)\,dV\right)_{NET} = \lim_{dt\to 0}\left\{\frac{1}{dt}\left[\left(\iiint_{dV(t+dt)-V(t)}\beta(t+dt)\,V\,n\delta tds\right)\right]+\left(\iiint_{dV(t)}\frac{\partial\beta}{\partial t}dV\right)\right\}$$

$$\Rightarrow \frac{d}{dt}\left(\iiint_{dV(t)}\beta(t)\,dV\right) = \lim_{dt\to 0}\left\{\left[\left(\oiiint_{S(t)}\beta(t+dt)\,V.nds\right)\right]+\left(\iiint_{dV(t)}\frac{\partial\beta}{\partial t}dV\right)\right\}$$

$$=\left\{\left[\left(\oiiint_{S(t)}\beta(t)\,V.nds\right)\right]+\left(\iiint_{dV(t)}\frac{\partial\beta}{\partial t}dV\right)\right\}$$

According to "**Divergence/Gauss theorem**", we have $\int_S a.nds = \int_V \nabla.adV$, where "a" is any vector and *n* is the unit normal.

So $$\oiiint_{S(t)}\beta(t)\,V.nds = \iiint_V \nabla.(\beta V)\,dV$$

Therefore, $$\Rightarrow \frac{d}{dt}\left(\iiint_{dV(t)}\beta(t)\,dV\right) = \iiint_{dV(t)}\nabla.(\beta V)\,dV + \iiint_{dV(t)}\frac{\partial\beta}{\partial t}dV$$

$$\Rightarrow \frac{d}{dt}\left(\iiint_{dV(t)}\beta(t)\,dV\right) = \iiint_{dV(t)}\left[\nabla.(\beta V)+\frac{\partial\beta}{\partial t}\right]dV = \iiint_{dV(t)}\left[\frac{\partial(\beta V_i)}{\partial x_i}+\frac{\partial\beta}{\partial t}\right]dV \quad (4.36)$$

Equation 4.36 forms the **Reynolds' Transport Theorem**, where LHS represents the Lagrangian Derivative and the RHS represents the Eulerian Derivatives.

If the flow is steady then the mass in flow is equal to the out flow and then Eq. (4.6) reduces to

$$\frac{d}{dt}\left(\iiint_{dV(t)} \beta(t)\,dV\right) = \iiint_{dV(t)} \left[\frac{\partial \beta}{\partial t}\right] dV$$

Example 4.29: Velocity vector of a flow field is given as $\bar{V} = 2xyi - x^2zj$. The vorticity vector at (1,1,1) will be **[GATE 2010]**

Solution: Given data: $\bar{V} = 2xyi - x^2zj; P(1,1,1)$

Vorticity is given by
$$\Omega = 2\bar{\omega} = \begin{vmatrix} i & j & k \\ \partial/\partial x & \partial/\partial y & \partial/\partial z \\ u & v & w \end{vmatrix} = \begin{vmatrix} i & j & k \\ \partial/\partial x & \partial/\partial y & \partial/\partial z \\ 2xy & -x^2z & 0 \end{vmatrix}$$

$$\Omega = \left[(0 - x^2)i - (0 - 0)j + (-2xz - 2x)k\right] = -x^2i - (2xz + 2x)k$$

Vorticity vector Ω at point $P(1,1,1)$ is $\Omega = -i - 4k$

Example 4.30: Determine the pressure variation of a fluid particle having streamline velocity $V = V_o\left(1 + \frac{r^3}{x^3}\right)$ over a sphere and tending towards the center of sphere.

Solution: Given data: $\bar{V} = V_o\left(1 + \frac{r^3}{x^3}\right)$

According to hydrostatic pressure, we have $\frac{-dp}{ds} = \rho g$

But as the acceleration of the stream particle towards the center is due to gravity and tangential acceleration of the particle exits while rotation of the particle but as the particle tends towards center, the normal acceleration is zero, so $a_s = g_s = V\frac{dV}{dx}$ and $a_n = 0$

So
$$\frac{-dp}{dx} = \rho V\frac{dV}{dx} = \rho V_o\left(1 + \frac{r^3}{x^3}\right)\frac{d}{dx}\left[V_o\left(1 + \frac{r^3}{x^3}\right)\right]$$

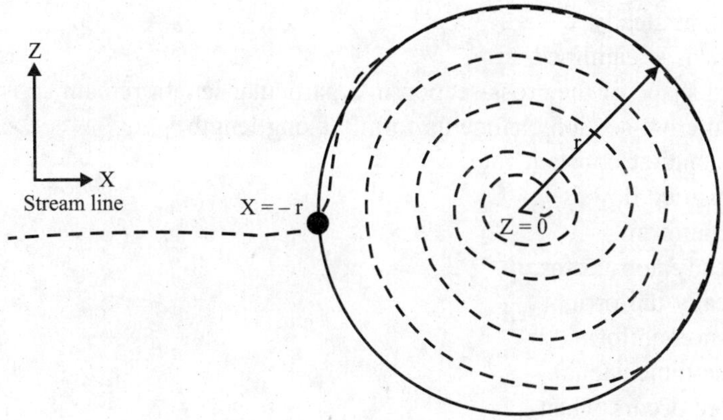

Figure P4.30

$$\Rightarrow \int_0^p dp = 3r^3\rho V_o^2 \int_{-\alpha}^{x}\left[\left(1+\frac{r^3}{x^3}\right)\right]\left(x^{-4}\right)dx$$

$$\Rightarrow P = 3r^3\rho V_o^2 \left|\left\{\left[-\frac{1}{3}x^{-3}\right]-\left[\frac{1}{6}r^3x^{-6}\right]\right\}\right|_{-\alpha}^{x} = -\rho V_o^2\left\{\left[\frac{r^3}{x^3}\right]+\left[\frac{1}{2}\frac{r^3}{x^6}\right]\right\}$$

EXERCISE

Objective

1. In the case of steady flow of fluid, the acceleration of any fluid particle is
   a) Constant
   b) Variable
   c) Zero
   d) Zero under limiting conditions
   e) Never zero.
2. Non-uniform flow occurs when
   a) The direction and magnitude of the velocity at all points are identical
   b) The velocity of successive fluid particles, at any point, is the same at successive periods of time
   c) The magnitude and direction of the velocity do not change from to point to point in the fluid
   d) The fluid particles move in plane or parallel planes and the streamline patterns are identical in each plane
   e) Velocity, depth, pressure, etc. change from point to point in the fluid flow.
3. Uniform flow occurs when

a) The flow is steady
b) The flow is streamlined
c) Size and shape of the cross-section in a particular length remain constant
d) Size and cross-section change uniformly along length
e) Flow occurs at constant rate.

4. Gradually varied flow is
a) Steady uniform
b) Non-steady non-uniform
c) Non-steady uniform
d) Steady non-uniform
e) True one-dimensional.

5. Steady flow occurs when
a) The direction and magnitude of the velocity at all points are identical
b) The velocity of successive fluid particles at any point, is the same at successive periods of time
c) The magnitude and direction of the velocity do not change from point to point in the fluid
d) The fluid particles move in plane or parallel planes and the streamline patterns are identical in the each plane
e) Velocity, depth, pressure, etc. change from point to point in the fluid flow.

6. Which of the following represents unsteady uniform flow?
a) Flow through an expanding tube at an increasing rate
b) Flow through an expanding tube at constant rate
c) Flow through a long pipe at decreasing rate
d) Flow through a long pipe at constant rate
e) None of the above

7. The flow in which each liquid particle has a definite path and their paths do not cross each other is called
a) One-dimensional flow
b) Uniform flow
c) Steady flow
d) Turbulent flow
e) Streamline flow.

8. The flow in which conditions do not change with time at any point, is known as
a) One-dimensional flow
b) Uniform flow
c) Steady flow
d) Turbulent flow
e) Streamline flow.

9. The flow in which particles of a fluid attain such velocities that varies from point to point in magnitude and direction as well as from instant to instant, is known as
a) One-dimensional flow
b) Uniform flow

c) Steady flow
d) Turbulent flow
e) Streamline flow.

10. A streamline is defined as the line
a) Parallel to central axis of flow
b) Parallel to outer surface of pipe
c) Of equal velocity in a flow
d) Along which the pressure drop is uniform
e) which occurs in all flows

11. The equation of continuity holds good when the flow
a) is steady
b) is one-dimensional
c) Velocity is uniform at all the cross-sections
d) All of the above
e) None of the above

12. When the flow parameters at any instant remain same at every point, then flow is said
a) Quasi static
b) Steady state
c) Laminar
d) Uniform
e) Static

13. A one-dimensional flow is one which
a) is uniform flow
b) is steady uniform flow
c) takes place in straight lines
d) involves zero transverse component of flow
e) takes place in one dimension

14. The continuity equation is connected with
a) Viscous unviscous fluids
b) Compressibility fluids
c) Conservation of mass
d) Steady unsteady flow
e) Open channel pipe flow

15. The equation of continuity of flow is based on the principle of conservation of
a) Flow
b) Mass
c) Momentum
d) Energy
e) Mass, momentum and energy

16. Choose the wrong statement about flow nets
a) Flow nets are drawn to indicate flow patterns in case of one-dimensional flow
b) Flow net consists of a system of streamlines so spaced that the rate of flow is the same between each successive pair of lines

c) Flow net consists of another system of lines normal to the streamlines and so spaced that the distance between the normal lines equals the distance between adjacent streamlines
d) An infinite number of streamline are required to describe completely the flow under given boundary condition
e) It is an usual practice to use a small number of such streamlines, as long as acceptable accuracy is obtained.

17. The general equation of continuity for three-dimensional flow of a compressible fluid for steady flow is

(a) $\frac{\partial u}{\partial x}+\frac{\partial v}{\partial y}+\frac{\partial w}{\partial z}=0$

(b) $\frac{\partial u}{\partial x}+\frac{\partial v}{\partial y}+\frac{\partial w}{\partial z}=0$

(c) $\frac{\partial u}{\partial x}+\frac{\partial v}{\partial y}+\frac{\partial w}{\partial z}=1$

(d) $\frac{\partial u}{\partial x}+\frac{\partial v}{\partial y}+\frac{\partial w}{\partial z}=u.v.w$

(e) None of the above

18. Consider steady, incompressible and irrotational flow through a reducer in a horizontal pipe where the diameter is reduced from 20 cm to 10 cm. The pressure in the 20 cm pipe just upstream of the reducer is 150 kPa. The fluid has a vapor pressure of 50 kPa and a specific weight of 5 kN/m3. Neglecting frictional effects, the maximum discharge (in m^3/s) that can pass through the reducer without causing cavitation is
(a) 0.05
(b) 0.16
(c) 0.27
(d) 0.38

19. You are asked to evaluate assorted fluid flows for their suitability in a given laboratory application. The following three flow choices, expressed in terms of the two-dimensional velocity fields in the xy-plane, are made available.
P. u = 2y, v = –3x
Q. u = 3xy, v = 0
R. u = –2x, v = 2y
Which flow(s) should be recommended when the application requires the flow to be incompressible and ir-rotational?
(a) P and R
(b) Q
(c) Q and R
(d) R

Answers

1. c	2. c	3. c	4. d	5. b	6. c
7. b	8. c	9.d	10. c	11. a	12. d
13. d	14. c	15. b	16. c	17. a	18. b
19. d					

Theory

1. When is the flow in pipe (a) unsteady and (b) non-uniform?
2. Define and distinguish between steady flow and uniform flow. Give two examples of each.
3. Differentiate between compressible and incompressible flows.
4. Define and distinguish laminar and turbulent flows.
5. Under what can one treat real fluid flow as irrotational.
6. Define the equation of continuity and obtain the equation of continuity equation.
7. Define the terms a) Velocity potential function and b) Stream function and vortex flow.
8. What is the condition for the flow to be irrotational?
9. Write different types of flow lines.
10. What do you mean by equipotential line and a line of constant stream function?
11. Explain the terms vorticity and circulation.
12. Define stream function and explain its significance.
13. Derive an expression for Bernoulli's equations for flow along a streamline.
14. Describe the uses and limitations of flow nets.

Problems

1. Determine the equation of a streamline at $t = t_0$ and passing through a point (x_0, y_0) for the velocity field $V = (1 + At)i + 2xj$.
2. Determine the equation of a streamline passing through a point $(1,0,1)$ for the velocity field $V = 2rxi + 2ryj - 4rzk$.
3. Find the magnitudes of velocity and acceleration for the flow field $V = 2x^3 i - 3x^2 yj$ at point $P(2,1,3,1)$.
4. For steady, incompressible flow, verify whether the following values of u and v are possible:

$$u = \frac{-x}{(x^2 + y^2)}, v = \frac{-y}{(x^2 + y^2)}$$

5. What the irrotational velocity field associated with potential $\phi = 3x^2 - 3x + 3y^2 + 16t^2 + 12zt$? Does it satisfy continuity incompressible equation?

QUICK REFERENCE OF FORMULAE TO REMEMBER

Property	Formulae	Remarks, if any
Velocity of particle (V)	$V = V_x i + V_y j + V_z k \quad V = \sqrt{u^2 + v^2 + w^2}$ $u = \dfrac{\partial x}{\partial t};\ v = \dfrac{\partial y}{\partial t},\ w = \dfrac{\partial z}{\partial t}$	u, v, w are velocity components along x, y and z directions resp.
Acceleration of particle (a)	$a = a_s + a_n \quad a_s = a_x i + a_y j + a_z k$ $a_s = \sqrt{{a_x}^2 + {a_y}^2 + {a_z}^2}$ $a_s = v\left(\dfrac{\partial v}{\partial s}\right) + \left(\dfrac{\partial v}{\partial t}\right);\ a_s = u\dfrac{\partial V}{\partial x} + v\dfrac{\partial V}{\partial y} + w\dfrac{\partial V}{\partial z} + \dfrac{\partial V}{\partial t}$ $a_n = \dfrac{v^2}{r}$	a_s, a_N are acceleration components along space (x, y, z) and radial directions resp.
Steady flow	$\dfrac{\partial \rho}{\partial t} = 0$	P is density, s is space (x, y, z axis)
Uniform flow	$\left(\dfrac{\partial v}{\partial s}\right)_t = 0$	
Incompressible flow	ρ = Constant	
Streamline	$\dfrac{dx}{u} = \dfrac{dy}{v} = \dfrac{dz}{w}$	
Continuity equation (3-D) Cartesian	$\dfrac{\partial(\rho u)}{\partial x} + \dfrac{\partial(\rho v)}{\partial y} + \dfrac{\partial(\rho w)}{\partial z} + \dfrac{\partial(\rho)}{\partial t} = 0$	
Continuity equation (1-D)	Discharge (Q) = area * Velocity	
Continuity equation (3-D) polar	$\dfrac{1}{r}(\rho V_r) + \dfrac{\partial(\rho V_r)}{\partial r} + \dfrac{1}{r}\dfrac{\partial(\rho V_\theta)}{\partial \theta} = 0$	
Continuity equation (3-D) spherical	$\dfrac{1}{r^2}\dfrac{\partial(r^2 V_r)}{\partial r} + \dfrac{1}{r\sin\theta}\dfrac{\partial(V_\theta \sin\theta)}{\partial \theta}$ $+ \dfrac{1}{r\sin\theta}\dfrac{\partial(V_w)}{\partial w} = 0$	

Contd.

Property	**Formulae**	**Remarks, if any**
Circulation (Γ)	$\Gamma = \left(\dfrac{\partial v}{\partial x} - \dfrac{\partial u}{\partial y}\right)\delta x \delta y$	
Vorticity (Ω)	$\Omega = \dfrac{\Gamma}{dA}, \quad \Omega = \left(\dfrac{\partial v}{\partial x} - \dfrac{\partial u}{\partial y}\right)$	"dA" is elemental area
Rotation velocity (ω)	$\omega = \omega_x i + \omega_y j + \omega_z k \omega = \dfrac{1}{2}\begin{vmatrix} i & j & k \\ \partial/\partial x & \partial/\partial y & \partial/\partial z \\ u & v & w \end{vmatrix}$ $\omega = \sqrt{\omega_x^2 + \omega_y^2 + \omega_z^2}\, \omega_x = \dfrac{1}{2}\left(\dfrac{\partial w}{\partial y} - \dfrac{\partial v}{\partial z}\right) = \dfrac{\Omega_x}{2}$ $\omega_y = \dfrac{1}{2}\left(\dfrac{\partial u}{\partial z} - \dfrac{\partial w}{\partial x}\right) = \dfrac{\Omega_y}{2}; \omega_z = \dfrac{1}{2}\left(\dfrac{\partial v}{\partial x} - \dfrac{\partial u}{\partial y}\right) = \dfrac{\Omega_z}{2}$	ω = Ω/2
Velocity potential (ϕ)	$u = -\dfrac{\partial \phi}{\partial x}; \ v = -\dfrac{\partial \phi}{\partial y}; \ w = -\dfrac{\partial \phi}{\partial z}; \ \dfrac{dy}{dx} = \left(\dfrac{u}{v}\right)$	*dy*/*dx* is the slpe of equipotential line
Stream function (ψ)	$u = \dfrac{\partial \psi}{\partial y}; \ v = -\dfrac{\partial \psi}{\partial x}; \ \dfrac{dy}{dx} = -\left(\dfrac{v}{u}\right)$	

Table 4.1: Important points to remember

If streamline are radial	$a_n = 0$
Rotational flow	$\omega \neq 0 \quad \Omega \neq 0$
Rotation	$\omega = \Omega / 2$
Irrotational flow	$\omega = 0; \ \Omega = 0$
Flow possibility	If the continuity equation is satisfied then the flow is possible.
$\omega = \dfrac{1}{2}(\nabla * V)$	Where $(\nabla * V)$ is lurl of velocity vector, $\Omega = (\nabla * V)$
Circulation (Γ) happens only	When vorticity Ω is present
Product of equipotential and stream function	$\left(\dfrac{dy}{dx}\right)_{\phi=c} * \left(\dfrac{dy}{dx}\right)_{\psi=c} = -1$
If ϕ = continues and satisfies Laplace equation	Flow is steady, incopressible, irrotational
If ψ = continues and satisfies Laplace equation	Flow is steady, incompressible, ir-rotational and 2-dimenstional
ϕ = constant *t* along a line	It is called as equipotential line
One any streamline	ψ = constant every where
Stream function ψ	Exists fan rotational and irrotational flow

CHAPTER 5

Fluid Dynamics

INTRODUCTION

Fluid dynamics is a science, which deals with the velocities and acceleration behavior of fluids, and the forces/energies causing them.

This chapter deals with three conservation equations commonly used in fluid mechanics: The mass equation, Bernoulli equation (Newton's second law of motion, also called as momentum conservation) and energy equation where the mass equation is an expression of the conservation of mass principle. The Bernoulli equation is concerned with the conservation of kinetic, potential, and flow energies of a fluid stream and their conversion to each other in regions of flow where net viscous forces are negligible and where other restrictive conditions apply. The energy equation is a statement of the conservation of energy principle. In fluid dynamics, the fluids assumed to be continuous (Continuum fluids, in addition to the application of conservation equations. In fluid mechanics, it is found convenient to separate mechanical energy from thermal energy and to consider the conversion of mechanical energy to thermal energy as a result of frictional effects as mechanical energy loss. Then the energy equation becomes the mechanical energy balance.

We start this chapter with an overview of conservation principles and the conservation of mass relation. This is followed by a discussion of various forms of energy and work. Then we derive the Bernoulli equation by applying Newton's second law to a fluid element along a streamline and demonstrate its use in a variety of applications like venturimeter, orifice meter and Pitot tube.

DIFFERENT TYPES OF ENERGIES OF A FLUID

a) Potential Energy: Potential energy is the energy possessed by a fluid due to the virtue of its position with respect to the datum line and is denoted by PE.

$$PE = mgZ = WZ \tag{5.0}$$

where W is the weight of liquid and Z is the height from the datum.

b) Kinetic Energy: Kinetic energy is the energy possessed by a fluid due to the virtue of its motion and denoted by KE.

$$KE = \frac{1}{2}mV^2 = \frac{W}{2g}V^2 \tag{5.1}$$

where W is the weight and V is the velocity of liquid respectively.

c) Pressure Energy/Pressure Head: Pressure energy is the energy created by a fluid, in rest, when contained in a vessel and denoted by h.

$$h = \frac{p}{\gamma} \tag{5.2}$$

where p is the pressure and γ is the specific weight of the liquid respectively .
Hence, pressure energy can also be defined as force per unit weight of fluid.

d) Total Energy: Total energy is the summation of all the energies above and denoted by "*TE*"

$$TE = \frac{p}{\gamma} + \frac{1}{2g}V^2 + Z \tag{5.3}$$

BERNOULLI'S EQUATION (from the concepts of total heads and work done)

Bernoulli stated that in an ideal and incompressible fluid, the total energy of a fluid is constant throughout a streamline, if the flow is steady and continuous.

In other words, "The total pressure which is summation of static and dynamic pressures is defined to be constant throughout a stream line", which indicates the principle of energy conservation.

i.e. Total head = Pressure head + Velocity head + Potential/Datum head = Constant

$$\Rightarrow \frac{p}{\gamma} + \frac{1}{2g}V^2 + Z = \text{Constant} \tag{5.4}$$

Proof: Observe a liquid continuous, ideal, incompressible passing through a non-uniform pipe of diameter, datum heights, pressures, velocities D_1, Z_1, P_1, V_1 and D_2, Z_2, P_2, V_2 at inlet section 1-1 and outlet section 2-2 respectively.

When liquid flows from one datum height to another datum height it losses or possess potential energy (depending upon the datum elevations), while having work done in moving the liquid from one position/section to another position/section. But according to conservation of energy principle, the energy can be neither created nor destroyed but only can be transformed from one form to another. So the energy due to work done due to motion and potentiality is converted into kinetic energy.

Hence, the gain in kinetic energy by the liquid is equal to the sum of the loss in potential energy and work done in moving the liquid. That is, **KE = (Loss of PE + Work done)**, provided the outlet datum height is less than the inlet datum height. Or else the gain in potential energy will be due to the loss of kinetic energy in doing work while moving the liquid from lower datum height to upper datum height. That is, **gain in PE = (Loss of KE + Work done)**

As the liquid is continuous, the weight of liquid W is constant at any section along the pipe length. i.e. $W = \gamma_1 A_1 dx_1 = \gamma_2 A_2 dx_2$

$$\text{since } \gamma = \frac{W}{volume}$$

$$\Rightarrow A_1 dx_1 = A_2 dx_2 = W/\gamma \quad \text{since } \gamma_1 = \gamma_2 = \gamma \tag{5.5}$$

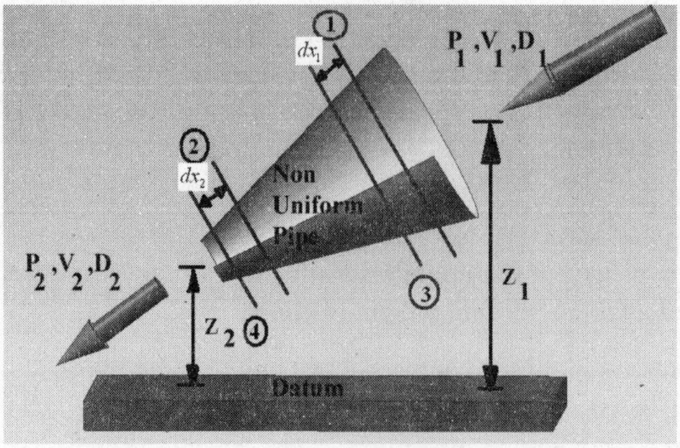

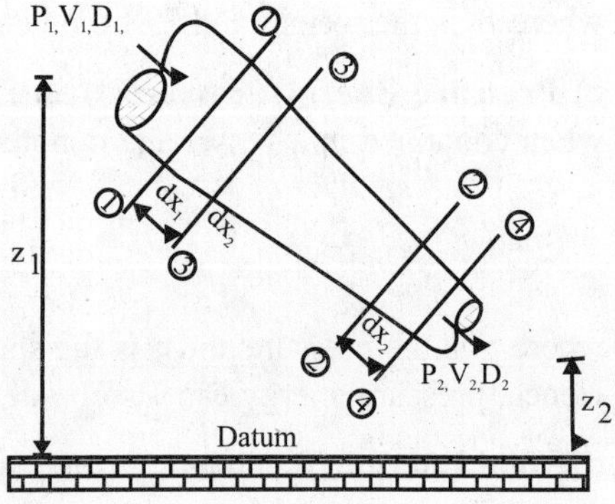

Figure 5.1: Bernoulli's experiment

Potential energies with respect to the two sections 1-1 and 2-2 are $PE_1 = mgh = WZ_1$ $PE_2 = mgh = WZ_2$ respectively.

As seen in Figure 5.1 there is a decrease in datum heights, Z_1, Z_1 while moving from section 1-1 to 2-2 and hence there will loss in potential energy.

Therefore, loss in

$$PE = W\left(Z_1 - Z_2\right) \tag{5.6}$$

Consider the sections 1-1, 2-2 be moved to new sections 3-3, 4-4 by a distance dx_1, dx_2 respectively. There is a net work done (*WD*) in moving the liquid from one position to new positions.

i.e. $$WD = WD_{1-3} - WD_{2-4}$$

But $WD = Force * Displacement$ and hence

$$WD_{1-3} = P_1 A_1 * dx_1\,,\; WD_{2-4} = P_2 A_2 * dx_2$$

$$\Rightarrow WD = A_1 dx_1\left(P_1 - P_2\right) \tag{5.7}$$

By substituting Eq:(5.5) in Eq:(5.7)

We get $$WD = \frac{W}{\gamma}\left(P_1 - P_2\right) \tag{5.8}$$

When the fluid moves from section 1-1 to 2-2 , it attains **kinetic energy** and hence gain in KE is

$$KE = \frac{W}{2g}(V_2^2 - V_1^2) \tag{5.9}$$

and is due to the loss in potential energy and net work done in moving the liquid from one section to the other section and so, **Gain in KE = (Loss in PE + Work done),**

i.e.
$$\frac{W}{2g}\left(V_2^2 - V_1^2\right) = \left(W\left(Z_1 - Z_2\right) + \frac{W}{\gamma}\left(P_1 - P_2\right)\right)$$

or $W(Z_1 - Z_2) = \frac{W}{2g}(V_2^2 - V_1^2) + \frac{W}{\gamma}(P_2 - P_1)$; if **gain in PE = (Loss in KE + Work done).**

By rearranging

$$\frac{p_1}{\gamma_1} + \frac{1}{2g}V_1^2 + Z_1 = \frac{p_2}{\gamma_2} + \frac{1}{2g}V_2^2 + Z_2 \tag{5.10}$$

$$\Rightarrow \frac{p}{\gamma} + \frac{1}{2g}V^2 + Z = \text{Constant} \tag{5.10a}$$

This proves **Bernoulli's equation for an ideal fluid.**

Note: Observing Eq. (5.10), we say that increase in velocity simultaneously decrease in pressure

i.e. $\frac{(p_1 - p_2)}{\gamma} = \frac{\left(V_2^2 - V_1^2\right)}{2g}$

*If the fluid is real, then there will be some losses occurring due to friction, etc. due to which total energy at outlet will be less by an amount of head loss h_L, so Bernoulli's equation for real fluids becomes

$$\boxed{\frac{p_1}{\gamma_1} + \frac{1}{2g}V_1^2 + Z_1 = \frac{p_2}{\gamma_2} + \frac{1}{2g}V_2^2 + Z_2 + h_L} \tag{5.10b}$$

*If the fluid is real and has a turbine at the outlet which develops energy, then there will be head loss he at exit and also energy developed by turbine per unit weight of fluid h_T, SoBernoulli's equation for real fluids with a turbine at outlet becomes

$$\boxed{\frac{p_1}{\gamma_1} + \frac{1}{2g}V_1^2 + Z_1 = \frac{p_2}{\gamma_2} + \frac{1}{2g}V_2^2 + Z_2 + h_L + h_T} \tag{5.10c}$$

*If the fluid is real and has a pump at the inlet which pumps, then there will be head losses h_L at the exit and also energy head added up to the pump per unit weight of the fluid h_p and so Bernoulli's equation for real fluids with a pump at the inlet becomes

$$\boxed{\frac{p_1}{\gamma_1} + \frac{1}{2g}V_1^2 + Z_1 + h_p = \frac{p_2}{\gamma_2} + \frac{1}{2g}V_2^2 + Z_2 + h_L} \tag{5.10d}$$

Example 5.1: Water is flowing through a pipe having diameter 0.35 m and 0.2 m at the upper end and bottom end having a pressure of 15 Pa and 25 Pa respectively. Determine the difference in the datum head, if the rate of flow through the pipe is 350 lit/sec.

Solution: Given data:

$$d_1 = 0.35\,\text{m}, d_2 = 0.2\,\text{m}, P_1 = 15\,\text{Pa}$$

$$P_2 = 25\,\text{Pa}, Q = 350\,\text{lit/s}, \gamma_{water} = 9810\,\text{N/m}^3$$

As per continuity equation, we have

$$Q = A_1V_1 = A_2V_2$$

$$\Rightarrow V_1 = Q/A_1 = \frac{0.35*4}{\pi*0.35^2} = 3.63\,\text{m/s}$$

$$\Rightarrow V_2 = Q/A_2 = \frac{0.35*4}{\pi*0.2^2} = 11.15\,\text{m/s}$$

As per Bernoulli's $\quad \frac{p_1}{\gamma_1} + \frac{1}{2g}V_1^2 + Z_1 = \frac{p_2}{\gamma_2} + \frac{1}{2g}V_2^2 + Z_2$

$$\Rightarrow Z_1 - Z_2 = \left(\frac{25-15}{9810}\right) + \frac{1}{2*9.81}\left(11.15^2 - 3.63^2\right) = 5.66\,\text{m}$$

Hence, the difference in datum head is 5.66 m.

Example 5.2: Find pressure and velocities within a contracting and expanding pipe having a fluid flowing with density $\rho = 1000\,\text{kg/m}^3$ steadily through it. The section diameters are $d_1 = 0.7\,\text{m}$ and $d_2 = 0.5\,\text{m}$. The gage pressure at 1 is $P_1 = 150\,\text{kPa}$. The velocity at 1 is $V_1 = 5\,\text{m/s}$. Assuming the tube to be horizontal, what is the gage pressure at section 2?

Solution: Given data:

$$d_1 = 0.7\,\text{m}, d_2 = 0.5\,\text{m}$$

$$P_1 = 150\,\text{kPa}, V_1 = 5\,\text{m/s}, Z_1 = Z_2$$

As per continuity equation, we have

$$Q = A_1V_1 = A_2V_2 \quad \Rightarrow V_2 = A_1V_1/A_2 = \frac{0.7^2*5}{0.5^2} = 9.8\,\text{m/s}$$

As per Bernoulli's equation

$$\frac{p_1}{\gamma_1} + \frac{1}{2g}V_1^2 + Z_1 = \frac{p_2}{\gamma_2} + \frac{1}{2g}V_2^2 + Z_2$$

$$\Rightarrow p_2 = \rho g\left(\frac{p_1}{\rho g} + \frac{1}{2g}\left(V_1^2 - V_2^2\right) + \left(Z_1 - Z_2\right)\right)$$

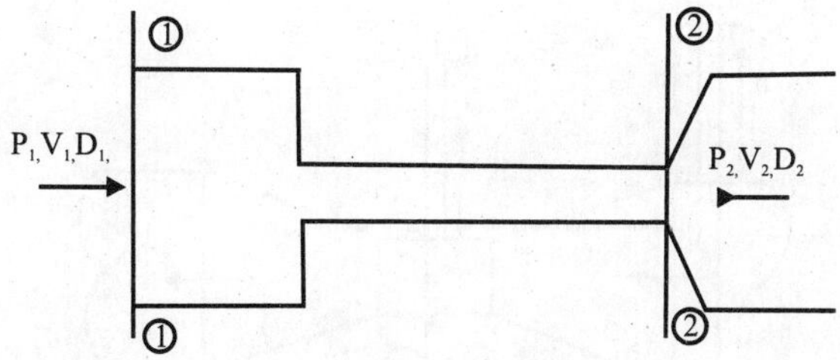

Figure P5.2

$$\Rightarrow p_2 = 1000 * 9.81\left(\frac{150000}{9810} + \frac{1}{2*9.81}\left(5^2 - 9.8^2\right)\right) = 114.4\,\text{kPa}$$

The gage pressure at section 2 is 114.4 kPa.

Applications of Bernoulli's Equation: The Bernoulli's equation is often combined with the Continuity equation to find velocities and pressures at points in the fluid flow connected by a streamline and is applicable to many situations not just the pipe flow but as well to flow measurement from tanks and open channels. It can be applied not only to incompressible fluids using work energy equations but also to compressible fluids like air using the pressure variation phenomena, such as in development of lift on aircraft wings, operation of air speed, indicators, etc.

The incompressible fluid flow measuring devices which use Bernoulli's equation are

1) Venturimeter
2) Orificemeter and Orifice
3) Pitot tube
4) Rotameter
5) Siphon
6) Sluice gate.

VENTURIMETER

Venturimeter is a device used to measure the rate of discharge in a pipeline and has been named after Italian engineer Venturi.

It consists of three parts, namely, a short converging part, a throat and a diverging part (as shown in Figure 5.2). The diameter of the converging and diverging portions has the same diameter as that of the pipeline, to which it is fitted.

Consider a venturimeter having D_1, P_1, V_1, A_1 and D_2, P_2, V_2, A_2 as diameter, pressure, velocity and areas at inlet and at the throat of venturimeter respectively.

If Bernoulli's equation is applied at the two sections like at inlet and throat, then the equation will be

$$\frac{p_1}{\gamma_1} + \frac{1}{2g}V_1^2 + Z_1 = \frac{p_2}{\gamma_2} + \frac{1}{2g}V_2^2 + Z_2$$

$$\Rightarrow \left(\frac{P_1 - P_2}{\gamma}\right) + (Z_1 - Z_2) = \frac{1}{2g}\left(V_2^2 - V_1^2\right)$$

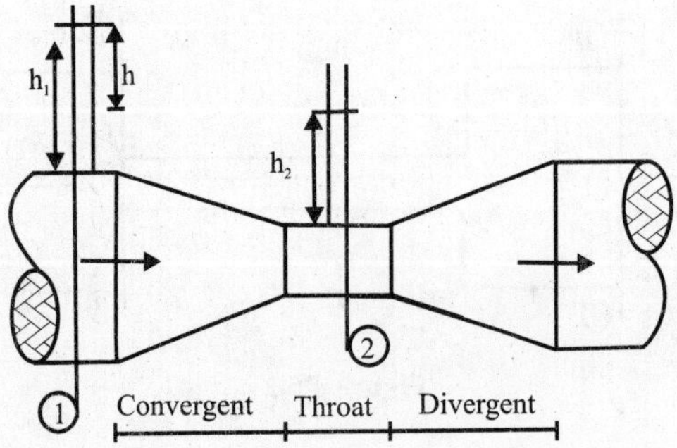

Figure 5.2: Convergent divergent-venturimeter

but $$h=\left(\frac{P_1-P_2}{\gamma}\right)=h_1-h_2$$;where h is the differential head

$$\therefore h=\left(\frac{P_1-P_2}{\gamma}\right)+(Z_1-Z_2)=\frac{1}{2g}\left(V_2^2-V_1^2\right) \tag{5.11}$$

Equation (5.11) is the **general equation for venturimeter** and is applicable for horizontal, vertical and inclined venturimeter and $\left(\frac{P}{\gamma}+Z\right)$ is called hydraulic gradient.

*If venturimeter is **horizontal**, then $Z_1=Z_2$ and Eq. (5.11) becomes

$$h=\left(\frac{P_1-P_2}{\gamma}\right)=\frac{1}{2g}\left(V_2^2-V_1^2\right) \tag{5.11a}$$

where $h=h_1-h_2$

According to **continuity equation,**

$$Q=A_1V_1=A_2V_2 \Rightarrow V_1=\frac{A_2V_2}{A_1} \tag{5.12}$$

Substitute Eq. (5.12) in Eq. (5.11a), to get V_2 in terms of A_1, A_1, h

$$\Rightarrow V_2=\frac{A_1}{\sqrt{A_1^2-A_2^2}}*\sqrt{2gh} \tag{5.12a}$$

But $Q=A_2V_2$

$$\Leftrightarrow Q=\frac{A_1A_2}{\sqrt{A_1^2-A_2^2}}*\sqrt{2gh} \tag{5.13}$$

$$\Rightarrow (Q_{Theoretical})_{venturi} = C * \sqrt{h} \tag{5.13a}$$

where constant of venturimeter $(C) = \dfrac{A_1 A_2}{\sqrt{A_1^2 - A_2^2}} * \sqrt{2g}$

*But in actual practice, the actual discharge $(Q_{Actual})_{venturi}$ will be less than the theoretical, due to some frictional losses and hence $(Q_{act})_{venturi} < (Q_{Th})_{venturi}$ and hence,

$$(Q_{Actual})_{venturi} = C_d * (Q_{Theoritical})_{venturi} \tag{5.13b}$$

where C_d coefficient of discharge < 1

*If Venturimeter is **Inclined** and having a **heavier manometric liquid** when compared to the liquid passing in the venturimeter, as shown in Figure.5.2a, then $Z_1 \neq Z_2$ and Eq. (5.11) becomes

$$h = y\left[\frac{S_{ML}}{S_{LP}} - 1\right] \tag{5.13c}$$

where y, S_{ML}, S_{LP} are differential gage reading, specific gravities of manometeric liquid and liquid in the pipe respectively.

*If venturimeter is **inclined** and having a **lighter manometeric liquid** when compared to the liquid passing in the venturimeter, as shown in Figure 5.2a, then and Eq. (5.11) becomes

$$h = y\left[1 - \frac{S_{LP}}{S_{ML}}\right] \tag{5.13d}$$

where y, S_{ML}, S_{LP} are differential gage reading, specific gravities of manometeric liquid and liquid in the pipe respectively.

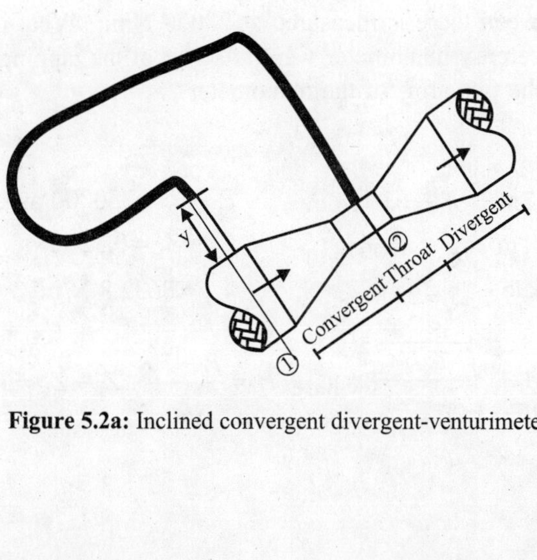

Figure 5.2a: Inclined convergent divergent-venturimeter

Example 5.3: A horizontal venturimeter with inlet diameter 20 cm and throat diameter 10 cm is used to measure the flow of water. The pressure at inlet is 17.5 Pa and the vaccum pressure at the throat is 30 cm of mercury. Find the discharge of water through venturimeter having $C_d = 0.95$.

Solution: Given data:

$$d_1 = 0.2 \text{ m}, d_2 = 0.1 \text{ m}, P_1 = 17.5 \text{ kPa}, C_d = 0.95$$
$$P_2 = 0.3 \text{ m } of\ mercury = 0.3 * 13.6 * 9810 = 40.0 \text{ kPa}$$

Area at the inlet and outlets

$$A_1 = 0.0314 \text{ m}^2$$
$$A_2 = 0.00785 \text{ m}^2 \text{ respectively.}$$

Discharge of water through the horizontal venturimeter is

$$Q_{actual} = C_d A_1 A_2 \sqrt{\frac{2gH}{A_1^2 - A_2^2}}$$

where $$H = \frac{P_1 - P_2}{\gamma}$$

$$\Rightarrow H = \frac{17.5 - 40.08}{9.81} = -2.296 \text{ m}$$

Negative indicates the vaccum condition

Therefore, discharge $Q_{actual} = 0.95 * 0.0314 * 0.00785 \sqrt{\dfrac{2 * 9.81 * 2.296}{0.0314^2 - 0.00785^2}} = 0.0517 \text{ m}^3/\text{s}$.

Example 5.4: A vertical venturimeter is being used to measure the flow of oil of specific gravity 0.88 in a 200 mm diameter pipe. The throat diameter of the venturimeter is 100 mm and the discharge coefficient is 0.96. Two tapping points at the throat and entrance are 320 mm apart and the pressure difference between these is measured at 28000 N/m². What is the discharge and the velocity in the pipe? If a mercury manometer were attached at the tapping points, what would be the difference in levels of the two arms of the manometer?

Solution: Given data:

$$d_1 = 0.2\text{m} \quad A_1 = 0.0314 \text{ m}^2 \quad P_1 - P_2 = 28000 \text{ Pascal}$$
$$d_2 = 0.1\text{m} \quad A_2 = 0.00785 \text{ m}^2 \quad Z_2 - Z_1 = 0.32 \text{ m}$$
$$C_d = 0.96 \quad S = 0.88 \quad \gamma_{oil} = 8632.8 \text{ N/m}^3$$

$$Q_{actual} = C_d A_1 A_2 \sqrt{\frac{2gH}{A_1^2 - A_2^2}} \quad \text{where } H = \frac{P_1 - P_2}{\gamma} + Z_1 - Z_2 = 3.56 \text{ m}$$

So actual discharge $Q_{actual} = 0.065 \text{ m}^3/\text{s}$ and velocity in the pipe $V = Q_{actual}/A_1 = 2.07 \text{ m/s}$.

If a differential mercury manometer attached at the two tapping points then

$$H = y\left[\frac{S_{ML}}{S_{LP}} - 1\right] = y\left[\frac{13.6}{0.88} - 1\right]$$

Hence, difference in levels of the two arms of the manometer will be $y = 2.46$ m.

Example 5.5: A venturimeter with an entrance diameter of 5/2 times the throat diameter. The inlet and throat are connected by a water filled tubes to a mercury U-tube manometer. The velocity of the flow along the pipe is to be $2.5\sqrt{H}$ m/s, where H is the manometer reading in meters of mercury. Determine the loss of head between inlet and throat of the venturi, when $H = 0.49\,\text{m}$.

Solution: Given data:

$$d_1 = \frac{5}{2}d_2, V_1 = 2.5\sqrt{H}, H = 0.49 \text{ m},$$

$$\rho_w = 1000 \text{ kg/m}^3, \rho_m = 13600 \text{ kg/m}^3$$

$$V_2 = \frac{A_1V_1}{A_2} = \frac{4*\pi*d_1^2*2.5\sqrt{H}}{4*\pi*d_2^2}$$

$$\Rightarrow V_2 = \frac{4*\pi*5^2*2.5\sqrt{0.49}}{4*\pi*2^2} = 10.93\,\text{m/s}$$

As per Bernoulli's

$$P_1 - P_2 = \frac{\rho_w V_2^2}{2} + \rho_w g Z_2 - \frac{\rho_w V_1^2}{2} - \rho_w g Z_1 + \rho_w g h_f$$

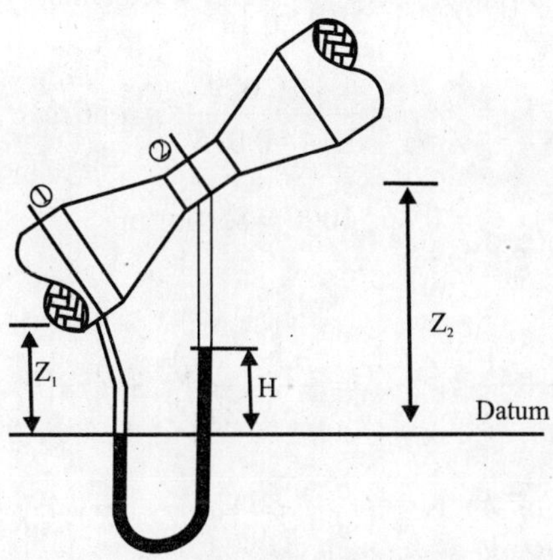

Figure P5.5

Using manometer equation, we have

$$P_1 - P_2 = \rho_w g Z_2 - \rho_w g Z_1 + \rho_w g H + \rho_m g H$$

Hence, from Bernoulli's and manometer equation,

We have $\rho_w g h_f = gH(\rho_m - \rho_w) - \frac{\rho_w}{2}\left(\frac{V_2^2}{2} - \frac{V_1^2}{2}\right)$

$$\Rightarrow h_f = \frac{gH(\rho_m - \rho_w) - \frac{\rho_w}{2}\left(\frac{V_2^2}{2} - \frac{V_1^2}{2}\right)}{\rho_w g}$$

So head loss $h_f = \frac{9.81*0.49(13600-1000) - \frac{1000}{2}\left(\frac{10.93^2}{2} - \frac{1.75^2}{2}\right)}{9810} = 0.233\,\text{m}$

Example 5.6: A venturimeter has its axis vertical, the inlet and throat diameters being 150 mm and 75 mm respectively. The throat is 300 mm above inlet, oil of specific gravity 0.85 flows up through the meter at a rate of Q = 0.029 m^3/s.If the pressure difference between the inlet and the throat is 20 kPa. Assume the losses in the pipe to be expressed as $(h)_f = k\frac{v^2}{2g}$. Find the value of *k*.

Solution: Given data:

$$d_i = 0.15\text{ m},\ d_t = 0.075\text{ m},\ Q = 0.029\text{ m}^3/\text{s}$$
$$S = 0.85,\ Z_2 - Z_1 = 0.3\text{ m},\ P_1 - P_2 = 20\text{ kPa}$$

As per continuity equation, we have $Q = A_1V_1 = A_2V_2$

$$\Rightarrow V_1 = Q/A_1 = \frac{4*0.029}{3.14*0.15^2} = 1.64\,\text{m/s}$$

$$\Rightarrow V_2 = Q/A_2 = \frac{4*0.029}{3.14*0.075^2} = 6.56\,\text{m/s}$$

Density of oil is $(\rho)_{oil} = S*(\rho)_{water} = 0.85*1000 = 850\,\text{kg/m}^3$

As per Bernoulli's equation

$$\frac{p_1}{\gamma_1} + \frac{1}{2g}V_1^2 + Z_1 = \frac{p_2}{\gamma_2} + \frac{1}{2g}V_2^2 + Z_2 + h_f$$

$$\Rightarrow \frac{p_1 - p_2}{\rho_{oil} g} + \frac{1}{2g}\left(V_1^2 - V_2^2\right) + (Z_1 - Z_2) = k\frac{V_2^2}{2g}$$

$$\Rightarrow k = 0.0192$$

Example 5.7: A venturimeter of inlet diameter 0.15 m and throat diameter 0.075 m is fitted in a 0.15 m diameter vertical pipe in which liquid of specific gravity 0.68 flows downwards. Pressure gages are fitted to the inlet and throat sections. The throat being 0.85 m below the inlet. Taking coefficient of discharge as 0.85, find the discharge (a) when the pressure gages read the same, (b) when the inlet gage reads 16 kPa higher than the throat gage.

Solution: Given data: $d_L = 0.15$ m, $d_1 = 0.075$ m, $C_d = 0.85$, $S = 0.68$, $Z_1 - Z_2 = 0.85$ m

As per continuity equation, we have $Q = A_1V_1 = A_2V_2$

$$\Rightarrow V_2 = {A_1V_1}/{A_2} = \frac{0.15^2 V_1}{0.075^2} = 4V_1$$

Density of liquid is

$$(\rho)_{oil} = S * (\rho)_{water} = 0.68 * 1000 = 680\,\text{kg/m}^3$$

1) $(P)_1 = (P)_2$ then

As per Bernoulli's $$\frac{p_1}{\gamma_1} + \frac{1}{2g}V_1^2 + Z_1 = \frac{p_2}{\gamma_2} + \frac{1}{2g}V_2^2 + Z_2$$

$$\Rightarrow \frac{p_1 - p_2}{\rho_{oil}g} + \frac{1}{2g}\left(V_1^2 - V_2^2\right) + \left(Z_1 - Z_2\right) = 0$$

$$\Rightarrow V_1 = \sqrt{\frac{0.85 * 2 * 9.81}{15}} = 1.054\,\text{m/s}$$

So discharge $Q = A_1V_1 \Rightarrow Q = A_1V_1 = \dfrac{\pi * 0.15^2}{4} * 1.054 = 0.0186\,\text{m}^3/\text{s}$

2) $P_1 - P_2 = 16000\,\text{Pa}$ then

As per Bernoulli's $$\frac{p_1}{\gamma_1} + \frac{1}{2g}V_1^2 + Z_1 = \frac{p_2}{\gamma_2} + \frac{1}{2g}V_2^2 + Z_2$$

$$\Rightarrow \frac{p_1 - p_2}{\rho_{oil}g} + \frac{1}{2g}\left(V_1^2 - V_2^2\right) + \left(Z_1 - Z_2\right) = 0$$

$$\Rightarrow V_1 = \sqrt{\frac{3.25 * 2 * 9.81}{15}} = 2.06\,\text{m/s}$$

So discharge $Q = A_1V_1 \Rightarrow Q = A_1V_1 = \dfrac{\pi * 0.15^2}{4} * 2.06 = 0.0364\,\text{m}^3/\text{s}$

ORIFICE-METER AND ORIFICE

Orifice is a small opening in a tank/reservoir, through which we can measure the discharge and orifice-meter is a device used to measure the discharge of fluid flow through a **pipeline** or in a **tank/reservoir**.

Orifice-meter is a flat circular plate having a hole with diameter D_0 at the throat and having sharp edges to minimize frictional losses by minimizing the contact between the hole and the liquid, through which the liquid flows and works on the same principal of venturimeter. As the orifice has sharp edges at the throat, the fluid coming out of the throat will tend to reduce its diameter of flow, till some length and then it expands to the diameter of the pipeline (as shown in Figure 5.3). The region where the fluid flow reduces its diameter is called **Vena contraction**.

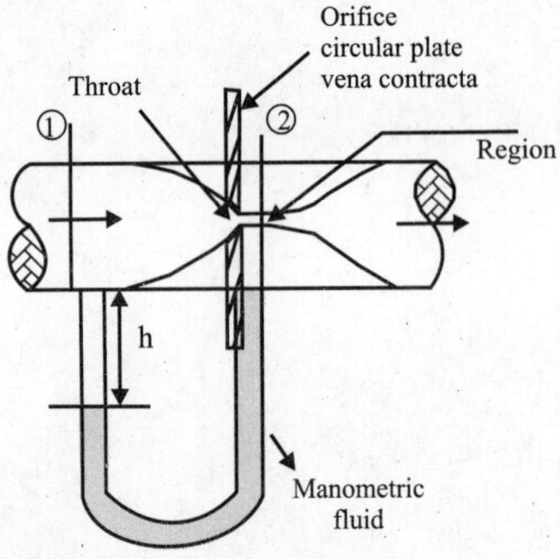

Figure 5.3: Orifice-meter fitted to a pipeline

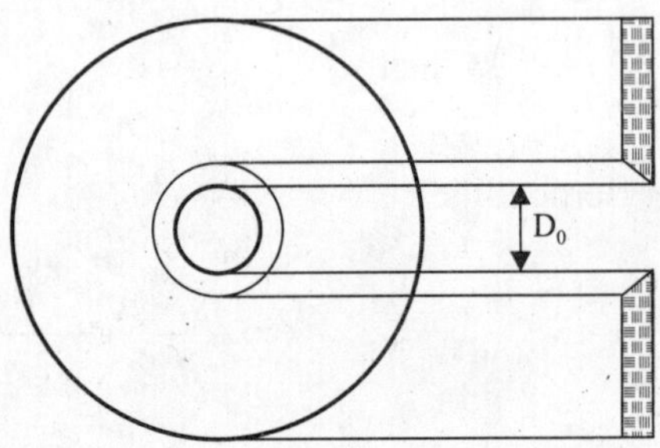

Figure 5.3a: Orifice disc

Discharge through an orifice meter fitted to a pipeline:

Consider the diameters of the pipeline, throat of the orifice and Vena-contraction to be D_1, D_0, D_V respectively.

Let P_1, V_1, A_1 and P_V, V_V, A_V be pressures, velocities and areas at inlet and Vena contraction of orifice, respectively.

And let

$$C_C = \frac{A_V}{A_0} \tag{5.14}$$

be coefficient of contraction.

If Bernoulli's equation is applied between sections at the inlet of the pipeline and Vena contraction region, as shown in Figure 5.3, the equation will be

$$\frac{p_1}{\gamma_1} + \frac{1}{2g}V_1^2 + Z_1 = \frac{p_V}{\gamma_V} + \frac{1}{2g}V_V^2 + Z_V$$

By rearranging

$$\left(\frac{p_1}{\gamma_1}+Z_1\right)-\left(\frac{p_V}{\gamma_V}+Z_V\right)=\frac{V_V{}^2-V_1^2}{2g}$$

where $\left(\frac{p}{\gamma}+Z\right)$ is **hydraulic Gradient.**

But $h=\left(\frac{p_1}{\gamma_1}\right)-\left(\frac{p_V}{\gamma_V}\right)=\frac{V_V{}^2-V_1^2}{2g}$ where h is differential head.

So
$$h=\frac{V_V{}^2-V_1^2}{2g} \Rightarrow V_V=\sqrt{2gh+V_1^2} \tag{5.15}$$

According to **continuity equation,**

$$Q=A_1V_1=A_VV_V \Rightarrow V_1=\frac{A_VV_V}{A_1}=\frac{A_0C_CV_V}{A_1} \tag{5.15a}$$

Substitute Eq. (5.15a) in Eq. (5.15) to get V_V in terms of A_1, A_0, h, C_C

$$\Rightarrow V_V=\frac{\sqrt{2gh}}{\sqrt{1-\left(\frac{A_0C_C}{A_1}\right)^2}}$$

Hence,

$$Q=A_VV_V=\frac{A_1A_V\sqrt{2gh}}{\sqrt{A_1^2-A_V{}^2}}$$

By rearranging
$$Q=\left[\frac{A_1A_0\sqrt{2gh}}{\sqrt{A_1^2-A_0^2}}\right]\left[\frac{A_V\sqrt{A_1^2-A_0^2}}{A_0\sqrt{A_1^2-A_V{}^2}}\right]=\left[\frac{A_1A_0\sqrt{2gh}}{\sqrt{A_1^2-A_0^2}}\right]\left[C_c\frac{\sqrt{A_1^2-A_0^2}}{\sqrt{A_1^2-A_V{}^2}}\right]$$

$$\Rightarrow (Q_{th})_{Orifice}=C_d\left[\frac{A_1A_0\sqrt{2gh}}{\sqrt{A_1^2-A_0^2}}\right]=C_d * K_o * A_0\sqrt{2gh} \tag{5.16}$$

where $C_d=C_C\left[\frac{\sqrt{A_1^2-A_0^2}}{\sqrt{A_1^2-A_V{}^2}}\right]$ and $K_o=\frac{A_1}{\sqrt{A_1^2-A_0^2}}$ = Flow/orifice coefficient

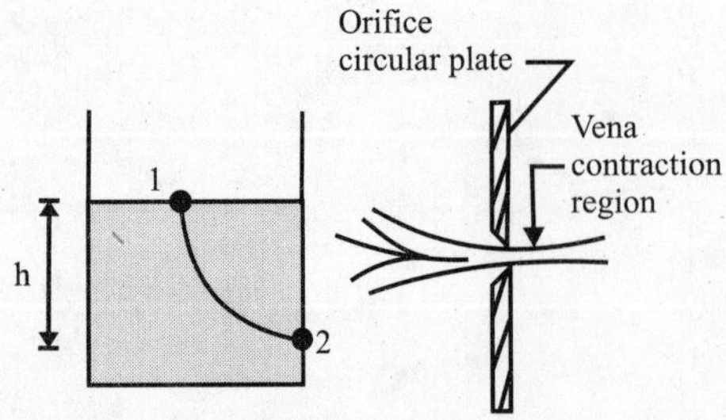

Figure 5.3a: Orifice-meter fitted to a tank

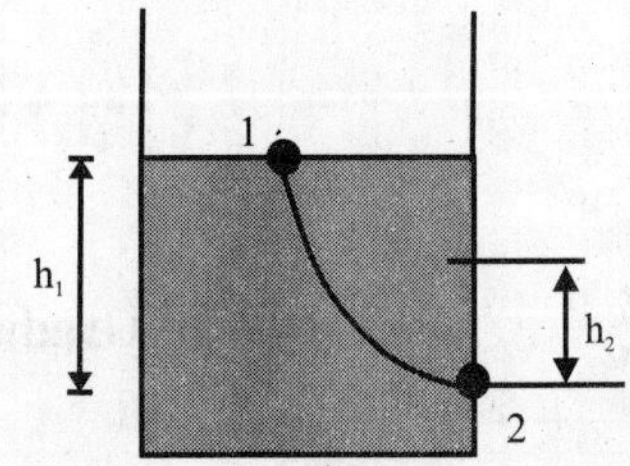

Figure 5.3b: Orifice-meter fitted to a tank, to empty

If an orifice-meter is substituted with a venturimeter of same inlet and throat diameters, then

$$(Q_{TH})_{venturi} \text{ becomes } (Q_{TH})_{Venturi} = \left[\frac{A_1 A_0 \sqrt{2gh}}{\sqrt{A_1^2 - A_0^2}}\right] \quad (5.17)$$

And by comparing Eq. (5.17) and Eq. (5.16), we have

$$(Q_{TH})_{Orifice} = C_d * (Q_{TH})_{Venturi} \quad (5.17a)$$

Note:

- Observing Eq. (5.16), it is clear that the coefficient of contraction C_C is less than the coefficient of discharge C_d, i.e. $(C_C) < (C_d)$.
- Observing Eq. (5.17a), it is cear that the Discharge through an Orifice meter is less than the Venturimeter discharge. i.e. $(Q_{TH})_{Orifice} < (Q_{TH})_{Venturi}$.

Discharge through an Orifice fitted to a Tank

Consider a tank/reservoir filled with a liquid and is open to atmosphere at the open end. And let a small orifice be fitted to one of the walls to the tank at a depth of "h" from the free surface, to let the liquid to flow out, through the orifice, as shown in Figure 5.3a.

Apply Bernoulli equation along the streamline joining point 1 on the free surface to point 2 at the center of the orifice and point 2 taken as datum, we have

$$\frac{p_1}{\gamma} + \frac{1}{2g}V_1^2 + Z_1 = \frac{p_2}{\gamma_2} + \frac{1}{2g}V_2^2 + Z_2 \quad (5.18)$$

As point 1 and 2 are subjected to atmosphere, the pressure heads are zero so, $\frac{p_2}{\gamma} = \frac{p_1}{\gamma} = V_1 = Z_2 = 0$ and $Z_1 = h$ and hence Eq. (5.18) becomes $h = \frac{1}{2g}V_2^2$.

$$\Rightarrow (V_2)_{Theoretical} = \sqrt{2gh} \quad (5.19)$$

$$\Rightarrow (V_2)_{Actual} = C_V (V_2)_{Theoretical} = C_V \sqrt{2gh} \tag{5.19a}$$

where C_v is coefficient of velocity.

Discharge through the orifice is $(Q)_{Actual} = (A)_{Actual} * (V)_{Actual}$; where (A) the area of the jet discharged is and $(V)_{Actual}$ jet velocity, i.e. $(V)_{Actual} = C_V \sqrt{2gh}$.

But $C_c = A_{jet} / A_{orifice}$

Hence, $$(Q)_{Actual} = (C_C * A_O) * (C_V \sqrt{2gh})$$

$$\Rightarrow (Q)_{Actual} = C_d * A_O \sqrt{2gh} \tag{5.20}$$

since $C_d = C_C * C_V$

Discharge through an orifice fitted to a tank/reservoir.

*Observing Eq. (5.20) it is clear that the discharge "Q" is directly proportional to the head "h".

Time of Emptying the Tank

Let δh be fall in the level of the liquid in the tank in time period of δt and as known $(Q)_{Actual} = (A)_{Actual} * (V)_{Actual}$ so Q = Area tank * velocity of discharge

$$\Rightarrow Q_{Actual} = -A * \frac{\delta h}{\delta t} \quad \text{since } V = -\frac{\delta h}{\delta t} \tag{5.21}$$

Substituting Eq. (5.20) in Eq. (5.21), we have

$$C_d * A_O \sqrt{2gh} = -A * \frac{\delta h}{\delta t}$$

$$\Rightarrow \delta t = -A * \frac{\delta h}{C_d * A_O \sqrt{2g} * \sqrt{h}} \tag{5.21a}$$

Integrating Eq. (5.21a) on both sides, we have

$$\int_0^t \delta t = \frac{-A}{C_d * A_O \sqrt{2g}} \int_{h_1}^{h_2} \frac{\delta h}{\sqrt{h}}$$

$$\Rightarrow t = \frac{2A}{C_d * A_O \sqrt{2g}} \left(\sqrt{h_1} - \sqrt{h_2}\right) \tag{5.22}$$

Time to empty a tank/reservoir with an orifice fitted.

Difference between venturimeter and orifice meter

Venturimeter	**Orifice meter**
1) Coefficient of discharge is more. 2) It can be fitted only in pipeline. 3) Flow velocity is known by pressure difference at inlet and throat.	Coefficient of discharge is less. It can be fitted in a pipeline as well as tank opening. Flow discharge measured by Pitot tube or by Trajectory method. Its PE converted to KE and liquid jet discharges to open atmosphere freely.

Example 5.8: An orifice-meter is to be fitted into a horizontal pipe 20 cm diameter, carrying oil of specific gravity 0.85 for the purpose of flow measurement. The differential head is to be indicated by U-tube manometer reading is not to exceed 0.2 m when the flow is $Q = 15$ kg/s, what should be the diameter of the orifice? Assume $C_d = 0.62$.

Solution: Given data: $A_1 = 0.0314\,\text{m}^2, C_d = 0.62, \overline{m} = 15\,\text{kg/s}, S_{oil} = 0.85, S_{Mercury} = 13.6, h \leq 0.2\,\text{m}$

Density of liquid is $(\rho)_{oil} = S*(\rho)_{water} = 0.85*1000 = 850\,\text{kg/m}^3$

Discharge through the orifice is $Q = \dfrac{\overline{m}}{\rho_{oil}} = \dfrac{15}{850} = 0.0176\ \text{m}^3/\text{s}$

Let the area of orifice be A_0, then as per the Eq. (5.16), we have

$$(Q_{th})_{Orifice} = C_d\left[\frac{A_1 A_0\sqrt{2gh}}{\sqrt{A_1^2 - A_0^2}}\right]$$

$$\Rightarrow 0.0176 = 0.62\left[\frac{0.0314 * A_0\sqrt{2*9.81*0.2}}{\sqrt{0.00098 - A_0^2}}\right]$$

$$\Rightarrow 0.00098 - A_0^2 = A_0^2\,(4.8)$$

$$\Rightarrow A_0 = 0.0129\,\text{m}^2$$

So the **diameter of orifice** is $d_0 = 0.128\,\text{m}$

Example 5.9: An orifice of 10 cm diameter discharges water under a head of 5 m. If the discharge is 0.015 m^3/s and has a vena contraction diameter of 0.08 m, determine the coefficient of discharge, coefficient of contraction, coefficient of velocity, assuming the entrance to the pipe is sharp-edge.

Solution: Given data: $d_0 = 0.1\,\text{m}, H = 5\,\text{m}, Q = 0.015\,\text{m}^3/\text{s}, d_V = 0.08\,\text{m}$

Refer Eq. (5.20), we have $(Q)_{Actual} = C_d * A_O\sqrt{2gh}$

$$\Rightarrow 0.015 = C_d * \frac{\pi * 0.1^2}{4}\sqrt{2*9.81*5}$$

$$\Rightarrow C_d = 0.19$$

We have $$C_c = \frac{A_V}{A_o} = \frac{0.08^2}{0.1^2} = 0.64$$

$$\Rightarrow C_V = \frac{C_d}{C_c} = \frac{0.19}{0.64} = 0.29$$

So the **coefficients of discharge, contraction, velocity** are 0.19, 0.64, 0.29 respectively.

Example 5.10: A swimming pool 12 m long and 7 m wide holds water to a depth of 2 m. If the water is discharged through an opening of area 0.2 m^2 at the bottom of the pool, find the time required to empty the tank. Take the coefficient of discharge as 0.6.

Solution: Given data: $L = 12$ m, $B = 7$ m, $H_1 = 2$ m, $A_o = 0.2$ m^2, $C_d = 0.6$

As the tank to be emptied the final height $H_2 = 0$

Area of the swimming pool is $A = (L * B) = (12 * 7) = 84\,\text{m}^2$

Time of emptying the tank will be

Refer Eq. (5.22), we have

$$t = \frac{2A}{C_d * A_O\sqrt{2g}}\left(\sqrt{h_1} - \sqrt{h_2}\right)$$

$$t = \frac{2*84}{0.6*0.2\sqrt{2*9.8}}\left(\sqrt{2} - 0\right) = 446.9 \text{ sec}$$

Hence, **time of emptying the tank** is 446.9 seconds.

Example 5.11: A circular swimming pool of 12 m diameter and 7 m depth holds water to a depth of 5 m. If the water is discharged through an opening of area 0.3 m^2 at the bottom of the pool, find the time required to to empty the tank and also to fall the water level by 5 m to 2 m. Take the coeffifcient of discharge as 0.6.

Solution: Given data: $D = 12\,\text{m}$, $H_1 = 5\,\text{m}$, $A_0 = 0.3\,\text{m}^2$, $C_d = 0.6$

1) As the tank to be emptied the final height $H_2 = 0$

Area of the swimming pool is $A = \frac{\pi}{4} * D^2 = \frac{\pi}{4} * 12^2 = 113\,\text{m}^2$

Time of emptying the tank will be

Refer Eq. (5.22), we have $t = \dfrac{2A}{C_d * A_O\sqrt{2g}}\left(\sqrt{h_1} - \sqrt{h_2}\right)$

$$t = \frac{2*113}{0.6*0.3\sqrt{2*9.8}}\left(\sqrt{5-0}\right) = 633.8 \text{ sec}$$

Hence, **time of emptying the tank** is 633.8 seconds.
2) As the tank level to be brought down from 5 m to 2 m, the final height H_2 = 2 m

Area of the swimming pool is $A = \dfrac{\pi}{4} * D^2 = \dfrac{\pi}{4} * 12^2 = 113\,\text{m}^2$

Time of emptying the tank will be

$$t = \frac{2A}{C_d * A_O\sqrt{2g}}\left(\sqrt{h_1} - \sqrt{h_2}\right)$$

$$t = \frac{2*113}{0.6*0.3\sqrt{2*9.8}}\left(\sqrt{5} - \sqrt{2}\right) = 233 \text{ sec}$$

Hence, **time of emptying the tank** for bringing the level down from 5 m to 2 m is 490.9 seconds.

Example 5.12: A circular swimming pool of 12 m diameter and 7 m depth holds water to a depth of 5 m. If the time required to fall the water level is 2 minutes. The water is being discharged through an opening of area 0.3 m^2 at the bottom of the pool, find the level of water above the opening after 2 minutes in the tank. Take the coefficient of discharge as 0.6.

Solution: Given data: $D = 12\,\text{m},\ H_1 = 5\,\text{m},\ A_0 = 0.3\,\text{m}^2,\ t = 120\,\text{s},\ C_d = 0.6$
As the tank level to being brought down in 2 minutes time, let the final height be H_2.

Area of the swimming pool is $A = \dfrac{\pi}{4} * D^2 = \dfrac{\pi}{4} * 12^2 = 113\,\text{m}^2$

Time of emptying the tank will be

$$t = \frac{2A}{C_d * A_O\sqrt{2g}}\left(\sqrt{h_1} - \sqrt{h_2}\right)$$

$$\Rightarrow 120 = \frac{2*113}{0.6*0.3\sqrt{2*9.8}}\left(\sqrt{5} - \sqrt{H_2}\right)$$

$$\Rightarrow H_2 = 3.28\,\text{m}$$

Hence, **the level of water above the opening after 2minutes in the tank** is 446.9 seconds.

Example 5.13: A cylindrical tank 3 m in diameter and 7 m high has an orifice 150 mm in diameter at the bottom center of the tank. A constant discharge of 85 lit/s of water is fed into the tank. At the same time, water is being discharged through the orifice. Determine the time taken to lower the water surface level in the tank from 5 m to 2.5 m above the orifice, take the coefficient of discharge as 0.72 and the top of the tank is left open to the atmosphere. **[UPSC Exam]**

Solution: Given data: $D = 3\,\text{m},\ H_1 = 5\,\text{m},\ H_2 = 2.5\,\text{m},\ d = 0.15\,\text{m},\ C_d = 0.72$

Area of the tank $$A = \frac{\pi}{4} * D^2 = \frac{\pi}{4} * 3^2 = 7.06\,\text{m}^2$$

Area of the orifice $$A_o = \frac{\pi}{4} * d^2 = \frac{\pi}{4} * 0.15^2 = 0.0176\,\text{m}^2$$

Let the discharge fed into the tank be $q = 0.085\ \text{m}^3/\text{s}$.
Let the discharge through the orifice be Q.
Consider a strip dh of water level in the tank at a height of h from the orifice, the volume of the tank water discharged through the orifice is equal to the difference in discharges of orifice and discharge fed, i.e. $Q_{left} = \left(Q_{orifice} - q_{fed}\right)$.

$$\Rightarrow -A * dh = \left(Q_{orifice} - q_{fed}\right) dT$$

$$\Rightarrow -A * dh = \left(C_d * A_o\sqrt{2gh} - q_{fed}\right) dT$$

Negative sign indicates the fall in the level.

$$\Rightarrow -dT = \frac{A * dh}{\left(C_d * A_o\sqrt{2gh} - q_{fed}\right)}$$

$$\Rightarrow -\int dT = \int_{H_1}^{H_2} \frac{A * dh}{\left(C_d * A_o\sqrt{2gh} - q_{fed}\right)}$$

$$\Rightarrow T = \int_{H_1}^{H_2} \frac{-A * dh}{\left(C_d * A_o\sqrt{2gh} - q_{fed}\right)} \qquad (1)$$

Let $B = \left(C_d * A_o\sqrt{2g}\right) = 0.0564$ and let $\left(B\sqrt{h} - \dot{q}_{fed}\right) = x$

$$\Rightarrow dh = \frac{2\sqrt{h}}{B} dx$$

So Eq. (1), becomes

$$\Rightarrow T = -\int_{H_1}^{H_2} \frac{A * 2\sqrt{h} * dx}{B\left(B\sqrt{h} - q_{fed}\right)}$$

$$\Rightarrow T = -\frac{2A}{B^2}\int_{H_1}^{H_2} \frac{B\sqrt{h}dx}{\left(B\sqrt{h} - q_{fed}\right)}$$

$$\Rightarrow T = -\frac{2A}{B^2}\int_{H_1}^{H_2} \frac{\left(x + q_{fed}\right)dx}{x} \quad \sin ce \left(B\sqrt{h} - q_{fed}\right) = x$$

$$\Rightarrow T = -\frac{2A}{B^2}\left[x + q_{fed}\log_e x\right]_{H_1}^{H_2}$$

$$\Rightarrow T = \frac{2A}{B^2}\left[B\left(\sqrt{H_1} - \sqrt{H_2}\right) + q_{fed}\log_e \frac{B\sqrt{H_1} - q_{fed}}{B\sqrt{H_2} - q_{fed}}\right]$$

$$\Rightarrow T = \frac{2 * 7.068}{0.0564^2}\left[0.0564\left(\sqrt{5} - \sqrt{2.5}\right) + 0.085 * \log_e \frac{0.0564\sqrt{5} - 0.085}{0.0564\sqrt{2.5} - 0.085}\right] = 1028\,\text{sec}$$

Hence, **the time period required** is 1028 seconds.

Example 5.14: How long does it take to draw the liquid from h=5 m to h=2 m, from a vessel in form a truncated cone having an orifice of diameter 0.2 m attached at the bottom base, with 1 m diameter at the bottom base and 60° vertex angle has a liquid filled in it. Take the coeffifcient of discharge as 0.85 of the orifice opening.

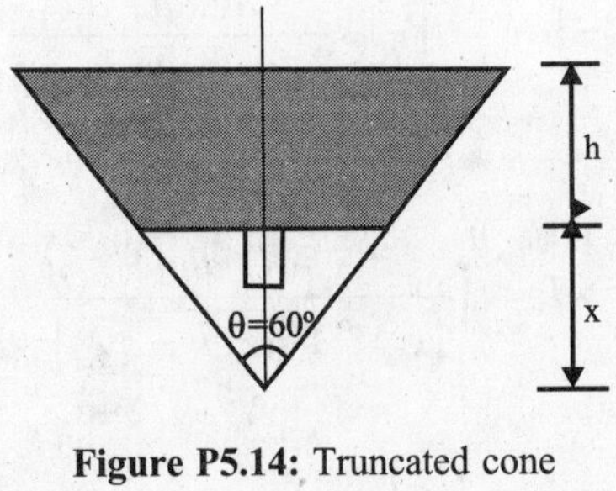

Figure P5.14: Truncated cone

Solution: Given data: $H_1 = 5\,\text{m},\ H_2 = 2\,\text{m},\ d = 0.2\,\text{m},\ C_d = 0.85,\ \theta = 60°,\ b = 1\,\text{m}$

Area of the orifice $A_o = \frac{\pi}{4} * d^2 = \frac{\pi}{4} * 0.2^2 = 0.0314\,\text{m}^2$

Based on the geometry of the cone, we have

$$\tan 30^\circ = \frac{b}{x} = \frac{0.5}{x} \quad \Rightarrow x = 0.866\,\text{m}$$

Let the radius of upper base be "r". Based on the geometry of the cone, we have

$$\tan 30^\circ = \frac{r}{x+h} \quad \Rightarrow r = \frac{(0.866+h)}{1.732}$$

Consider a strip *dh* of liquid level in the cone vessel at a height of *h* from the orifice, the volume of the liquid discharged through the orifice is equal to the discharge left the vessel,

i.e. $$Q_{left} = \left(Q_{orifice}\right) \Rightarrow -A * dh = \left(Q_{orifice}\right) dT$$

$$\Rightarrow -\pi r^2 * dh = \left(C_d * A_O \sqrt{2gh}\right) dT$$

$$\Rightarrow -3.14 r^2 * dh = \left(0.85 * 0.0314 \sqrt{2 * 9.81 h}\right) dT$$

Negative sign indicates the fall in the level.

$$\Rightarrow -dT = 8.85 * \frac{(0.866+h)^2 * dh}{\left(\sqrt{h}\right)} \quad \text{since } r = \frac{(0.866+h)}{1.732}$$

The time required to empty the cone from h=5 m to h=2 m,

$$\Rightarrow T = 8.85 \int_{H_1}^{H_2} \frac{(0.866+h)^2 * dh}{\left(\sqrt{h}\right)}$$

$$\Rightarrow T = 8.85 \left[1.5\sqrt{h} + \frac{2}{5} h^{5/2} + 1.732 * \frac{2}{3} * h^{3/2}\right]_2^5$$

$$\Rightarrow T = 8.85 \left[1.5\left(\sqrt{5} - \sqrt{2}\right) + \frac{2}{5}\left(5^{5/2} - 2^{5/2}\right) + 1.732 * \frac{2}{3} * \left(5^{3/2} - 2^{3/2}\right)\right] = 304.6\,\text{seconds}$$

Hence, the **time required to bring down the level** of liquid from h=5 m to h=2 m is 304.6 seconds.

PITOT TUBE

A uniform velocity flow hitting a solid blunt body has streamlines similar to the below shown Figure 5.4a. Some streamlines move to the left and some to the right and the central one hit the blunt body and stops. At this point "2" velocity is zero and the fluid does not move at this one point. This point is known as the **stagnation point**. Pitot tube is a **"simple velocity measuring tube in a flow, by converting kinetic energy of flow into pressure energy, at that point where the velocity becomes zero"**.

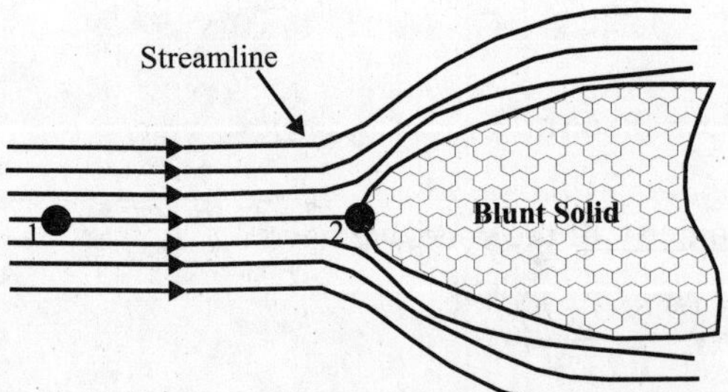

Figure 5.4a: Streamlines over a blout body

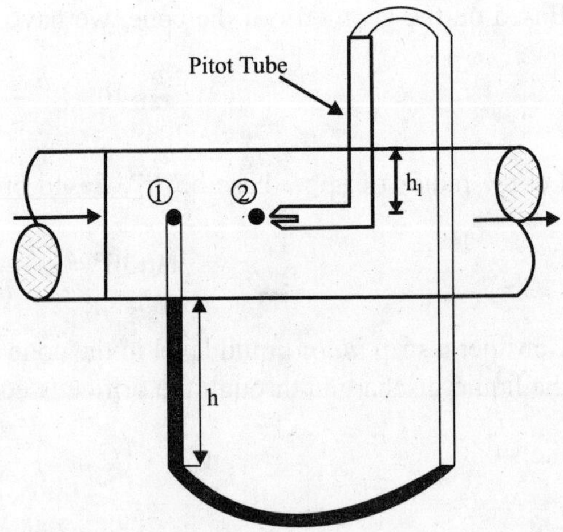

Figure 5.4b: Pitot tube

It consists of a tube bent in right angle as shown in Figure 5.4b, with a small opening, where the flow velocity becomes zero i.e. $V_2 = 0$ and the kinetic energy at this point 2 is thus converted to pressure energy and the fluid raises into the tube up to some level above the free surface of fluid flow.

Henri Pitot a French hydraulic engineer, invented Pitot tube consisting of two coaxial tubes bent in right angles, as shown in the Figure 5.4b. The interior tube has an opening perpendicular to the flow where the flow velocity becomes zero, i.e. $V_2 = 0$ and the kinetic energy at this point 2 is thus converted to stagnation pressure, while the exterior tube has an opening parallel to the flow which indicates the static pressure.

Fluid raise is indicated in the manometer/pressure gage indicator connected to the Pitot end and thus the velocity of the fluid flow or flow rate is determined using Bernoulli's equation between the flow and the stagnation point.

Consider a Pitot tube inserted in a fluid flow velocity V_1. Assume two points 1, 2; point 1 before Pitot tube (static pressure position P_1) and point 2 at the mouth (Stagnation pressure point P_2) of the Pitot tube, as shown in figure. The pressure head at point 1 will be

$$\frac{P_1}{\gamma_1} = h_1 \tag{5.23}$$

But pressure head at point 2 is due to stagnation, which is combined effect of static and dynamic pressure

$$\frac{P_2}{\gamma_2} = h_1 + h \tag{5.23a}$$

Apply Bernoulli's equation between point 1 and point 2,

we have $$\frac{p_1}{\gamma_1} + \frac{1}{2g}V_1^2 + Z_1 = \frac{p_2}{\gamma_2} + \frac{1}{2g}V_2^2 + Z_2$$

As $Z_1 = Z_2$

$$\Rightarrow V_1 = \sqrt{\frac{2(p_2 - p_1)}{\rho}} \tag{5.24}$$

In terms of pressure heads $h_1 + \frac{1}{2g}V_1^2 = h + h_1 + 0$ since $Z_1 = Z_2$ & $V_2 = 0$

$$\Rightarrow (V_1)_{th} = \sqrt{2gh} \tag{5.24a}$$

So, actual velocity is

$$(V_1)_{act} = C_V\sqrt{2gh} \tag{5.25}$$

where c_V is the coefficient of Pitot tube.

Note:
If Pitot tube has a differential manometer attached to it then, Eq. (5.25) becomes

$$(V_1)_{act} = C_V\sqrt{2g\left[\frac{S_{ML}}{S_{LP}} - 1\right]} \tag{5.25a}$$

Example 5.15: The velocity of water flowing in venturimeter is measured using a Pitot-static tube located centrally in the flow. If the height measured on the attached mercury manometer is 150 mm, determine the velocity of the oil.

Solution: Given data: Differential gage reading $y = 0.15$ m, $S_{ML} = 13.6$, $S_{LL} = 1$.

Equation of Pitot tube $\quad (V_1)_{act} = \sqrt{2gH} = \sqrt{2gy\left[\frac{S_{ML}}{S_{LP}} - 1\right]}$

$$\Rightarrow (V_1)_{act} = \sqrt{2*9.8*0.15\left[\frac{13.6}{s1} - 1\right]} = 6.08 \text{ m/s}$$

Example 5.16: A Pitot-static tube located centrally in the flow of water, in a pipe. If the differential pressure noted across Pitot tapings is 2.8 kPa, determine the velocity of water in the pipe, assuming 0.85 of coefficient of Pitot tube.

Solution: Given data: $\Delta P = 28000\,\text{Pa}, C_V = 0.85$

$$\frac{\Delta P}{\rho g} = h = \frac{28000}{1000*9.81} = 2.85\,\text{m}$$

Equation of Pitot tube

$$(V_1)_{act} = C_V\sqrt{2gH} = 0.85*\sqrt{2*9.81*2.85} = 6.36\,\text{m/s}$$

EULER'S EQUATION—CONSERVATION OF MOMENTUM

Euler's equation is based on conservation of momentum and stated by Newton's second law, i.e. the force applied is equal to the change in momentum.

$$F = m\frac{dV}{dt} \quad \Rightarrow F = ma \tag{5.26}$$

Consider a fluid particle flowing in a streamline having dA, dS, dW as cross-section area, length and weight, respectively and at a datum level dz.

When the particle moves, it is subjected to three forces, two pressure forces on either ends of the particle and the third is due to its weight, as shown in Figure 5.5. Hence, net forces acting on the fluid particle is summation of pressure forces on the either ends, i.e. F_{AB}, F_{CD} and weight force F_w.

If the forces along stream flow direction are taken as positive, then pressure force on one end will be positive and the other pressure force will be negative, so as to keep the fluid in equilibrium and the force due to weight will be negative.

Hence,

$$\Rightarrow F = \{[F_{AB} - F_{CD}] - F_W\}$$

$$\Rightarrow F = \{[P * dA - (P + dP)dA] - \rho g * dA * ds * \cos\theta\}$$

$$\Rightarrow F = \{-[(dP)dA] - \rho g * dA * ds * \cos\theta\}$$

But
$$\cos\theta = \frac{dz}{ds};$$

$$\Rightarrow F = \{-[(dP)dA] - \rho g * dA * dz\} \tag{5.27}$$

Mass of the fluid particle is
$$m = \rho(dA * ds) \tag{5.28}$$

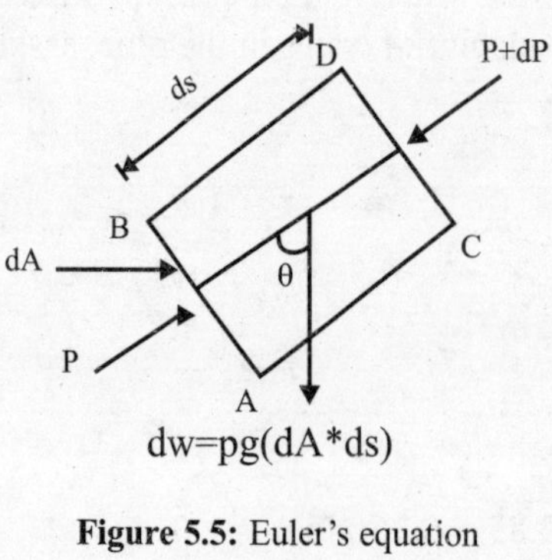

Figure 5.5: Euler's equation

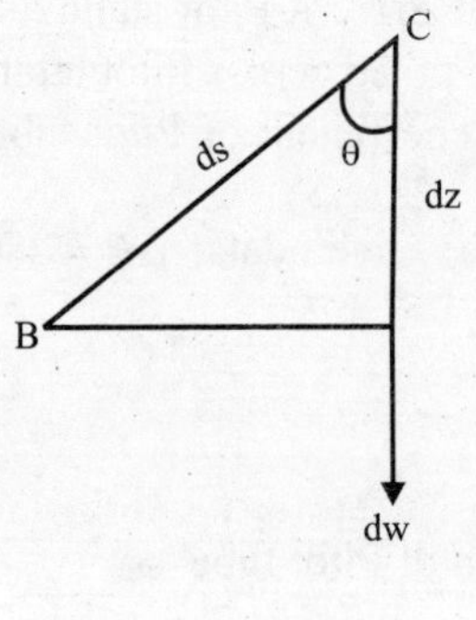

Figure 5.5a

Acceleration of the particle is $$a = V\frac{dV}{ds} \tag{5.29}$$

Substitute Eq. (5.27), Eq. (5.28), Eq. (5.29) in Eq. (5.26)

$$\Rightarrow \{-[(dP)dA] - \rho g * dA * dz\} = \rho(dA * ds) * V\frac{dV}{ds}$$

$$\Rightarrow \{-[(dP)dA] - \rho g * dA * dz\} = \rho(dA) * VdV \tag{5.30}$$

Taking $\rho * dA$ as common term on both sides of Eq. (5.30), we have

$$\boxed{\frac{dP}{\rho} + gdz + VdV = 0} \tag{5.31}$$

This is the **Euler's Equation of Motion** and is differential form of Bernoulli's equation.

VORTEX MOTION

It is defined as the rotation of fluid in a container, about a fixed axis. When a container containing liquid at rest, is rotated along a fixed axis (z) at a speed (N), then the liquid particles rotate in a circular direction, along the boundary of the container, forming a parabolic curve with decrease in normal free surface level of the liquid, as shown in Figure 5.6.

The liquid particles, while rotating in circular curve has their pressure depending upon two parameters like radius from the fixed axis and the height from the bottom base, i.e.

$$P = f(r, z) \tag{5.32}$$

and hence the partial derivative of pressure becomes

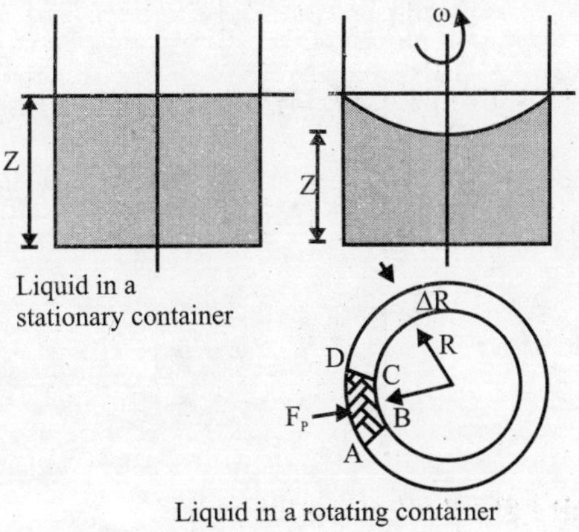

Figure 5.6: Vortex motion

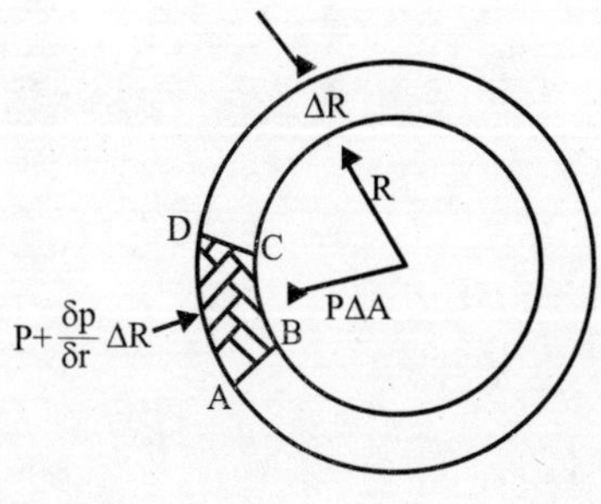

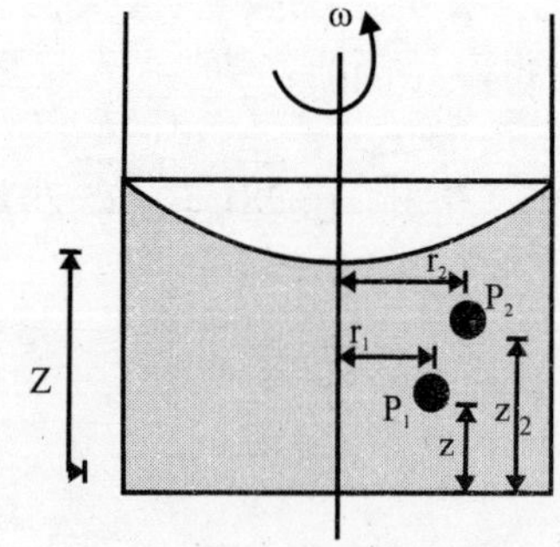

Figure 5.6a

$$dP = \frac{\partial P}{\partial r} dr + \frac{\partial P}{\partial z} dz \tag{5.32a}$$

and also have two force components, like centrifugal force and normal force along the radius, which are to be balanced for the fluid particle to be in equilibrium. And hence the centrifugal force (*F*) is equal to the pressure force (*F*) i.e.

$$F_C = F_N \tag{5.33}$$

where

$$F_C = \frac{mV^2}{r} \tag{5.33a}$$

and

$$F_N = [F_{AD} - F_{BC}]$$

$$\Rightarrow F_N = \left[\left(P + \frac{\partial P}{\partial r}\Delta r\right)\Delta A - (P)\Delta A\right] \tag{5.33b}$$

Substitute Eq. (5.33a), Eq. (5.33b) in Eq. (5.33)

$$\Rightarrow \frac{mV^2}{r} = \left[\left(P + \frac{\partial P}{\partial r}\Delta r\right)\Delta A - (P)\Delta A\right] = \frac{(\rho \Delta A \Delta r)V^2}{r}$$

$$\boxed{\Rightarrow \frac{\partial P}{\partial r} = \frac{\rho V^2}{r}} \tag{5.34}$$

where $\frac{\partial P}{\partial r}$ is called **pressure gradient** along the radius.

According to hydrostatic law

$$\frac{\partial P}{\partial z} = -\gamma = -\rho g \tag{5.35}$$

{refer Eq. (2.7a)}
Substitute Eq. (5.35), Eq. (5.34) in Eq. (5.32a), we have

$$dP = \frac{\rho V^2}{r} dr - \rho g dz \tag{5.36}$$

Eq. (5.36) is the **general equation for variation of pressure** of a rotating fluid about a fixed plane.

Vortex motion can be classified as

1) Forced vortex motion.
2) Free vortex motion.

Forced Vortex Flow

A forced vortex flow is one, which has the fluid mass rotated by some external means, thereby causing the whole mass of fluid to rotate at constant angular velocity ω. And so there is always a constant external torque requirement, to be applied to the fluid mass resulting in an expenditure of energy, in a forced vortex. And hence **forced vortex motion** has

$$V = r\omega$$

Examples of forced vortex motion are:

1) Stirring of water in a glass with a spoon.
2) Rotation of liquid inside the impeller of a centrifugal pump.
3) Rotation of water through the runner of a turbine.

Case 1: If two points are taken inside the liquid with r_1,r_2 and Z_1,Z_2 as radius and heights respectively (as shown in figure 5.6a).

Substitute $V = r\omega$ and integrate Eq. (5.36), with the limits, which gives

$$\int_{P_1}^{P_2} \partial P = \rho\omega^2 \int_{r_1}^{r_2} r\partial r - \int_{z_1}^{z_2} \rho g dz$$

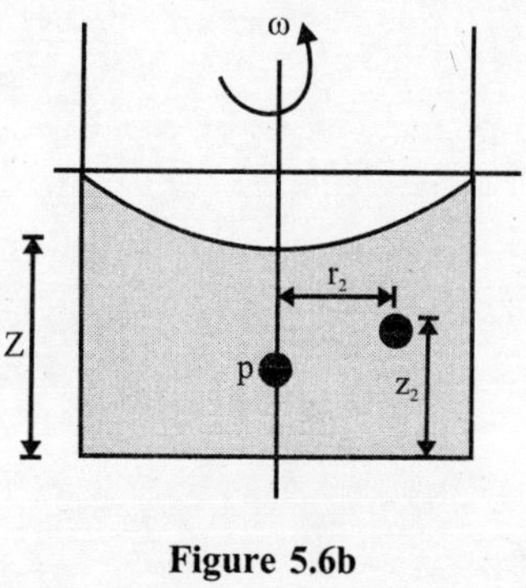

Figure 5.6b

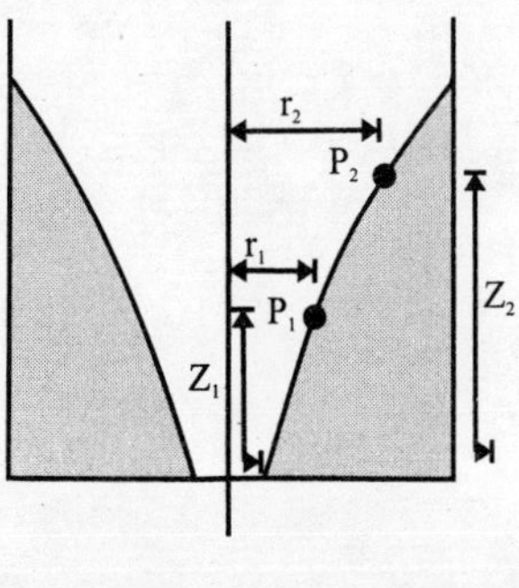

Figure 5.6c: Free vortex motion

$$\Rightarrow (P_2 - P_1) = \frac{\rho(V_2^2 - V_1^2)}{2} - \rho g(Z_2 - Z_1) \tag{5.37}$$

Case 2: If two points are taken on the free surface of the liquid with radius r_1, r_2 and Z_1, Z_2 as radius and heights respectively. Then $P_2 = P_1$ and substitute $V = r\omega$ and integrate Eq. (5.36), with the limits, which gives

$$\int_{P_1}^{P_2} \partial P = \rho\omega^2 \int_{r_1}^{r_2} r\partial r - \int_{z_1}^{z_2} \rho g dz$$

$$\Rightarrow (Z_2 - Z_1) = \frac{(V_2^2 - V_1^2)}{2g} \tag{5.38}$$

Case 3: If one point taken on axis of rotation and other on the free surface of the liquid with radius r_2 and Z_1, Z_2 as heights respectively (as shown in Figure 5.6b). Then $P_2 = P_1$ $r_1 = 0, V_1 = 0$ and substitute $V = r\omega^2$ and integrate Eq. (5.36), with the limits, which gives

$$\int_{P_1}^{P_2} \partial P = \rho\omega^2 \int_{r_1}^{r_2} r\partial r - \int_{z_1}^{z_2} \rho g dz$$

$$Z = \frac{V_2^2}{2g} = \frac{r_2^{\ 2}\omega^2}{2g} = \frac{P}{\gamma} \tag{5.39}$$

And hence, Eq. (5.39) shows that the free surface profile is a parabola.

Free Vortex Flow: In free vortex motion, $V \propto \frac{1}{r}$

$$\Rightarrow V = \frac{C}{r} \tag{5.40}$$

Substituting Eq. (5.40) in Eq. (5.36), we get

$$\int_{P_1}^{P_2} \partial P = \rho C \int_{r_1}^{r_2} r^{-3} \partial r - \int_{z_1}^{z_2} \rho g dz$$

$$\Rightarrow (P_2 - P_1) = \rho C \left| \frac{r^{-2}}{-2} \right|_{r_1}^{r_2} - \rho g(Z_2 - Z_1)$$

$$\Rightarrow (P_2 - P_1) = \frac{-\rho C}{2}\left(r_2^{-2} - r_1^{-2}\right) - \rho g\left(Z_2 - Z_1\right)$$

$$\Rightarrow (P_2 - P_1) = \frac{-\rho}{2}\left(V_2^2 - V_1^2\right) - \rho g\left(Z_2 - Z_1\right) \qquad (5.41)$$

Eq. (5.41) in the Bernoulli's equation and so in case of free vortex, Bernoulli's equation is applicable. Examples of free vortex motion are:

1) A whirlpool in a river.
2) Flow around a circular bend.
3) Rotation of sea water due to depression.

PRESSURE VARIATION FOR FLOW OF FLUID OVER A CURVED BODY

Consider a sphere over which a fluid particle is flowing. When considered a fluid particle passing over the sphere it is subjected to inertia force and according to Newton's second law, we have

$$dF = \delta m * a_s$$

As the fluid particle has some weight, the force causing the motion comprises two parts like force due to fluid pressure and other is due to its own weight and so total force acting on the fluid particle

$$dF = dF_P + dF_g = -\frac{\partial P}{\partial s} * \delta \text{ volume} - \gamma\delta \text{ volume } \sin\theta \qquad (5.42)$$

So Eq. (5.42) reduces to $-\frac{\partial P}{\partial s} * \delta volume - \gamma * \delta volume * \sin\theta = \rho * \delta volume * V\frac{\partial V}{\partial s}$

$$\Rightarrow -\frac{\partial P}{\partial s} - \gamma * \sin\theta = \rho * V\frac{\partial V}{\partial s} \qquad (5.43)$$

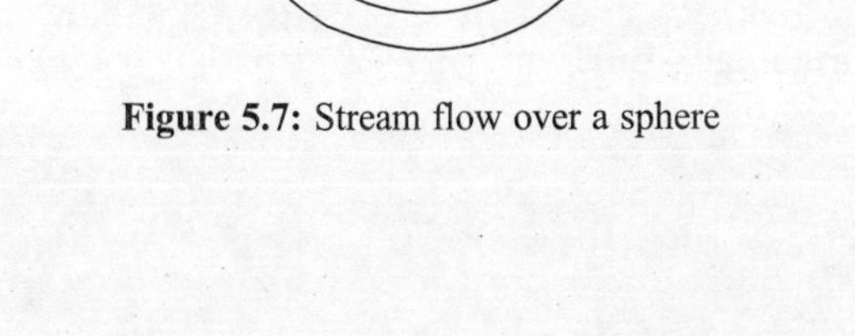

Figure 5.7: Stream flow over a sphere

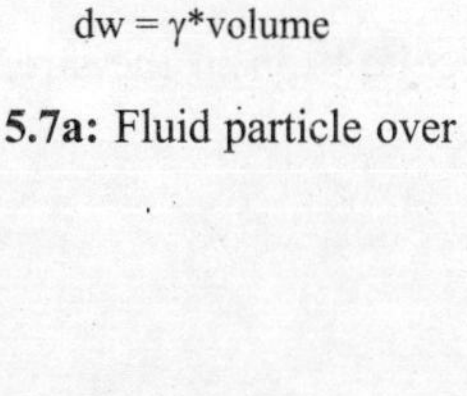

Figure 5.7a: Fluid particle over a sphere

Example 5.17: Determine the orifice diameter that will result in 150 kPa pressure drop in a 12 mm pipe with a flow rate of 80 mL/s. Assume the orifice coefficient to be 0.6. Also find the ratio of orifice diameter to pipe diameter?

If the smallest pressure differential that can accurately be measured with the pressure sensor is 1 kPa, what is the smallest flow that can accurately be measured using this orifice?

Solution: Given data : $\Delta P = 150$ kPa, $d_1 = 0.012$ m, $Q = 80 * 10^{-6}\,\text{m}^3/\text{s}$, $K_o = 0.6$

a) Orifice diameter: $Q = K_o * A_0\sqrt{2gh} = K_{ori} * \dfrac{\pi d_o^2}{4} * \sqrt{\dfrac{2\Delta P}{\rho}}$

$$d_o = \sqrt{\frac{4Q}{\pi K_{orifice}\sqrt{\frac{2\Delta p}{\rho}}}} = \sqrt{\frac{4\left(80\times10^{-6}\right)}{\pi(0.6)\sqrt{\frac{2(150000)}{1000}}}} = 3.13 \text{ mm}$$

b) Ratio of orifice diameter to pipe diameter: $\dfrac{d_o}{d_1} = \dfrac{3.13}{12} = 0.26$

c) Smallest flow that can accurately be measured for ΔP = 1 kPa:

$$Q = K_{ori} * \frac{\pi d_o^2}{4} * \sqrt{\frac{2\Delta P}{\rho}} = 0.6 * \frac{\pi * 0.0013^2}{4} * \sqrt{\frac{2 * 1000}{1000}} = 1.125 \text{ mL/s}$$

Example 5.18: Determine the pressure variation of a fluid particle having streamline velocity $V = V_o\left(1 + \dfrac{r^3}{x^3}\right)$ over a sphere and tending towards the center of sphere.

Solution: Given data: $V = V_o\left(1 + \dfrac{r^3}{x^3}\right)$

According to pressure variation of fluid flow over a curved surface, we have

$$-\frac{\partial P}{\partial s} - \gamma * \sin\theta = \rho * V\frac{\partial V}{\partial s}$$

But as the acceleration of the stream particle is horizontal $\theta = 0°$, $-\dfrac{\partial P}{\partial s} = \rho * V\dfrac{\partial V}{\partial s}$

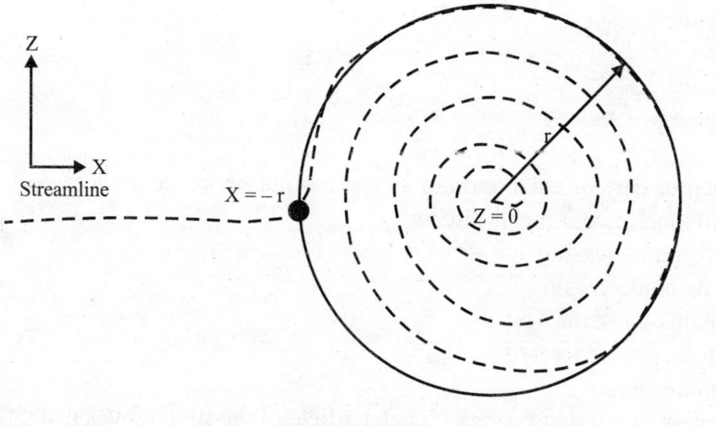

Figure P5.18

So $$\frac{-dp}{dx} = \rho V \frac{dV}{dx} = \rho V_o \left(1+\frac{r^3}{x^3}\right)\frac{d}{dx}\left[V_o\left(1+\frac{r^3}{x^3}\right)\right]$$

$$\Rightarrow \int_0^p dp = 3r^3 \rho V_o^2 \int_{-\alpha}^{x} \left[\left(1+\frac{r^3}{x^3}\right)\right]\left(x^{-4}\right)dx$$

$$\Rightarrow P = 3r^3 \rho V_o^2 \left\{\left[-\frac{1}{3}x^{-3}\right] - \left[\frac{1}{6}r^3 x^{-6}\right]\right\}\Bigg|_{-\alpha}^{x} = -\rho V_o^2 \left\{\left[\frac{r^3}{x^3}\right] + \left[\frac{1}{2}\frac{r^3}{x^6}\right]\right\}$$

EXERCISE

Objective

1. In a free vortex motion the radial component of velocity everywhere is
   a) Maximum
   b) Minimum
   c) Zero
   d) Non-zero and finite
   e) Unpredictable
2. In a free vortex the velocity of flow everywhere with the fluid is
   a) Maximum
   b) Minimum
   c) Zero
   d) Non-zero and finite
   e) Unpredictable
3. Pitot tube is used for measurement of

a) Pressure
b) Flow
c) Velocity
d) Discharge
e) Viscosity

4. The total energy of each particle at various places in the case of perfect incompressible fluid flowing in continuous stream
   a) Keeps on increasing
   b) Keeps on decreasing
   c) Remains constant
   d) May increase/decrease
   e) Unpredictable
5. The motion of a fluid is vortex if each particle of the fluid moves in a circular path with the speed which
   a) Is constant
   b) Is directly proportional to distance from center
   c) Varies as square of distance from center
   d) Varies inversely as the distance from center
   e) Varies inversely as square of distance from center.
6. Time required to empty rectangular tank is proportional to its
   a) H
   b) $\sqrt{H}$
   c) H2
   d) $H^{3/2}$
7. The contraction of area for flow through orifice in tank depends on
   a) Shape of orifice
   b) Size of orifice
   c) Head in tank
   d) All of the above
   e) None of the above
8. The discharge through an orifice fitted to a tank can be increased by
   a) Fitting a short length of pipe to the outside
   b) Sharpening the edges of orifice
   c) Fitting a long length of pipe to the outside
   d) Fitting a long length of pipe to the inside
   e) All of the above
9. When a liquid rotates at constant angular velocity about a vertical axis as a rigid body, the pressure
   a) Increase linearly as its radial distance
   b) Varies inversely as the altitude along any vertical line
   c) Varies as the square of the radial distance
   d) Decrease as the square of radial distance
   e) None of the above

10. When venturimeter is inclined, then for given flow it will show
    a) Less reading
    b) More reading
    c) Same reading
    d) Inaccurate reading
11. The value of coefficient of velocity compared to coefficient of discharge
    a) is less
    b) is more
    c) has no relation
    d) is same
    e) none of the above
12. Bernoulli's theorem is applicable for
    a) Streamline flow
    b) Steady flow
    c) Turbulent flow
    d) Perfect incompressible fluid flowing in continuous streams
13. Euler's equation in the different form for motion of liquids is given by

    (a) $\frac{dp}{\rho} - gdz + vdv = 0$

    (b) $\frac{dp}{\rho} - gdz - gdz = 0$

    (c) $\frac{dp}{\rho} + gdz + vdv = 0$

    (d) $pdp + vdv = 0$

    (e) None
14. The flow in venturimeter takes place at
    a) Atmosphere pressure
    b) At pressure greater than atmospheric pressure
    c) Vacuum
    d) High pressure
    e) Any pressure
15. A fluid flow taking place continuously round a curved path about a fixed axis of rotation, is known as
    a) Vortex flow
    b) Radial flow
    c) Circular flow
    d) Unsteady flow
    e) Avoid interruption in the flow.
16. In a free vortex motion
    a) Rotation of fluid, moving as a solid, takes place about an axis

b) Each particle moves in a circular path with a speed varying inversely as the distance from the center
c) Velocity decrease with the radius
d) Velocity remains constant
e) None

17. In a forced vortex motion
a) Rotation of fluid, moving as a solid, takes place about an axis
b) Each particle moves in a circular path with a speed varying inversely as the distance from the center
c) Velocity decreases with the radius
d) Velocity remains constant
e) None

18. In a free vortex motion, the tangential velocity of the water particles is proportional to
a) Distance from the center (r)
b) r^2
c) 1/r
d) $1/r^2$
e) $1/\sqrt{r}$

19. A right circular cylinder open at top is filled with water and rotated about its vertical axis at such speed that half the liquid spills out. The pressure at center of bottom is
a) One half its value when cylinder was full
b) One-fourth its value when cylinder was full
c) Zero
d) Can't be determined due to insufficient data
e) None of the above

Answers

1. a	2.b	3. b	4. c	5.d	6.b
7. d	8. a	9. d	10. c	11. a	12.d
13. c	14. c	15. a	16. b	17. a	18. c
19. c					

Theory

1. Differentiate between free and forced vortex flow.
2. Define stagnation point.
3. Derive an expression for the depth of paraboloid formed by the surface of a liquid contained in a cylindrical tank, which is rotated at constant angular velocity *w* about its vertical axis.
4. Derive from the first principles, the condition for irrotational flow. Prove that, for potential flow, both the stream and velocity function satisfy Laplace equation.
5. Derive Euler's equation of motion for a fluid flow.
6. Define Bernoulli's equation.
7. Explain orifice-meter in detail with diagram. Also derive an expression for finding out the actual discharge from a given orifice meter.

8. Derive generalized equation for vortex flow and deduce it to free vortex flow.

Problems

1. A venturimeter is to be fitted in a horizontal pipe of 0.15 m diameter to measure a flow of water, which may be anything up to 240 m^3/h. The pressure head at the inlet for this flow is 18 m above the atmospheric and the pressure head at the throat must be lower than 7 m below atmospheric. Between the inlet and throat, there is an estimated frictional loss of 10% of the difference in pressure head between these points. Calculate the minimum allowable diameter for the throat.
2. Starting with the Bernoulli and continuity equations, show that the following expression gives the discharge measured by a venturimeter.

$$Q = C_d A_1 A_2 \sqrt{\frac{2g\left(\frac{P_1 - P_2}{\rho g} + z_1 - z_2\right)}{A_1^2 - A_2^2}}$$

   And that when a manometer is attached the discharge may be given by

$$Q_{actual} = C_d A_1 A_2 \sqrt{\frac{2gh\left(\frac{\rho_{mano}}{\rho} - 1\right)}{A_1^2 - A_2^2}}$$

3. A venturimeter has its axis vertical, the inlet and throat diameters being 150 mm and 75 mm respectively. The throat is 225 mm above inlet, petrol of specific gravity 0.78 flows up through the meter at a rate of 0.029 m^3/s. Find the pressure difference between the inlet and the throat.
4. A venturimeter has its axis vertical, the outlet and throat diameters being 150 mm and 75 mm respectively. The throat is 2.5 m below outlet, and when the flow rate down the pipe is 0.02 m^3/s. The pressure at the throat is 14715 Pa greater than that of outlet pressure. Assuming the losses in the pipe as $(h)_f = k\frac{v^2}{2g}$, where v is the velocity at the outlet, find the value of k and also if a differential mercury manometer is connected in between the two points, determine the level in difference.
5. A 150 mm × 75 mm venturimeter with C_d = 0.98 is to be replaced by an orifice meter having a value of C_d = 0.6. If both the meters are to give the same differential mercury manometer reading for a discharge of 100 lit/s and the inlet diameter to remain 150 mm, what should be the diameter of orifice?
6. A venturimeter is used for measuring the flow of petrol in a pipeline inclined at 35°C to horizontal. The specific gravity of the petrol is 0.81 and throat area ratio is 4. If the difference in mercury levels in the gage is 50mm, calculate the flow in liters per hour if the pipe diameter is 0.3 m. Take C_d = 0.975.

7. Determine the diameter of the throat of a venturimeter used for measuring the flow of petrol (specific gravity of the petrol is 0.81) in a pipeline of 0.6 m diameter horizontal, so that the difference in mercurry level in the gage is 50 cm, Assume coefficient of discharge $C_d = 0.975$.
8. A pipe 15 cm diameter 15 metres long is connected to the bottom of a tank,15 meters long by 12 meters wide. The original head over the open end of the pipe is 5 meters. Find the time of emptying the tank, assuming the entrance to the pipe is sharp-edge.
9. An orifice meter is to be fitted into a horizontal pipe 20 cm diameter, carrying oil of specific gravity 0.85 for the purpose of flow measurement. The differential head is to be indicated by a U-tube manometer containing mercury (specific gravity = 13.6). If the manometer reading is not to exceed 0.2 m when the flow is 15 kg/sec, what should be the diameter of the orifice? Assume $C_d = 0.62$.
10. A pipeline 60 cm in diameter bifurcates at a Y-junction into two branches 40 cm and 30 cm diameter. If the rate of flow in main pipe is 1.5 m^3/s and mean velocity of flow in 30 cm diameter pipe is 7.5 m/s, determine the rate of flow in the 40 cm diameter pipe.
11. The pressure difference measured between two tapings of a Pitot static tube, one taping pointing upstream and the other perpendicular to the flow placed in the center of the pipeline of diameter 40 cm is 10 cm of water. The mean velocity of the pipe is 0.75 cm times the central velocity. Find the discharge through the pipe. Take coefficient of Pitot tube as 0.98.
12. A 300 mm × 150 mm diameters venturimeter is provided in a pipeline to carry oil of specific gravity 0.9, flow being upwards. The difference in elevation of the throat section and inlet section is 300 mm. The differential U-Tube mercury manometer shows a gage reading of 250 mm. Calculate discharge, change in pressure. Take $C_d = 0.9$ and $S_{Hg} = 13.6$.
13. A closed cylinder vessel of diameter 30 cm and height 100 cm contains water up to a depth of 80 cm. The air above the water surface is at a pressure of 5.886N/cm^2. The vessel is rotated at a speed of 250 rpm about its vertical axis. Find the pressure head at the bottom of the vessel at the center and at the edge.
14. A venturimeter with an entrance diameter of 03 m and a throat diameter of 0.2 m is used to measure the volume of a gas flowing through a pipe. The discharge coefficient of the meter is 0.96. Assuming the specific weight of the gas to be a constant at 19.62 N/m^3, calculate the volume flowing when the pressure difference between the entrance and the throat is measured as 0.06 m on a water U-tube manometer.
15. What is the inflow of the shown (Figure E5.15) control volume of a tank shape?

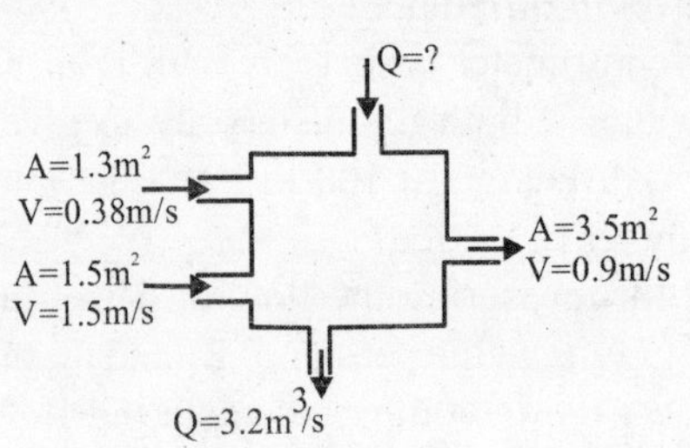

Figure E5.15: Tank having inflow/out flow

QUICK REFERENCE OF FORMULAE:

Property	Formulae	Remarks, if any	
Potential energy (PE)	$PE = mgZ = WZ$		
Kinetic energy (KE)	$KE = \frac{1}{2}mV^2 = \frac{W}{2g}V^2$		
Pressure energy (h)	$h = \frac{P}{\gamma}$		
Total energy (TH)	$TH = \frac{p}{\gamma} + \frac{1}{2g}V^2 + Z$		
Bernoulli's equation	$\frac{p}{\gamma} + \frac{1}{2g}V^2 + Z = \text{Constant}$		
General equations of a venturimeter	$h = \left(\frac{P_1 - P_2}{\gamma}\right) + (Z_1 - Z_2) = \frac{1}{2g}\left(V_2^2 - V_1^2\right)$ and $(Q)_{Theoretical} = \frac{A_1 A_2}{\sqrt{A_1^2 - A_2^2}} * \sqrt{2gh}$		h is differential head Q is discharge. A_1 is area at inlet A_2 is area at throat.
Equation of a horizontal venturimeter	$h = \left(\frac{P_1 - P_2}{\gamma}\right) = \frac{1}{2g}\left(V_2^2 - V_1^2\right)$	$h = h_1 - h_2$	
Equation of a inclined and vertical venturimeter	$h = y\left[\frac{S_{ML}}{S_{LP}} - 1\right]$ or $h = y\left[1 - \frac{S_{LL}}{S_{mL}}\right]$		
Discharge through an orifice meter fitted to a pie line (Q_{th})	$(Q_{th})_{Orifice} = C_d\left[\frac{A_1 A_0 \sqrt{2gh}}{\sqrt{A_1^2 - A_0^2}}\right]$		h is differential head A_1 is area at inlet A_0 is area at throad
Discharge through an orifice meter fitted to a Tank (Q_{Actual})	$(V_2)_{Actual} = C_V * \sqrt{2gh}$ and $(Q)_{Actual} = C_d * A_O\sqrt{2gh}$ or $Q_{Actual} = -A * \frac{\delta h}{\delta t}$	$C_d = C_C - C_V$ δh is fall of liquid level. δt is time of fall of liquid level. A is area at tank.	

Contd.

Property	**Formulae**	**Remarks, if any**	
Time of emptying a tank fitted with an orifice-meter.	$t = \frac{2A}{C_d * A_O\sqrt{2g}}\left(\sqrt{h_1 - h_2}\right)$		
Actual velocity through a Pitot tube	$(V)_{actual} = C_V\sqrt{2gh}$		h is differential head
Euler's Equation of Motion	$\frac{dP}{\rho} + gdz + VdV = 0$		
General equation of pressure variation in a vortex motion.	$dP = \frac{\rho V^2}{r}dr - \rho gdz$	$\partial P/\partial z = - y = - \rho g$ is hydrostatic law.	$\partial P/\partial r = \rho V^2/r$ is pressure gradient.
Velocity of forced and free vortex motion	$V = r\omega$ and $V = \frac{C}{r}$		ω rev/min

Points to Remember

For turbines	$\frac{p_1}{\gamma_1} + \frac{1}{2g}V_1^2 + Z_1 = \frac{p_2}{\gamma_2} + \frac{1}{2g}V_2^2 + Z_2 + H_L + H_T$	H_L is head loss. H_T is power developed by turbine per unit weight of liquid.
For pumps	$\frac{p_1}{\gamma_1} + \frac{1}{2g}V_1^2 + Z_1 + H_P = \frac{p_2}{\gamma_2} + \frac{1}{2g}V_2^2 + Z_2 + H_L$	H_t is head loss. H_P is energy added by pump per unit weight of liquid.
For real fluids	$\frac{p_1}{\gamma_1} + \frac{1}{2g}V_1^2 + Z_1 = \frac{p_2}{\gamma_2} + \frac{1}{2g}V_2^2 + Z_2 + H_L$	H_L is head loss.
Power required and developed by pump P_p and turbine P_T	$P_P = \frac{\gamma QH_P}{\eta_O}$ and $P_T = \gamma Q\eta_T H_T$	η_0 is overall efficiency. Q is discharge. η_T is turbine efficiency. γ specific weight of liquid.

CHAPTER 6

Viscous Flow

INTRODUCTION

Low velocity fluid flow does have the Reynolds number so low that the viscous effect dominates the inertial effect, mainly at the boundary of the flow and surface contact. In continuation to this, the viscous effect is eliminated by increase in velocity of the flow, which in turn increases the Reynolds number.

In this chapter, we discuss differential equations of viscid fluid flow fields, considering the flow to be one dimensional, wherein viscous effect and pressure effects are only predominant, like equations developed by Couette and Hagen Poiseulle, basing on equilibrium state of fluid particle in a flow field. And in continuation, we discuss fluid particles subjected to two-dimensional forces like inertial, viscous and gravitational forces been discussed in deriving Navier Stokes equation, basing on several assumptions and eliminations.

And moreover considering conservation of mass (the continuity equation), momentum equations and conservation of energy (Bernoulli equation), the viscous flow through closed conduits (pipes–generally circular cross-sections) and the losses raised due to viscous effect have been discussed by taking into account Fanning and Darcy Weisbach factors, in pipes. Pipe flow is an important study in domestic water supply, flow between reservoirs, flow of oils through pipelines, etc. considering the losses for effective flow potential.

COUETTE FLOW

Couette flow derives equation for velocity and shear profiles between wide gap plates like water channels and states that, **"If any viscous fluid flows in between two parallel plates, one plate fixed and other plate moving, has parabolic velocity profile, then the fluid is said to have Couette flow"**.

Consider a fluid particle in one-dimensional flow having dimensions ($\Delta x, \Delta y, 1$), moving at a height of "y" distance from the fixed plate, in between two plates (i.e. one plate fixed and other plate moving), as shown Figure 6.1.

The fluid particle has four force components acting on it, like two pressures forces (F_P) on both ends of the fluid particle and two shear forces (F_S) one on either face of the fluid particle, as shown in Figure 6.1a.

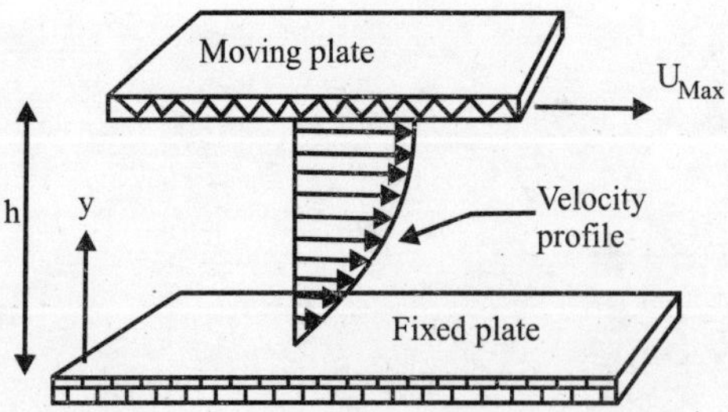

Figure 6.1

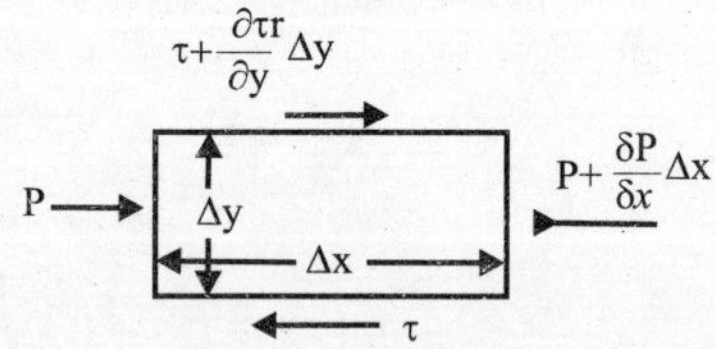

Figure 6.1a

If a particle to be in equilibrium, the summation of all the forces acting on the particle along the direction of flow must be zero, i.e. $\sum(F_P + F_S) = 0$ and so

$$\left[(P)\Delta y * 1\right] - \left[\left(P + \frac{dp}{dx}\Delta x\right)\Delta y * 1\right] + \left[\left(\tau + \frac{d\tau}{dy}\Delta y\right)\Delta x * 1 - (\tau)\Delta x * 1\right] = 0$$

$$\Rightarrow \left[\left(\frac{dp}{dx}\right)\right] = \left[\left(\frac{d\tau}{dy}\right)\right] \qquad (6.0)$$

where $\left(\frac{dp}{dx}\right)$ and $\left(\frac{d\tau}{dy}\right)$ is **pressure and shear gradient** respectively.

Equation 6.0 says that the "**pressure gradient along the flow is same as that of the shear gradient across the flow**".

$$\Rightarrow \tau = \frac{dp}{dx} * y + C \qquad (6.0a)$$

According to Newton's law of viscosity

$$\tau = \mu \frac{du}{dy} \qquad (6.0b)$$

Equate Eq. (6.0a) and Eq. (6.0b) and by rearranging, we have

$$du = \frac{1}{\mu}\left(\frac{dp}{dx}\right) y * dy + C * dy$$

and integrating both sides, we have

$$u = \frac{1}{2\mu}\left(\frac{dp}{dx}\right) y^2 + Cy + D \qquad (6.1)$$

Apply Boundary condition in Eq. (6.1), i.e. at $y = 0, u = 0$

$$\Rightarrow D = 0$$

Similarly, at $y = h, u = U_{MAX}$ $\Rightarrow C = \frac{U_{MAX}}{h} - \frac{h}{2\mu}\left(\frac{dp}{dx}\right)$

Substitute *C*, *D* in Eq. (6.1)

$$\Rightarrow u = \frac{U_{MAX} * y}{h} + \frac{1}{2\mu}\left(\frac{dp}{dx}\right)\left[y^2 - hy\right] \quad (6.2)$$

Velocity profile will be

$$\Rightarrow \frac{u}{U_{MAX}} = \frac{y}{h} + \frac{h^2}{2\mu U_{MAX}}\left(\frac{dp}{dx}\right)\left[\frac{y^2}{h^2} - \frac{y}{h}\right] \quad (6.2a)$$

And **shear stress** (τ) profile,

$$\tau = \mu\left(\frac{du}{dy}\right) = \mu\left(\frac{U_{MAX}}{h} - \frac{1}{2\mu}\left(\frac{dp}{dx}\right)[h - 2y]\right) \quad (6.3)$$

And **discharge** (*Q*)

$$Q = AV = \int_0^h u(dy * 1) \Rightarrow Q = AV = \int_0^h \left\{\frac{U_{MAX} * y}{h} + \frac{1}{2\mu}\left(\frac{dp}{dx}\right)\left[y^2 - hy\right]\right\} dy$$

$$\Rightarrow Q - \frac{U_{MAX} h}{2} - \frac{h^3}{12\mu}\left(\frac{dp}{dx}\right) \quad (6.4)$$

Equations (6.2), (6.3) and (6.4) are the **general equations for velocity, shear stress and discharge in a Couette flow**.

From Eqs. (6.2) and (6.4) the velocity and shear profiles are dependent on only the pressure gradient $\left(\frac{dp}{dx}\right)$ and the pressure gradient may either be zero or negative or positive, hence we need to analyze the three cases.

Case 1: Zero pressure Gradient/Plane Couette Flow $\left(\frac{dp}{dx} = 0\right)$

Apply boundary conditions in Eq. (6.2) and Eq. (6.3) with zero pressure gradient, i.e. at

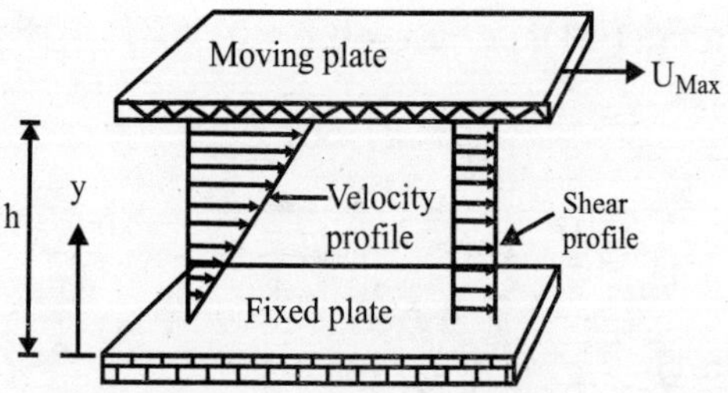

Figure 6.2: Plane couette flow

$$y = 0 \quad u = 0 \text{ and } \tau = \mu \frac{U_{MAX}}{h} \tag{6.5a}$$

at $$y = h\,,\ u = U_{MAX} \text{ and } \tau = \mu \frac{U_{MAX}}{h} \tag{6.5b}$$

Figure 6.2 shows the velocity profile and shear profiles of zero pressure gradient based on the velocity and shear stress values obtained by applying boundary conditions.

Case 2: Negative pressure gradient/favorable pressure gradient $\left(\dfrac{dp}{dx} = -ve\right)$

Apply boundary conditions in Eq. (6.2) and Eq. (6.3) with negative pressure gradient,

i.e. at $$y = 0, u = 0 \text{ and } \tau_{y=0} = \mu \left[\frac{U_{MAX}}{h} + \frac{h}{2\mu}\left(\frac{dp}{dx}\right)\right] \tag{6.6a}$$

At $$y = h\,,\ u = U_{MAX} \text{ and } \tau_{y=h} = \mu \left[\frac{U_{MAX}}{h} - \frac{h}{2\mu}\left(\frac{dp}{dx}\right)\right] \tag{6.6b}$$

At $$y = h/2\,,\ u = (U_{MAX})\frac{1}{2} - \left[\frac{h^2}{8\mu}\left(\frac{dp}{dx}\right)\right]$$

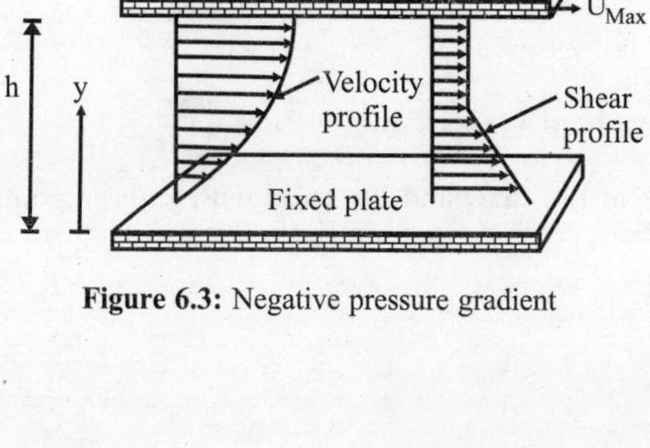

Figure 6.3: Negative pressure gradient

and $$\tau_{y=h/2} = \mu \frac{U_{MAX}}{h} \tag{6.6c}$$

Figure 6.3 shows the velocity profile and shear profiles of negative pressure gradient based on the velocity and shear stress values obtained by applying boundary conditions.

By observing Eq. (6.6a) and Eq. (6.6b), $\tau_{y=0} > \tau_{y=h}$ and which shows that the shear stress at the moving plate is less than the fixed plate and hence less force is needed to drag the moving plate with respect to the fixed plate, which is a favorable position and so-called **favorable pressure gradient**. And moreover this condition can arise only when the fixed plate is also moving, which means the two plates is moving in same direction.

Case 3: Positive pressure gradient/adverse pressure gradient $\left(\frac{dp}{dx} = +\text{ve}\right)$.

Apply boundary conditions in Eq. (6.2) and Eq. (6.3) with positive pressure gradient,

i.e. At $$y = 0, u = 0 \text{ and } \tau_{y=0} = \mu\left[(U_{MAX})\frac{1}{h} - \left[\frac{1}{2\mu}\left(\frac{dp}{dx}\right)(h)\right]\right] \tag{6.7a}$$

At $$y = h, u = U_{MAX} \text{ and } \tau_{y=h} = \mu\left[(U_{MAX})\frac{1}{h} + \left[\frac{h}{2\mu}\left(\frac{dp}{dx}\right)\right]\right] \tag{6.7b}$$

$$y = h/2,\ u = (U_{MAX})\frac{1}{2} + \left[\frac{h^2}{8\mu}\left(\frac{dp}{dx}\right)\right] \text{ and } \tau_{y=h/2} = \mu\frac{U_{MAX}}{h} \tag{6.7c}$$

Figure 6.4 shows the velocity profile and shear profiles of positive pressure gradient based on the velocity and shear stress values obtained by applying boundary conditions. By observing Eq: (6.7a), Eq. (6.7b) and Eq. (6.7c), i.e. $\tau_{y=0}, \tau_{y=h/2}, \tau_{y=h}$ are some finite values, which shows that the shear stress at the fixed plate is zero and has some value less than $\tau_{y=h/2}$ and hence the fixed plate is also moving in opposite direction to the moving plate and in turn generating additional shear stress to the moving plate at the top and hence more force is needed to drag the moving plate with respect

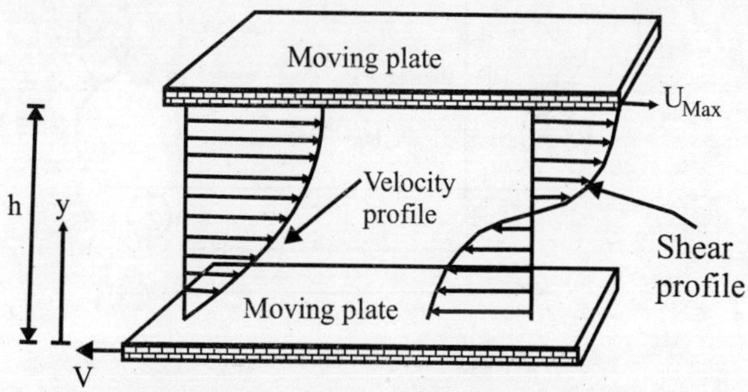

Figure 6.4: Positive pressure gradient

to the fixed plate, which is an adverse position and so called **adverse pressure gradient. And this condition can arise only when the fixed plate is also moving with some velocity –*V*, which means the two plates are moving in opposite directions.**

HAGEN POISEULLE LAW —IN A CIRCULAR PIPE

Reynolds number justify by a numeric value of the flow in circular pipe, to be laminar or turbulent. If the flow has Reynolds number less than 2000, then the flow is said to be laminar and in between 2000 to 4000, the flow is treated as transient flow and Reynolds number more than 4000 is treated as turbulent flow.

Hagen treating the fluid flow in a circular pipe to be Laminar (i.e. Reynolds number less than 2000) has given velocity and shear stress profiles in a circular pipe basing on two assumptions, he made:

a) Fluid follows Newton's law of viscosity.
b) The fluid has no slip of particles at the boundaries i.e. ($V_w = 0$).

Consider a pipe having radius "R" and length "L", in which a cylindrical fluid particle with area ($a = \Pi r^2$) and length (Δx) flowing.

The fluid particle is subjected to three force components like two pressure forces (F_{P1}, F_{P2}) on both ends and a shear force on the surface of the particle (F_τ) as shown in Figure 6.5. And hence in equilibrium state, the sum of forces along the flow direction will be zero. i.e. $F_{P1} - F_{P2} - F_\tau = 0$

where $\quad F_{P1} = P*\left(\Pi r^2\right),$

$$F_{P2} = \left[P + \frac{dp}{dx}\Delta x\right]*\Pi r^2,$$

$$F_\tau = \tau*(2\Pi r)*\Delta x$$

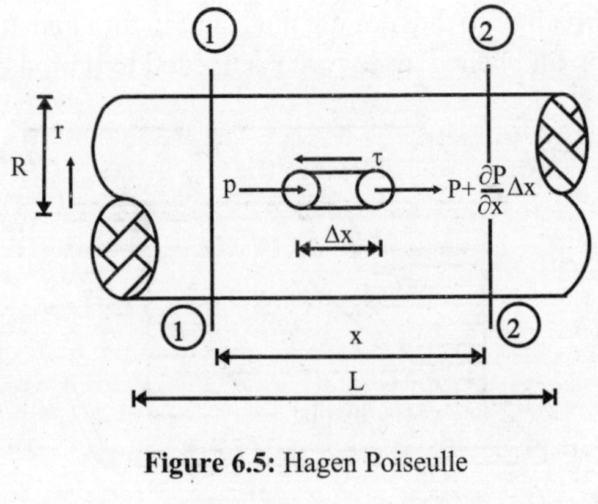

Figure 6.5: Hagen Poiseulle

$$\Rightarrow \tau = \left(\frac{-dp}{dx}\right)\frac{r}{2} \tag{6.8}$$

where $\left(\frac{dp}{dx}\right)$ is the **pressure gradient** acting along the flow direction.

As per the Newton's law of viscosity, $\tau = \mu\left(\frac{du}{dy}\right)$

but $y = R - r \Rightarrow dy = -dr$ therefore $\tau = -\mu\left(\frac{du}{dr}\right)$ (6.9)

Equating Eq. (6.9) and Eq. (6.8) $\Rightarrow -\mu\left(\frac{du}{dr}\right) = \left(\frac{-dp}{dx}\right)\frac{r}{2}$

Integrate on both sides $\Rightarrow u = \frac{1}{4\mu}\left(\frac{dp}{dx}\right)r^2 + C$ (6.10)

Apply boundary condition as

$$r = R,\ u = o \ \Rightarrow C = \frac{-1}{4\mu}\left(\frac{dp}{dx}\right)R^2$$

and by substituting C value in Eq. (6.10) we get Eq. (6.10) as

$$\Rightarrow u = \frac{-1}{4\mu}\left(\frac{dp}{dx}\right)\left[R^2 - r^2\right] \tag{6.11}$$

Apply boundary conditions in Eq. (6.11), then at $r = 0$ and $u = U_{MAX}$

$$U_{MAX} = \frac{-1}{4\mu}\left(\frac{dp}{dx}\right)\left[R^2\right] \tag{6.12}$$

$$\Rightarrow \frac{u}{U_{MAX}} = \left[1 - \left(\frac{r}{R}\right)^2\right] \tag{6.12a}$$

And discharge $Q = Area * Velocity = \int_0^R u * (2\Pi r * dr)$

$$\Rightarrow Q = \int_0^R U_{MAX}\left[1-\left(\frac{r}{R}\right)^2\right]*(2\Pi r*dr) \Rightarrow Q = U_{MAX}\frac{\Pi}{2}R^2$$

But average or mean velocity u_{AV} is

$$u_{AV} = \frac{Q}{A}$$

$$\Rightarrow u_{AV} = \frac{U_{MAX}\frac{\Pi}{2}R^2}{\Pi R^2} = \frac{U_{MAX}}{2} \quad (6.13)$$

Substitute Eq. (6.12) in Eq. (6.13)

$$u_{AV} = \frac{-1}{8\mu}\left(\frac{dp}{dx}\right)\left[R^2\right]$$

Integrate both sides with the limits, we have

$$\int_0^L \frac{8\mu u_{AV} * dx}{R^2} = -\int_{P_1}^{P_2} dp$$

$$(P_1 - P_2) = \frac{32\mu u_{AV} L}{D^2} \quad (6.14)$$

This is the **equation of Hagen Poiseulle**

Dividing both sides by γ,

$$h_f = \frac{(P_1 - P_2)}{\gamma} = \frac{32\mu u_{AV} L}{\gamma D^2} \quad (6.15)$$

where h_f is **head loss due to friction**

In general, if a circular pipe is substituted by a square pipe of same dimension, as shown in Figure 6.5b, then

$$\frac{D_h}{4} = \frac{A_w}{P_w} \quad (6.16)$$

and A_w/P_w is called as Hydraulic mean depth HMD" or Hydraulic radius, where D_h is hydraulic diameter, A_w is the wetted area and P_w is perimeter wetted. Hence, by substituting Eq. (6.16) in Eq. (6.15), we get

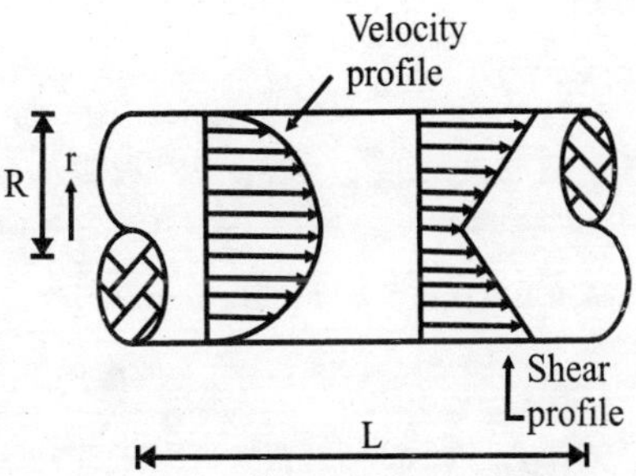

Figure 6.5a: Velocity and shear profiles of Hagen Poiseulle

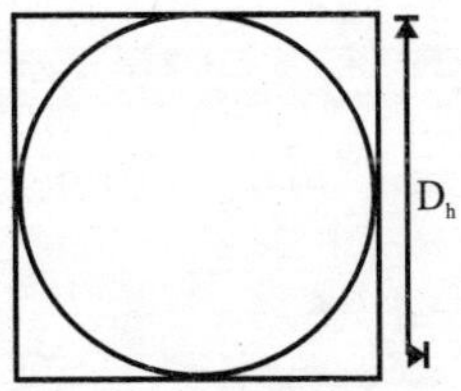

Figure 6.5b

$$h_f = \frac{(P_1 - P_2)}{\gamma} = \frac{8\mu u_{AV} L}{\gamma D_h} \tag{6.17}$$

Example 6.1: A liquid whose coefficient of viscosity is "m" flows below the critical velocity for laminar flow in a circular pipe of diameter "d" and with mean velocity "u". Show that the pressure loss in a length of pipe is $32\mu u / d^2$. And also calculate pressure loss in a length of 120 m, if oil of viscosity 0.05 kg/ms flows through a pipe diameter of 0.1 m with a velocity of 0.6 m/s.

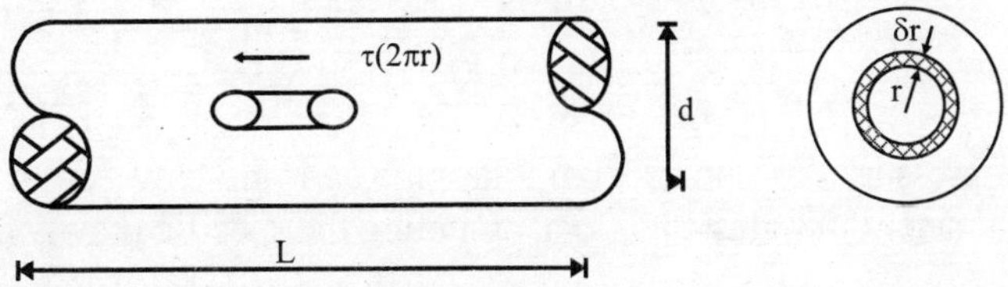

Figure P6.1

Solution:

Given data: $L = 120\,\text{m},\ d = 0.1\,\text{m},\ u = 0.6\,\text{m/s},\ \mu = 0.05\,\text{kg/ms}$.

In equilibrium conditions, we have shear force equal to pressure forces and hence

$$\tau(2\pi r L) = \Delta P \pi r^2 \quad \Rightarrow \tau = \frac{\Delta P r}{2L}$$

As per Newton's law, we have

$$\tau = -\mu \frac{du}{dr}$$

so

$$-\mu \frac{du}{dr} = \frac{\Delta P r}{2L} \quad \Rightarrow u = -\frac{\Delta P r}{2\mu L} \int_o^R r\,dr \quad \Rightarrow u = -\frac{\Delta P r^2}{4\mu L} + C$$

Applying boundary conditions, we have at $r = 0, u = U_{Max}$ and $r = R, u = 0$.

Hence,

$$\Rightarrow u = \frac{\Delta P}{4\mu L}\left(R^2 - r^2\right)$$

The flow rate in a strip of thickness will be $\delta Q = u\delta A$ where $\delta A = \pi\left(r + \delta r\right)^2 - \pi r^2 \approx 2\pi r\delta r$.
Hence,

$$\delta Q = u\left(2\pi r\delta r\right) = \frac{\Delta P}{4\mu L}\left(R^2 - r^2\right)\left(2\pi r\delta r\right)$$

Total flow rate will be

$$Q = \int_0^R \frac{\Delta P}{4\mu L}\left(R^2 - r^2\right)\left(2\pi r dr\right) = \frac{\Delta P \pi d^4}{128\mu L}$$

$$\Rightarrow \Delta P = \frac{32L\mu u}{d^2} \text{ per unit length}$$

If substituted the given values, we get

$$\Delta P = \frac{32uL\mu}{d^2} = \frac{32 * 0.05 * 120 * 0.6}{0.1^2} = 11.52\,\text{kPa}$$

Example 6.2: If the pressure drop in a 120 m long pipe carrying crude oil of 8.5 poise through a horizontal pipe of diameter 0.08 m is 1500 kPa, assuming the specific gravity of the crude oil to be 0.75 determine

a) Discharge of the oil.
b) Reynolds number and determine whether the flow is laminar or turbulent
c) Center line velocity.
d) Total frictional drag over entire length and power required to maintain the flow.
e) The velocity gradient at the wall and velocity at 0.01 m from the wall.
f) Shear stress at a distance of 0.01 m from the wall.
g) Power required by a pump to pump oil per 40 m of length to maintain the flow.

Solution: Given data: $L = 120\,\text{m}$, $D = 0.08\,\text{m}$, $\Delta P = 1500\,\text{kPa}$, $S = 0.75$, $\mu = 0.85\,\text{Ns/m}^2$
Area of the pipe be

$$A = \frac{\pi * D^2}{4} = \frac{\pi * 0.08^2}{4} = 0.005024\,\text{m}^2$$

a) **Discharge through the pipe:**
According to Hagen Poiseulle law, we have

$$\left(P_1 - P_2\right) = \frac{32\mu u_{AV} L}{D^2}$$

$$\Rightarrow u_{AV} = 2.94\,\text{m/s}$$

Hence, the discharge will be $\Rightarrow Q = Au_{AV} = 0.005024 * 2.94 = 0.0147\,\text{m}^3/\text{s}$

b) **Reynolds number:**

$$R_e = \frac{\rho u_{AV} D}{\mu} = \frac{0.75*1000*2.94*0.08}{0.85} = 207.6 < 2000,$$

hence the flow in the circular pipe is laminar.

c) **Center line velocity:** As the center line velocity in a pipe will be maximum, refer Eq. (6.13) we have

$$\Rightarrow U_{MAX} = 2u_{AV} = 2*2.94 = 5.88\,\text{m/s}$$

d) **Total frictional drag:** We have $\tau = \left(\frac{-dp}{dx}\right)\frac{r}{2}$

where $$-\frac{\Delta P}{dx} = \left(\frac{P_1 - P_2}{L}\right) \Rightarrow \tau = \left(\frac{P_1 - P_2}{L}\right)\frac{r}{2}$$

As the friction exists at the boundary, we have taken the value of "r" as "R" and hence the equation becomes

$$\tau_w = \left(\frac{P_1 - P_2}{L}\right)\frac{R}{2} \Rightarrow \tau_w = \left(\frac{1500*1000}{120}\right)*\frac{0.04}{2} = 250\,\text{Pa}$$

So the **frictional drag over entire length** is $F_D = \tau_w * \pi DL = 0.754\,\text{kN}$

Power required will be $P = F_D * u_{Av} = 2.216\,\text{kW}$

e) The **velocity gradient** at the wall and velocity at 0.01 m from the wall:
As per the Newton's law of viscosity, we have

$$\tau_w = \mu\left(\frac{du}{dy}\right)_{y=o} \Rightarrow \left(\frac{du}{dy}\right)_{y=o} = \frac{\tau_w}{\mu} = \frac{250}{0.85} = 294.12\,\text{s}^{-1}$$

So velocity gradient at the wall is 294.

Velocity at distance of 0.01 m from the wall:

Refer Eq. (6.11) $u = \frac{-1}{4\mu}\left(\frac{dp}{dx}\right)\left[R^2 - r^2\right]$

But $r = R - y = 0.04 - 0.01 = 0.03\,\text{m}$ $\Rightarrow u_{y=0.1} = \frac{1}{4*0.85}\left(\frac{1500*1000}{120}\right)\left[0.04^2 - 0.03^2\right] = 2.57$ m/s

f) **Shear stress** at a distance of 0.01 m from the wall.

The shear profile of the flow in a pipe will follow $\frac{\tau_y}{r} = \frac{\tau_w}{R}$.

Hence,

$$\frac{\tau_{y=0.1}}{0.03} = \frac{250}{0.04}$$
$$\Rightarrow \tau_{y=0.1} = 187.5\,\text{Pa}$$

g) **Power required** per 40 m length:

Head loss $h_f = \frac{\Delta P}{\gamma} = \frac{1500*1000}{0.75*1000*9.81} = 203.8$ m

Power required per 40 m of pipe length will be

$$40*Power = 40*\gamma Q h_f = 40*0.75*1000*9.81*0.0147*203.8 = 881.7\text{ kW}$$

Example 6.3: A pump having 80% efficiency required 70 kW power to pump an oil (specific gravity of 0.75) at a rate of 890 lit/sec through a pipe of diameter 0.25 m and 300 m long, determine:

a) Dynamic viscosity of the oil
b) Reynolds number and state whether the flow is laminar or turbulent.

Solution: Given data: $L = 300\text{m}$, $D = 0.25\text{m}$, $\eta = 80\%$, $S = 0.75$, $Q = 0.89\text{m}^3/\text{s}$, $Power = 70\text{kW}$

Area of the pipe $$A = \frac{\pi * D^2}{4} = \frac{\pi * 0.25^2}{4} = 0.049\,\text{m}^2$$

Given discharge will be

$$Q = Au_{AV}$$
$$\Rightarrow u_{AV} = 0.89/0.049 = 18.14\,\text{m/s}$$

a) **Dynamic viscosity of the oil**:

Power required is given as $Power = \frac{\gamma Q h_f}{\eta}$

$$\Rightarrow Power = \gamma Q h_f / \eta = 0.75*1000*9.81*0.89*h_f/0.8 = 70*1000\,\text{W}$$
$$\Rightarrow h_f = 8.55\,\text{m}$$

But head loss $h_f = \frac{\Delta P}{\gamma} \Rightarrow \Delta P = h_f * \gamma = 8.55*0.75*1000*9.81 = 62.9\,\text{kPa}$

According to Hagen Poiseulle law, we have

$$(P_1 - P_2) = \frac{32\mu u_{AV} L}{D^2} \quad \Rightarrow \mu = 0.0226\,\text{Ns/m}^2$$

b) **Reynolds number:**

$$R_e = \frac{\rho u_{AV} D}{\mu} = \frac{0.75*1000*18.14*0.25}{0.0226} = 150497 > 4000$$

And as the Reynolds number is exceeding 4000 value for a flow in a pipe and so the flow is turbulent.

Example 6.4: A plunger of 0.8 m and length 0.13 m has four holes of diameter 5/1600 m drilled through it in the direction of its length. The plunger is a close fit inside a cylinder, containing oil, such that no oil is assumed to pass between the plunger and the cylinder. If the plunger is subjected to a vertical downward force of 45 N (including its own weight) and it is assumed that the upward flow through the four small holes is laminar, determine the speed of the fall of the plunger. The viscosity of the oil is 0.2 kg/ms.

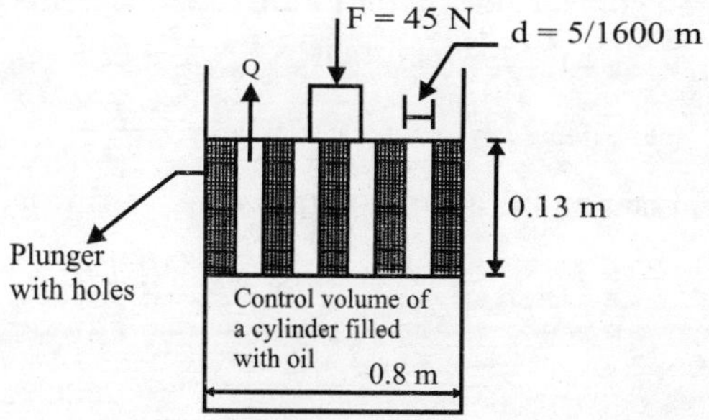

Figure P6.4

Solution:
Given data:

$$L = 0.13 \text{ m}, \; D = 0.8 \text{ m}, \; d = 5/1600 \text{ m}, \; F = 45 \text{ N}, \mu = 0.2 \text{ kg/ms}$$

Let area of the plunger hole be $a = \frac{\pi * d^2}{4}$ and area of the cylinder be $A = \frac{\pi * D^2}{4}$.

$$F = \Delta P(A - 4a) = \Delta P\left(\pi\left(\frac{0.8^2}{4}\right) - 4\pi\left(\frac{5/1600}{2}\right)^2\right) = 45\,\text{N}$$

$$\Rightarrow \Delta P = 90 \text{ Pa}$$

According to Hagen Poiseulle law, we have $Q = \frac{\Delta P * A * D^2}{32L\mu}$, but the number of holes is 4 and hence the discharge through a hole will be $Q_{hole} = Q/4 = \frac{\Delta P * A * D^2}{128L\mu} = 34.78\ \text{m}^3/\text{s}$

As we are aware discharge is equal to the product of the area and velocity of the plunger, so

$$Q = A * V_p \Rightarrow V_p = \frac{Q}{A} = 277 \text{ m/s}$$

Head loss (h_f) due to Fanning and Darcy Weisbach friction factors in circular pipe:

Fluid cylindrical element in the center line of the flow in a pipeline will be subjected to turbulent and laminar effects on either side, along its length and hence the head losses at the fluid element due to turbulent and laminar flow will be same, i.e. $h_{f_{Laminar}} = h_{f_{Turbulent}}$. But it is treated that the flow in a pipeline is turbulent due to presence of the roughness of the pipe material. In turbulence, the velocity of each fluid particles will be of no same, as like in case of laminar flow and hence the pressure force is not uniform throughout the flow by every fluid elements and hence the pressure force is called as **Propelling Force** F_{PP}, such that

$$F_{PP} = (P_1 - P_2)A_w \tag{6.18}$$

where A_w is the wetted area of the fluid elements

Similarly, the shear stres at the walls of fluid elements are not uniform and vary as a function of **velocity square on the wetted surface perimeter of the fluid elements and is called as Frictional Resistance** $F_{\tau P}$. That is, $F_{\tau P} \alpha (P_W L)V^2$, which makes

$$F_{\tau P} = f'(P_W L)V^2 \tag{6.19}$$

where P_W is the wetted perimeter and f' is a non-dimensional perimeter.

Hence, under equilibrium $F_{PP} = F_{\tau P} \quad \Rightarrow (P_1 - P_2)A_W = f'(P_W L)V^2$

By rearranging, we have
$$(P_1 - P_2) = \frac{f'(P_W L)V^2}{A_W} \tag{6.20}$$

Dividing by specific weight of the fluid γ and rearranging, we have Friction losses

$$\frac{(P_1 - P_2)}{\gamma} = \left(\frac{2gf'}{\gamma}\right)\left(\frac{P_W}{A_w}\right)\left(\frac{LV^2}{2g}\right)$$

But $\frac{D_h}{4} = \frac{A_W}{P_W} = \textit{Hydraulic diameter}$ and $\lambda = \frac{2gf'}{\gamma} = \textit{Fanning Friction factor}$

$$\Rightarrow h_f = \frac{(P_1 - P_2)}{\gamma} = \frac{2\lambda L u_{AV}^2}{gD}$$

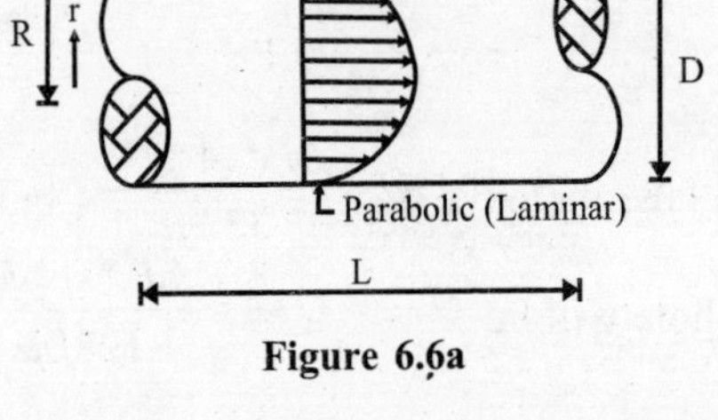

Figure 6.6a

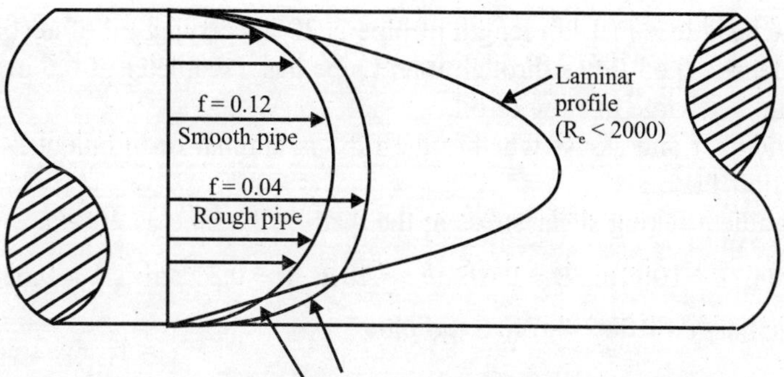

Fig. 6.6b: Velocity profile of laminar and turbulant flow.

So **head loss due to fanning friction factor** is

$$h_f = \frac{2\lambda L u_{AV}^2}{gD} \tag{6.21}$$

Darcy Weisbach friction factors $f = 4\lambda$,

Hence,

$$h_f = \frac{f L u_{AV}^2}{2gD} \tag{6.22}$$

From Hagen Poiseulle equation

$$h_f = \frac{32\mu u_{AV} L}{\gamma D^2} \tag{6.23}$$

Equate Eq. (6.23) and Eq. (6.22)

$$\Rightarrow \frac{f L u_{AV}^2}{2gD} = \frac{32\mu u_{AV} L}{\rho g D^2}$$

$$\Rightarrow \begin{pmatrix} f = 64\left(\dfrac{\mu}{\rho u_{AV} D}\right) = \left(\dfrac{64}{R_e}\right) \\ \lambda = 16\left(\dfrac{\mu}{\rho u_{AV} D}\right) = \left(\dfrac{16}{R_e}\right) \end{pmatrix} \tag{6.24}$$

where R_e is Reynolds number.

The entrance length for laminar is $\frac{l_e}{D} = 0.06 R_e$ and for turbulent $\frac{l_e}{D} = 4.4 R_e^{1/6}$

Example 6.5: If the head loss in 1 km length of pipe is 20 m carrying oil of absolute viscosity 3.0 poise and relative density 0.65 flows through it and pipe has a diameter of 0.6 m, then determine:

a) Velocity and discharge of flow of oil.
b) Reynolds number and assess whether the flow is laminar or turbulent.
c) Darcy friction factor.
d) Pressure gradient taking shear stress at the wall of the pipe as 200 Pa.

Solution: Given data: $L = 1000\,\text{m}$, $D = 0.6\,\text{m}$, $h_f = 20\,\text{m}$, $\mu = 0.3\,\text{Ns/m}^2$, $S = 0.65$

a) **Velocity and discharge** of flow through the pipe:

According to Hagen Poiseulle law, we have

$$(P_1 - P_2) = \frac{32\mu u_{AV} L}{D^2}$$

$$\Rightarrow (P_1 - P_2) = \rho_{oil} g h_f = \frac{32\mu u_{AV} L}{D^2}$$

$$\Rightarrow u_{AV} = 4.78\,\text{m/s}$$

So velocity of flow is 4.78 m/s.

Discharge will be $\Rightarrow Q = A * u_{AV} = \frac{\pi}{4} * D^2 * u_{AV} = 1.35\ \text{m}^3/\text{s}$

b) **Reynolds number:**

$$R_e = \frac{\rho_{oil} * u_{AV} * D}{\mu} = \frac{S * \rho_{water} * u_{AV} * D}{\mu} = 6214$$

Hence, the flow is turbulent.

c.) **Darcy friction factor:**

$$f = \frac{64}{R_e} = \frac{64}{6214} = 0.0103$$

d.) **Pressure gradient** taking shear stress at the wall of the pipe as 200 Pa:

We have

$$\tau_w = \left(\frac{-dp}{dx}\right)\frac{R}{2} \Rightarrow \left(\frac{-dp}{dx}\right) = \frac{2\tau_w}{R} = \frac{4 * 200}{0.6} = -1333.3\,\text{N/m}^3$$

Example 6.6: A liquid of specific gravity 2.8 and a viscosity 0.8 poise flows through a smooth pipe of unknown diameter, resulting in a pressure drop of 800 Pa in 2 km length of the pipe. What is the pipe diameter if the mass flow rate is 2500 kg/h.

Solution: Given data: $L = 2000\,\text{m}, \Delta p = 800\,\text{Pa}, \mu = 0.08\,\text{Ns/m}^2, S = 2.8,\ \bar{m} = 0.69\ \text{kg/s}$

We know mass flow rate $\overline{m} = \rho_{liq} * A_{pipe} * u_{AV}$

$$\Rightarrow u_{AV} = \frac{4*0.69}{2.8*1000*3.14*D^2} = \frac{3.16\times10^{-4}}{D^2} \quad (1)$$

According to Hagen Poiseulle law, we have

$$\left(P_1 - P_2\right) = \frac{32\mu u_{AV} L}{D^2} \quad (2)$$

Substituting Eq. (1) in Eq. (2), we get

$$\Delta P = \frac{32\mu * 3.16\times10^{-4} * L}{D^4}$$

$$\Rightarrow D = 0.212\,\text{m}$$

So **diameter of the unknown pipe** is 0.212 m.

Example 6.7: If oil has absolute viscosity of 30 poise and relative density of 0.65 flows through a 35 cm diameter and 250 m long pipe, then determine

a) What is the maximum flow that will ensure laminar flow (m^3/s)?
b) What would be the pressure drop for this flow?
c) What is the head loss?

Solution: Given data: $L = 250\,\text{m}$, $D = 0.35\,\text{m}$, $\mu = 0.3\,\text{Ns/m}^2$, $S = 0.65$

a) To ensure the laminar flow in a pipe, the Reynolds number should be equal to 2000 and hence as we know, Reynolds number

$$R_e = \frac{\rho u_{AV} D}{\mu} = \frac{S * \rho_{water} * u_{AV} * D}{\mu} \Rightarrow u_{AV} = 2.64\,\text{m/s}$$

b) Pressure drop:
According to Hagen Poiseulle law, we have

$$\left(P_1 - P_2\right) = \frac{32\mu u_{AV} L}{D^2} = 51.67\,\text{kPa}$$

c) Head loss due to pressure drop is

$$h_L = \frac{\left(P_1 - P_2\right)}{\rho_{oil} g} = \frac{\left(P_1 - P_2\right)}{S * \rho_{water} * g} = 8.1\,\text{m}$$

Example 6.8: A pipe of diameter 100 mm and length 1 km is used to pump oil of viscosity 8.5 poise and specific gravity 0.92 at the rate of 1.2 m^3/min. The first 300 m of pipe is laid along the ground sloping upwards 10° to the horizontal and the remaining pipe is laid on the ground sloping upwards at 15° to the horizontal.

a) State whether the flow is laminar or turbulent.
b) Determine pressure to be developed by the pump and the power of the driving motor if the pump has 65% efficiency.

Assume suitable data for friction factor "f", if required. **[UPSC]**

Solution: Given data:

$$L = 1000\,\text{m}, L_1 = 300\,\text{m}, L_2 = 700\,\text{m}, D = 0.1\,\text{m}, S = 0.92, \theta_1 = 10^\circ$$
$$\theta_2 = 15^\circ, \eta_P = 65\%, \mu = 0.85\,\text{Ns/m}^2, Q = 0.02\,\text{m}^3/\text{s}$$

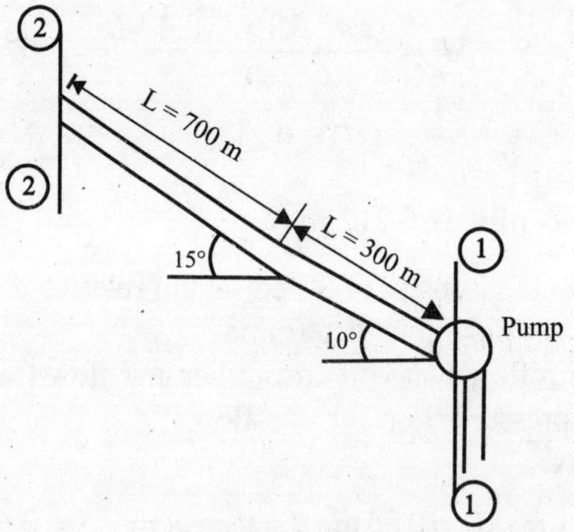

Figure P6.8

a) As per continuity equation, we have

$$Q = AV = \frac{\pi}{4}D^2 * V \Rightarrow V = \frac{4Q}{\pi D^2} = \frac{4*0.02}{3.14*0.1^2} = 2.548\text{ m/s}$$

To ensure the flow to be laminar or turbulent in the pipe, then the Reynolds number

$$R_e = \frac{\rho VD}{\mu} = \frac{S*\rho_{water}*u_2 D}{\mu}$$

$$\Rightarrow R_e = \frac{0.92*1000*2.548*0.1}{0.85} = 275.78$$

As the Reynolds number is less than 2000, the flow is treated as **laminar**.

And the friction factor will be

$$f = \frac{64}{R_e} = 0.232$$

b) Applying Bernoulli's equation at sections 1-1 and 2-2, as shown in figure, we have

$$\frac{P_1}{\gamma_{oil}}+\frac{V_1^2}{2g}+Z_1=\frac{P_2}{\gamma_{oil}}+\frac{V_2^2}{2g}+Z_2+h_L$$

$$\Rightarrow \frac{P_1}{\gamma_{oil}}+\frac{V^2}{2g}+Z_1=\frac{P_2}{\gamma_{oil}}+\frac{V^2}{2g}+Z_2+h_{f1}+h_{f1} \text{ since } \begin{matrix} Q \text{ is constant} \\ h_L=h_{f1}+h_{f2} \end{matrix}$$

$$\Rightarrow \frac{P_1}{\gamma_{oil}}+0=0+\left[(L_1\sin 10°)+(L_2\sin 15°)\right]+\frac{fL_1V^2}{2gD}+\frac{fL_2V^2}{2gD}$$

$$\Rightarrow P_1=9033 \text{ kPa}$$

Hence, the pressure at the pump is 9033 kPa.

The power required by the pump is

$$Power=\gamma QH_P/\eta_P=\gamma QP/\gamma\eta_P=QP/\eta_P$$

$$\Rightarrow Power=QP/\eta_P=0.02*9033*1000/0.65=278 \text{ kW}$$

Hence, the **power required by the pump** is 266 kW.

Example 6.9: What would be the power required by a pump having overall efficiency of 80% to pump an oil having specific gravity of 0.9 and absolute viscosity of 5 poise, from a sump and is to be lifted to a height of 10 m at free end with discharge rate at 0.05 m^3/s, with the help of a pipe of diameter 0.3m and length 100 m. Take frictional losses into account and assume friction coefficient as 0.05.

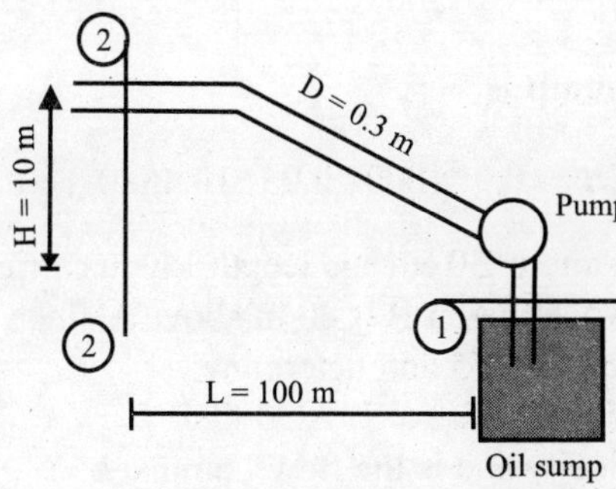

Figure P6.9

Solution: Given data:

$$L=100\,\text{m},\ D=0.3\,\text{m},\ S=0.9,\ f=0.05$$

$$Q=0.05\,\text{m}^3/\text{s}, \eta_o=80\%, H=10\,\text{m}, \mu=0.5\,\text{Ns/m}^2$$

As per continuity equation, we have

$$Q = AV_2 = \frac{\pi}{4}D^2 * V_2 \Rightarrow V_2 = \frac{4Q}{\pi D^2} = \frac{4*0.05}{3.14*0.3^2} = 0.707\,\text{m/s}$$

Applying Bernoulli's equation, we have

$$\frac{P_1}{\gamma_{oil}} + \frac{V_1^2}{2g} + Z_1 + H_P = \frac{P_2}{\gamma_{oil}} + \frac{V_2^2}{2g} + Z_2 + h_L$$

$$0 + 0 + 0 + H_P = 0 + \frac{0.707^2}{2*9.81} + 10 + h_L \qquad (1)$$

Head loss due to friction is

$$h_L = \frac{fLV_2^2}{2gD} = \frac{0.05*100*0.707^2}{2*9.81*0.3} = 0.425\,\text{m}$$

So Eq. (1) becomes $H_P = \dfrac{0.707^2}{2*9.81} + 10 + 0.425 \Rightarrow H_P = 10.45\,\text{m}$

Hence, the energy added to the pump is 10.45 m.

To ensure the flow to be in the pipe, then the Reynolds number should be less than or equal to 2000 and hence as we know, Reynolds number

$$R_e = \frac{\rho u_2 D}{\mu} = \frac{S * \rho_{water} * u_2 * D}{\mu} = 382$$

Hence, the flow is **Laminar**.

The **power required by the pump** is

$$Power = \gamma Q H_P / \eta_o = 0.9*1000*0.05*10.45/0.8 = 0.587\,\text{kW}$$

Example 6.10: A pipeline AB of diameter 30 cm and length 140 m carries oil at the rate of 50 lit/sec. The flow takes place from A to B where point B is 30 m above A. Take f = 0.08, absolute viscosity as 3 poise and specific gravity of oil as 0.65 and determine:

a) Find the pressure at A if the pressure at B is 25 kPa.
b) What is the center line velocity and is the flow Laminar?
c) What is the power of the pump required for pumping oil, assuming overall efficiency as 80%?
d) Shear stress at the boundary?

Solution: Given data:

a) Pressure at A if the pressure at B is 25 kPa:
Applying Bernoulli's equation, we have

$$L = 140\,\text{m},\ D = 0.3\,\text{m},\ \mu = 0.3\,\text{Ns/m}^2,\ S = 0.65$$

$$f = 0.08,\ Z_B = 30\,\text{m},\ Q = 0.005\,\text{m}^3/\text{s},\ \eta_o = 80\%$$

$$\Rightarrow \frac{P_A}{S*\rho_{water}g} + 0 = \frac{P_B}{S*\rho_{water}g} + 30 + h_L \text{ since } Q \text{ is constant} \quad (1)$$

As per continuity equation, we have $Q = AV_B = \frac{\pi}{4}D^2 * V_B$

$$\Rightarrow V_B = \frac{4Q}{\pi D^2} = \frac{4*0.005}{3.14*0.3^2} = 0.07\,\text{m/s}$$

So head loss due to friction is

$$h_L = \frac{fLV_B^2}{2gD} = \frac{0.08*140*0.07^2}{2*9.81*0.3} = 0.0093\,\text{m}$$

So from Eq. (1) the pressure at section A is 216.69 kPa.

b) Center line velocity and is the flow Laminar:

As per continuity equation, we have

$$Q = AV_B = \frac{\pi}{4}D^2 * V_B$$

$$\Rightarrow V_B = u_{AV} = \frac{4Q}{\pi D^2} = \frac{4*0.005}{3.14*0.3^2} = 0.07\,\text{m/s}$$

So the average velocity is 0.07 m/s.

To ensure the flow to be in the pipe, then the Reynolds number should be less than or equal to 2000 and hence as we know, Reynolds number

$$R_e = \frac{\rho u_{AV} D}{\mu} = \frac{S*\rho_{water}*u_{AV}*D}{\mu} = 45.5$$

Hence, the flow is **Laminar**.

c) Power required by the pump is

$$Power = Q*\Delta P/\eta_o = 0.005*(216.49 - 25)/0.8 = 1.196\,\text{kW}$$

d) Shear stress at the boundary:

$$\tau_w = \left(\frac{-dp}{dx}\right)\frac{R}{2} \Rightarrow \tau_w = -\left(\frac{25 - 216.49}{140}\right) * \frac{0.15}{2} = 0.10\,\text{Pa}$$

TYPES OF LOSSES

When a liquid flows from one point to another point, it comes across different types of losses depending on the shape and geometry of the container through which it is traveling in. Hence, a real fluid flow is to take care of all the head losses a rising while in motion like losses due to friction, Eddies, curved bends, etc. And the major head loss is due to friction and remaining other losses can be treated as minor.

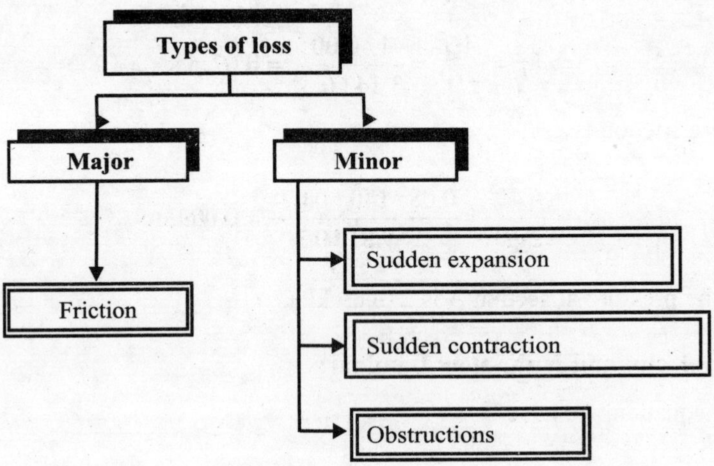

Figure 6.7: Types of lossess

Forces and Head Loss (h_E) due to Sudden Expansion of Pipe

Consider a pipe with diameter (*D*) having a fluid flow through it and is suddenly expanded in diameter to (*D*) as shown in Figure 6.8. When the liquid is passing from small diameter to larger diameter, it decelerates and attains the larger diameter after passing through some length of larger diameter. Hence, there will be an area $(A_2 - A_1)$ in between the sudden enlargement point to the section where it attains the larger diameter. In this area, there will be some losses like Eddie flow occur, which reduces the pressure head and is considered head loss due to sudden expansion.

Apply Bernoulli's equation at section 1 – 1 and 2 – 2

$$\frac{p_1}{\gamma_1} + \frac{1}{2g}V_1^2 + Z_1 = \frac{p_2}{\gamma_2} + \frac{1}{2g}V_2^2 + Z_2 + h_E$$

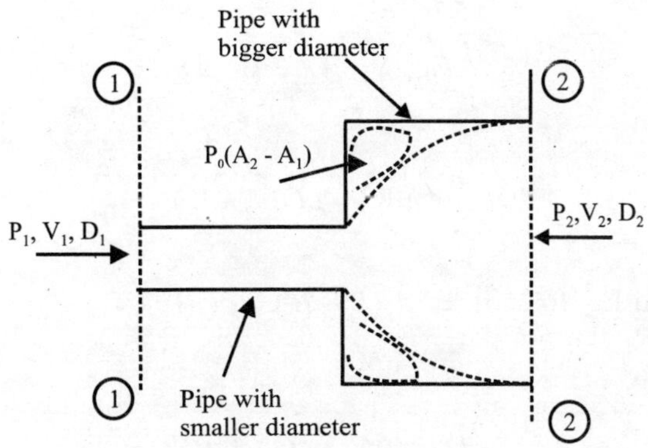

Figure 6.8: Sudden expansion

By rearranging, we get

$$\Rightarrow h_E = \left(\frac{P_1 - P_2}{\gamma}\right) + \frac{1}{2g}\left(V_2^2 - V_1^2\right) \tag{6.25}$$

since $Z_1 = Z_2$

According to the conservation of momentum, the total net force is equal to the rate of change of momentums, i.e.

$$F = \frac{d}{dt}(mV) \tag{6.26}$$

Under equilibrium state, there will be three components of forces like pressure forces (F_{p1}) at inlet, (F_{p2}) at outlet and at Eddie area (F_{p0}) along the flow direction. And hence net force $F = F_{p1} + F_{p0} - F_{p2}$

$$\Rightarrow F = P_1 A_1 + P_0 (A_2 - A_1) - P_2 A_2$$

If $P_1 = P_0$ then $$F = A_2 (P_1 - P_2) \tag{6.26a}$$

$\frac{d}{dt}(mV)$ at inlet and outlet are $\bar{M}_1 = \rho_1 A_1 V_1^2$ and $\bar{M}_2 = \rho_2 A_2 V_2^2$ respectively (since $m = \rho AV$) and

hence **total change in momentum** is $\frac{d}{dt}(mV) = (\bar{M}_2 - \bar{M}_1)$.

$$\Rightarrow \frac{d}{dt}(mV) = (\rho_2 A_2 V_2^2 - \rho_1 A_1 V_1^2)$$

$$\Rightarrow \frac{d}{dt}(mV) = \rho A_2 V_2 (V_2 - V_1)$$

$$\Rightarrow \frac{d}{dt}(mV) = \rho Q (V_2 - V_1) \tag{6.26b}$$

So equate Eq. (6.26a) and Eq. (6.26b) $\Rightarrow A_2(P_1 - P_2) = \rho Q(V_2 - V_1)$

Dividing both sides by ρg,

$$\Rightarrow \frac{(P_1 - P_2)}{\rho g} = \frac{(V_2^2 - V_2 V_1)}{g} \tag{6.27}$$

Substitute Eq. (6.27) in Eq. (6.25), we get

$$h_E = \frac{(V_2^2 - V_2 V_1)}{g} + \frac{1}{2g}\left(V_2^2 - V_1^2\right)$$

$$\boxed{h_E = \frac{(V_1 - V_2)^2}{2g}} \tag{6.28}$$

This is the **head loss due to sudden expansion**.

Example 6.11: If the hydraulic gradient rise is 20 mm in a sudden enlarged water main pipe from 300 mm to 500 mm diameter, then calculate the rate of flow.

Solution: Given data:

$$d_1 = 0.3\,\text{m}, d_2 = 0.5\,\text{m}$$

Hydraulic gradient $\left\{\left(\frac{P_2}{\gamma} + z_2\right) - \left(\frac{P_2}{\gamma} + z_2\right) = 0.02\,\text{m}\right\}$

Area of smaller cross-section pipe $A_1 = \frac{\pi * d_1^2}{4} = \frac{\pi * 0.3^2}{4} = 0.07065\,\text{m}^2$

Area of bigger cross-section pipe $A_2 = \frac{\pi * d_2^2}{4} = \frac{\pi * 0.5^2}{4} = 0.196\,\text{m}^2$

Applying Bernoulli's equation, we have $\frac{p_1}{\gamma_1} + \frac{1}{2g}V_1^2 + Z_1 = \frac{p_2}{\gamma_2} + \frac{1}{2g}V_2^2 + Z_2 + h_E$

h is head loss due to sudden enlargement $\Rightarrow h_E = \left(\frac{P_1 - P_2}{\gamma}\right) + \frac{1}{2g}\left(V_2^2 - V_1^2\right)$

And we have $h_E = \frac{(V_1 - V_2)^2}{2g}$

As per continuity equation, we have $V_1 = \frac{A_2 V_2}{A_1} = \frac{0.196V_2}{0.07065} = 2.77V_2$

Hence, h becomes $h_E = \frac{(V_1 - V_2)^2}{2g} = \frac{3.15V_2^2}{2g}$

And hence Bernoulli's equation becomes

$$\left\{\left(\frac{p_2}{\gamma} + z_2\right) - \left(\frac{p_1}{\gamma} + z_1\right)\right\} = \frac{7.67}{2g}V_2^2 - \frac{1}{2g}V_2^2 - \frac{3.15V_2^2}{2g}$$

$$\Rightarrow V_2 = 0.333\,\text{m/s}$$

So rate of flow will be $\Rightarrow Q = A_2V_2 = 0.196 * 0.333 = 0.0654\,\text{m}^3/\text{s}$.

Example 6.12: For sudden expansion in a pipe flow, work out the optimum ratio between the diameter of the pipe before expansion and the diameter of the pipe after expansion, so that the pressure rise is maximum? Also find the maximum pressure rise. **[UPSC]**

Solution: Given data: $d_1, d_2, z_1 = z_2$

Applying Bernoulli's equation, we have

$$\frac{p_1}{\gamma_1} + \frac{1}{2g}V_1^2 = \frac{p_2}{\gamma_2} + \frac{1}{2g}V_2^2 + h_E$$

And we have

$$h_E = \frac{(V_1 - V_2)^2}{2g}$$

So Bernoulli's equation becomes

$$\left\{\frac{p_2}{\gamma} - \frac{p_1}{\gamma}\right\} = \frac{1}{2g}V_1^2 - \frac{1}{2g}V_2^2 - \frac{(V_1 - V_2)^2}{2g}$$

$$V_1 = \frac{A_2V_2}{A_1} = \frac{d_2^2V_2}{d_1^2}$$

As per continuity equation, we have

$$\Rightarrow V_2 = \left(\frac{d_1}{d_2}\right)^2 V_1$$

So
$$\Rightarrow \left\{\frac{p_2}{\gamma} - \frac{p_1}{\gamma}\right\} = \frac{\Delta P}{\gamma} = \frac{1}{2g}V_1^2 \left\{1 - \left(\frac{d_1}{d_2}\right)^2 - \left[1 - \left(\frac{d_1}{d_2}\right)^2\right]^2\right\}$$

To get the optimum pressure ratio, we need to differentiate the above equation with diameter ratio and equate to zero and hence the equation becomes

$$\frac{d}{d\left(d_1/d_2\right)}\Delta P = 0$$

$$\Rightarrow \frac{d}{d\left(d_1/d_2\right)}\Delta P = \frac{d}{d\left(d_1/d_2\right)}\left[\frac{\gamma}{2g}V_1^2\right]\left\{1 - \left(\frac{d_1}{d_2}\right)^2 - \left[1 - \left(\frac{d_1}{d_2}\right)^2\right]^2\right\} = 0$$

$$\Rightarrow \left(\frac{d_1}{d_2}\right) = \frac{1}{\sqrt{2}}$$

So for optimum pressure ratio the diameter must be in ratio of

$$\boxed{\left(\frac{d_1}{d_2}\right) = \frac{1}{\sqrt{2}}}$$

Maximum pressure rise occurs at the diameter ratio and hence the maximum pressure ratio will be

$$\Rightarrow \Delta P_{\max} = \frac{\gamma}{2g}V_1^2 \left\{1 - \left(\frac{1}{\sqrt{2}}\right)^2 - \left[1 - \left(\frac{1}{\sqrt{2}}\right)^2\right]^2\right\}$$

$$\Rightarrow \Delta P_{\max} = \frac{\gamma}{4g}V_1^2$$

So optimum pressure ratio will be

$$\boxed{\Delta P_{\max} = \frac{\gamma}{4g}V_1^2}$$

Example 6.13: In a pipe of 80 mm diameter an oil of specific gravity 0.8 is flowing at the rate of 0.0125 m³/s. A sudden enlargement takes place into a second pipeline of such diameter that maximum pressure rise is obtained. Find

a) Loss of energy in sudden enlargement.
b) Differential gage length indicated by an oil-mercury connected between the two pipes.

[UPSC]

Solution: Given data:

$$d_1 = 0.08\,\text{m},\ S_{oil} = 0.8,\ Q = 0.0125\ \text{m}^3/\text{s}$$

Area of smaller cross-section pipe $A_1 = \frac{\pi * d_1^2}{4} = \frac{\pi * 0.08^2}{4} = 0.00502\,\text{m}^2$

Velocity in the smaller pipe will be $V_1 = \frac{Q}{A_1} = \frac{0.0125}{0.00502} = 2.49\,\text{m/s}$

The pressure rise will be maximum when $\frac{d_1}{d_2} = \frac{1}{\sqrt{2}}$

Then diameter of bigger pipe will be $d_2 = \sqrt{2} * d_1 = \sqrt{2} * 0.08 = 0.113\,\text{m}$

Area of bigger cross-section pipe $A_2 = \frac{\pi * d_2^2}{4} = \frac{\pi * 0.113^2}{4} = 0.01\,\text{m}^2$

Velocity in the bigger pipe will be $V_2 = \frac{Q}{A_2} = \frac{0.0125}{0.01} = 1.24\,\text{m/s}$

And we have $h_E = \frac{(V_1 - V_2)^2}{2g} = \frac{(2.49 - 1.24)^2}{2 * 9.81} = 0.079\,\text{m}$

Hence, the loss of head due to sudden expansion is 0.079 m of oil.

(b) Applying Bernoulli's equation, we have

$$\Rightarrow \left(\frac{p_2}{\gamma} - \frac{p_1}{\gamma}\right) = \frac{1}{2g}V_1^2 - \frac{1}{2g}V_2^2 - h_E$$

But when a differential gage is attached then the pressure head becomes

$$\left(\frac{p_2}{\gamma} - \frac{p_1}{\gamma}\right) = h\left(\frac{S_{mercury}}{S_{oil}} - 1\right) \text{ refer chatper 5}$$

$$\left(\frac{p_2}{\gamma}-\frac{p_1}{\gamma}\right)=h\left(\frac{13.6}{0.8}-1\right)=16h$$

so $$16*h=\frac{1}{2g}V_1^2-\frac{1}{2g}V_2^2-h_E$$

$$\Rightarrow h=9.875\,\text{mm}$$

Hence, **differential gage length** indicated by an oil-mercury connected between the two pipes is 9.875 mm.

Forces and Head Loss (h_c) due to Sudden Contraction of Pipe

Consider a pipe with diameter *(D)* having a fluid flow through it and is suddenly contraction in diameter to *(D)*, as shown in Figure 6.9.

When the liquid is passing from larger diameter to smaller diameter there exists a region called **Vena-contraction** of diameter *(D)* before attaining the diameter of smaller pipe, as seen in case of orifice-meter, is shown in figure. Hence, there exists a coefficient of contraction C,

$$C_C=\frac{A_0}{A_2} \tag{6.29}$$

Apply Bernoulli's equation at section 1-1 and 2-2

i.e. $$\frac{p_1}{\gamma_1}+\frac{1}{2g}V_1^2+Z_1=\frac{p_2}{\gamma_2}+\frac{1}{2g}V_2^2+Z_2+h_C$$

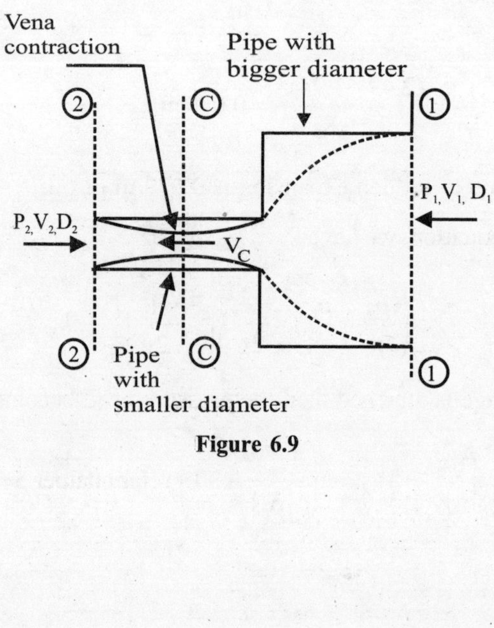

Figure 6.9

The total head loss h_C due to contraction comprises head loss due to contraction h_{C1} between section 1-1 and $c-c$ and head loss due to sudden expansion h_E between section $c-c$ and 2-2. So $h_C = h_{C1} + h_E$.

But the head loss h_{C1} is relatively small in quantity and hence it can be neglected.

$$\Rightarrow h_C = h_E = \frac{(V_0 - V_2)^2}{2g} \tag{6.30}$$

$$\left(\text{since } h_E = \frac{(V_0 - V_2)^2}{2g}\right)$$

As per the continuity equation

$$V_0 = \frac{A_2 V_2}{A_0} = \frac{V_2}{C_C} \tag{6.30a}$$

Substitute Eq. (6.30a) in Eq. (6.30), we have

$$h_C = \frac{\left(\frac{V_2}{C_C} - V_2\right)^2}{2g} \Rightarrow h_C = \frac{V_2^2}{2g}\left(\frac{1}{C_C} - 1\right)^2$$

$$h_C = K\frac{(V_2)^2}{2g} \tag{6.31}$$

where $K = \left(\frac{1}{C_C} - 1\right)^2$ is loss coefficient.

This is the **head Loss due to sudden contraction.**
The approximate value of K is 0.5, hence

$$h_C = 0.5\frac{(V_2)^2}{2g} \tag{6.31a}$$

Example 6.14: In a pipe of 200 mm diameter an oil of specific gravity 0.8 is flowing at the rate of 0.04 m^3/s. A sudden contraction of pipe has taken place thereby the diameter reduced to 120 mm. Find:

a) Loss of energy in sudden contraction.
b) Coefficient of contraction if differential gauge length indicated by an oil-mercury connected between the two pipes is 40 mm.

Solution: Given data: $d_1 = 0.2\,\text{m},\ d_2 = 0.12\,\text{m},\ S_{oil} = 0.8, Q = 0.04\,\text{m}^3/\text{s},\ h = 40$ mm

Area of bigger cross-section pipe $A_1 = \frac{\pi * d_1^2}{4} = \frac{\pi * 0.2^2}{4} = 0.0314\,\text{m}^2$

Velocity in the bigger pipe will be $V_1 = \dfrac{Q}{A_1} = \dfrac{0.04}{0.0314} = 1.27\,\text{m/s}$

Area of bigger cross-section pipe $A_2 = \dfrac{\pi * d_2^{\,2}}{4} = \dfrac{\pi * 0.12^2}{4} = 0.0113\,\text{m}^2$

Velocity in the bigger pipe will be $V_2 = \dfrac{Q}{A_2} = \dfrac{0.04}{0.0113} = 3.54\,\text{m/s}$

a) Referring Eq. (6.31a), we have $h_C = \dfrac{0.5(V_2)^2}{2g} = \dfrac{0.5 * (3.54)^2}{2 * 9.81} = 0.319\,\text{m}$

Hence, the loss of head due to sudden contraction is 0.319 m of oil.

b) Applying Bernoulli's equation, we have

$$\left(\frac{p_1}{\gamma} - \frac{p_2}{\gamma}\right) = -\frac{1}{2g}V_1^2 + \frac{1}{2g}V_2^{\,2} + h_C$$

But when a differential gage is attached then the pressure head becomes

$$\left(\frac{p_1}{\gamma} - \frac{p_2}{\gamma}\right) = h\left(\frac{S_{mercury}}{S_{oil}} - 1\right) \quad \textit{refer Chapter 5}$$

$$\left(\frac{p_1}{\gamma} - \frac{p_2}{\gamma}\right) = 0.04 * \left(\frac{13.6}{0.8} - 1\right) = 0.64$$

Referring Eq. (6.31), we have $h_C = \left[\dfrac{1}{C_c} - 1\right]^2 \dfrac{(V_2)^2}{2g}$

So $$0.64 = -\frac{1}{2g}V_1^2 + \frac{1}{2g}V_2^{\,2} + \left[\frac{1}{C_c} - 1\right]^2 \frac{(V_2)^2}{2g}$$

$$\Rightarrow \left[\frac{1}{C_c} - 1\right]^2 = 0.01302$$

$$\Rightarrow C_c = 0.736$$

Hence, coefficient of contraction at differential gage length of 40 mm is 0.736.

Example 6.15: In a pipe of 200 mm diameter pipe carrying water at the rate of 0.04 m³/s. A sudden contraction of pipe has taken place thereby the diameter reduced to 120 mm. Find the pressure change in the pipes due to sudden contraction, if the coefficient of contraction is taken as 0.736.

Solution: Given data: $d_1 = 0.2\,\text{m}$, $d_2 = 0.12\,\text{m}$, $Q = 0.04\,\text{m}^3/\text{s}$, $C_C = 0.736$

Area of bigger cross-section pipe $A_1 = \dfrac{\pi * d_1^2}{4} = \dfrac{\pi * 0.2^2}{4} = 0.0314\,\text{m}^2$

Velocity in the bigger pipe will be $V_1 = \dfrac{Q}{A_1} = \dfrac{0.04}{0.0314} = 1.27\,\text{m/s}$

Area of bigger cross-section pipe $A_2 = \dfrac{\pi * d_2^2}{4} = \dfrac{\pi * 0.12^2}{4} = 0.0113\,\text{m}^2$

Velocity in the bigger pipe will be $V_2 = \dfrac{Q}{A_2} = \dfrac{0.04}{0.0113} = 3.54\,\text{m/s}$

We have $h_C = \left[\dfrac{1}{C_c} - 1\right]^2 \dfrac{(V_2)^2}{2g} = 0.082\ \text{m}$

Applying Bernoulli's equation, we have

$$\left(\frac{p_2}{\gamma} - \frac{p_1}{\gamma}\right) = \frac{1}{2g}V_1^2 - \frac{1}{2g}V_2^2 - h_C$$

$$\Rightarrow \Delta P = 9.81 * \left(\frac{1}{2*9.81}1.27^2 - \frac{1}{2*9.81}*3.54^2 - 0.082\right) = -6.27\,\text{kPa}$$

Hence, the negative sign indicates that the pressure change is a loss and is of 6.27 kPa in value.

Example 6.16: If the coefficient of contraction is assumed to be 0.65, for a pipeline having a sudden contraction in diameter from 300 mm diameter to 150 mm and a pressure change of 125 kPa to 70 kPa. Calculate the water flow rate.

Following this if there is sudden enlargement from 150 mm diameter to 300 m diameter and if the pressure at the 150 mm diameter section is 70 kPa, what is the pressure at the 300 m section enlarged portion?

Solution: Given data: $d_1 = 0.3\,\text{m}$, $d_2 = 0.15\,\text{m}$, $P_1 = 125\,\text{kPa}$, $P_2 = 70\,\text{kPa}$, $C_C = 0.65$, $P_3 = 70$ kPa

Area of bigger cross-section pipe $A_1 = \dfrac{\pi * d_1^2}{4} = \dfrac{\pi * 0.3^2}{4} = 0.07065\,\text{m}^2$

Area of smaller cross-section pipe $A_2 = \dfrac{\pi * d_2^{\,2}}{4} = \dfrac{\pi * 0.15^2}{4} = 0.0177\,\text{m}^2$

a) Head loss due to sudden contraction is $h_C = \left[\dfrac{1}{C_c} - 1\right]^2 \dfrac{(V_2)^2}{2g}$

$$h_C = \left[\frac{1}{0.65} - 1\right]^2 \frac{(V_2)^2}{2 * 9.81} = 0.0147 V_2^{\,2}$$

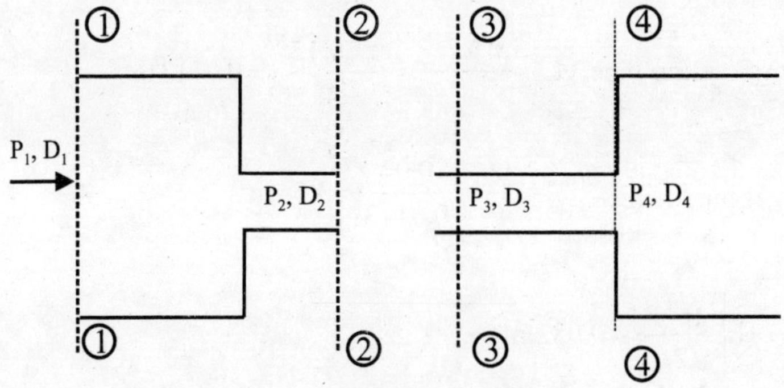

Figure P6.16

As per continuity equation, we have $V_1 = \dfrac{A_2 V_2}{A_1} = 0.25 V_2$

Applying Bernoulli's equation, we have

$$\left(\frac{p_2}{\gamma} - \frac{p_1}{\gamma}\right) = \frac{1}{2g} V_1^2 - \frac{1}{2g} V_2^{\,2} - h_C$$

$$\Rightarrow V_2 = 12.4 \text{ m/s}$$

Hence, the rate of flow $Q = A_2 V_2 = 0.219$ m^3/s

b) Applying Bernoulli's equation in the second portion, we have

$$\frac{p_3}{\gamma_3} + \frac{1}{2g} V_3^{\,2} + Z_3 = \frac{p_4}{\gamma_4} + \frac{1}{2g} V_4^{\,2} + Z_4 + h_E$$

As known $$V_3 = V_2, V_4 = V_1 = \frac{V_2}{4}$$

$$\Rightarrow P_4 = 41.1\text{kPa}$$

Hence, pressure at the 300 mm enlarged pipe will be 41.1 kPa.

Forces Due to Pipe Bend

In general, the pipes may or may not be straight and it may have bends either. When it has a bend, the fluid flowing through the bend will experience some force exerted by the walls of the pipe, while change in flow direction takes place, i.e. the force exerted by the walls of pipe along the direction of flow may be along x-axis and y-axis, and hence the forces acting on the fluid flow will have force components along x-axis and y-axis.

Consider two sections 1-1 and 2-2, one at inlet and other at outlet of the pipe bend, having pressures, diameter and velocity P_1, V_1, D_1 and P_2, V_2, D_2 respectively.

The pressure forces at the inlet and outlet sections are F_{p1}, F_{p2} respectively. Where

$$F_{p1} = P_1 * A_1$$

$$F_{p2} = P_2 * A_2 \qquad (6.32)$$

When the two sections are displaced to new positions 3-3 and 4-4 due to inertial force, as shown in Figure 6.9, there will be displacement ds_1, ds_2 respectively.
where

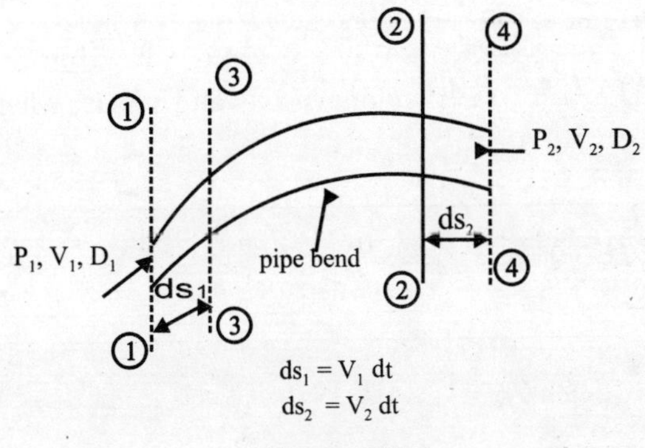

Figure 6.10

$$ds_1 = V_1 dt \qquad (6.32a)$$

$$ds_2 = V_2 dt \qquad (6.32b)$$

According to Newton's second law, $F_i = \frac{d}{dt}(mV)$

$$\Rightarrow F_i dt = d(mV) \qquad (6.33)$$

Change in momentum at inlet and outlet are $M_1 = m_1 * V_1$ and $M_2 = m_2 * V_2$ respectively.

$$M_1 = (\rho_1 A_1 ds_1) * V_1 = (\rho_1 A_1 V_1 dt) * V_1 \tag{6.34a}$$

$$M_2 = (\rho_2 A_2 ds_2) * V_2 = (\rho_2 A_2 V_2 dt) * V_2 \tag{6.34b}$$

Hence, total change in momentum is $d(mV) = (M_2 - M_1)$

$$\Rightarrow d(mV) = \rho Q(V_2 - V_1)dt \tag{6.34c}$$

Substituting Eq. (6.34) in Eq. (6.33) $F_i dt = \rho Q(V_2 - V_1)dt$

$$\Rightarrow F_i = \frac{\gamma Q}{g}(V_2 - V_1) \tag{6.35}$$

where $\frac{\gamma Q}{g}$ is mass flux.

The three forces (F_{p1}, F_{p2}, F_i) acting on the fluid flow, in a bend pipe can be resolved along the x-axis and y-axis, i.e. the velocity components along x-axis and y-axis will be $V_1 = (V_1 \cos\theta_{1x})\left(\frac{\pi}{2} - \theta\right)$ $V_2 = (V_2 \cos\theta_{2x})$ and $V_1 = (V_1 \sin\theta_{1y})$ $V_2 = (V_2 \sin\theta_{2y})$ respectively.

Net force $F = \left(\frac{\gamma Q}{g}(V_2 - V_1) + P_1 A_1 - P_2 A_2\right)$ along x-axis and y-axis, where $F = (F_i + F_{p1} - F_{p2})$

We get

$$F_X = \frac{\gamma Q}{g}(V_2 \cos\theta_{2x} - V_1 \cos\theta_{1x}) + p_1 A_1 \cos\theta_{1x} - p_2 A_2 \cos\theta_{2x} \tag{6.36a}$$

$$F_Y = \frac{\gamma Q}{g}(V_2 \sin\theta_{2y} - V_1 \sin\theta_{1y}) + p_1 A_1 \sin\theta_{1y} - p_2 A_2 \sin\theta_{2y} \tag{6.36b}$$

Hence, resultant force

$$F_R = \sqrt{F_X^2 + F_Y^2} \tag{6.36c}$$

and inclination of F_R is $\theta = Tan^{-1}\left(\frac{F_Y}{F_X}\right)$.

Example 6.17: The discharge of water through a 40° bend, shown in Figure, is 30 liters/s. The bend is lying in the horizontal plane and the diameters at the entrance and exit are 200 mm and 100

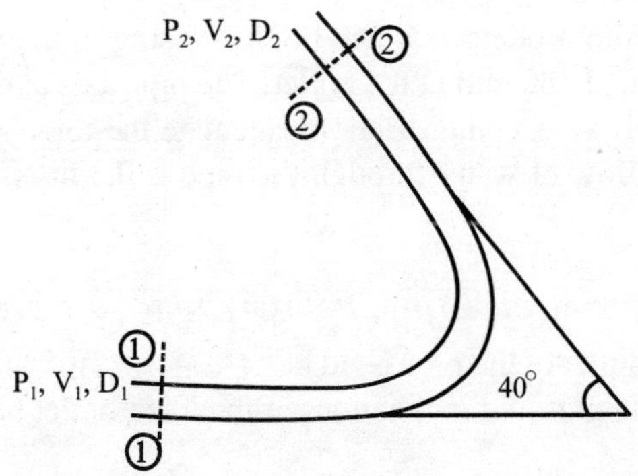

Figure P6.17: Pipe bend

mm respectively. The pressure measured at the entrance is 100 kPa, what is the magnitude and direction of the force exerted by the water on the bend?

Solution: Given data: $d_1 = 0.2\,\text{m}$, $d_2 = 0.1\,\text{m}$, $P_1 = 100\,\text{KN/m}^2$, $\gamma = 9810\,\text{N/m}^3$, $Q = 30\,\text{lit/s}$

Let the areas at the inlet and outlet of the pipe bend be $A_1 = 0.0314\ \text{m}^2, A_2 = 0.00785\ \text{m}^2$ respectively.

Let the velocity, datum heights and inclinations at the inlet, outlet be

$$V_1 = Q/A_1 = \frac{0.030}{0.0314} = 0.955\,\text{m/s} \qquad Z_1 = Z_2$$

$$\&$$

$$V_2 = Q/A_2 = \frac{0.030}{0.00785} = 3.82\,\text{m/s} \qquad \theta_1 = 0, \theta_2 = 40$$

respectively.

Applying Bernoulli's equation,

we have
$$\frac{P_2 - P_1}{\rho g} = \frac{\left(V_1^2 - V_2^2\right)}{2g} + (Z_1 - Z_2)$$

So substituting the values, we have $P_2 = 93160$ Pa

$$F_X = \frac{\gamma Q}{g}(V_2 \cos\theta_{2x} - V_1 \cos\theta_{1x}) + p_1 A_1 \cos\theta_{1x} - p_2 A_2 \cos\theta_{2x} = 2638.9\ \text{N}$$

$$F_Y = \frac{\gamma Q}{g}(V_2 \sin\theta_{2y} - V_1 \sin\theta_{1y}) + p_1 A_1 \sin\theta_{1y} - p_2 A_2 \sin\theta_{2y} = -396.4\ \text{N}$$

The resultant force $F_R = \sqrt{F_X^2 + F_Y^2}$ and so $F_R = \sqrt{(2638.9)^2 + (396.4)^2} = 2668.5\,\text{N}$

And the inclination is
$$\theta = Tan^{-1}\left(\frac{F_Y}{F_X}\right) = Tan^{-1}\left(\frac{-396.4}{2638.9}\right) = -8.54°$$

Example 6.18: The outlet from a pump is a bend of 45° rising in the vertical plane. The bend is 150 mm diameter at its inlet and 300 mm at the outlet. The pipe axis at the inlet is horizontal and at the outlet it is 1.0 m higher. By neglecting friction, calculate the force and its direction if the inlet pressure is 100 kPa and the flow of water through the pipe is 0.3 m³/s. The volume of the pipe is 0.075 m³.

Solution: Given data: $d_1 = 0.15\,\text{m},\ d_2 = 0.3\,\text{m},\ P_1 = 100\,\text{KN/m}^2,\ \gamma = 9810\,\text{N/m}^3,\ Q = 0.3\,\text{m}^3/\text{s}$

Let the areas at the inlet and outlet of the pipe bend be $A_1 = 0.0177\,\text{m}^2, A_2 = 0.0707\,\text{m}^2$ respectively.

Let the velocity, datum heights and inclinations at the inlet, outlet be

$$V_1 = Q/A_1 = \frac{0.3}{0.0177} = 16.98\,\text{m/s} \qquad Z_1 = 0, Z_2 = 1.0\,\text{m}$$

$$\&$$

$$V_2 = Q/A_2 = \frac{0.3}{0.0707} = 4.24\,\text{m/s} \qquad \theta_1 = 0, \theta_2 = 45^\circ$$

respectively.

Applying Bernoulli's equation, we have $\dfrac{P_2 - P_1}{\rho g} = \dfrac{\left(V_1^2 - V_2^2\right)}{2g} + \left(Z_1 - Z_2\right)$

So substituting the values, we have $P_2 = 225361.4$ N

$$F_X = \frac{\gamma Q}{g}(V_2 \cos\theta_{2x} - V_1 \cos\theta_{1x}) + p_1 A_1 \cos\theta_{1x} - p_2 A_2 \cos\theta_{2x} = -10990\text{ N}$$

$$F_Y = \frac{\gamma Q}{g}(V_2 \sin\theta_{2y} - V_1 \sin\theta_{1y}) + p_1 A_1 \sin\theta_{1y} - p_2 A_2 \sin\theta_{2y} = -10366.9\text{ N}$$

The resultant force $F_R = \sqrt{F_X^2 + F_Y^2}$ and so $F_R = \sqrt{(-10990)^2 + (-10366)^2} = 15108\,\text{N}$

And the inclination is $\theta = \tan^{-1}\left(\dfrac{F_Y}{F_X}\right) = \tan^{-1}\left(\dfrac{-10366.93}{-10990}\right) = 43^\circ.32'$

Example 6.19: Water flows horizontally along a 200 mm pipeline fitted with a 90° bend that moves the water vertically upwards. The diameter at the outlet of the bend is 125 mm and it is 0.8 m above the center line of the inlet. If the flow through the bend is 125 lit/s and assuming no losses due to friction, calculate the magnitude and direction of the resultant force the bend support must withstand. The volume of the bend is 0.02 m³ and the pressure at the inlet is 120 kPa.

Solution: Given $d_1 = 0.2\,\text{m},\ d_2 = 0.125\,\text{m},\ P_1 = 120\,\text{kPa},\ \gamma = 9810\,\text{N/m}^3,\ Q = 125\,\text{lit/s}$

Let the areas at the inlet and outlet of the pipe bend be $A_1 = 0.0314\text{ m}^2, A_2 = 0.01226\text{ m}^2$ respectively.

Let the velocity, datum heights and inclinations at the inlet, outlet be

$$V_1 = Q/A_1 = \frac{0.125}{0.0314} = 3.98\,\text{m/s} \qquad Z_1 = 0, Z_2 = 0.8\,\text{m}$$
&
$$V_2 = Q/A_2 = \frac{0.125}{0.0122} = 10.19\,\text{m/s} \qquad \theta_1 = 0, \theta_2 = 90^\circ \quad \text{respectively}$$

Applying Bernoulli's equation, we have $\dfrac{P_2 - P_1}{\gamma} = \dfrac{\left(V_1^2 - V_2^2\right)}{2g} + (Z_1 - Z_2)$

So substituting the values, we have $P_2 = 68.515\,\text{kPa}$

$$F_X = \frac{\gamma Q}{g}(V_2 \cos\theta_{2x} - V_1 \cos\theta_{1x}) + p_1 A_1 \cos\theta_{1x} - p_2 A_2 \cos\theta_{2x} = 3767.4\ \text{N}$$

$$F_Y = \frac{\gamma Q}{g}(V_2 \sin\theta_{2y} - V_1 \sin\theta_{1y}) + p_1 A_1 \sin\theta_{1y} - p_2 A_2 \sin\theta_{2y} = -530.18\ \text{N}$$

The resultant force $F_R = \sqrt{F_X^2 + F_Y^2}$ and so $F_R = \sqrt{(3767)^2 + (530)^2} = 3804\,\text{N}$

And the inclination is $\theta = \tan^{-1}\left(\dfrac{F_Y}{F_X}\right) = \tan^{-1}\left(\dfrac{-530.18}{3767.4}\right) = -8°.01'$

Forces and Head Losses Due to Porous Medium in a Pipe Based on Darcy Equation

Whenever a pipe is occupied by a porous material like sand or a bundle of rods, etc., which can be treated as porous (a porous material is one which is loosely packed), then the floor area for the flow of the fluid through the pipe is reduced. And hence the effective diameter of the pipe will be modified to a new diameter called hydraulic diameter, which is nothing but

$$\frac{D_h}{4} = \frac{A_W}{P_W} \tag{6.37}$$

where A_W is the wetted area by the fluid flow and P_W is the perimeter wetted by the fluid flow.

Let the volume of the pipe with no porous material, volume of the porous material in the pipe and volume of the voids in between the porous material be V, V_p, V_V respectively.

Volume of the pipe with no porous material

$$V = \left(V_V + V_p\right) \tag{6.37a}$$

then **porosity** of the porous material will be

$$\varepsilon = \frac{V_V}{V} \tag{6.38}$$

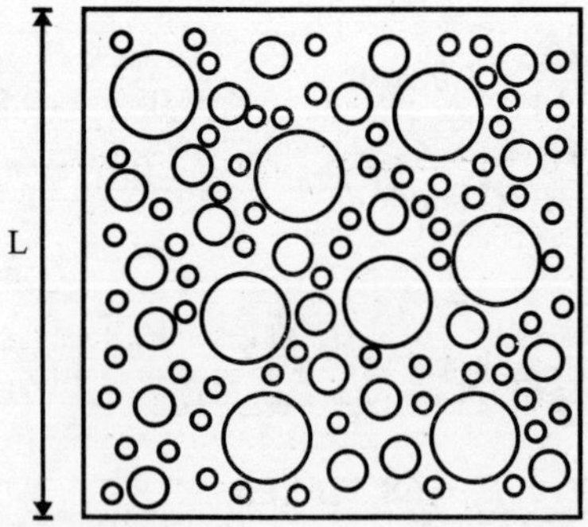

Figure 6.11: Porous material packed in a pipe

Darcy proposed a relation for flow of fluid in a porous medium as $u = -\frac{K}{\mu}\left(\frac{dp}{dx}\right)$ where K is the permeability.

$$\Rightarrow \frac{u\mu}{K} = -\left(\frac{dp}{dx}\right) \tag{6.39}$$

We know the **Fanning friction factor** $h_f = \frac{P_1 - P_2}{\rho g} = \frac{4\lambda L u_{AV}^2}{2gD}$

By rearranging $$\Rightarrow -\left(\frac{dp}{L}\right) = \frac{4\lambda \rho u_{AV}^2}{2D} \tag{6.40}$$

Equate Equation (6.39) and Eq. (6.40), i.e. $\frac{u\mu}{K} = \frac{4\lambda \rho u_{AV}^2}{2D}$

$$\Rightarrow \lambda = \frac{u\mu D}{2K\rho u_{AV}^2} \quad \Rightarrow \lambda = \frac{u\nu D}{2K u_{AV}^2} \tag{6.40a}$$

since $\nu = \frac{\mu}{\rho}$

According to continuity equation $Q = A_{Void} u_{AV} = Au$

$$\Rightarrow \frac{A_{Void}}{A} = \frac{u}{u_{Void}} = \varepsilon \quad \left(\text{since } \varepsilon = \frac{V_V}{V} = \frac{A_V * L}{A * L} = \frac{A_V}{A}\right)$$

$$\Rightarrow \lambda R_e = \frac{\varepsilon D^2}{2K} = C \tag{6.41}$$

$$\left(\text{since } R_e = \frac{u_{Av}D}{\nu}\right)$$

So the value of "*C*" is based on the size and geometry of the solid porous material.

By rearranging
$$h_f = \frac{4\lambda L u_{AV}^2}{2gD} = (\lambda)\left(\frac{\rho L u_{AV}}{\mu}\right)\left(\frac{2\mu u_{AV}}{\rho g D}\right) = h_f = \left(\frac{2\lambda R_{e\mu u_{AV}}}{\gamma D}\right) \tag{6.41a}$$

Substituting Eq. (6.41) in Eq. (6.41a), we get

$$h_f = \left(\frac{\varepsilon D \mu u_{AV}}{K\gamma}\right) \tag{6.42}$$

Eq. (6.42) is the **head loss due to porous material.**

Flow Potential and Flow Resistance

In general when a fluid stored in a reservoir at some elevation has potential energy to flow to downward elevation which can be termed **flow potential**. Due to flow potential, the liquids in reservoirs are able to flow from one place to the other place, which are at the downward elevation and while flowing there exists several kinds of losses like frictional, contraction, expansion, bends, etc. and so flow from one reservoir point to another reservoir point through a pipe(s) passage connected in between them, as shown in Figure 6.12, have their flow effected by **head loss** h_L like due to sudden contraction h_C from the tank "1" to the pipe, frictional losses h_f in the pipe and sudden expansion h_E from the pipe to tank "2".

Hence,
$$h_L = h_C + h_f + h_E \tag{6.42a}$$

Consider a tank "1" at an elevation z_A and tank "2" at an elevation z_B from the datum, connected by a pipe passage of diameter, length, velocity of flow *D, L, V* respectively, as shown in Figure 6.11.

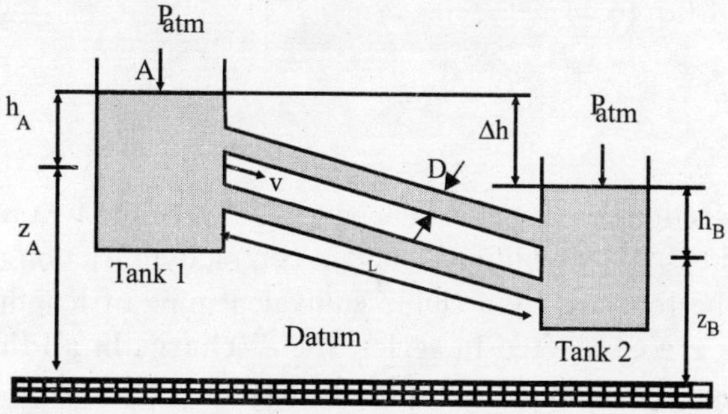

Figure 6.12: Flow potential and flow resistance.

Apply Bernoulli's equation at the points A and B at the free surfaces of the two tanks, then

$$\frac{P_{atm}}{\rho g}+\frac{V_A^2}{2g}+\left(h_A+Z_A\right)=\frac{P_{atm}}{\rho g}+\frac{V_B^2}{2g}+\left(h_B+Z_B\right)+h_L$$

where h_A, h_B is the pressure head.

$$\Rightarrow h_L=\left(h_A+Z_A\right)-\left(h_B+Z_B\right)=\Delta h \tag{6.43}$$

since $V_A = V_B = 0$

Consider Eq. (6.42), i.e.

$$h_L=\left(0.5\frac{V^2}{2g}\right)+\frac{fLV^2}{2gD}+\left(\frac{V^2}{2g}\right)$$

$$\Rightarrow h_L=\left(1.5+\frac{fL}{D}\right)\left(\frac{V^2}{2g}\right) \tag{6.43a}$$

As per continuity equation,

$$Q=aV \Rightarrow V^2={Q^2}/{a^2} \tag{6.43b}$$

where a is the area of the pipe and substituting Eq. (6.43b) in Eq. (6.43a), we have

$$h_L=RQ^2 \tag{6.44}$$

where

$$R=\left(1.5+\frac{fL}{D}\right)\left(\frac{1}{2a^2g}\right)$$

is resistance of flow.

Observing Eq. (6.43) and Eq. (6.44)

$$h_L=\Delta h=RQ^2 \tag{6.45}$$

Note

If more number of pipe sections/passages is attached in between the two reservoirs, then the head losses will be further increased, by additional sudden contraction, friction and sudden expansions, hence all the pipes can be replaced by a single equivalent pipe of length and diameter L_{eq}, D_{eq} respectively, if the **pipes are connected in series, the discharge in all the pipes are equal and so** $Q = Q_1 = Q_2 = Q_3$ using the relation of **Dupit's Equation**

$$h_{f_{eq}} = h_{f1} + h_{f2} + h_{f3} + ... \Rightarrow \frac{fL_{eq}V_{eq}^2}{2gD_{eq}} = \frac{fL_1V_1^2}{2gD_1} + \frac{fL_2V_2^2}{2gD_2} + \frac{fL_3V_3^2}{2gD_3} + ...$$

$$\Rightarrow \frac{L_{eq}}{D_{eq} * D_{eq}^4} = \frac{L_1}{D_1 * D_1^4} + \frac{L_2}{D_2 * D_2^4} + \frac{L_3}{D_3 * D_3^4} + ...$$

Since $$Q = Q_1 = Q_2 = Q_3 = \dots$$

$$h_{f_{eq}} = h_{f1} + h_{f2} + h_{f3} + .. \& \frac{L_{eq}}{D_{eq}^5} = \frac{L_1}{D_1^5} + \frac{L_2}{D_2^5} + \frac{L_3}{D_3^5} + \dots \quad (6.45a)$$

And if the more number of **pipes are connected in Parallel,** then discharge in the main pipe will be distributed equally to the branches of the pipe and hence

$$Q = Q_1 + Q_2 + Q_3 + \dots \& h_{f1} = h_{f2} = h_{f3} \cdots \quad (6.45b)$$

$$\Rightarrow h_{f_{eq}} = \frac{fL\,V_{eq}^2}{2gD_{eq}} = \frac{fL\,V_1^2}{2gD_1} = \frac{fL\,V_2^2}{2gD_2} = \frac{fL\,V_3^2}{2gD_3} + ...$$

Example 6.20: Two pipes of 25 m and 15 m length are in diameters of 0.15 m, 0.3 m respectively are connected in series to a tank at one end and the other end is left to atmosphere. The first pipe has sudden enlargement of diameter and the height of the water level in the tank is 8 m above the center of the pipe. Considering all the losses of head which occur, determine the rate of flow.

Take coefficient of friction for all the pipes as 0.01.

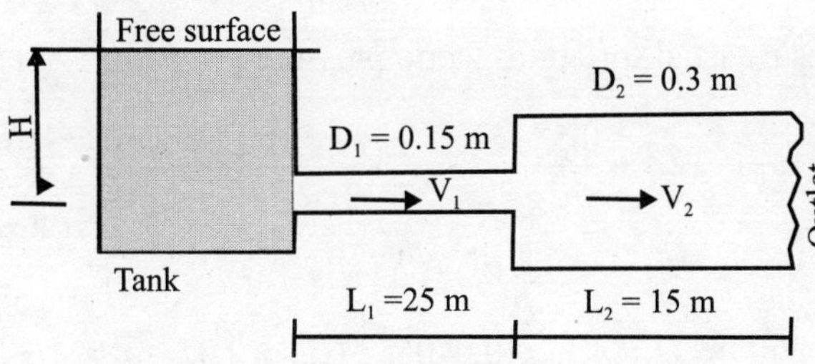

Figure P6.20

Solution:

Given data: $d_1 = 0.15\,\text{m},\ d_2 = 0.3\,\text{m},\ L_1 = 25\,\text{m},\ L_2 = 15\,\text{m},\ H = 8\,\text{m},\ f = 0.01$

Area of smaller cross-section pipe $A_1 = \dfrac{\pi * d_1^2}{4} = \dfrac{\pi * 0.15^2}{4} = 0.0177\,\text{m}^2$

Area of bigger cross-section pipe $A_2 = \dfrac{\pi * d_2^2}{4} = \dfrac{\pi * 0.3^2}{4} = 0.07065\,\text{m}^2$

Applying Bernoulli's equation, we have $\dfrac{p_1}{\gamma_1} + \dfrac{1}{2g}V_1^2 + Z_1 = \dfrac{p_2}{\gamma_2} + \dfrac{1}{2g}V_2^2 + Z_2 + h_L$

where total head losses $h_L = h_C + h_{f1} + h_E + h_{f2}$

So Bernoulli's equation becomes $\dfrac{p_1}{\gamma_1} + \dfrac{1}{2g}V_1^2 = \dfrac{p_2}{\gamma_2} + \dfrac{1}{2g}V_2^2 + h_C + h_{f1} + h_E + h_{f2}$ (1)

But we know, Head loss due to contraction is $h_C = \dfrac{0.5}{2g}V_1^2$

Head loss due to expansion is $h_E = \dfrac{(V_1 - V_2)^2}{2g}$

Head loss due to friction is $h_f = \dfrac{4fL}{2gd}V^2$

So Eq. (1) becomes $\dfrac{p_1}{\gamma_1} + \dfrac{1}{2g}V_1^2 = \dfrac{p_2}{\gamma_2} + \dfrac{1}{2g}V_2^2 + \dfrac{0.5}{2g}V_1^2 + \dfrac{4fL_1}{2gd_1}V_1^2 + \dfrac{(V_1 - V_2)^2}{2g} + \dfrac{4fL_2}{2gd_2}V_2^2$

As per continuity equation, we have $V_1 = \left(\dfrac{d_2}{d_1}\right)^2 V_2 = \left(\dfrac{0.3}{0.15}\right)^2 V_2 = 4V_2$

$$\Rightarrow \frac{p_1}{\gamma_1} + \frac{16}{2g}V_2^2 = \frac{p_2}{\gamma_2} + \frac{1}{2g}V_2^2 + \frac{0.5*16}{2g}V_2^2 + \frac{4fL_1*16}{2gd_1}V_2^2 + \frac{(4V_2 - V_2)^2}{2g} + \frac{4fL_2}{2gd_2}V_2^2$$

As there will be no pressure head for atmospheric pressure

$$(H) + \frac{16}{2g}V_2^2 = 0 + \frac{1}{2g}V_2^2 + \frac{0.5*16}{2g}V_2^2 + \frac{4fL_1*16}{2gd_1}V_2^2 + \frac{(4V_2 - V_2)^2}{2g} + \frac{4fL_2}{2gd_2}V_2^2$$

$$\Rightarrow V_2 = 1.11\,\text{m/s}$$

Hence, the **rate of flow** will be $Q = A_2V_2 = 0.07065 * 1.11 = 0.078\,\text{m}^3/\text{s}$

Example 6.21: Water is being pumped at the rate of 0.02 m^3/s to an overhead tank through a 150 mm diameter 300 m long delivery pipe. In the tank, the pipe discharges freely at height of 15 m above the pump. If the Darcy-Weisbach friction factor is equal to 0.03 for the pipe, determine:

a) The pressure developed by the pump on its delivery side
b) The power delivered to water by the pump. **[UPSC Civil Exam]**

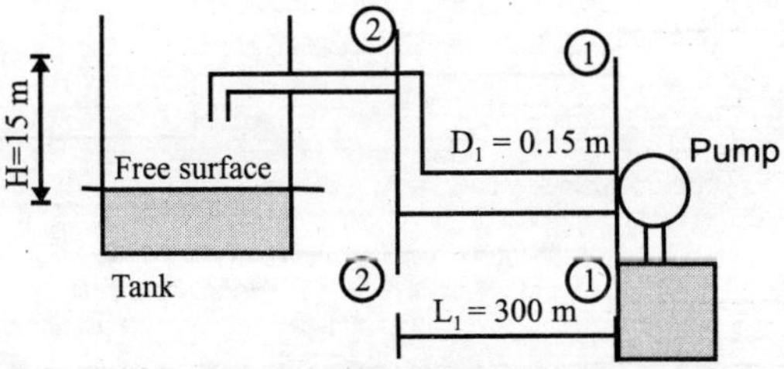

Figure P6.21

Solution:

Given data:

$$Q = 0.02\,\text{m}^3/\text{s}, d_1 = 0.15\,\text{m}$$
$$f = 0.03, L_1 = 300\,\text{m}, H = 15\,\text{m}$$

Area of cross-section of pipe $A_1 = \dfrac{\pi * d_1^2}{4} = \dfrac{\pi * 0.15^2}{4} = 0.0177\,\text{m}^2$

Velocity in the pipe is $\quad V_1 = V_2 = \dfrac{Q}{A_1} = \dfrac{0.02}{0.0177} = 1.129\,\text{m/s}$

a) Applying Bernoulli's equation we have

$$\frac{p_1}{\gamma_1} + \frac{1}{2g}V_1^2 + Z_1 = \frac{p_2}{\gamma_2} + \frac{1}{2g}V_2^2 + Z_2 + h_L$$

$$\left(\text{since} \quad \begin{matrix} z_2 = 15 \\ \textit{atmosphere head zero} \end{matrix}\right)$$

$$\Rightarrow p_1 = 18.9 * 9.81 = 185\,\text{kPa}$$

Hence, the **pressure developed by the pump** on delivery side is 185 kPa.

b) **Power developed to water by the pump** is $P = \gamma Q h_f = 9.81 * 0.02 * .918 = 0.768\,\text{kW}$

Example 6.22: A compound piping system consists of 1800 m of 50 cm diameter, 1200 m of 40 cm diameter and 600 m of 30 cm diameter pipes of same material connected in series. What is the equivalent length of a 40 cm diameter pipe of the same material? State clearly the assumptions made. **[Nagarjuna University]**

Solution: Given data:

$$L_1 = 1800\,\text{m},\ d_1 = 0.5\,\text{m},\ L_2 = 1200\,\text{m},\ d_2 = 0.4\,\text{m},\ L_3 = 600\,\text{m},\ d_3 = 0.3\,\text{m},\ d_e = 0.4\,\text{m}$$

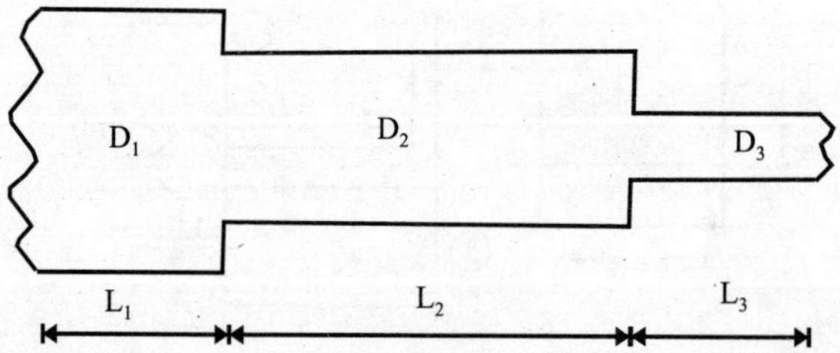

Figure P6.22

Refer Eq. (6.42b), we have

$$\frac{L_{eq}}{D_{eq}^5} = \frac{L_1}{D_1^5} + \frac{L_2}{D_2^5} + \frac{L_3}{D_3^5}$$

$$\Rightarrow \frac{L_{eq}}{0.4^5} = \frac{1800}{0.5^5} + \frac{1200}{0.4^5} + \frac{600}{0.3^5} \Rightarrow L_{eq} = 4318.2\,\text{m}$$

Hence, the **equivalent length** is 4318.2 m for 0.4 m diameter pipe of same material.

Assumptions made are:

a) Friction is constant through all the pipes.

b) The head losses due to sudden contraction are neglected.

Example 6.23: A pipe 150 mm in diameter and 15 m long is connected to the bottom of a tank 15 m long and 12 m wide. The original head over the open end of the pipe is 5 m. Find the time of emptying the tank, assuming the entrance of the pipe is sharp edged and take friction factor $f = 0.01$. **[UPSC Exam]**

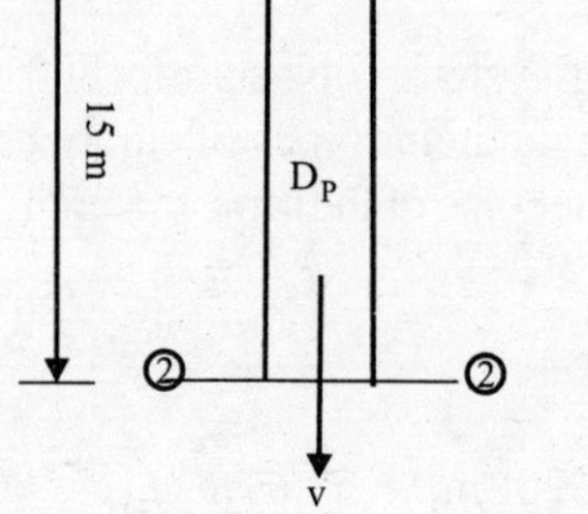

Figure P.6.23

Solution: Given data: $L_t = 5\,\text{m},\ W_t = 12\,\text{m},\ L_P = 15\,\text{m},\ D_P = 0.15\,\text{m},\ f = 0.01$

Apply Bernoulli's equation at section 1-1 and 2-2, thereby we get

$$\frac{p_1}{\gamma_1} + \frac{1}{2g}V_1^2 + Z_1 = \frac{p_2}{\gamma_2} + \frac{1}{2g}V_2^2 + Z_2 + h_f + h_C$$

$$\Rightarrow (h+15) + 0 + 0 = 0 + \frac{1}{2g}V^2 + \frac{fL_P V^2}{2gD_P} + \frac{0.5}{2g}V^2$$

$$\Rightarrow V = \sqrt{\frac{2g(h+15)}{2.5}}$$

The volume of liquid leaving the tank must be equal to the discharge through the sharp edge pipe and hence $A_t * dh = Q * dt$.

$$.\Rightarrow -L_t * W_t * dh = A_P * V * dt$$

$$\Rightarrow -12 * 15 * dh = \frac{\pi * D_P^2}{4} * V * dt$$

$$\Rightarrow -180 * dh = \frac{\pi * 0.15^2}{4} * \sqrt{\frac{2g(h+15)}{2.5}} * dt$$

Integrating on both sides, which give

$$-180 * \int_5^0 dh = \frac{\pi * 0.15^2}{4} * \sqrt{\frac{2g(h+15)}{2.5}} * \int_0^T dt$$

$$-\frac{180}{0.0495} * \int_5^0 \sqrt{\frac{1}{(h+15)}} * dh = \int_0^T dt \quad T = 4357 \text{ seconds}$$

Hence, the **time of emptying** the tank is 1.21 hours.

Example 6.24: To increase the discharge of a pipeline of 600 mm diameter of 2000 m long, another pipeline of the same diameter is introduced parallel to the first pipe in the second half of the length. If the friction factor $f = 0.06$ and head at the inlet is 200 mm, calculate the increase in discharge. Neglect minor losses.

Solution: Given data:

$$L_m = 2000\,\text{m},\ D_m = D_P = 0.6\,\text{m},$$
$$L_P = 1000\,\text{m},\ f = 0.06,\ h = 0.2\,\text{m}$$

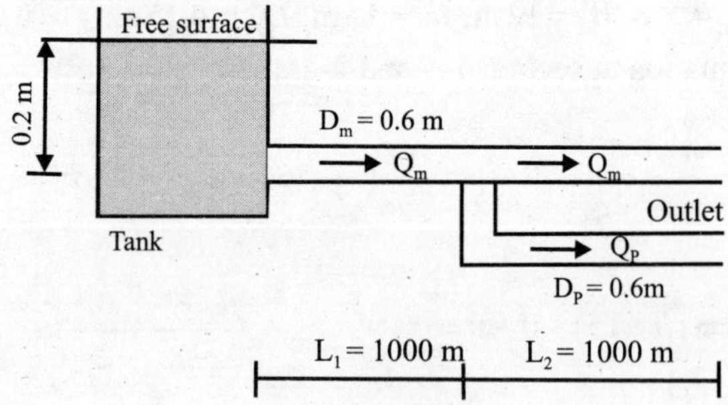

Figure P6.24

a) Head loss through a single pipe of length L = 2000 m and diameter D = 0.6 m will be

$$h_f = h = \frac{fLV^2}{2gD} \Rightarrow V = 0.14 \text{ m/s}$$

So discharge "Q" through a single pipe of L = 2000 m and diameter D = 0.6 m will be

$$Q = AV = \frac{\pi D_m^2}{4} * 0.14 = 0.0395 \text{ m}^3/\text{s}$$

b) When the discharge is divided into two pipes which are of same diameter and lengths, then the discharge at the inlet will be the sum of the divided discharges and also the individual discharges will be half of the main discharge.

i.e. $$Q_m = Q_1 + Q_P \quad \text{and} \quad Q_m/2 = Q_1 = Q_P = 0.0395/2 = 0.0198$$

Refer Eq. (6.42b), we have head loss in series pipes as $h_f = h_{f1} + h_{fm}$

$$\Rightarrow 0.2 = \frac{fLQ_1^2}{2gdA_1^2} + \frac{fLQ_m^2}{2gdA_m^2}$$

$$\Rightarrow Q_m = 0.05 \text{ m}^3/\text{s}$$

Hence, **increase in discharge** is $Q_m - Q = 0.05 - 0.0395 = 0.0106 \text{ m}^3/\text{s}$

Example 6.25: Two tanks are connected by two sharp ended pipes of diameters 50 mm and 100 mm respectively, each of lengths 300 mm respectively, which are in parallel between the two tanks and the two tanks have a difference of levels of 30 m. If the friction factor for each pipe is f = 0,12 calculate:

a) Discharge for each pipe.
b) The equivalent pipe diameter of 300 m long, which would give same discharge, as that of the combination of two pipes.

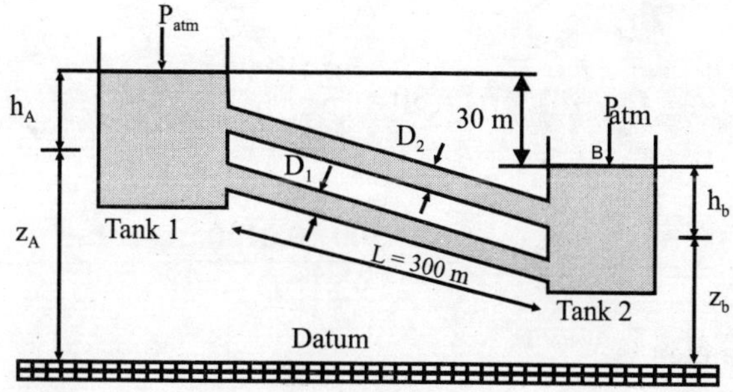

Figure P6.25

Solution: Given data: $L_1 = L_2 = 300\,\text{m}$, $D_1 = 0.05\,\text{m}$, $D_2 = 0.1\,\text{m}$, $f = 0.12$, $\Delta h = 30\,\text{m}$

a) **Flow rate in each pipes:**

For pipe 1,
$$h_{f1} = 30 = \frac{fLV_1^2}{2gD_1}$$

$$30 = \frac{0.12 * 300 * V_1^2}{2 * 9.81 * 0.05} \Rightarrow V_1 = 0.904\ \text{m/s}$$

So discharge through pipe 1 is $Q = A_1V_1 = \dfrac{\pi * D_1^2}{4} * V_1 = 0.00178\ \text{m}^3/\text{s}$

For pipe 2,
$$h_{f2} = 30 = \frac{fLV_2^2}{2gD_2}$$

$$30 = \frac{0.12 * 300 * V_2^2}{2 * 9.81 * 0.1} \Rightarrow V_2 = 1.278\ \text{m/s}$$

So discharge through pipe 2 is

$$Q = A_2V_2 = \frac{\pi * D_2^2}{4} * V_2 = 0.01004\ \text{m}^3/\text{s}$$

b) **Diameter of equivalent pipe:**

Refer Eq. (6.45b), we have $h_f = h_{f1} = h_{f2}$ & $Q = Q_1 + Q_2$ for parallel pipes and hence the equivalent discharge will be $Q = Q_1 + Q_2 = 0.0118\,\text{m}^3/\text{s} \Rightarrow V = Q/A = 0.01505/D_{eq}^2$

And head loss will be

$$h_f = 30 = \frac{fLV^2}{2gD_{eq}}$$

$$\Rightarrow 30 = \frac{0.12 * 300 * 0.01505^2}{2 * 9.81 * D_{eq}^5}$$

Hence, **equivalent diameter** is

$$D_{eq} = 0.1067\,\text{m}$$

Example 6.26: A 250 mm diameter, 3 km long straight pipe runs between two reservoirs of surface elevations 135 m and 60 m. A 1.5 km long, 300 mm diameter pipe is laid parallel to the 250 mm diameter pipe from its mid-point to the lower reservoir. Neglecting all minor losses and assuming friction factor of 0.02 for both the pipes, find the increase in discharge caused by addition of 300mm diameter pipe.

Solution: Given data:

$$h_A + Z_A = 135\,\text{m},\ h_B + Z_B = 60\,\text{m},\ L = 3000\,\text{m}$$
$$L_1 = L_p = 1500\,\text{m},\ D_m = D_1 = 0.25\,\text{m},\ D_p = 0.3\,\text{m}$$

Let the areas of the pipes D, D_m, D_1, D_p (shown in the figure) be respectively

$$A = A_m = A_1 = \frac{\pi * D^2}{4} = \frac{\pi * 0.25^2}{4} = 0.049\,\text{m}^2$$

$$A_p = \frac{\pi * D_p{}^2}{4} = \frac{\pi * 0.3^2}{4} = 0.07065\,\text{m}^2$$

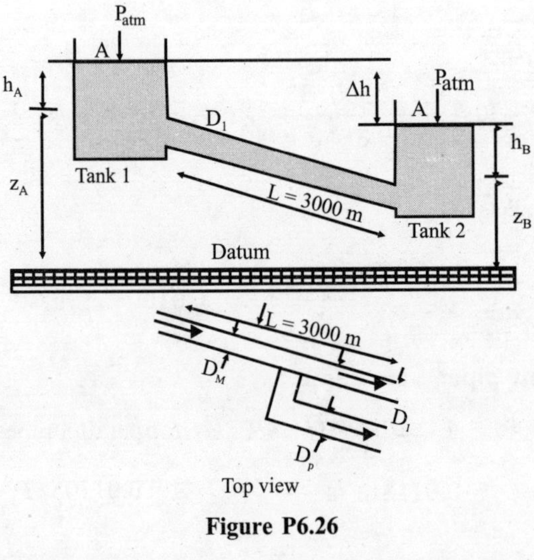

Figure P6.26

a) Flow rate single pipe of diameter 250 mm:

$$\Delta h = h_f = 135 - 60 = \frac{fLV^2}{2gD}$$

$$\Rightarrow V = 2.476\,\text{m/s}$$

So discharge through the pipe is $Q = AV = 0.049 * 2.476 = 0.1215\ \text{m}^3\text{/s}$

b) Discharge through individual pipes (see top view):

Refer Eq. (6.42c), we have $h_{fm} = h_{f1} = h_{fp}$ and $Q_m = Q_1 + Q_p$ for parallel pipes

As $h_{f1} = h_{fp}$
$$\Rightarrow \frac{fL_1V_1^2}{2gD_1} = \frac{fL_pV_p^{\,2}}{2gD_p}$$

$$\Rightarrow \frac{0.02 * 1500 * V_1^2}{2 * 9.81 * 0.25} = \frac{0.02 * 1500 * V_p^{\,2}}{2 * 9.81 * 0.3}$$

$$\Rightarrow V_p = 1.095V_1$$

The total discharge is $Q_m = Q_1 + Q_p = 0.00721\,\text{m}^3\text{/s}$

since
$$Q = AV \Rightarrow A_mV_m = A_1V_1 + A_pV_p$$

$$\Rightarrow 0.049 * V_m = 0.049 * V_1 + 0.0706 * 1.095 * V_1$$

$$\Rightarrow V_m = 2.577V_1$$

As the pipes of diameter D_m, D_1 are is series, we have $\Delta h = h_{fm} + h_{f1}$

$$\Rightarrow 75 = \frac{fL_mV_m^{\,2}}{2gD_m} + \frac{fL_1V_1^2}{2gD_1}$$

$$\Rightarrow 75 = \frac{0.02 * 1500 * V_m^{\,2}}{2 * 9.81 * 0.25} + \frac{0.02 * 1500 * 0.388V_m^{\,2}}{2 * 9.81 * 0.25}$$

$$\Rightarrow V_m = 3.26\,\text{m/s}$$

So discharge due to distributed pipes is $Q_m = A_mV_m = 0.049 * 3.26 = 0.16\,\text{m}^3\text{/s}$

Hence, **increase in discharge due to distribution of pipeline** is

$$Q_{increase} = Q_m - Q = 0.16 - 0.1215 = 0.0385\,\text{m}^3/\text{s}$$

Example 6.27: A 0.8 m diameter, 3 km long straight pipe runs between two reservoirs of surface elevations 100 m and 70 m. In order to increase the rate of flow between the two reservoirs by 20%, it is decided to lay an additional 0.8 m diameter pipeline to the upper reservoir. The second pipeline is to be parallel to the original pipeline and is to be connected to the latter at some suitable point. Determine the point of connection, neglecting all minor losses and assuming friction factor of 0.04 for both the pipes. **[UPSC]**

Solution: Given data:

$$h_A + Z_A = 100\,\text{m},\ h_B + Z_B = 70\,\text{m},\ L = 3000\,\text{m}$$
$$L_m + L_1 = 3000\,\text{m},\ D = 0.8\,\text{m} = D_1 = D_p = D_m$$

Let the areas of the pipes D, D_m, D_1, D_p shown in the figure be respectively

$$A = A_m = A_1 = A_p = \frac{\pi * D^2}{4}$$

$$\Rightarrow A = \frac{\pi * 0.8^2}{4} = 0.5024\,\text{m}^2$$

a) Flow rate single pipe of diameter 250 mm:

$$\Delta h = h_f = 100 - 70 = \frac{fLV^2}{2gD}$$

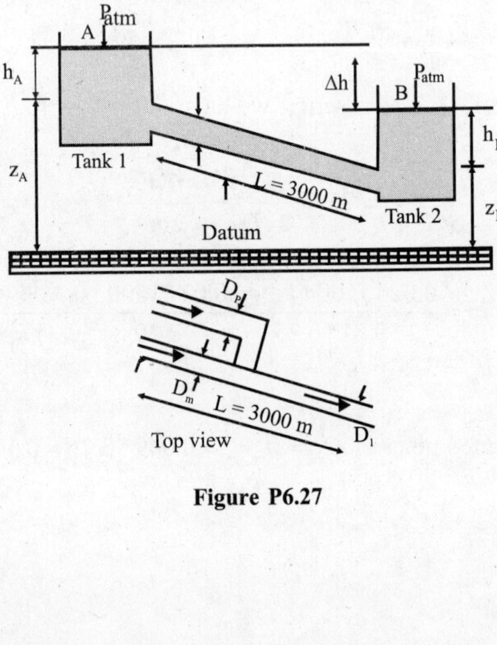

Figure P6.27

$$\Rightarrow V = 1.98\,\text{m/s}$$

So discharge through the pipe is $Q = AV = 0.5024 * 1.98 = 0.99\,\text{m}^3\text{/s}$.

b) Discharge through individual pipes (see top view): As there is increase of 20% of discharge due to the addition of additional pipe of diameter 0.8 m and length L_m, we have $Q_{increase} = Q_m - Q$

$$\Rightarrow 0.2 = Q_m - 0.99$$
$$\Rightarrow Q_m = 1.19\,\text{m}^3\text{/s}$$

Hence, discharge in each sub-pipe will be half of Q_m, i.e. $Q_m / 2 = 1.19/2 = 0.595\,\text{m}^3\text{/s}$.

So $V_p = V_m = V_1 = Q_m / A = 0.595/0.5024 = 1.18\,\text{m}^3\text{/s}$ since area is same.

As the pipes of diameter D_m, D_1 are in series, we have $\Delta h = h_{fm} + h_{f1}$

$$\Rightarrow 70 = \frac{fL_m V_m^{\,2}}{2gD_m} + \frac{f(L - L_m)V_1^2}{2gD_1}$$

$$\Rightarrow 70 = \frac{0.04 * L_m * 1.184^2}{2 * 9.81 * 0.8} + \frac{0.04 * (3000 - L_m) * 1.184^2}{2 * 9.81 * 0.8}$$

$$\Rightarrow L_m = 1213.87\text{m}$$

Hence, increase of 20% in discharge due to distribution of pipeline will be raised if the **second pipe located at a distance** of 1213.87 m from the upper reservoir.

Example 6.28: A farmer wishes to connect two pipes of different lengths and diameters to a common header supplied with 8 + 10^{-3} m^3/s of water from a pump. One pipe is 100 m long and 5 cm diameter. The other pipe is 800 m long. Determine the diameter of the second pipe such that both pipes have the same rate of flow. Assume the pipes to be laid on level ground and friction factor of 0.02. Also determine the head loss in meters of water in the pipes. **[GATE]**

Solution: Given data: $Q = 8 \times 10^{-3}\,\text{m}^3\text{/s},\ L_1 = 100\,\text{m},\ L_2 = 800\,\text{m},\ D_1 = 0.05\,\text{m},\ f = 0.02$

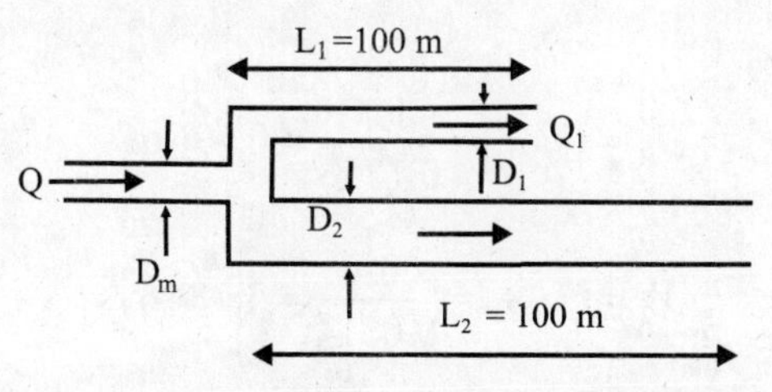

Figure P6.28

a) Discharge through individual pipes:

Refer Eq. (6.42c), we have $h_f = h_{f1} = h_{f2}$ & $Q = Q_1 + Q_2$ for parallel pipes

As $h_{f1} = h_{fp}$ $$\Rightarrow \frac{fL_1V_1^2}{2gD_1} = \frac{fL_pV_2^2}{2gD_2}$$

$$\Rightarrow \frac{L_1Q_1^2}{D_1^5} = \frac{L_2Q_2^2}{D_2^5}$$

$$\Rightarrow \frac{Q_1^2}{Q_2^2} = \frac{800 * D_1^5}{100 * D_2^5}$$

But as given $Q_1 = Q_2 = Q/2 = 4\times10^{-3}\,\text{m}^3/\text{s}$ and so $\Rightarrow 800 * D_1^5 = 100 * D_2^5$

$$\Rightarrow D_2 = 7.578 \text{ cm}$$

As the pipes of diameter D_1, D_2 are is parallel, we have

$$h_{f1} = h_{f2} = \frac{32\, fL_1Q_1^2}{\pi^2 * g * D_1^5} = 33.84 \text{ m}$$

Hence, **head loss in the pipes** is 33.84 m.

Example 6.29: find the loss head incurred when two pipes of diameter 400 mm and 200 mm are 300 mm long each are connected in series and has discharge through the pipeline as 0.1 m³/s.

And also determine the loss of head in the system to pass the same total discharge when the pipes are connected in parallel? Take friction factor as 0.0075 for each pipe.

Solution: Given data: $D_1 = 0.4\,\text{m}$, $L_1 = 300\,\text{m}$, $D_2 = 0.2\,\text{m}$, $L_2 = 300\,\text{m}$, $f = 0.0075$, $Q = 0.1\,\text{m}^3/\text{s}$

Area of first pipe is $$A_1 = \frac{\pi}{4}D_1^2 = \frac{\pi}{4} * 0.4 = 0.1256\,\text{m}^2$$

Area of second pipe is $$A_2 = \frac{\pi}{4}D_2^2 = \frac{\pi}{4} * 0.2 = 0.0314\,\text{m}^2$$

Velocity in first pipe is $$V_1 = Q/A_1 = \frac{0.1}{0.1256} = 0.796\,\text{m/s}$$

Velocity in second pipe is $$V_2 = Q/A_2 = \frac{0.1}{0.0314} = 3.184\,\text{m/s}$$

a) **Pipes are connected in series**:

The head loss will be $H = h_{f1} + h_C + h_{f2}$

$$\Rightarrow H = \frac{fL_1V_1^2}{2gD_1} + \frac{0.5V_2^2}{2g} + \frac{fL_2V_2^2}{2gD_2} = 6.16 \text{ m}$$

Hence, **head loss due to series connection of pipes** is 6.16 m.

b) **Pipes are connected in parallel:**

The discharge will be $Q = Q_1 + Q_2$.

$$\Rightarrow Q = A_1V_1 + A_2V_2$$

$$\Rightarrow 0.1 = 0.1256V_1 + 0.0314V_2 \quad (1)$$

The head loss will be same in both the pipes, when connected in parallel and so

$$\Rightarrow \frac{fL_1V_1^2}{2gD_1} = \frac{fL_2V_2^2}{2gD_2}$$

$$\Rightarrow V_1 = 1.414V_2 \quad \text{since} \quad \begin{matrix} f_1 = f_2 \\ L_1 = L_2 \end{matrix} \quad (2)$$

So from Eq. (1) and Eq. (2), we have $V_1 = 0.677 m/s, 0.97 \text{ m/s}$

if $V_1 = 0.677 \text{ m/s}$ then $\Rightarrow Q_1 = A_1V_1 = 0.1256 * 0.67 = 0.085 \text{m}^3/\text{s}$

if $V_1 = 0.97 \text{ m/s}$ then $\Rightarrow Q_1 = A_1V_1 = 0.1256 * 0.97 = 0.1219 \text{m}^3/\text{s}$

From the above values, $V_1 = 0.677 \text{ m/s}$ can only be considered since the value of discharge at that velocity is less than the actual discharge, i.e. $V_1 = 0.677 \text{ m/s}$.

Head loss in parallel pipes is

$$\Rightarrow H = \frac{fL_1V_1^2}{2gD_1} = \frac{0.0075 * 300 * 0.67^2}{2 * 9.81 * 0.4} = 0.1312 \text{m}$$

Example 6.30: The pipes of diameter "D" and "d"of equal length "L" are arranged in parallel, the loss of head for either pipe for a flow of "Q" is "h". If the pipes are arranged in series and the same quantity "Q" flows through them, the head loss is "H". If d = 0.5 D, find the percentage of total flow through each pipe when placed in parallel and the ratio of H to h neglecting minor losses and assuming friction coefficient to be constant. **[UPSC]**

Solution: Given data: $d = 0.5D, L_1 = L_2, f_1 = f_2$

Area of first pipe is $$A_1 = \frac{\pi}{4} D^2$$

Area of second pipe is $$A_2 = \frac{\pi}{4} d^2$$

Velocity in first pipe is $$V_1 = Q / A_1 = \frac{0.1}{0.1256} = 0.796 \text{ m/s}$$

Velocity in second pipe is $$V_2 = Q / A_2 = \frac{0.1}{0.0314} = 3.184 \text{ m/s}$$

a) Discharge through individual pipes for parallel pipes: Given: *head lost* = *h*

Refer Eq. (6.42c), we have $h = h_{f1} = h_{f2}$ & $Q = Q_1 + Q_2$ for parallel pipes

As $h_{f1} = h_{f2}$ $$\Rightarrow \frac{fL_1 V_1^2}{2gD} = \frac{fL_2 V_2^2}{2gd}$$

$$\Rightarrow \frac{32 fL_1 Q_1^2}{\pi^2 * g * D^5} = \frac{32 fL_2 Q_2^2}{\pi^2 * g * d^5}$$

$$\Rightarrow Q_1 = 5.567 Q_2$$

But $Q = Q_1 + Q_2$ and so $\Rightarrow Q_2 = 0.15Q$ & $Q_1 = 0.85Q$

And head loss in individual pipes are $h = h_{f1} = h_{f2} = \frac{32 fL_1 Q_1^2}{\pi^2 * g * D^5} = \frac{32 fL_2 Q_2^2}{\pi^2 * g * d^5}$

b) Pipes are connected in series:

Velocity in the individual pipes will be $V_1 = Q / A_1$ & $V_2 = Q / A_2$ since $Q = Q_1 = Q_2$

The head loss will be $H = h_{f1} + h_{f2}$

$$\Rightarrow H = \frac{fL_1 V_1^2}{2gD} + \frac{fL_2 V_2^2}{2gd}$$

$$\Rightarrow H = \frac{32 fLQ^2}{\pi^2 g D^5} + \frac{32 fLQ^2}{\pi^2 g d^5}$$

but $$\frac{h}{Q_1^2} = \frac{32 f L_1}{\pi^2 * g * D^5} \; \& \; \frac{h}{Q_2^2} = \frac{32 f L_2}{\pi^2 * g * d^5}$$

And so head loss becomes $$H = Q^2 * \frac{h}{Q_1^2} + Q^2 * \frac{h}{Q_2^2}$$

$$\frac{H}{h} = \frac{Q^2}{Q_1^2} + \frac{Q^2}{Q_2^2}$$

$$\frac{H}{h} = \frac{Q^2}{(0.85Q)^2} + \frac{Q^2}{(0.15Q)^2} \quad \text{since } Q_2 = 0.15Q \,\&\, Q_1 = 0.85Q \quad \frac{H}{h} = 45.83$$

Hence, **head loss ratio between series and parallel connection arrangements** is 45.83.

Example 6.31: Determine the diameter of new pipe lay beside an existing pipeline of diameter 500 mm, so that the two pipes work parallel and reduce the friction loss of 30 m to 8.5 m arising by a pumping plant while pumping the water from the old pipeline. Assume that the two pipes may work in parallel for the entire length and has friction factor and material of same.

Solution: Given data: $h_f = 30\,\text{m}, h_{f1} = h_{fnew} = 8.5\,\text{m}, L = L_1 = L_{new}, f = f_1 = f_{new}, D_m = 0.5\,\text{m}$

Head loss due to a single pipe existing: $h_f = \dfrac{fLV^2}{2gD_m}$

$$\Rightarrow 30 = \frac{32 fLQ^2}{\pi^2 * g * D_m^5}$$

$$\Rightarrow 90.67 = \frac{fLQ_m^2}{D_m^5}$$

$$\Rightarrow Q_m = \frac{90.67 D_m^5}{fL} \qquad (1)$$

b) As the pipes of diameter $D_m = D, D_{new} 90^\circ$ are in parallel, we have

Refer Eq. (6.42c), we have $Q = Q_1 + Q_{new}$ for parallel pipes

And as $h_{f1} = h_{fnew}$ $$\Rightarrow 8.5 = \frac{fLV_1^2}{2gD_m} = \frac{fL_{new}V_{new}^2}{2gd_{new}}$$

$$\Rightarrow 8.5 = \frac{32 fLQ_1^2}{\pi^2 * g * D^5} = \frac{32 fL_2 Q_2^2}{\pi^2 * g * d_{new}^5}$$

$$\Rightarrow Q_1 = \frac{25.69 * D^5}{fL} \quad (2)$$

Similarly, for second new pipe, we have $Q_{new} = \frac{25.69 * D_{new}^5}{fL_{new}}$ (3)

Taking the ratio Eq. (1) and Eq. (2), we get $Q_1 = 0.283 Q_m$

but $Q = Q_1 + Q_{new}$ and so substituting Eq. (1), Eq. (2) and Eq. (3), we get

$$Q = 0.283Q + Q_{new}$$
$$\Rightarrow Q_{new} = 0.7167Q$$

And head loss in individual pipes are $h_{f1} = h_{fnew}$

$$\Rightarrow \frac{32 fL_1 Q_1^2}{\pi^2 * g * D^5} = \frac{32 fL_{new} Q_{new}^2}{\pi^2 * g * D_{new}^5}$$

$$\Rightarrow D^5 = 0.156 D_{new}^5 \quad (4)$$

And also we get for $Q = Q_1 + Q_{new}$ and so substituting Eq. (1), Eq. (2) and Eq. (3), we get

$$\frac{90.67 D^5}{fL} = \frac{25.69 * D^5}{fL} + \frac{25.69 * D_{new}^5}{fL_{new}}$$

$$\Rightarrow 90.67 D^5 = 25.69 * D^5 + 25.69 * D_{new}^5$$

$$\Rightarrow D^5 = 0.395 D_{new}^5 \quad (5)$$

So from Eq. (4) and Eq. (5), we get $D_{new} = 1.204\,\text{m}$

Hence, the appropriate diameter pipe of same length that needed to reduce the head loss from 30 m to 8.5 m is 1.204 m.

Example 6.32: The water levels in the two reservoirs A and B are 104.5 m and 100 m respectively above the datum. A pipe joins each to a common point D, where pressure is 98.1 kPa gauge and height 83.5 m above datum. Another pipe connects D to another tank C. What will be the height of the water level in the tank C, assuming the same value of "F" for all pipes. Take friction coefficient as 0.0075. The diameters of the pipes AD, BD and CD are 300 mm, 450 mm, 600 mm respectively and their lengths are 240 m, 270 m and 300 m respectively. **[IIT Delhi]**

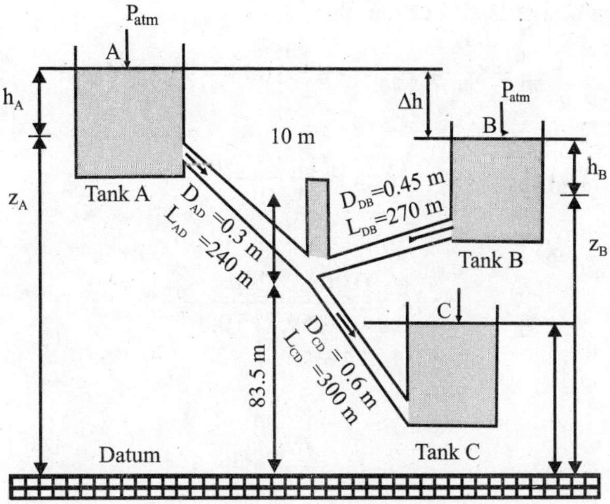

Figure P6.32

Solution: Given data:

$$D_{AD} = 0.3\,\text{m},\ D_{DB} = 0.45\,\text{m},\ D_{CD} = 0.6\,\text{m},\ L_{AD} = 240\,\text{m},\ L_{DB} = 270\,\text{m},\ L_{CD} = 300\,\text{m},\ f = 0.0075$$

$$h_A + Z_A = 104.5\,\text{m},\ h_B + Z_B = 100\,\text{m},\ Z_D = 83.5\,\text{m},\ P_D = 98.1\,\text{kPa}$$

Areas of pipe AD, DB, CD will be respectively

$$A_{AD} = \frac{\pi}{4}D_{AD}^2 = \frac{\pi}{4}*0.3^2 = 0.07065\,\text{m}^2$$

$$A_{DB} = \frac{\pi}{4}D_{DB}^2 = \frac{\pi}{4}*0.45^2 = 0.1589\,\text{m}^2$$

$$A_{CD} = \frac{\pi}{4}D_{CD}^2 = \frac{\pi}{4}*0.6^2 = 0.283\,\text{m}^2$$

Pressure head at the point D will be

$$h_D = \frac{P_D}{\gamma} = \frac{98.1}{9.81} = 10\,\text{m of water}$$

Hence, the total head at point D will be

$$h_{fD} = h_D + Z_D = 10 + 83.5 = 93.5\,\text{m}$$

Total head loss between point A and D will be

$$h_{fAD} = h_{AD} + Z_{AD} - h_{fD}$$

$$\Rightarrow h_{fAD} = 104.5 - 93.5 = 11.0\,\text{m}$$

Total head loss between points B and D will be

$$h_{fBD} = h_{BD} + Z_{BD} - h_{fD} = 100.0 - 93.5 = 6.5\,\text{m}$$

For pipe AD, we have head loss as $\quad h_{fAD} = \dfrac{fL_{AD}V_{AD}^2}{2gD_{AD}}$

$$\Rightarrow 1.1 = \frac{0.0075 * 240 * V_{AD}^2}{2 * 9.81 * 0.3}$$
$$\Rightarrow V_{AD} = 3\,\text{m/s}$$

For pipe BD, we have head loss as $\quad h_{fBD} = \dfrac{fL_{BD}V_{BD}^2}{2gD_{BD}}$

$$\Rightarrow 6.5 = \frac{0.0075 * 270 * V_{BD}^2}{2 * 9.81 * 0.45}$$
$$\Rightarrow V_{BD} = 2.66\,\text{m/s}$$

As we know $\quad Q_{AD} + Q_{BD} = Q_{CD}$

$$\Rightarrow A_{AD} * V_{AD} + A_{BD} * V_{BD} = A_{CD} * V_{CD}$$
$$\Rightarrow 0.07065 * 3 + 0.1589 * 2.66 = 0.283 * V_{CD}$$
$$\Rightarrow V_{CD} = 2.24\,\text{m/s}$$

So velocity in the pipe CD is 2.24 m/s and the head loss in tank C will be

$$h_{fCD} = \frac{fL_{CD}V_{CD}^2}{2gD_{CD}}$$
$$\Rightarrow h_{fCD} = \frac{0.0075 * 300 * 2.24^2}{2 * 9.81 * 0.6} = 3.84\,\text{m}$$

Hence, the **level of the tank C from the datum** will be

$$h_C + Z_C = h_{fD} - h_{fCD} = 93.5 - 3.84 = 89.66\,\text{m}$$

Example 6.33: Three reservoirs A, B and C are connected by a pipe system shown in the figure. Find the discharge into or from the reservoir B and C if the rate "Q" from reservoir A is 60 lit/sec. Find the height of water level in the reservoir C. f = 0.006 for all pipes.

Solution: Given data:

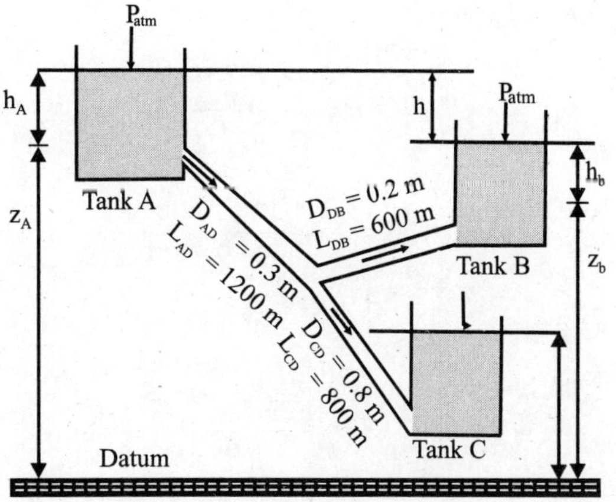

Figure P6.33

$D_{AD} = 0.3\,\text{m}$, $D_{DB} = 0.2\,\text{m}$, $D_{CD} = 0.8\,\text{m}$, $L_{AD} = 1200\,\text{m}$, $L_{DB} = 600\,\text{m}$, $L_{CD} = 800\,\text{m}$

$f = 0.006$, $h_A + Z_A = 40\,\text{m}$, $h_B + Z_B = 38\,\text{m}$, $Q_{AD} = 0.06\,\text{m}^3/\text{s}$

Areas of pipes AD, DB, CD will be respectively

$$A_{AD} = \frac{\pi}{4} D_{AD}^2 = \frac{\pi}{4} * 0.3^2 = 0.07065\,\text{m}^2$$

$$A_{DB} = \frac{\pi}{4} D_{DB}^2 = \frac{\pi}{4} * 0.2^2 = 0.0314\,\text{m}^2$$

$$A_{CD} = \frac{\pi}{4} D_{CD}^2 = \frac{\pi}{4} * 0.8^2 = 0.5024\,\text{m}^2$$

For pipe AD: We know as per continuity equation

$$Q_{AD} = A_{AD} * V_{AD}$$

$$\Rightarrow 0.06 = 0.07065 * V_{AD}$$

$$\Rightarrow V_{AD} = 0.849\,\text{m/s}$$

Head loss
$$h_{fAD} = \frac{fL_{AD} V_{AD}^2}{2gD_{AD}}$$

$$\Rightarrow h_{fAD} = \frac{0.006 * 1200 * 0.849^2}{2 * 9.81 * 0.3} = 0.88\text{ m}$$

For pipe BD, we have head loss as

$$\Delta h = h_{fBD} = \frac{fL_{BD}V_{BD}^2}{2gD_{BD}}$$

$$\Rightarrow 40 - 38 = 2 = \frac{0.006 * 600 * V_{BD}^2}{2 * 9.81 * 0.2}$$

$$\Rightarrow V_{BD} = 1.476 \text{ m/s}$$

As we know $Q_{AD} + Q_{BD} = Q_{CD}$

$$\Rightarrow A_{AD} * V_{AD} + A_{BD} * V_{BD} = A_{CD} * V_{CD}$$
$$\Rightarrow 0.07065 * 0.849 + 0.0314 * 1.476 = 0.5024 * V_{CD}$$
$$\Rightarrow V_{CD} = 0.212 \text{ m/s}$$

So discharge in the pipe CD is 0.106 m^3/s

And similarly discharge in the pipe BD is 0.0463 m^3/s.

Hence, the surface level of the tank C to the surface level of tank B, will be

$$\Delta h = \sum RQ^2 = \left\{\left(1.5 + \frac{fL_{BD}}{D_{BD}}\right)\left(\frac{1}{2A_{BD}{}^2 g}\right)Q_{BD}{}^2 + \left(1.5 + \frac{fL_{CD}}{D_{CD}}\right)\left(\frac{1}{2A_{CD}{}^2 g}\right)Q_{CD}{}^2\right\}$$

$$\Rightarrow \Delta h = 2.14 \text{ m}$$

And hence the **surface level of tank C from the datum** will be

$$(h_C + Z_C) = (h_B + Z_B) - \Delta h$$
$$(h_C + Z_C) = 38 - 2.14 = 35.8 \text{ m}$$

Example 6.34: Three reservoirs A, B and C are connected by a pipe system shown in the figure. Find the diameter of Tank C, if the discharge into tank B and Tank C is same from the Tank A. Take f = 0.006 for all pipes.

Solution: Given data:

$$D_{AD} = 0.3 \text{ m}, D_{DB} = 0.2 \text{ m}, L_{AD} = 1200 \text{ m}, L_{DB} = 600 \text{ m}, L_{CD} = 800 \text{ m}$$
$$f = 0.006, h_A + Z_A = 40 \text{ m}, h_B + Z_B = 38 \text{ m}, Q_{BD} = Q_{CD}, h_C + Z_C = 35 \text{ m}$$

Areas of pipes AD, DB, CD will be respectively

$$A_{AD} = \frac{\pi}{4}D_{AD}^2 = \frac{\pi}{4} * 0.3^2 = 0.07065 \text{ m}^2$$

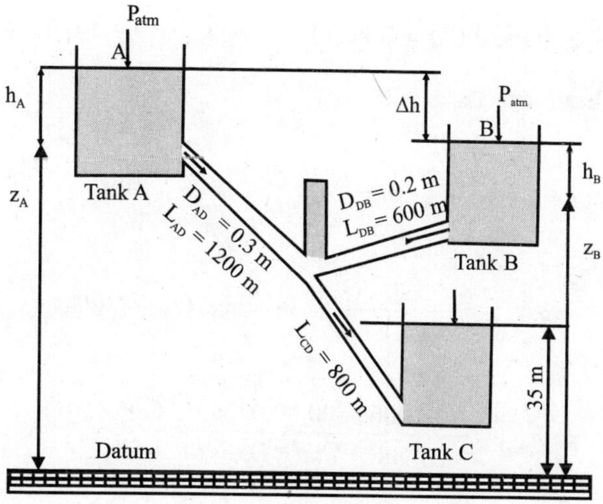

Figure P6.34

$$A_{DB} = \frac{\pi}{4} D_{DB}^2 = \frac{\pi}{4} * 0.2^2 = 0.0314\,\text{m}^2$$

Head loss in pipe AD,

$$h_{fAD} = \frac{fL_{AD} V_{AD}^2}{2gD_{AD}}$$

$$\Rightarrow h_{fAD} = \frac{0.006 * 1200 * Q_{AD}^2}{2 * 9.81 * 0.3 * 0.07065^2} = 245 Q_{AD}^2$$

For pipe BD, we have head loss as

$$h_{fBD} = \frac{fL_{BD} V_{BD}^2}{2gD_{BD}}$$

$$\Rightarrow h_{fBD} = \frac{fL_{BD} Q_{AD}^2}{2gD_{BD} * 4 * A_{BD}^2} \quad \text{since } Q_{AD} = 2Q_{BD}$$

$$\Rightarrow h_{fBD} = \frac{0.006 * 600 * Q_{AD}^2}{2 * 9.81 * 0.2 * 4 * 0.0314^2} = 2326 Q_{AD}^2$$

But the head loss for the pipe ADB, we have $\Delta h = h_{fAD} + h_{fBD}$

$$\Rightarrow 2 = 245 Q_{AD}^2 + 2326 Q_{AD}^2 = 2571.2 Q_{AD}^2$$

$$\Rightarrow Q_{AD} = 0.0279\ \text{m}^3/\text{s}$$

So discharge through pipe BD and CD will be $Q_{AD}/2 = Q_{BD} = Q_{CD} = 0.0139\,\text{m}^3/\text{s}$

For pipe CD, we have head loss as

$$h_{fCD} = \frac{fL_{CD}V_{CD}^2}{2gD_{CD}}$$

$$\Rightarrow h_{fCD} = \frac{fL_{CD}Q_{AD}^2}{2gD_{CD} * 4 * A_{CD}^2} \quad \text{since } Q_{AD} = 2Q_{BD}$$

$$\Rightarrow h_{fCD} = \frac{4 * 0.006 * 800 * 0.0279^2}{2 * 9.81 * 4 * \pi * D_{CD}^{\ 3}} = \frac{6.06 \times 10^{-5}}{D_{CD}^{\ 3}}$$

But the head loss for the pipe ADC, we have $\Delta h = h_{fAD} + h_{fCD}$

$$\Rightarrow 40 - 35 = 5 = 245Q_{AD}^2 + \frac{6.06 \times 10^{-5}}{D_{CD}^{\ 3}} = 245 * 0.0279^2 + \frac{6.06 \times 10^{-5}}{D_{CD}^{\ 3}}$$

$$\Rightarrow D_{CD} = 0.0232 \text{ m}$$

Hence, the diameter of pipe to tank C is 0.0232 m.

Example 6.35: Figure P6.35 shows a network in which "Q" and "h_f" refer to discharges and pressure drops respectively. Subscripts 1, 2, 3, 4 and 5 designated respective values in pipe lengths AC, BC, CD and AC. Subscripts A, B, C and D designate discharge entering or leaving the junction points A, B, C and D respectively. By sticking the values given in the figure, find the following discharges Q_B, Q_2, Q_5, & Q_5, and pressure drops h_{f4} & h_{f5} and give these computed values at their respective places on a neat sketch of the network along with the following directions. **[GATE]**

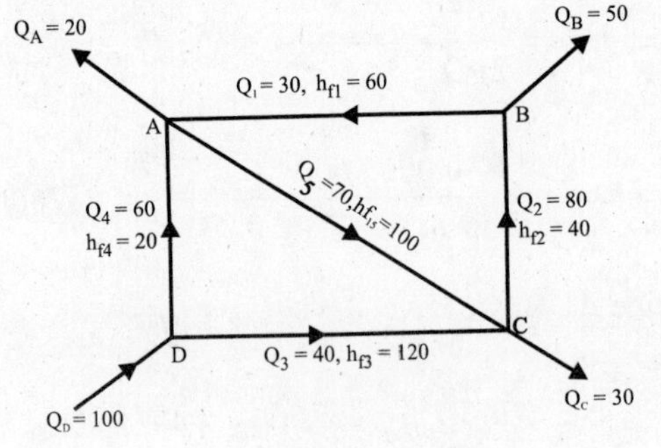

Figure P6.35a

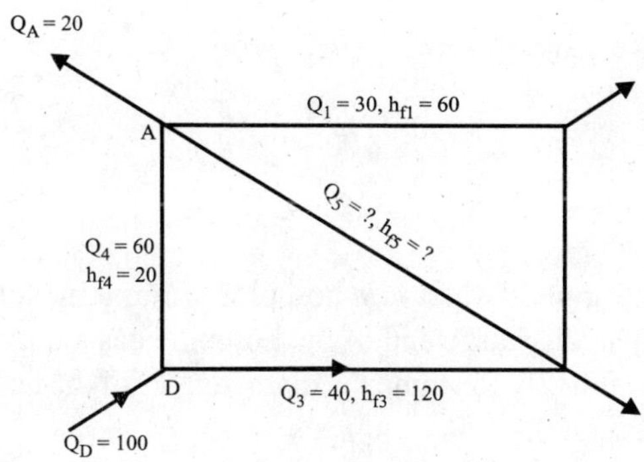

Figure P6.35b

Solution: Given data:

$$Q_A = 20\,\text{m}^3/\text{s},\ Q_c = 30\,\text{m}^3/\text{s},\ Q_D = 100\,\text{m}^3/\text{s},\ h_{f1} = 60\,\text{m},\ h_{f2} = 40\,\text{m},\ h_{f3} = 120\,\text{m}$$

As observed in the figure, we can take four junction points like A, B, C and D for equating the discharges entering is equal to the discharges leaving from each junctions and so

At junction D, we have $Q_D = Q_3 + Q_4$

$$\Rightarrow 100 = 40 + Q_4$$
$$\Rightarrow Q_4 = 60\,\text{m}^3/\text{s}$$

So the discharge at pipe 4 is leaving.

Similarly, at junction A, we have (assuming discharge in pipe 1 to be entering) $Q_1 + Q_5 + Q_4 = Q_A$

$$\Rightarrow 30 + Q_5 + 60 = 20 \ \Rightarrow Q_5 = -70\,\text{m}^3/\text{s}$$

So the discharge at pipe 5 is leaving.

Similarly, at junction C, we have $Q_2 + Q_5 + Q_3 - Q_C$

$$\Rightarrow Q_2 + 70 + 40 = 30$$
$$\Rightarrow Q_2 = -80\,\text{m}^3/\text{s}$$

So the discharge at pipe 2 is leaving.

Similarly, at junction B, we have $Q_2 = Q_B + Q_1$

$$\Rightarrow 80 = Q_B + 30$$
$$\Rightarrow Q_B = 50\,\text{m}^3/\text{s}$$

For each closed circuit of pipe network, the sum of head losses is zero and hence

For circuit ABC, we have $h_{f2} + h_{f1} + h_{f5} = 0$

$$\Rightarrow 40 + 60 + h_{f5} = 0$$
$$\Rightarrow h_{f5} = -100\,\text{m}$$

Similarly, for circuit ADC, we have $h_{f4} + h_{f5} - h_{f3} = 0$.

$$\Rightarrow h_{f4} - 100 - 120 = 0$$

$$\Rightarrow h_{f4} = 20\,\text{m}$$

Example 6.36: A fire engine supplies water to a hosepipe, 75 m long and 0.075 m in diameter, at a pressure of 294 kPa (gage). The discharge end of the hosepipe has a nozzle of diameter "d" fixed to it. Taking the friction factor as 0.032, determine the diameter "d" of the nozzle, so that the momentum of the issuing jet may be maximum. **[UPSC Exams]**

Solution: Given data: $D_1 = 0.75\,\text{m},\ L_1 = 75\,\text{m},\ P_1 = 294\,\text{kPa(gage)},\ d,\ f = 0.032$

Applying Bernoulli's equation in between in the pipe and nozzle end, we have

$$\frac{P_1}{\gamma} + \frac{V_1^2}{2g} + Z_1 = \frac{P_{atm}}{\gamma} + \frac{V_2^2}{2g} + Z_2 + h_f$$

As per continuity equation, we have $Q = A_1V_1 = A_2V_2$

$$\Rightarrow V_2 = \left(\frac{A_1}{A_2}\right)V_1$$

$$\Rightarrow V_2 = \left(\frac{D^2_1}{d^2}\right)V_1 = \left(\frac{0.075}{d}\right)^2 V_1 \qquad (1)$$

Hence, the Bernoulli's equation becomes $\frac{P_1}{\gamma} + \frac{V_1^2}{2g} = \frac{P_{atm}}{\gamma} + \left(\frac{0.075}{d}\right)^4 \frac{V_1^2}{2g} + \frac{fL_1V_1^2}{2gD_1}$

$$\frac{294}{9.81} + \frac{V_1^2}{2*9.81} = \frac{0}{9.81} + \left(\frac{0.075}{d}\right)^4 * \frac{V_1^2}{2g} + \frac{0.032*75*V_1^2}{2*9.81*0.075}$$

$$\Rightarrow V_1 = \sqrt{\frac{588.6}{(31) + \left(\left(\frac{0.075}{d}\right)^4\right)}} \qquad (2)$$

Hence, velocity at the nozzle end will be

$$V_2 = \left(\frac{0.075}{d}\right)^2 V_1 = \left(\frac{0.075}{d}\right)^2 \sqrt{\frac{588.6}{(31) + \left(\left(\frac{0.075}{d}\right)^4\right)}}$$

Momentum of the jet from the nozzle will be $M = \rho Q_2 V_2 = 1000 * \frac{\pi}{4} d^2 * V_2^2$

$$\Rightarrow M = 1000 * \frac{\pi}{4} d^2 * \left(\frac{0.075}{d}\right)^4 * \frac{588.6}{(31) + \left(\left(\frac{0.075}{d}\right)^4\right)}$$

$$\Rightarrow M = \frac{14.6}{d^2\left[31 + \left(\left(\frac{0.075}{d}\right)^4\right)\right]} \qquad (3)$$

To get the maximum momentum, we have to differentiate the momentum with respect to the diameter of the nozzle and equate to zero, i.e

$$\frac{dM}{dd} = 0$$

$$\frac{dM}{dd} = \frac{d}{dd}\left\{\frac{14.6}{d^2\left[31 + \left(\left(\frac{0.075}{d}\right)^4\right)\right]}\right\} = 0$$

$$d = 31.8\,\text{mm}$$

So the **diameter of nozzle at the end** is 31.8 mm to get maximum momentum of issuing jet.

HYDRAULIC GRADIENT

Any water distribution transports system should supply water, without impairing its **adequate quantities** and at **sufficient pressures** to meet system requirements like domestic, commercial and industrial uses. To meet the demand of water supply a service reservoir/tank stores the water and supplies it at the required pressure to the farthest point in the area.

A water supply system needs to guarantee a minimum pressure even at the most remote point in the area of supply. Therefore, it is essential that the **hydraulic gradient line** (i.e., sum of pressure head p/γ and potential head z) should always be above the required pressure. This section briefs the various aspects of supply of water through pipelines maintaining no less than minimum flow rate, energies and pressures.

Till this section, the losses in pipe flow have been studied, individually. But in this section we deal with water flowing in a pipeline from one reservoir/tank to another reservoir/tank taking in account all losses same time like loss at the entry, exit, at bends, valves, fittings, sudden expansion, sudden contraction and friction.

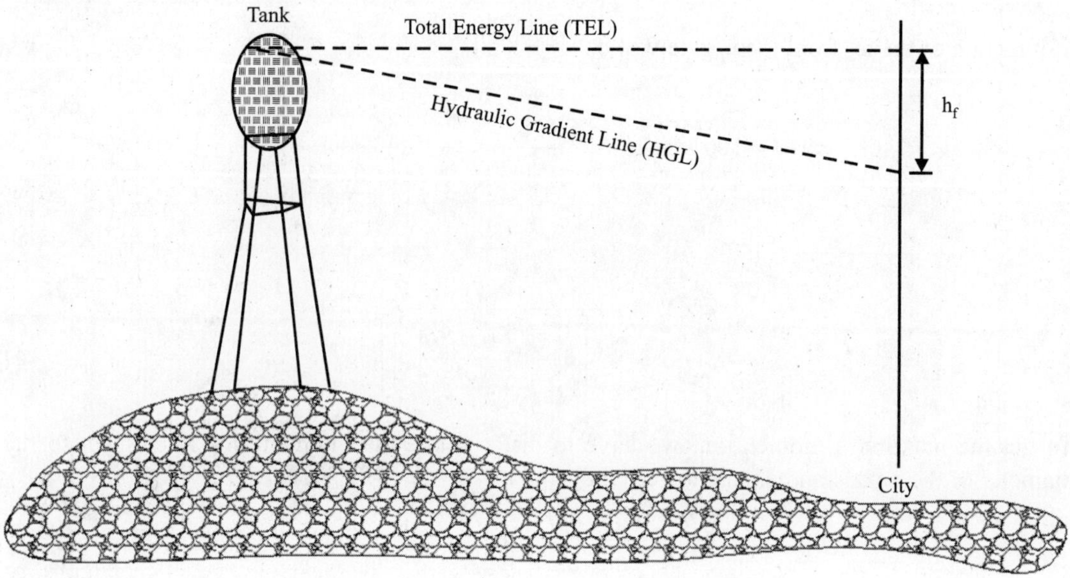

Figure 6.13a: TEL and HGL

When water flows through a **uniform pipeline** from a high level reservoir/tank to a lower level reservoir/tank, it has a total energy comprising the potential head, pressure head and velocity head and so the **total energy** becomes $TE = \frac{p}{\gamma} + \frac{V^2}{2g} + z$. If the total energies at different points along the pipeline are plotted, generate a horizontal line termed "**Total Energy line (TEL or EGL)**", which will be a horizontal line at the level of the supply reservoir/tank. But as the major loss in a pipeline is due to friction, there will be a reduction of pressure head (h_f) only and hence the available head from the starting to the exit or lower level reservoir/tank has deviation in the total energy line (horizontal line) to a different line termed "**hydraulic gradient line (HGL)**" which will be below the total energy line, as shown in Figure 6.13a. The hydraulic gradient line always lies below the total energy line but still it fluctuates with respect to the pressure changes.

When the losses are due to **variance of pipeline dimensions** at all point/sections and existence of friction in pipeline are taken into consideration, then there arise loss of quantity flow (i.e. input quantity of water will be more than the exit quantity) from the starting to the end point. And so the hydraulic gradient line (which is a straight line) due to friction consideration only, changes its shape when all losses (like entry loss ($0.5*V^2/2g$), friction ($fLV^2/2gd$), sudden contraction , sudden expansion, exit losses etc) are included at a time, as shown in Figure 6.13b.

The velocity head at the different salient points in the pipe are also calculated and these are subtracted from the Energy Line to give the **Hydraulic Gradient** line.

Normally when a pipeline is laid, it should be below or at the level of hydraulic gradient line or else an air airlock is formed.

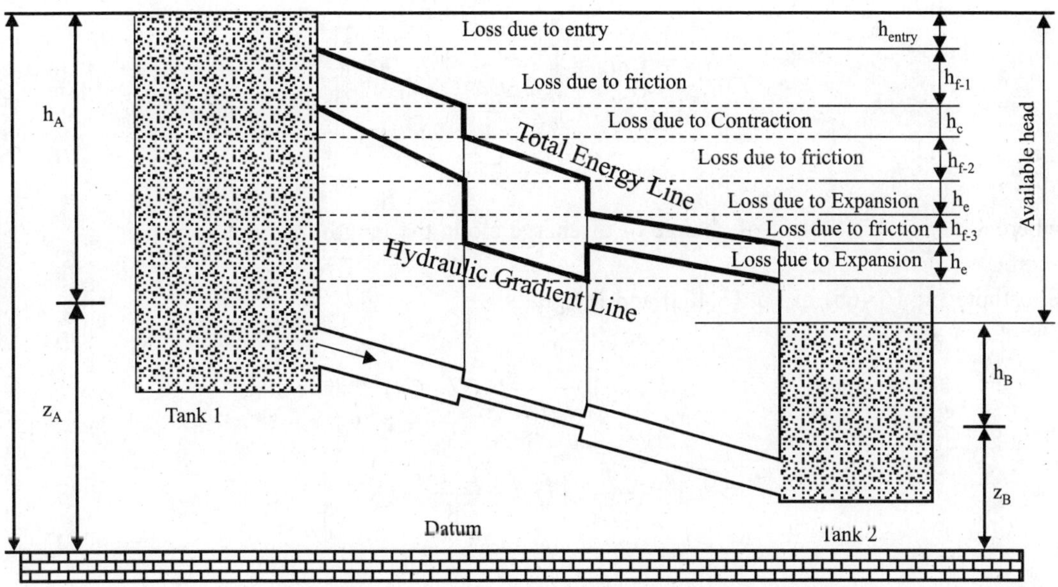

Figure 6.13b: Generating TEL and HGL lines

Flow through Perforated Pipes

In household water supply, the discharge quantity "*Q*" of water from a larger diameter (D) pipe (called **trunk pipe**) is connected with the branched pipes (called **perforated pipes**) of same smaller diameters (*d*). These perforated pipes absorb equal quantity of water "q" from the trunk pipe such that all the quantity of water "*Q*" supplied in the trunk is absorbed, when moved from inlet to outlet positions of length "*L*", i.e. $Q = 0$ at the end ($x = L$) of the trunk pipe, as shown in Figure 6.14.

The head loss in the trunk due to a single perforated pipe be

$$dh_f = \frac{fV^2}{2gD} * dx = \frac{f}{2gDa^2} * q^2 * dx \tag{6.46}$$

Then total head loss in the trunk due to several perforated pipes will be

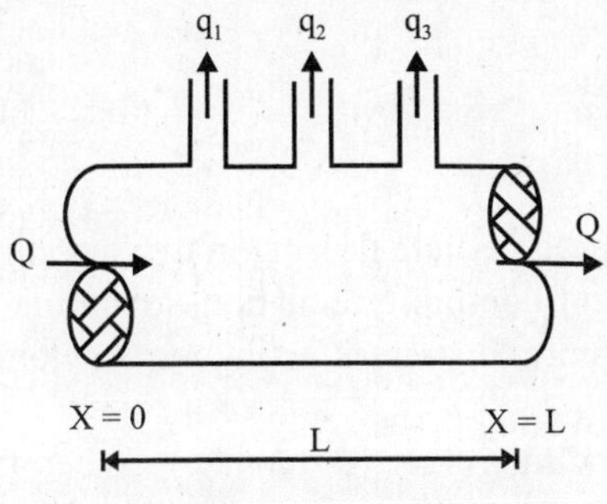

Figure 6.14: Perforated pipes

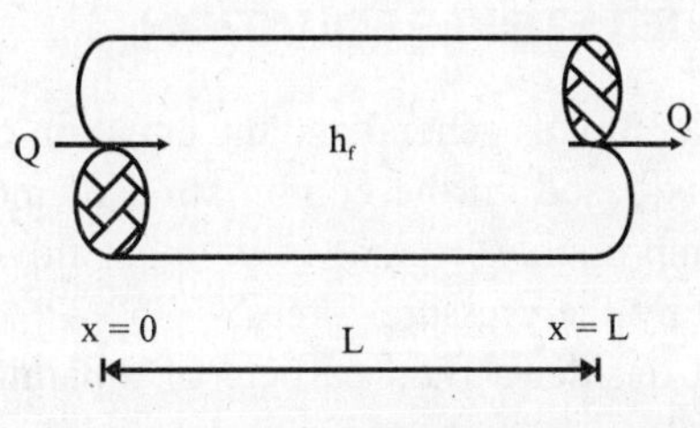

Figure 6.14a: Pipe with no perforated pipes

$$h_{f1} = \int_0^L dh_f = \int_0^L \left[\frac{f}{2gDa^2} * q^2 * dx \right] \tag{6.46a}$$

but

$$q = Q - Cx \tag{6.46b}$$

where $C = \frac{dQ}{dx} = \frac{Q}{L}$ is rate of change of discharge along the length.

Substitute Eq. (6.46b) in Eq. (6.46a) and integrate

$$h_{f1} = \int_0^L dh_f = \int_0^L \left[\frac{f}{2gDa^2} * (Q - Cx)^2 * dx \right]$$

$$\Rightarrow h_{f1} = \frac{f}{2gDa^2} \left[Q^2 L + \frac{C^2 L^3}{3} - QCL^2 \right]$$

$$\Rightarrow h_{f1} = \frac{fC^2 L^3}{3 * 2gDa^2} \text{ (since } Q = CL)$$

$$\boxed{\Rightarrow h_{f1} = \frac{1}{3} \left[\frac{fQ^2 L}{2gDa^2} \right]} \tag{6.47}$$

Equation (6.47) gives the **head loss in a trunk pipe having perforated pipes.**

If h_f is the head loss of trunk pipe (D) without perforated pipes, as shown in Figure 6.13a, then

$$h_f = \frac{fLQ^2}{2gDa^2} \tag{6.48}$$

Observing Eq. (6.47) and Eq. (6.48), we can say that

$$\boxed{h_{f1} = \frac{1}{3} h_f} \tag{6.49}$$

NAVIER STOKES EQUATION

Navier Stokes generalized the equation of forces acting on a fluid particle flowing in two directions and is based on the combination of momentum equation and continuity equation, assuming an incompressible laminar Newtonian flow and it replaces the normal stresses acting on Newtonian fluid by the pressure forces.

Consider a fluid particle of a parallelepiped shape with dimensions $(\Delta x, \Delta y, \Delta z)$. When this particle is in motion due to inertial force (F_I), it is subjected to resistance forces like pressure

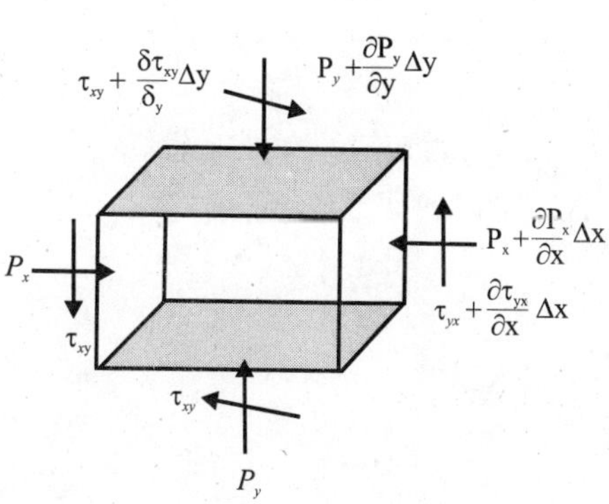

Figure 6.15: Navier Stokes equation

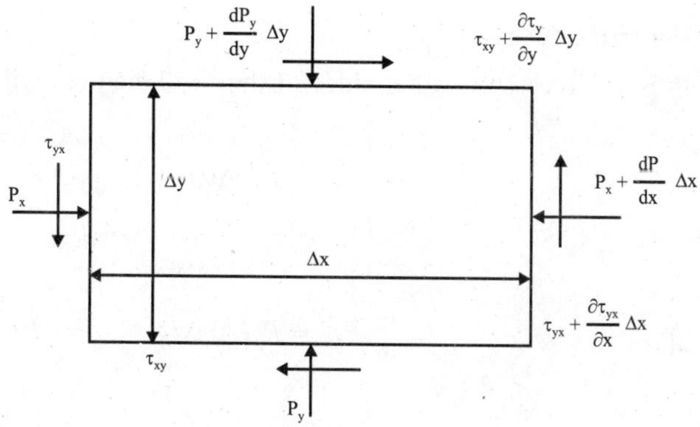

Figure 6.15a: Navier Stokes equation

forces (F_P), shear forces (F_τ) on either sides of the fluid particle and gravity forces (F_G), as shown in the Figures 6.15 and 6.15a.

To have the motion to the fluid particle, the particle must have inertial force equal or greater to the sum of all the three resisting forces, in the flow direction as well as across the flow direction and hence

$$\begin{pmatrix} F_{I_x} = F_{P_x} + F_{\tau_x} + F_{G_x} \\ F_{I_y} = F_{P_y} + F_{\tau_y} + F_{G_y} \end{pmatrix} \tag{6.50}$$

Inertia forces (F_I)

As per Newton's second law of motion $F_i = Mass(m) * Acceleration$ hence

Inertia force along the flow direction

$$F_{I_x} = Mass * Acceleration = (\rho \Delta x \Delta y \Delta z)\frac{du}{dt} \tag{6.51a}$$

Similarly, inertia force across the flow direction

$$F_{I_y} = Mass * Acceleration = (\rho \Delta x \Delta y \Delta z)\frac{dv}{dt} \tag{6.51b}$$

Gravity force (F_G)

As the gravity force is the product of gravity and mass, so gravity forces along the flow direction

$$F_{G_x} = G_x * mass = G_x * (\rho \Delta x \Delta y \Delta z) \tag{6.52a}$$

Similarly, gravity force across the flow direction

$$F_{G_y} = G_y * mass = G_y * (\rho \Delta x \Delta y \Delta z) \tag{6.52b}$$

Pressure forces (F_P)

Pressure forces on either sides of the fluid element will

$$F_{P_{x1}} = P_x(\Delta y \Delta z), F_{P_{x2}} = \left(P_x + \frac{dp}{dx}\Delta x\right)(\Delta y \Delta z)$$

and $$F_{P_{y1}} = P_y(\Delta x \Delta z), F_{Py2} = \left(P_y + \frac{dp}{dy}\Delta y\right)(\Delta x \Delta z)$$

in flow direction and flow across direction respectively.

So net pressure force along the flow direction

$$F_{P_x} = F_{P_{x1}} - F_{P_{x2}}$$

$$\Rightarrow F_{P_x} = P_x(\Delta y \Delta z) - \left(P_x + \frac{dp}{dx}\Delta x\right)(\Delta y \Delta z)$$

$$\Rightarrow F_{P_x} = -\left(\frac{dp}{dx}\Delta x\right)(\Delta y * 1) \tag{6.53a}$$

Similarly, net pressure force across the flow direction $F_{P_y} = F_{P_{y1}} - F_{P_{y2}}$

$$\Rightarrow F_{P_y} = -\left(\frac{dp}{dy}\Delta y\right)(\Delta x \Delta z) \tag{6.53b}$$

Shear forces (F_τ)

Shear forces on either sides of the fluid element will be

$$F_{\tau_{xy1}} = \tau_{xy}(\Delta x \Delta z), F_{\tau_{xy2}} = \left(\tau_{xy} + \frac{d\tau_{xy}}{dy}\Delta y\right)(\Delta x \Delta z)$$

and $$F_{\tau_{yx1}} = \tau_{yx}(\Delta y \Delta z),\ F_{\tau_{yx2}} = \left(\tau_{yx} + \frac{d\tau_{yx}}{dx}\Delta x\right)(\Delta y \Delta z)$$

in flow direction and flow across direction respectively

Shear forces along the flow direction $F_{\tau_{xy}} = F_{\tau_{xy1}} - F_{\tau_{xy2}}$

$$\Rightarrow F_{\tau_{xy}} = \left(\frac{d\tau_{xy}}{dy}\Delta y\right)(\Delta x \Delta z) \tag{6.54a}$$

Shear forces across the flow direction $F_{\tau_{yx}} = F_{\tau_{yx1}} - F_{\tau_{yx2}}$

$$\Rightarrow F_{\tau_{yx}} = \left(\frac{d\tau_{yx}}{dx}\Delta x\right)(\Delta y \Delta z) \tag{6.54b}$$

As per Newton's law of viscosity $\tau_{yx} = \tau_{xy} = \mu\left[\frac{du}{dx} + \frac{dv}{dy} + \frac{dw}{dz}\right]$

$$\Rightarrow \frac{d\tau_{yx}}{dy} = \frac{d\tau_{xy}}{dx} = \mu\left[\frac{d}{dx}\left(\frac{du}{dx}\right) + \frac{d}{dy}\left(\frac{dv}{dy}\right) + \frac{d}{dz}\left(\frac{dw}{dz}\right)\right]$$

$$\Rightarrow \frac{d\tau_{yx}}{dy} = \frac{d\tau_{xy}}{dx} = \mu\left[\left(\frac{d^2u}{dx^2}\right) + \left(\frac{d^2v}{dy^2}\right)\right] \tag{6.54c}$$

Substitute Eq. (6.54c) in Eq. (6.54a & 6.54b), we get shear forces along the flow direction as

$$F_{\tau_{xy}} = \mu\left[\left(\frac{d^2u}{dx^2}\right) + \left(\frac{d^2v}{dy^2}\right)\right](\Delta y)(\Delta x \Delta z) \tag{6.55a}$$

Similarly, shear forces across the flow direction

$$F_{\tau_{yx}} = \mu\left[\left(\frac{d^2u}{dx^2}\right) + \left(\frac{d^2v}{dy^2}\right)\right](\Delta x)(\Delta y \Delta z) \tag{6.56a}$$

Substituting Eq. (6.51a), Eq. (6.52a), Eq. (6.53a) and Eq. (6.56a) in Eq. (6.50), we get **Navier Stokes equation along the flow direction** of a two-dimensional flow as

$$-\frac{dp_x}{dx}(\Delta x)(\Delta y \Delta z) + \mu\left[\left(\frac{d^2u}{dx^2}\right) + \left(\frac{d^2v}{dy^2}\right)\right](\Delta y)(\Delta x \Delta z) + G_x * (\rho \Delta x \Delta y \Delta z) = (\rho \Delta x \Delta y \Delta z)\frac{du}{dt}$$

$$\Rightarrow -\frac{dp_x}{dx} + \mu\left[\left(\frac{d^2u}{dx^2}\right) + \left(\frac{d^2v}{dy^2}\right)\right] + G_x * (\rho) = (\rho)\frac{du}{dt} \tag{6.57a}$$

Similarly, substituting Eq. (6.51a), Eq. (6.52a), Eq. (6.53a) and Eq. (6.56a) in Eq. (6.50), we get **Navier Stokes equation along the flow direction** of a two-dimensional flow as

$$\Rightarrow -\frac{dp_y}{dy} + \mu\left[\left(\frac{d^2u}{dx^2}\right) + \left(\frac{d^2v}{dy^2}\right)\right] + G_y * (\rho) = (\rho)\frac{dv}{dt} \qquad (6.57b)$$

So from equations Eq. (6.57a) and Eq. (6.57b) **generalized equation for Navier Stokes equation** can be written as

$$\rho\frac{dV}{dt} = -\nabla(P_s) + \rho G_s + \mu\nabla^2(V) \qquad (6.58)$$

where $\nabla = \left(\frac{d}{dx} + \frac{d}{dy} + \frac{d}{dz}\right)$ is a Laplace operator and V is the velocity vector.

Generalized equation for Navier Stokes equation including continuity equation, can be written as

$$\rho\left(\frac{dV}{dt} + V\frac{dV}{ds}\right) = -\nabla(P_s) + \rho G_s + \mu\nabla^2(V) \qquad (6.59)$$

Example 6.37: Derive an expression for the pressure force along the flow at a section in a gradually decreasing cross-section pipe, where the area is A_x connected to a reservoir and the other end is left to atmosphere. Assume uniform, invisicid and incompressible flow and also neglect body forces. (b) Comment on the force along the flow direction, if the pipe has gradually decrease in cross-section (i.e. $A_2 < A_1$) and also when gradually increase in cross-section (i.e. $A_2 > A_1$).

Solution: Given data:

ρ = constant, P_{atm}, A_1, A_x, A_2

Gradually decreasing cross section pipe

Consider a gradually decrease in cross-section pipe with inlet (section 1-1) and outlet (section 2-2) diameters as A_1, A_2 respectively. And consider a section at some location at a distance of "x" (named section x-x) in the pipe and apply Bernoulli's equation in between sections x-x and 2-2:

So we get
$$\frac{P_x}{\rho g} + \frac{V_x^2}{2g} = \frac{P_{atm}}{\rho g} + \frac{V_2^2}{2g}$$

$$\Rightarrow \Delta P_x = (P_x - P_{atm}) = \frac{1}{2}\rho\left(V_2^2 - V_x^2\right) = \frac{1}{2}\rho Q^2\left(\frac{1}{A_2^2} - \frac{1}{A_x^2}\right)$$

We know pressure force $F_x = \int_{A_1}^{A_2} \Delta P_x * dA = \frac{1}{2}\rho Q^2 \int_{A_1}^{A_2} \left(\frac{1}{A_2^2} - \frac{1}{A_x^2} \right) dA$

As there is gradual decrease in area $A_2 < A_1$ and so

$$F_x = -\frac{1}{2}\rho Q^2 \int_{A_1}^{A_2} \left(\frac{1}{A_2^2} - \frac{1}{A_x^2} \right) dA = \frac{1}{2}\rho Q^2 \int_{A_2}^{A_1} \left(\frac{1}{A_2^2} - \frac{1}{A_x^2} \right) dA$$

$$\Rightarrow F_x = \frac{1}{2}\rho Q^2 \int_{A_1}^{A_2} \left(\frac{(A_1 - A_2)^2}{A_1 A_2^2} \right)$$

So the pressure force along the flow direction at section "x" will be $F_x = \frac{1}{2}\rho Q^2 \int_{A_1}^{A_2} \left(\frac{(A_1 - A_2)^2}{A_1 A_2^2} \right)$

b) If $A_2 < A_1$, $\Delta P_x = +ve$, then and pressure inside the pipe is more than surrounding pressure and hence the pressure force acts on the inner walls of the pipe along the flow direction.

If $A_2 < A_1$, then $\Delta P_x = -ve$ and pressure inside the pipe is less than surrounding pressure and hence the pressure force acts on the outer walls of the pipe along the flow direction.

Example 6.38: Water is flowing in a horizontal concrete pipe of circular-cross section. The diameter of the pipe is 30 m and the length is 2000 m. The average velocity in the pipe is 48 m/s. The equivalent roughness of the pipe $\varepsilon = 0.01$ m. Find the pressure drop in the pipe.

Solution: Given data:

$$\rho = 1000\,\text{kg/m}^3,\ \mu = 1.12\times10^{-3},\ \varepsilon = 0.01\,\text{m},\ L = 2000\,\text{m},\ d = 30\,\text{m},\ V = 48\,\text{m/s}$$

From the Moody chart, taking the $\frac{\varepsilon}{d} = 0.00033$ and $R_e = \frac{\rho\mu V}{d} = 1.79$, we have friction factor as $f = 0.026$.

According to Darcy equation, we have

$$(P_1 - P_2) = \gamma * h_f = \gamma * \frac{fLV^2}{2gd} = \frac{9810 * 0.026 * 2000 * 48^2}{2 * 9.81 * 30} = 1996.8\,\text{KPa}$$

Hence, the drop in pressure is 1996.8 KPa.

Example 6.39: Two tanks A and B are connected by a pipe of 2000 m long with 3 m diameter and then suddenly reduced to 2 m in diameter for the remaining 1000 m. The difference of levels in the tanks is constant at 30 m. Assume friction factor as 0.005 and coefficient of contraction as 0.58.

Find all the losses including at the sharp edge pipe entry at A in terms of velocity V_2 and determine the flow in m^3/min. Draw the hydraulic gradient diagram.

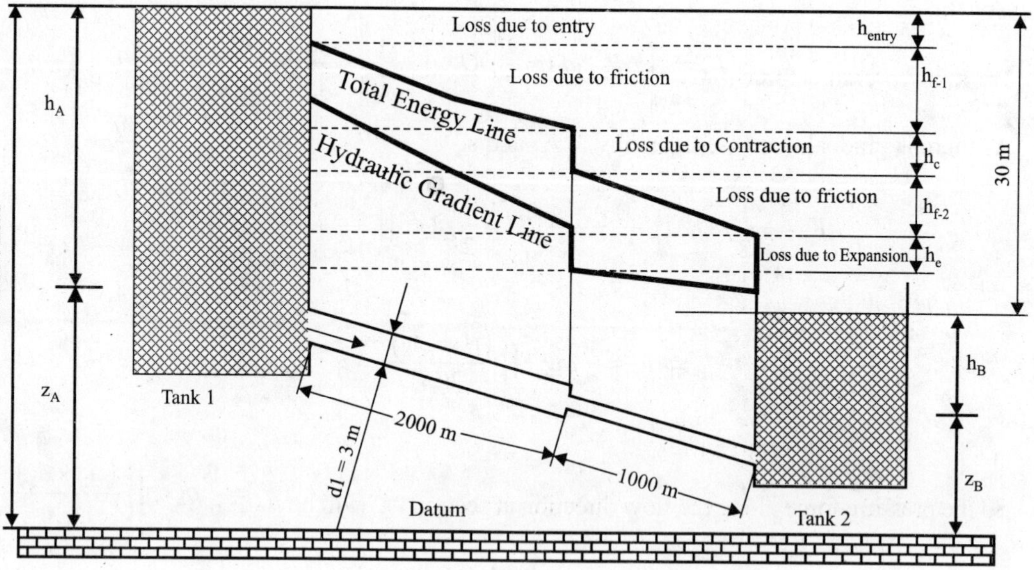

Figure P6.39: TEL and HGL lines

Solution: Given data:

$$\rho = 1000\,\text{kg/m}^3,\ d_1 = 3\,\text{m},\ d_2 = 2\,\text{m},\ L_1 = 2000\,\text{m},\ L_2 = 1000\,\text{m},\ f = 0.005,\ C_c = 0.58$$

For pipes connected in series have the flow rate equal, $Q = Q_1 = Q_2$ and so $V_1 = \left(\frac{d_2}{d_1}\right)^2 V_2 = 0.44\,V_2$

As we know the loss at entry is $h_{entry} = 0.5 * \frac{V_1^2}{2g} = 0.025\,V_1^2 = 0.00484\,V_2^2$

Frictional loss in pipe 1 will be $h_{f_1} = \frac{fLV^2}{2gd} = 0.169\,V_1^2 = 0.0327\,V_2^2$

The loss due to contraction is $h_{entry} = \left(\frac{1}{C_c} - 1\right)^2 * \frac{V_1^2}{2g} = 0.0267\,V_1^2 = 0.0052\,V_2^2$

Frictional loss in pipe 2 will be $h_{f_2} = \frac{fLV^2}{2gd} = \frac{0.005 * 1000 * V_2^2}{2 * 9.81 * 2} = 0.1275\,V_2^2$

The loss due to Expansion is $h_{\text{expansion}} = \frac{V_2^2}{2g} = 0.05\,V_2^2$

Flow rate is $Q = A_1 * V_1 = 0.0518 * V_2$ m³/min.

EXERCISE

Objective

1. For pipes, laminar flow occurs when Reynolds number is
   a) Less than 2000
   b) Between 2000 and 4000
   c) More than 4000
   d) Less than 4000
   e) None of the above
2. At the centre line of a pipe flowing under pressure where the velocity gradient is zero, the shear stress will be
   a) Minimum
   b) Maximum
   c) Zero
   d) Negative value
   e) Could be any value.
3. The friction head loss due to the flow of a viscous fluid through a circular pipe of length L and diameter d with a velocity v, and Fanning pipe friction 'f' is

   (a) $\dfrac{4\lambda LV^2}{2gd}$ (b) $\dfrac{4fLV^2}{g\pi d^2}$

   (c) $\dfrac{4fLV^2}{gd^2}$ (d) None
4. Two pipe systems can be said to be equivalent, when the following quantities are same
   a) Friction loss and flow
   b) Length and diameter
   c) Flow and length
   d) Friction factor and diameter
   e) Velocity and diameter
5. Hydraulic diameter used in place of diameter for non-circular ducts is equal to

   (a) $\dfrac{A_W}{P_W}$ (b) $\dfrac{2A_W}{P_W}$ (c) $\dfrac{4A_W}{P_W}$ (d) $\dfrac{P_W}{A_W}$

   where A = area wetted, P = perimeter wetted.
6. The losses due to contraction are expressed by

   (a) $\dfrac{V_1^2 - V_2^2}{2g}$ (b) $\dfrac{V_2^2 - V_1^2}{2g}$ (c) $\dfrac{V_1^2 - V_2^2}{g}$ (d) $\dfrac{0.5V_2^2}{2g}$
7. The losses due to sudden expansion are expressed by

   (a) $\dfrac{(V_1 - V_2)^2}{2g}$ (b) $\dfrac{V_2^2 - V_1^2}{2g}$ (c) $\dfrac{V_1^2 - V_2^2}{g}$ (d) $\dfrac{0.5V_2^2}{2g}$

   e) none

8. The velocity distribution for flow between two fixed parallel plates
   a) Is constant over the cross-section
   b) Is zero at the plates and increases linearly to the mid-plane
   c) Varies parabolically across the section
   d) is zero in middle and increases linearly towards the plates
   e) None
9. According to Darcy's formula, loss of head due to friction in the pipe is

   (a) $\frac{4flv^2}{gd}$ (b) $\frac{flv^2}{gd}$ (c) $\frac{flv^2}{2gd}$ (d) $\frac{4flv}{gd}$ (e) $\frac{4flv^2}{d}$

   where f = Darcy's friction coefficient, l = length of pipe, v = velocity of liquid of flow, d = diameter of pipe.
10. Friction factor for pipes depends on
    a) Rate of flow
    b) Fluid density
    c) Viscosity
    d) Pipe roughness
    e) All of the above
11. In series–pipe applications
    a) The head losses through each pipe are added to obtain the total head loss
    b) The head loss is same through each pipe
    c) Friction factors are assumed for each pipe
    d) Flow increases
    e) None of the above
12. In the case flow through parallel pipes
    a) Flow in each pipe is same
    b) Head loss in each pipe is same
    c) Head loss depends upon flow conditions
    d) Total head loss is sum of the head losses in individual pipes
    e) None of the above
13. If l_1, l_2, l_3 ... be the lengths and d_1, d_2, d_3 ... be the diameter of the parts of a compound pipe, then length L_{eq} and diameter D_{eq} of a uniform equivalent pipe will be related an under

    a) $\frac{L_{eq}}{D_{eq}} = \frac{l_1}{d_1} + \frac{l_2}{d_2} + \frac{l_3}{d_3} + \ldots$ b) $\frac{L_{eq}}{D_{eq}^2} = \frac{l_1}{d_{1^2}} + \frac{l_2}{d_{2^2}} + \frac{l_3}{d_{3^2}} + \ldots$

    c) $\frac{L_{eq}}{D_{eq}} = \frac{l_1}{d_{1^3}} + \frac{l_2}{d_{2^3}} + \frac{l_3}{d_{3^3}} + \ldots$ d) $\frac{L_{eq}}{D_{eq}} = \frac{l_1}{d_{1^4}} + \frac{l_2}{d_{2^4}} + \frac{l_3}{d_{3^4}} + \ldots$

    e) $\frac{L_{eq}}{D_{eq}} = \frac{l_1}{d_1^5} + \frac{l_2}{d_2^5} + \frac{l_3}{d_3^5} + \ldots$

14. The hydraulic radius is given by
    a) Wetted perimeter divided by area
    b) Area divided by squared of wetted perimeter
    c) Wetted area divided by wetted perimeter
    d) Square root of area
    e) none
15. The maximum velocity of a one-dimensional incompressible fully developed viscous flow, between two fixed parallel plates is 6 m/s. The mean velocity (in m/s) of the flow is
    a) 2
    b) 3
    c) 4
    d) 5
16. Can the Darcy-Weisbach equation and Moody diagram be used for fluids other than water?
    a) no
    b) cannot say
    c) yes
17. Do smooth pipes have head loss and is it possible to decrease the head loss in a pipe by installing a smooth liner?
    a) no
    b) yes
    c) cannot say
18. Water at 25°C is flowing through a 1.0 km long G.I. pipe of 200 mm diameter at the rate of 0.07 m^3/s. If value of Darcy friction factor for this pipe is 0.02 and density of water is 1000 kg/m^3, the pumping power (in kW) required to maintain the flow is
    a) 1.8
    b) 17.4
    c) 20.5
    d) 41.0
19. The hydraulic gradient line always lies below the total energy line
    a) False
    b) True
    c) Cannot say
20. The hydraulic line do not fluctuates with respect to the pressure changes.
    a) False
    b) True
    c) Cannot say

Answers

1. a	2. c	3. a	4. a	5. a	6. d
7. a	8. c	9.c	10. e	11. a	12. b
13. e	14. c	15. b	16. c	17. b	18.b
19. b	20. a				

Theory

1. Explain why the negative pressure gradient is favorable.
2. Explain why the positive pressure gradient is adverse.
3. Derive Hazen-Poiseulle equation for laminar flow in circular pipes and extend it to derive Friction factor.
4. Explain different types of losses.
5. Derive the head losses due to sudden enlargement.
6. Derive the head losses due to sudden contraction.
7. Derive the forces due to pipe bends.
8. Derive the head losses due to porous medium.
9. Write a note on potential flow and derive the head loss equation.
10. Explain perforated pipe flows.
11. Write a note in Navier Stokes equation and solutions.

Problems

1. Gasoline at 20°C temperature is pumped at 0.2 m^3/s through a 15 km long cast iron pipe of diameter 0.18 m. Compute the power required in kilowatts if the pumps are 80 percent efficient. Take friction factor as 0.0216.
2. Crude oil is pumped through a refinery pipeline at a rate of 2,50,000 m^3/day. The diameter of the pipeline is 120 mm. The pressure at the just outlet of the pump is 8.5 kPa and the pressure outside the pipeline is 340 kPa. Determine the length of the pipeline that the pump can pump the crude oil, for these conditions, if the specific gravity of the oil is 0.9 and viscosity is 0.017 Ns/m^2. And also calculate the power required by the pump if it has a efficiency of 80 percent.
3. A steady on the plunger causes a flow rate of 0.4 cm^3/s of fluid flow in the rigid cylinder, through a needle opening. The diameters of the cylinder and the needle opening are 10 mm, 0.25 mm respectively. Determine the force required to maintain the flow push, if the fluid has a density of 900 kg/m^3 and viscosity of 0.002 kg/m-s.
4. A 30 cm diameter pipe is bifurcated into two nozzles at a Y-junction having diameter of 7.5 cm and 10cm diameter respectively. The 7.5 cm diameter pipe bifurcates with an angle of 25° to the horizontal, whereas the 10 cm diameter pipe bifurcates with an angle of 35° to the horizontal. The velocity of the flow at the exit of the two nozzles is 10 m/s. The junction is in single plane. Neglecting friction, determine the resultant force acting and its direction.
5. A non-uniform pipe of diameter 30 cm and 15 cm at the inlet and outlet respectively bends from inlet to outlet by an angle of 120°. The axis at the inlet is horizontal and 1.5 m above the outlet section. Rate of flow of fluid in this pipe bend is 0.9 m^3/s. Neglecting friction, determine the magnitudes and directions of the resultant force acting, when fluid is water and has inlet pressure of 0.15 MPa.
6. Three reservoirs A, B and C are connected by a pipe system. Find the discharge in all the pipes, if the tank A is 70 m, Tank B is 30 m and Tank C 15 m above from the ground respectively. Take all the pipes are of lengths 1500 m, 0 cm diameter and friction factor f = 0.006 for all pipes.

7. Consider the flow of glycerine between two parallel plates having a separation of 36 cm. The upper plate moves to the right with a velocity of 10 m/s, and the lower plate moves to the left with a velocity of 5 m/s. Assume that there is no pressure gradient and ignore gravity.

   Assume that the flow is uniform in x, and that there is no dependence on the coordinate perpendicular to the picture below.

   (a) Write down the x-component of the steady two-dimensional Navier Stokes equations, and eliminate all the terms that are zero in this particular flow. Explain each elimination. After eliminating the zero terms solve for the x-component of velocity.

   (b) Use your solution to find the force per unit area required to pull the upper plate.

   (c) Find the average velocity in the region between the plates.

8. Two tanks A and B are 1000 m apart and are connected by two pipes of 2 m and 3 m in diameters. The height of water in tank A is maintained at 250 m above the tank B. Assume the friction factor for both the pipes as 0.008 and coefficient of contraction at the entry as 0.585.

   Calculate the rate of flow in each pipe and draw the hydraulic gradient. If the pipes to be replaced by one larger pipe, calculate the diameter of the pipe required for the total rate of flow from A and B, to remain unchanged.

QUICK REFERENCE OF FORMULAE

Property	Formulae	Remarks, if any
General equation of **Couette flow**	$u = \frac{U_{MAX} * y}{h} + \frac{1}{2\mu}\left(\frac{dp}{dx}\right)[y^2 - hy]$ $\tau = \mu\left(\frac{du}{dy}\right) = \mu\left(\frac{U_{MAX}}{h} - \frac{1}{2\mu}\left(\frac{dp}{dx}\right)[h - 2y]\right)$ $Q = \frac{U_{MAX}h}{2} - \frac{h^3}{12\mu}\left(\frac{dp}{dx}\right)$	***u***, U_{Max} are velocity and maximum respectively. *y* is distance of fluid particle from fixed plate. *h* is distance between two plates. *Q* is discharge, μ is viscosity.
Hagen Poiseulle equations for a circular pipe	$\tau = \left(\frac{-dp}{dx}\right)\frac{r}{2}, U_{MAX} = \frac{-1}{4\mu}\left(\frac{dp}{dx}\right)[R^2]$ $\frac{u}{U_{MAX}} = \left[1 - \left(\frac{r}{R}\right)^2\right], u_{AV} = \frac{U_{MAX}}{2}$ $h_f = \frac{(P_1 - P_2)}{\gamma} = \frac{32\mu u_{AV}L}{\gamma D^2}; h_f = \frac{(P_1 - P_2)}{\gamma} = \frac{d\mu u_{AV}L}{\gamma D_h}$	τ is shear stress, U_{AV} is average velocity of velocity, μ is viscosity, h_f is head loss due to friction, D_h is hydraulic diameter.
Fanning friction factor λ and Darcy friction factor ($f = 4\lambda$)	$f = \left(\frac{64}{R_e}\right), \lambda = \left(\frac{16}{R_e}\right)$	$R_e = \frac{\rho u_{av}L}{\mu} = \frac{\rho u_{av}D}{\mu}$ is Reynolds number.
Head loss due to friction h_f	$h_f = \frac{(P_1 - P_2)}{\gamma} = \frac{fLV^2}{2gD}; h_f = \frac{(P_1 - P_2)}{\gamma} = \frac{2\lambda LV^2}{gD}$	($f = 4\lambda$)
Head loss due to sudden expansion h_E	$h_E = \frac{(V_1 - V_2)^2}{2g}$	
Head loss due to sudden contraction h_C	$h_C = K\frac{(V_2)^2}{2g}; K = \left(\frac{1}{C_C} - 1\right)^2; C_C = \frac{A_0}{A_2}$	where *K* is loss coefficient and C_C, A_0, A_2 is coefficient of contraction, areas at throat and outlet.
Forces in pipe bend	$F_X = \frac{\gamma Q}{g}(V_2 \cos\theta_{2x} - V_1 \cos\theta_{1x}) + p_1 A_1 \cos\theta_{1x} - p_2 A_2 \cos\theta_{2x}$ $F_Y = \frac{\gamma Q}{g}(V_2 \sin\theta_{2y} - V_1 \sin\theta_{1y}) + p_1 A_1 \sin\theta_{1y} - p_2 A_2 \sin\theta_{2y}$ $F_R = \sqrt{F_X^2 + F_Y^2}; \theta = Tan^{-1}\left(\frac{F_Y}{F_X}\right)$	F_x, F_y, F_R, θ are force components along X-axis and Y-axis and resultant force and its inclination. $\frac{\gamma Q}{g}$ is mass flux.

Contd.

QUICK REFERENCE OF FORMULAE

Property	**Formulae**	**Remarks, if any**
Head loss due to porous material h_p	$\varepsilon = \frac{V_V}{V}; \lambda R_e = \frac{\varepsilon D^2}{2K}; h_p = \left(\frac{\varepsilon D \mu u_{AV}}{K\gamma}\right)$	ε, V_v, V, K are Porosity, volume of voids, volume of pipe and permeability.
Fow potential and flow resistance	$R = \left(1.5 + \frac{fL}{D}\right)\left(\frac{1}{2a^2 g}\right); h_L = \Delta h = RQ^2$ $\Rightarrow h_L = (h_A + Z_A) - (h_B + Z_B) = \Delta h$	R, Δh, f, h_A, h_B are resistance, difference in elevation of two reservoirs, Darcy friction factor and pressure heads respectively.
Flow through pipes connected in **series**.	$Q = Q_1 = Q_2 = Q_3 = \ldots$ $h_{feq} = h_{f1} + h_{f2} + h_{f3} + \ldots$ $\frac{L_{eq}}{D_{eq}^5} = \frac{L_1}{D_1^5} + \frac{L_2}{D_2^5} + \frac{L_3}{D_3^5} + \ldots$	h_{feq}, L_{eq}, D_{eq} are Frictional head loss, length and diameter of equivalent pipe.
Flow through pipes in **parallel**.	$Q = Q_1 + Q_2 + Q_3 + \ldots \,\&\, h_{f1} = h_{f2} = h_{f3} \ldots$	
Generalized Navier Stockes equation.	$\rho\left(\frac{dV}{dt} + V\frac{dV}{ds}\right) = -\nabla(P_s) + \rho G_s + \mu\nabla^2(V)$	$\nabla = \left(\frac{d}{dx} + \frac{d}{dy} + \frac{d}{dz}\right)$ Laplace operator

Points to Remember

Hydraulic diameter D_h	$\frac{D_h}{4} = \frac{A_w}{P_w}$	A_w, P_w are wetted area and wetted perimeter.
If branched pipes are parallel	$h_{f1} = h_{f2} = h_{f3} = \ldots$	h_{f1}, h_{f2}, ... are head friction losses in the branched parallel pipes.
Head loss due to perforated pipes h_{f1}	$h_{f1} = \frac{1}{3} h_f$	h_f is head loss due to pipe without perforated pipes.

CHAPTER 7

Boundary Layer Equations

INTRODUCTION

Fluid molecule layers flowing over a flat surface with no-slip condition possess maximum velocity at the free surface layers. The molecular layers at the contact of the flat surface do adhere as per the viscous theory, which give large shear gradient and extends up to some vertical distance and therefrom the free surface stream layers exists. Hence, flow past over a flat surface have two layer regions, one having high viscous layers which is of our interest and called "**boundary layer region**" and other layer with free viscous layers, as shown in Figure 7.0. Defining the boundary layer, the region where velocity gradients are high producing significant amount of viscous forces. And the boundary layer region thickness can be found out using displacement, energy or momentum of boundary layers methods, which have been discussed in this chapter.

The boundary layer region may be further classified into laminar and turbulent boundary sub-layers basing on the momentum and drag effects in the boundary layer region and is given by Von-Karman.

If considered fluid layers in closed conduits, the viscous effect does have predominance over the entire cross-section in a symmetric fashion and prevails over the whole cross-sectional area and length, give no chance for free stream region formation and is called as "**fully developed boundary layers**", as shown in Figure 7.0a. The laminar or turbulence depends on the Reynolds number.

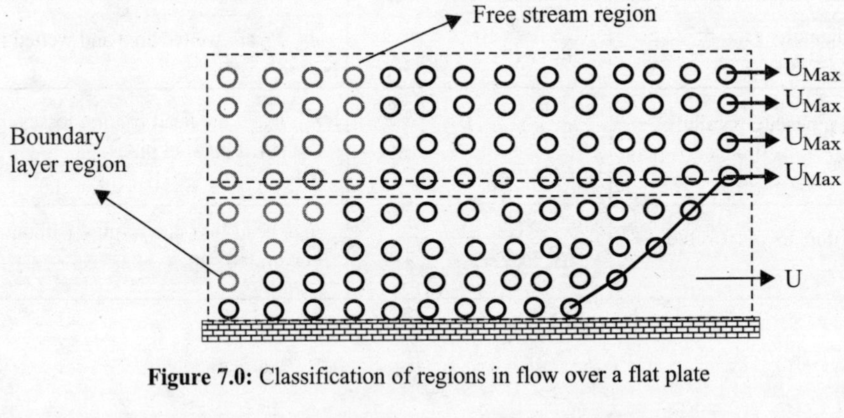

Figure 7.0: Classification of regions in flow over a flat plate

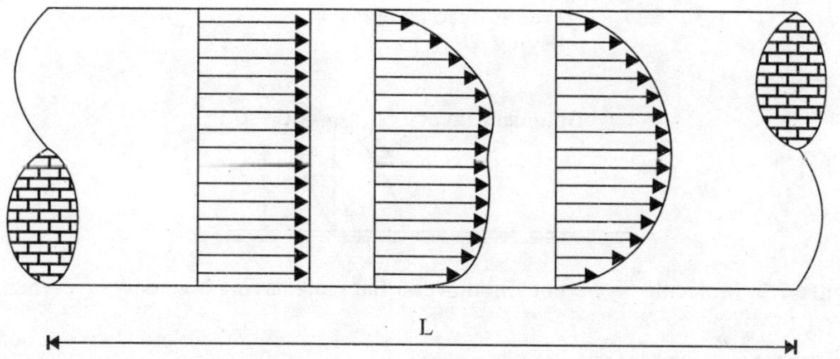

Figure 7.0a: Fully developed boundary layer region in pipe flow

BOUNDARY LAYER THICKNESS

When a viscid fluid flow is led to flow over a flat surface with a velocity U_{Max} the free surface molecule layers have same velocity as that of the dragged velocity U_{Max} up to some thickness, as the inertia dominates the viscous effects. The layers below this thickness region have strong tendency to pull back the above consecutive layers due to viscosity and these region layers do have little lesser velocity u . This phenomenon prevails over this entire region thickness and is called as **boundary layer thickness (δ)**, due to the increase in viscous effect domination over the inertial, as shown in Figures 7.1 and 7.1a and hence the boundary layer thickness (δ) prevails in the region having some thickness wherein the layers do possess high shear stress.

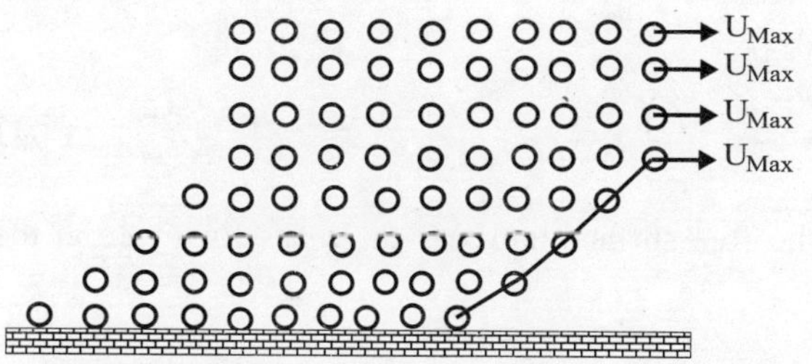

Figure 7.1: Molecule layers in motion over a flat plate

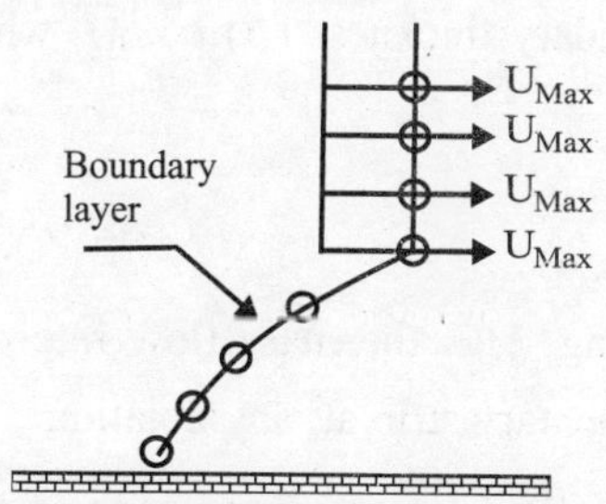

Figure 7.1a: Molecule layers in motion over a flat plate

If the movements of each layer forwarded in the boundary layer region are traced, then it forms a curve called boundary layer curve, as shown in Figures 7.1a & 7.2. And boundary layer thickness (δ) is defined as **"the distance of the velocity of viscous layer from the boundary surface, whose velocity value tends to 0.99 times the velocity of free surface velocity"**.

The boundary layer thickness can be found out in three methods, like

1) Displacement thickness method (δ_d)
2) Momentum thickness method (δ_M)
3) Energy thickness method (δ_E)

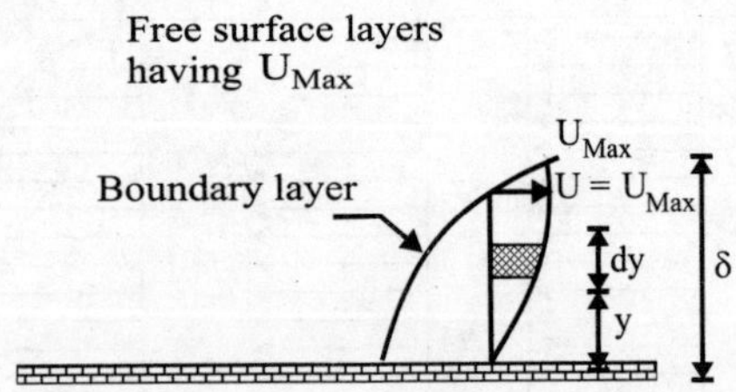

Figure 7.2: Molecule layers in motion over a flat plate having boundary layer thickness

Displacement Thickness Method (δ_d)

Displacement method is based on the velocity of displaced layers and is defined as "**the perpendicular distance of the viscous layer from the boundary surface, whose velocity value tends to 0.99 times the velocity of free surface velocity**".

Once the fluid layers move from one position to the other, in the boundary layer region, there arise change in mass flow rate $\bar{M}$ between fluid layers along the boundary layer thickness, in perpendicular distance from the flat plate and this change in mass flow rate is a loss of mass flow rate $\bar{M}_L$ due to viscous effect present in the boundary layer up to a thickness of (δ_d) and hence

$$\bar{M}_L = \bar{M} \tag{7.0}$$

Consider a fluid elementary strip thickness of dimension (*dy*, 1) at a distance of "y" from the flat plate, as shown in Figure 7.3.

Change in total mass flow rate occurs during the displacement of molecule layers, over entire boundary thickness (δ) be $\bar{M}$ where

$$\bar{M} = \int_0^{\delta} \left(\bar{M}_2 - \bar{M}_1\right) \tag{7.1}$$

having $\bar{M}_2$ the mass flow rate occurred at the free surface and $\bar{M}_1$ the mass flow rate at the elementary strip at, any position.

Mass flow rate occurred at the free surface

$$\bar{M}_2 = \rho\left(dy * 1\right)U_{\max} \tag{7.1a}$$

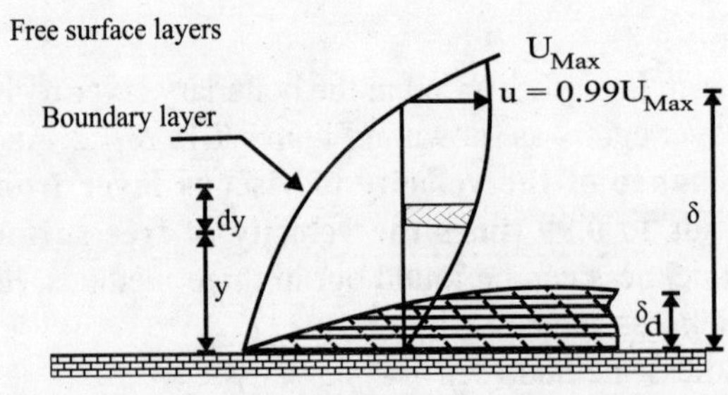

Figure 7.3: Displacement boundary layer thickness

mass flow rate in the boundary layer

$$\bar{M}_1 = \rho(dy * 1)u \tag{7.1b}$$

Substituting Eq. (7.1b), Eq. (7.1a) in Eq. (7.1)

$$\bar{M} = \int_0^\delta \rho(U_{max} - u)\,dy \tag{7.2}$$

Total loss of mass flow rate of fluid is

$$\bar{M}_L = \rho AV = \rho(\delta_d * 1)U_{max} \tag{7.3}$$

Observing Eq. (7.0), we have Eq. (7.3) is equal to Eq. (7.2)

$$\Rightarrow \rho\delta_d U_{max} = \int_0^\delta \rho(U_{max} - u)\,dy$$

$$\boxed{\delta_d = \int_0^\delta \left(1 - \frac{u}{U_{max}}\right)dy} \tag{7.4}$$

This is the **equation for displacement boundary layer thickness.**

Momentum Thickness Method (δ_M)

This method is based on the momentum of displaced layers and is defined as **"the perpendicular distance of the viscous layer from the boundary surface, whose momentum flux corresponding to the free stream velocity is the loss in momentum due to the boundary layer formation"**.

Once the fluid layers move from one position to the other, in the boundary layer region, there arise change in momentum flow rate $\bar{M}V$ between fluid layers along the boundary layer thickness, in perpendicular distance from the flat plate and this change in momentum flow rate is a loss of mass flow rate $\bar{M}_v$ due to viscous effect present in the boundary layer up to a thickness of (δ_M) and hence

$$\bar{M}V_m = \bar{M}V \tag{7.5}$$

Consider a fluid elementary strip thickness of dimension (*dy*, 1) at a distance of "y" from the boundary plate, as shown in Figure 7.3.

Change in total momentum flow rate occurs during the movement of molecule layers, over entire boundary thickness (δ) be $\bar{M}V$ where

$$\bar{M} = \int_0^\delta \left(\bar{M}V_2 - \bar{M}_1\right) \tag{7.6}$$

having $\bar{M}V_2$ the momentum flow rate occurred at the free surface and $\bar{M}V_1$ the momentum flow rate at the elementary strip at, any position.

Momentum rate at the free surface

$$\bar{M}V_2 = (\rho AV)V = \rho u(dy*1)U_{\max} \tag{7.6a}$$

Momentum rate in the boundary layer

$$\bar{M}V_1 = (\rho AV)V = \rho(dy*1)u^2 \tag{7.6b}$$

Substituting Eq. (7.6b) and Eq. (7.6a) in Eq.(7.6),

$$\bar{M}V = \int_0^{\delta} \rho u(U_{\max} - u)\,dy \tag{7.7}$$

Total loss of momentum rate of fluid is

$$\bar{M}V_m = (\rho AV)V = \rho\delta_M U_{\max}^2 \tag{7.8}$$

Observing Eq. (7.5), we have Eq. (7.8) equal to Eq. (7.7)

$$\Rightarrow \rho\delta_M U_{\max}^2 = \int_0^{\delta} \rho u(U_{\max} - u)\,dy$$

$$\delta_M = \int_0^{\delta} \frac{u}{U_{\max}}\left(1 - \frac{u}{U_{\max}}\right) dy \tag{7.9}$$

This is the **equation for momentum boundary layer thickness.**

Energy Thickness Method (δ_E)

This method is based on the energy of motion, i.e. kinetic energy of displaced layers and is defined as **"the amount of distance by which the surface would have to be moved, so that there is no boundary layer, and the kinetic energy is same as that of actual free flow".**

Once the fluid layers move with kinetic energy from one position to the other, in the boundary layer region, there arise change in kinetic energy KE between fluid layers along the boundary layer thickness, in perpendicular distance from the flat plate and this change in kinetic energy is a loss of energy KE_L due to viscous effect present in the boundary layer up to a thickness of (δ_M) and hence

$$KE_L = KE \tag{7.10}$$

Consider a fluid elementary strip thickness of dimension (*dy*, 1) at a distance of "*y*" from the boundary plate, as shown in Figure 7.3.

Change in total kinetic energy occurs during the movement of molecule layers, over entire boundary thickness (δ) be *KE* where

$$KE = \int_0^{\delta} (KE_2 - KE_1) \tag{7.11}$$

having KE_2 as the kinetic energy occurred at the free surface and KE_1 as the kinetic energy at the elementary strip at, any position.
Kinetic energy at free surface

$$KE_2 = \frac{1}{2}mV^2 = \frac{1}{2}\left[\rho u(dy*1)\right]U^2_{\max} \tag{7.11a}$$

Kinetic energy in the boundary layer

$$KE_1 = \frac{1}{2}(m)V^2 = \frac{1}{2}\left[\rho u(dy*1)\right]u^2 \tag{7.11b}$$

Substituting Eq. (7.11b), Eq. (7.11a) in Eq. (7.11)

$$KE = \int_0^\delta \frac{\rho u}{2}\left(U^2_{\max} - u^2\right)dy \tag{7.12}$$

Total loss of kinetic energy of fluid is

$$KE_L = \frac{1}{2}(m)V^2 = \frac{1}{2}(\rho\delta_E U_{MAX})U^2_{\max} \tag{7.13}$$

Observing Eq. (7.10), we have Eq. (7.13) equal to Eq. (7.12)

$$\Rightarrow \frac{1}{2}(\rho\delta_E U_{MAX})U^2_{\max} = \int_0^\delta \frac{\rho u}{2}\left(U^2_{\max} - u^2\right)dy$$

$$\delta_E = \int_0^\delta \frac{u}{U_{\max}}\left(1 - \frac{u^2}{U^2_{\max}}\right)dy \tag{7.14}$$

This is the **equation for energy boundary layer thickness.**

Example 7.1: For the given velocity profile $\frac{u}{U_{Max}} = 2\left(\frac{y}{\delta}\right) - \left(\frac{y}{\delta}\right)^2$, determine

a) Displacement thickness.
b) Momentum thickness.
c) Energy thickness.
d) Shape factor $\left(\delta_d / \delta_M\right)$

Solution: Given data: velocity profile

$$\frac{u}{U_{Max}} = 2\left(\frac{y}{\delta}\right) - \left(\frac{y}{\delta}\right)^2$$

a) We have **displacement thickness** as

$$\delta_d = \int_0^{\delta}\left(1 - \frac{u}{U_{\max}}\right)dy$$

Substituting the velocity profile in the above equation, we get

$$\delta_d = \int_0^{\delta}\left(1 - 2\left(\frac{y}{\delta}\right) + \left(\frac{y}{\delta}\right)^2\right)dy$$

$$\Rightarrow \delta_d = \left| y - \frac{2}{2}\left(\frac{y^2}{\delta}\right) + \left(\frac{y^3}{3\delta^2}\right)\right|_0^{\delta} = \frac{\delta}{3}$$

b) We have **momentum thickness** as

$$\delta_M = \int_0^{\delta}\frac{u}{U_{\max}}\left(1 - \frac{u}{U_{\max}}\right)dy$$

Substituting the velocity profile in the above equation, we get

$$\delta_M = \int_0^{\delta} 2\left(\frac{y}{\delta}\right) - \left(\frac{y}{\delta}\right)^2\left(1 - 2\left(\frac{y}{\delta}\right) + \left(\frac{y}{\delta}\right)^2\right)dy$$

$$\Rightarrow \delta_M = \left|\frac{2}{2}\left(\frac{y^2}{\delta}\right) - \left(\frac{y^3}{3\delta^2}\right) - \frac{4}{3}\left(\frac{y^3}{\delta^2}\right) + \frac{2}{4}\left(\frac{y^4}{\delta^3}\right) + \frac{2}{4}\left(\frac{y^4}{\delta^3}\right) - \frac{1}{5}\left(\frac{y^5}{\delta^4}\right)\right|_0^{\delta} = \frac{2}{15}\delta$$

c) We have **energy thickness** as

$$\delta_e = \int_0^{\delta}\frac{u}{U_{\max}}\left(1 - \frac{u^2}{U^2_{\max}}\right)dy$$

Substituting the velocity profile in the above equation, we get

$$\delta_e = \int_0^{\delta} 2\left(\frac{y}{\delta}\right) - \left(\frac{y}{\delta}\right)^2\left(1 - \left[2\left(\frac{y}{\delta}\right) - \left(\frac{y}{\delta}\right)^2\right]^2\right)dy$$

$$\Rightarrow \delta_e = \left| \frac{2}{2}\left(\frac{y^2}{\delta}\right) - \left(\frac{y^3}{3\delta^2}\right) - \frac{4}{4}\left(\frac{y^4}{\delta^3}\right) + \frac{8}{5}\left(\frac{y^5}{\delta^4}\right) - \left(\frac{y^6}{\delta^5}\right) + \frac{1}{7}\left(\frac{y^7}{\delta^6}\right) \right|_0^\delta = \frac{43}{105}\delta$$

d) Shape factor: $\left(\frac{\delta_d}{\delta_M}\right) = \frac{1}{3} * \frac{15}{2} = \frac{5}{2}$

Example 7.2: For the given velocity profile $\frac{u}{U_{Max}} = \frac{3}{2}\left(\frac{y}{\delta}\right) - \frac{1}{2}\left(\frac{y}{\delta}\right)^3$, determine

a) Displacement thickness.
b) Momentum thickness.
c) Energy thickness.
d) Shape factor $\left(\delta_d / \delta_M\right)$

Solution: Given data: velocity profile

$$\frac{u}{U_{Max}} = \frac{3}{2}\left(\frac{y}{\delta}\right) - \frac{1}{2}\left(\frac{y}{\delta}\right)^3$$

a) We have **displacement thickness** as

$$\delta_d = \int_0^\delta \left(1 - \frac{u}{U_{\max}}\right) dy$$

Substituting the velocity profile in the above equation, we get

$$\delta_d = \int_0^\delta \left(1 - \frac{3}{2}\left(\frac{y}{\delta}\right) + \frac{1}{2}\left(\frac{y}{\delta}\right)^3\right) dy$$

$$\Rightarrow \delta_d = \left| y - \frac{3}{2*2}\left(\frac{y^2}{\delta}\right) + \frac{1}{2*4}\left(\frac{y^4}{\delta^3}\right) \right|_0^\delta = \frac{3\delta}{8}$$

b) We have **momentum thickness** as $\delta_M = \int_0^\delta \frac{u}{U_{\max}}\left(1 - \frac{u}{U_{\max}}\right) dy$

Substituting the velocity profile in the above equation, we get

$$\delta_M = \int_0^\delta \left(\frac{3}{2}\left(\frac{y}{\delta}\right) - \frac{1}{2}\left(\frac{y}{\delta}\right)^3\right)\left(1 - \frac{3}{2}\left(\frac{y}{\delta}\right) + \frac{1}{2}\left(\frac{y}{\delta}\right)^3\right) dy$$

$$\Rightarrow \delta_M = \left| \frac{3}{4}\left(\frac{y^2}{\delta}\right) - \frac{1}{8}\left(\frac{y^4}{\delta^3}\right) - \frac{9}{12}\left(\frac{y^3}{\delta^2}\right) + \frac{3}{20}\left(\frac{y^5}{\delta^4}\right) + \frac{3}{20}\left(\frac{y^5}{\delta^4}\right) - \frac{1}{28}\left(\frac{y^7}{\delta^6}\right) \right|_0^\delta = \frac{39\delta}{280}$$

c) We have **energy thickness** as $\delta_e = \int_0^\delta \frac{u}{U_{\max}}\left(1 - \frac{u^2}{U^2_{\max}}\right) dy$

Substituting the velocity profile in the above equation, we get

$$\delta_e = \int_0^\delta \frac{3}{2}\left(\frac{y}{\delta}\right) - \frac{1}{2}\left(\frac{y}{\delta}\right)^3 \left(1 - \left[\frac{3}{2}\left(\frac{y}{\delta}\right) - \frac{1}{2}\left(\frac{y}{\delta}\right)^3\right]^2\right) dy$$

$$\Rightarrow \delta_e = \left| \frac{3}{2*2}\left(\frac{y^2}{\delta}\right) - \frac{31}{8*4}\left(\frac{y^4}{\delta^3}\right) + \frac{27}{8*6}\left(\frac{y^6}{\delta^5}\right) - \frac{9}{8*8}\left(\frac{y^8}{\delta^7}\right) + \frac{1}{8*10}\left(\frac{y^{10}}{\delta^9}\right) \right|_0^\delta = \frac{1323}{8640}\delta$$

d) Shape factor: $\left(\frac{\delta_d}{\delta_M}\right) = \frac{3}{8} * \frac{280}{39} = \frac{35}{13}$

Example 7.3: For the given velocity profile $\frac{u}{U_{Max}} = \left(\frac{y}{\delta}\right)$,

determine

a) Displacement thickness.
b) Momentum thickness.
c) Energy thickness.
d) Shape factor $\left(\delta_d / \delta_M\right)$

Solution: Given data: velocity profile $\frac{u}{U_{Max}} = \left(\frac{y}{\delta}\right)$

a) We have **displacement thickness** as $\delta_d = \int_0^\delta \left(1 - \frac{u}{U_{\max}}\right) dy$

Substituting the velocity profile in the above equation, we get

$$\delta_d = \int_0^\delta \left(1 - \left(\frac{y}{\delta}\right)\right) dy$$

$$\Rightarrow \delta_d = \left| y - \frac{1}{2}\left(\frac{y^2}{\delta}\right) \right|_0^\delta = \frac{\delta}{2}$$

b) We have **momentum thickness** as $\delta_M = \int_0^{\delta} \frac{u}{U_{\max}}\left(1-\frac{u}{U_{\max}}\right)dy$

Substituting the velocity profile in the above equation, we get

$$\delta_M = \int_0^{\delta}\left(\frac{y}{\delta}\right)\left(1-\left(\frac{y}{\delta}\right)\right)dy$$

$$\Rightarrow \delta_M = \left|\frac{1}{2}\left(\frac{y^2}{\delta}\right)-\frac{1}{3}\left(\frac{y^3}{\delta^2}\right)\right|_0^{\delta} = \frac{1}{6}\delta$$

c) We have **energy thickness** as $\delta_e = \int_0^{\delta} \frac{u}{U_{\max}}\left(1-\frac{u^2}{U^2_{\max}}\right)dy$

Substituting the velocity profile in the above equation, we get

$$\delta_e = \int_0^{\delta}\left(\frac{y}{\delta}\right)\left(1-\left[\left(\frac{y}{\delta}\right)\right]^2\right)dy$$

$$\Rightarrow \delta_e = \left|\frac{1}{2}\left(\frac{y^2}{\delta}\right)-\frac{1}{4}\left(\frac{y^4}{\delta^3}\right)\right|_0^{\delta} = \frac{1}{2}\delta$$

d) Shape factor: $\left(\frac{\delta_d}{\delta_M}\right) = \frac{1}{2} * \frac{6}{1} = 3$

Example 7.4: For the given velocity profile $\frac{u}{U_{Max}} = 1+\left(\frac{y}{\delta}\right)$, determine

a) Displacement thickness.
b) Momentum thickness.
c) Energy thickness.
d) Shape factor $\left(\delta_d/\delta_M\right)$

Solution: Given data: velocity profile $\frac{u}{U_{Max}} = 1+\left(\frac{y}{\delta}\right)$

a) We have **displacement thickness** as $\delta_d = \int_0^{\delta}\left(1-\frac{u}{U_{\max}}\right)dy$

Substituting the velocity profile in the above equation, we get

$$\delta_d = \int_0^{\delta}\left(1-1-\left(\frac{y}{\delta}\right)\right)dy$$

$$\Rightarrow \delta_d = \left|-\frac{1}{2}\left(\frac{y^2}{\delta}\right)\right|_0^{\delta} = -\frac{\delta}{2}$$

b) We have **momentum thickness** as $\delta_M = \int_0^{\delta}\frac{u}{U_{\max}}\left(1-\frac{u}{U_{\max}}\right)dy$

Substituting the velocity profile in the above equation, we get

$$\delta_M = \int_0^{\delta}\left(1+\frac{y}{\delta}\right)\left(1-1-\left(\frac{y}{\delta}\right)\right)dy$$

$$\Rightarrow \delta_M = \left|-\frac{1}{2}\left(\frac{y^2}{\delta}\right)-\frac{1}{3}\left(\frac{y^3}{\delta^2}\right)\right|_0^{\delta} = \frac{-5}{6}\delta$$

c) We have **energy thickness** as $\delta_e = \int_0^{\delta}\frac{u}{U_{\max}}\left(1-\frac{u^2}{U^2_{\max}}\right)dy$

Substituting the velocity profile in the above equation, we get

$$\delta_e = \int_0^{\delta}\left(1+\frac{y}{\delta}\right)\left(1-\left[1+\left(\frac{y}{\delta}\right)\right]^2\right)dy$$

$$\Rightarrow \delta_e = \left|-\frac{2}{2}\left(\frac{y^2}{\delta}\right)-\frac{3}{3}\left(\frac{y^3}{\delta^2}\right)-\frac{1}{4}\left(\frac{y^4}{\delta^3}\right)\right|_0^{\delta} = \frac{-9}{4}\delta$$

d.) Shape factor: $\left(\frac{\delta_d}{\delta_M}\right) = \frac{1}{2} * \frac{6}{5} = \frac{3}{5}$

Example 7.5: For the given velocity profile $\frac{u}{U_{Max}} = \left(\frac{y}{\delta}\right)^{1/7}$, determine

a) Displacement thickness.
b) Momentum thickness.
c) Energy thickness.
d) Shape factor $\left(\delta_d / \delta_M\right)$

Solution: Given data: velocity profile $\frac{u}{U_{Max}} = \left(\frac{y}{\delta}\right)^{1/7}$

a) We have **displacement thickness** as $\delta_d = \int_0^\delta \left(1 - \frac{u}{U_{\max}}\right) dy$

Substituting the velocity profile in the above equation, we get

$$\delta_d = \int_0^\delta \left(1 - \left(\frac{y}{\delta}\right)^{1/7}\right) dy$$

$$\Rightarrow \delta_d = \left| y - \frac{7}{8}\left(\frac{y^{8/7}}{\delta^{1/7}}\right) \right|_0^\delta = \frac{\delta}{8}$$

b) We have **momentum thickness** as $\delta_M = \int_0^\delta \frac{u}{U_{\max}}\left(1 - \frac{u}{U_{\max}}\right) dy$

Substituting the velocity profile in the above equation, we get

$$\delta_M = \int_0^\delta \left(\frac{y}{\delta}\right)^{1/7}\left(1 - \left(\frac{y}{\delta}\right)^{1/7}\right) dy$$

$$\Rightarrow \delta_M = \left| \frac{7}{8}\left(\frac{y^{8/7}}{\delta^{1/7}}\right) - \frac{7}{9}\left(\frac{y^{9/7}}{\delta^{2/7}}\right) \right|_0^\delta = \frac{7}{72}\delta$$

c) We have **energy thickness** as $\delta_e = \int_0^\delta \frac{u}{U_{\max}}\left(1 - \frac{u^2}{U^2_{\max}}\right) dy$

Substituting the velocity profile in the above equation, we get

$$\delta_e = \int_0^\delta \left(\frac{y}{\delta}\right)^{1/7}\left(1 - \left[\left(\frac{y}{\delta}\right)^{1/7}\right]^2\right) dy$$

$$\Rightarrow \delta_e = \left| \frac{7}{8}\left(\frac{y^{8/7}}{\delta^{1/7}}\right) - \frac{7}{10}\left(\frac{y^{10/7}}{\delta^{3/7}}\right) \right|_0^\delta = \frac{7}{40}\delta$$

d) Shape factor: $\left(\dfrac{\delta_d}{\delta_M}\right) = \dfrac{1}{8} * \dfrac{72}{7} = \dfrac{9}{7}$

Example 7.6: For the given velocity profile $\dfrac{u}{U_{Max}} = 1 + \left(\dfrac{y}{\delta}\right) - 2\left(\dfrac{y}{\delta}\right)^2$

a) Displacement thickness.
b) Momentum thickness.
c) Shape factor $\left(\delta_d / \delta_M\right)$

Solution: Given data: velocity profile $\dfrac{u}{U_{Max}} = 1 + \left(\dfrac{y}{\delta}\right) - 2\left(\dfrac{y}{\delta}\right)^2$

a) We have **displacement thickness** as $\delta_d = \int_0^{\delta} \left(1 - \dfrac{u}{U_{\max}}\right) dy$

Substituting the velocity profile in the above equation, we get

$$\delta_d = \int_0^{\delta} \left(1 - 1 - \left(\frac{y}{\delta}\right) + 2\left(\frac{y}{\delta}\right)^2\right) dy$$

$$\Rightarrow \delta_d = \left| -\frac{1}{2}\left(\frac{y^2}{\delta}\right) + \frac{2}{3}\left(\frac{y^3}{\delta^2}\right) \right|_0^{\delta} = \frac{\delta}{6}$$

b) We have **momentum thickness** as $\delta_M = \int_0^{\delta} \dfrac{u}{U_{\max}} \left(1 - \dfrac{u}{U_{\max}}\right) dy$

Substituting the velocity profile in the above equation, we get

$$\delta_M = \int_0^{\delta} 1 + \left(\frac{y}{\delta}\right) - 2\left(\frac{y}{\delta}\right)^2 \left(1 - 1 - \left(\frac{y}{\delta}\right) + 2\left(\frac{y}{\delta}\right)^2\right) dy$$

$$\Rightarrow \delta_M = \left| \frac{-1}{2}\left(\frac{y^2}{\delta}\right) + \frac{1}{3}\left(\frac{y^3}{\delta^2}\right) + \frac{4}{4}\left(\frac{y^4}{\delta^3}\right) - \frac{4}{5}\left(\frac{y^5}{\delta^4}\right) \right|_0^{\delta} = \frac{1}{30}\delta$$

d) Shape factor: $\left(\dfrac{\delta_d}{\delta_M}\right) = \dfrac{1}{8} * \dfrac{72}{7} = \dfrac{9}{7}$

Example 7.7: Find the shape factor for a study laminar flow in a pipe.

Solution: Given data: Study laminar flow in a pipe.
Velocity profile for a laminar flow through a pipe is given in Hagen Poiseulle equation of viscous flow chapter and hence referring Eq. (6.12a), we have

$$\frac{u}{U_{MAX}} = \left[\left(\frac{R^2 - r^2}{R}\right)\right]$$

but $y = R - r$ and hence Eq. (6.12a) becomes

$$\frac{u}{U_{MAX}} = \left[\left(\frac{y}{R}\right)^2\right]$$

Refer Eq. (7.4), we have **displacement thickness** as

$$\delta_d = \int_0^{\delta}\left(1 - \frac{u}{U_{\max}}\right)dy$$

Substituting the velocity profile in the above equation, we get

$$\delta_d = \int_0^{R}\left(\left[1 - \left(\frac{y}{R}\right)^2\right]\right)dy$$

$$\delta_d = \frac{1}{R^2}\int_0^{R}\left[\left(R^2 - y^2\right)\right]dy$$

$$\Rightarrow \delta_d = \frac{1}{R^2}\left|R^2 y - \left(\frac{y^3}{3}\right)\right|_0^R = \frac{2R}{3}$$

Refer Eq. (7.9), we have **momentum thickness** as $\delta_M = \int_0^{\delta}\frac{u}{U_{\max}}\left(1 - \frac{u}{U_{\max}}\right)dy$

Substituting the velocity profile in the above equation, we get

$$\delta_M = \int_0^{R}\left[\left(\frac{y}{R}\right)^2\right]\left(\left[1 - \left(\frac{y}{R}\right)^2\right]\right)dy$$

$$\Rightarrow \delta_M = \left|\left(\frac{y^3}{R^2}\right) - \frac{1}{9}\left(\frac{y^5}{R^4}\right)\right|_0^R = \frac{4}{5}R$$

Shape factor: $\left(\frac{\delta_d}{\delta_M}\right) = \frac{2}{3} * \frac{5}{4} = \frac{5}{6}$

LAMINAR TURBULENT TRANSITION

The laminar flow in a fluid exists until the inertia force equals the viscous force. And as the fluid starts moving with higher velocities (i.e. inertia force exceeds viscous force), the viscous effect starts decreasing due to which the fluid particles flowing starts jumping on to one another and the flow starts splitting. Once the velocity of flow becomes very high, the viscous effect is eliminated and the flow becomes completely uncontrolled and hence called **turbulent flow**.

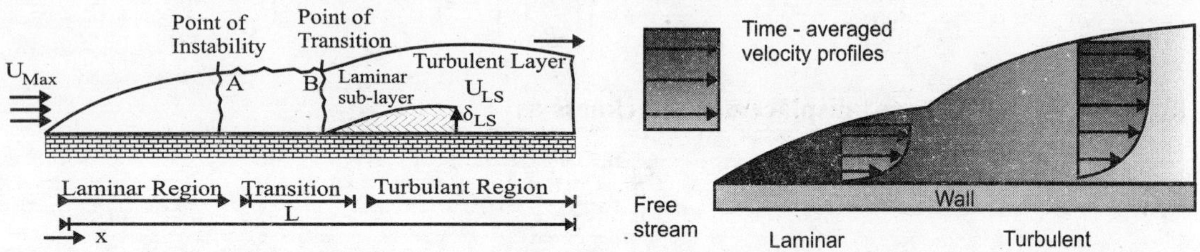

Figure 7.4: Boundary layer transitions over a flat plate

The fluid flowing has laminar flow existence at low velocities up to some extent, over a flat plate (i.e. up to a Reynolds number $R_e < 3\times10^5$) and as the velocity starts increasing beyond the limit called **point of instability** (Point "A", in the Figure 7.4), wherein the fluid flow will be in the inter phase of laminar flow and turbulent flow, and is named **transition zone** (i.e. up to Reynolds number $3\times10^5 < R_e < 5\times10^5$). And once the inertia is very much high, the flow will be in the turbulent zone, as the viscous force starts eliminated (i.e. up to Reynolds number $R_e < 10^7$). Once the Reynolds number $R_e > 10^7$, the fluid flow becomes fully developed turbulent.

The turbulent zone does contain a sub-layer which is of laminar flow, up to some length called critical length L_C, wherein the flow will be acting as laminar flow and will be holding the fluid flow from contact with the walls of the surface. This sub-layer of laminar flow is called as **laminar sub-layer**, beyond which the flow becomes fully developed turbulent flow.

The velocity profile of laminar flow follows a parabolic curve $U \approx \left(\frac{y}{\delta}\right)$ and a logarithmic curve for a turbulent flow $U \approx \log\left(\frac{y}{\delta}\right)$, $\Rightarrow \frac{U}{U_{MAX}} = \left(\frac{y}{\delta}\right)^{1/n}$ and where in it is seen experimentally that the power index n=7.

Examples of laminar turbulent transition from a gas stove flame, cigarette smoke and smoke from chimney are shown below respectively.

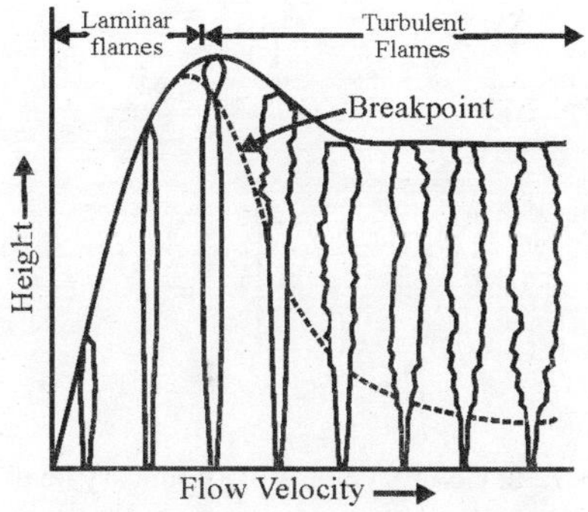

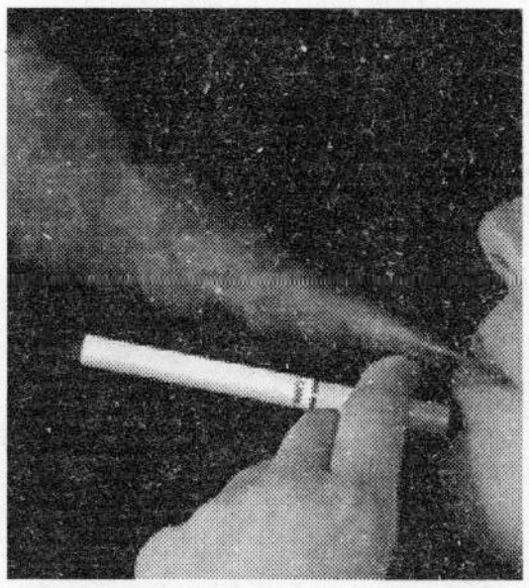

LAMINAR BOUNDARY THEORY

Any fluid flow having its fluid particles moving parallel to the boundary of the surface is called **Laminar flow** and the fluid particles have consistent paths. Whenever a laminar viscous flow is flowing over a flat plate it has shear stress τ_w at the contact of the wall of the plate and fluid and due to this wall shear stress, there arises a backward pull/force due to the friction between the object and the fluid and is called **frictional drag force** (F_{Df}) and moreover in addition to the frictional drag, there exists separation of fluid flow over the object which creates a **lift force component** to the smooth flow of fluid and hence creates additional drag force called **form force** (F_L). So the **Total drag force** $F_D = F_{Df} + F_L$ and **coefficient of drag** $C_D = {}^*C_f + C_L$ where ${}^*C_f$ is coefficient of drag due to friction and C_L coefficient due to form drag. But the shearing effect takes place in the direction of stream flow and hence the lift component will be comparatively smaller, i.e. $F_L <<< F_{Df}$. Any designer and analyst need to take into account the drag force whilst constructions of any structural design and stability, which play a significant role in collapsing of heavy structures due to drag force created by heavy flowing winds and storms over them.

LAMINAR MOMENTUM INTEGRAL EQUATION (VON KARMAN EQUATION)

This equation based on net momentum effect acting in the boundary layer, for finding the shear stress at the wall τ_w along the length (L) of a unit width (Th=1) flat plate and also the skin friction coefficient/drag (C_D) equation.

Consider a smooth flat plate of length (L) and a unit width and thickness, having a free stream velocity U_M and shear stress at the wall τ_w due to the predominance of viscous effect in the boundary layer thickness δ. Consider a fluid strip (*ABCD*), as shown in Figure 7.5, at a height of (dy) and length (dx).

According to Momentum equation: Due to the stream flow, there exists momentum rate of fluid mass entering into the closed control volume *ABCD* through faces *AB* and *CD*, while the fluid mass leaves from face *BC*, as shown in Figure 7.5a.

The fluid momentum rate entering through the face (AD) will be with velocity (u) and hence momentum rate of fluid entering through the face (AD) is

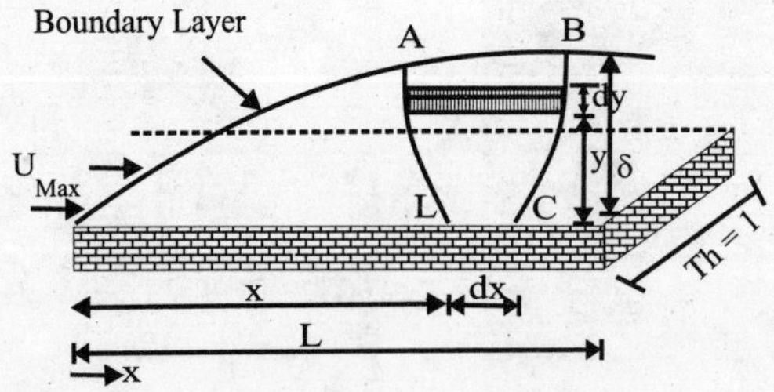

Figure 7.5: Boundary laminar layer over a thin flat plate

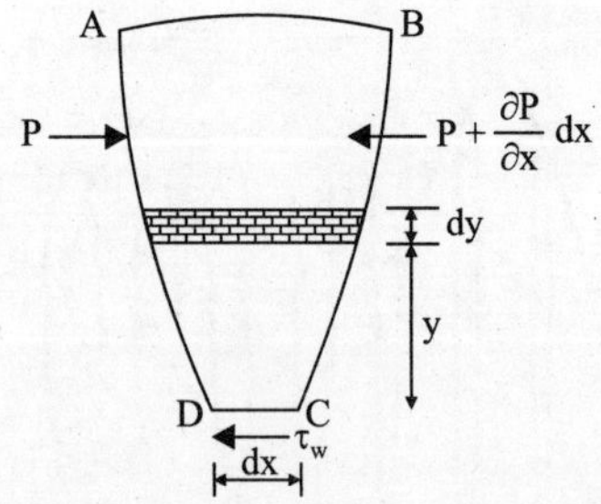

Figure 7.5a: Control volume of a boundary level

$$\bar{m}V_{AD} = \frac{d}{dt}(mV_{AD})$$

$$\Rightarrow \frac{d}{dt}(mV_{AD}) = \int_0^\delta (\rho Au)(u) = \int_0^\delta (\rho Au^2)$$

$$\Rightarrow \bar{m}V_{AD} = \int_0^\delta (\rho(dy*1)u)(u) = \int_0^\delta (\rho * dy * u^2) \tag{7.15a}$$

Momentum rate of fluid entering through the face (AB) will be velocity (U_{Max}) and hence momentum rate of fluid entering through the face (AD) is $\bar{m}V_{AB} = \frac{d}{dt}(mV_{AB})$.

$$\Rightarrow \frac{d}{dt}(mV_{AB}) = \int_0^\delta (\rho Au)(U_{Max}) = \int_0^\delta (\rho Au * U_{Max})$$

$$\Rightarrow \frac{d}{dt}(mV_{AB}) = \frac{d}{dx}\int_0^\delta (\rho(1*dy)u)(U_M) * dx = U_M \frac{d}{dx}\left[\int_0^\delta [\rho * dy * u]\right]dx \tag{7.15b}$$

Momentum rate of fluid leaving through the face (CB) will be velocity (u) and hence momentum rate of fluid leaving through the face (CB) is $\bar{m}V_{CB} = \frac{d}{dt}(mV_{CB})$.

$$\frac{d}{dt}mV_{CB} = \bar{m}V_{AD} + \frac{d}{dx}(\bar{m}V_{AD})dx$$

$$\Rightarrow \frac{d}{dt}mV_{CB} = \int_0^\delta (\rho * dy * u^2) + \frac{d}{dx}\left[\int_0^\delta (\rho * dy * u^2)\right]dx \tag{7.15c}$$

Net momentum rate of fluid passing through the control volume $(ABCD)$ be $\bar{m}V$, then

$$\bar{m}V = \bar{m}V_{CB} - \bar{m}V_{AD} - \bar{m}V_{AB} \tag{7.16}$$

and substituting Eq. (7.15c), Eq. (7.15b), Eq. (7.15a) in Eq. (7.16), we get

$$\bar{m}V = \frac{d}{dx}\left[\int_0^{\delta}\left(\rho u^2 * dy\right) - \int_0^{\delta}\left(\rho u U_M * dy\right)\right]dx$$

$$\bar{m}V = \rho\frac{d}{dx}\left[\int_0^{\delta}\left(u^2 - uU_M\right)dy\right]dx \tag{7.17}$$

As this change in momentum rate Eq. (7.17) is due to the shear stress τ_w in opposing direction to the flow and net pressure force, so the change in momentum rate must be in equilibrium to the sum of shear stress acting throughout the contact surface and net pressure force.

i.e.
$$-\tau_w\left(dx^*1\right) - \frac{\partial P}{\partial x}dx^*y = p\frac{d}{dx}\left[\int_0^{\delta}\left(u^2 - uU_M\right)dy\right]dx$$

$$\Rightarrow \tau_w + \frac{dP}{dx}\int_0^{\delta}dy = \rho\frac{d}{dx}\left[\int_0^{\delta}uU_M dy\right] - \rho\frac{d}{dx}\int_0^{\delta}\left(u^2\right)dy \tag{7.17a}$$

To evaluate pressure gradient $\frac{dP}{dx}$, use Bernoulli's equation in the free stream region.

i.e.
$$\frac{P}{\gamma} + \frac{U_{Max}^2}{2} = C \Rightarrow \frac{P}{\rho} = C - \frac{U_{Max}^2}{2}$$

Differentiating we get Bernoulli's equation as $\Rightarrow \frac{dP}{dx} = -\rho U_{Max}\frac{dU_{Max}}{dx}$

Substituting the above pressure gradient value in Eq. (7.17a), we have.

$$\tau_w - \rho U_{Max}\frac{dU_{Max}}{dx}\int_0^{\delta}dy = \rho\frac{d}{dx}\left[\int_0^{\delta}uU_M dy\right] - \rho\frac{d}{dx}\int_0^{\delta}\left(u^2\right)dy \tag{7.17b}$$

But as per the chain rule of differentiation $\frac{d}{dx}uU_M = u\frac{d}{dx}U_M + U_M\frac{d}{dx}u$, so Eq. (7.17b) becomes

$$\tau_w - \rho U_{Max}\frac{dU_{Max}}{dx}\int_0^{\delta} dy = \rho\left\{\frac{dU_{Max}}{dx} * \int_0^{\delta} u dy + U_M * \frac{d}{dx}\left[\int_0^{\delta} u dy\right]\right\} - \rho\frac{d}{dx}\int_0^{\delta} u^2 dy$$

$$\Rightarrow \frac{\tau_w}{\rho} = U_{Max}\frac{dU_{Max}}{dx}\int_0^{\delta} dy + \left\{\frac{dU_{Max}}{dx} * \int_0^{\delta} u dy + U_M * \frac{d}{dx}\left[\int_0^{\delta} u dy\right]\right\} - \frac{d}{dx}\int_0^{\delta} u^2 dy$$

$$\Rightarrow \frac{\tau_w}{\rho} = \frac{dU_{Max}}{dx}\int_0^{\delta}\left(U_{Max} - u\right)dy + \frac{d}{dx}\int_0^{\delta}\left(uU_{Max} - u^2\right)dy \qquad (7.18)$$

Equation (7.18) is the **momentum integral equation for both laminar and turbulent flow.**

If laminar flow is considered, then the $U_{\max} = C \,\&\, \frac{dU_{Max}}{dx} = 0$ and hence we have shear stress at the wall for a laminar flow as

$$\frac{\tau_w}{\rho U_M^2} = \frac{d}{dx}\left[\int_0^{\delta}\frac{u}{U_M}\left(1 - \frac{u}{U_M}\right)dy\right]$$

$$\Rightarrow \frac{\tau_w}{\rho U_M^2} = \frac{d}{dx}\left[\delta_M\right] \qquad (7.18a)$$

Equation (7.18a) is also called **Von Karman equation** for laminar.
If considered the boundary layer conditions, like

a) At the leading edge of the plate $y = 0 \,\&\, u = 0$

b) At the leading edge of the plate, no shear stress $y = 0 \,\&\, \frac{du}{dy} = Finite\ Value \,\&\, \frac{d^2u}{dy^2} = 0$

c) At the tail edge of the plate $y = \delta \,\&\, u = U_{Max}$

d) At the tail edge of the plate $y = \delta \,\&\, \frac{du}{dy} = 0$

The velocity profile equation can be written in the form of a polynomial of y having equal number of terms as that of the boundary conditions, i.e.

$$u = a + by + cy^2 + dy^3 \qquad (7.19)$$

Applying boundary conditions in Eq. (7.19), we have $a = 0 = c, b = \frac{3U_{Max}}{2\delta}, d = -\frac{U_{Max}}{2\delta^3}$ and hence Eq. (7.19) becomes

$$\frac{u}{U_{Max}} = \frac{3}{2}\left(\frac{y}{\delta}\right) - \frac{1}{2}\left(\frac{y}{\delta}\right)^3 \qquad (7.19a)$$

Substituting Eq. (7.19a) in Eq. (7.18a) and integrating, we have

$$\tau_w = \rho U_{Max}^2 \frac{d}{dx}\left(\frac{39}{280}\delta\right) \tag{7.19b}$$

For a laminar Newtonian fluid $\tau_w = \mu\left(\frac{du}{dy}\right)_{y=0}$

$$\Rightarrow \tau_w = \frac{3}{2}\frac{\mu U_{Max}}{\delta} \tag{7.19c}$$

Equate Eq. (7.19b) and Eq. (7.19c)

$$\Rightarrow \frac{3}{2}\frac{\mu U_{Max}}{\delta} = \rho U_{Max}^2 \frac{d}{dx}\left(\frac{39}{280}\delta\right)$$

$$\Rightarrow \delta = \left(\frac{4.64x}{\sqrt{R_{e_x}}}\right) \tag{7.20}$$

Skin friction coefficient/drag coefficient: Skin friction coefficient is defined as the ratio of shear stress at the walls and $\frac{1}{2}\rho U_M^2$, and hence ${}^*C_f = \frac{\tau_w(x)}{\frac{1}{2}\rho U_M^2}$ and substituting Eq. (7.19c), we get

$${}^*C_f = \frac{3\mu}{\rho U_{Max}\delta}$$

$$\Rightarrow {}^*C_f = 3\left(\frac{\mu}{\rho x U_{Max}}\right)\left(\frac{x}{\delta}\right)$$

$$\Rightarrow {}^*C_f = \frac{3}{4.64}\left(R_{e_x}\right)^{-0.5}$$

$$\Rightarrow {}^*C_f = \frac{0.646}{\sqrt{R_{e_x}}} \tag{7.21}$$

Average skin friction coefficient $\bar{C}_f = \frac{1}{L}\int_0^L {}^*C_f(x)dx$ and by taking integration constant $C = 4.64$ for a third degree velocity profile, we have

$$\Rightarrow \bar{C}_f = \frac{1.328}{\sqrt{R_{e_L}}} \tag{7.22}$$

The **drag force** (F_D) will be in opposite direction to the shear stress τ_w and as force = stress * area we have $F_D = \int -\tau_w (A)$

$$\Rightarrow F_D = \int_0^L \tau_w (dx * th) = \int_0^L \tau_w (dx * 1)$$

$$F_D = \rho U_M^2 \left(\frac{0.664}{\sqrt{R_{e_L}}} \right) A \tag{7.23}$$

hence we can observe that $$F_D = \frac{\rho U_M^2}{2} * \bar{C}_f * A \tag{7.23a}$$

In similar procedure to the above, various velocity profiles were formulated by **Prandtl** and **H. Blasius**, by varying number of boundary conditions for various velocity profiles and the corresponding values of boundary layer thickness, skin friction/drag coefficient, average skin friction coefficients/drag coefficients $\delta, {}^*C_f, \bar{C}_f$ respectively, for a flat plate have been developed, as detailed below in the table.

Table 7.1: Boundary layer thickness, skin friction coefficient for various velocity profiles

Name	Velocity profile	Boundary layer thickness δ	Skin friction coefficient $^*C_f$	Average Skin friction coefficient $\bar{C}_f$
	$\frac{u}{U_{Max}} = \sin\left(\frac{\Pi}{2}\frac{y}{\delta}\right)$	$\delta = \frac{4.795x}{\sqrt{R_{e_x}}}$	${}^*C_f = \frac{0.654}{\sqrt{R_{e_x}}}$	$\bar{C}_f = \frac{1.31}{\sqrt{R_{e_L}}}$
Prandtl	$\frac{u}{U_{Max}} = 2\left(\frac{y}{\delta}\right) - \left(\frac{y}{\delta}\right)^2$	$\delta = \frac{5.48x}{\sqrt{R_{e_x}}}$	${}^*C_f = \frac{0.73}{\sqrt{R_{e_x}}}$	$\bar{C}_f = \frac{1.46}{\sqrt{R_{e_L}}}$
	$\frac{u}{U_{Max}} = \frac{3}{2}\left(\frac{y}{\delta}\right) - \frac{1}{2}\left(\frac{y}{\delta}\right)^3$	$\delta = \frac{4.64x}{\sqrt{R_{e_x}}}$	${}^*C_f = \frac{0.646}{\sqrt{R_{e_x}}}$	$\bar{C}_f = \frac{1.292}{\sqrt{R_{e_L}}}$
	$\frac{u}{U_{Max}} = 2\left(\frac{y}{\delta}\right) - 2\left(\frac{y}{\delta}\right)^3 + \left(\frac{y}{\delta}\right)^4$	$\delta = \frac{5.84x}{\sqrt{R_{e_x}}}$	${}^*C_f = \frac{0.686}{\sqrt{R_{e_x}}}$	$\bar{C}_f = \frac{1.372}{\sqrt{R_{e_L}}}$
H. Blasius		$\delta = \frac{5x}{\sqrt{R_{e_x}}}$	${}^*C_f = \frac{0.664}{\sqrt{R_{e_x}}}$	$\bar{C}_f = \frac{1.328}{\sqrt{R_{e_L}}}$

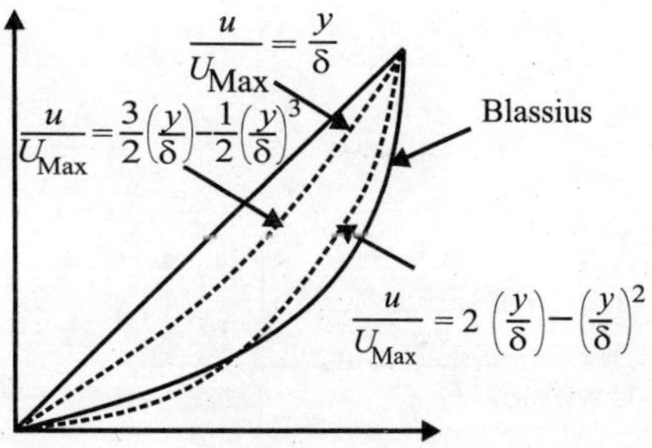

Figure 7.6: Comparison of velocity profile polynomials from linear to higher order with exact Blassius velocity profiles

Example 7.8: Applying appropriate boundary conditions, determine the velocity profile over a flat plate having velocity over it as $u = a\sin(by) + c$

Solution: Given data: velocity over plate $u = a\sin(by) + c$

The prevailing boundary conditions over a flat plate are

a) At the leading edge of the plate $y = 0$ and $u = 0$

b) At the leading edge of the plate, no shear stress $y = 0$ and $\frac{du}{dy} =$ finite Values and $\frac{d^2u}{dy^2} = 0$

c) At the tail edge of the plate $y = \delta$ and $u = U_{Max}$

d) At the tail edge of the plate $y = \delta$ and $\frac{du}{dy} = 0$

And so applying boundary condition "a", we get $c = 0$ and hence the velocity becomes

$$u = a\sin(by) \tag{1}$$

Similarly, apply Boundary conditions "c", we get

$$U_{MAX} = a\sin(b\delta) \tag{2}$$

Dividing Eq. (1) and Eq. (2), we get $$\frac{u}{U_{MAX}} = \frac{\sin(by)}{\sin(b\delta)} \tag{3}$$

Similarly, apply boundary conditions "d", we get $\left(\frac{du}{dy}\right)_{y=\delta} = ab\cos(by)$

$$\Rightarrow \left(\frac{du}{dy}\right)_{y=\delta} = ab\cos(b\delta) = 0 \tag{4}$$

$$= ab\cos(b\delta) = 0$$

$$\Rightarrow b = \frac{\pi}{2\delta} \tag{4a}$$

Substituting, Eq. (4a) in Eq. (3), we get $\dfrac{u}{U_{MAX}} = \dfrac{\sin\left(\dfrac{\pi y}{2\delta}\right)}{\sin\left(\dfrac{\pi\delta}{2\delta}\right)}$

$$\Rightarrow \frac{u}{U_{MAX}} = \frac{\sin\left(\dfrac{\pi y}{2\delta}\right)}{\sin\left(\dfrac{\pi}{2}\right)} = \frac{\sin\left(\dfrac{\pi y}{2\delta}\right)}{1}$$

$$\Rightarrow \frac{u}{U_{MAX}} = \sin\left(\frac{\pi y}{2\delta}\right)$$

Hence, the **velocity profile** is $\dfrac{u}{U_{Max}} = \sin\left(\dfrac{\Pi}{2}\dfrac{y}{\delta}\right)$.

Example 7.9: For the given velocity profile $\dfrac{u}{U_{Max}} = \sin\left(\dfrac{\Pi}{2}\dfrac{y}{\delta}\right)$

Determine a) Derive the boundary layer thickness.
b) Wall shear stress.
c) Coefficient of drag.
d) Drag force and average drag coefficient.

Solution: Given data: velocity profile $\dfrac{u}{U_{Max}} = \sin\left(\dfrac{\Pi}{2}\dfrac{y}{\delta}\right)$

a) Boundary layer thickness:

Refer Eq. (7.9), we have momentum thickness as $\delta_M = \displaystyle\int_0^\delta \frac{u}{U_{\max}}\left(1 - \frac{u}{U_{\max}}\right)dy$

Substituting the velocity profile in the above equation, we get

$$\delta_M = \int_0^\delta \sin\left(\frac{\Pi}{2}\frac{y}{\delta}\right)\left(1 - \sin\left(\frac{\Pi}{2}\frac{y}{\delta}\right)\right)dy$$

$$\Rightarrow \delta_M = \int_0^\delta \left\{\sin\left(\frac{\Pi}{2}\frac{y}{\delta}\right) - \sin^2\left(\frac{\Pi}{2}\frac{y}{\delta}\right)\right\}dy$$

$$\Rightarrow \delta_M = \int_0^{\delta} \left\{ \sin\left(\frac{\Pi}{2}\frac{y}{\delta}\right) - \frac{1-\cos\left(\frac{\pi y}{\delta}\right)}{2} \right\} dy \quad \text{since} \quad 2\sin^2\theta = 1-\cos 2\theta$$

$$\Rightarrow \delta_M = \left| \frac{-\cos\left(\frac{\pi y}{2\delta}\right)}{\frac{\pi}{2\delta}} - \frac{y}{2} + \frac{\sin\left(\frac{\pi y}{\delta}\right)}{\frac{\pi}{\delta}} \right|_0^{\delta} = \left(\frac{4-\pi}{2\pi}\right)\delta$$

Refer Eq. (7.18), we have $\Rightarrow \dfrac{\tau_w}{\rho U_M^2} = \dfrac{d}{dx}[\delta_M]$

$$\Rightarrow \tau_w = \rho U_{Max}^2 \frac{d}{dx}\left(\frac{4-\pi}{2\pi}\right)\delta \tag{1}$$

From the velocity profile, we have $u = U_{Max}\left\{\sin\left(\dfrac{\Pi}{2}\dfrac{y}{\delta}\right)\right\}$

Hence, the velocity gradient will be $\dfrac{du}{dy} = U_{Max} * \dfrac{\pi}{2\delta}\left\{\cos\left(\dfrac{\Pi}{2}\dfrac{y}{\delta}\right)\right\}$

For a laminar Newtonian fluid $\tau_w = \mu\left(\dfrac{du}{dy}\right)_{y=0}$

$$\tau_w = \mu\left(\frac{du}{dy}\right)_{y=0} = \mu U_{Max}\left\{\frac{\pi}{2\delta}\left\{\cos\left(\frac{\Pi}{2}\frac{y}{\delta}\right)\right\}\right\}_{y=0}$$

$$\Rightarrow \tau_w = \frac{\pi\mu U_{Max}}{2\delta} \tag{2}$$

Equate Eq. (1) and Eq. (2) $\Rightarrow \dfrac{\pi\mu U_{Max}}{2\delta} = \rho U_{Max}^2 \dfrac{d}{dx}\left(\dfrac{4-\pi}{2\pi}\right)\delta$

$$\Rightarrow \frac{\pi\mu U_{Max}}{2\rho U_{Max}^2} * \left(\frac{2\pi}{4-\pi}\right)dx = \delta d\delta$$

Integrating both sides, we get **boundary layer thickness**

$$\boxed{\delta = \frac{4.795x}{\sqrt{R_{e_x}}}} \tag{3}$$

b) Wall shear stress:

Substitute Eq. (3) in Eq. (2), then we get wall shear stress as

$$\Rightarrow \tau_w = 0.327\frac{\mu/U_{Max}}{x}\sqrt{R_{e_x}} \tag{4}$$

c) Skin friction coefficient/drag coefficient:

Skin friction coefficient is defined as the ratio of shear stress at the walls and $\frac{1}{2}\rho U_M^2$ and hence

$$^*C_f = \frac{\tau_w(x)}{\frac{1}{2}\rho U_M^2} \tag{5}$$

and substituting Eq:(4) in Eq:(5), we get

$$^*C_f = \frac{2*0.327\frac{\mu U_{Max}}{x}\sqrt{R_{e_x}}}{\rho U^2{}_{Max}}$$

$$\Rightarrow {}^*C_f = \left(\frac{0.654\mu}{\rho x U_{Max}}\right)\left(\sqrt{R_{e_x}}\right)$$

$$\Rightarrow {}^*C_f = \frac{0.654}{\sqrt{R_{e_x}}} \tag{6}$$

d) Drag force and average drag coefficient:

The **drag force** (F_D) will be in opposite direction to the shear stress τ_w and as $force = stress * area$ we have $FD = \int -\tau_w(A)$

$$\Rightarrow F_D = \int_0^L \tau_w(dx * th) = \int_0^L \tau_w(dx * th)$$

$$\Rightarrow F_D = \int_0^L \tau_w(dx * th) = \int_0^L 0.327\frac{\mu U_{Max}}{x}\sqrt{R_{e_x}} * dx * th$$

$$\Rightarrow F_D = 0.327\mu U_{Max} * th * \sqrt{\frac{\rho U_{Max}}{\mu}}\int_0^L \frac{1}{x} * dx$$

$$\Rightarrow F_D = 0.327\mu U_{Max} * th * \sqrt{\frac{\rho U_{Max}}{\mu}}\left|\frac{x^{1/2}}{1/2}\right|_0^L$$

$$\Rightarrow F_D = 0.654\mu U_{Max} * th * \sqrt{\frac{\rho U_{Max} L}{\mu}}$$

$$\Rightarrow F_D = \rho U_M \left(\frac{0.654}{\sqrt{R_{e_L}}} \right) Area \tag{7}$$

Average skin friction coefficient is defined as the drag force and $\frac{1}{2}\rho A U_M^2$, and hence

$$F_D = \frac{\rho U_M^2}{2} * \bar{C}_f * A \tag{8}$$

and substituting Eq.(7) in Eq.(8), we get

$$\bar{C}_f = \frac{F_D}{\frac{\rho U_M^2}{2} * A} = \frac{\rho U_M^2 \left(\frac{0.654}{\sqrt{R_{e_L}}} \right) A}{\frac{\rho U_M^2}{2} * A} = \frac{1.31}{\sqrt{R_{e_L}}}$$

Example 7.10: Oil (specific gravity 0.85 and kinematic viscosity of 2.6 × 10^{-4} m^2/s) is led to flow over a flat plate of 0.5 m long and 0.2 m wide, with a velocity of 7 m/s. Determine the skin friction coefficient, drag force, boundary layer thickness and also the shear stress at the trailing edge.

Solution: Given data: velocity profile $S = 0.85, \nu = 2.6\times10^{-4}\,\text{m}^2/\text{s}, L = 0.5\text{m}, B = 0.2\text{m}, U_{Max} = 7\,\text{m/s}$

Reynolds number will be

$$R_e = \frac{U_{MAX} * L}{\nu} = \frac{7 * 0.5}{2.6\times10^{-4}} = 1.3\times10^4$$

As the Reynolds number is less than 3×10^5, hence the flow is laminar over the flat plate and so the **drag force**

$$F_D = \frac{\rho U_M^2}{2} * \bar{C}_f * A \text{ where } \bar{C}_f = \frac{1.328}{\sqrt{R_{e_L}}} \text{ according to Blasius}$$

$$\Rightarrow F_D = \frac{\rho U_M^2}{2} * \frac{1.328}{\sqrt{R_{e_L}}} * A$$

$$\Rightarrow F_D = \frac{0.85 * 1000 * 7^2}{2} * \frac{1.328}{\sqrt{1.3\times10^4}} * 0.5 * 0.2 = 24.26\text{N}$$

Boundary layer thickness will be

$$\delta = \frac{5x}{\sqrt{R_{ex}}} = \frac{5*0.5}{\sqrt{1.3*10^4}} = 0.0219\,\text{m}$$

Wall shear stress is

$$\tau_w = \frac{\rho U_M^2}{2} * {}^*C_f \quad \text{where} \quad {}^*C_f = \frac{0.664}{\sqrt{R_{ex}}} \quad \text{according to Blasius}$$

$$\Rightarrow \tau_w = \frac{\rho U_M^2}{2} * \frac{0.664}{\sqrt{R_{ex}}} = \frac{850*49}{2} * \frac{0.664}{\sqrt{1.3*10^4}} = 121.28\,\text{Pa}$$

Example 7.11: For the given velocity profile $\frac{u}{U_{Max}} = 2\left(\frac{y}{\delta}\right) - \left(\frac{y}{\delta}\right)^2$

Determine a) Derive the boundary layer thickness.

b) Wall shear stress.

c) Coefficient of drag.

d) Drag force and Average drag coefficient.

Solution: Given data: velocity profile $\frac{u}{U_{Max}} = 2\left(\frac{y}{\delta}\right) - \left(\frac{y}{\delta}\right)^2$

a) Boundary layer thickness:

Refer Eq. (7.9), we have momentum thickness as $\delta_M = \int_0^{\delta} \frac{u}{U_{\max}}\left(1 - \frac{u}{U_{\max}}\right) dy$

Substituting the velocity profile in the above equation, we get

$$\delta_M = \int_0^{\delta} 2\left(\frac{y}{\delta}\right) - \left(\frac{y}{\delta}\right)^2 \left(1 - 2\left(\frac{y}{\delta}\right) + \left(\frac{y}{\delta}\right)^2\right) dy$$

$$\Rightarrow \delta_M = \int_0^{\delta} \left\{\left(\frac{2y}{\delta}\right) - \left(\frac{y}{\delta}\right)^2 - 4\left(\frac{y}{\delta}\right)^2 + 2\left(\frac{y}{\delta}\right)^3 + 2\left(\frac{y}{\delta}\right)^3 - \left(\frac{y}{\delta}\right)^4\right\} dy$$

$$\Rightarrow \delta_M = \left|\frac{2}{2}\left(\frac{y^2}{\delta}\right) - \left(\frac{y^3}{3\delta^2}\right) - \frac{4}{3}\left(\frac{y^3}{\delta^2}\right) + \frac{2}{4}\left(\frac{y^4}{\delta^3}\right) + \frac{2}{4}\left(\frac{y^4}{\delta^3}\right) - \frac{1}{5}\left(\frac{y^5}{\delta^4}\right)\right|_0^{\delta} = \frac{2}{15}\delta$$

We have

$$\Rightarrow \frac{\tau_w}{\rho U_M^2} = \frac{d}{dx}[\delta_M]$$

$$\Rightarrow \tau_w = \rho U_{Max}^2 \frac{d}{dx}\left(\frac{2}{15}\delta\right) \quad (1)$$

From the velocity profile, we have $u = U_{Max}\left(2\left(\frac{y}{\delta}\right) - \left(\frac{y}{\delta}\right)^2\right)$

Hence, the velocity gradient will be $\frac{du}{dy} = U_{Max}\left(2\left(\frac{1}{\delta}\right) - \left(\frac{2y}{\delta^2}\right)\right)$.

For a laminar Newtonian fluid $\tau_w = \mu\left(\frac{du}{dy}\right)_{y=0}$

$$\tau_w = \mu\left(\frac{du}{dy}\right)_{y=0} = \mu\left\{U_{Max}\left(2\left(\frac{1}{\delta}\right) - \left(\frac{2y}{\delta^2}\right)\right)\right\}_{y=0}$$

$$\Rightarrow \tau_w = 2\frac{\mu U_{Max}}{\delta} \quad (2)$$

Equate Eq. (1) and Eq. (2)

$$\Rightarrow 2\frac{\mu U_{Max}}{\delta} = \rho U_{Max}^2 \frac{d}{dx}\left(\frac{2}{15}\delta\right)$$

$$\Rightarrow 2 * \frac{15}{2} * \frac{\mu U_{Max}}{\rho U_{Max}^2} dx = \delta(d\delta)$$

Integrating both sides, we get **boundary layer thickness**

$$\boxed{\delta = \frac{5.48x}{\sqrt{R_{e_x}}}} \quad (3)$$

b) Wall shear stress:
Substitute Eq. (3) in Eq. (2), then we get wall shear stress as

$$\boxed{\Rightarrow \tau_w = 0.365\frac{\mu / U_{Max}}{x}\sqrt{R_{e_x}}} \quad (4)$$

c) Skin friction coefficient/drag coefficient:

Skin friction coefficient is defined as the ratio of shear stress at the walls and $\frac{1}{2}\rho U_M^2$, and hence

$$^{*}C_f = \frac{\tau_w(x)}{\frac{1}{2}\rho U_M^2} \quad (5)$$

and substituting Eq. (4) in Eq. (5), we get

$$^{*}C_f = \frac{2*0.365*\frac{\mu U_{Max}}{x}\sqrt{R_{e_x}}}{\rho U^2_{Max}}$$

$$\Rightarrow {}^{*}C_f = \left(\frac{0.73\mu}{\rho x U_{Max}}\right)\left(\sqrt{R_{e_x}}\right)$$

$$\boxed{\Rightarrow {}^{*}C_f = \frac{0.73}{\sqrt{R_{e_x}}}} \quad (6)$$

d) Drag force and average drag coefficient:

The **drag force** (F_D) will be in opposite direction to the shear stress τ_w and as force = stress*area we have $F_D = \int -\tau_w(A)$

$$\Rightarrow F_D = \int_0^L \tau_w(dx*th) = \int_0^L \tau_w(dx*th)$$

$$\Rightarrow F_D = \int_0^L \tau_w(dx*th) = \int_0^L 0.365\frac{\mu U_{Max}}{x}\sqrt{R_{e_x}}*dx*th$$

$$\Rightarrow F_D = 0.365\mu U_{Max}*th*\sqrt{\frac{\rho U_{Max}}{\mu}}\int_0^L \frac{1}{x}*dx$$

$$\Rightarrow F_D = 0.365\mu U_{Max}*th*\sqrt{\frac{\rho U_{Max}}{\mu}}\left|\frac{x^{1/2}}{1/2}\right|_0^L$$

$$\Rightarrow F_D = 0.73\mu U_{Max}*th*\sqrt{\frac{\rho U_{Max}L}{\mu}}$$

$$\boxed{\Rightarrow F_D = \rho U_M\left(\frac{0.73}{\sqrt{R_{e_L}}}\right)Area} \quad (7)$$

Average skin friction coefficient is defined as the drag force and $\frac{1}{2}\rho A U_M^2$, and hence

$$F_D = \frac{\rho U_M^2}{2} * \bar{C_f} * A \tag{8}$$

and substituting Eq. (7) in Eq. (8), we get

$$\bar{C_f} = \frac{F_D}{\frac{\rho U_M^2}{2} * A} = \frac{\rho U_M \left(\frac{0.73}{\sqrt{R_{e_L}}}\right) A}{\frac{\rho U_M^2}{2} * A} = \frac{1.46}{\sqrt{R_{e_L}}}$$

Example 7.12: For the given velocity profile $\frac{u}{U_{Max}} = \frac{3}{2}\left(\frac{y}{\delta}\right) - \frac{1}{2}\left(\frac{y}{\delta}\right)^3$

Determine

a) Derive the boundary layer thickness.
b) Wall shear stress.
c) Coefficient of drag.
d) Drag force and average drag coefficient.

Solution: Given data: velocity profile $\frac{u}{U_{Max}} = \frac{3}{2}\left(\frac{y}{\delta}\right) - \frac{1}{2}\left(\frac{y}{\delta}\right)^3$

a) Boundary layer thickness:

Refer Eq. (7.9), we have momentum thickness as $\delta_M = \int_0^\delta \frac{u}{U_{\max}}\left(1 - \frac{u}{U_{\max}}\right) dy$

Substituting the velocity profile in the above equation, we get

$$\delta_M = \int_0^\delta \frac{3}{2}\left(\frac{y}{\delta}\right) - \frac{1}{2}\left(\frac{y}{\delta}\right)^3 \left(1 - \frac{3}{2}\left(\frac{y}{\delta}\right) + \frac{1}{2}\left(\frac{y}{\delta}\right)^3\right) dy$$

$$\Rightarrow \delta_M = \int_0^\delta \left\{\frac{3}{2}\left(\frac{y}{\delta}\right) - \frac{1}{2}\left(\frac{y}{\delta}\right)^3 - \frac{9}{4}\left(\frac{y}{\delta}\right)^2 + \frac{3}{4}\left(\frac{y}{\delta}\right)^4 + \frac{3}{4}\left(\frac{y}{\delta}\right)^4 - \frac{1}{4}\left(\frac{y}{\delta}\right)^6\right\} dy$$

$$\Rightarrow \delta_M = \left|\frac{3}{2*2}\left(\frac{y^2}{\delta}\right) - \frac{1}{2*4}\left(\frac{y^4}{\delta^3}\right) - \frac{9}{4*3}\left(\frac{y^3}{\delta^2}\right) + \frac{3}{4*5}\left(\frac{y^5}{\delta^4}\right) + \frac{3}{4*5}\left(\frac{y^5}{\delta^4}\right) - \frac{1}{4*7}\left(\frac{y^7}{\delta^6}\right)\right|_0^\delta = \frac{39}{280}\delta$$

Refer Eq. (7.18), we have

$$\Rightarrow \frac{\tau_w}{\rho U_M^2} = \frac{d}{dx}[\delta_M]$$

$$\Rightarrow \tau_w = \rho U_{Max}^2 \frac{d}{dx}\left(\frac{39}{280}\delta\right) \quad (1)$$

For a laminar Newtonian fluid $\tau_w = \mu\left(\frac{du}{dy}\right)_{y=0}$

From the velocity profile, we have $u = U_{Max}\left(\frac{3}{2}\left(\frac{y}{\delta}\right) - \frac{1}{2}\left(\frac{y}{\delta}\right)^3\right)$

Hence, the velocity gradient will be $\frac{du}{dy} = U_{Max}\left(\frac{3}{2}\left(\frac{1}{\delta}\right) - \frac{1}{2}\left(\frac{3y^2}{\delta^3}\right)\right)$

For a laminar Newtonian fluid $\tau_w = \mu\left(\frac{du}{dy}\right)_{y=0}$

$$\tau_w = \mu\left(\frac{du}{dy}\right)_{y=0} = \mu\left\{U_{Max}\left(\frac{3}{2}\left(\frac{1}{\delta}\right) - \frac{1}{2}\left(\frac{3y^2}{\delta^3}\right)\right)\right\}_{y=0}$$

$$\Rightarrow \tau_w = \frac{3}{2}\frac{\mu U_{Max}}{\delta} \quad (2)$$

Equate Eq. (1) and Eq. (2)

$$\Rightarrow \frac{3}{2}\frac{\mu U_{Max}}{\delta} = \rho U_{Max}^2 \frac{d}{dx}\left(\frac{39}{280}\delta\right)$$

$$\Rightarrow \frac{3}{2} * \frac{280}{39} * \frac{\mu U_{Max}}{\rho U_{Max}^2} dx = \delta(d\delta)$$

Boundary layer thickness $$\delta = \left(\frac{4.64x}{\sqrt{R_{e_x}}}\right) \quad (3)$$

b) Wall shear stress:

Substitute Eq. (3) in Eq. (2), then we get wall shear stress as

$$\tau_w = \frac{3}{2}\frac{\mu U_{Max}}{4.64x}\sqrt{R_{e_x}}$$

$$\boxed{\Rightarrow \tau_w - 0.323\frac{\mu U_{Max}}{x}\sqrt{R_{e_x}}} \qquad (4)$$

c) Skin friction coefficient/Drag coefficient:

Skin friction coefficient is defined as the ratio of shear stress at the walls and $\frac{1}{2}\rho U_M^2$, and hence

$$^*C_f = \frac{\tau_w(x)}{\frac{1}{2}\rho U_M^2} \qquad (5)$$

and substituting Eq. (4) in Eq. (5), we get

$$^*C_f = \frac{2*0.323\frac{\mu U_{Max}}{x}\sqrt{R_{e_x}}}{\rho U^2_{Max}}$$

$$\Rightarrow {}^*C_f = \left(\frac{0.646\mu}{\rho x U_{Max}}\right)\left(\sqrt{R_{e_x}}\right)$$

$$\boxed{\Rightarrow {}^*C_f = \frac{0.646}{\sqrt{R_{e_x}}}} \qquad (6)$$

d) Drag force and average drag coefficient:

The **drag force** (F_D) will be in opposite direction to the shear stress τ_w and as *force = stress * area* we have $F_D = \int -\tau_w(A)$

$$\Rightarrow F_D = \int_0^L \tau_w(dx * th) = \int_0^L \tau_w(dx * th)$$

$$\Rightarrow F_D = \int_0^L \tau_w(dx * th) = \int_0^L 0.323\frac{\mu U_{Max}}{x}\sqrt{R_{e_x}} * dx * th$$

$$\Rightarrow F_D = 0.323\mu U_{Max} * th * \sqrt{\frac{\rho U_{Max}}{\mu}}\int_0^L \frac{1}{x} * dx$$

$$\Rightarrow F_D = 0.323\mu U_{Max} * th * \sqrt{\frac{\rho U_{Max}}{\mu}} \left| \frac{x^{1/2}}{1/2} \right|_0^L$$

$$\Rightarrow F_D = 0.646\mu U_{Max} * th * \sqrt{\frac{\rho U_{Max} L}{\mu}}$$

$$\Rightarrow F_D = \rho U_M \left(\frac{0.664}{\sqrt{R_{e_L}}} \right) Area \tag{7}$$

Average skin friction coefficient is defined as the drag force and $\frac{1}{2}\rho A U_M^2$, and hence

$$F_D = \frac{\rho U_M^2}{2} * \bar{C_f} * A \tag{8}$$

and substituting Eq. (7) in Eq. (8), we get

$$\bar{C_f} = \frac{F_D}{\frac{\rho U_M^2}{2} * A} = \frac{\rho U_M \left(\frac{0.664}{\sqrt{R_{e_L}}} \right) A}{\frac{\rho U_M^2}{2} * A} = \left(\frac{1.292}{\sqrt{R_{e_L}}} \right)$$

Example.7.13: For the given velocity profile $\frac{u}{U_{Max}} = 2\left(\frac{y}{\delta}\right) - 2\left(\frac{y}{\delta}\right)^3 + \left(\frac{y}{\delta}\right)^4$.

Determine a) Derive the boundary layer thickness.
b) Wall shear stress.
c) Coefficient of drag.
d) Drag force and average drag coefficient.

Solution: Given data: velocity profile $\frac{u}{U_{Max}} = 2\left(\frac{y}{\delta}\right) - 2\left(\frac{y}{\delta}\right)^3 + \left(\frac{y}{\delta}\right)^4$

a) Boundary layer thickness:

Refer Eq. (7.9), we have momentum thickness as $\delta_M = \int_0^{\delta} \frac{u}{U_{max}} \left(1 - \frac{u}{U_{max}}\right) dy$

Substituting the velocity profile in the above equation, we get

$$\delta_M = \int_0^{\delta} 2\left(\frac{y}{\delta}\right) - 2\left(\frac{y}{\delta}\right)^3 + \left(\frac{y}{\delta}\right)^4 \left(1 - 2\left(\frac{y}{\delta}\right) + 2\left(\frac{y}{\delta}\right)^3 - \left(\frac{y}{\delta}\right)^4\right) dy$$

$$= \int_0^{\delta} \left\{ 2\left(\frac{y}{\delta}\right) - 2\left(\frac{y}{\delta}\right)^3 - 4\left(\frac{y}{\delta}\right)^2 + 7\left(\frac{y^4}{\delta^4}\right) - 4\left(\frac{y}{\delta}\right)^5 - 2\left(\frac{y^6}{\delta^6}\right) + 4\left(\frac{y^7}{\delta^7}\right) - \left(\frac{y^8}{\delta^8}\right) \right\} dy$$

$$\Rightarrow \delta_M = \left| \frac{2}{2}\left(\frac{y^2}{\delta}\right) - \frac{2}{4}\left(\frac{y^4}{\delta^3}\right) - \frac{4}{3}\left(\frac{y^3}{\delta^2}\right) + \frac{7}{5}\left(\frac{y^5}{\delta^4}\right) + \frac{4}{6}\left(\frac{y^6}{\delta^5}\right) - \frac{2}{7}\left(\frac{y^7}{\delta^6}\right) + \frac{4}{8}\left(\frac{y^8}{\delta^7}\right) - \frac{1}{9}\left(\frac{y^9}{\delta^8}\right) \right|_0^{\delta}$$

$$\Rightarrow \delta_M = \frac{37}{315}\delta$$

Refer Eq. (7.18), we have

$$\Rightarrow \frac{\tau_w}{\rho U_M^2} = \frac{d}{dx}[\delta_M]$$

$$\Rightarrow \tau_w = \rho U_{Max}^2 \frac{d}{dx}\left(\frac{37}{315}\delta\right) \tag{1}$$

From the velocity profile, we have

$$u = U_{Max}\left\{ 2\left(\frac{y}{\delta}\right) - 2\left(\frac{y}{\delta}\right)^3 + \left(\frac{y}{\delta}\right)^4 \right\}$$

Hence, the velocity gradient will be

$$\frac{du}{dy} = U_{Max}\left\{ 2\left(\frac{1}{\delta}\right) - 6\left(\frac{y^2}{\delta^3}\right) + 4\left(\frac{y^5}{\delta^4}\right) \right\}$$

For a laminar Newtonian fluid

$$\tau_w = \mu\left(\frac{du}{dy}\right)_{y=0}$$

$$\tau_w = \mu\left(\frac{du}{dy}\right)_{y=0} = \mu U_{Max}\left\{ 2\left(\frac{1}{\delta}\right) - 6\left(\frac{y^2}{\delta^3}\right) + 4\left(\frac{y^5}{\delta^4}\right) \right\}_{y=0}$$

$$\Rightarrow \tau_w = 2\frac{\mu U_{Max}}{\delta} \tag{2}$$

Equate Eq. (1) and Eq. (2)

$$\Rightarrow 2\frac{\mu U_{Max}}{\delta} = \rho U_{Max}^2 \frac{d}{dx}\left(\frac{37}{315}\delta\right)$$

$$\Rightarrow 2 * \frac{315}{37} * \frac{\mu U_{Max}}{\rho U_{Max}^2} dx = \delta\left(d\delta\right)$$

Integrating both sides, we get **boundary layer thickness**

$$\delta = \frac{5.84x}{\sqrt{R_{e_x}}} \tag{3}$$

b) Wall shear stress:

Substitute Eq. (3) in Eq. (2), then we get wall shear stress as

$$\Rightarrow \tau_w = 0.343\frac{\mu U_{Max}}{x}\sqrt{R_{e_x}} \tag{4}$$

c) Skin friction coefficient/drag coefficient:

Skin friction coefficient is defined as the ratio of shear stress at the walls and $\frac{1}{2}\rho U_M^2$, and hence,

$$^*C_f = \frac{\tau_w(x)}{\frac{1}{2}\rho U_M^2} \tag{5}$$

and substituting Eq. (4) in Eq. (5), we get

$$^*C_f = \frac{2 * 0.343\frac{\mu U_{Max}}{x}\sqrt{R_{e_x}}}{\rho U^2_{Max}}$$

$$\Rightarrow {}^*C_f = \left(\frac{0.686\mu}{\rho x U_{Max}}\right)\left(\sqrt{R_{e_x}}\right)$$

$$\Rightarrow {}^*C_f = \frac{0.686}{\sqrt{R_{e_x}}} \tag{6}$$

d) Drag force and average drag coefficient:

The **drag force** $\left(F_D\right)$ will be in opposite direction to the shear stress τ_w and as *force* = *stress* * *area*

we have $F_D = \int -\tau_w\left(A\right)$

$$\Rightarrow F_D = \int_0^L \tau_w (dx * th) = \int_0^L \tau_w (dx * th)$$

$$\Rightarrow F_D = \int_0^L \tau_w (dx * th) = \int_0^L 0.343 \frac{\mu U_{Max}}{x} \sqrt{R_{e_x}} * dx * th$$

$$\Rightarrow F_D = 0.343 \mu U_{Max} * th * \sqrt{\frac{\rho U_{Max}}{\mu}} \int_0^L \frac{1}{x} * dx$$

$$\Rightarrow F_D = 0.343 \mu U_{Max} * th * \sqrt{\frac{\rho U_{Max}}{\mu}} \left| \frac{x^{1/2}}{1/2} \right|_0^L$$

$$\Rightarrow F_D = 0.686 \mu U_{Max} * th * \sqrt{\frac{\rho U_{Max} L}{\mu}}$$

$$\Rightarrow F_D = \rho U_M \left(\frac{0.686}{\sqrt{R_{e_L}}} \right) \text{Area} \qquad (7)$$

Average skin friction coefficient is defined as the drag force and $\frac{1}{2}\rho A U_M^2$, and hence,

$$F_D = \frac{\rho U_M^2}{2} * \bar{C}_f * A \qquad (8)$$

and substituting Eq. (7) in Eq. (8), we get

$$\bar{C}_f = \frac{F_D}{\frac{\rho U_M^2}{2} * A} = \frac{\rho U_M \left(\frac{0.686}{\sqrt{R_{e_L}}} \right) A}{\frac{\rho U_M^2}{2} * A} = \frac{1.372}{\sqrt{R_{e_L}}}$$

Example 7.14: For the given velocity profile $\frac{u}{U_{Max}} = 1 + \left(\frac{y}{\delta}\right)$

Determine a) Derive the boundary layer thickness.
b) Wall shear stress.
c) Coefficient of drag.
d) Drag force and average drag coefficient.

Solution: Given data: velocity profile $\frac{u}{U_{Max}} = 1 + \left(\frac{y}{\delta}\right)$

a) Boundary layer thickness:

Refer Eq. (7.9), we have momentum thickness as

$$\delta_M = \int_0^{\delta} \frac{u}{U_{\max}}\left(1 - \frac{u}{U_{\max}}\right) dy$$

Substituting the velocity profile in the above equation, we get

$$\delta_M = \int_0^{\delta}\left(1 + \frac{y}{\delta}\right)\left(1 - 1 - \left(\frac{y}{\delta}\right)\right) dy$$

$$\delta_M = \int_0^{\delta}\left(1 + \frac{y}{\delta}\right)\left(-\left(\frac{y}{\delta}\right)\right) dy$$

$$\delta_M = \int_0^{\delta}\left(-\left(\frac{y}{\delta}\right) - \left(\frac{y}{\delta}\right)^2\right) dy$$

$$\Rightarrow \delta_M = \left| -\frac{1}{2}\left(\frac{y^2}{\delta}\right) - \frac{1}{3}\left(\frac{y^3}{\delta^2}\right)\right|_0^{\delta} = \frac{-5}{6}\delta$$

Refer Eq. (7.18), we have $\Rightarrow \frac{\tau_w}{\rho U_M^2} = \frac{d}{dx}[\delta_M]$

$$\Rightarrow \tau_w = -\rho U_{Max}^2 \frac{d}{dx}\left(\frac{5}{6}\delta\right) \quad (1)$$

From the velocity profile, we have

$$u = U_{Max}\left\{1 + \left(\frac{y}{\delta}\right)\right\}$$

Hence, the velocity gradient will be

$$\frac{du}{dy} = U_{Max}\left\{\left(\frac{1}{\delta}\right)\right\}$$

For a laminar Newtonian fluid

$$\tau_w = \mu\left(\frac{du}{dy}\right)_{y=0} = \mu U_{MAX}\left\{\left(\frac{1}{\delta}\right)\right\} \Rightarrow \tau_w = \frac{\mu U_{Max}}{\delta} \quad (2)$$

Equate Eq. (1) and Eq. (2)

$$\Rightarrow \frac{\mu U_{Max}}{\delta} = \rho U_{Max}^2 \frac{d}{dx}\left(\frac{-5}{6}\delta\right) \Rightarrow \frac{6}{5} * \frac{\mu U_{Max}}{\rho U_{Max}^2} dx = -\delta(d\delta)$$

Integrating both sides, we get **boundary layer thickness**

$$\delta = \frac{1.2x}{\sqrt{R_{e_x}}} \tag{3}$$

b) Wall shear stress:

Substitute Eq. (3) in Eq. (2), then we get wall shear stress as

$$\Rightarrow \tau_w = 0.8333 \frac{\mu U_{Max}}{x} \sqrt{R_{e_x}} \tag{4}$$

c) Skin friction coefficient/drag coefficient:

Skin friction coefficient is defined as the ratio of shear stress at the walls and $\frac{1}{2}\rho U_M^2$, and hence,

$$^*C_f = \frac{\tau_w(x)}{\frac{1}{2}\rho U_M^2} \tag{5}$$

and substituting Eq. (4) in Eq. (5), we get

$$^*C_f = \frac{2 * 0.8333 \frac{\mu U_{Max}}{x} \sqrt{R_{e_x}}}{\rho U^2_{Max}}$$

$$\Rightarrow {}^*C_f = \left(\frac{1.666\mu}{\rho x U_{Max}}\right)\left(\sqrt{R_{e_x}}\right)$$

$$\Rightarrow {}^*C_f = \frac{1.666}{\sqrt{R_{e_x}}} \tag{6}$$

d) Drag force and average drag coefficient:

The **drag force** (F_D) will be in opposite direction to the shear stress τ_w and as *force = stress * area* we have

$$F_D = \int -\tau_w(A) \Rightarrow F_D = \int_0^L \tau_w(dx * th) = \int_0^L \tau_w(dx * th)$$

$$\Rightarrow F_D = \int_0^L \tau_w \left(dx * th\right) = \int_0^L 0.8333 \frac{\mu U_{Max}}{x} \sqrt{R_{e_x}} * dx * th$$

$$\Rightarrow F_D = 0.833 \mu U_{Max} * th * \sqrt{\frac{\rho U_{Max}}{\mu}} \int_0^L \frac{1}{x} * dx$$

$$\Rightarrow F_D = 0.033 \mu U_{Max} * th * \sqrt{\frac{\rho U_{Max}}{\mu}} \left| \frac{x^{1/2}}{1/2} \right|_0^L$$

$$\Rightarrow F_D = 1.666 \mu U_{Max} * th * \sqrt{\frac{\rho U_{Max} L}{\mu}}$$

$$\Rightarrow F_D = \rho U_M \left(\frac{1.666}{\sqrt{R_{e_L}}} \right) \text{Area} \tag{7}$$

Average skin friction coefficient is defined as the drag force and $\frac{1}{2}\rho A U_M^2$, and hence,

$$F_D = \frac{\rho U_M^2}{2} * \bar{C}_f * A \tag{8}$$

and substituting Eq. (7) in Eq. (8), we get $\bar{C}_f = \dfrac{F_D}{\dfrac{\rho U_M^2}{2} * A} = \dfrac{\rho U_M \left(\dfrac{1.666}{\sqrt{R_{e_L}}} \right) A}{\dfrac{\rho U_M^2}{2} * A} = \dfrac{3.333}{\sqrt{R_{e_L}}}$

Example 7.15: For the given velocity profile $\dfrac{u}{U_{Max}} = 1 + \left(\dfrac{y}{\delta}\right) - 2\left(\dfrac{y}{\delta}\right)^2$

Determine: a) Derive the boundary layer thickness.
b) Wall shear stress.
c) Coefficient of drag.
d) Drag force and average drag coefficient.

Solution: Given data: velocity profile $\dfrac{u}{U_{Max}} = 1 + \left(\dfrac{y}{\delta}\right) - 2\left(\dfrac{y}{\delta}\right)^2$

a) Boundary layer thickness:

Refer Eq. (7.9), we have momentum thickness as $\delta_M = \int_0^{\delta} \frac{u}{U_{\max}}\left(1-\frac{u}{U_{\max}}\right)dy$

Substituting the velocity profile in the above equation, we get

$$\delta_M = \int_0^{\delta} 1+\left(\frac{y}{\delta}\right)-2\left(\frac{y}{\delta}\right)^2\left(1-1-\left(\frac{y}{\delta}\right)+2\left(\frac{y}{\delta}\right)^2\right)dy$$

$$\Rightarrow \delta_M = \int_0^{\delta}\left[1+\left(\frac{y}{\delta}\right)-2\left(\frac{y}{\delta}\right)^2\right]\left(-\left(\frac{y}{\delta}\right)+2\left(\frac{y}{\delta}\right)^2\right)dy$$

$$= \int_0^{\delta}\left\{-\left(\frac{y}{\delta}\right)-\left(\frac{y}{\delta}\right)^2+2\left(\frac{y}{\delta}\right)^3+2\left(\frac{y}{\delta}\right)^2+2\left(\frac{y}{\delta}\right)^3-4\left(\frac{y}{\delta}\right)^4\right\}dy$$

$$\Rightarrow \delta_M = \left|\frac{-1}{2}\left(\frac{y^2}{\delta}\right)+\frac{1}{3}\left(\frac{y^3}{\delta^2}\right)+\frac{4}{4}\left(\frac{y^4}{\delta^3}\right)-\frac{4}{5}\left(\frac{y^5}{\delta^4}\right)\right|_0^{\delta} = \frac{1}{30}\delta$$

Refer Eq. (7.18), we have $\Rightarrow \frac{\tau_w}{\rho U_M^2} = \frac{d}{dx}[\delta_M]$

$$\Rightarrow \tau_w = \rho U_{Max}^2 \frac{d}{dx}\left(\frac{1}{30}\delta\right) \quad (1)$$

From the velocity profile, we have $u = U_{Max}\left\{1+\left(\frac{y}{\delta}\right)-2\left(\frac{y}{\delta}\right)^2\right\}$

Hence, the velocity gradient will be $\frac{du}{dy} = U_{Max}\left\{\left(\frac{1}{\delta}\right)-4\left(\frac{y}{\delta^2}\right)\right\}$

For a laminar Newtonian fluid $\tau_w = \mu\left(\frac{du}{dy}\right)_{y=0}$

$$\tau_w = \mu\left(\frac{du}{dy}\right)_{y=0} = \mu U_{Max}\left\{\left(\frac{1}{\delta}\right)\right\}$$

$$\Rightarrow \tau_w = \frac{\mu U_{Max}}{\delta} \qquad (2)$$

Equate Eq. (1) and Eq. (2)

$$\Rightarrow \frac{\mu U_{Max}}{\delta} = \rho U_{Max}^2 \frac{d}{dx}\left(\frac{1}{30}\delta\right)$$

$$\Rightarrow \frac{30}{1} * \frac{\mu U_{Max}}{\rho U_{Max}^2} dx = \delta(d\delta)$$

Integrating both sides, we get **boundary layer thickness**

$$\delta = \frac{30x}{\sqrt{R_{e_x}}} \qquad (3)$$

b) Wall shear stress:
Substitute Eq. (3) in Eq. (2), then we get wall shear stress as

$$\Rightarrow \tau_w = 0.0333 \frac{\mu U_{Max}}{x} \sqrt{R_{e_x}} \qquad (4)$$

c) Skin friction coefficient/drag coefficient:

Skin friction coefficient is defined as the ratio of shear stress at the walls and $\frac{1}{2}\rho U_M^2$, and hence,

$$^*C_f = \frac{\tau_w(x)}{\frac{1}{2}\rho U_M^2} \qquad (5)$$

and substituting Eq. (4) in Eq. (5), we get

$$^*C_f = \frac{2 * 0.0333 \frac{\mu U_{Max}}{x} \sqrt{R_{e_x}}}{\rho U^2_{Max}}$$

$$\Rightarrow {}^*C_f = \left(\frac{0.0666\mu}{\rho x U_{Max}}\right)\left(\sqrt{R_{e_x}}\right)$$

$$\Rightarrow {}^*C_f = \frac{0.0666}{\sqrt{R_{e_x}}} \qquad (6)$$

d) Drag force and average drag coefficient:

The **drag force** (F_D) will be in opposite direction to the shear stress τ_w and as $force = stress * area$ we have $F_D = \int -\tau_w(A)$

$$\Rightarrow F_D = \int_0^L \tau_w(dx * th) = \int_0^L \tau_w(dx * th)$$

$$\Rightarrow F_D = \int_0^L \tau_w(dx * th) = \int_0^L 0.0333 \frac{\mu U_{Max}}{x} \sqrt{R_{e_x}} * dx * th$$

$$\Rightarrow F_D = 0.033 \mu U_{Max} * th * \sqrt{\frac{\rho U_{Max}}{\mu}} \int_0^L \frac{1}{x} * dx$$

$$\Rightarrow F_D = 0.033 \mu U_{Max} * th * \sqrt{\frac{\rho U_{Max}}{\mu}} \left| \frac{x^{1/2}}{1/2} \right|_0^L$$

$$\Rightarrow F_D = 0.0666 \mu U_{Max} * th * \sqrt{\frac{\rho U_{Max}}{\mu}} \left| \frac{x^{1/2}}{1/2} \right|_0$$

$$\Rightarrow F_D = pU_M \left(\frac{0.0666}{\sqrt{R_{eL}}} \right) \text{Area} \tag{7}$$

Average skin friction coefficient is defined as the drag force and $\frac{1}{2}\rho A U_M^2$, and hence,

$$F_D = \frac{\rho U_M^2}{2} * \bar{C}_f * A \tag{8}$$

and substituting Eq. (7) in Eq. (8), we get

$$\bar{C}_f = \frac{F_D}{\frac{\rho U_M^2}{2} * A} = \frac{\rho U_M \left(\frac{0.0666}{\sqrt{R_{e_L}}} \right) A}{\frac{\rho U_M^2}{2} * A} = \frac{0.1333}{\sqrt{R_{e_L}}}$$

Example 7.16: Oil having kinematic viscosity of 2.6×10^{-4} m²/s flows past over a flat plate 5 m, at a velocity of 30 m/s. Determine the boundary layer thickness at the length of 2 m from the leading edge.

Solution: Given data: $\nu = 2.6 \times 10^{-4} \text{m}^2/\text{s}$, $U_{Max} = 3 \text{m/s}$, $L = 5\text{m}, x = 2\text{m}$

We know Reynolds number at $x = 2$ m, will be

$$R_e = \frac{xU_{Max}}{\nu} = \frac{2*3}{2.6 \times 10^{-4}} = 23079$$

And according to Blasius, we have $\delta = \dfrac{5x}{\sqrt{R_{e_x}}}$

$$\Rightarrow \delta = \frac{5*2}{\sqrt{23079}} = 0.0658 \text{ m}$$

Hence, the **boundary layer thickness** is 0.0658 m, from the leading edge of the plate.

TURBULENT FLOW THEORY

Turbulent, the other name of irregular fluctuations like eddies, which reduce the energy of the flow, in form of inter-rubbing action of fluid molecules due to which the energy possessed due to kinetic energy is dissipated in form of heat.

In nature, there exist no perfect laminar flows, when viewed microscopically. There exist small disturbances in between fluid molecules in motion, at microscopic level. But if these disturbances can be neglected in macroscopic level, then the flow is as "laminar flow", but if they are appreciable in macroscopic level, then the flow is known as "**Turbulent flow**".

At lower velocity (low Reynolds number), the viscous effect in the flow overcome the fluctuations and so the flow looks like laminar, but once the velocity increases high (high Reynolds number) the fluctuations (eddies) momentum increases, which makes the flow turbulent and the properties of the fluid flow changes considerably, when compared with laminar fluid. And hence it is of much importance as that of laminar flow study.

The turbulence can be due to any kind of disturbances like low viscosity flow, flow over non-aerodynamic shapes, sudden obstructions, vibrations, etc. and hence there is no mathematical relation with completeness for establishing turbulent flow and so **turbulent flow cannot be defined but only can be described basing on some approximations** and incorporating them in Navier-Stokes equation.

Momentum Equation for Turbulent Flow and Reynold Stresses

As per the continuity equation for an incompressible fluid steady state, the curl of velocity is zero, i.e.

$$\frac{du}{dx}+\frac{dv}{dy}+\frac{dw}{dz}=0 \tag{7.24}$$

In a turbulent flow, the flow will be in a steady state and hence each fluid particle has its own velocity and hence it is necessary to take the average and also include the fluctuation components, as shown in Figure 7.7.

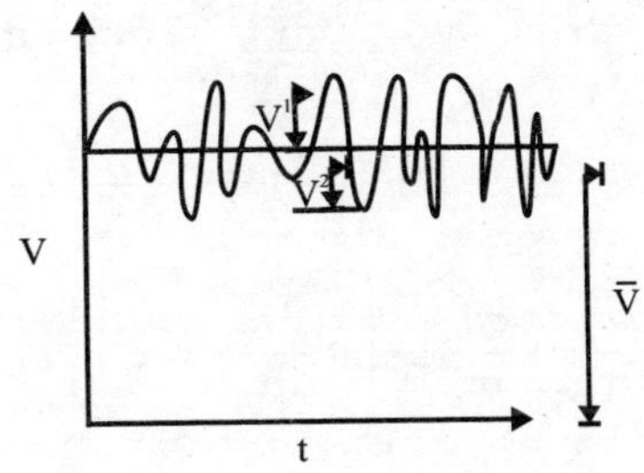

Figure 7.7: Variation of mean and fluctuating velocities with time "t" in turbulent flow

$$\begin{aligned}
u(x,y,z,t)&=\bar{u}(x,y,z,t)+u'(x,y,z,t)\\
v(x,y,z,t)&=\bar{v}(x,y,z,t)+v'(x,y,z,t)\\
w(x,y,z,t)&=\bar{w}(x,y,z,t)+w'(x,y,z,t)\\
P(x,y,z,t)&=\bar{P}(x,y,z,t)+P'(x,y,z,t)
\end{aligned} \tag{7.24a}$$

Where $\bar{V}$, V' are average and fluctuation components of velocity.

Hence, substituting Eq. (7.24a) in Eq. (7.24), we have

$$\left(\frac{d\bar{u}}{dx}+\frac{d\bar{v}}{dy}+\frac{d\bar{w}}{dz}\right)+\left(\frac{du'}{dx}+\frac{dv'}{dy}+\frac{dw'}{dz}\right)=0 \tag{7.24b}$$

But the differentials of fluctuation components will be near to zero, as the fluctuations are very small and hence

$$\left(\frac{du'}{dx}+\frac{dv'}{dy}+\frac{dw'}{dz}\right)\approx 0 \tag{7.24b}$$

and this reduces Eq. (7.24b) as

$$\left(\frac{d\bar{u}}{dx}+\frac{d\bar{v}}{dy}+\frac{d\bar{w}}{dz}\right)=0 \tag{7.24c}$$

Consider Navier Stokes equation expressed in terms of momentum and continuity (refer Eq. (6.59)) for a incompressible steady state fluid flow,

i.e.
$$\rho\left(\frac{dV}{dt}+V\frac{dV}{ds}\right)=-\nabla\left(P_s\right)+\rho G_s+\mu\nabla^2\left(V\right)$$

Along x-axis

$$\Rightarrow \rho\left(u\frac{\partial u}{\partial x}+v\frac{\partial u}{\partial y}\right)=-\frac{\partial p}{\partial x}+\mu\left(\frac{\partial^2 u}{\partial x^2}+\frac{\partial^2 u}{\partial y^2}\right) \tag{7.25}$$

Substituting the fluctuation components in Eq. (7.24a) in Eq. (7.25), we have

$$\rho\left\{\left(\bar{u}\frac{\partial \bar{u}}{\partial x}+\bar{v}\frac{\partial \bar{u}}{\partial y}\right)+\left(\bar{u}\frac{\partial u'}{\partial x}+\bar{v}\frac{\partial u'}{\partial y}\right)+\left(u'\frac{\partial \bar{u}}{\partial x}+v'\frac{\partial \bar{u}}{\partial y}\right)+\left(u'\frac{\partial u'}{\partial x}+v'\frac{\partial u'}{\partial y}\right)\right\}=$$
$$=-\left(\frac{\partial \bar{p}}{\partial x}+\frac{\partial p'}{\partial x}\right)+\mu\left\{\left(\frac{\partial^2 \bar{u}}{\partial x^2}+\frac{\partial^2 \bar{u}}{\partial y^2}\right)+\left(\frac{\partial^2 u'}{\partial x^2}+\frac{\partial^2 u'}{\partial y^2}\right)\right\} \tag{7.25a}$$

$$\rho\left\{\left(\bar{u}\frac{\partial \bar{u}}{\partial x}+\bar{v}\frac{\partial \bar{u}}{\partial y}\right)+\cancel{\left(\bar{u}\frac{\partial u'}{\partial x}+\bar{v}\frac{\partial u'}{\partial y}\right)}+\cancel{\left(u'\frac{\partial \bar{u}}{\partial x}+v'\frac{\partial \bar{u}}{\partial y}\right)}+\left(u'\frac{\partial u'}{\partial x}+v'\frac{\partial u'}{\partial y}\right)\right\}=$$
$$=-\left(\frac{\partial \bar{p}}{\partial x}+\cancel{\frac{\partial p'}{\partial x}}\right)+\mu\left\{\left(\frac{\partial^2 \bar{u}}{\partial x^2}+\frac{\partial^2 \bar{u}}{\partial y^2}\right)+\cancel{\left(\frac{\partial^2 u'}{\partial x^2}+\frac{\partial^2 u'}{\partial y^2}\right)}\right\}$$

$$\Rightarrow \rho\left\{\left(\bar{u}\frac{\partial \bar{u}}{\partial x}+\bar{v}\frac{\partial \bar{u}}{\partial y}\right)+\left(u'\frac{\partial u'}{\partial x}+v'\frac{\partial u'}{\partial y}\right)\right\}=-\left(\frac{\partial \bar{p}}{\partial x}\right)+\mu\left\{\left(\frac{\partial^2 \bar{u}}{\partial x^2}+\frac{\partial^2 \bar{u}}{\partial y^2}\right)\right\} \tag{7.25b}$$

Multiply Eq. (7.24b) with $\rho u'$ for two-dimensional and add this to the second term in L.H.S. in Eq. (7.25b), without change in value, then it becomes

$$\left\{\left(\frac{\partial (u')^2}{\partial x}+\frac{\partial u'v'}{\partial y}\right)\right\} \tag{7.25c}$$

By replacing Eq. (7.25c) in Eq. (7.25b) and taking time average of the inertia terms, we have

$$\rho\left\{\left(\bar{u}\frac{\partial \bar{u}}{\partial x}+\bar{v}\frac{\partial \bar{u}}{\partial y}\right)+\left(\frac{\partial}{\partial x}\left(\overline{u'}\right)^2+\frac{\partial}{\partial y}\left(\overline{u'v'}\right)\right)\right\}=-\left(\frac{\partial \bar{p}}{\partial x}\right)+\mu\left\{\left(\frac{\partial^2 \bar{u}}{\partial x^2}+\frac{\partial^2 \bar{u}}{\partial y^2}\right)\right\} \tag{7.25d}$$

Equation 7.26 gives the **momentum equation of a turbulent flow.**

As per the laminar viscosity $\tau = u\frac{\partial u}{\partial y}$ but once the flow becomes turbulent the viscosity has the effect of fluctuations and hence the viscosity becomes $\tau = \mu\frac{\partial u}{\partial y} - \rho\overline{u}'^2$ and so the Eq. (7.25d) becomes

$$\left\{\left(\bar{u}\frac{\partial \bar{u}}{\partial x} + \bar{v}\frac{\partial \bar{u}}{\partial y}\right)\right\} = -\left(\frac{\partial \bar{p}}{\partial x}\right) + \frac{\partial}{\partial x}\left(\mu\frac{\partial \bar{u}}{\partial y} - \rho\overline{u'^2}\right) + \frac{\partial}{\partial y}\left(\mu\frac{\partial \bar{u}}{\partial y} - \rho\overline{u'v'}\right) \tag{7.26}$$

The terms $-\rho\overline{u'^2}$ and $-\rho\overline{u'v'}$ are the components due to turbulence and they are known as "**Reynold stresses**" and if written in three-dimensional form, we get Reynold stresses as

$$\begin{pmatrix} -\rho\overline{u'^2} & -\rho\overline{u'v'} & -\rho\overline{u'w'} \\ -\rho\overline{u'v'} & -\rho\overline{v'^2} & -\rho\overline{u'w'} \\ -\rho\overline{u'w'} & -\rho\overline{w'v'} & -\rho\overline{w'^2} \end{pmatrix}$$

TURBULENT FLOW THROUGH A PIPE

Consider a pipe having diameter "D" and length "L" and with a fluid flowing in it. The velocity profile will be a smooth parabolic curve with maximum velocity at the center. But as the boundary of the pipe may not be smooth, due to which the velocity profile of the flow will be having a laminar parabolic profile on one side of the pipe and logarithmic turbulent profile

$$\frac{u}{U_{MAX}} = \left[\frac{y}{R}\right]^{1/n} \tag{7.27}$$

on the other side of the pipe, as shown in Figures 7.8a and b.

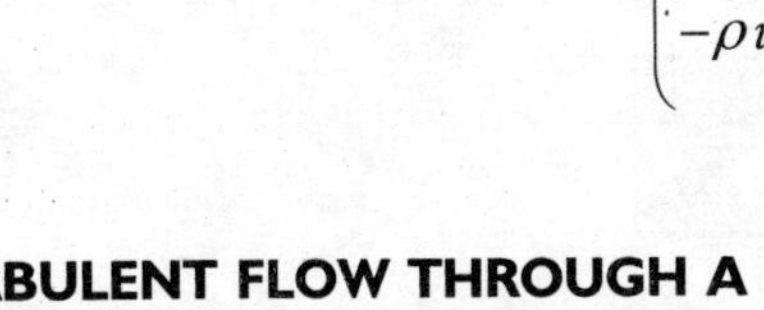

Figure 7.8a

Figure 7.8b

As the turbulent and laminar layers mix up with each other at the center, the frictional losses increase due to this turbulence effect for the same flow rate and same pipe diameter with laminar flow.

Consider a fluid cylindrical element in the center line of the flow, so that the element is subjected to turbulent and laminar effects on either side, along its length and hence the head losses at the fluid element due to turbulent and laminar flow will be same, i.e. $h_{f_{Lamiinar}} = h_{f_{Turbulent}}$.

Laminar Region

A pressure force F_P will be driving the fluid element from the inlet to outlet of the pipe. And while moving, due to turbulence effect, there exists a shear force F_τ at the walls of the fluid element, opposite in direction to the flow of fluid element.

Under equilibrium conditions, the pressure force F_P will be equal to shear force F_τ, i.e. $F_P = F_\tau$ where $F_P = (P_1 - P_2)A$ and $F_\tau = \tau_w(2\Pi rL)$.

So
$$(P_1 - P_2)A = \tau_w(2\Pi rL)$$

$$\Rightarrow (P_1 - P_2) = \frac{4\tau_w L}{D}$$

$$\Rightarrow h_{f_{Laminar}} = \frac{(P_1 - P_2)}{\gamma} = \frac{4\tau_w L}{\gamma D} \qquad (7.28)$$

Turbulent Region

In turbulence the velocity of each fluid particle will be not same, as like in case of laminar flow and hence the pressure force is not uniform throughout the flow by every fluid element and so this pressure force is called **propelling force** F_{PP}, such that $F_{PP} = (P_1 - P_2)A_W$, where A_W is the wetted area of the fluid elements. Similarly, the shear stress at the walls of fluid elements are not uniform and vary as a function of velocity square on the wetted perimeter of the fluid elements and is called **frictional resistance** $F_{\tau P}$ and so $F_{\tau P} \alpha (P_W L)V^2$, which makes $F_{\tau P} = f'(P_W L)V^2$ where P_W is the wetted perimeter and f' is a non-dimensional parameter.

Hence, under equilibrium $F_{PP} = F_{\tau P}$

$$\Rightarrow (P_1 - P_2)A_W = f'(P_W L)V^2$$

By rearranging, we have

$$(P_1 - P_2) = \frac{f'(P_W L)V^2}{A_W}$$

Dividing by specific weight of the fluid γ and rearranging, we have friction losses

$$h_{f_{Turbulent}} = \frac{(P_1 - P_2)}{\gamma} = \left(\frac{2gf'}{\gamma}\right)\left(\frac{P_W}{A_W}\right)\left(\frac{LV^2}{2g}\right) = \frac{4\lambda LV^2}{2gD_h} \qquad (7.29)$$

since $\frac{D_h}{4} = \frac{A_W}{P_W}$ = *Hydraulic diameter* and $\lambda = \frac{2gf'}{y}$ = *Fanning friction factor*

but $h_{f_{Laminar}} = h_{f_{Turbulent}}$ so equate Eq. (7.28) and Eq. (7.29),

i.e.
$$\frac{fLV^2}{2gD_h} = \frac{4\tau_w L}{\gamma D}$$

$$\Rightarrow f = \frac{8\tau_w}{\rho V^2} = \frac{8\tau_w}{\rho u_{av}^2} \tag{7.30}$$

since $f = 4\lambda$ = Darcy friction factor

H. Blasius has established the experimental value of Darcy friction factor as

$$f = 0.3164(R_e)^{0.25} \tag{7.30a}$$

for $R_e \leq 10^5$ in a pipe flow. And experimentally the power index $n = 7$ satisfy the Reynolds number $R_e \leq 5\times10^5$ to $R_e \leq 10^7$, and hence the turbulent flow is taken as Seventh power law, i.e.

$$\frac{u}{U_{MAX}} = \left[\frac{y}{R}\right]^{1/7}$$

Discharge

Discharge Q = area*velocity and hence discharge through the pipe will be $Q = \Pi R^2 * u_{av}$ but if considered the fluid element then discharge $Q = \int_0^R 2\Pi r * u * dr$ and hence

$$Q = \Pi R^2 * u_{av} = \int_0^R 2\Pi r * u * dr \tag{7.31}$$

as $r = R - y$ and $\frac{u}{U_{MAX}} = \left[\frac{y}{R}\right]^{1/n}$, if substituted this values in the Eq. (7.31), we get

$$\Pi R^2 * u_{av} = -\int_0^R 2\Pi(R-y) * U_{MAX}\left[\frac{y}{R}\right]^{1/n} * dy$$

$$\frac{u_{av}}{U_{MAX}} = \left(\frac{2n^2}{(n+1)(2n+1)}\right) \Rightarrow u_{av} = 0.816U_{MAX} \tag{7.31a}$$

If n = 7

If substituted Eq. (7.30a) and Eq. (7.31a) in Eq. (7.30), we get

$$\frac{\tau_w}{\rho} = 0.0226U_M^2\left(\nu/_{RU_M}\right)^{1/4} = 0.0226U_M^2\left(\mu/_{\rho U_M\delta}\right)^{1/4} \quad (7.32)$$

Equation 7.32 gives the **shear stress at the walls of a pipe having turbulent flow.**

The **total friction force** acting on the fluid due to the pipe over a length L is given by

$$F = \int_0^L \tau_w (2\pi R)dz = -8\pi\mu L U_{Max} \quad (7.32a)$$

Example 7.17: A smooth pipe of diameter 8 cm and 800 m long, carries water at the rate of 0.48 m³/minute. Calculate the center line velocity, head loss, wall shear and total friction force. Take $f = 0.0791(R_e)^{-0.25}$ and $\nu = 0.015$ stokes

Solution: Given data:

$$D = 0.08\,\text{m}, L = 800\,\text{m}, Q = 0.48/60 = 0.008\,\text{m}^3/\text{s}$$

$$f = 0.0791\left(R_e\right)^{-0.25}, \nu = 0.015\times10^{-4}\,\text{m}^2/\text{s}$$

We know discharge $Q = Au_{AV} = \frac{\pi}{4}D^2 * u_{AV}$

$$\Rightarrow u_{AV} = \frac{4Q}{\pi * D^2} = \frac{4*0.008}{\pi*0.08^2} = 1.592\,\text{m/s}$$

So Reynolds number will be

$$R_e = \frac{u_{AV} * D}{\nu} = \frac{1.592*0.08}{0.015\times10^{-4}} = 84925$$

As the Reynolds number is greater than 4000 and hence the flow is **turbulent** in the pipe.

a) **Center line velocity:**

$$U_{MAX} = 2*u_{AV} = 2*1.592 = 3.184\,\text{m/s}$$

b) **For head loss:** $h_f = \frac{fLu_{AV}^2}{2gD}$

$$\Rightarrow h_f = \frac{0.0791*\left(R_e\right)^{-0.25}*800*1.592^2}{2*9.81*0.08} = \frac{0.0791*\left(84925\right)^{-0.25}*800*1.592^2}{2*9.81*0.08} = 5.98\,\text{m}$$

c) For **wall shear stress:** Refer Eq.(7.32), we have

$$\tau_w = \rho 0.0226 U_M^2 \left(\nu / RU_M \right)^{1/4}$$

$$\Rightarrow \tau_w = \rho 0.0226 U_M^2 \left(\nu / RU_M \right)^{1/4} = 1000 * 0.0226 * 3.184^2 \left(0.015\times10^{-4} / 0.04 * 3.184 \right)^{1/4} = 13.42\,\text{Pa}$$

d) **Total friction force:** Refer Eq.(7.32a), we have

$$F = \int_0^L \tau_w (2\pi R) dz = -8\pi\mu L U_{Max}$$

$$\Rightarrow F = -8\pi\nu\rho L U_{Max} = -8 * 3.14 * 0.015\times10^{-4} * 1000 * 800 * 3.184 = 95.98\,\text{N}$$

Negative sign indicates that the friction force is against shear.

TURBULENT FLOW OVER A FLAT PLATE WITH VISCOUS SUB-LAYER

The turbulent flow has some laminar sub-region up to the critical length before it fully develops and this laminar has shear stress τ_w at the wall and it has boundary layer thickness (δ_{LS}) with a maximum velocity U_{MLS}

As seen in Eq. (7.32) the shear stress

$$\frac{\tau_w}{\rho} = 0.0226 U_M^2 \left(\nu / \delta U_M \right)^{1/4} \tag{7.33a}$$

for flow over a flat plate, as the radius of pipe substitutes the boundary layer thickness.

According to **Von-Kármán** equation $\frac{\tau_w}{\rho U_M^2} = \frac{d}{dx}[\delta_M]$ refer Eq. (7.18) and evaluating

$$\delta_M = \int_0^\delta \frac{u}{U_{max}} \left(1 - \frac{u}{U_{max}} \right) dy$$

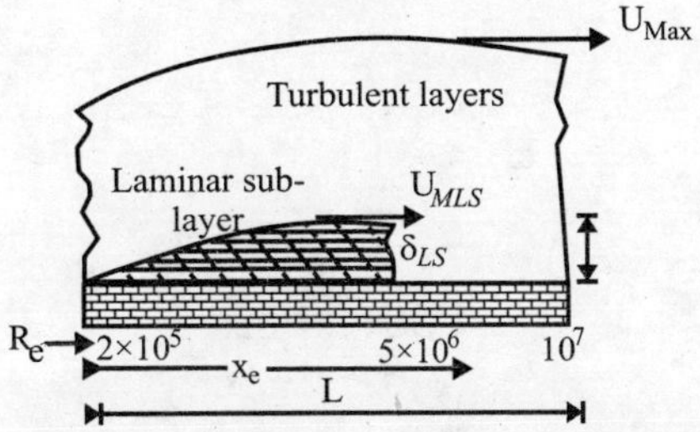

Figure 7.9: Turbulent flow over a flat plate with viscous layers

with the seventh power law

$$\text{(i.e. } \frac{u}{U_{\max}} = \left[\frac{y}{R}\right]^{1/n} \text{),}$$

we get $\delta_M = \dfrac{7}{72}\delta$, Von-Kármán equation reduces to

$$\frac{\tau_w}{\rho U_M^2} = \frac{7}{72}\frac{d\delta}{dx} \tag{7.33b}$$

Equating Eq. (7.33a) and Eq. (7.33b), we have

$$0.0226U\left(\nu/\delta U_M\right)^{1/4} = \frac{7}{72}\frac{d\delta}{dx}$$

$$\Rightarrow \delta = \frac{0.371x}{\left(R_{e_x}\right)^{1/5}} \tag{7.34}$$

Substituting Eq. (7.34) in Eq. (7.33a), we have

$$\tau_w = \left(\frac{\rho U_M^2}{2}\right)\left(\frac{0.0576}{\left(R_{e_x}\right)^{1/5}}\right) \ \& \ {}^*C_f = \left(\frac{0.0576}{\left(R_{e_x}\right)^{1/5}}\right)$$

$$\overline{C}_f = \left(\frac{0.074}{\left(R_{e_L}\right)^{1/5}}\right) \ \& \ F_D = \left(\frac{\rho U_M^2}{2}\right)\left(\frac{0.074}{\left(R_{e_L}\right)^{1/5}}\right)A \tag{7.35}$$

$${}^*C_f = \frac{\tau_w}{\dfrac{\rho U_M^2}{2}}$$

since

$$\overline{C}_f = \frac{1}{L}\int_0^L {}^*C_f * dx$$

$$F_D = \left(\frac{\rho U_M^2}{2}\right)\left(\overline{C}_f\right)A$$

Example 7.18: For the given velocity profile $\frac{u}{U_{Max}} = \left(\frac{y}{\delta}\right)^{1/7}$, determine:

a) Derive the boundary layer thickness.
b) Wall shear stress.
c) Coefficient of drag.
d) Drag force and average drag coefficient.

Solution: Given data: velocity profile $\frac{u}{U_{Max}} = \left(\frac{y}{\delta}\right)^{1/7}$

a) Boundary layer thickness

Refer Eq. (7.9), we have momentum thickness as

$$\delta_M = \int_0^\delta \frac{u}{U_{max}}\left(1 - \frac{u}{U_{max}}\right) dy$$

Substituting the velocity profile in the above equation, we get

$$\delta_M = \int_0^\delta \left(\frac{y}{\delta}\right)^{1/7}\left(1 - \left(\frac{y}{\delta}\right)^{1/7}\right) dy$$

$$\delta_M = \int_0^\delta \left(\left(\frac{y}{\delta}\right)^{1/7} - \left(\frac{y}{\delta}\right)^{2/7}\right) dy$$

$$\Rightarrow \delta_M = \left|\frac{7}{8}\left(\frac{y^{8/7}}{\delta^{1/7}}\right) - \frac{7}{9}\left(\frac{y^{9/7}}{\delta^{2/7}}\right)\right|_0^\delta = \frac{7}{72}\delta$$

Refer Eq. (7.18), we have

$$\Rightarrow \frac{\tau_w}{\rho U_M^2} = \frac{d}{dx}[\delta_M]$$

$$\Rightarrow \tau_w = \rho U_{Max}^2 \frac{d}{dx}\left(\frac{7}{72}\delta\right) \qquad (1)$$

From the velocity profile, we have

$$u = U_{Max}\left\{\left(\frac{y}{\delta}\right)^{1/7}\right\}$$

Hence, the velocity gradient will be

$$\frac{du}{dy} = U_{Max} * \frac{1}{7}\left\{\left(\frac{y^{8/7}}{\delta^{1/7}}\right)\right\}$$

For a laminar Newtonian fluid $\tau_w = \mu\left(\frac{du}{dy}\right)_{y=0}$

$$\tau_w = \mu\left(\frac{du}{dy}\right)_{y=0} = \mu U_{Max}\left\{\frac{1}{7}\left\{\left(\frac{y^{8/7}}{\delta^{1/7}}\right)\right\}\right\}_{y=0} = \text{infinite}$$

And hence use shear stress given by H. Blasius

$$\boxed{\frac{\tau_w}{\rho} = 0.0226U_M^2\left(\mu/\rho U_M\delta\right)^{1/4}} \quad \text{refer Eq. (7.32)}$$

$$\Rightarrow \tau_w = 0.0226\rho U_M^2\left(\mu/\rho U_M\delta\right)^{1/4} \tag{2}$$

Equate Eq. (1) and Eq. (2)

$$\Rightarrow 0.0226\rho U_M^2\left(\mu/\rho U_M\delta\right)^{1/4} = \rho U_{Max}^2\frac{d}{dx}\left(\frac{7}{72}\delta\right)$$

$$\Rightarrow 0.0226\rho U_M^2 * \frac{72}{7\rho U_{Max}^2} * \left(\mu/\rho U_M\right)^{1/4} dx = \delta^{1/4} d\delta$$

Integrating both sides, we get **boundary layer thickness**

$$\delta^{5/4} = 0.9298 * \left(\mu/\rho U_M\right)^{1/4} * x$$

$$\boxed{\Rightarrow \delta = \frac{0.371x}{\sqrt[5]{R_{e_x}}}} \tag{3}$$

b) Wall shear stress:

Substitute Eq. (3) in Eq. (2), then we get wall shear stress as

$$\boxed{\Rightarrow \tau_w = \frac{\rho U^2{}_{Max}}{2} * \frac{0.0576}{\sqrt[5]{R_{e_x}}}} \tag{4}$$

c) Skin friction coefficient/drag coefficient:

Skin friction coefficient is defined as the ratio of shear stress at the walls and $\frac{1}{2}\rho U_M^2$, and hence

$$^*C_f = \frac{\tau_w(x)}{\frac{1}{2}\rho U_M^2} \tag{5}$$

and substituting Eq. (4) in Eq. (5), we get

$$^*C_f = \frac{2*\frac{\rho U^2{}_{Max}}{2}*\frac{0.0576}{\sqrt[5]{R_{e_x}}}}{\rho U^2{}_{Max}}$$

$$\Rightarrow {}^*C_f = \frac{0.0576}{\sqrt[5]{R_{e_x}}} \tag{6}$$

d) Drag force and average drag coefficient:

The drag force (F_D) will be in opposite direction to the shear stress τ_w and as $force = stress * area$ we have $F_D = \int -\tau_w(A)$

$$\Rightarrow F_D = \int_0^L \tau_w(dx * th) = \int_0^L \frac{\rho U^2{}_{Max}}{2} * \frac{0.0576}{\sqrt[5]{R_{e_x}}} * dx * th$$

$$\Rightarrow F_D = 0.0576 * \frac{\rho U^2{}_{Max}}{2} \int_0^L \frac{1}{x * \sqrt[5]{R_{e_x}}} * dx$$

$$\Rightarrow F_D = \rho U_M^2 \left(\frac{0.036}{\sqrt[5]{R_{e_L}}}\right) Area \tag{7}$$

Average skin friction coefficient is defined as the drag force and $\frac{1}{2}\rho A U_M^2$, and hence

$$F_D = \frac{\rho U_M^2}{2} * \bar{C}_f * A \tag{8}$$

and substituting Eq.(7) in Eq. (8), we get

$$\bar{C}_f = \frac{F_D}{\frac{\rho U_M^2}{2} * A} = \frac{\rho U_M^2 \left(\frac{0.036}{\sqrt[5]{R_{e_L}}}\right) A}{\frac{\rho U_M^2}{2} * A} = \frac{0.072}{\sqrt[5]{R_{e_L}}}$$

And is valid for $5\times10^5 \le R_e < 10^7$

TOTAL TURBULENT FLOW OVER A FLAT PLATE

Prandtl suggested that the mean velocity distribution of a turbulent flow can be approximated by the seventh power law (i.e. n = 7), where $5\times10^5 < R_e < 10^7$ and hence the turbulent boundary layer follows $\frac{u}{U_{MAX}} = \left[\frac{y}{R}\right]^{1/7}$ velocity profile between the said Reynolds number range. But once the Reynolds number exceeds 10^7, the turbulent flow becomes fully developed and the seventh power law is violated and hence Prandtl had experimentally given the value of **average skin friction coefficient** value in the range of $10^7 < R_e < 10^9$ as

$$\bar{C}_f = \left(\frac{0.445}{\left(\log_{10} R_{e_L}\right)^{2.58}}\right) \tag{7.36}$$

In Figure 7.10, we can see two regions in the turbulent flow, one region with turbulent flow having viscous sub-layer and other region with fully developed turbulence.

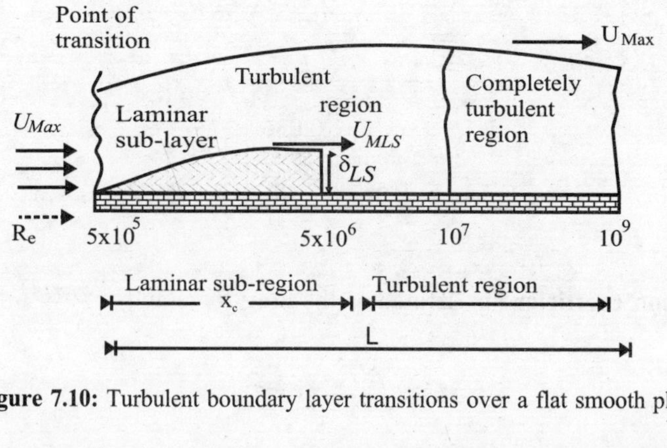

Figure 7.10: Turbulent boundary layer transitions over a flat smooth plate

So if needed to calculate the **actual drag force** $(F_D)_{Actual\ Turb}$ due to turbulence that is existing in a flow; then we need to subtract the drag force due to turbulent flow with viscous layer $(F_D)_{Turb\ viscous}$ from the fully developed turbulent flow $(F_D)_{Fully\ Turb}$ which is from the leading edge of the plate and so

$$(F_D)_{Actual\ Turb} = (F_D)_{Fully\ Turb} - (F_D)_{Turb\ viscous} \tag{7.37}$$

Fully developed turbulent flow which is from the leading edge of the plate $(F_D)_{Fully\ Turb}$

As we know the drag force

$$F_D = \left(\frac{\rho U_M^2}{2}\right)(\bar{C}_f)A$$

hence as given by Prandtl the value of Average skin friction coefficient as

$$\bar{C}_f = \left(\frac{0.445}{(\log_{10} R_{e_L})^{2.58}}\right)$$

and hence the drag force becomes

$$(F_D)_{Fully\ Turb} = \left(\frac{\rho U_M^2}{2}\right)\left(\frac{0.445}{(\log_{10} R_{e_L})^{2.58}}\right)L * B \tag{7.37a}$$

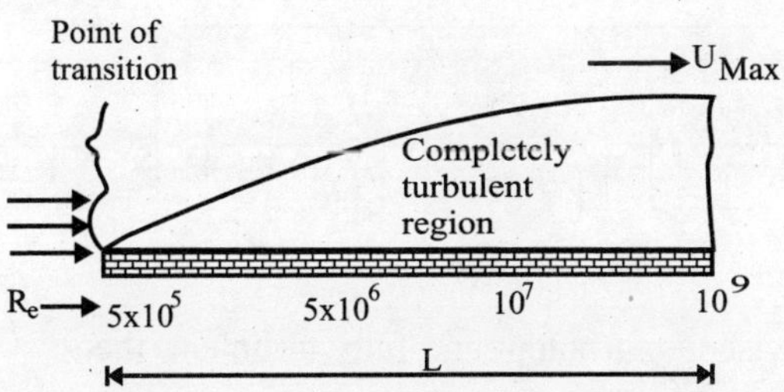

Figure 7.10a: Turbulent boundary layer from the leading edge over a flat smooth plate

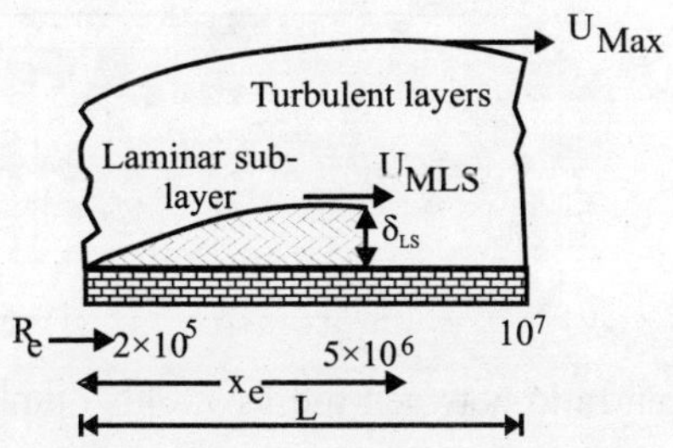

Figure 7.10b: Turbulent flow over a flat plate with viscous layers

Drag force due to turbulent flow with viscous layer $(F_D)_{Turb\ viscous}$

As the drag force

$$F_D = \left(\frac{\rho U_M^2}{2}\right)(\bar{C}_f)A$$

hence as derived in turbulent flow with viscous over a flat plate and is given in Eq. 7.37 as

$$\bar{C}_f = \left(\frac{0.0.72}{\left(R_{e_{xc}}\right)^{0.2}}\right)$$

and hence the drag force becomes

$$\left(F_D\right)_{Turb\ Viscous} = \left(\frac{\rho U_M^2}{2}\right)\left(\frac{0.072}{\left(R_{e_{x_c}}\right)^{0.2}}\right) x_c * B \tag{7.37b}$$

Substituting Eqs. (7.37a) and (7.37b) in Eq. (7.37), we have

$$\left(F_D\right)_{Actual\ Turb} = \left(\frac{\rho U_M^2}{2}\right)\left(\frac{0.445}{\left(\log_{10} R_{e_L}\right)^{2.58}}\right) L * B - \left(\frac{\rho U_M^2}{2}\right)\left(\frac{0.072}{\left(R_{e_{x_c}}\right)^{0.2}}\right) x_c * B$$

$$\Rightarrow \left(F_D\right)_{Actual\ Turb} = \left(\frac{B\rho U_M^2}{2}\right)\left[\left(\frac{0.445L}{\left(\log_{10} R_{e_L}\right)^{2.58}}\right) - \left(\frac{0.072x_c}{\left(R_{e_{x_c}}\right)^{0.2}}\right)\right] \tag{7.38}$$

but

$$\left(F_D\right)_{Total\ Turb} = \left(F_D\right)_{Fully\ Turb} - \left(F_D\right)_{Turb\ viscous} + \left(F_D\right)_{Laminar\ Sub-Layer} \tag{7.38a}$$

$$\Rightarrow \left(F_D\right)_{Total\ Turb} = \left(\frac{B\rho U_M^2}{2}\right)\left[\left(\frac{0.445L}{\left(\log_{10} R_{e_L}\right)^{2.58}}\right) - \left(\frac{0.072x_c}{\left(R_{e_{x_c}}\right)^{0.2}}\right) + \left(\frac{1.328x_c}{\left(R_{e_{x_c}}\right)^{0.5}}\right)\right] \tag{7.39}$$

If taken ratio between the Reynolds numbers of viscous turbulent and fully turbulent, then

$$\frac{\left(R_e\right)_{x_c}}{\left(R_e\right)_L} = \frac{\rho U_M x_c}{\mu} * \frac{\mu}{\rho U_M L}$$

$$\Rightarrow x_c = \frac{L * \left(R_e\right)_{x_c}}{\left(R_e\right)_L} \tag{7.39a}$$

And substituting this value in Eq. (7.39), we have

$$\left(F_D\right)_{Total\ Turb} = \left(\frac{\rho U_M^2}{2}\right)\left[\left(\frac{0.445}{\left(\log_{10} R_{e_L}\right)^{2.58}}\right) - \left(\frac{0.072\left(R_{e_{x_c}}\right)^{0.8}}{\left(R_{e_L}\right)}\right) + \left(\frac{1.328\left(R_{e_{x_c}}\right)^{0.5}}{\left(R_{e_L}\right)}\right)\right]A \quad (7.40)$$

As the transition of viscous turbulent flow to fully developed turbulent starts from Reynolds number $\left(R_e\right)_{x_c} = 5\times10^5$, so if substituted this value in Eq. 7.38, we get

$$\left(F_D\right)_{Total\ Turb} = \left(\frac{\rho U_M^2}{2}\right)\left[\left(\frac{0.445}{\left(\log_{10} R_{e_L}\right)^{2.58}}\right) - \left(\frac{1700}{\left(R_{e_L}\right)}\right)\right]A \quad (7.41)$$

and

$$\left(\bar{C}_f\right)_{Total\ Turb} = \left[\left(\frac{0.445}{\left(\log_{10} R_{e_L}\right)^{2.58}}\right) - \left(\frac{1700}{\left(R_{e_L}\right)}\right)\right] \quad (7.42)$$

Observe the plots between Reynolds number and the average skin friction coefficient of a turbulent flow and its transition to fully developed turbulent, basing on the increase in Reynolds number, in Figure 7.10c.

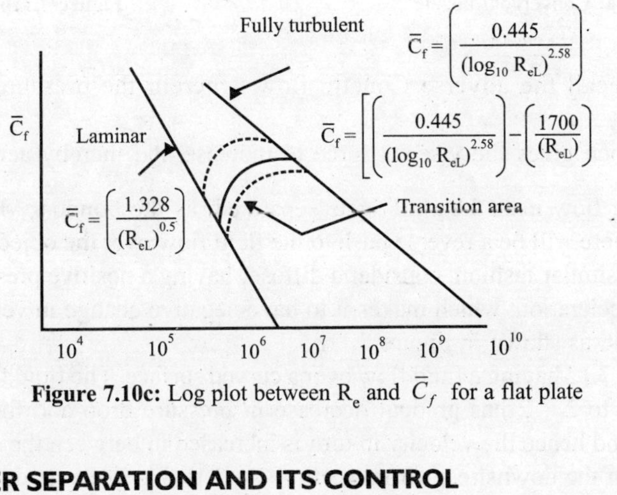

Figure 7.10c: Log plot between R_e and $\bar{C}_f$ for a flat plate

BOUNDARY LAYER SEPARATION AND ITS CONTROL

When the fluid flowing is unable to follow an object's shape it separates and hence **"separation"** is a phenomenon to get separated by the boundary layer of the flow from the surface of the object. The separations of boundary layers of the flow from the object moving give rise to development of eddies. The one way to reduce the formation of eddies in between object moving a fluid flow and the boundary layer of the flow is to have aerodynamic shape and geometry. The eddies thus formed due to irregular shape of the objects create a back flow of the fluid over the object, thereby enhancing object fluctuations, increase in drag, developing heat dissipation due to friction between flow and object, etc.

The separation phenomenon mostly acts adversely in case high speed moving objects like airplanes wings, jets, projectiles, etc.

Recall the favorable Couette flow, wherein the pressure gradient is negative (i.e. $\frac{dP}{dx} = -ve$), which gives the viscous force to reduce and thereby act as additional push to the fluid flow over the object. In similar fashion consider a convergent nozzle having a negative pressure gradient, do have only increase in acceleration, which makes it to have positive change in velocity and stabilized fluid flow through it, as shown in Figure 7.11a. Hence, the velocity may decrease with increase in pressure gradient along flow directions.

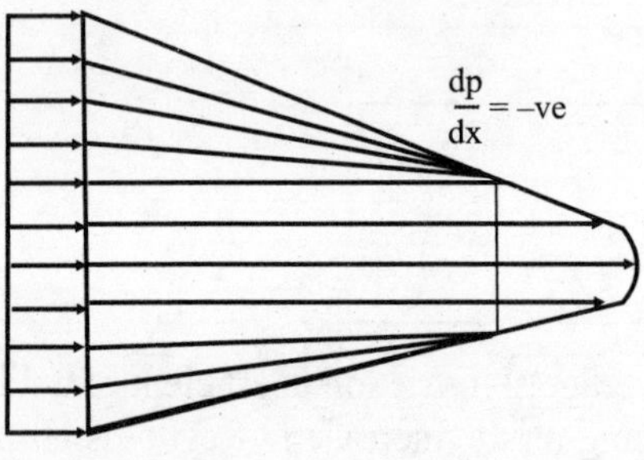

Figure 7.11a: Convergent nozzle

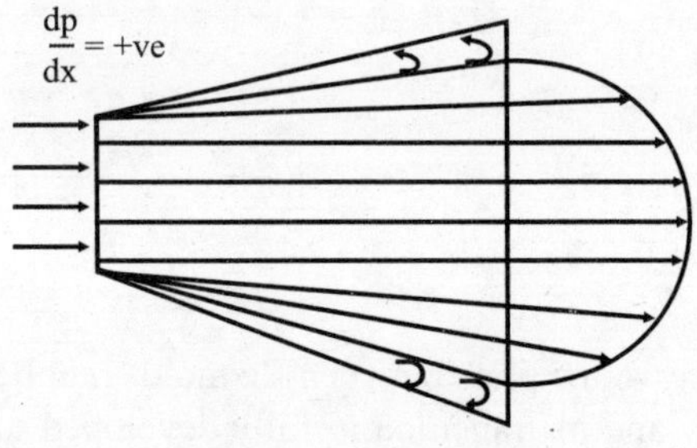

Figure 7.11b: Diffuser

Similarly, if recall the adverse Couette flow, wherein the pressure gradient is positive (i.e. $\frac{dP}{dx} = +ve$), which gives the viscous force to increase and thereby act shearing at the walls increase and so the flow near the walls gets separated, as the boundary layer at the walls gets thicker and hence there will be a reverse push to the fluid flow over the object, thereby reducing the velocity of flow. In similar fashion, consider a diffuser having a positive pressure gradient; do have only decrease in acceleration, which makes it to have negative change in velocity and unstabilized fluid flow through it, as shown in Figure 7.11b.

Observe Figure 7.12 having a fluid flow over a curved surface. The fluid flowing in the upstream from section 1 – 1 to 2 – 2 has gradual decrease in pressure drop and thereby reduces viscous effect at the walls and hence the velocity in turn is increased in between the sections 1 – 1 to 2 – 2. The fluid flowing in the downstream from section 2 – 2 to 3 – 3 has gradual increase in pressure drop and thereby starts increasing viscous effect at the walls and hence the velocity in turn is increased in between the sections 2 – 2 and 3 – 3, but the viscous effect does not overcome the velocity and hence the flow remains stable.

The fluid flowing in the downstream from section 3 – 4 to 4 – 4 has vast increase in pressure drop and thereby increasing viscous effect at the walls rapidly and hence the boundary layer becomes unstable and separation of flow emerges in between the surface and fluid flow and so the boundary layer shifts to new position 4 to 4^1, as shown in Figure 7.12, wherein the eddies form and gives a

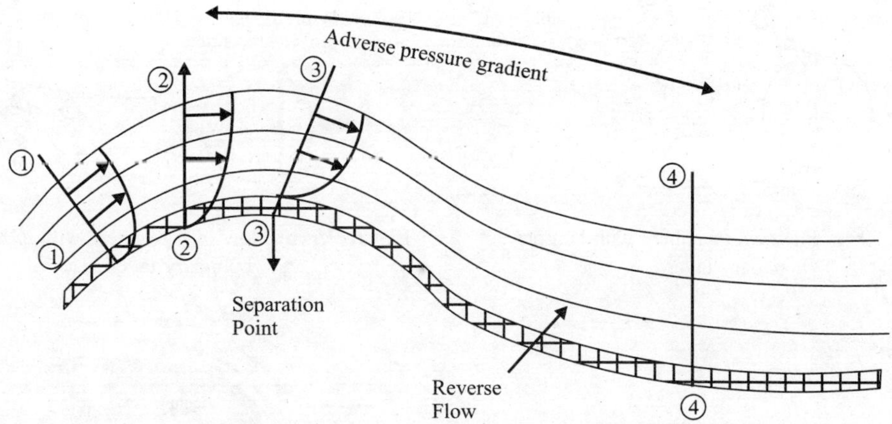

Figure 7.12: Flow over a curved surface

reverse flow which reduces the velocity of the flow in between the sections 3 – 4 and 4 – 4. And so the flow in the downstream is treated as adverse flow conditions.

The adverse flow conditions depend on the point of separation (section 3) and the separation point can be positioned based on the roughness of the surface, nature of boundary layer, geometry of the surface Reynolds number.

Laminar boundary layers tend to be thinner than turbulent boundary layers and separation occurs both in laminar and turbulent flows but as the turbulent flow has random movement of fluid molecules, it possesses more energy and resists more to the separation than a laminar flow.

Methods to Control Separation

There are five methods to control separation or to shift separation point away from the surface so that eddies form in a narrow gap of the boundary layer separation, thereby reducing the separation effects.

The methods of separation are being shown in the below classification:

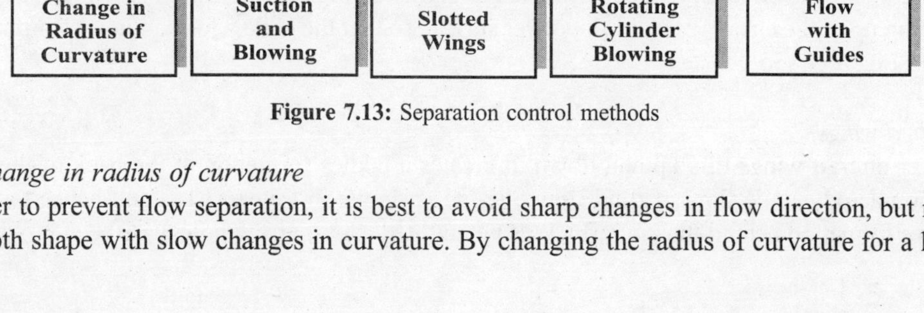

Figure 7.13: Separation control methods

1) Change in radius of curvature

In order to prevent flow separation, it is best to avoid sharp changes in flow direction, but making a smooth shape with slow changes in curvature. By changing the radius of curvature for a laminar

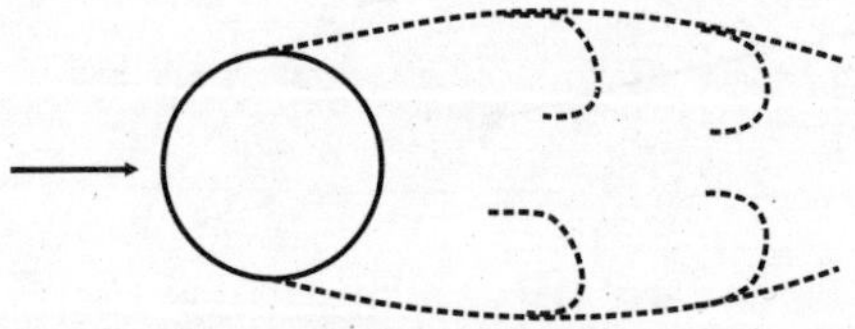

Figure 7.14a: Flow over a sphere with laminar boundary layer

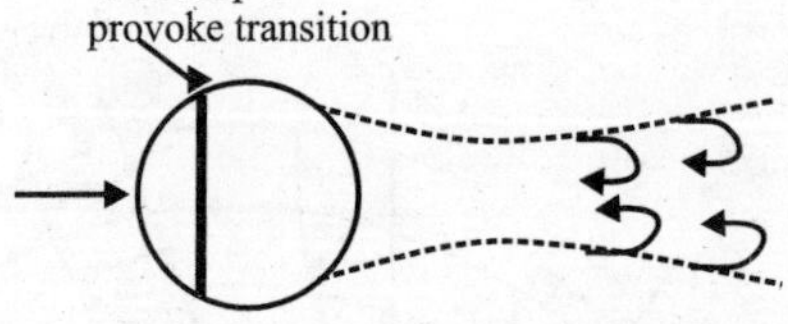

Figure 7.14b: Flow over a sphere with turbulent boundary layer

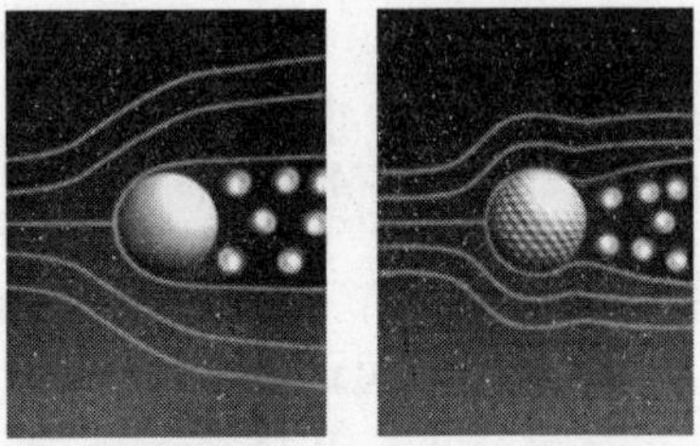

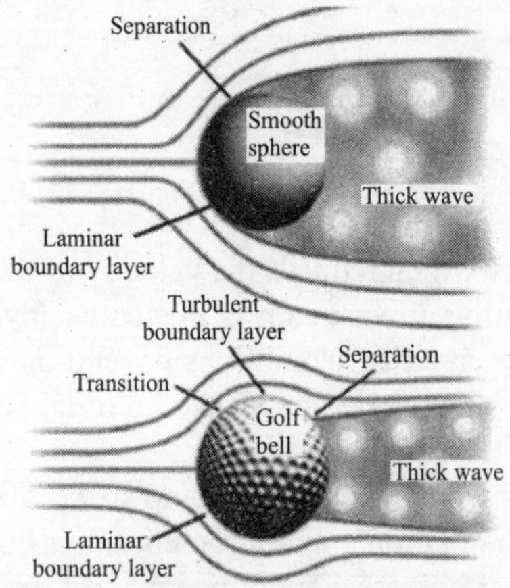

flow over sphere, do shifts the separation point away downwards so that the eddies form far away from the surface, as shown in Figure 7.14a and similarly by placing a disturbance like wire strip which develops turbulence, as shown in Figure 7.14b and as turbulence has strong energy to flow which thereby mix up the fluid molecules to the surface for more distance, thereby narrowing the eddies gap formation.

Turbulent boundary layers may cause more skin friction drag, but they tend to be thicker and separate later. So for objects in which the pressure drag is dominant, it can be beneficial to artificially trigger a boundary layer to go turbulent to reduce the flow separation effect. The most well-known example of this is the golf ball, as shown in the above pictures.

2) Suction and blowing

Suction is the process of sucking in the eddies that forms in the flow and the surface of the objects, thereby reducing the effect of eddies formation, as shown in Figure 7.15a. And blowing is a process of speeding up of the energy of fluid flow over surface so that the eddies formation is minimized, as shown in Figure 7.15b.

3) Slotted wings

Whenever aircraft wings tilts up and down, it creates a eddies formation at the back of the wing, due to non-aerodynamic flow of fluid flow over it, as shown in Figure 7.16a and hence to reduce the eddies formation behind the aircraft wings when tilted, slot (narrow holes) are been provided so

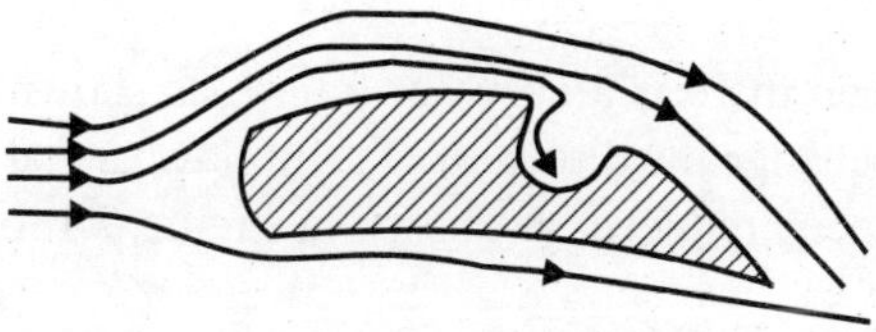

Figure 7.15a: Suction

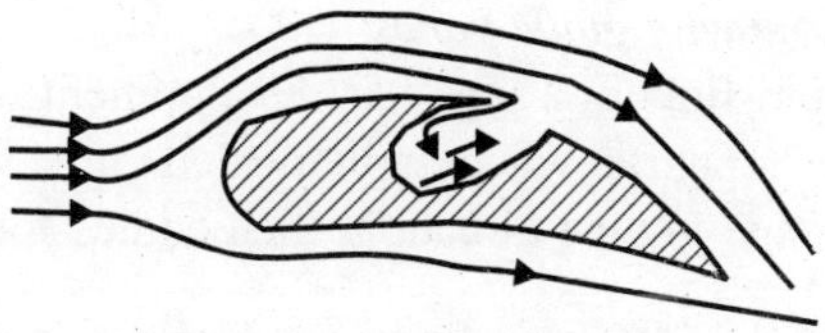

Figure 7.15b: Blowing

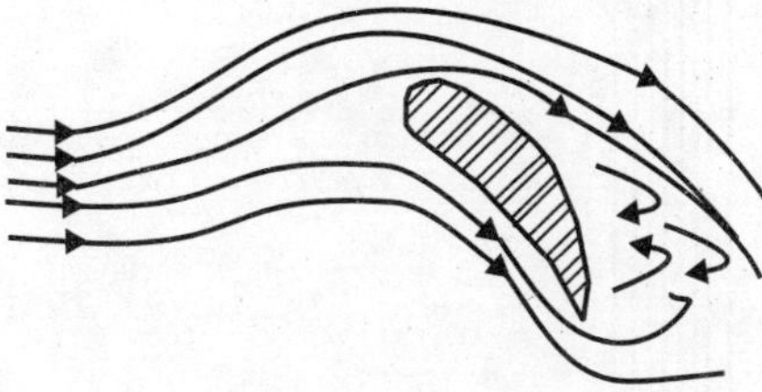

Figure 7.16a: Aerofoil with no slots

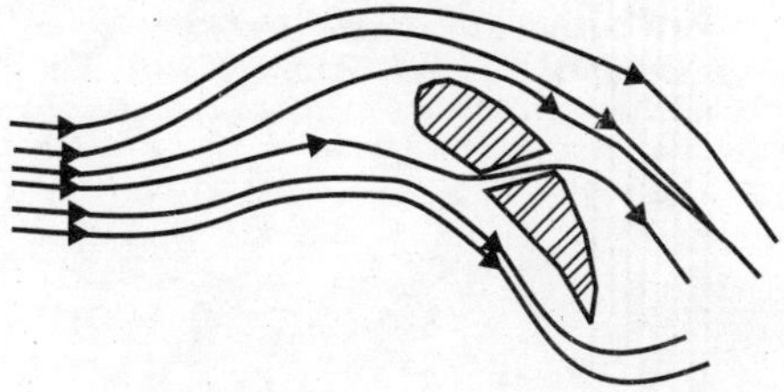

Figure 7.16b: Aerofoil with slots

that the flow below the aircraft wing energize the flow and reduces the formation of eddies, as shown in Figure 7.16b.

Note: The process of suction, blowing, slotting is been implemented in design of aircraft wings.

4) Rotating cylinder at the leading edge

As observed in Figure 7.17a, the eddies form on the surface, but when the cylinder in front of the leading edge is rotated slightly which gives the fluid flow over the surface to follow the surface shape profile, as shown in Figure 7.17b.

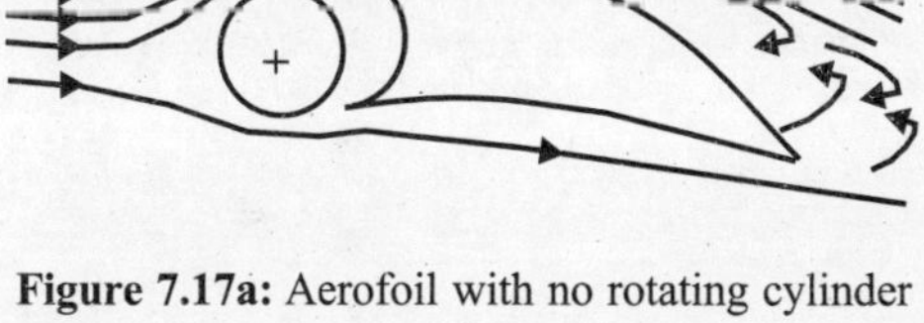

Figure 7.17a: Aerofoil with no rotating cylinder

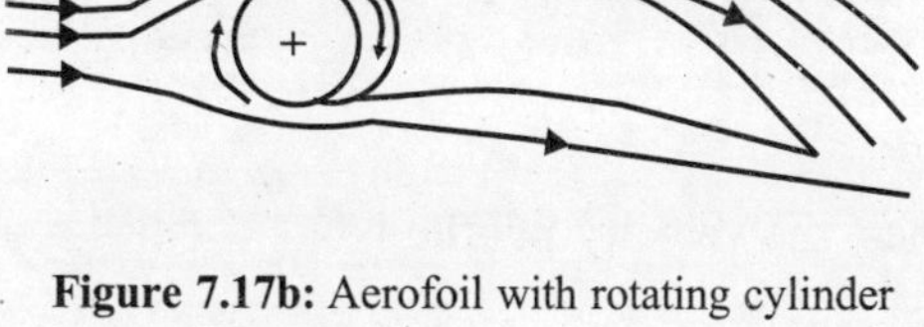

Figure 7.17b: Aerofoil with rotating cylinder

5) Put a flap on the end of the wing and tilt it, as shown in Figure 7.17c.

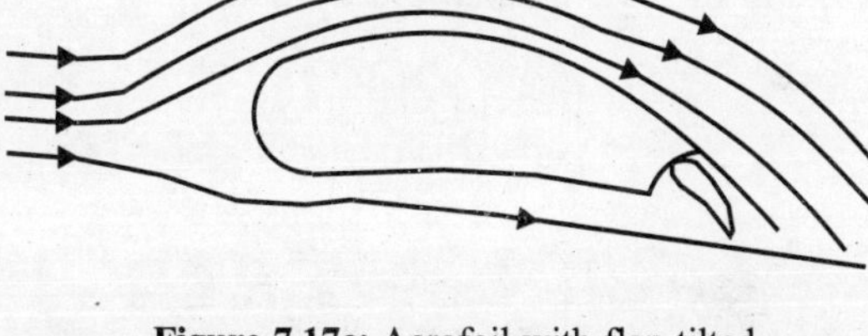

Figure 7.17c: Aerofoil with flap tilted

6) Flow with guide vanes

The fluid flow in a pipe has development of eddies, when there is a curve or a turn, as shown in Figure 7.18a. But when implemented guide vanes which change the direction of fluid flow as that of the change in pipe direction, then eddies formation are been reduced, as shown in Figure 7.18b.

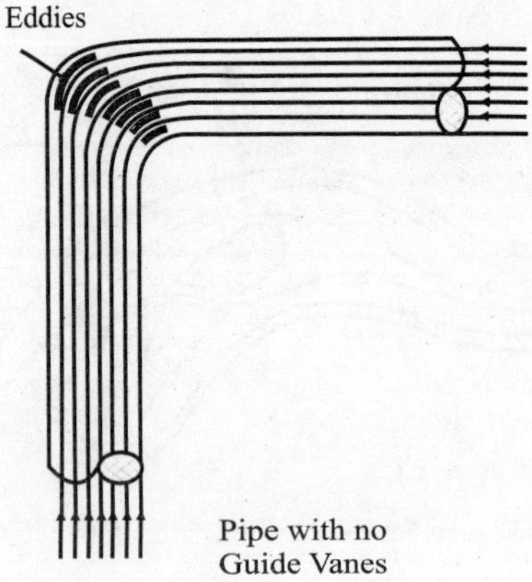

Figure 7.18a: Pipe with no Guide Vanes

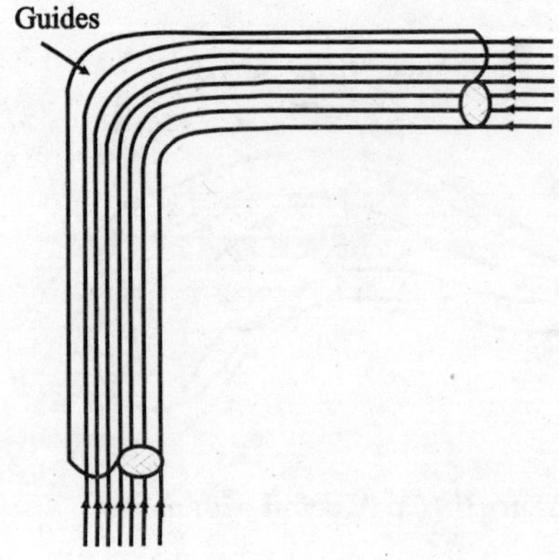

Figure 7.18b: Pipe with Guide Vanes

Example 7.19: Find the velocity profile, wall shear stress (τ_w), Boundary layer thickness (δ), Local drag coefficient and drag force, when the velocity profile is $u = a + by + cy^2$.

Solution: Given data: $u = a + by + cy^2$

The prevailing boundary conditions over a flat plate are

a. At the leading edge of the plate $y = 0$ & $u = 0$

b. At the leading edge of the plate, no shear stress $y = 0 \,\&\, \frac{du}{dy} = Finite\ Value \,\&\, \frac{d^2u}{dy^2} = 0$

c. At the tail edge of the plate $y = \delta$ & $u = U_{max}$

Applying boundary conditions, we get

$$a = 0,\ c = 0\ \&\ U_{MAX} = b\delta$$

Hence, the **velocity profile** is $\frac{u}{U_{max}} = \left(\frac{y}{\delta}\right)$

Boundary layer thickness: We have momentum thickness as $\delta_M = \int_0^{\delta} \frac{u}{U_{max}}\left(1 - \frac{u}{U_{max}}\right) dy$

Substituting the velocity profile in the above equation, we get

$$\delta_M = \int_0^{\delta} \left(\frac{y}{\delta}\right)\left(1 - \left(\frac{y}{\delta}\right)\right) dy$$

$$\Rightarrow \delta_M = \left|\frac{1}{2}\left(\frac{y^2}{\delta}\right) - \frac{1}{3}\left(\frac{y^3}{\delta^2}\right)\right|_0^{\delta} = \frac{\delta}{6}$$

Refer Von Karman equation, we have $\dfrac{\tau_w}{\rho U_M^2} = \dfrac{d}{dx}\left(\dfrac{1}{6}\delta\right)$ (1)

For a Laminar Newtonian fluid $\tau_w = \mu\left(\dfrac{du}{dy}\right)_{y=0} = \dfrac{\mu U_{Max}}{\delta}$ (2)

Equate Eq. (1) and Eq. (2) $\Rightarrow \dfrac{uU_{\max}}{\delta} = \rho U_{\max}^2 \dfrac{d}{dx}\left(\dfrac{1}{6}\delta\right)$

$$\Rightarrow 6 * \frac{\mu U_{Max}}{\rho U_{Max}^2} dx = \delta(d\delta)$$

Integrating both sides, we get **boundary layer thickness** $\delta = \dfrac{2.45x}{\sqrt{R_{e_x}}}$ (3)

Wall shear stress: Substitute Eq. (3) in Eq. (2), then we get wall shear stress as

$$\Rightarrow \tau_w = 0.408 \frac{\mu U_{Max}}{x}\sqrt{R_{e_x}} \quad (4)$$

Skin friction coefficient/Drag coefficient: Skin friction coefficient is defined as the ratio of shear stress at the walls and $\frac{1}{2}\rho U_M^2$, and hence

$$^*C_f = \frac{\tau_w(x)}{\frac{1}{2}\rho U_M^2} \quad (5)$$

and substituting Eq. (4) in Eq. (5), we get

$$^*C_f = \frac{2 * 0.408 \frac{\mu U_{Max}}{x}\sqrt{R_{e_x}}}{\rho U^2_{Max}} = \left(\frac{0.816}{\sqrt{R_{e_x}}}\right) \quad (6)$$

Drag force and average drag coefficient: Average skin friction coefficient is defined as the

$$\bar{C}_f = \frac{1}{L}\int_0^L {}^*C_f = \frac{1.224}{\sqrt{R_{e_L}}} \quad (7)$$

and drag force $F_D = \rho U_M^2 \left(\bar{C}_f\right) Area$ (8)

Substituting Eq:(7) in Eq:(8), we get drag force as $F_D = \rho U_M^2 \left(\dfrac{1.224}{\sqrt{R_{e_L}}}\right) A$

EXERCISE

Objective

1. The region between the separation streamline and boundary surface of the solid body is known as
   a) Wake
   b) Drag
   c) Lift
   d) Boundary layer
   e) Aerofoil section
2. Velocity distribution in the turbulent boundary layer follows the following law
   a) Linear
   b) Square
   c) Parabola
   d) Logarithmic
   e) Cubic
3. Reynolds number is significant in
   a) Supersonics, as with projectiles and jet propulsion
   b) Full immersion or completely enclosed flow, as with pipes, aircraft wings, nozzles, etc.
   c) Simultaneous motions through fluid where there surface of discontinuity, gravity forces, and wave making effect, as with ship's hulls
   d) All of the above
   e) None of the above
4. The laminar sub-layer acts as
   a) An insulating medium
   b) Good conductor of heat
   c) Refractory substance
   d) Heat absorber
   e) Heat generator
5. Laminar flow occurs in pipes, when Reynolds number
   a) Lies between 2000–3000
   b) Lies between 3000–4000
   c) Is more than 2000
   d) Is less than 2000
   e) None of the above
6. Separation occurs when
   a) The velocity of sound is reached
   b) The boundary layer comes to rest
   c) The cross-section of a channel is reduced
   d) The pressure reaches a minimum
   e) None
7. Head loss in turbulent flow in a pipe
   a) Varies directly as velocity
   b) Varies inversely as square of velocity

c) Varies approximately as square of velocity
d) Depends upon orientation of pipe
e) Varies inversely as velocity.

8. Turbulent flow generally occurs for cases involving
   a) Very slow motions
   b) Very viscous fluids
   c) Very narrow passage
   d) All of the above
   e) None
9. In turbulent flow
   a) The shear stress is generally larger than in a similar laminar flow
   b) Fluid particles move in an orderly manner
   c) Momentum transfer is on a molecular scale only
   d) Momentum is more effective than momentum transfer in causing shear stress
   e) None
10. A body is said to bc provided optimum amount of streamlining when
   a) Friction drag is minimum
   b) Pressure drag is minimum
   c) Profile drag (i.e. sum of friction drag and pressure drag) is minimum.
   d) Product of friction and pressure drag is minimum.
   e) Friction drag is minimum and pressure drag is max.
11. Drag is defined as the force component exerted on an immersed object, the component acting in direction
   a) Normal to flow direction
   b) Parallel to flow direction
   c) At resultant angle
   d) Radial to flow direction
   e) Opposite to flow direction

Answers

1.a	2.d	3.d	4.b	5.d	6.d
7.c	8.e	9. d	10.c	11.e	

Theory

1. What is the physical significance of displacement thickness of boundary layer theory?
2. Define boundary layer thickness.
3. What boundary conditions must be satisfied by the velocity distribution in laminar boundary layer over a flat plate?
4. Derive head losses in a pipe due to turbulent flow.
5. The velocity distribution in the boundary layer was found to fit the equation (u/U) = (y/d)1/7. Find the displacement thickness.

6. Explain Reynolds number and its significance in detail. What is the significance of upper and lower critical Reynolds numbers.
7. Derive Von-Karman equation for a laminar flow.
8. Write a note on laminar turbulent transition.
9. Derive Von-Karman momentum integral equation for boundary later flows.
10. Explain the mechanisms of growth of the boundary layers including how, why and where separation occurs. You should also be able to suggest what might be done to prevent separation.

Problems

1. A thin flat plate, 0.05 tall and 3.0 m wide, forms the leading edge of a banner towed by an aircraft at 44 m/s on a 294 K day. How far from the leading edge of the plate does the laminar portion of boundary layer extend? What is the boundary layer thickness at the downstream end of the laminar boundary layer? Find the drag force on the plate contributed by the laminar boundary layer, and the corresponding drag coefficient.
2. A box shaped body 2.5 m wide, 3 m high and 7 m long is traveling at a velocity of 20 m/s in 20°C. Calculate the contributions to the total drag of the truck from the sides and top of the body. Assume that boundary layer flow over it is of complete turbulent. Also find the wall shear stress, boundary layer thickness along the top of the box and maximum value of the wall shear stress and boundary layer thickness along the top side of the box.
3. The length of hull of a submarine is 80 m and its surface area is 3000 m^2. Determine the frictional drag if a) The boundary layer is turbulent over entire surface. b) The turbulent is preceded by laminar. Take the velocity of submarine as 4.17 m/s.
4. A smooth wind tunnel 1.5 m long, 0.75 dia is used for testing the model of an airplane. If the velocity of air is 70 m/s and the pressure drop in the tunnel is 0.981 kPa. Determine the drag on the model. Take density of air as 1.2 kg/m^3 and viscosity as 1.854×10^4 Ns/m^2. Determine the frictional drag on an airship 92 m long, average diameter 18 m, being propelled at 120 km/hour. Take density of air as 1.3 kg/m^3 and kinematic viscosity of 1.5×10^{-5} m^2/s.
5. Determine the frictional drag on the airship 92 m long, average diameter 18 m, being propelled at 120 km/hour. Take density of air as 1.3 kg/m^3 and kinematic viscosity as 1.5 $\times$ 10^{-5} m^2/s.

QUICK REFERENCE OF FORMULAE

Property	**Formulae**			**Remarks, if any**
Displacement boundary layer thickness	$\delta_d = \int_0^{\delta}\left(1-\dfrac{u}{U_{\max}}\right)dy$			
Momentum boundary layer thickness	$\delta_M = \int_0^{\delta}\dfrac{u}{U_{\max}}\left(1-\dfrac{u}{U_{\max}}\right)dy$			
Energy boundary layer thickness	$\delta_E = \int_0^{\delta}\dfrac{u}{U_{\max}}\left(1-\dfrac{u^2}{U_{\max}^2}\right)dy$			
Momentum integral equation–Von Kármán equation	$\dfrac{\tau_w}{\rho U_M^2} = \dfrac{d}{dx}[\delta_M]$			
	Boundary layer thickness δ	Skin friction coefficient $^*C_f$	Average skin friction coefficient $\bar{C}_f$	Velocity profile
Prandtl velocity profiles	$\delta = \dfrac{4.795x}{\sqrt{R_{e_x}}}$	$^*C_f = \dfrac{0.654}{\sqrt{R_{e_x}}}$	$\bar{C}_f = \dfrac{1.31}{\sqrt{R_{e_L}}}$	$\dfrac{u}{U_{Max}} = \sin\left(\dfrac{\Pi}{2}\dfrac{y}{\delta}\right)$
	$\delta = \dfrac{5.48x}{\sqrt{R_{e_x}}}$	$^*C_f = \dfrac{0.73}{\sqrt{R_{e_x}}}$	$\bar{C}_f = \dfrac{1.46}{\sqrt{R_{e_L}}}$	$\dfrac{u}{U_{Max}} = 2\left(\dfrac{y}{\delta}\right)-\left(\dfrac{y}{\delta}\right)^2$
	$\delta = \dfrac{4.64x}{\sqrt{R_{e_x}}}$	$^*C_f = \dfrac{0.646}{\sqrt{R_{e_x}}}$	$\bar{C}_f = \dfrac{1.292}{\sqrt{R_{e_L}}}$	$\dfrac{u}{U_{Max}} = \dfrac{3}{2}\left(\dfrac{y}{\delta}\right)-\dfrac{1}{2}\left(\dfrac{y}{\delta}\right)^3$
	$\delta = \dfrac{5.84x}{\sqrt{R_{e_x}}}$	$^*C_f = \dfrac{0.686}{\sqrt{R_{e_x}}}$	$\bar{C}_f = \dfrac{1.372}{\sqrt{R_{e_L}}}$	$\dfrac{u}{U_{Max}} = 2\left(\dfrac{y}{\delta}\right)-2\left(\dfrac{y}{\delta}\right)^3+\left(\dfrac{y}{\delta}\right)^4$
H. Blasius	$\delta = \dfrac{5x}{\sqrt{R_{e_x}}}$	$^*C_f = \dfrac{0.664}{\sqrt{R_{e_x}}}$	$\bar{C}_f = \dfrac{1.328}{\sqrt{R_{e_L}}}$	
Turbulent flow in a pipe	$f = \dfrac{2\tau_w}{\rho V^2} = \dfrac{2\tau_w}{\rho u_{av}^2}$			f Darcy friction factor
Shear stress at the walls in a pipe having turbulent flow	$\Rightarrow f = \dfrac{0.3164}{(R_e)^{1/4}}$ $\Rightarrow \dfrac{\tau_w}{\rho} = 0.0226U_M^2\left(\mu/\rho U_M\delta\right)^{1/4}$			Given by H. Blasius

Contd.

Property	Formulae	Remarks, if any
Turbulent flow over a flat plate, with viscous layers	$\delta = \frac{0.371x}{(R_{e_x})^{1/5}}$ $\tau_w = \left(\frac{\rho U_M^2}{2}\right)\left(\frac{0.0576}{(R_{e_x})^{1/5}}\right)$ & $^*C_f = \left(\frac{0.0576}{(R_{e_x})^{1/5}}\right)$ $\bar{C}_f = \left(\frac{0.074}{(R_{e_L})^{1/5}}\right)$ & $F_D = \left(\frac{\rho U_M^2}{2}\right)\left(\frac{0.074}{(R_{e_L})^{1/5}}\right)A$	Boundary layer thickness, shear stress, local and average skin friction coefficient, drag force.
Total turbulent flow over a flat plate	$(F_D)_{Total\ Turb} = \left(\frac{\rho U_M^2}{2}\right)\left[\left(\frac{0.445}{(\log_{10} R_{e_L})^{2.58}}\right) - \left(\frac{1700}{(R_{e_L})}\right)\right]A$	Total drag force in turbulent flow.
	$(\bar{C}_f)_{Total\ Turb} = \left[\left(\frac{0.445}{(\log_{10} R_{e_L})^{2.58}}\right) - \left(\frac{1700}{(R_{e_L})}\right)\right]$	Total skin friction coefficient in turbulent flow.

Points to Remember

Property	Formulae	
Used formulae	$^*C_f = \frac{\tau_w}{\frac{\rho U_M^2}{2}}, \bar{C}_f = \frac{1}{L}\int_0^L {}^*C_f * dx, F_D = \left(\frac{\rho U_M^2}{2}\right)\left(\bar{C}_f\right)A$	$\tau_w, {}^*C_f, \bar{C}_f, F_D$ Are shear stress, local, average skin friction coefficient, drag force, respectively,
Total turbulent flow over a flat plate	$\frac{U}{U_{MAX}} = \left(\frac{y}{\delta}\right)^{1/n}, n = 7$	Total drag force in turbulent flow.
Total Turbulent flow over a flat plate	$f = 0.3164(R_e)^{-0.25}$	for $R_e \leq 10^5$, where f is Darcy friction factor
Fully developed turbulent flow as given by Prandtl	$\bar{C}_f = \left(\frac{0.445}{(\log_{10} R_{e_L})^{2.58}}\right)$	$\bar{C}_f$ Average skin friction coefficient in fully developed turbulent flow.
Relation between critical length and total length of the plate	$x_c = \frac{L*(R_e)_{x_c}}{(R_e)_L}$	

CHAPTER 8

Dimensional Analysis and Model Similitude

INTRODUCTION

Dimensional analysis is a tool to analyze the physical quantities in terms of its dimensions like mass (M), length (L), time (T), temperature (θ) and electric current (I) and luminous intensity (J), for solving engineering problems.

Dimensional analysis is a process to show how dependent variable varies with a variety of independent variables and moreover dimensional analysis predicts the actual independent and dependent variables, depending on the requirement and usage. But it does not surely predict the exact number of dependent and independent variables, so as to minimize the process of dimensional analysis. Hence, dimensional analysis is a process of obtaining required information of a particular problem in the form of dimensionless groups. Dimensional analysis is considered a powerful tool for engineers and scientists, which combines dimensional dependent variables with dimensional independent variables, non-dimensional variables, and dimensional constants and is applied to several practical and real time problems to illustrate both its usage and drawbacks.

Any physical quantity should possess' only unique dimensions irrespective of its environmental usage and hence dimensional analysis is a method to compare and solve the quantities of same dimensions of any engineering applications. But this principle of formulation arise only difficulty of including and excluding of independent variables and parameters.

In this chapter, we first recall dimensions, units and their homogeneity and apply it to develop dimensionless numbers. In later half of this chapter, we discuss the concept of similarity between a model and a prototype.

FUNDAMENTAL AND DERIVED DIMENSIONS OF QUANTITIES

Representing any physical quantity in terms of M-L-T-θ is termed fundamental quantities. Any physical fundamental quantities used in engineering application must possess dimensional homogeneity (i.e. irrespective of the usage of the physical quantity, in any engineering subject, its fundamental dimensional system M-L-T-θ should be the same).

E.g.: when physical quantity like force used in solid mechanics or fluid mechanics or engineering mechanics, its fundamental dimensional system will remain the same, i.e. MLT^{-2}. Hence, irrespective of the physical quantity usage in different subjects of engineering application, the fundamental derived dimensions do not change and is called as **Dimensional Homogeneity**.

We have basically three types of derived fundamental quantities in fluid mechanics:

1. **Geometric quantities**: These represent the size and shapes of the physical quantity like area, volume, and moment of inertia.
2. **Kinematic quantities**: These quantities represent the motion of an object or physical quantity like velocity, acceleration, angular velocity, discharge, gravity, kinematic viscosity, vorticity, circulation, etc.
3. **Dynamic quantities**: These quantities represent the forces causing motion of an object or physical quantity like forces, pressure, viscosity, shear stress, work, torque, power, and density, etc.

Derived Quantity	Physical Quantity	M-L-T- θ SYSTEM
Geometric	Area	$M^0L^2T^0$
	Volume	$M^0L^3T^0$
Kinematic	Velocity	$M^0L^1T^{-1}$
	Acceleration	$M^0L^1T^{-2}$
	Discharge	$M^0L^3T^{-1}$
	Stream function	$M^0L^2T^{-1}$
	Angular velocity	$M^0L^0T^{-1}$
Dynamic	Force	$M^1L^1T^{-2}$
	Pressure	$M^1L^{-1}T^{-2}$
	Work	$M^1L^2T^{-2}$
	Shear stress	$M^1L^{-1}T^{-2}$
	Viscosity	$M^1L^{-1}T^{-1}$
	Density	$M^1L^{-3}T^0$

Uses and Advantages of Dimensional Analysis

Uses:

1. To test dimensional homogeneity of any equation of fluid motion.

2. To derive equation expressed in terms of non-dimensional parameters to show the relative significance of each parameter.
3. To derive rational formulae for a flow phenomenon.
4. To make possibility of analyzing a large experiment of fluid flow phenomenon.

Advantages:

1. To express the functional relationship between the variables in flow phenomenon, in dimensional terms.
2. It enables to derive a simplified dimensional theoretical equation.
3. It facilitates to convert the units of different variables from one unit system to other unit system.
4. It provides solution to complex mathematical problems.
5. With proper selection of variables, logical deductions can be made using dimensionless parameters, for a problem.

METHODS OF DIMENSIONAL ANALYSIS

There are two types of methods for dimensional analysis, and they are

a) Rayleigh method
b) Buckingham's π-method.
c) Matrix-tensor method.
d) Bridgman's method.
e) Visual inspection of variables involved.

In this context, only the first two methods are being dealt with.

RALEIGH METHOD

This method gives the expressing of the dependent variables within terms of the independent variables but it restricts the independent variables to three or four and becomes complex in solving if more number of independent variables are involved.

Procedure:

Step 1: Express the given quantity Z as a function of its independent physical quantities.

$$Z = f\left(A, B, C, ...\right)$$

Step 2: Give powers X1, X2, and X3 … to the independent variables, in the formulated function.

$$Z = f\left(A^{X1}, B^{X2}, C^{X3}, ...\right)$$

Step 3: Express function "f" as a non-dimensional constant "C".

$$Z = K\left(A^{X1}.B^{X2}.C^{X3} ....\right)$$

Step 4: Write M-L-T- θ system on both sides of the function, corresponding to the physical quantities.

$$M^Y L^X T^W \theta^U = K\left((MLT\theta)^{X1}.(MLT\theta)^{X2}.(MLT)^{X3}....\right)$$

Step 5: Compare the 'M', 'L', 'T', 'θ' powers of the dependent quantity with the summation of 'M', 'L , 'T', 'θ' powers of the independent variables, so as to get four equations in the form of powers a, b, c.

E.g. for "M" powers $Y = X1 + X2 + X3 + ...$

for "L" powers $X = X1 + X2 + X3 + ...$

for "T" powers $W = X1 + X2 + X3 + ...$

for "θ" powers $U = X1 + X2 + X3 + ...$

Step 6: Solve these power equations to get the values of the powers ($X1, X2, X3,......$). If the numbers of power unknowns are more than the number of equations, then express all the powers in terms of any one of the power.

Step 7: Substitute these values of powers obtained, in the function of Step 3.

E.g. If let $X1 = 3, X2 = 2, X3 = 4,........$ then $Z = K\left(A^3.B^2.C^4........\right)$

E.g. If let $X1 = 3 + X2, X3 = 4 + X2,..$ then $Z = K\left(A^3.B^{X2}.C^4.A^{X2}.C^{X2}...\right)$

Step 8: Bring all the known power quantities outside the function and express the dependent function in terms of known quantities and other unknown function.

E.g. If the powers are solvable, then $Z = A^3.B^2.C^4........$

E.g. If the powers are not solvable

i.e. $X1 = 3 - X2, X3 = 4 - X2,....$ then $Z = A^3 C^4 \psi\left(\frac{B}{A.C}\right)^{X2}$

Example 8.1: Show that drag force $F = \rho V^2 d^2 \phi\left(\frac{1}{R_e}\right)$ using Rayleigh method.

Solution: Given data: $F = f(\rho, V, d, \mu)$

Step 1:
$$F = f\left([\rho]^{x1}[V]^{x2}[d]^{x3}[\mu]^{x4}\right)$$

Step 2:
$$M^1 L^1 T^{-2} = f\left[ML^{-3}\right]^{x1}\left[LT^{-1}\right]^{x2}[L]^{x3}\left[M^1 L^{-1} T^{-1}\right]^{x4}$$

Step 3: Equate the powers

Power of M: $$1 = (x1) + x4$$

Power of L: $$1 = -3(x1) + x2 + x3 - (x4) \Rightarrow x2 + x3 = 4 - 2x4$$

Power of T: $$-2 = -1(x2) - 1(x4)$$

So $$(x2) = (x3) = 2 - (x4) \text{ and } (x1) = 1 - (x4)$$

Step 4: $$F = f\left([\rho]^{(1-x4)}[V]^{(2-x4)}[d]^{(2-x4)}[\mu]^{x4}\right)$$

Step 5: $$F = \rho V^2 d^2 \phi\left(\frac{\mu}{\rho V d}\right)^{x4} \Rightarrow F = \rho V^2 d^2 \phi\left(\frac{1}{R_e}\right)$$

Example 8.2: Show that pressure drop $\Delta P = \frac{\mu V}{l}\phi\left(\frac{d}{l}\right)$ using Rayleigh method.

Solution: Given data: $\Delta P = f(l, V, d, \mu)$

Step 1: $$\Delta P = f\left([l]^{x1}[V]^{x2}[d]^{x3}[\mu]^{x4}\right)$$

Step 2: $$M^1L^{-1}T^{-2} = f[L]^{x1}\left[LT^{-1}\right]^{x2}[L]^{x3}\left[M^1L^{-1}T^{-1}\right]^{x4}$$

Step 3: Equate the powers: $M - L - T$

So we get $$(x2) = (x4) = 1 \text{ and } (x1) = -1 - (x3)$$

Step 4: $$\Delta P = f\left([l]^{(-1-x3)}[V]^{(1)}[d]^{(x3)}[\mu]^{1}\right)$$

Step 5: $$\Delta P = \frac{\mu V}{l}\phi\left(\frac{d}{l}\right)^{x3} \Rightarrow \Delta P = \frac{\mu V}{l}\phi\left(\frac{d}{l}\right)$$

Example 8.3: Show that resistance $R = \mu V D \phi\left(\frac{\rho V D}{\mu}\right)$ using Rayleigh method:

Solution: Given data: $R = f(\rho, V, D, \mu)$

Step 1: $$R = f\left([D]^{x1}[V]^{x2}[\rho]^{x3}[\mu]^{x4}\right)$$

Step 2: $$M^1L^1T^{-2} = f[L]^{x1}\left[LT^{-1}\right]^{x2}\left[ML^{-3}\right]^{x3}\left[M^1L^{-1}T^{-1}\right]^{x4}$$

Step 3: Equate the powers: $M-L-T$

So we get $$(x2) = (x1) = 1 + x3 \text{ and } (x4) = 1 - (x3)$$

Step 4: $$R = f\left([D]^{(1+x3)}[V]^{(1+x3)}[\rho]^{(x3)}[\mu]^{(1-x3)}\right)$$

Step 5: $$R = D\mu V * \left(\frac{DV\rho}{\mu}\right)^{x3} \Rightarrow R = D\mu V\phi\left(\frac{DV\rho}{\mu}\right)$$

Exmaple 8.4: Show that discharge $Q = d^2\sqrt{P/\rho}\,\varphi\left(\frac{d\sqrt{P\rho}}{\mu}\right)$ using Rayleigh method:

Solution: Given data: $Q = f(d, \rho, P, \mu)$

Step 1: $$Q = f\left([d]^{x1}[\rho]^{x2}[P]^{x3}[\mu]^{x4}\right)$$

Step 2: $$M^0L^3T^{-1} = f\left([L]^{x1}\left[ML^{-3}\right]^{x2}\left[ML^{-1}T^{-2}\right]^{x3}\left[M^1L^{-1}T^{-1}\right]^{x4}\right)$$

Step 3: Equate the powers: $M-L-T$

So we get $$(x2) = x3 - 1 \text{ and } (x1) = 1 + 2(x3), (x4) = 1 - 2(x3)$$

Step 4: $$Q = f\left([d]^{(1+2(x3))}[\rho]^{(x3-1)}[P]^{(x3)}[\mu]^{(1-2(x3))}\right)$$

Step 5: $$Q = \frac{d\mu}{\rho}\phi\left(\frac{d^2\rho P}{\mu^2}\right)^{x3}$$

Step 6: Multiplying by $\frac{d\sqrt{P\rho}}{\mu}$, we get $Q = d^2\sqrt{P/\rho}\,\varphi\left(\frac{d\sqrt{P\rho}}{\mu}\right)$

BUCKINGHAM π-THEOREM

It states, "the difference of summation of all the independent variables and dependent variable (n) and the fundamental quantities (m) are to be expressed in terms of non-dimensional parameters called π terms".

It overcomes the limit of number of independent variables as in the case of Rayleigh method, and hence Buckingham π-theorem has become more advantageous than Rayleigh method.

Procedure:

Step 1: Express the given quantity Z as a function of its independent physical quantities.

$$Z = f(A, B, C, D, E........)$$

Step 2: Merge the dependent quantity with the independent quantities, so that the function formulates as a function of dependent and independent quantities. $f(Z, A, B, C, D, E........) = 0$

So that the total number of quantities "n" becomes sum of dependent quantity and independent quantities.

Step 3: Express function "f" as a function of dimensionless "π" terms, such that the π terms are equal to (n-m) terms, where "m" is the number of fundamental dimensions (m=3, i.e. $M-L-T$)'

$$f(\pi_1, \pi_2, \pi_3, \pi_4, ..., \pi_{n-m}) = 0$$

Step 4: Take any three known independent quantities such that one quantity represents geometric similarity, second quantity represents kinematic similarity, third quantity represents dynamic similarity and repeat them in all the π terms with powers incorporated to the quantities, where π terms are functions of these three repeating quantities and remaining left over quantities such that every π terms has (m+1) quantities in its functions.

$$\begin{aligned}
\pi_1 &= A^{x1} \quad B^{y1} \quad C^{w1} \quad Z \\
\pi_2 &= A^{x2} \quad B^{y2} \quad C^{w2} \quad D \\
\pi_3 &= A^{x3} \quad B^{y3} \quad C^{w3} \quad E \\
&\ \ \vdots \\
\pi_{n-m} &= A^{x(n-m)} \quad B^{y(n-m)} \quad C^{w(n-m)} \quad X
\end{aligned}$$

Step 5: As the π terms are dimensionless, write M-L-T-θ system on both sides of the π term function and its corresponding physical quantities in the Π term function.

$$M^0 L^0 T^0 \theta^0 = \left((MLT\theta)^{x1} . (MLT\theta)^{y1} . (MLT)^{w1} ....\right)$$

And get the power values and write the π term in a single quantity form. Like

$$\pi_1 = \frac{A^2 B}{CD^3 E^2}$$

Step 6: Repeat Step 5 for all the π term functions and get all the π term functions.

Like $$\pi_1 = \frac{A^2 Z}{CD^3 E^2}, \pi_2 = \frac{AB^3}{C^3 D^2 E}$$

Step 7: Substitute all the π term functions obtained, in the function of step 3.

E.g. $$f\left(\frac{A^2 Z}{CD^3 E^2}, \frac{AB^3}{C^3 D^2 E}, ...\right) = 0$$

Step 8: Bring π term function containing the dependent quantity, outside the function f and express the dependent function in terms of other quantities in the function. $Z = A^3 . B^2 . C^4 ....$

Difference between Rayleigh method and Buckingham π-theorem:

Rayleigh Method	Buckingham π-theorem
1. It can accommodate only 3 or 4 independent variables.	1. It can accommodate more number of independent variables.
2. It does interpret the number of dimensionless quantities, in advance.	2. It does not interpret the number of dimensionless quantities, in advance.
	3. Designated dimensionless quantities like π terms are used for solving.

Note: The below mentioned modifications/changes are permitted with variables in a function occurring during solving problems:

a) Addition, reciprocal, multiplications with constants, square roots of any group in the function is permitted.
b) A group may be raised to any power, in the function.
c) Expressing as a function of the other groups is permitted.

Example 8.5: Show that $\Delta P = \rho V^2 \phi\left(\frac{L}{d}, R_e, \frac{t}{L}\right)$ using Buckingham method when ΔP is a function of density ρ, length L, height of turbulence "t", velocity V and viscosity μ.

Solution: Given data: $\Delta P = f(\rho, V, L, d, t, \mu)$

Step 1: $$0 = f(\Delta P, \rho, V, L, d, t, \mu)$$

Step 2: Number of dependent variables are 1 and number of independent variables are 6, so total number of variables n=7 and fundamental variables m=3.

Step 3: So number of dimensionless π terms are (7–3=4).

Step 4: Write the function in terms of π terms, i.e. $f(\pi_1,\pi_2,\pi_3,\pi_4)=0$

Step 5:

$$\pi_1 = L^{x1}\ V^{y1}\ \rho^{w1}\ \Delta P$$
$$\pi_2 = L^{x2}\ V^{y2}\ \rho^{w2}\ d$$
$$\pi_3 = L^{x3}\ V^{y3}\ \rho^{w3}\ \mu$$
$$\pi_4 = L^{x4}\ V^{y4}\ \rho^{w4}\ t$$

where $LV\rho$ are repeating variables

Step 6: Equate the powers of π_1 term, so that

$$M^0L^0T^0 = [L]^{x1}\left[LT^{-1}\right]^{y1}\left[ML^{-3}\right]^{w1}\left[M^1L^1T^{-2}\right]$$

Equate the powers: Power of M: $-1 = w1$

Power of L: $0=(x1)+y1-3(w1)+1$

Power of T: $0=-y1-2 \Rightarrow (y1)=-2$

And so $(x1=2),(y1=-2),(w1=-1)$

Therefore, π_1 term becomes $\pi_1 = L^2V^{-2}\rho^{-1}\Delta P = \dfrac{\Delta P * L^2}{V^2\rho}$

Repeat step 6 to get the powers of π_2 term, so that

$$M^0L^0T^0 = [L]^{x2}\left[LT^{-1}\right]^{y2}\left[ML^{-3}\right]^{w2}\left[L^1\right]$$

Equate the powers: Power of M: $0 = w2$

Power of L: $0=(x2)+y2-3(w2)$

Power of T: $0=-y2 \Rightarrow (y2)=0$

And so $(x2=0),(y2=0),(w2=0)$

Therefore, π_2 term becomes $\pi_2 = d$

Similarly, π_4 term becomes $\pi_4 = t$

Repeat step 6 to get the powers of π_3 term, so that

$$M^0L^0T^0 = [L]^{x3}\left[LT^{-1}\right]^{y3}\left[ML^{-3}\right]^{w3}\left[ML^{-1}T^{-1}\right]$$

Equate the powers: Power of M: $-1 = w3$

Power of L: $0 = (x3) + y3 - 3(w3) - 1$

Power of T: $0 = -y3 - 1 \Rightarrow (y3) = -1$

And so $(x3 = -1), (y3 = -1), (w3 = -1)$

Therefore, π_3 term becomes $\pi_3 = L^{-1}V^{-1}\rho^{-1}\mu = \dfrac{\mu}{LV\rho}$

Step 7: So the function $f\left(\dfrac{\Delta P * L^2}{V^2\rho}, d, \dfrac{\mu}{LV\rho}, t\right) = 0$

Step 8: $\Delta P = \rho V^2 \phi\left(\dfrac{d}{L^2}, \dfrac{1}{R_e}, \dfrac{t}{L^2}\right)$

Step 9: By rearranging, multiplying, dividing and reciprocal of individual π terms, we get

$$\Rightarrow \Delta P = \rho V^2 \phi\left(\frac{L}{d}, R_e, \frac{t}{L}\right)$$

Example 8.6: Show that $F = \rho V^2 d^2 \phi\left(\dfrac{Nd}{V}, \dfrac{1}{R_e}\right)$ using Buckingham method.

Solution: Given data: $F = f(\rho, V, N, d, \mu)$

Step 1: $0 = f(F, \rho, V, N, d, \mu)$

Step 2: Number of dependent variables is 1 and numbers of independent variables are 5, so total number of variables n=6 and fundamental variables m=3.

Step 3: So number of dimensionless π terms are (6–3=3).

Step 4: Write the function in terms of π terms, i.e. $f(\pi_1, \pi_2, \pi_3) = 0$

Step 5:

$$\pi_1 = d^{x1}\ V^{y1}\ \rho^{w1}\ F$$

$$\pi_2 = d^{x2}\ V^{y2}\ \rho^{w2}\ N$$

$$\pi_3 = d^{x3}\ V^{y3}\ \rho^{w3}\ \mu$$

where $dV\rho$ are repeating variables.

Step 6: Equate the powers of π_1 term, so that

$$M^0L^0T^0 = [L]^{x1}\left[LT^{-1}\right]^{y1}\left[ML^3\right]^{w1}\left[M^1L^1T^{-2}\right]$$

Equate the powers: $M - L - T$

So we get $(x1 = -2), (y1 = -2), (w1 = -1)$

Therefore, π_1 term becomes $\pi_1 = d^{-2}V^{-2}\rho^{-1}F = \dfrac{F*}{d^2V^2\rho}$

Repeat step 6 to get the powers of π_2 term, so that

$$M^0L^0T^0 = [L]^{x2}\left[LT^{-1}\right]^{y2}\left[ML^{-3}\right]^{w2}\left[T^{-1}\right]$$

Equate the powers: $M - L - T$

And so we get $(x2 = 1), (y2 = -1), (w2 = 0)$

Therefore, π_2 term becomes $\pi_2 = dV^{\ 1}N = \dfrac{Nd}{V}$

Repeat step 6 to get the powers of π_3 term, so that

$$M^0L^0T^0 = [L]^{x3}\left[LT^{-1}\right]^{y3}\left[ML^{-3}\right]^{w3}\left[ML^1T^{-1}\right]$$

Equate the powers: Power of M: $M - L - T$

And so $(x3 = -1), (y3 = -1), (w3 = -1)$ Therefore, π_3 term becomes

$$\pi_3 = d^{-1}V^{-1}\rho^{-1}\mu = \frac{\mu}{dV\rho}$$

Step 7: So the function $$f\left(\frac{F}{d^2V^2\rho},\frac{Nd}{V},\frac{\mu}{dV\rho}\right)=0$$

Step 8: $$F=\rho V^2 d^2\phi\left(\frac{Nd}{V},\frac{1}{R_e}\right)$$

Example 8.7: Can it be shown that the resistance to motion, R, of an object moving in a fluid can be given by the following expression:

$$R=\rho V^2 d^2\varphi\left(\frac{\rho Vd}{\mu}\right)$$

Solution: Given data: $R=f(\rho,V,d,\mu)$

Step 1: $$0=f(R,\rho,V,d,\mu)$$

Step 2: Number of dependent variables is 1 and number of independent variables are 4, so total number of variables n=5 and fundamental variables m=3.

Step 3: So number of dimensionless π terms are (5–3=2).

Step 4: Write the function in terms of π terms, i.e. $f(\pi_1,\pi_2)=0$

Step 5: $$\pi_1=d^{x1}\ V^{y1}\ \rho^{w1}\ R$$
$$\pi_2=d^{x2}\ V^{y2}\ \rho^{w2}\ \mu$$

where d,V,ρ are repeating variables.

Step 6: Equate the powers of π_1 term, so that

$$M^0L^0T^0=[L]^{x1}\left[LT^{-1}\right]^{y1}\left[ML^{-3}\right]^{w1}\left[M^1L^1T^{-2}\right]$$

Equate the powers: $M-L-T$

And so we get $$(x1=-2),(y1=-2),(w1=-1)$$

Therefore, π_1 term becomes $$\pi_1=d^{-2}V^{-2}\rho^{-1}R=\frac{R}{d^2V^2\rho}$$

Repeat step 6 to get the powers of π_2 term, so that

$$M^0L^0T^0=[L]^{x2}\left[LT^{-1}\right]^{y2}\left[ML^{-3}\right]^{w2}\left[ML^{-1}T^{-1}\right]$$

Equate the powers: $M-L-T$

And so we get $(x2=-1),(y2=-1),(w2=-1)$

Therefore, π_3 term becomes $\pi_2 = d^{-1}V^{-1}\rho^{-1}\mu = \dfrac{\mu}{dV\rho}$

Step 7: So the function $f\left(\dfrac{R}{d^2V^2\rho}, \dfrac{\mu}{dV\rho}\right) = 0$

Step 8: $R = \rho V^2 d^2 \phi\left(\dfrac{dV\rho}{\mu}\right)$

Exmaple 8.8: Show that $Q = \dfrac{d^2\sqrt{P}}{\sqrt{\rho}}\varphi\left(\dfrac{d\sqrt{P\rho}}{\mu}\right)$.

Solution: Given data: $Q = f(\rho, P, d, \mu)$

Step 1: $0 = f(Q, \rho, P, d, \mu)$

Step 2: Number of dependent variables is 1 and number of independent variables are 4, so total number of variables n=5 and fundamental variables m=3.

Step 3: So number of dimensionless π terms are (5 3−2).

Step 4: Write the function in terms of π terms, i.e. $f(\pi_1, \pi_2) = 0$

Step 5:

$$\pi_1 = d^{x1}\ P^{y1}\ \rho^{w1}\ Q$$
$$\pi_2 = d^{x2}\ P^{y2}\ \rho^{w2}\ \mu$$

where d, P, ρ are repeating variables.

Step 6: Equate the powers of π_1 term, so that

$$M^0L^0T^0 = [L]^{x1}\left[M^1L^{-1}T^{-2}\right]^{y1}\left[ML^{-3}\right]^{w1}\left[L^3T^{-1}\right]$$

Equate the powers: $M-L-T$

And so we get $(x1=-2),(y1=-\frac{1}{2}),(w1=\frac{1}{2})$

Therefore, π_1 term becomes $\pi_1 = d^{-2}P^{-\frac{1}{2}}\rho^{\frac{1}{2}}Q = \frac{Q*\sqrt{\rho}}{\sqrt{P}*d^2}$

Repeat step 6 to get the powers of π_2 term, so that

$$M^0L^0T^0 = [L]^{x2}\left[ML^{-1}T^{-2}\right]^{y2}\left[ML^{-3}\right]^{w2}\left[ML^{-1}T^{-1}\right]$$

Equate the powers: Power of M: $0 = y2 + (w2) + 1$

Power of L: $0 = (x2) - y2 - 3(w2) - 1$

Power of T: $0 = -2(y2) - 1 \Rightarrow (y2) = -\frac{1}{2}$

And so $(x2=-1),(y2=-\frac{1}{2}),(w2=-\frac{1}{2})$

Therefore, π_2 term becomes $\pi_2 = d^{-1}P^{-\frac{1}{2}}\rho^{-\frac{1}{2}}\mu = \frac{\mu}{d\sqrt{P\rho}}$

Step 7: So the function $f\left(\frac{Q*\sqrt{\rho}}{\sqrt{P}*d^2}, \frac{\mu}{d\sqrt{P\rho}}\right) = 0$

Step 8: $Q = \frac{d^2\sqrt{P}}{\sqrt{\rho}}\phi\left(\frac{\mu}{d\sqrt{P\rho}}\right)$

Step 9: By taking reciprocal of the function and multiplying by "d^3", we get

$$Q = \frac{d^2\sqrt{P}}{\sqrt{\rho}}\phi\left(\frac{d\sqrt{P\rho}}{\mu}\right)$$

Example 8.9: Show that pressure drop $V = \sqrt{gH}\phi\left(\frac{d}{H}, \frac{\mu}{\rho VH}\right)$ using Buckingham's method.

Solution: Given data:

$$V = \sqrt{gH}\,\phi\left(\frac{d}{H}, \frac{\mu}{\rho VH}\right)$$

Step 1: $$0 = f(V, g, H, \rho, d, \mu)$$

Step 2: Number of dependent variables is 1 and number of independent variables are 5, so total number of variables n=6 and fundamental variables m=3.

Step 3: So number of dimensionless π terms are (6 – 3 = 3).

Step 4: Write the function in terms of π terms, i.e. $f(\pi_1, \pi_2, \pi_3) = 0$

Step 5:

$$\pi_1 = d^{x1}\ g^{y1}\ \rho^{w1}\ V$$
$$\pi_2 = d^{x2}\ g^{y2}\ \rho^{w2}\ \mu$$
$$\pi_3 = d^{x3}\ g^{y3}\ \rho^{w3}\ H$$

where d, g, ρ are repeating variables.

Step 6: Equate the powers of π_1 term, so that

$$M^0L^0T^0 = [L]^{x1}\left[LT^{-2}\right]^{y1}\left[ML^{-3}\right]^{w1}\left[LT^{-1}\right]$$

Equate the powers: $M - L - T$

And so we get $$\left(x1 = -\frac{1}{2}\right), \left(y1 = -\frac{1}{2}\right), (w1 = 0)$$

Therefore, π_1 term becomes $$\pi_1 = d^{-3/2} g^{-1/2} V = \frac{V}{\sqrt{g} * d^{3/2}}$$

Repeat step 6 to get the powers of π_2 term, so that

$$M^0L^0T^0 = [L]^{x2}\left[LT^{-2}\right]^{y2}\left[ML^{-3}\right]^{w2}\left[ML^{-1}T^{-1}\right]$$

Equate the powers: $M - L - T$

And so we get $\left(x2 = -\frac{3}{2}\right), \left(y2 = -\frac{1}{2}\right), (w2 = -1)$. Therefore, π_2 term becomes

$$\pi_2 = d^{-3/2} g^{-1/2} \rho^{-1} \mu = \frac{\mu}{\rho d^{3/2} \sqrt{g}}$$

Repeat step 6 to get the powers of π_3 term, so that

$$M^0 L^0 T^0 = [L]^{x3} \left[L T^{-2}\right]^{y3} \left[ML^{-3}\right]^{w3} [L]$$

Equate the powers: $M - L - T$

And so we get $(x3 = -1), (y3 = 0), (w3 = 0)$

Therefore, π_3 term becomes $\pi_3 = d^{-1} H = \frac{H}{d}$

Step 7: So the function $f\left(\frac{V}{\sqrt{gd}}, \frac{\mu}{\rho d^{3/2}\sqrt{g}}, \frac{H}{d}\right) = 0$

Step 8: By multiplying term 1 by $\sqrt{d/H}$ and term 2 by $d^{3/2}\sqrt{g}$ and by rearranging, we get

$$V = \sqrt{gH}\,\phi\left(\frac{\mu}{\rho VH}, \frac{d}{H}\right)$$

Example 8.10: The discharge Q of a convergent pump depends upon the mass density of the fluid ρ, speed of the pump N, diameter of the impeller D, manometric head H and the viscosity of the fluid μ. Show that $Q = ND^3 \varphi\left(\frac{gHN^2}{D^2}, \frac{\mu}{\rho ND^2}\right)$ using Buckingham's method.

Solution: Given data: $Q = f(D, H, N, \rho, \mu, g)$

Step 1: $0 = f(D, H, N, \rho, \mu, g, Q)$

Step 2: Number of dependent variables is 1 and number of independent variables are 6, so total number of variables n = 7 and fundamental variables m = 3.

Step 3: So number of dimensionless π terms are (7 – 3 = 4).

Step 4: Write the function in terms of π terms, i.e. $f(\pi_1, \pi_2, \pi_3, \pi_4) = 0$

Step 5:

$$\pi_1 = D^{x1}\ g^{y1}\ \rho^{w1}\ Q$$
$$\pi_2 = D^{x2}\ g^{y2}\ \rho^{w2}\ H$$
$$\pi_3 = D^{x3}\ g^{y3}\ \rho^{w3}\ \mu \text{ where } D, g, \rho \text{ are repeating variables}$$
$$\pi_4 = D^{x4}\ g^{y4}\ \rho^{w4}\ N$$

Step 6: Equate the powers of π_1 term, so that

$$M^0 L^0 T^0 = [L]^{x1}\left[LT^{-2}\right]^{y1}\left[ML^{-3}\right]^{w1}\left[L^3T^{-1}\right]$$

Equate the powers: $M - L - T$

And so $$\left(x1 = -\frac{5}{2}\right), (y1 = -\frac{1}{2}), (w1 = 0)$$

Therefore, π_1 term becomes $$\pi_1 = D^{-5/2} g^{-1/2} Q = \frac{Q}{\sqrt{g} * D^{5/2}}$$

Repeat step 6 to get the powers of π_2 term, so that

$$M^0 L^0 T^0 = [L]^{x2}\left[LT^{-2}\right]^{y2}\left[ML^{-3}\right]^{w2}[L]$$

Equate the powers: $M - L - T$

And so $(x2 = -1), (y2 = 0), (w2 = 0)$

Therefore, π_2 term becomes $\pi_2 = D^{-1}H = \dfrac{H}{D}$

Repeat step 6 to get the powers of π_3 term, so that

$$M^0 L^0 T^0 = [L]^{x3}\left[LT^{-2}\right]^{y3}\left[ML^{-3}\right]^{w3}\left[ML^{-1}T^{-1}\right]$$

Equate the powers: $M - L - T$

And so $$\left(x3 = -\frac{3}{2}\right), (y3 = -\frac{1}{2}), (w3 = -1)$$

Therefore, π_3 term becomes $$\pi_3 = D^{-3/2} g^{-1/2} \rho^{-1} \mu = \frac{\mu}{\rho D^{3/2}\sqrt{g}}$$

Repeat step 6 to get the powers of π_4 term, so that

$$M^0L^0T^0 = [L]^{x4}\left[LT^{-2}\right]^{y4}\left[ML^{-3}\right]^{w4}\left[T^{-1}\right]$$

Equate the powers: $M - L - T$

And so $\left(x4 = \frac{1}{2}\right), (y4 = -\frac{1}{2}), (w4 = 0)$

Therefore, π_4 term becomes $\pi_4 = D^{1/2} g^{-1/2} N = \frac{N\sqrt{D}}{\sqrt{g}}$

Step 7: So the function $f\left(\frac{Q}{\sqrt{g} * D^{5/2}}, \frac{H}{D}, \frac{\mu}{\rho D^{3/2}\sqrt{g}}, \frac{N\sqrt{D}}{\sqrt{g}}\right) = 0$

Step 8: $Q = D^{5/2} * \sqrt{g}\phi\left(\frac{H}{D}, \frac{\mu}{\rho D^{3/2}\sqrt{g}}, \frac{N\sqrt{D}}{\sqrt{g}}\right)$

Step 9: Rewriting $Q = D^3 D^{-1/2} * \sqrt{g}\phi\left(\frac{H}{D}, \frac{\mu}{\rho D^{3/2}\sqrt{g}}, \frac{N\sqrt{D}}{\sqrt{g}}\right)$

Step 10: Multiply "$\pi_4 - term$ ", we get $Q = ND^3\phi\left(\frac{H}{D}, \frac{\mu}{\rho D^{3/2}\sqrt{g}}\right)$

Step 11: Multiply "π_2 by π_4^2 & g^2/D^2", we get $Q = ND^3\varphi\left(\frac{gH}{N^2D^2}, \frac{\mu}{\rho ND^2}\right)$

Example 8.11: The drag force F_d on a body in supersonic flow is considered to depend on the length L, velocity V, diameter D, mass density of the fluid ρ, viscosity of the fluid μ and bulk modulus K. Show that $F_d = \rho V^2 l^2 \varphi\left(\frac{D}{l}, \frac{\mu}{\rho Vl}, \frac{K}{\rho V^2}\right)$.

Solution: Given data: $F_d = \rho V^2 l^2 \varphi\left(\frac{D}{l}, \frac{\mu}{\rho V l}, \frac{K}{\rho V^2}\right)$

Step 1: $$0 = f(D, V, l, \rho, \mu, K, F_d)$$

Step 2: Number of dependent variables is 1 and numbers of independent variables are 6, so total number of variables n=7 and fundamental variables m=3.

Step 3: So number of dimensionless π terms are (7–3 = 4).

Step 4: Write the function in terms of π terms, i.e. $f(\pi_1, \pi_2, \pi_3, \pi_4) = 0$

Step 5:
$$\begin{aligned}
\pi_1 &= D^{x1}\ V^{y1}\ \rho^{w1}\ F_d \\
\pi_2 &= D^{x2}\ V^{y2}\ \rho^{w2}\ l \\
\pi_3 &= D^{x3}\ V^{y3}\ \rho^{w3}\ \mu \\
\pi_4 &= D^{x4}\ V^{y4}\ \rho^{w4}\ K
\end{aligned}$$
where D, V, ρ are repeating variables

Step 6: Equate the powers of π_1 term, so that

$$M^0 L^0 T^0 = [L]^{x1}\left[LT^{-1}\right]^{y1}\left[ML^{-3}\right]^{w1}\left[MLT^{-2}\right]$$

Equate the powers: $M - L - T$

And so $$(x1 = -2), (y1 = -2), (w1 = -1)$$

Therefore, π_1 term becomes $$\pi_1 = D^{-2} V^{-2} \rho^{-1} F_d = \frac{F_d}{\rho V^2 D^2}$$

Repeat step 6 to get the powers of π_2 term, so that

$$M^0 L^0 T^0 = [L]^{x2}\left[LT^{-1}\right]^{y2}\left[ML^{-3}\right]^{w2}[L]$$

Equate the powers: $M - L - T$

And so $$(x2 = -1), (y2 = 0), (w2 = 0)$$

Therefore, π_2 term becomes $\pi_2 = D^{-1}l = \dfrac{l}{D}$

Repeat step 6 to get the powers of π_3 term, so that

$$M^0L^0T^0 = [L]^{x3}\left[LT^{-1}\right]^{y3}\left[ML^{-3}\right]^{w3}\left[ML^{-1}T^{-1}\right]$$

Equate the powers: $M - L - T$

And so $(x3 = -1), (y3 = -1), (w3 = -1)$

Therefore, π_3 term becomes $\pi_3 = D^{-1}V^{-1}\rho^{-1}\mu = \dfrac{\mu}{DV\rho}$

Repeat step 6 to get the powers of π_4 term, so that

$$M^0L^0T^0 = [L]^{x4}\left[LT^{-1}\right]^{y4}\left[ML^{-3}\right]^{w4}\left[ML^{-1}T^{-2}\right]$$

Equate the powers: $M - L - T$

And so $(x4 = 0), (y4 = -2), (w4 = -1)$

Therefore, π_4 term becomes $\pi_4 = V^{-2}\rho^{-1}K = \dfrac{K}{\rho\sqrt{V}}$

Step 7: So the function $f\left(\dfrac{F_d}{\rho V^2 D^2}, \dfrac{l}{D}, \dfrac{\mu}{VD\rho}, \dfrac{K}{\rho\sqrt{V}}\right) = 0$

Step 8: $F_d = \rho V^2 D^2 \phi\left(\dfrac{l}{D}, \dfrac{\mu}{VD\rho}, \dfrac{K}{\rho\sqrt{V}}\right)$

Step 9: Multiply l^2 / D^2 , we get $F_d = \rho V^2 l^2 \phi\left(\dfrac{l}{D}, \dfrac{\mu}{VD\rho}, \dfrac{K}{\rho\sqrt{V}}\right)$

Step 10: Multiply "$\begin{matrix}\pi_3 & by & D/l \\ \pi_4 & by & 1/V^{-3/2}\end{matrix}$" we get $F_d = \rho V^2 l^2 \varphi\left(\dfrac{D}{l}, \dfrac{\mu}{\rho V l}, \dfrac{K}{\rho V^2}\right)$

Example 8.12: A small sphere of density ρ and diameter D settles at a terminal velocity V in a liquid of density ρ_l and dynamic viscosity μ. Gravity g is known to be a parameter. Show that

$$V = \sqrt{gD}\varphi\left(\frac{\rho}{\rho_l}, \frac{\mu}{\rho D\sqrt{gD}}\right)$$

Solution: Given data:

$$V = \sqrt{gD}\varphi\left(\frac{\rho}{\rho_l}, \frac{\mu}{\rho D\sqrt{gD}}\right)$$

Step 1: $0 = f(V, g, D, \rho, \rho_l, \mu)$

Step 2: Number of dependent variables is 1 and number of independent variables are 5, so total number of variables n=6 and fundamental variables m=3.

Step 3: So number of dimensionless π terms are (6 – 3 = 3).

Step 4: Write the function in terms of π terms, i.e. $f(\pi_1, \pi_2, \pi_3) = 0$

Step 5:

$$\pi_1 = d^{x1}\ g^{y1}\ \rho^{w1}\ V$$
$$\pi_2 = d^{x2}\ g^{y2}\ \rho^{w2}\ \mu$$
$$\pi_3 = d^{x3}\ g^{y3}\ \rho^{w3}\ \rho_l$$

where d, g, ρ are repeating variables

Step 6: Equate the powers of π_1 term, so that

$$M^0L^0T^0 = [L]^{x1}\left[LT^{-2}\right]^{y1}\left[ML^{-3}\right]^{w1}\left[LT^{-1}\right]$$

Equate the powers: $M - L - T$

And so $\left(x1 = -\frac{1}{2}\right), (y1 = -\frac{1}{2}), (w1 = 0)$

Therefore, π_1 term becomes $\pi_1 = d^{-1/2} g^{-1/2} V = \frac{V}{\sqrt{gd}}$

Repeat step 6 to get the powers of π_2 term, so that

$$M^0L^0T^0 = [L]^{x2}\left[LT^{-2}\right]^{y2}\left[ML^{-3}\right]^{w2}\left[ML^{-1}T^{-1}\right]$$

Equate the powers: $M - L - T$

Power of M: $0 = (wz) + 1$

Power of L: $$0 = (x2) + y2 - 3(w2) - 1$$
$$\Rightarrow (x2) + y2 = -2$$

Power of T: $$0 = -2(y2) - 1 \Rightarrow (y2) = -\frac{1}{2}$$

And so $$\left(x2 = -\frac{3}{2}\right), (y2 = -\frac{1}{2}), (w2 = -1)$$

Therefore, π_2 term becomes $$\pi_2 = d^{-3/2} g^{-1/2} \rho^{-1} \mu = \frac{\mu}{\rho d^{3/2} \sqrt{g}}$$

Repeat step 6 to get the powers of π_3 term, so that

$$M^0 L^0 T^0 = [L]^{x3} \left[LT^{-2}\right]^{y3} \left[ML^{-3}\right]^{w3} \left[ML^{-3}\right]$$

Equate the powers: $M - L - T$

And so $$(x3 = 0), (y3 = 0), (w3 = -1)$$

Therefore, π_3 term becomes $$\pi_3 = \rho^{-1} \rho_l = \frac{\rho_l}{\rho}.$$

Step 7: So the function $$f\left(\frac{V}{\sqrt{gd}}, \frac{\mu}{\rho d^{3/2} \sqrt{g}}, \frac{\rho_l}{\rho}\right) = 0.$$

Step 8: $$V = \sqrt{gd}\phi\left(\frac{\mu}{\rho d^{3/2} \sqrt{g}}, \frac{\rho_l}{\rho}\right).$$

Step 9: By rearranging, we get

$$V = \sqrt{gd}\varphi\left(\frac{\rho}{\rho_l}, \frac{\mu}{\rho d \sqrt{gd}}\right)$$

Example 8.13: Show that the power developed in a turbine is

$$P = \rho N^3 D^5 \varphi\left(\frac{D}{B}, \frac{\rho N D^2}{\mu}, \frac{ND}{\sqrt{gH}}\right)$$

where D, B are the diameter and width of the turbine runner, H is the operating head P is the power, mass density of the fluid ρ , viscosity of the fluid μ .

Solution: Given data:

$$P = \rho N^3 D^5 \varphi\left(\frac{D}{B}, \frac{\rho N D^2}{\mu}, \frac{ND}{\sqrt{gH}}\right)$$

Step 1: $$0 = f(P, D, B, N, \rho, \mu, g, H)$$

Step 2: Number of dependent variables is 1 and number of independent variables are 7, so total number of variables n=8 and fundamental variables m=3.

Step 3: So number of dimensionless π terms are (8 – 3 = 5).

Step 4: Write the function in terms of π terms, i.e. $f(\pi_1, \pi_2, \pi_3, \pi_4, \pi_5) = 0$

Step 5:

$$\begin{aligned}
\pi_1 &= D^{x1}\ g^{y1}\ \rho^{w1}\ P \\
\pi_2 &= D^{x2}\ g^{y2}\ \rho^{w2}\ B \\
\pi_3 &= D^{x3}\ g^{y3}\ \rho^{w3}\ \mu \\
\pi_4 &= D^{x4}\ g^{y4}\ \rho^{w4}\ N \\
\pi_5 &= D^{x5}\ g^{y5}\ \rho^{w5}\ H
\end{aligned}$$

where D, g, ρ are repeating variables

Step 6: Equate the powers of π_1 term, so that

$$M^0 L^0 T^0 = [L]^{x1}\left[LT^{-2}\right]^{y1}\left[ML^{-3}\right]^{w1}\left[ML^2T^{-3}\right]$$

Equate the powers: $$M - L - T$$

And so $$\left(x1 = -\frac{7}{2}\right), \left(y1 = -\frac{3}{2}\right), (w1 = -1)$$

Therefore, π_1 term becomes $$\pi_1 = D^{-7/2} g^{-3/2} \rho^{-1} P = \frac{P}{\rho g^{3/2} D^{7/2}}$$

Repeat step 6 to get the powers of π_2 term, so that

$$M^0 L^0 T^0 = [L]^{x2}\left[LT^{-2}\right]^{y2}\left[ML^{-3}\right]^{w2}[L]$$

Equate the powers: $$M - L - T$$

And so $(x2=-1),(y2=0),(w2=0)$

Therefore, π_2 term becomes $\pi_2 = D^{-1}B = \dfrac{B}{D}$

Repeat step 6 to get the powers of π_3 term, so that

$$M^0L^0T^0 = [L]^{x3}\left[LT^{-2}\right]^{y3}\left[ML^{-3}\right]^{w3}\left[ML^{-1}T^{-1}\right]$$

Equate the powers: $M-L-T$

And so $\left(x3=-3/2\right),(y3=-\dfrac{1}{2}),(w3=-1)$

Therefore, π_3 term becomes $\pi_3 = D^{-3/2}g^{-1/2}\rho^{-1}\mu = \dfrac{\mu}{\rho D\sqrt{gD}}$

Repeat step 6 to get the powers of π_4 term, so that

$$M^0L^0T^0 = [L]^{x4}\left[LT^{-2}\right]^{y4}\left[ML^{-3}\right]^{w4}\left[T^{-1}\right]$$

Equate the powers: $M-L-T$

And so $\left(x4=\dfrac{1}{2}\right),(y4=-\dfrac{1}{2}),(w4=0)$

Therefore, π_4 term becomes $\pi_4 = D^{1/2}g^{-1/2}N = \dfrac{N\sqrt{D}}{\sqrt{g}}$

Repeat step 6 to get the powers of π_5 term, so that

$$M^0L^0T^0 = [L]^{x5}\left[LT^{-2}\right]^{y5}\left[ML^{-3}\right]^{w5}[L]$$

Equate the powers: Power of M: $M-L-T$

And so $(x5=-1),(y5=0),(w5=0)$

Therefore, π_5 term becomes $\pi_5 = D^{-1}H = \dfrac{H}{D}$

Step 7: So the function $f\left(\frac{P}{\rho g^{3/2} D^{7/2}}, \frac{B}{D}, \frac{\mu}{\rho D\sqrt{gD}}, \frac{N\sqrt{D}}{\sqrt{g}}, \frac{H}{D}\right) = 0$

Step 8: $P = \rho g^{3/2} D^{7/2} \phi\left(\frac{B}{D}, \frac{\mu}{\rho D\sqrt{gD}}, \frac{N\sqrt{D}}{\sqrt{g}}, \frac{H}{D}\right)$

Step 9: Multiply by $N^3 D^{3/2} / g^{3/2}$, we get $P = \rho N^3 D^5 \varphi\left(\frac{D}{B}, \frac{\rho N D^2}{\mu}, \frac{ND}{\sqrt{gH}}\right)$

Step 10: Taking reciprocals of π_2, π_3 and multiply "π_3 *by* $\sqrt{D}/\sqrt{H}$", we get

$$P = \rho N^3 D^5 \varphi\left(\frac{D}{B}, \frac{\rho N D^2}{\mu}, \frac{ND}{\sqrt{gH}}\right)$$

MODEL SIMILITUDE/ANALYSIS

Model similitude is the study of model and prototype of a machine/component, so as to ascertain the behavioral aspects like performance, efficiencies, movement, appearance, designing feasibility, flow conditions, etc. when the actual machine/component works in the actual working conditions. This analysis helps in predicting the evil effects of the actual machine/components, if any, after manufacturing and can be rectified by taking the changes in design stage in the model itself and analyzing it and these changes can be incorporated at the time of manufacture of the actual machine/components.

Physical modeling of scaled components/objects is predominantly growing in use like in the heavy industrial manufacturing (aeronautical, shipbuilding, missiles and torpedo, etc.), geotechnical structures (foundations, dams, retaining walls, irrigation projects), etc.

Model similitude is an approach to functional testing of an object/component by correlating the behavior of a model and the actual object to be produced by coupling between material properties, geometric shape, kinematics and dynamics of the model and actual object that produce failure of the actual object, at the homologous points. E.g: Each and every point on the prototype has a scaling factor on its corresponding point on the model and hence by correlating the appropriate scale factors, we can predict the behaviors and performance of the actual machine/object.

Simulation of scaled objects are built in laboratories and tested under same conditions as that of the actual object that operates, in a controlled and carefully monitored conditions. Physical modeling has many benefits like relatively little cost in producing a scaled model, relatively less space requirement, feasibility to produce and transport. Modeling also has limitations, which principally involves issues related to boundary conditions and similitude (like simultaneous scaling of multiple phenomena, scale effects of instrumentation, side boundary friction, etc.).

Model: A model is the replica of actual machine/components with reduced scale dimensions and of same shape, but may not work in actual working conditions.

Prototype: A prototype is the reduced dimensional machine/components that works in the actual working conditions and is of same shape and material of actual machine/components.

The models can be made as replica of actual machine/components and used in analysis of the actual prototype and this can be done only if the model and prototype has compatibility of comparing various terms of non-dimensional terms of same similarities in between them, like

- Geometric similarities
- Kinematic similarities
- Dynamic similarities

Established laws of similitude are based on the similarity between the model and prototype with respect to geometry, kinematic and dynamics. While there is a sound theoretical basis for these scaling factors, experimental verification is nevertheless required to fully validate established similitude relationships.

GEOMETRIC SIMILARITIES

Geometric similarity exists between a model and a prototype when they both have geometrical dimensions like length, area, volume, inertia, etc. in a proper linear scale ratio (S_L) at any corresponding points on them and are of same shape.

E.g.: Consider a prototype and a model of cube having dimensions L_M, B_M, h_M and L_P, B_P, h_P of prototype and model respectively and also prototype and model of a ball having diameter "D_P" and "D_M" respectively, as shown in Figures 8.1a and 8.1b.

Then

$$\frac{L_M}{L_P} = \frac{B_M}{B_P} = \frac{h_M}{h_P} = \frac{D_M}{D_P} = S_L \tag{8.0}$$

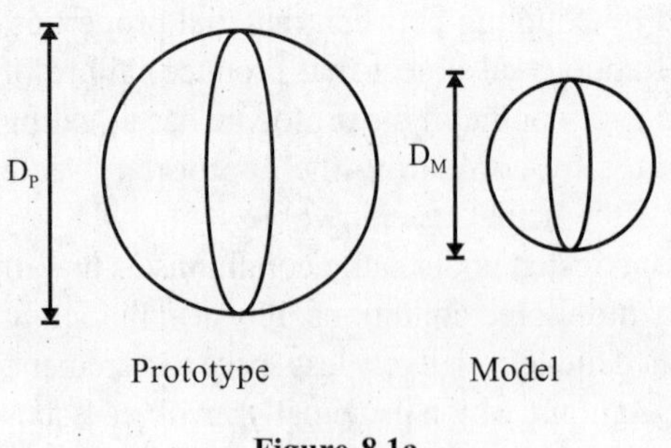

Figure 8.1a

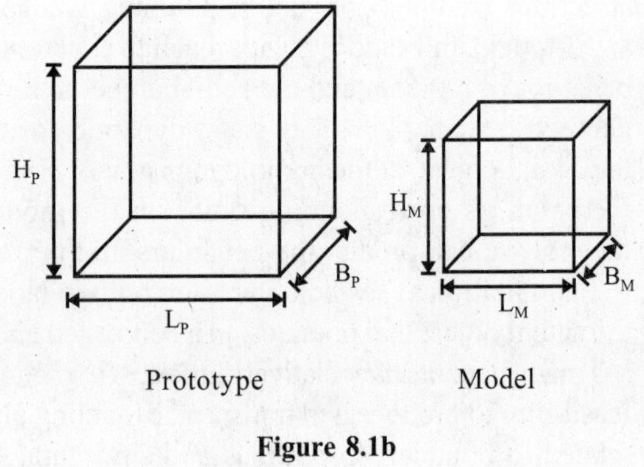

Figure 8.1b

$$\frac{(Area)_M}{(Area)_P} = \frac{L_M * B_M}{L_P * B_P} = \frac{(D_M)^2}{(D_P)^2} = (S_L)^2$$

$$\frac{(Volume)_M}{(Volume)_P} = \frac{L_M * B_M * h_M}{L_P * B_P * h_P} = \frac{(D_M)^3}{(D_P)^3} = (S_L)^3$$

KINEMATIC SIMILARITIES

Kinematic similarity exists between a model and a prototype when they both have geometrical similarity and are subjected to same kinematic conditions like velocity, acceleration, discharge, etc. to a proper scale ratio, at any corresponding points on them, then

$$(Velocity_M) = (Velocity_P) \Rightarrow \frac{(Velocity_M)}{(Velocity_P)} = \frac{(L_M / T_M)}{(L_P / T_P)} = \frac{L_M * T_P}{L_P * T_M} = \frac{S_L}{S_T}$$

where (S_T) is the time scale ratio.

$$\frac{(Q_M)}{(Q_P)} = \frac{(L_M^3 / T_M)}{(L_P^3 / T_P)} = \frac{L_M^3 * T_P}{L_P^3 * T_M} = (S_L^3) \tag{8.1}$$

DYNAMIC SIMILARITIES

Dynamic similarity exists between a model and a prototype when they both have geometrical similarity, kinematic similarity and are subjected to same dynamic conditions like inertia force, gravity force, pressure force, shear force, elastic force, etc. in a proper linear scale ratio, at any corresponding points on them.

E.g.: Consider a prototype and a model of a fluid particle moving on a aerofoil having forces subjected on it like inertia force F_I, shear force F_τ, pressure force F_P, gravitational force F_G, surface tension force F_σ, as shown in Figure 8.2.

Then $$\frac{\sum(F_I)_{Model}}{\sum(F_I)_{Prototype}} = \frac{(Ma)_{Model}}{(Ma)_{Prototype}} \Rightarrow \frac{\sum(F_I)_{Model}}{\sum(F_I)_{Prototype}} = \frac{(F_\tau + F_P + F_E + F_G + F_\sigma)_{Model}}{(F_\tau + F_P + F_E + F_G + F_\sigma)_{Prototype}}$$

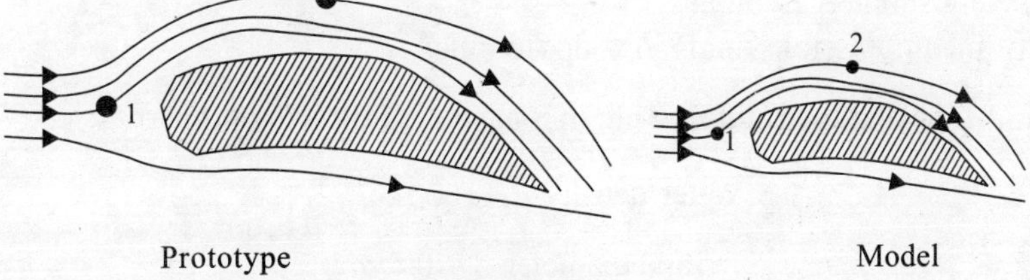

Figure 8.2

Similarly, $$\frac{(F_I)_M}{(F_I)_P} = \frac{(F_\tau)_M}{(F_\tau)_P} = \frac{(Ma)_M}{(Ma)_M} = \frac{(\rho * L^3 * a)_M}{(\rho * L^3 * a)_P} = \frac{S_L^4}{S_T^2} \tag{8.2}$$

if $\rho_M = \rho_P$

Similarly, $$\frac{(F_I)_M}{(F_I)_P} = \frac{(F_P)_M}{(F_P)_P} = \frac{(F_G)_M}{(F_G)_P} = \frac{(F_\sigma)_M}{(F_\sigma)_P} = \frac{(F_E)_M}{(F_E)_P}$$

DIMENSIONLESS GROUPS AND THEIR SIGNIFICANCE

Whenever there is a motion of fluid or object, it is due to the inertia force MLT^{-2} , as per the Newton's second law. But the inertia force is not achieved completely, when a fluid/object flows and it's due to the presence of other resisting forces like viscous force, gravity force, pressure force, surface tension force, elastic/compressive forces, etc.

In order to take into account of these resisting forces in model analysis having dynamic similarities, we use dimensionless quantities like Reynolds number, Froude number, Euler's number, Weber number and Mach number, to accommodate these resisting forces while in motion.

Reynolds Number: Reynolds number is the ratio of **inertia force** and **viscous force**. And it indicates the overcome of inertia force over the viscous force and signifies in knowing the type of flow, i.e. laminar or turbulent.

i.e. $$R_e = \frac{Inertia\ Force}{Viscous\ Force} = \frac{m * a}{\tau * A}$$

$$\Rightarrow R_e = \frac{m * a}{\tau * A} = \frac{\rho A V^2}{\mu \frac{V}{L} A}$$

$$\Rightarrow R_e = \frac{\rho V L}{\mu} \tag{8.3}$$

Reynolds number can be correlated for a model and prototype in the situation of problems where the viscous effect is more predominant than the other resisting forces and can be used for correlation as below mentioned case, like

a) Low velocity condition problems.
b) Completely submerged bodies.
c) Flow of incompressible fluids in a closed pipe.

Example 8.14: A pipe and the fluid flowing in it have the following properties:

Water density $\rho = 1000\,\text{kg/m}^3$

Pipe diameter $d = 0.7$ m

Dynamic viscosity, $\mu = 490\,\text{Ns/m}^2$

What is the MAXIMUM velocity when flow is laminar i.e. Re = 1800?

Solution: Given data: Water density $\rho = 1000\ \text{kg/m}^3$, Pipe diameter $d = 0.5$ m,

Dynamic viscosity, $\mu = 490\ \text{Ns/m}^2$

We know Reynolds number $R_e = \dfrac{\rho V d}{\mu}$

Hence, maximum velocity in the pipe is $\Rightarrow V = \dfrac{R_e * \mu}{\rho * d} = \dfrac{1800 * 490}{1000 * 0.7} = 1260\,\text{m/s}$

Example 8.15: An underwater missile, diameter 2 m and length 10 m is tested in a water tunnel to determine the forces acting on the real prototype. A 1/20 scale is to be used. If the maximum allowable speed of the prototype missile is 10 m/s, what should be the speed of the water in the tunnel to achieve dynamic similarity?

Solution: Given data

Scale		Model	Prototype
$D_m/D_P = 1/20$	Medium Velocity	Water $U_{max} = ?$	Water $U_{max} = 10$ m/s $D = 2$ m

According to dynamic similarity, Reynolds numbers are equal and hence

$$\left(\frac{\rho u d}{\mu}\right)_m = \left(\frac{\rho u d}{\mu}\right)_P \quad \text{but } \mu_m = \mu_P \ \&\ \rho_m = \rho_P$$

so $$u_m = u_P\left(\frac{d_P}{d_m}\right) = 200\,\text{m/s}$$

Example 8.16: An experiment is being designed for a sphere balloon of 1 m traveling in air so that it has dynamic similarity with a sphere balloon of diameter 0.01 m traveling in water at 2 m/s. What speed must the air be traveling?

Solution: For dynamic similarity, the Reynolds numbers must be equal in air and water

$$\left(\frac{\rho u d}{\mu}\right)_{air} = \left(\frac{\rho u d}{\mu}\right)_{water}$$

$$\left(\frac{1.25 * u * 1.0}{1.7\times10^{-5}}\right) = \left(\frac{1000 * 2 * 0.01}{1.0\times10^{-6}}\right)$$

$$u_{air} = 272\text{m/s}$$

Example 8.17: A pipe of diameter 1.7 m is required to transport an oil of specific gravity 0.9 and viscosity 3.5×10^{-2} poise at the rate of 3000 lit/sec. Tests were conducted on a 17 cm diameter pipe using water at 20°C. Find the velocity and rate of flow in the model. Viscosity of water at 20°C is 0.01poise.

Solution: Given data:

Scale		Model	Prototype
$\frac{D_m}{D_P} = \frac{0.17}{1.7} = \frac{1}{10}$	Medium	Water $\left(\rho = 1000\,\text{kg/m}^3\right)$	Oil
		$d = 0.17$ m	$d = 1.7$ m
		$\mu = 0.01\,\text{poise}$	$\mu = 3.5\times10^{-2}\,\text{poise}$
		$V_m = Q_m = ?$	$S_P = 0.9$
			$Q = 3\,\text{m}^3/\text{s}$

Velocity of prototype is $V_P = \frac{Q_P}{A_P} = \frac{3*4}{\pi * 1.7^2} = 1.32\,\text{m/s}$

Density of oil is $S_P = \frac{\rho_{oil}}{\rho_{water}} \Rightarrow \rho_{oil} = \rho_{water} * S_P = 1000 * 0.9 = 900\,\text{kg/m}^3$

According to dynamic similarity, Reynolds numbers are equal and hence

$$\left(\frac{\rho V d}{\mu}\right)_{oil} = \left(\frac{\rho V d}{\mu}\right)_{water}$$

$$V_m = 3.39\,\text{m/s}$$

So discharge through the model is $Q_m = A_m V_m = \frac{\pi}{4} * 0.17^2 * 3.39 = 0.077\,\text{m}^2/\text{s}$

Froude Number: Froude number is the ratio of inertia force and gravity force. It indicates the overcome of inertia force over the gravity force and comes into account when the flow is having free surface waves, like motion of sea water.

$$F_r = \sqrt{\frac{Inertia\ Force}{Gravity\ Force}}$$

$$\Rightarrow F_r = \sqrt{\frac{\rho A V^2}{m * g}} = \sqrt{\frac{\rho A V^2}{\rho * AL * g}}$$

$$\Rightarrow F_r = \frac{V}{\sqrt{gL}} \qquad (8.4)$$

Froude number can be correlated for a model and prototype in the situation of problems where the gravitational effect is more predominant than the other resisting forces and can be used for correlation as below mentioned case, like

a) Spillways of dams and irrigation projects.
b) Open channels having wave jumping.
c) Flow out of nozzles.
d) Motion of ships in a turbulent condition, which arise large wave jumping's.

Example 8.18: A model of a ship built at a 1/15 scale and is tested in a fresh water and operating at a velocity of 3 m/s, at which the resistance was measured as 60 N. Determine the velocity of the prototype if the prototype is 50 m long works out in sea water and also determine the resisting force observed by the prototype. Take density of sea water as $(\rho = 1100\,\text{kg/m}^3)$

Solution: Given data:

Scale		Model	Prototype
$\frac{L_m}{L_P} = \frac{1}{15}$	Medium	Water $(\rho = 1100\,\text{kg/m}^3)$ $F_m = 60N$ $V_m = 3\,\text{m/s}$	Sea water. $(\rho = 1100\,\text{kg/m}^3)$ $L_P = 50$ m

Problems related with ships, do have gravity force predominance and hence the Froude number is to be taken into account and so

$$\left(\frac{V}{\sqrt{gL}}\right)_m = \left(\frac{V}{\sqrt{gL}}\right)_P \Rightarrow \left(\frac{3\sqrt{15}}{\sqrt{L_P}}\right) = \left(\frac{V_P}{\sqrt{L_P}}\right)$$

So **velocity of prototype** is $V_P = 3\sqrt{15} = 11.62\,\text{m/s}$

According to dynamic similarity, Froude number equals and as resistance force =mass × acceleration, and hence equating the forces of model and prototype, we have $(F)_m = (F)_P$

$$\Rightarrow \frac{(F)_m}{(F)_P} = \frac{\rho_{water} * L_m^3 * a_m}{\rho_{Seawater} * L_P^3 * a_P}$$

$$\Rightarrow \frac{(F)_m}{(F)_P} = \frac{\rho_{water} * L_m^4}{\rho_{Seawater} * L_P^4}$$

$$\Rightarrow F_P = 3341.25\,\text{kN}$$

So resisting force on the prototype is 3341.75 kN.

Example 8.19: A 1:20 model of a flying boat is towed through water. The prototype is moving in sea water of specific weight of 1100 kg/m^3 at a velocity of 15 m/sec. Find the corresponding speed of the model. Also determine the resistance due to wave on model, if the resistance due to wave of prototype is 5000 kN.

Solution: Given data:

Scale		Model	Prototype
$\frac{L_m}{L_P} = \frac{1}{20}$	Medium	Water	Sea water $\left(\rho = 1100\,\text{kg/m}^3\right)$
		$\left(\rho = 1000\,\text{kg/m}^3\right)$	$F_P = 5000$ kN, $V_P = 15\,\text{m/s}$

Problems related with ships, do have gravity force predominance and hence the Froude number is to be taken into account and so

$$\left(\frac{V}{\sqrt{gL}}\right)_m = \left(\frac{V}{\sqrt{gL}}\right)_P \Rightarrow \left(\frac{V_m\sqrt{20}}{\sqrt{L_P}}\right) = \left(\frac{15}{\sqrt{L_P}}\right)$$

So **velocity of model** is $V_P = 15/\sqrt{20} = 3.354\,\text{m/s}$

According to dynamic similarity, Froude number equals and as Resistance force =mass x acceleration, and hence equating the forces of model and prototype, we have $(F)_m = (F)_P$

$$\Rightarrow \frac{(F)_m}{(F)_P} = \frac{\rho_{water} * L_m^3 * a_m}{\rho_{Sea\,water} * L_P^3 * a_P}$$

$$\Rightarrow \frac{(F)_m}{(F)_P} = \frac{\rho_{water} * L_m^4}{\rho_{Seawater} * L_P^4} \quad \text{since} \quad \frac{T_m}{T_P} = 1$$

$$\Rightarrow F_m = 28.4 \text{ N}$$

So **resisting force** on the model is 28.4 N.

Example 8.20: A model is built at a 1/15 scale and is to be tested in a wind tunnel operating at a pressure of 15 times the atmosphere pressure. The aeroplane flies at 500 km/h. At what speed should the wind tunnel operate to give dynamic similarity between the model and the prototype? If the drag force measured on the model as 335 N, what will be the drag on the plane? Take the relation of drag (F) as $F = \rho l^2 u^2$.

Solution: Given data:

Scale		Model	Prototype
	Medium	Air	Air
$\frac{L_m}{L_P} = \frac{1}{15}$		$P_m = 15 * 100 = 1500\,Pa$	$V_P = 500 * 1000/3600 = 138.9\,\text{m/s}$
		$F_m = 335\,\text{N}$	

Using Froude number, we can get the velocity of the model, as the scale ratio is given and hence referring Froude number

$$\left(\frac{V}{\sqrt{gL}}\right)_m = \left(\frac{V}{\sqrt{gL}}\right)_P \Rightarrow \left(\frac{V_m\sqrt{15}}{\sqrt{L_P}}\right) = \left(\frac{138.9}{\sqrt{L_P}}\right)$$

So **velocity of model** is $V_m = 138.9/\sqrt{15} = 35.86\,\text{m/s}$

According to dynamic similarity given in the problem, we have drag force $F = \rho l^2 u^2$ and hence equating the forces of model and prototype, we have $(F)_m = (F)_P$

$$\Rightarrow \frac{(F)_m}{(F)_P} = \frac{\rho_{Air} * L_m^2 * V_m^2}{\rho_{Air} * L_P^2 * V_P^2}$$

$$\Rightarrow F_P = 1130.87 \text{ kN}$$

So **drag force** on the prototype is 1130.87 kN.

Euler's Number: Euler number is the ratio of inertia force and pressure force. It indicates the overcome of inertia force over the pressure force and comes into account when the flow has pressure gradients.

$$E_r = \sqrt{\frac{\textit{Inertia Force}}{\textit{Pressure Force}}}$$

$$\Rightarrow E_r = \sqrt{\frac{\rho A V^2}{P * A}}$$

$$E_r = \frac{V}{\sqrt{P/\rho}} \qquad (8.5)$$

Euler's number can be correlated for a model and prototype in the situation of problems where the pressure force effect is more predominant than the other resisting forces and can be used for correlation as below mentioned case, like

a) Flow through pipe problems.
b) Water hammer problems.
c) Sudden closure of pipe valves and cavitation based problems.

Weber Number: Weber number is the ratio of inertia force and surface tension force. It indicates the overcome of inertia force over the surface tension force and comes into account when the capillary movement of the flow of fluid is important, like water in soil and blood in veins.

$$W_r = \sqrt{\frac{Inertia\ Force}{Surface\ Tension\ Force}}$$

$$\Rightarrow E_r = \sqrt{\frac{\rho AV^2}{\sigma * L}} = \sqrt{\frac{\rho L^2 V^2}{\sigma * L}}$$

$$\Rightarrow W_e = \frac{V}{\sqrt{\sigma / \rho L}} \quad (8.6)$$

Weber number can be correlated for a model and prototype in the situation of problems where the surface tension or capillary effects is more predominant than the other resisting forces and can be used for correlation as below mentioned case, like

a) Capillary flow of blood in human body.
b) Capillary effect of ink in a pen.
c) Motion of underground water in soil problems.
d) Fluid flow over a sheet.

Mach number: Mach number is the ratio of inertia force and elastic/compressive force. It indicates the overcome of inertia force over the elastic/compressive force and comes into account when the compressibility factor and the acoustic velocities of the flow of fluid is important, like movement of airplanes, projectiles, bullets, etc.

$$M = \sqrt{\frac{Inertia\ Force}{Elastic / Compressive\ Force}}$$

$$\Rightarrow M = \sqrt{\frac{\rho AV^2}{K * A}}$$

where K is bulk modulus.

$$\Rightarrow M = \frac{V}{\sqrt{K / \rho}} = \frac{V}{a} \quad (8.7)$$

where a is velocity of sound in fluids.

Mach number can be correlated for a model and prototype in the situation of problems where the compressive forces effects is more predominant than the other resisting forces and can be used for correlation as below mentioned case, like

a) High velocity related problems.
b) Problems related with compressible flow.
c) Aerodynamic testing problems.
d) Torpedo related problems.

Note: The type of dimensionless number/group that is to be used in dynamic similarities depends upon the problem given. i.e. E.g.: If the problem has only free surface waves or gravitational effect then we use dimensionless Froude number F_r for correlating a model and prototype

i.e. $$(F_r)_{model} = (F_r)_{Prototype} \Rightarrow \frac{V_m}{\sqrt{L_m g_m}} = \frac{V_p}{\sqrt{Lpg_p}} \Leftrightarrow \frac{V_m}{\sqrt{L_m}} = \frac{V_p}{\sqrt{Lp}}$$

Similarly, if the problem has only viscous effects, then we use dimensionless Reynolds number R_e for correlating a model and prototype.

i.e. $$(R_e)_{model} = (R_e)_{Prototype} \Rightarrow \left(\frac{\rho VL}{\mu}\right)_m = \left(\frac{\rho VL}{\mu}\right)_p \Leftrightarrow \left(\frac{VL}{\nu}\right)_m = \left(\frac{VL}{\nu}\right)_p$$

Similarly, if the problem has only pressure forces or frictional loss effects, then we use dimensionless Euler's number E_r for correlating a model and prototype

i.e. $$(E_r)_{model} = (E_r)_{Prototype} \Rightarrow \frac{V_m}{\sqrt{P_m/\rho_m}} = \frac{V_p}{\sqrt{P_p/\rho_p}}$$

Similarly, if the problem has only surface tension forces then we use dimensionless Weber number W_r for correlating a model and prototype

i.e. $$(W_r)_{model} = (W_r)_{Prototype} \Rightarrow \frac{V_m}{\sqrt{\sigma_m/\rho_m L_m}} = \frac{V_p}{\sqrt{\sigma_p/\rho_p L_m}}$$

Similarly, if the problem has only elastic/compressive forces then we use dimensionless Mach number M for correlating a model and prototype

i.e. $$(M)_{model} = (M)_{Prototype} \Rightarrow \frac{V_m}{\sqrt{K_m/\rho_m}} = \frac{V_p}{\sqrt{K_p/\rho_p}}$$

Example 8.21: What should be the speed and thrust on the model of a sea-plane scaled to 1/15th of the prototype. The prototype has a takeoff speed of 33m/s and a thrust of 4 kN. Assume similar working conditions.

Solution: Given data:

Scale	Model	Prototype
L_m/L_P=1/15	Seawater $u_m = F_m = ?$	Seawater u_P = 33 m/s, F_p = 4000 N

As the sea-plane towed on the surface of the sea, it is subjected to gravitational/wave resistance and hence, we need to use Froude number.

According to Dynamic similarity, Froude numbers are equal and hence,

$$\left(\frac{V}{\sqrt{Lg}}\right)_m = \left(\frac{V}{\sqrt{Lg}}\right)_P$$

So $$u_m = u_P\left(\sqrt{\frac{L_m}{L_p}}\right) = 8.5 \text{ m/s}$$

Similarly Thrust force on the model will be $F_m = F_P\left(\frac{L_m}{L_p}\right)^4 = 0.079 \text{ N}$

Example 8.22: Find the build in pressure (P) in the prototype of an air-compressor, when tested its 1/25 scaled model in water which is 40 times more viscous than air. Assume the build in the model compressor as 3kPa and density of air as 1.2 kg/m³.

Solution: Given data:

Scale	Model	Prototype
$L_m/L_P = 1/25$	Water ρ_m = 1000 kg/m³	Air ρ_p = 1.2 kg/m³
	ρ_m = 3 kPa, $\mu_m = 40\ \mu_p$	P_p = ?

According to Dynamic similarity, Reynolds numbers are equal and hence,

$$\left(\frac{\rho VL}{\mu}\right)_m = \left(\frac{\rho VL}{\mu}\right)_P$$

So $$\frac{u_m}{u_P} = \left(\frac{\mu_m}{\mu_p}\right)\left(\frac{L_p}{L_m}\right)\left(\frac{\rho_p}{\rho_m}\right) = 0.12$$

As there need to measure the pressure, we use Dynamic similarity of Euler's numbers and hence

$$P_p = P_m\left(\frac{u_m}{u_p}\right)^2\left(\frac{\rho_m}{\rho_p}\right) = 250 \text{ Pa}$$

Example 8.23: 1/10th model of an air wing is tested in a wind tunnel to analyze the vibrations caused of an airplane at a frequency of 10 Hz, which appeared when moving at speed f 33 m/s. The pressure in the wind tunnel adjusted to 10 times that of atmosphere and the model is analyzed at frequency of 30 Hz.

Neglect the temperature variations, viscosity. Ensure whether the vibration is due to the aerofoil vorticity or not.

Solution: Given data:

Scale	Model	Prototype
$L_m/L_P = 1/10$	Air $P_m = 10\ P_{atm}$ $f_m = 30$ Hz, $\mu_m = \mu_p$	Air $u_p = 33$ m/s $f_p = 10$ Hz, P_{atm}

As the pressure in the model is 10 times that of the prototype, we have $\left(\rho_p/\rho_m\right) = 10$

According to Dynamic similarity, Reynolds numbers are equal and hence,

$$\left(\frac{\rho VL}{\mu}\right)_m = \left(\frac{\rho VL}{\mu}\right)_P$$

So $$u_m = u_P\left(\frac{\mu_m}{\mu_p}\right)\left(\frac{L_p}{L_m}\right)\left(\frac{\rho_p}{\rho_m}\right) = 33 \text{ m/s}$$

As there need to measure the frequency, we have f = Velocity/Length

So $$f_p = f_m\left(\frac{u_p}{u_m}\right)\left(\frac{L_m}{L_p}\right) = 3 \text{ Hz} < 10 \text{ Hz}$$

As the vibration frequcncy of the prototype is within the limit, the aerofoil vorticity is not the cause.

Example 8.24: Show that wall shear stress τ_w at the wall of a pipe can be given as $\tau_w = \rho U_m^2 * \phi(R_e)$ where U_m is center line max velocity, μ is viscosity of liquid flowing in the pipe, D is the diameter of the pipe and ρ density of the liquid flowing in the pipc.

Solution: Given data: $\tau_w = f(\rho, U_m, D, \mu)$

Step 1: $0 = f(\tau_w, \rho, U_m, D, \mu)$

Step 2: Number of dependent variables is 1 and numbers of independent variables are 4, so total number of variables n = 5 and fundamental variables m = 3.

Step 3: So number of dimensionless π terms are (5 – 3 = 2).

Step 4: Write the function in terms of terms, i.e. $f(\pi_1, \pi_2) = 0$

Step 5:

$$\pi_1 = D^{x1}\ U_m^{y1}\ \rho^{w1}\ \tau_w$$

$$\pi_2 = D^{x2}\ U_m^{y2}\ \rho^{w2}\mu$$

Step 6: Equate the powers of π_1 term, so that

$$M^0L^0T^0 = [L]^{x1}\left[LT^{-1}\right]^{y1}\left[ML^{-3}\right]^{w1}\left[ML^{-1}T^{-2}\right]$$

Equate the powers: M – L – T, we get $(x1 = 0), (y1 = -2), (w1 = -1)$

Therefore, π_1 term becomes $\pi_1 = D^0U_m^{-2}\rho^{-1}\tau_w = \dfrac{\tau_w}{U_m^2\rho}$

Repeat step 6 to get the powers of π_2 term, so that

$$M^0L^0T^0 = [L]^{x2}\left[LT^{-1}\right]^{y2}\left[ML^{-3}\right]^{w2}\left[ML^{-1}T^{-1}\right]$$

Equate the powers: $M - L - T$, we get $(x2 = -1), (y2 = -1), (w2 = -1)$

Therefore π_2 term becomes $\pi_2 = D^1U_m^{-1}\rho^{-1}\mu = \dfrac{\mu}{DU_m\rho\mu}$

Step 7: So the function $f\left(\dfrac{\tau_w}{U_m^2\rho}, \dfrac{\mu}{DU_m\rho\mu}\right) = 0$

Step 8: $\tau_w = U_m^2\rho\phi\left(\dfrac{\mu}{DU_m\rho\mu}\right) = \rho U_m^2 * \phi(R_e)$

EXERCISE

Objective

1. Mach number is significant in
   a) Supersonics, as with projectiles and jet propulsion
   b) Full immersion or completely enclosed flow, as with pipes, aircraft wings, nozzles, etc.
   c) Simultaneous motion through two fluids where there is a surface of discontinuity, gravity forces, and wave making effects as with ship's hulls
   d) All of the above
   e) None of the above
2. Froude number is significant in
   a) Supersonics, as with projectiles and jet propulsion
   b) Full immersion or completely enclosed flow, as with pipes, aircraft wings, nozzles, etc.
   c) Simultaneous motions through fluid where there surface of discontinuity, gravity forces, and wave making effect, as with ship's hulls
   d) All of the above
   e) None of the above
3. Principle of similitude forms the basis of
   a) Comparing two identical equipment
   b) Designing models so that the result can be converted to prototypes
   c) Comparing similarity between design and actual equipment

d) Hydraulic designs
e) Performing acceptance tests

4. For similarity in addition to models being geometrically similar to prototype the following in both cases should also be equal
a) Ratio of inertial force to force due to viscosity
b) Ratio of inertial force to force due to gravitation
c) Ratio of inertial force to force due to surface tension
d) All the four ratios of inertial force to force due to viscosity, gravitation , surface tension, and elasticity
e) None
5. Froude number is the ratio of inertial force to
a) Gravitation force
b) Surface tension
c) Elasticity
d) Viscosity
e) None of the above
6. The nondimensional factor governing viscous or frictional resistance is
a) Reynolds number
b) Weber number
c) Froude number
d) Mach number
e) None of the above
7. Euler's dimensionless number relates
a) Pressure force and inertia force
b) Pressure force or viscous force
c) Inertia force and gravity force
d) Buoyancy force and viscous force
e) Inertia force and viscous force
8. Mach number is the ratio of inertia force to
a) Gravitation force
b) Surface tension
c) Elasticity
d) Viscosity
e) None of the above
9. Reynolds number is the ratio of inertial force to
a) Gravitational force
b) Surface tension
c) Elasticity
d) Viscosity
e) None
10. Weber number is the ratio of inertia force to
a) Gravitational force
b) Surface tension
c) Elasticity

d) Viscosity
e) None

11. The dimensions of bulk modulus is

a) $M^1L^{-1}T^{-2}$ b) $L^{-1}T^{-2}$ c) $M^2L^{-1}T^{-2}$ d) $M^{-1}L^{-1}T^{-2}$ e) None

12. A phenomenon is modeled using "*n*" dimensional variables with "*k*" primary dimensions. The number of non-dimensional variables is

a) k
b) n
c) $n - k$
d) $n + k$

13. Match the following:

P: Compressible flow	U: Reynolds number
Q: Free surface flow	V: Mach number
R: Boundary layer flow	W: Weber number
S: Pipe flow	X: Froude number
T: Surface tension flow	Y: Skin friction coefficient

a) $P - V; Q - X; R - Y; S - W; T - U$
b) $P - V; Q - X; R - Y; S - U; T - W$
c) $P - X; Q - V; R - W; S - U; T - Y$
d) $P - X; Q - V; R - Y; S - U; T - W$

Answers

1.a	2.c	3. b	4. d	5.a	6.a
7. a	8.c	9. d	10. b	11. a	12. c
13. b					

Theory

1. State Buckingham's π-theorem. Why this is considered superior over the Rayleigh's method for dimensional analysis?
2. Deduce the basic dimensions of dynamic viscosity.
3. What do you mean by repeating variables? How are repeating variables selected for dimensional analysis?
4. What are the uses of model similitude?
5. Write fundamental derived formula for viscosity and force.
6. Explain what is meant by *dimensional homogeneity* and describe how it may be used.
7. What is a *non-dimensional* or *dimensionless* number? Some common non-dimensional numbers are given names. State three of these.
8. Describe the term *similarity* and what is required to achieve *geometric similarity*, *kinematic similarity* and *dynamic similarity*.

Problems

1. The efficiency η of a fan depends on the density ρ, the dynamic viscosity μ of the fluid, the angular velocity ω, diameter D of the rotor and discharge "Q". Express η in terms of dimensionless parameters.

2. The resistance force "R" of a supersonic plane during flight can be considered as dependent upon the length of the aircraft "l", velocity "V", air viscosity μ, air density ρ and bulk modulus "K". Express the functional relationship between these variables and the resulting force.
3. Derive expression for the discharge of an oil having viscosity μ, density ρ, through an orifice. The orifice diameter is "D", working under a head "H" and the oil is discharged freely into atmosphere.
4. The pressure drop ΔP in a pipe of diameter "D" and length "l" depends on mass density ρ and viscosity μ of the fluid flowing with a mean velocity "V" and a average height "k" of roughness projections on the pipe surface. Obtain a dimensionless expression for ΔP and show that $h_f = \dfrac{flV^2}{2gD}$, where $h_f = \dfrac{\Delta P}{\gamma}$ and called as head loss due to friction.
5. Obtain the equation of specific speed for turbine as $N_s = \dfrac{N\sqrt{P}}{H^{5/4}}$, if the turbine works under a head of "H", speed "N", power to be developed "P", density of fluid ρ and acceleration due to gravity "g".
6. A geometrically similar model of an air-duct is built to 1/25 scale and tested with water which is 50 times more and 800 times than air. When tested under dynamically similarly conditions, the pressure drop is 2 bar in the model. Find the corresponding pressure drop in the full scale prototype. [102.4 Pa]

QUICK REFERENCE OF FORMULAE

Property	Formulae	Units	Remarks, if any
Reynolds number	$R_e = \dfrac{\rho VL}{\mu} = \dfrac{\rho VD}{\mu}$	Dimensionless	$R_e = \dfrac{\textit{Inertia Force}}{\textit{Viscous Force}}$
Froude number	$F_r = \dfrac{V}{\sqrt{Lg}}$	Dimensionless	$F_r = \sqrt{\dfrac{\textit{Inertia Force}}{\textit{Gravity Force}}}$
Euler number	$E_r = \dfrac{V}{\sqrt{P/\rho}}$	Dimensionless	$E_r = \sqrt{\dfrac{\textit{Inertia Force}}{\textit{Pressure Force}}}$
Weber number	$W_e = \dfrac{V}{\sqrt{\sigma/\rho L}}$	Dimensionless	$F_r = \sqrt{\dfrac{\textit{Inertia Force}}{\textit{Surface Tension Forece}}}$
Mach number	$M = \dfrac{V}{\sqrt{K/\rho}} = \dfrac{V}{a}$	Dimensionless	$M = \sqrt{\dfrac{\textit{Inertia Force}}{\textit{Compressive Force}}}$ a is the velocity of sound.

CHAPTER-9

Compressible Flow

INTRODUCTION

In case of liquids, the changes in pressure and velocity vary the compressibility of the liquid which makes them compressible fluids, but still if the variation range is very less then it can be treated as incompressible and hence liquids are treated as incompressible fluids at low range of pressure and velocity variations. But in case of gases, slight variation in pressure and velocity makes vast difference in compressibility due to which gases are treated as compressible fluids.

Till now the fluids have been assumed as incompressible (ρ = constant) and the effect of temperature has not been taken into account. In this chapter, the fluid flows are considered as having a significant change in density (i.e., fluids are compressible) like in case of gases and also the effect of temperature has been included, for the necessity for analyzing the theory of compressible fluids.

In this chapter, we develop the general relations associated with compressible flows for an ideal gas with constant specific heats.

We start this chapter by introducing the concepts of stagnation state, speed of sound, and Mach number for compressible flows. The relationships between the static and stagnation fluid properties are developed for isentropic flows of ideal gases, and these are expressed as functions of specific heat ratios and the Mach number. The effects of area changes for one-dimensional isentropic, subsonic and supersonic flows are discussed. These effects are illustrated by considering the isentropic flow through converging and converging-diverging nozzles. The concept of shock waves and the variation of flow properties across normal and oblique shock waves are discussed. Finally, we consider the effects of friction and heat transfer on compressible flows and develop relations for property changes.

ONE-DIMENSIONAL BASIC EQUATIONS OF A COMPRESSIBLE FLOW

The basic equations of any fluid flow are based on

1) Continuity equation (mass equation)
2) Bernoulli's energy equation
3) Momentum equation.

General Thermodynamic Processes and Relations related with compressible fluids

Process	Equations	Power Index	Relations
Perfect gases	$P * Vol = mRT$ $P = \rho RT$		$\frac{P_1V_1}{T_1} = \frac{P_2V_2}{T_2}$
	$\frac{dP}{P} = \frac{d\rho}{\rho} + \frac{dT}{T}$ Differential form		$\rho = \frac{Mass}{Volume}$
Polytropic Process	$Pv^n = C$	n	$\frac{T_2}{T_1} = \left(\frac{P_2}{P_1}\right)^{\frac{n-1}{n}} = \left(\frac{V_1}{V_2}\right)^{n-1}$
Adiabatic Process	$Pv^\gamma = C$	$n = \gamma$	$\frac{T_2}{T_1} = \left(\frac{P_2}{P_1}\right)^{\frac{\gamma-1}{\gamma}} = \left(\frac{V_1}{V_2}\right)^{\gamma-1} = \left(\frac{\rho_2}{\rho_1}\right)^{\gamma-1}$
Isothermal	$Pv = C$	$n = 1$	$P_1V = P_2V_2$
Isobaric Process	$P = C$	$n = 0$	$\frac{T_2}{T_1} = \left(\frac{V_2}{V_1}\right)$
Isochoric Process	Volume $= C$	$n = \alpha$	$\frac{T_2}{T_1} = \left(\frac{P_2}{P_1}\right)$
	$C_P - C_V = R$	$\frac{C_P}{C_V} = \gamma$	$\gamma = 1.4$ for Air

Continuity Equation

It has been studied in fluid kinematics that the mass flow rate *Density*(ρ)* *Area*(A)* *Velocity*(V) = *Constant* based on conservation of mass. But if the fluid is incompressible then p = constant and hence AV = constant but when the fluid is compressible $\rho \neq$ constant and so

$$\rho AV = \text{Constant} \tag{9.1}$$

Taking logarithm on both sides and differentiating, we get

$$\Rightarrow \log(\rho) + \log(A) + \log(V) = \log(C)$$

$$\frac{d\rho}{\rho} + \frac{dA}{A} + \frac{dV}{V} = 0 \tag{9.2}$$

Differential form of Continuity equation for a compressible flow

Bernoulli's Equation

Stated by Bernoulli's "in an ideal and incompressible fluid, the total energy of a fluid is constant throughout a streamline, if the flow is steady and continuous".

As the density varies in case of a compressible fluid, the differential form of Bernoulli's equation (Euler's equation) is required, and hence

$$\Rightarrow \frac{dp}{\rho} + VdV + gdZ = 0 \tag{9.3}$$

Integrating both sides, we have

$$\frac{1}{g}\int \frac{dp}{\rho} + \frac{V^2}{2g} + Z = C \tag{9.4}$$

General equation of Bernoulli for a compressible flow

The compressible fluid while flowing may have different thermal processes, as mentioned in below cases:

Case 1: Bernoulli's equation with adiabatic process:

For **adiabatic process** $Pv^{\gamma} = Constant$

$$\frac{P}{\rho^{\gamma}} = Constant \tag{9.5}$$

$$\Rightarrow \int \frac{dp}{\rho} = \int \frac{dp}{\left(P/C\right)^{1/\gamma}} = \frac{\gamma}{\gamma - 1}\left(\frac{P}{\rho}\right) \tag{9.5a}$$

Substituting Eq. (9.5a) in Eq. (9.4), we get

$$\frac{\gamma}{\gamma - 1}\left(\frac{P}{\rho g}\right) + \frac{V^2}{2g} + Z = C \tag{9.6}$$

or

$$\left(\frac{\gamma}{\gamma - 1}\right) RT + \frac{V^2}{2} + gZ = C \tag{9.6a}$$

Equations (9.6) and (9.6a) are the **equations for Bernoulli's equation with adiabatic process.**

If the process is polytropic, then

$$\frac{n}{n - 1}\left(\frac{P}{\rho g}\right) + \frac{V^2}{2g} + Z = C \tag{9.6b}$$

Case 2: Bernoulli's equation with isothermal process:
For **Isothermal process** $Pv = Constant$

$$\frac{P}{\rho} = Constant \tag{9.7}$$

$$\int \frac{dp}{\rho} = \int \frac{dp}{\left(P/C\right)} = \left(\frac{P}{\rho}\right)\log(P) \tag{9.8}$$

Substituting Eq. (9.8) in Eq. (9.4), we get

$$\left(\frac{P}{\rho g}\right)\log(P) + \frac{V^2}{2g} + Z = C \tag{9.9}$$

Equation. (9.9) is the **equation for Bernoulli's equation with isothermal process.**

Case 3: Bernoulli's equation with Isobaric process:
For **Isobaric process**

$$P = Constant \tag{9.10}$$

$$\int \frac{dp}{\rho} = 0 \tag{9.11}$$

Substituting Eq. (9.11) in Eq. (9.4)

$$\frac{V^2}{2g} + Z = C \tag{9.12}$$

Equation (9.12) is the **equation for Bernoulli's equation with Isobaric process.**

Example 9.1: A gas with velocity of 300 m/s is flowing through a horizontal pipe at a section where pressure is 78 kN/m^2 (*abs*) and temperature 40°C. The pipe changes in diameter and at this section the pressure is 117 kN/m^2 (*abs*). Find the velocity of the gas at this section if the flow is adiabatic. Take $R = 287$ J/kg $- K$ & $\gamma = 1.4$

Solution:
Given data:

$$P_1 = 78 \text{ kPa}(abs), T_1 = 313\text{K}, V_1 = 300 \text{ m/s}, P_2 = 117 \text{ kPa}(abs), R = 287 \text{ J/kg} - \text{K}, \gamma = 1.4$$

As per Bernoulli's equation for adiabatic process

$$\left(\frac{\gamma}{\gamma-1}\right)RT_1 + \frac{V_1^2}{2} + gZ_1 = \left(\frac{\gamma}{\gamma-1}\right)RT_2 + \frac{V_2^2}{2} + gZ_2$$

$$\Rightarrow \left(\frac{\gamma}{\gamma-1}\right)RT_1 + \frac{V_1^2}{2} = \left(\frac{\gamma}{\gamma-1}\right)RT_2 + \frac{V_2^2}{2}$$

We know for adiabatic process $\dfrac{T_2}{T_1} = \left(\dfrac{P_2}{P_1}\right)^{\frac{\gamma-1}{\gamma}}$

$$\Rightarrow T_2 = T_1\left(\frac{P_2}{P_1}\right)^{\frac{\gamma-1}{\gamma}} = 313 * \left(\frac{117000}{78000}\right)^{\frac{1.4-1}{1.4}} = 351\text{K}$$

Substituting the values in Eq. (9.6a), we get

$$\Rightarrow V_2 = 115\,\text{m/s}$$

Hence, the velocity of gas at the section is 115 m/s.

Example 9.2: A gas flowing through a horizontal pipe at a section where pressure is 6×10^4 N/m^2 (*abs*) and temperature 40°C. The pipe changes in diameter and at this section the pressure is 9×10^4 N/m^2 (*abs*). Find the velocity of the gas at the inlet if the velocity at that section is half the velocity at the exit. Assume the flow to be isothermal. Take $R = 287$ J/kg – K & $\gamma = 1.4$

Solution: Given data:

$$P_1 = 6\times10^4\,\text{Pa}(abs),\ T_1 = T_2 = 313\text{ K},\ 2V_1 = V_2$$
$$P_2 = 9\times10^4\,\text{Pa}(abs),\ R = 287\,\text{J/kg} - \text{K}, \gamma = 1.4$$

We know, as per ideal gas equation, we have

$$\frac{P_1}{\rho_1} = RT_1 = 287 * 313 = 89.831$$

$$\Rightarrow \rho_1 = \frac{60000}{89.831} = 668\text{ kg/m}^3$$

Similarly, at other section, we have $\dfrac{P_2}{\rho_2} = RT_1 = 287 * 313 = 89.831$

$$\Rightarrow \rho_2 = \frac{90000}{89.831} = 1001.9\text{ kg/m}^3$$

As per Bernoulli's equation for isothermal process

$$\left(\frac{P_1}{\rho_1 g}\right)\log(P_1)+\frac{V_1^2}{2}+gZ_1=\left(\frac{P_2}{\rho_2 g}\right)\log(P_2)+\frac{V_2^2}{2}+gZ_2$$

$$\left(\frac{P_1}{\rho_1 g}\right)\log(P_1)+\frac{V_2^2}{4*2}=\left(\frac{P_2}{\rho_2 g}\right)\log(P_2)+\frac{V_2^?}{4*2}$$

Substituting the values, we get

$$\Rightarrow V_2 = 230.9 \text{ m/s}$$

Hence, the velocity of gas at the section is 200 m/s.

Momentum Equation

Consider a steady compressible fluid flow, through a non-uniform circular duct. As the duct has gradual increase in cross-section, there arise increase in pressure from one section to the other in the circular duct and also as the cross-section is not uniform, there arise a normal pressure force along the surface of the duct, as shown in Figure 9.0 and hence the pressure force along the length will be sum of the pressure forces along the length and the pressure force along the normal surface of the duct.

As per momentum equation $$F=\frac{d}{dt}(mV)$$

$$\Rightarrow F=(\bar{m}dV)=(\rho AV)dV \tag{9.13}$$

Net pressure force along the length and normal surface will be

$$F_{xL}=PA-[(P+dP)A]$$

$$\Rightarrow F_{xL}=-dPA \tag{9.14a}$$

$$F_x=F_{xL}+F_{xN}=PA-\{[(P+dP)(A+dA)]+F_{xN}\}$$

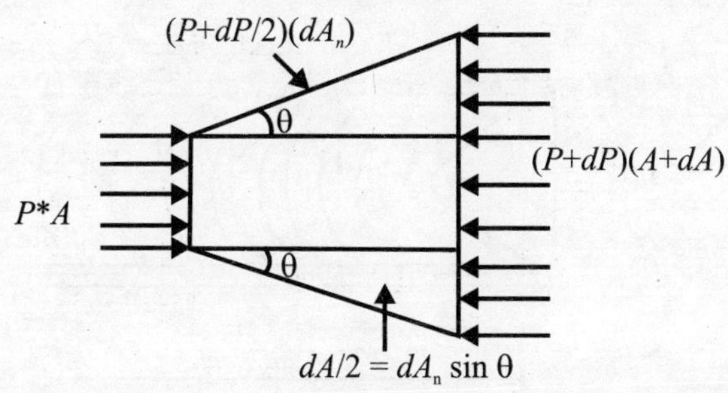

Figure 9.0: Non-uniform circular duct.

$$\Rightarrow F_x = PA - \left\{\left[(P+dP)(A+dA)\right] + \left(P+\frac{dP}{2}\right)\frac{dA}{2}\right\}$$

$$\Rightarrow F_x = -dPA \tag{9.14b}$$

Equating Eq. (9.1) and Eq. (9.14b), we have

$$-dPA = (\rho AV)\,dV$$

$$\Rightarrow \frac{dP}{\rho} + (V)\,dV = 0 \tag{9.15a}$$

$$\Rightarrow \int\frac{dP}{\rho} + \frac{V^2}{2} = C \tag{9.15b}$$

If gravitational effect is considered

$$\frac{dP}{\rho} + (V)\,dV + gdZ = 0 \tag{9.16a}$$

$$\Rightarrow \int\frac{dP}{\rho} + \frac{V^2}{2} + gZ = C \tag{9.16b}$$

Case 1: Momentum equation with adiabatic process:

For adiabatic process $Pv^\gamma = Constant$ i.e. $P = \rho^\gamma C$

Differentiating, we get $dP = C\gamma\rho^{\gamma-1}d\rho$

Then $$\int\frac{dp}{\rho} = \int\frac{dp}{\left(P/C\right)^\gamma} = \frac{\gamma}{\gamma-1}\left(\frac{P}{\rho}\right) \tag{9.17}$$

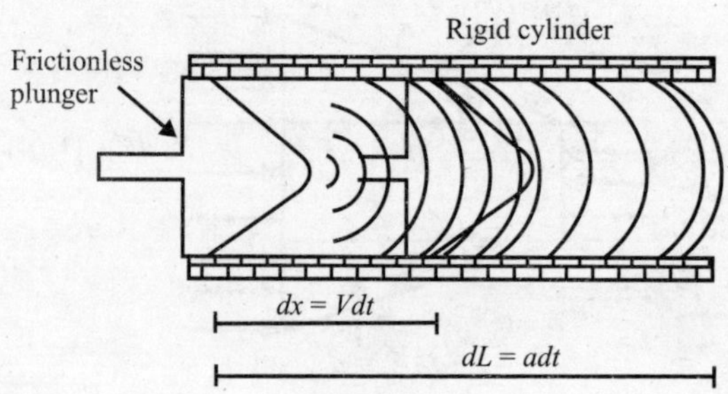

Figure 9.1

$$\frac{\gamma}{\gamma-1}\left(\frac{P}{\rho g}\right)+\frac{V^2}{2g}+Z=C \tag{9.18}$$

$$\left(\frac{\gamma}{\gamma-1}\right)RT+\frac{V^2}{2}+gZ=C \tag{9.18a}$$

Eq. (9.18) and Eq. (9.18a) are the equation for momentum equation with adiabatic process.

VELOCITY OF SOUND

Consider a rigid cylinder or duct of uniform cross-section area (A) and length dL filled with a fluid having pressure "P" and have a frictionless plunger in it, at rest. So the initial mass of the fluid in the pipe is

$$m_1 = \rho * volume = \rho * AdL \tag{9.19a}$$

When the plunger is given a sudden impulse with a velocity V in a time dt in a time $dx(dx = Vdt)$, then the fluid pressure existing in the cylinder gets disturbed and rises with some amount (dP) and the volume of the fluid reduces, which in turn reduces the mass to

$$m_2 = (\rho + d\rho)A(dL - dx) \tag{9.19b}$$

This disturbance of pressure, propagates as a wave along the length dL and is termed **Pressure Wave** and it travels with a velocity of sound "a" and hence $dL = adt$.

As per continuity equation $m_2 = m_1$

$$\Rightarrow (\rho + d\rho)A(dL - dx) = \rho AdL$$

$$\Rightarrow (\rho + d\rho)A(a - V) = \rho Aa$$

$$[\text{since } dx = Vdt \ \& \ dL = adt]$$

$$\Rightarrow \rho a + ad\rho - Vd\rho - \rho V = \rho a$$

But $V <<< a$ & also $d\rho$ is very small and hence we can neglect $Vd\rho \approx 0$

$$V = \frac{ad\rho}{\rho} \tag{9.20}$$

Applying impulse momentum equation for the compressed fluid, we have

$$F = \frac{d}{dt}(m\Delta V) \tag{9.21}$$

and $$m = (\rho * A * dL) \; \& \; \Delta V = (V - 0)$$

$$\Rightarrow [(P + dP) - P]A = m\frac{d\Delta V}{dt} = \left(\rho * A\frac{dL}{dt}\right)(V - 0)$$

But $$a = \frac{dL}{dt} \Rightarrow dP * A = \rho * A * a * (V - 0)$$

$$\Rightarrow a = \frac{dP}{\rho V} \tag{9.22}$$

Substituting Eq. (9.20) in Eq. (9.22), we get sonic velocity "*a*" as

$$a = \sqrt{\frac{dP}{d\rho}} \tag{9.23}$$

Equation (9.23) is the **general equation of velocity of sound.**

Case 1: Sonic velocity of sound in terms of Bulk modulus:

Bulk modulus $$K = \frac{-dP}{\left(\text{change in Volume}/\text{original Volume}\right)}$$

$$\Rightarrow -dVol/Vol = \frac{dP}{K} \tag{9.24}$$

As the volume decreases for a compressible fluid, the density increases and hence

$$Volume \;\alpha\; \frac{1}{\rho}$$

$$\Rightarrow Vol * \rho = constant \tag{9.24a}$$

Differentiating Eq. (9.24a) on both sides, we get $Vd\rho + \rho dV = 0$

$$\Rightarrow \frac{dP}{\rho} = \frac{-dV}{V} \tag{9.24b}$$

From Eq. (9.24b) in Eq. (9.24), we have

$$\Rightarrow \frac{dP}{d\rho} = \frac{K}{\rho} \tag{9.24c}$$

Substituting Eq. (9.24c) in Eq. (9.23), we get **Sonic Velocity** as

$$a = \sqrt{\frac{K}{\rho}} \tag{9.25}$$

Case 2: Sonic velocity of sound in terms of adiabatic process:

$$PV^{\gamma} = constant \quad \Rightarrow \frac{P}{\rho^{\gamma}} = constant$$

$$\Rightarrow P\rho^{-\gamma} = constant \tag{9.26}$$

Differentiating Eq. (9.26) and dividing by $\rho^{-\gamma}$, we get $\rho^{-\gamma}dP - P\gamma\left(\rho^{-\gamma-1}d\rho\right) = 0$

$$\Rightarrow \frac{dP}{d\rho} = \frac{\gamma P}{\rho} = \gamma RT \tag{9.26a}$$

Substituting Eq. (9.26a) in Eq. (9.23),we get **sonic velocity** as

$$a = \sqrt{\gamma\left(\frac{P}{\rho}\right)} = \sqrt{\gamma RT} \tag{9.27}$$

Case 3: Sonic velocity of sound in terms of Isothermal process PV = Constant i.e., $\gamma = 1$

Then **sonic velocity** becomes $$a = \sqrt{\left(\frac{P}{\rho}\right)} = \sqrt{RT} \tag{9.28}$$

Example 9.3: Find the sonic velocity for the following fluids:

a) Crude oil of specific gravity 0.8 and bulk modulus 1.5 GPa.
b) Mercury having a bulk modulus of 27 GPa.

Solution:

(a) Given data: $S = 0.8$, $K = 1.5\times10^{9}$ Pa

Density of crude oil is $\rho_{oil} = S * \rho_{water} = 0.8 * 1000 = 800$ kg/m^3

Sonic velocity $a = \sqrt{\dfrac{1.5\times10^{9}}{800}} = 1369$ m/s

(b) Given data: $S = 13.6$, $K = 27 \times 10^9$ Pa

Density of mercury is $\rho_{Mer} = S * \rho_{water} = 13.6 * 1000 = 13600$ kg/m^3

Sonic velocity $a = \sqrt{\dfrac{27 \times 10^9}{13600}} = 1409$ m/s

ELASTIC WAVE PROPAGATIONS IN A COMPRESSIBLE FLOW

When an object/projectile moves with a velocity "V" in a compressible fluid which is at rest, then the disturbance created by the object in the compressible fluid generates as a pressure wave, which propagates in the form of concentric circles, having the object as the centre and with a speed of sound "a", in the outward direction.

The concentric wave propagates as a function of change in time "dt" and velocity of sound "a", as shown in Figure 9.2.

E.g.: When an object is dropped in a bucket full of water gradually with a velocity, then the disturbance caused due to immersion of the object in the water is generated as a wave in form of concentric circles (assume four concentric circles are generated).

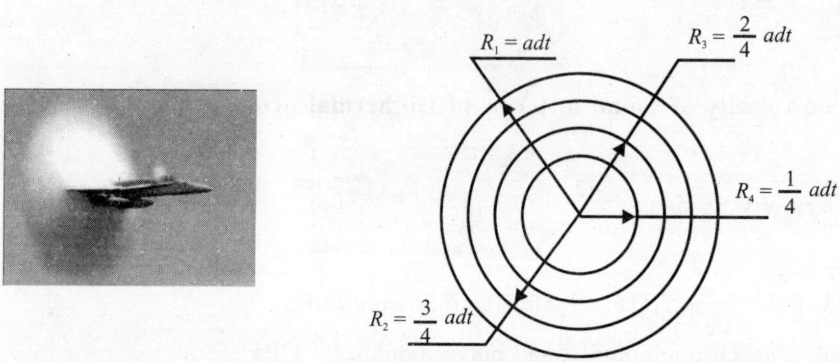

Figure 9.2: Pressure wave propagation in form of concentric circles

As the object/projectile moves with different types of sonic speeds like Sub-Sonic, Sonic, Super-Sonic etc then the pressure wave distribution circles vary with respect to the type of sonic velocity, with which the object is moving.

Case 1: When velocity "V" is subsonic velocity (i.e. $M < 1$ or $V < a$)
In this case, the velocity of object is less than the sonic velocity and the wave advances from point to point as a function of time and sonic velocity. The object cannot cross the wave propagation circles, wherever it advances and it will be enclosed in the pressure wave circles only, as shown in Figure 9.3.

Let the object propagate waves at point "C" in form of circles with radius $R_1 = adt$, $R_2 = \frac{3}{4}adt$, $R_2 = \frac{2}{4}adt$, $R_2 = \frac{1}{4}adt$, etc. These circles attain their maximum radius in some change in time periods "dt" and by this time change, the object advances to new positions $C_1\left(CC_1 = dx_1 = Vdt\right)$, C_2, C_1 and so on. But the maximum circle radius R_1 is attained before object comes to new position C_1 since $V < a$ and $R_1 > dx$, so the object displacement will not cross the first circle. Similarly the maximum circle radius R_2 is attained before object comes to new position C_2 since $V < a$ and $R_2 > dx_2$, so the object displacement will not cross the second circle. And this process repeats for number of circles. Hence, while the object advances from point "C" to newer positions C_1, C_2, C_3, etc. with $V < a$ its displacement position will be placed in the wave circles only, as shown in Figure 9.3.

E.g. A bullet from a gun fired travelling with subsonic velocity can be seen, passing inside a distance of the pressure wave propagation circle, from an observer, by the time he hears the sound of its motion, as shown in the Figure 9.3a.

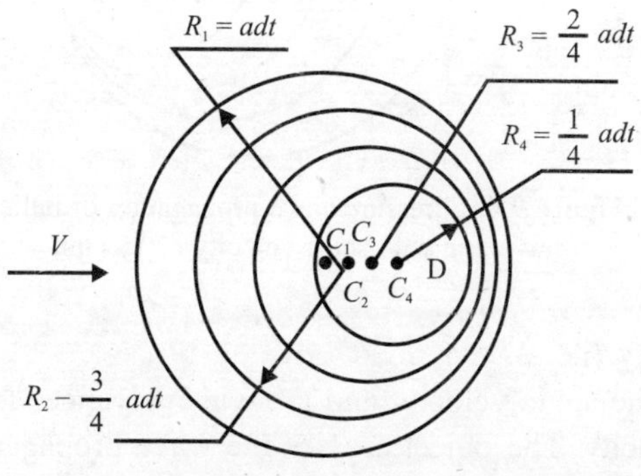

Figure 9.3: Pressure wave propagation of subsonic velocity

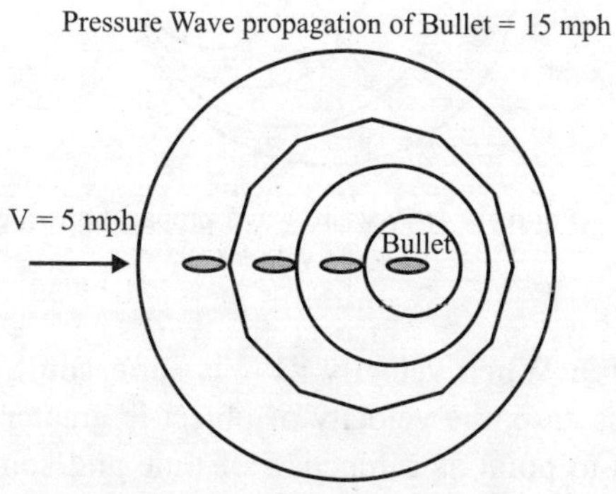

Figure 9.3a: Pressure wave propagation of bullet travelling with sub-sonic velocity of 5 mph

Case 2: When velocity "V" is sonic velocity (i.e. $M = 1$ or $V = a$) In this case the velocity of object is equal to the sonic velocity and the wave advances from point to point as a function of time and sonic velocity. The object remains on the circumferences of the wave propagation circles, wherever it advances and it will be on the common circumference point of the pressure wave circles only, as shown in Figure 9.4.Let the object propagate waves at point "C" in form of circles with radius $R_1 = adt$, $R_2 = \frac{3}{4}adt$, $R_2 = \frac{2}{4}adt$, $R_2 = \frac{1}{4}adt$, etc. These circles attain their maximum radius in some change in time periods "dt" and by this time change, the object advances to new positions $C_1\left(CC_1 = dx_1 = Vdt = adt\right)$, C_2, C_3, and so on. But the maximum circle radius R_1 is attained at the same instant of object, when comes to new position C_1 since $V = a$ and $R_1 = dx_1$, so the object displacement will lie on the circumference of the first circle. Similarly, the maximum circle radius

R_2 is attained at the same instant of object, when comes to new position C_2 since $V = a$ and $R_2 = dx_2$, so the object displacement will lie on the circumference of the second circle. And this process repeats for number of circles. Hence, while the object advances from point "*C*" to newer positions C_1, C_2, C_3 etc. with $V = a$ its displacement position will be placed on the common point of all the circumferences of the wave circles, as shown in Figure 9.4.

E.g. A bullet from a gun fired travelling with sonic velocity can be seen, passed a distance of the pressure wave propagation circle, from an observer, by the time he hears the sound of its motion, as shown in the Figure 9.4a.

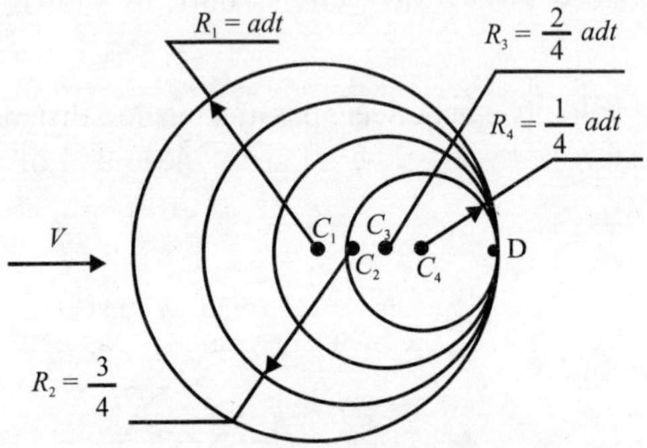

Figure 9.4: Pressure wave propagation of sonic velocity

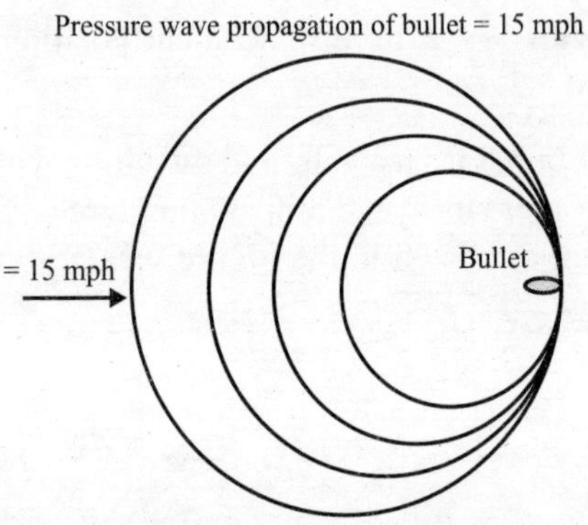

Figure 9.4a: Pressure wave propagation of bullet travelling with sonic velocity of 15 mph

Case 3: When velocity "*V*" is supersonic velocity (i.e. $M > 1$ or $V > a$)
In this case, the velocity of object is greater than the sonic velocity and the wave advances from point to point as a function of time and sonic velocity. The object crosses the wave propagation circles, wherever it advances and it will not be enclosed in the pressure wave circles, as shown in Figure 9.5.

Let the object propagate waves at point "*C*" in form of circles with radius $R_1 = adt$, $R_2 = \frac{3}{4}adt$, $R_2 = \frac{2}{4}adt$, $R_2 = \frac{1}{4}adt$, etc. These circles attain their maximum radius in some change in time periods "*dt*" and by this time change, the object advances to new positions $C_1\left(CC_1 = dx_1 = Vdt\right)$, C_2, C_3, and so on. But the maximum circle radius R_1 is attained after object comes to new position C_1 since $V > a$ and $R_1 < dx_1$, so the object displacement will cross the first circle. Similarly, the maximum circle radius R_2 is attained after the object comes to new position C_2 since $V > a$ and $R_2 < dx_2$, so the object displacement will cross the second circle. And this process repeats for number of circles. Hence, while the object advances from point "*C*" to newer positions C_1, C_2, C_3, etc. with $V > a$ its displacement position will be placed outside the wave circles, as shown in Figure 9.5.

If two common tangents to all these pressure wave circles are drawn, they meet at a point and which makes a cone-like structure. This cone structure is called **Mach Cone** and the angle made by the displacement line with the common tangent is called **Mach Angle** α where

$$\sin\alpha = \frac{adt}{Vdt} = \frac{a}{V} = \frac{1}{M} \tag{9.28a}$$

E.g. A bullet from a gun fired travelling with supersonic velocity can be seen, passed over a long distance from an observer, by the time he hears the sound of its motion, as shown in the Figure 9.5a.

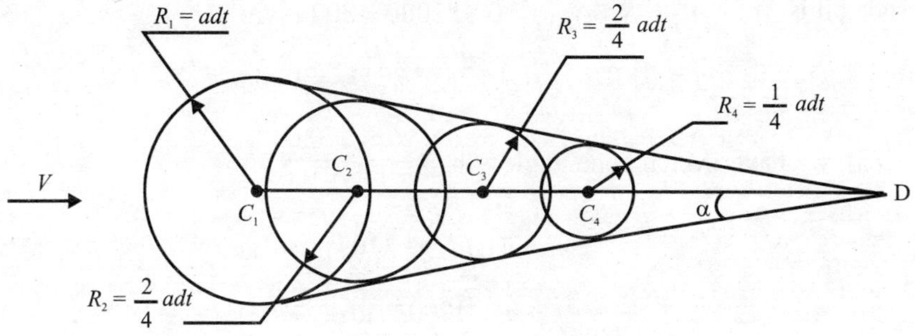

Figure 9.5: Pressure wave propagation of supersonic velocity

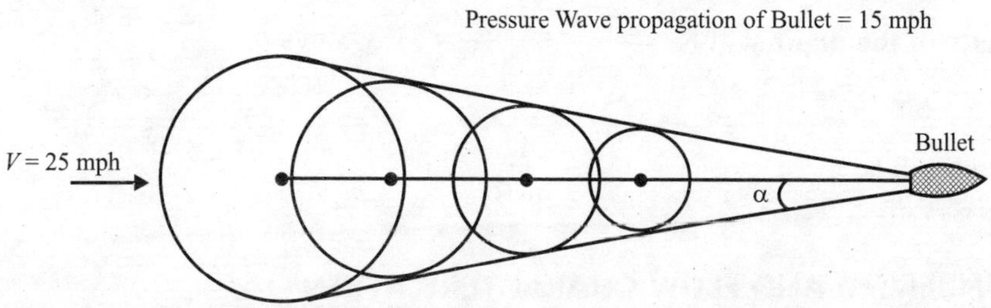

Figure 9.5a: Pressure wave propagation of bullet travelling with supersonic velocity of 25 mph

Example 9.4: A projectile is travelling in air having pressure and temperature as 88.3 kPa and -2°C. If the mach angle is 40°, find the velocity of the projectile. Take $R = 287$ J/kg – K, $\gamma = 1.4$.

Solution: Given data: $\alpha = 40°, T = 271$ K, $P = 88.3$ kPa, $R = 287$ J/kg – K, $\gamma = 1.4$

Density of crude oil is $\rho_{oil} = S * \rho_{water} = 0.8 * 1000 = 800$ kg/m^3

Sonic velocity $a = \sqrt{\gamma RT} = \sqrt{1.4 * 287 * 271} = 330$ m/s

Mach cone angle $\sin\alpha = \frac{a}{V} = \frac{330}{V}$

$$\Rightarrow V = \frac{330}{\sin 40°} = 513.4 \text{ m/s}$$

Example 9.5: An aeroplane is flying at 1000 m above in a still air having a pressure of 98.1 kPa (absolute) and temperature of 271 K. The sound of the airplane is heard by an observer after 3 seconds, while it passed over his head. Determine the velocity and Mach number of the aeroplane.

Solution: Given data: $t = 3$ sec, $T = 271$ K, $P = 98.1$ kPa, $R = 287$ J/kg – K, $\gamma = 1.4, h = 1000$ m

Density of crude oil is $\rho_{oil} = S * \rho_{water} = 0.8 * 1000 = 800$ kg/m³

Sonic velocity $a = \sqrt{\gamma RT} = \sqrt{1.4 * 287 * 271} = 330$ m/s

Refer Eq. (9.28a), we have Mach cone angle $\sin\alpha = \dfrac{a}{V} = \dfrac{330}{V}$

$$\Rightarrow a = V\sin\alpha = 330 \qquad (1)$$

But
$$\tan\alpha = \frac{h}{Vt} = \frac{1000 * \sin\alpha}{3 * 330}$$
$$\Rightarrow \alpha = 8.1°$$

So the **velocity of the plane** will be $\Rightarrow V = \dfrac{330}{\sin 8.1°} = 2339.3$ m/s

Mach number will be $M = \dfrac{1}{\sin\alpha} = 7.09$

MACH NUMBER (M) AND FLOW CHARACTERIZATION

Mach number is the ratio of inertia force over an elastic force in a compressible fluid flow. i.e.

$$M = \sqrt{\frac{inertia\ \ force}{elastic\ \ force}} = \sqrt{\frac{\rho AV^2}{KA}} = \frac{V}{\sqrt{K/\rho}} = \frac{V}{a}$$

where "a" is the velocity of sound and hence, velocity of sound of a compressible fluid is dependent on the Mach number and is useful to find the type of sonic speeds of the compressible fluid.

If $M < 1$ then $V < a$ this shows that the velocity of object is less than the velocity of sound and is called **Subsonic Velocity** and the disturbance created by the object in the compressible fluid travel upstream along the aerofoil shape of the object, as shown in Figure 9.6a.

If $M = 1$ then $V = a$ this shows that the velocity of object is equal to the velocity of sound and is called **Sonic Velocity,** where the normal shock waves emerges at the leading edges, as shown in Figure 9.6b.

If $M > 1$ then $V > a$ this shows that the velocity of object is greater than the velocity of sound and is called **Supersonic Velocity**. Like, where in the velocity of object moving in a compressible

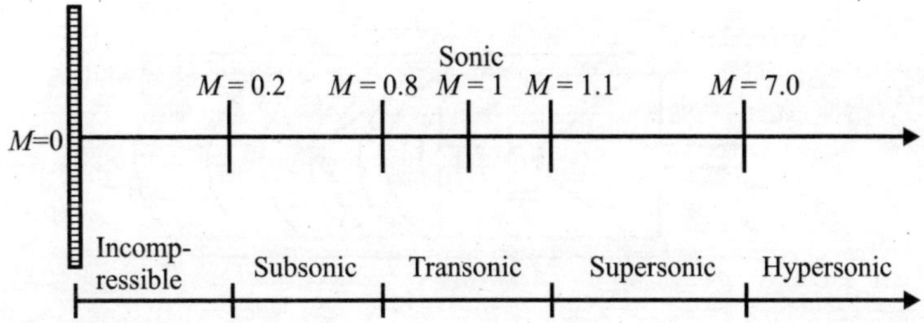

Figure 9.6: Relation graph between different types of speeds and Mach number

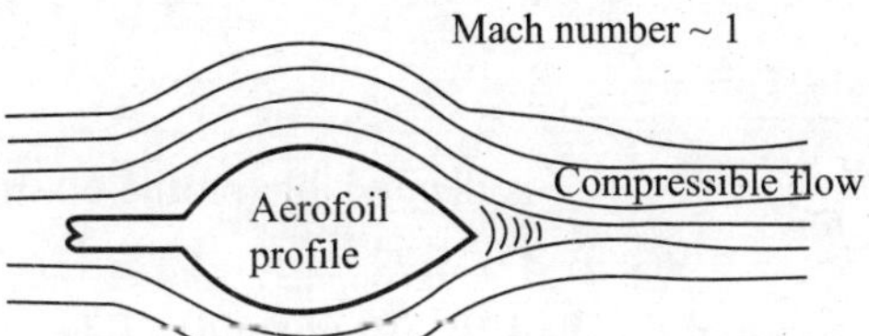

Figure 9.6a: Pressure waves passing over head of aerofoil for compressible flow at Mach < 1

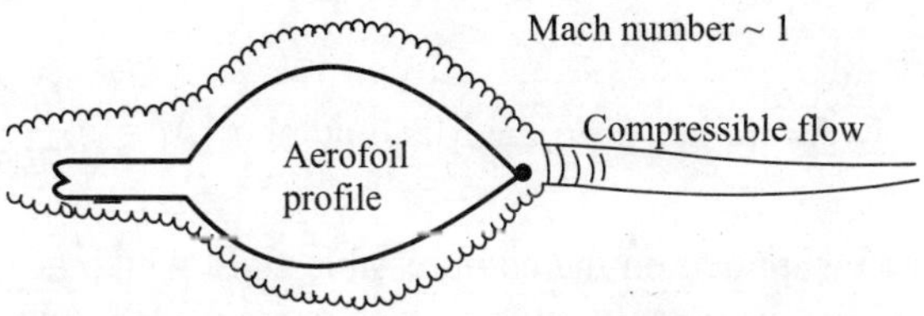

Figure 9.6b: Shock waves emerging at leading edge for compressible flow at Mach ~1

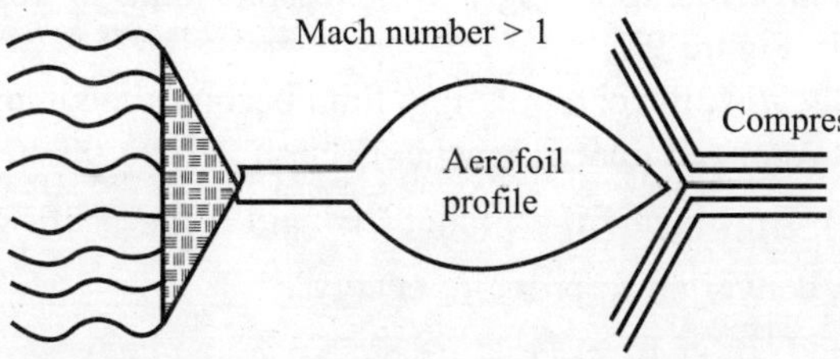

Figure 9.6c: Oblique shock waves emerging at leading edges for compressible flow at Mach > 1

fluid are predominant, as in cases of high speed moving objects like airplanes, projectiles, missiles, bullets, jet-planes, satellite launching etc.

The disturbance created by the object moving with sonic or supersonic velocities in the compressible fluid creates oblique shock waves, as shown in Figure 9.6c.

LIMITS IN INCOMPRESSIBILITY

Any compressible fluid can be compressed up to some amount of its volume reduction, which gives change in the density $d\rho$ for compressible fluids. But if the compressible fluids are still further compressed, they (compressible fluids) start opposing the change in its density and become incompressible.

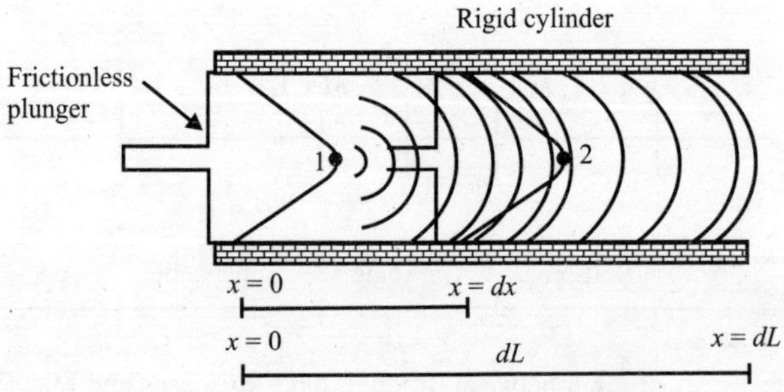

Figure 9.7: Limits in incompressibility

As we had seen the Mach number $M = \sqrt{\dfrac{Inertia\ force}{Elastic\ force}} = \dfrac{V}{\sqrt{K/\rho}}$ is dependable on the elastic force applied on the compressible fluid which acts in changing the density $d\rho$ of fluid and hence the elastic force applied on any compressible fluid can be applied only up to some limit, thereby the fluid becomes stagnated and incompressible. And hence the Mach number which signifies the compressibility forces has a definite value beyond which it has no significance.

Consider a frictionless plunger compressing a compressible fluid up to a length "dL" from section 1-1 to 2-2, as shown in Figure 9.7.

Assume, beyond the length "dL", the compressible fluid becomes incompressible from section 2-2 and hence the plunger gets stopped once it reaches section 2-2, so the velocity at this section will be zero (i.e. $V_2 = V_S \approx 0$). And hence the plunger is said to be at the "**stagnation point**", where-in the kinetic energy is converted to pressure energy.

Apply Bernoulli's equation between section 1-1 and 2-2, we have

$$\frac{dP}{\rho} + VdV = 0 \tag{9.29}$$

If integrated from initial stage to stagnation stage, then we get

$$\Rightarrow \int_{P_1}^{P_S} \frac{dP}{\rho} = -\int_{V_1}^{V_S} VdV \Rightarrow \frac{(P_S - P_1)}{\rho} = \frac{dP}{\rho} = \frac{1}{2}V^2 \tag{9.29a}$$

But $a^2 d\rho = dP$ and substituting it in Eq. (9.29a), we get

$$\frac{a^2 d\rho}{\rho} = \frac{1}{2}V^2 \Rightarrow \frac{d\rho}{\rho} = \frac{1}{2}\frac{V^2}{a^2} \Rightarrow \frac{d\rho}{\rho} = \frac{1}{2}M^2 \quad \left(\text{since } M = \frac{V}{a}\right)$$

By taking the density change $\left({d\rho}/{\rho}\right)$ of a compressible fluid not to exceed 2%, we get

$$\frac{1}{2}M^2 \le 0.02 \Rightarrow M \le 0.2$$

And hence the compressible fluids remain incompressible till Mach number is equal or less than 0.2.

And so from Eq. (9.29a), we can have

$$P_S = P_1 + \frac{1}{2}\rho V^2 \quad (9.30)$$

And if two points are taken not on the stream line having stagnation point, then

$$P_2 = P_1 + \frac{1}{2}\rho\left(V_1^2 - V_2^2\right) \Rightarrow \frac{P_2 - P_1}{\frac{1}{2}\rho V_1^2} = 1 - \frac{A_1^2}{A_2^2} \quad \left(\text{since } A_1V_1 = A_2V_2\right) \quad (9.30a)$$

Once the Mach number exceeds beyond (i.e. $M > 0.2$), the fluid becomes compressible. And if compressibility is considered, then

$$P_S = P_1\left[1 + \left(\frac{\gamma - 1}{2}\right)M_1^2\right]^{\gamma/\gamma-1} \quad (9.30b)$$

STAGNATION POINT AND ITS RELATIONSHIP WITH MACH NUMBER AND OTHER FLOW VARIABLES

When a frictionless compressible fluid flows over a bluff body, there arises a point on the body where the velocity of the compressible fluid flow is zero (i.e. $V_S \approx 0$) and this velocity is converted to pressure energy. This point is called "**stagnation point**".

Consider a compressible fluid flow having isentropic flow PV^γ = constant, pass over a body as shown in Figure 9.8 and let P_1, γ_1, V_1, Z_1 & P_S, γ_S, V_S, Z_S be pressure, specific weight of compressible fluid, velocity and datum height of stream line at point 1 and stagnation point "S" respectively.

As per Bernoulli's equation in between points "*S*" and "1", we have

$$\frac{p_1}{\gamma_1} + \frac{1}{2g}V_1^2 + Z_1 = \frac{p_S}{\gamma_S} + \frac{1}{2g}V_S^{\,2} + Z_S$$

Applying the integration

$$\Rightarrow \int\frac{p_1}{\rho_1 g} + \frac{1}{2g}V_1^2 + Z_1 = \int\frac{p_S}{\rho_S g} + \frac{1}{2g}V_S^{\,2} + Z_S$$

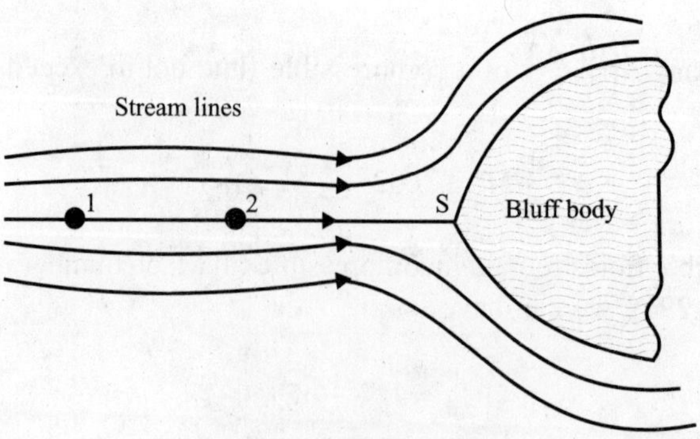

Figure 9.8: Stagnation properties

$$\Rightarrow \int \frac{p_1}{\rho_1 g} + \frac{1}{2g} V_1^2 = \int \frac{p_S}{\rho_S g} \tag{9.31}$$

since $Z_1 = Z_S$ & $V_S = 0$

As the compressible flow is isentropic.
(i.e. P/ρ^γ = constant)

we have
$$\int \frac{dp}{\rho} = \int \frac{dp}{\left(P/C\right)^\gamma} = \frac{\gamma}{\gamma-1}\left(\frac{P}{\rho}\right) \tag{9.31a}$$

Substituting Eq. (9.31a) in Eq. (9.31), we have

$$\frac{\gamma}{\gamma-1}\left(\frac{P_1}{\rho_1 g}\right) + \frac{1}{2g}V_1^2 = \frac{\gamma}{\gamma-1}\left(\frac{P_S}{\rho_S g}\right)$$

By rearranging
$$\frac{\gamma}{\gamma-1}\left(\frac{P_1}{\rho_1}\right)\left[1-\left(\frac{P_S}{P_1}\right)\left(\frac{\rho_1}{\rho_S}\right)\right] = -\frac{V_1^2}{2} \tag{9.31b}$$

But $\left(\frac{\rho_1}{\rho_S}\right) = \left(\frac{P_1}{P_S}\right)^{1/\gamma}$ from P/ρ^γ = constant and substituting this in Eq. (9.31b), we have

$$\frac{\gamma}{\gamma-1}\left(\frac{P_1}{\rho_1}\right)\left[1-\left(\frac{P_S}{P_1}\right)^{\gamma-1/\gamma}\right] = -\frac{V_1^2}{2}$$

By rearranging, we have
$$P_S = P_1\left[1+\left(\frac{\gamma-1}{\gamma}\right)\left(\frac{V_1^2}{2}\right)\left(\frac{\rho_1}{P_1}\right)\right]^{\gamma/\gamma-1} \tag{9.31c}$$

But $a=\sqrt{\gamma\left(\frac{P}{\rho}\right)}$ i.e. $\left(\frac{\rho_1}{P_1}\right)=\frac{\gamma}{a^2}$ and substituting this in Eq. (9.31c) and rearranging, we have

$$P_S = P_1\left[1+\left(\frac{\gamma-1}{2}\right)M_1^2\right]^{\gamma/\gamma-1} \quad \text{since } M=\left(\frac{V_1}{a}\right) \tag{9.32}$$

Similarly,
$$T_S = T_1\left[1+\left(\frac{\gamma-1}{2}\right)M_1^2\right] \quad \text{since } \frac{T_S}{T_1}=\left(\frac{P_S}{P_1}\right)^{\gamma/\gamma-1} \tag{3.33}$$

Similarly,
$$\rho_S = \rho_1\left[1+\left(\frac{\gamma-1}{2}\right)M_1^2\right]^{1/\gamma-1} \quad \text{since } \frac{P_S}{P_1}=\left(\frac{\rho_1}{\rho_S}\right)^{\gamma} \tag{9.34}$$

If considered flow variables like pressure, temperatures and densities with their local Mach numbers (i.e., **relationship of flow variables with Mach number between two local point**), we can deduce by considering two points at two different sections "1" and "2" on the isentropic compressible flow streamline, as shown in Figure 9.8.

According to Eq. (9.33), we have

$$T_S = T_1\left[1+\left(\frac{\gamma-1}{2}\right)M_1^2\right] \;\&\; T_S = T_2\left[1+\left(\frac{\gamma-1}{2}\right)M_2^2\right]$$

So
$$\frac{T_1}{T_2}=\left[\left(\frac{2+(\gamma-1)M_2^2}{2+(\gamma-1)M_1^2}\right)\right] \tag{9.34a}$$

Similarly
$$\frac{P_1}{P_2}=\left[\left(\frac{1+\left(\frac{\gamma-1}{2}\right)M_2^2}{1+\left(\frac{\gamma-1}{2}\right)M_1^2}\right)\right]^{\gamma/\gamma-1} \tag{9.34b}$$

since
$$\left(\frac{T_1}{T_2}\right)=\left(\frac{P_1}{P_2}\right)^{\gamma-1/\gamma}=\left(\frac{\rho_1}{\rho_2}\right)^{\gamma-1}$$

Similarly
$$\frac{\rho_1}{\rho_2}=\left[\left(\frac{1+\left(\frac{\gamma-1}{2}\right)M_2^2}{1+\left(\frac{\gamma-1}{2}\right)M_1^2}\right)\right]^{1/\gamma-1} \tag{9.34c}$$

As per continuity equation $\dfrac{A_2}{A_1} = \dfrac{\rho_1 V_1}{\rho_2 V_2}$

$$\Rightarrow \frac{A_2}{A_1} = \frac{\rho_1 M_1 \sqrt{\gamma R T_1}}{\rho_2 M_2 \sqrt{\gamma R T_2}} = \left(\frac{\rho_1}{\rho_2}\right)\left(\frac{M_1}{M_2}\right)\left(\frac{T_1}{T_2}\right)^{1/2} \qquad \text{since } V = M\sqrt{\gamma RT}$$

Substituting Eq. (9.34a) and Eq. (9.34c), we have

$$\frac{A_2}{A_1} = \left(\frac{M_1}{M_2}\right)\left(\left[\left(\frac{1+\left(\dfrac{\gamma-1}{2}\right)M_1^2}{1+\left(\dfrac{\gamma-1}{2}\right)M_2^2}\right)\right]\right)^{\gamma+1/2(\gamma-1)} \tag{9.34d}$$

Or in other terms, at the stagnation point we can deduce for an isentropic incompressible flow by considering the enthalpy. The **Enthalpy** (ΔH) (measure of energy per unit mass of the fluid/gas) is sum of **Internal energy** (*U*) (Energy possessed due to motion of fluid at molecular level like chemical or electrical energy) and **work done** is given by $\Delta H = \Delta U + pV$ $\Rightarrow (h) = (C_p T_1) = C_v T_1 + RT_1$ as $pV = RT_1$.

And hence the enthalpy at the stagnation point (h_s)will be the sum of enthalpy at any point (h) and the mechanical energy (potential and kinetic energies). The body weight (potential energy) of the compressible fluid can be neglected and if kinetic energy is only included then

$$(h_s - h) = K.E. \quad \Rightarrow C_p T_s = C_v T_1 + RT_1 + V_1^2/2$$

$$\Rightarrow T_s = T_1\left(\frac{1}{\gamma} + 1 - \frac{1}{\gamma}\right) + V_1^2/2C_p \qquad \text{since} \quad C_p - C_v = R$$

$$\Rightarrow T_S = T_1 + \frac{V_1^2}{2 * C_p}$$

or

$$\Rightarrow T_S = T_1\left(1 + \frac{\gamma-1}{2}M^2\right) \tag{9.35}$$

Note:

If a Pitot tube is placed at the stagnation point to determine the stagnation variables like air flow, then the actual air flow velocity will be higher than that of the theoretical due to the reason that the **compressibility correction factor (CF)** of air adds up to the differential head indication. So According to Eq. (9.32), we have $P_S = P_1\left[1 + \left(\dfrac{\gamma-1}{2}\right)M_1^2\right]^{\gamma/\gamma-1}$

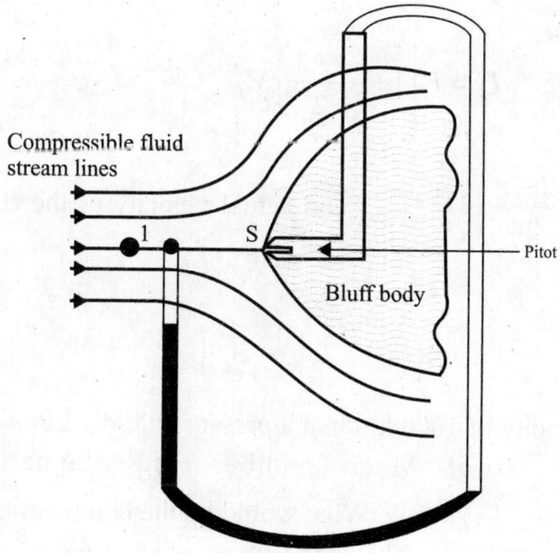

Figure 9.8: Stagnation properties in pitot tube

So the differential pressure head in the Pitot tube will be

$$\Delta P = P_S - P_1 = P_1\left[1+\left(\frac{\gamma-1}{2}\right)M_1^2\right]^{\gamma/\gamma-1} - P_1$$

$$\Rightarrow \Delta P = P_1\left\{\left[\left(\frac{\gamma-1}{2}\right)M_1^2\right]^{\gamma/\gamma-1}\right\}$$

But as $$P_1 = \left(\frac{\rho * a_1^2}{\gamma}\right) = \left(\frac{\rho V_1^2}{2}\right)\left(\frac{2}{\gamma M_1^2}\right) \Rightarrow \Delta P = \left(\frac{\rho V_1^2}{2}\right) * \left(\frac{2}{\gamma M_1^2}\right)\left\{\left[\left(\frac{T_s}{T_1}\right)\right]^{\gamma/\gamma-1} - 1\right\} \quad (9.36)$$

Where $$CF = \left(\frac{2}{\gamma M_1^2}\right)\left\{\left[\left(\frac{T_s}{T_1}\right)\right]^{\gamma/\gamma-1} - 1\right\} \quad (9.36a)$$

When Mach number "$M < 0.2$" and actual velocity of the compressible flow will be $V_a = \dfrac{V}{\sqrt{CF}}$.

And if the Mach number exceeds "$M > 0.2$", then the compressibility correction factor follows binomial expansion like

$$P_S = P_1\left[1+\left(\frac{\gamma-1}{2}\right)M_1^2 * CF\right]^{\gamma/\gamma-1} \tag{9.36b}$$

where $CF = \left(1+\left(\frac{M_1^4}{4}\right)+\left(\frac{2-\gamma}{24}\right)M_1^4+\ldots\right)$ and actual velocity of the compressible flow will be $V_a = V/\sqrt{CF}$.

Example 9.6: Air has a velocity of 1000 km/h at a pressure of 9.81 kPa vacuum and a temperature of 47°C. Compute its local Mach number and stagnation properties. Take $P_{atm} = 98.1\text{ kPa}, R = 287\text{ j/kg}-K, \gamma = 1.4$. What would be the compressibility correction factor for a Pitot static tube to measure the velocity at a Mach number of 0.8 and also calculate the actual velocity of the air flow velocity.

Solution: Given data:

$$P_{atm} = 98.1\text{ kPa},\ P_1 = 9.81\text{ kPa}(vaccum),\ V_1 = 277.8\text{ m/s}$$

$$\text{T}_1 = 320\text{ K}, R = 287\text{J/kg}-\text{K},\ \gamma = 1.4$$

a) Sonic velocity $a_1 = \sqrt{\gamma RT_1} = \sqrt{1.4*287*320} = 358.6\text{ m/s}$

Mach number
$$M_1 = \frac{V_1}{a_1} = \frac{278.8}{358.6} = 0.775 > 0.2$$

Absolute pressure $P_1(abs) = P_{atm} - P_1(vacuum) = 98.1 - 9.81 = 88.29\text{ kPa}$

$$P_1 = \rho_1 RT_1$$

Density
$$\Rightarrow \rho_1 = \frac{P_1}{RT_1} = \frac{88.29}{287*320} = 0.096\text{ kg/m}^3$$

Stagnation pressure

$$P_S = P_1\left[1+\left(\frac{\gamma-1}{2}\right)M_1^2\right]^{\gamma/\gamma-1} = 131.3\text{ kPa}$$

Stagnation temperature

$$T_S = T_1\left[1+\left(\frac{\gamma-1}{2}\right)M_1^2\right] = 358.4\text{ K}$$

Stagnation density $\rho_s = \dfrac{P_s}{RT_s} = 1.276 \text{ kg/m}^3$

b) Compressibility factor $CF = 1 + \dfrac{M_1^2}{4} + \left(\dfrac{2-\gamma}{24}\right) M_1^4 \Rightarrow CF = 1 + \dfrac{0.8^2}{4} + \left(\dfrac{2-1.4}{24}\right) 0.8^4 = 1.17$

c) Actual air flow velocity $V_a = {V_1}/{\sqrt{CF}} = {277.8}/{\sqrt{1.17}} = 25.68 \text{ m/s}$

Example 9.7: An aeroplane is flying at 230 km/h through a still air having a pressure of 98.1 kPa (absolute) and temperature of 320 K. Compute the stagnation properties on the nose of the plane:

a) Stagnation pressure
b) Stagnation temperature
c) Stagnation density
d) Include the compressibility correction factor and calculate the actual velocity of air flow, when a Pitot tube is used at the nose of the plane.

Take $P_{atm} = 101.1 \text{ kPa}, R = 287 \text{ J/kg} - \text{K}, \gamma = 1.4$

Solution: Given data:

$$P_{atm} = 101.1 \text{ kPa}, P_1 = 9.81 \text{ kPa}(absolute), T_1 = 320 \text{ K}, R = 287 \text{ J/kg} - \text{K}, \gamma = 1.4$$

Velocity of the air plane $V_1 = 230 * 1000/60 * 60 = 63.8 \text{ m/s}$

a) Sonic velocity $a_1 = \sqrt{\gamma R T_1} = \sqrt{1.4 * 287 * 320} = 358.6 \text{ m/s}$

Mach number $M_1 = \dfrac{V_1}{a_1} = \dfrac{63.88}{358.6} = 0.178 < 0.2$

$$P_1 = \rho_1 R T_1$$

Density $\Rightarrow \rho_1 = \dfrac{P_1}{RT_1} = \dfrac{98.1*1000}{287*320} = 1.068 \text{ kg/m}^3$

Stagnation pressure $P_S = P_1 \left[1 + \left(\dfrac{\gamma-1}{2}\right) M_1^2\right]^{\gamma/\gamma-1} = 100.25 \text{ kPa}$

b) Stagnation temperature $T_S = T_1 \left[1 + \left(\dfrac{\gamma-1}{2}\right) M_1^2\right] = 322 \text{ K}$

c) Stagnation density $\rho_S = \rho_1\left[1+\left(\frac{\gamma-1}{2}\right)M_1^2\right]^{1/\gamma-1} = 1.085$ kg/s

d) Compressibility factor $CF = \left(\frac{2}{\gamma M_1^2}\right)\left\{\left[\left(\frac{T_s}{T_1}\right)\right]^{\gamma/\gamma-1} - 1\right\} = 0.994$

and $V_a = V/\sqrt{CF} = 63.8/\sqrt{0.994} = 63.99$ m/s

Example 9.8: In an aircraft at an altitude where the pressure was 35 kPa and temperature –38°C Stagnation pressure was measured as 65.4 kPa. Calculate the speed of the aircraft. Take molecular weight of air as 28. **[UPSC]**

Solution: Given data:

$$P_S = 65400 \text{ Pa}, P_1 = 35000 \text{ Pa}, T_1 = 235 \text{ K}, M = 28, R_0 = 8314 \text{ Nm/mole} - \text{K}$$

Density

$$P_1 = \rho_1\left(\frac{R_0}{M}\right)T_1$$

$$\Rightarrow \rho_1 = \frac{P_1 * M}{R_0 T_1} = \frac{35000 * 28}{8314 * 235} = 0.5 \text{ kg/m}^3$$

Using the relation of Eq. (9.30), we have

$$P_S = P_1 + \frac{1}{2}\rho V^2$$

$$\Rightarrow 65400 = 35000 + \frac{1}{2} * 0.5 * V_a^2$$

$$\Rightarrow V_a = 348.7 \text{ m/s}$$

Hence, the **velocity of aircraft** is 348.7 m/s.

Example 9.9: The pressure leads from a Pitot static tube mounted on an aircraft were connected to a pressure gage in the cockpit. The dial of the pressure gage is calibrated to read the aircraft speed in m/s. The calibration is done on the ground by applying a known pressure across the gage and calculating the equivalent velocity using the incompressible Bernoulli's equation and assuming that the density is 1.224 kg/m^3.

The gage having been calibrated in this way the aircraft is flown at 9200 m, where the density is 0.454 kg/m^3 and ambient pressure is 30 kPa. The gage indicates a velocity of 152 m/s. What is the true speed of the aircraft? **[UPSC]**

Solution: Given data:

$$P_1 = 30 \text{ kPa}, \ V_1 = 152 \text{ m/s}, \ \rho_1 = 1.224 \text{ kg/m}^3$$

Stagnation pressure in a static Pitot tube is

$$P_S = P_1 + \frac{1}{2}\rho V^2 = 44.14 \text{ kPa}$$

Neglecting compressibility factor, the speed of the aircraft when the density is 0.454 kg/m³ is

$$P_S = P_1 + \frac{1}{2}\rho V^2$$

$$\Rightarrow V = 249.6 \text{ m/s}$$

Sonic velocity

$$a = \sqrt{\gamma\left(\frac{P_1}{\rho}\right)} = \sqrt{1.4 * \left(\frac{30000}{0.454}\right)} = 304.16 \text{ m/s}$$

Mach number is $$M = \frac{V}{a} = \frac{249.6}{304.16} = 0.82$$

Compressibility factor $$CF = 1 + \frac{M^2}{4}$$

$$\Rightarrow CF = 1 + \frac{0.82^2}{4} = 1.168$$

True speed of the aircraft is $$V_a = \frac{V}{\sqrt{CF}} = \frac{249.6}{\sqrt{1.168}} = 230.9 \text{ m/s}$$

Hence, the **true speed of the aircraft** is 230.9 m/s.

AREA, VELOCITY, PRESSURE RELATIONS OF A COMPRESSIBLE FLUID—NOZZLES AND DIFFUSERS

Any object like aeroplanes, projectiles, missiles, etc. moving in compressible fluid will be having its motion depending on the Newton's third law (i.e.) for every action, there is equal and opposite reaction).

These objects will be moving in compressible fluid at very high velocity like with supersonic velocity. And to attain these supersonic velocities, they take in the compressible fluid at supersonic velocity (i.e., as fluid medium having Mach number $M > 1$) into a device and decrease the velocity of the compressible fluid to sonic velocity $M = 1$ and then further decreased to subsonic velocity $M < 1$ and this fluid at subsonic velocity is led into another device to increase this velocity to supersonic velocity with decrease in pressure. Once the supersonic velocity with less pressure is attained, the compressible fluid is then released out into atmosphere with this supersonic velocity which give in turn an opposite thrust force to push the object forward with super sonic velocity.

To achieve this complete action, devices called **diffusers** and **nozzles** are being used.

A **diffuser** is a device which has varying cross-section area and reduces velocity of the fluid in spite of increase or decrease in cross-section area and at the same time it increases the pressure, as shown in Figure 9.10. The diffusers catch the compressible fluid flow at supersonic velocity and reduces their velocity to subsonic as well increases its pressure so that the subsonic compressible fluid can be still further compressed in compressors and can be used in high-speed engines for combustion with fuel mixed.

A **nozzle** is a device of varying cross-section having converging section, throat section and divergent section, as shown in Figure 9.11. When a nozzle fitted with a divergent section in form of a venturimeter, then the divergent section can is termed a diffuser. If the inlet pressure is maintained constant and the outlet pressure at the exit of the divergent portion is reduced in stages, at some point the pressure at the throat reaches sonic velocity. And when further reduced the throat pressure do not fall below the critical chocked pressure and in turn the flow expand with high velocities at the exit and so the main purpose of a nozzle is to increase the thrust force liberated by the high speed engines to the atmosphere with super sonic velocity, so as to achieve equal and opposite thrust force on the object for its super sonic speed.

The thrust power developed in the engine is being sent with subsonic velocity to the convergent portion of the nozzle, wherein the velocity is increased from subsonic velocity $M < 1$ to sonic velocity $M = 1$ by reduction in cross-section area and thereby this fluid with sonic velocity is sent to divergent portion of the nozzle, wherein velocity of fluid is increased to supersonic velocity $M > 1$ by increase in cross-section area and reduce in pressure.

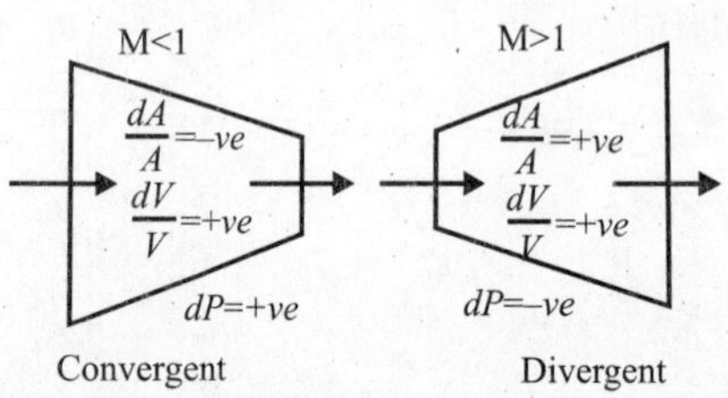

Figure 9.9: Flow through Nozzles

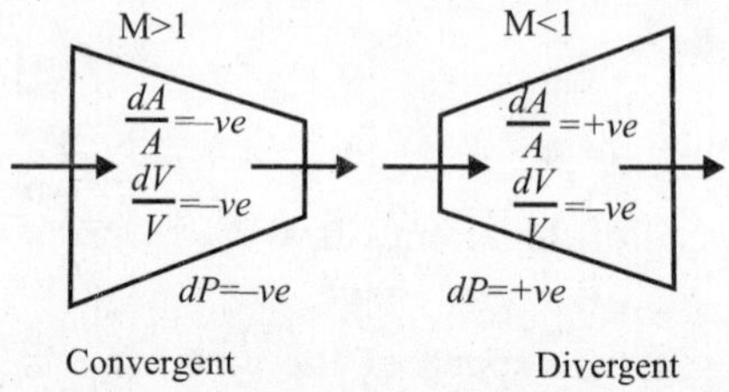

Figure 9.10: Flow through diffusers

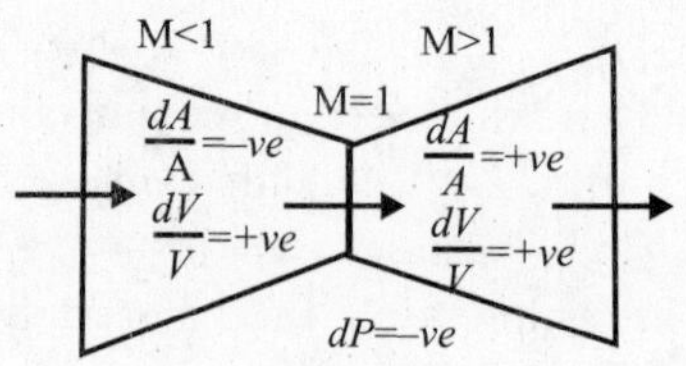

Figure 9.11: Flow through convergent-divergent nozzle

As per continuity equation of compressible fluids

$$\rho AV = \text{Constant} \tag{9.35}$$

$$\Rightarrow \frac{d\rho}{\rho} = -\left(\frac{dA}{A} + \frac{dV}{V}\right) \tag{9.36}$$

As per Bernoulli's equation for compressible fluid

$$\frac{dP}{\rho} + VdV + gdZ = 0$$

$$\Rightarrow \frac{dP}{\rho} = -VdV \Rightarrow \left[\left(\frac{dP}{d\rho}\right)\left(\frac{d\rho}{\rho}\right)\right] = -VdV \tag{9.37}$$

Substituting $a^2 = \frac{dP}{d\rho}$ and Eq. (9.36) in Eq. (9.37), we have

$$-a^2\left[\left(\frac{dV}{V} + \frac{dA}{A}\right)\right] = -VdV$$

$$\Rightarrow \frac{dA}{A} = \left[\left(\frac{dV}{V}\right)(M^2 - 1)\right] = 0 \tag{9.38}$$

Equation (9.38) gives **relation between area and velocity for compressible fluid for a convergent-divergent nozzle (also called Laval nozzle).**
And if the fluid is of incompressible $dP = 0$ and so $M = 0$ and hence

$$\frac{dA}{A} = -\left[\left(\frac{dV}{V}\right)\right] \tag{9.38a}$$

From Eq. (9.38) it is clear that for a nozzle or diffuser having convergent portion ($M < 1$), the change in area is inversely proportional to change in velocity and hence in convergent portion the dA/A and dV/V will be having opposite signs, as shown in Figures 9.9 and 9.10.

Similarly, a nozzle or diffuser having divergent portion ($M > 1$), the change in area is directly proportional to change in velocity and hence in divergent portion the dA/A and dV/V will be having same signs, as shown in Figures 9.09 and 9.10.

From Eq. (9.37), we have $\quad \frac{dP}{\rho} = -VdV \Rightarrow dP = -\rho V^2 \frac{dV}{V}$

But from Eq. (9.38), we have

$$\Rightarrow \frac{dA}{A}\left(\frac{1}{(M^2-1)}\right) = \left(\frac{dV}{V}\right)$$

$$dP = \frac{dA}{A}\left[\rho V^2\left(\frac{1}{1-M^2}\right)\right] \quad (9.39)$$

From Eq. (9.38) and Eq. (9.39) it is clear that for a nozzle or diffuser having throat portion ($M = 1$), the change in area, change in velocity and change in pressure are zero $\left({}^{dA}\!/_{A} = {}^{dV}\!/_{V} = dP = 0\right)$, as shown in Figure 9.11. (i.e. if the area is sufficiently reduced; the velocity at the throat reaches sonic velocity) and if the convergent portion is made longer and area at exit is further reduced, then also the velocity tends to be sonic by reducing the mass flow rate.

Hence, $\qquad \bar{m}_{throat} = \bar{m}_{exit} \Rightarrow (\rho AV)_{throat} = (\rho AV)_{exit}$

$$\Rightarrow \frac{A_{throat}}{A_{exit}} = M_{exit}\left(\frac{P_{exit}}{P_{throat}}\right)^{1+\gamma/2\gamma} \quad (3.39a)$$

EFFECT OF BACK PRESSURE ON THE FLOW IN LAVAL NOZZLE

The very purpose of including a convergent-divergent nozzle (Laval nozzle) to a reservoir having high pressure and temperature (P_0, T_0) is to have supersonic acceleration of fluid at the outlet **without** shock wave formation, as shown in Figure 9.12 and this condition can be possible if

$$P_S = P_{throat}\left[1 + \left(\frac{\gamma-1}{2}\right)M^2_{throat}\right]^{\gamma/\gamma-1} \Rightarrow {}^{P_{throat}}\!/_{P_S} = 0.528\,.$$

When a laval nozzle is connected to high pressure (P_o) and temperature (T_o) reservoir, the pressure variation in the convergent and divergent portions may be varying depending on the length of the nozzle and can be listed as in Figure 9.12a. The pressure P_C represents the critical pressure at the throat and the curve P_o to P_1 is the desired design condition of a Laval nozzle. With increase in pressure from normal design condition in divergent portion due to back pressure, there exists modification of flow in divergent portion but the isentropic expansion do not get disturbed, like:

a) Curve "D" which starts at somewhere point "j" below the critical pressure, giving raise to shock waves in the divergent portion at supersonic decelerated velocities, once the critical pressure is reached.
b) Curve "C" which starts directly form the critical pressure at the throat and follows back to a subsonic isentropic diffusion flow, if the divergent portion has subsonic conditions.

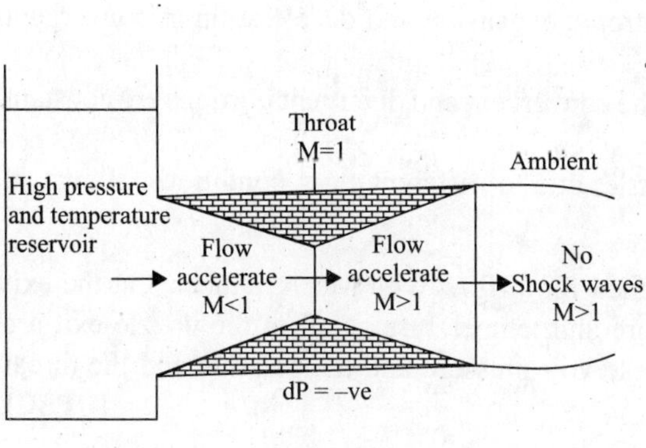

Figure 9.12: Flow through Laval nozzle with desired design condition

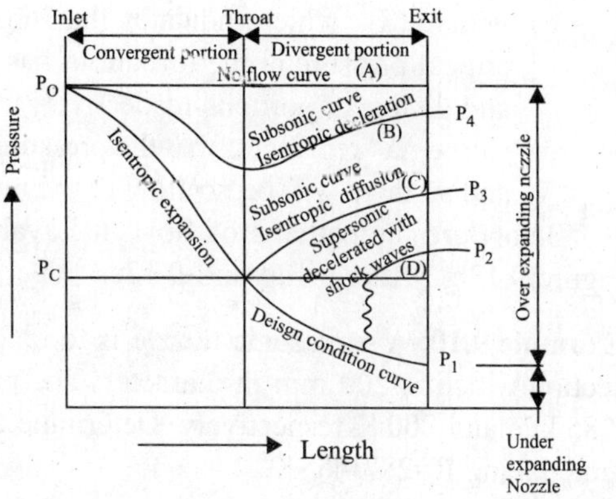

Figure 9.12a: Pressure distribution along the Laval nozzle

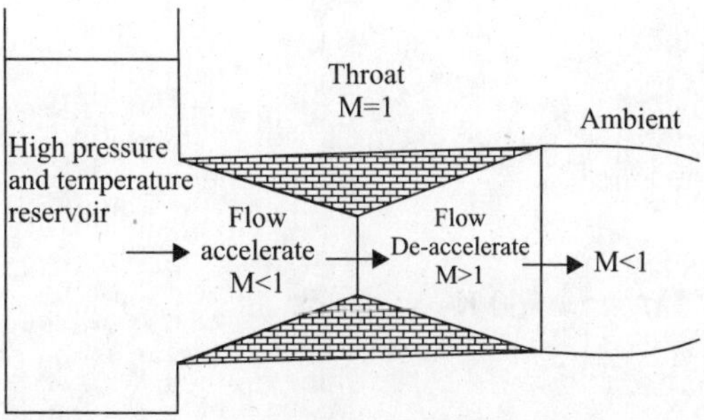

Figure 9.12b: Flow through Laval nozzle with subsonic flow

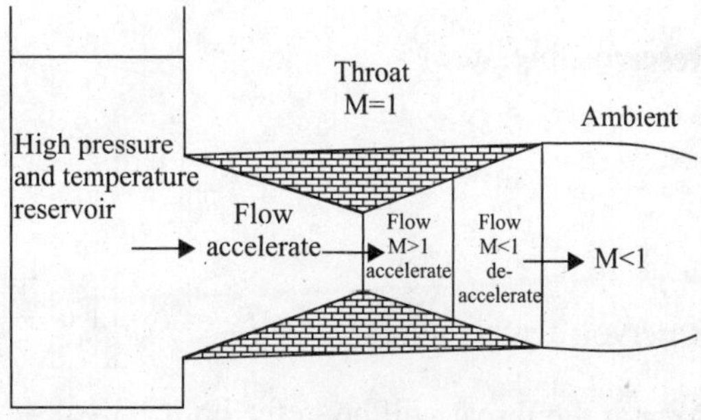

Figure 9.12c: Flow through Laval nozzle with shock in nozzle

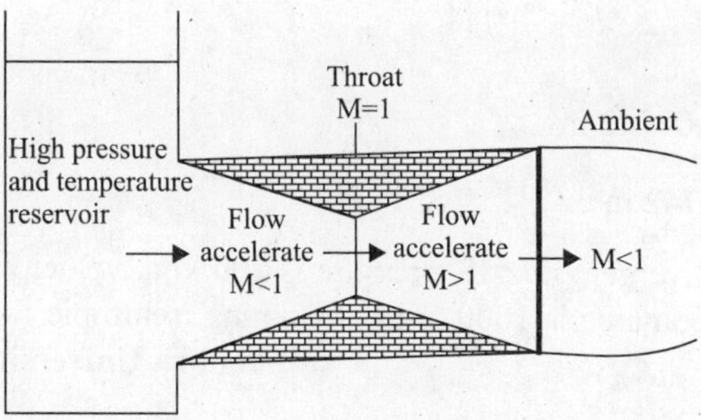

Figure 9.12d: Flow through Laval nozzle with shock at exit

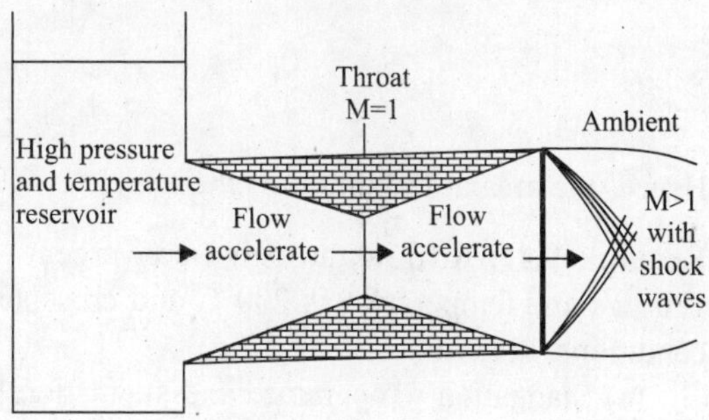

Figure 9.12e: Flow through Laval nozzle with supersonic shock waves

c) Curve "B" which indicates the subsonic condition throughout the nozzle, do not attain critical pressure at the throat but has isentropic expansion and deceleration in convergent and divergent portions respectively.

d) Curve "A" represents that the pressure in the convergent and divergent portions are constant and so there will be no flow in the nozzle.

Below are the patterns of flow in Laval nozzle, due to different flow conditions, shown in Figure 9.12b, 9.12c, 9.12d and 9.12e.

Example 9.10: A supersonic nozzle is to be designed for airflow with Mach number 3 at the exit section which is 200 mm in diameter. The pressure and temperature of air at the nozzle exit are 7.85 kPa and 200 K respectively. Determine the reservoir pressure and temperature and the throat area, taking R=287J/kg-K. **[UPSC]**

Solution: Given data: $M = 3, d = 0.2$ m, $P_{exit} = 7.85$ kPa, $T_{exit} = 200$ K

For supersonic nozzle (M>1) a **divergent nozzle section** and hence area at the exit is

$$A_{exit} = \frac{\pi d^2}{4} = 0.0314 \text{ m}^2$$

Reservoir pressure

$$P_{res} = P_{exit}\left[1+\left(\frac{\gamma-1}{2}\right)M_{exit}^2\right]^{\gamma/\gamma-1} = 288.35 \text{ kPa}$$

Reservoir temperature $\quad T_{res} = T_{exit}\left[1+\left(\frac{\gamma-1}{2}\right)M_{exit}^2\right] = 560 \text{ K}$

Area at the throat will be, refer Eq. (9.3.4d)

$$\frac{A_{exit}}{A_{throat}} = \left(\frac{M_{throat}}{M_{exit}}\right)\left(\left[\frac{1+\left(\frac{\gamma-1}{2}\right)M_{throat}^2}{1+\left(\frac{\gamma-1}{2}\right)M_{exit}^2}\right]\right)^{\gamma+1/2(\gamma-1)}$$

$$\Rightarrow A_{throat} = 0.00742 \text{ m}^2$$

Hence, the area at the throat of the nozzle is 0.00742 m^2.

Example 9.11: Airflow through a convergent – divergent nozzle, pressure is 200 kPa, velocity is 170 m/s and temperature is 200°C and cross-section area is 1000 mm^2. Assuming isentropic flow conditions, determine **[Andhra University]**

(a) Stagnation temperature and stagnation pressure

(b) Sonic velocity and Mach number at this section

(c) Velocity, Mach number and flow area at outlet section where pressure is 110 kPa.

(d) Pressure, temperature, velocity and flow area at throat of the nozzle.

Taking $R = 287$ J/kg – K, $C_p = 1.0$ kJ/kg – K, $\gamma = 1.4$

Solution: Given data:

$$P_x = 200 \text{ kPa}, V_x = 170 \text{ m/s}, T_x = 473 \text{ K}$$

$$A = 1000 \text{ mm}^2, R = 287 \text{ J/kg} - \text{K}, C_P = 1.0 \text{ kJ/kg} - \text{K}, \gamma = 1.4$$

a) Stagnation temperature $T_s = T_x + \dfrac{V_x^2}{2C_P} = 473 + \dfrac{170^2}{2*1000} = 487.45 \text{ K}$

Stagnation pressure, as per isentropic relation, we have

$$\left(\frac{P_s}{P_x}\right) = \left(\frac{T_s}{T_x}\right)^{\gamma/\gamma-1} \Rightarrow \left(\frac{P_s}{200}\right) = \left(\frac{487.45}{473}\right)^{1.4/1.4-1}$$

$$\Rightarrow P_s = 222.2 \text{ kPa}$$

b) Sonic velocity and Mach number

Sonic velocity $\quad a_x = \sqrt{\gamma R T_x} = \sqrt{1.4 * 287 * 473} = 435.9 \text{ m/s}$

Mach number $\quad M_x = \dfrac{V_x}{a_x} = \dfrac{170}{435.9} = 0.39$

c) Velocity at the outlet where $P_2 = 110$ kPa

Stagnation pressure $\quad P_S = P_2\left[1 + \left(\dfrac{\gamma - 1}{2}\right) M_2^{\,2}\right]^{\gamma/\gamma-1}$

$$\Rightarrow 222.2 = 110 * \left[1 + \left(\frac{1.4-1}{2}\right) * M_2^2\right]^{1.4/1.4-1}$$

$$\Rightarrow M_2 = 1.05$$

As per isentropic relation, we have $\left(\dfrac{P_s}{P_2}\right) = \left(\dfrac{T_s}{T_2}\right)^{\gamma/\gamma-1} \Rightarrow \left(\dfrac{222.2}{110}\right) = \left(\dfrac{487.45}{T_2}\right)^{1.4/1.4-1}$

$$\Rightarrow T_2 = 398.7 \text{ K}$$

Sonic velocity $\quad a_2 = \sqrt{\gamma R T_2} = \sqrt{1.4 * 287 * 398.7} = 400.2 \text{ m/s}$

Hence, velocity at the outlet is

$$M_2 = \frac{V_2}{a_2} = \frac{V_2}{400.2} = 1.05$$

$$\Rightarrow V_2 = 420.26 \text{ m/s}$$

As per continuity equation, we have

$$\rho_1 A_1 V_1 = \rho_2 A_2 V_2$$

$$\Rightarrow \frac{P_1}{RT_1} A_1 V_1 = \frac{P_2}{RT_2} A_2 V_2$$

$$\Rightarrow A_2 = 619.6 \text{ mm}^2$$

Hence, the area at the outlet of the nozzle is 619.6 mm^2.

d) At the throat Mach number is unit.

Stagnation pressure $P_S = P_t\left[1 + \left(\frac{\gamma - 1}{2}\right) M_t^2\right]^{\gamma/\gamma-1}$

$$\Rightarrow 222.2 = P_t * \left[1 + \left(\frac{1.4 - 1}{2}\right) * 1^2\right]^{1.4/1.4-1}$$

$$\Rightarrow P_t = 117.32 \text{ kPa}$$

As per isentropic relation, we have $\left(\frac{P_s}{P_t}\right) = \left(\frac{T_s}{T_t}\right)^{\gamma/\gamma-1} \Rightarrow \left(\frac{222.2}{117.32}\right) = \left(\frac{487.45}{T_t}\right)^{1.4/1.4-1}$

$$\Rightarrow T_t = 406.2K$$

Velocity $V_t = a_t = \sqrt{\gamma R T_t} = \sqrt{1.4 * 287 * 406.2} = 404 \text{ m/s}$

As per continuity equation, we have $\rho_1 A_1 V_1 = \rho_t A_t V_t$

$$\Rightarrow \frac{P_1}{RT_1} A_1 V_1 = \frac{P_t}{RT_t} A_t V_t$$

$$\Rightarrow \frac{200}{287 * 473} * 0.001 * 170 = \frac{117.32}{287 * 406.2} * A_2 * 404$$

$$\Rightarrow A_2 = 616 \text{ mm}^2$$

Hence, the area at the throat of the nozzle is 616 mm^2.

MASS FLOW RATE OF A COMPRESSIBLE FLOW THROUGH A NOZZLE

Consider a nozzle having convergent portion fitted to a reservoir/tank having a compressible fluid filled, as shown in Figure 9.13 and let the compressible fluid, flow through the nozzle adiabatically (PV^γ = constant).

Applying Bernoulli's equation at the two sections 1-1 and 2-2 as shown in Figure 9.13, we have

$$\left(\frac{\gamma}{\gamma-1}\right)\left(\frac{P_1}{\rho_1 g}\right)+\frac{V_1^2}{2g}+Z_1=\left(\frac{\gamma}{\gamma-1}\right)\left(\frac{P_2}{\rho_2 g}\right)+\frac{V_2^2}{2g}+Z_2$$

But $Z_1 = Z_2$ and assuming velocity at inlet as zero (i.e. $V_1 = 0$), we have

$$\left(\frac{\gamma}{\gamma-1}\right)\left(\frac{P_1}{\rho_1 g}\right)=\left(\frac{\gamma}{\gamma-1}\right)\left(\frac{P_2}{\rho_2 g}\right)+\frac{V_2^2}{2g}$$

By rearranging and substituting $\left(\frac{\rho_1}{\rho_2}\right)=\left(\frac{P_1}{P_2}\right)^{1/\gamma}=\left(\frac{P_2}{P_1}\right)^{-1/\gamma}$ for adiabatic flow, we have

$$\Rightarrow\left(\frac{\gamma}{\gamma-1}\right)\left(\frac{P_1}{\rho_1}\right)\left[1-\left(\frac{P_2}{P_1}\right)\left(\frac{\rho_1}{\rho_2}\right)\right]=\frac{V_2^2}{2}$$

$$\Rightarrow V_2^2=\left(\frac{2\gamma}{\gamma-1}\right)\left(\frac{P_1}{\rho_1}\right)\left[1-\left(\frac{P_2}{P_1}\right)\left(\frac{P_2}{P_1}\right)^{-1/\gamma}\right]$$

$$\Rightarrow V_2=\sqrt{\left(\frac{2\gamma}{\gamma-1}\right)\left(\frac{P_1}{\rho_1}\right)\left[1-\left(\frac{P_2}{P_1}\right)^{\gamma-1/\gamma}\right]} \tag{9.40}$$

Mass flow rate at the outlet section 2-2 be $\overline{m}$, then

$$\overline{m}=\rho_2 A_2 V_2 \tag{9.40a}$$

and substituting Eq. (9.40) in Eq. (9.40a), we have

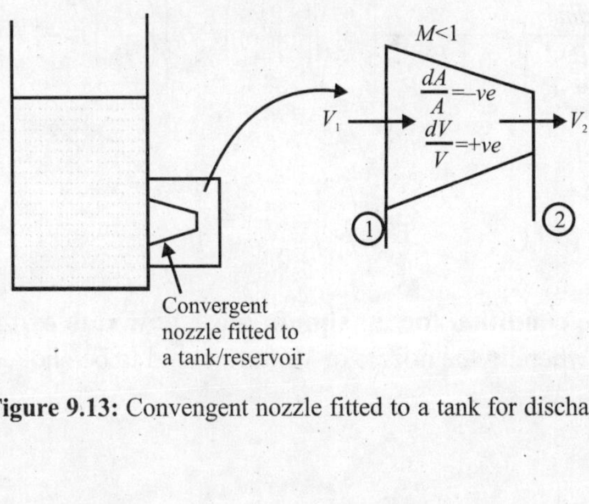

Figure 9.13: Convengent nozzle fitted to a tank for discharge

$$\bar{m} = \rho_2 A_2 \sqrt{\left(\frac{2\gamma}{\gamma-1}\right)\left(\frac{P_1}{\rho_1}\right)\left[1-\left(\frac{P_2}{P_1}\right)^{\gamma-1/\gamma}\right]}$$

$$\Rightarrow \bar{m} = A_2 \sqrt{\left(\frac{2\gamma}{\gamma-1}\right)\left(\frac{P_1}{\rho_1} * \rho_2^2\right)\left[1-\left(\frac{P_2}{P_1}\right)^{\gamma-1/\gamma}\right]}$$

By rearranging and substituting $\left(\frac{\rho_1}{\rho_2}\right) = \left(\frac{P_1}{P_2}\right)^{1/\gamma}$ for adiabatic flow, we have

$$\Rightarrow \bar{m} = A_2 \sqrt{\left(\frac{2\gamma}{\gamma-1}\right) P_1 \rho_1 \left[\left(\frac{P_2}{P_1}\right)^{2/\gamma} - \left(\frac{P_2}{P_1}\right)^{\gamma+1/\gamma}\right]} \qquad (9.41)$$

Condition for Maximum Mass Flow Rate/Discharge

To have the maximum mass flow rate through a nozzle for a compressible fluid flow, we need to differentiate Eq. (9.41) with respect to P_2/P_1 and equate to zero.

i.e.

$$\frac{d\bar{m}}{d\left(P_2/P_1\right)} = 0$$

$$\Rightarrow \frac{d\bar{m}}{d\left(P_2/P_1\right)} = \frac{d}{d\left(P_2/P_1\right)}\left[A_2 \sqrt{\left(\frac{2\gamma}{\gamma-1}\right) P_1 \rho_1 \left[\left(\frac{P_2}{P_1}\right)^{2/\gamma} - \left(\frac{P_2}{P_1}\right)^{\gamma-1/\gamma}\right]}\right]$$

$$\Rightarrow \left(\frac{P_2}{P_1}\right) = \left(\frac{2}{\gamma+1}\right)^{\gamma/\gamma-1} \qquad (9.42)$$

Hence, Eq. (9.42) is the **condition for maximum mass flow rate or discharge**, which is termed **critical pressure ratio** where in the nozzle or Venturi is said to be choked when attained maximum mass flow rate.

Hence, substituting Eq. (9.42) in Eq. (9.40) and Eq. (9.41), gives the maximum velocity at the outlet section and maximum mass flow rate through the nozzle

i.e.
$$V_{2_{max}} = \sqrt{\left(\frac{2\gamma}{\gamma+1}\right)\left(\frac{P_1}{\rho_1}\right)} \tag{9.43}$$

Similarly, for maximum mass flow rate/discharge

$$\bar{m}_{max} = A_2\sqrt{\left(\frac{2\gamma}{\gamma-1}\right)P_1\rho_1\left[\left(\frac{2}{\gamma+1}\right)^{2/\gamma-1} - \left(\frac{\gamma-1}{\gamma+1}\right)^{\gamma+1/\gamma-1}\right]} \tag{9.44}$$

If value of γ is replaced by value γ = 1.4 for compressible fluid like **air**, then $(P_2/P_1) = 0.528$ and if maximum flow rate will me

$$\bar{m}_{max} = 0.685 A_2\sqrt{P_1\rho_1} \tag{5.44a}$$

Substituting $\rho_1 = \rho_2\left(\frac{P_1}{P_2}\right)^{1/\gamma} = \left(\frac{P_2}{P_1}\right)$ and $\left(\frac{P_2}{P_1}\right) = \left(\frac{2}{\gamma+1}\right)^{\gamma/\gamma-1}$ in Eq. (9.40), we have

$$\Rightarrow V_2 = \sqrt{\left(\frac{2\gamma}{\gamma-1}\right)\left(\frac{\rho_1}{\rho_2}\right)^{\gamma} P_2\left[1-\left(\frac{2}{\gamma+1}\right)\right]\frac{1}{\rho_2}\left(\frac{P_2}{P_1}\right)^{1/\gamma}}$$

$$\Rightarrow V_2 = \sqrt{\frac{\gamma P_2}{\rho_2}} = a_2 \tag{9.45}$$

Hence, the velocity at the exit of the nozzle is sonic velocity.

Example 9.12: Find the maximum flow rate of air, taking $R = 287$ J/kg – K, $P_{atm} = 100$ kPa, $\gamma = 1.4$ for a large tank containing air at 284kPa (gage) and 297 K temperature. The air flows from the tank to the atmosphere through a convergent nozzle and the diameter at the outlet of the nozzle is 20 mm.

Solution: Given data:

$$R = 287 \text{ J/kg} - \text{K},\ P_{atm} = 100 \text{ kPa},\ P_{gauge} = 284 \text{ kPa},\ T_1 = 297 \text{ K},\ \gamma = 1.4,\ d_{outlet} = 0.02 \text{ m}$$

Pressure in the tank (absolute) is $P_{abs} = P_{atm} + P_{gauge} = 100 + 284 = 384$ kPa

Area at the outlet of the nozzle is $A_{outlet} = \frac{\pi d^2}{4} = \frac{3.14 * 0.02^2}{4} = 0.000314 \text{ m}^2$

Consider two points, one at the tank and other at the outlet of the nozzle, then density of the air in the tank is

$$\rho_1 = \frac{P_{abs}}{RT_1} = \frac{384}{287 * 297} = 4.5 \text{ kg/m}^3$$

So as per the Eq. (9.44a), we have $\bar{m}_{\max} = 0.685 A_2 \sqrt{P_1 \rho_1}$

$$\Rightarrow \bar{m}_{\max} = 0.685 * 0.000314 \sqrt{384000 * 4.5} = 0.283 \text{ kg/s}$$

Hence, the maximum flow rate at the exit of the nozzle is 0.283 kg/s.

Example 9.13: Assuming adiabatic flow in a tank fitted with a convergent divergent nozzle having air in it at temperature of 305 K, the diameter of nozzle outlet is 45 mm. Find the mass rate of flow of air through the convergent divergent nozzle into the atmosphere, when the pressure in the tank is

a) 165 kPa (absolute)
b) 320 kPa (absolute)

Take $R = 287$ J/kg – K, $\gamma = 1.4, P_{atm} = 101$ kPa

Solution: Given data: $d_o = 0.45$ m, $P_{atm} = 101$ kPa, $T_1 = 305$ K, $R = 287$ J/kg – K, $\gamma = 1.4$

Area at the outlet is $A_o = \frac{\pi d_o^2}{4} = \frac{3.14 * 0.45^2}{4} = 0.158 \text{ m}^2$

a) As per ideal gas equation $\left(\frac{P_1}{\rho_1}\right) = RT_1 \quad \Rightarrow \rho_1 = \frac{P_1}{RT_1} = \frac{165000}{287 * 305} = 1.88 \text{ kg/m}^3$

As per the relation of pressure ratio, we have $\left(\frac{P_{atm}}{P_1}\right) = \frac{101000}{165000} = 0.612$

Refer Eq. (9.41), for mass flow rate through a nozzle

$$\Rightarrow \bar{m} = A_o \sqrt{\left(\frac{2\gamma}{\gamma - 1}\right) P_1 \rho_1 \left[\left(\frac{P_{atm}}{P_1}\right)^{2/\gamma} - \left(\frac{P_{atm}}{P_1}\right)^{\gamma+1/\gamma}\right]} = 59.2 \text{ kg/s}$$

b) As per ideal gas equation $\left(\frac{P_1}{\rho_1}\right) = RT_1 \quad \Rightarrow \rho_1 = \frac{P_1}{RT_1} = \frac{320000}{287 * 305} = 3.655 \text{ kg/m}^3$

As per the relation of pressure ratio, we have $\left(\frac{P_{atm}}{P_1}\right) = \frac{101000}{320000} = 0.315$

Refer Eq. (9.41), for mass flow rate through a nozzle

$$\Rightarrow \bar{m} = A_o\sqrt{\left(\frac{2\gamma}{\gamma-1}\right)P_1\rho_1\left[\left(\frac{P_{atm}}{P_1}\right)^{2/\gamma}-\left(\frac{P_{atm}}{P_1}\right)^{\gamma+1/\gamma}\right]} = 105.01 \text{ kg/s}$$

Hence, the mass flow rate through the nozzle is 3.177 kg/s.

Example 9.14: Having a pressure at the outlet of the nozzle having diameter of 0.03 m has 1500 kPa (absolute), which is connected to a container containing air at 3000 kPa and temperature of 300 K. Find the velocity of air flowing out of the nozzle and also the mass rate of flow of air through the nozzle. Take $R = 287$ J/kg – K and $\gamma = 1.4$

Solution: Given data:

$d_o = 0.3$ m, $P_{atm} = 101$ kPa, $T_1 = 300$ K, $P_1 = 3000$ kPa, $P_o = 1500$ kPa, $R = 287$ J/kg – K, $\gamma = 1.4$

Area at the outlet is $A_o = \dfrac{\pi d_o^2}{4} = \dfrac{3.14 * 0.3^2}{4} = 0.07068 \text{ m}^2$

As per ideal gas equation $\left(\dfrac{P_1}{\rho_1}\right) = RT_1 \Rightarrow \rho_1 = \dfrac{P_1}{RT_1} = \dfrac{3000000}{287 * 300} = 34.9 \text{ kg/m}^3$

Refer Eq. (9.40), we have velocity at outlet

$$V_2 = \sqrt{\left(\frac{2\gamma}{\gamma-1}\right)\left(\frac{P_1}{\rho_1}\right)\left[1-\left(\frac{P_o}{P_1}\right)^{\gamma-1/\gamma}\right]}$$

$$\Rightarrow V_2 = \sqrt{\left(\frac{2*1.4}{1.4-1}\right)\left(\frac{3000000}{34.9}\right)\left[1-\left(\frac{1500000}{3000000}\right)^{1.4-1/1.4}\right]} = 328.79 \text{ m/s}$$

As per the relation of pressure ratio, we have $\left(\dfrac{P_o}{P_1}\right) = \dfrac{150000}{300000} = 0.5$

Refer Eq. (9.41), for mass flow rate through a nozzle

$$\Rightarrow \bar{m} = A_o\sqrt{\left(\frac{2\gamma}{\gamma-1}\right)P_1\rho_1\left[\left(\frac{P_o}{P_1}\right)^{2/\gamma}-\left(\frac{P_o}{P_1}\right)^{\gamma+1/\gamma}\right]} = 494.28 \text{ kg/s}$$

VARIATION OF MASS FLOW RATE WITH RESPECT TO PRESSURE RATIO

When a fluid is flowing in a tube/nozzle and it has to leave to the atmosphere outside, then the pressure at the inlet needs to be more than the outlet conditions and hence if the inlet and outlet pressure be P_1, P_2 respectively as shown in Figure 9.13a. Then $P_2 < P_1$ and hence the pressure ratio $\left({}^{P_2}\!/_{P_1} \right)$ will be ranging from 0 to 1 (i.e. $0 < P_2 < 1$).

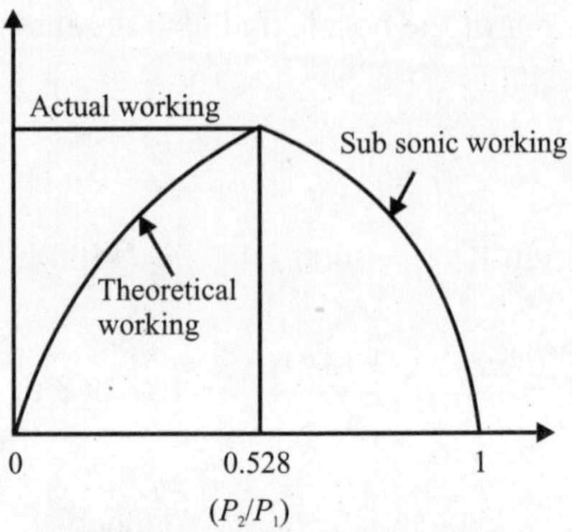

Figure 9.13a: Variation of mass flow rate per unit outlet area and pressure ratio

So if seen theoretical, by substituting different values of pressure ratio in Eq. (9.41), then at the mass flow rate per unit outlet area is zero at pressure ratio zero and one, i.e. $\left({}^{P_2}\!/_{P_1} = 0 \;\&\; {}^{P_2}\!/_{P_1} = 1 \right)$ and attains a maximum value at the pressure ratio 0.528, i.e. $\left({}^{P_2}\!/_{P_1} = 0.528 \right)$ and so the theoretical relation graph has been shown in thin lines in Figure 9.13a.

If seen actual/practical, by substituting different values of pressure ratio in Eq. (9.41), then at the mass flow rate per unit outlet area is zero at pressure ratio one, i.e. $\left({}^{P_2}\!/_{P_1} = 1 \right)$, since there will be no mass flow from inlet to outlet and as the outlet pressure is decreased, the mass flow rate attains a maximum value at the pressure ratio 0.528, i.e. $\left({}^{P_2}\!/_{P_1} = 0.528 \right)$ and so the actual/practical relation graph has been shown in thick lines in Figure 9.13. But after crossing below the maximum pressure ratio to zero pressure ratio, the mass flow rate do not tend to become zero but else as there will be no pressure obstruction to the mass flow, the complete mass flow rate entered will be let out to the surroundings with maximum quantity and hence the graph line of mass flow rate will be a horizontal line, as shown in Figure 9.13a.

COMPRESSIBLE FLUID FLOW THROUGH A VENTURIMETER

Consider a Venturimeter having convergent-throat-divergent portions, as shown in figure. Let a compressible fluid flow through it adiabatically (PV^{γ} = constant)

Applying Bernoulli's equation at the two sections like inlet section 1-1 and throat section 2-2 as shown in Figure 9.14, we have

$$\left(\frac{\gamma}{\gamma-1}\right)\left(\frac{P_1}{\rho_1 g}\right)+\frac{V_1^2}{2g}+Z_1=\left(\frac{\gamma}{\gamma-1}\right)\left(\frac{P_2}{\rho_2 g}\right)+\frac{V_2^2}{2g}+Z_2$$

$$\Rightarrow\left(\frac{\gamma}{\gamma-1}\right)\left(\frac{P_1}{\rho_1}\right)\left[1-\left(\frac{P_2}{\rho_2}\right)\left(\frac{\rho_1}{P_1}\right)\right]=\frac{V_2^2}{2}-\frac{V_1^2}{2}$$

By rearranging and substituting $\left(\frac{\rho_1}{\rho_2}\right)=\left(\frac{P_1}{P_2}\right)^{1/\gamma}$ for adiabatic flow, we have

$$\left(\frac{\gamma}{\gamma-1}\right)\left(\frac{P_1}{\rho_1}\right)\left[1-\left(\frac{P_2}{P_1}\right)\left(\frac{\rho_1}{\rho_2}\right)^{1/\gamma}\right]=\frac{V_2^2}{2}-\frac{V_1^2}{2}$$

$$\Rightarrow\left(\frac{\gamma}{\gamma-1}\right)\left(\frac{P_1}{\rho_1}\right)\left[1-\left(\frac{P_2}{P_1}\right)^{\gamma-1/\gamma}\right]=\frac{V_2^2}{2}-\frac{V_1^2}{2} \tag{9.46}$$

According to continuity equation $\rho_1 A_1 V_1=\rho_2 A_2 V_2$ for compressible flow

$$\Rightarrow V_1=\frac{\rho_2 A_2 V_2}{\rho_1 A_1} \tag{9.46a}$$

Substituting Eq. (9.46a) in Eq. (9.46), we have

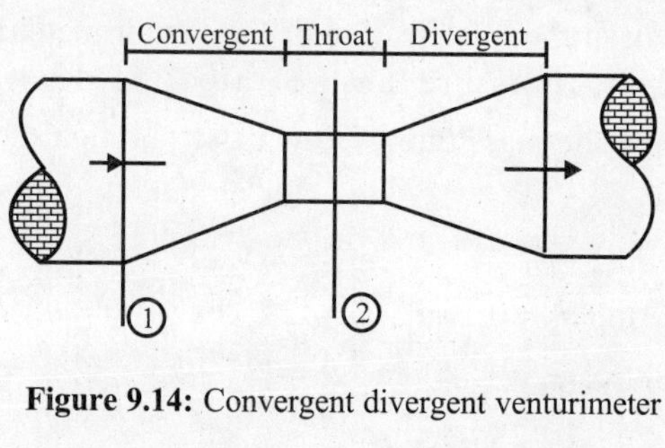

Figure 9.14: Convergent divergent venturimeter

$$\left(\frac{\gamma}{\gamma-1}\right)\left(\frac{P_1}{\rho_1}\right)\left[1-\left(\frac{P_2}{P_1}\right)^{\gamma-1/\gamma}\right]=\frac{V_2^2}{2}\left[1-\left(\frac{\rho_2}{\rho_1}\right)^2\left(\frac{A_2}{A_1}\right)^2\right]$$

$$\Rightarrow\left(\frac{\gamma}{\gamma-1}\right)\left(\frac{P_1}{\rho_1}\right)\left[1-\left(\frac{P_2}{P_1}\right)^{\gamma-1/\gamma}\right]=\frac{V_2^2}{2}\left[1-\left(\frac{P_2}{P_1}\right)^{2/\gamma}\left(\frac{A_2}{A_1}\right)^2\right] \text{ since } \left(\frac{\rho_1}{\rho_2}\right)=\left(\frac{P_1}{P_2}\right)^{1/\gamma}$$

$$\Rightarrow V_2=\sqrt{\frac{\left(\frac{2\gamma}{\gamma-1}\right)\left(\frac{P_1}{\rho_1}\right)\left[1-\left(\frac{P_2}{P_1}\right)^{\gamma-1/\gamma}\right]}{\left[1-\left(\frac{P_2}{P_1}\right)^2\left(\frac{A_2}{A_1}\right)^2\right]}} \tag{9.47}$$

Mass flow rate at the throat section 2-2 be $\bar{m}$, then $\bar{m}=\rho_2 A_2 V_2=\rho_2 A_2 V_2 C_d$ and substituting Eq. (9.46) in Eq. (9.46a), we have

$$\bar{m}=\rho_2 A_2\sqrt{\frac{\left(\frac{2\gamma}{\gamma-1}\right)\left(\frac{P_1}{\rho_1}\right)\left[1-\left(\frac{P_2}{P_1}\right)^{\gamma-1/\gamma}\right]}{\left[1-\left(\frac{P_2}{P_1}\right)^2\left(\frac{A_2}{A_1}\right)^2\right]}} \tag{9.48}$$

Note:
In case of a nozzle, the inlet velocity can be assumed to be negligible because it can be connected to a reservoir having inlet velocity zero but in case of Venturimeter it is not possible as there will be some inlet velocity in the pipe to which the Venturi is connected.

Example 9.15: Venturimeter having inlet diameter 300 mm and throat diameter 150 mm. The pressure and temperature of air at the inlet section of the Venturimeter are 137 kPa (absolute) and 288 K respectively. And the pressure at the throat is 127 kPa (absolute). Taking $R=287$ J/kg – K, $\gamma=1.4$. Find the mass rate of flow of air through the Venturimeter.

Solution: Given data:

$$d_i=0.3\text{m},\ d_t=0.15\text{m},\ P_i=137\text{ kPa}$$
$$T_i=288\text{ K},\ P_t=127\text{ kPa},\ R=287\text{ J/kg}-\text{K},\ \gamma=1.4$$

Area at the inlet is $A_i = \frac{\pi d_i^2}{4} = \frac{3.14 * 0.3^2}{4} = 0.07068 \text{ m}^2$

Area at the throat is $A_t = \frac{\pi d_t^2}{4} = \frac{3.14 * 0.15^2}{4} = 0.01767 \text{ m}^2$

As per ideal gas equation $\left(\frac{P_i}{\rho_i}\right) = RT_i \Rightarrow \rho_i = \frac{P_i}{RT_i} = \frac{137000}{287 * 288} = 1.657 \text{ kg/m}^3$

As per the relation of adiabatic expansion, we have

$$\left(\frac{\rho_t}{\rho_i}\right) = \left(\frac{P_2}{P_i}\right)^{1/\gamma}$$

$$\Rightarrow \left(\frac{\rho_t}{1.657}\right) = \left(\frac{127}{137}\right)^{1/1.4}$$

$$\Rightarrow \rho_t = 1.57 \text{ kg/m}^3$$

Refer Eq. (9.48), for mass flow rate through a Venturimeter

$$\bar{m} = \rho_t A_t \sqrt{\frac{\left(\frac{2\gamma}{\gamma - 1}\right)\left(\frac{P_i}{\rho_i}\right)\left[1 - \left(\frac{P_t}{P_i}\right)^{\gamma - 1/\gamma}\right]}{\left[1 - \left(\frac{P_t}{P_i}\right)^2\left(\frac{A_t}{A_i}\right)^2\right]}} = 3.177 \text{ kg/s}$$

Hence, the mass flow rate through the Venturimeter is 3.177 kg/s.

Example 9.16: Calculate the air speed, if a Pitot tube is pointed into an airstream having 135 kPa pressure and temperature 45°c. The differential pressure reading of the Pitot tube is 10 kPa.

Solution: Given data:

$$\Delta P = 10 \text{ kPa},\ P_1 = 135 \text{ kPa},\ T_1 = 45 + 273 = 318 \text{ K},\ \gamma = 1.4,\ R = 287 \text{ J/kg–K}$$

Differential Pitot tube in an airstream has differential pressure $P_S = \Delta P + P_1 = 10 + 135 = 145 kPa$

We know
$$\frac{P_S}{P_1} = \left[1 + \left(\frac{\gamma - 1}{2}\right) M_1^2\right]^{\gamma/\gamma - 1}$$

So
$$\Rightarrow M_1 = 0.32 = \frac{V}{a} = \frac{V_1}{\sqrt{\gamma R T_1}}$$
$$\Rightarrow V_1 = 114.38 \text{ m/s}$$

Hence, the air speed is 114.38 m/s.

Example 9.17: Determine the throat area if air of 300g/s must flow through a nozzle. The inlet pressure and temperature being 2 bar, 383 K respectively. Neglect the inlet velocity and assume the coefficient of discharge as 0.97 and also assume that the flow is choked.

Solution: Given data:

$$\bar{m} = 0.3 \text{ kg/s}, P_1 = 2 \text{ bar}, T_1 = 383 \text{ K}, \gamma = 1.4, R = 287 \text{ J/kg-K}, C_d = 0.97$$

As per ideal gas equation $\left(\frac{P_1}{\rho_1}\right) = RT_1 \;\; \Rightarrow \rho_1 = \frac{P_1}{RT_1} = \frac{200000}{287 * 383} = 1.819 \text{ kg/m}^3$

As per the condition of choked, we have $\left(\frac{P_2}{P_1}\right) = \left(\frac{2}{\gamma+1}\right)^{\gamma/\gamma-1}$

$$\Rightarrow P_2 = 200000\left(\frac{2}{1.4+1}\right)^{3.5} = 105.65 \text{ kPa}$$

So the throat/exit pressure is 105.65 kPa and the exit density will be

$$\rho_2 = \rho_1\left(\frac{P_2}{P_1}\right)^{1/\gamma} = 1.153 \text{ kg/m}^3$$

We know $\bar{m} - C_d A_2 \sqrt{\left(\frac{2\gamma}{\gamma-1}\right) P_1 \rho_1 \left[\left(\frac{P_2}{P_1}\right)^{2/\gamma} - \left(\frac{P_2}{P_1}\right)^{\gamma+1/\gamma}\right]} \Rightarrow A_2 = 7.4895 * 10^{-4}$

Hence, the throat area is 0.00074895m^2.

EXERCISE

Objective

1. Fire hose nozzle is generally made of
   a) Divergent shape
   b) Convergent shape

c) Convergent – divergent shape
d) Cylindrical shape
e) None

2. Velocity of pressure waves due to pressure disturbances imposed in a fluid is equal to

a) $\sqrt{\dfrac{K}{\rho}}$ b) $\sqrt{K\rho}$ c) $\sqrt{\rho/K}$ d) $\sqrt{\dfrac{1}{\rho K}}$ e) None of the above

3. If air ($\gamma = 1.4, R = 287\text{J/kg} - \text{K}$) at 20°C is expanded isentropically, the maximum velocity that can be achieved is
a) 342.5 m/s
b) 766 m/s
c) 1000 m/s
d) 518 m/s
e) None.

4. Air ($\gamma = 1.4, R = 287\ \text{J/kg} - \text{K}$) flows at a Mach number of 1.5 in a duct. If the ambient temperature is 7°C, the stagnation temperature is
a) 133°C
b) 10.5°C
c) 406°C
d) 256°C.

5. In isentropic flow of a perfect gas
a) The velocity always decreases in conduits of increasing area.
b) The velocity is always critical at the throat of a nozzle.
c) If the flow is subsonic, through a convergent nozzle, the maximum velocity is always at the throat.
d) If the flow is in a convergent-divergent nozzle, the maximum velocity is always at the throat.
e) None.

6. An aeroplane is cruising at a speed of 800 km/h at an altitude where the air temperature is 0°C. The flight Mach number at this speed is nearly
a) 1.33
b) 0.67
c) 0.25
d) 0.24

7. The maximum Mach number for which the flow of air can be considered incompressible within 1% error is
a) 0.1
b) 0.2
c) 0.4
d) 0.6

8. Identify the incorrect statement.
For isentropic flow through a duct, the maximum mass flow possible is
a) Proportional to the stagnation pressure.

b) Proportional to the stagnation temperature.
c) Inversely proportional to the square root of stagnation temperature.
d) Proportional to the area of the throat.
e) None

9. In a supersonic flow, a diverging passage results in
   a) Increase in velocity and pressure.
   b) Pressure, density and temperature increase.
   c) Velocity, pressure and density increase.
   d) Pressure, density and momentum flux increase.
   e) None.
10. In an isentropic flow of air ($\gamma = 1.4, R = 287$ J/kg – K), the stagnation temperature and temperature at any Mach number are related as
    a) $(1 + 0.2M^2)$
    b) $(1 + 0.2M^2)^{2.5}$
    c) $1 + 0.2M^2)^{3.5}$
    d) $1 + 0.2M^2)^{-0.5}$
    e) None.

Answers

1. b	2. a	3. a	4. b	5. b	6. b
7. a	8. a	9. a	10.a		

Theory

1. Describe and derive equation for velocity of sound for compressible flow and extend it for the adiabatic.
2. Write a note on elastic pressure wave propagation of sounds in compressible flows.
3. What is Mach number? Why this parameter is so important for the study of flow of compressible fluid?
4. State the Bernoulli's theorem for compressible flow. Derive an expression for Bernoulli's equation when the process is isothermal.
5. Write a note variation of pressure ratio and mass flow.
6. Write a note on limits in incompressibility.
7. Derive an expression for the velocity of sound for compressible fluid when the process is assumed as (i) isothermal (ii) adiabatic.
8. Find an expression for mass rate of flow of compressible fluid through an orifice or nozzle fitted to a tank. What is the condition for maximum rate of flow?
9. Derive equation for a Venturimeter discharge in an incompressible flow.
10. Define stagnation pressure and stagnation temperature.

Problems

1. A cylinder contains 10 cubic metres of air at 50°C and pressure of 294.3 kPa (absolute). The air is compressed to a volume of 2 cubic meters. Assuming the process to be adiabatic, calculate the final pressure and temperature. Take $\gamma = 1.4$.

2. Oxygen of mass 5 kg at t = 20°, is compressed isentropically to 50% its original volume. Find the final temperature, pressure, the work required and change in internal energy and enthalpy. Take initial pressure as 100 kPa (absolute), $C_V = 649$ J/kg – K, $C_P = 909$ J/kg – K
3. Calculate the speed of sound in dry air at sea level, $t = 20°, \gamma = 1.4, R = 287$ J/kg – K.
4. A gas flows isothermally from a pipe of 15 cm to a pipe of 10 cm diameters respectively. While flowing, there is a decrease in pressure from 600 kPa to 450 kPa (absolute) respectively. Calculate the velocity of the gas flow in the 10 cm diameter. Take $t = 20°$, $\gamma =$ 1.4, R = 287 J/kg – K.
5. A supersonic plane flies at 200km/h in air having a pressure of 30 kPa (absolute) and density of 0.5 kg/m^3. Calculate the stagnation temperature, pressure and density at the nose of the plane. Take oxygen of mass 5 kg at t = 20°, is compressed isentropically to 50% its original volume. Find the final temperature, pressure, the work required and change in internal energy and enthalpy. Take initial pressure as 100 kPa (absolute), $\gamma = 1.4, R = 287$ J/kg – K.
6. A tank containing air at a pressure of 3000 kPa (absolute) and 40°C is discharged through a convergent nozzle attached to the tank. The velocity of the airflow from the nozzle exit is 250 m/s. Calculate the Mach number at the exit of the nozzle. Take $\gamma = 1.4$, $R =$ 287 J/kg – K.
7. A rocket travels at 1800 km/h in air of pressure 35.6 kPa (absolute) and temperature –37°C. Find the mach number and mach angle. Take $\gamma = 1.4$, $R = 287$ J/kg – K.

QUICK REFERENCE OF FORMULAE

Property	**Formulae**	**Remarks, if any**
1-D Continuity equation	$\frac{d\rho}{\rho} + \frac{dA}{A} + \frac{dV}{V} = 0$	
1-D General Bernoulli's equation and Bernoulli's equation in terms of adiabatic, isothermal, isobaric	$\frac{1}{g}\int\frac{dp}{\rho} + \frac{V^2}{2g} + Z = 0$; $\frac{\gamma}{\gamma-1}\left(\frac{P}{\rho g}\right) + \frac{V^2}{2g} + Z = C$ or $\frac{\gamma}{\gamma-1}(RT) + \frac{V^2}{2g} + Z = C$; $\left(\frac{P}{\rho g}\right)\log(P) + \frac{V^2}{2g} + Z = C$; $\frac{V^2}{2g} + Z = C$	For compressible flow.
General equation of velocity of sound & velocity of sound in terms of bulk modulus, adiabatic flow, isothermal.	$a = \sqrt{\frac{dP}{d\rho}}$; $a = \sqrt{\frac{K}{\rho}}$; $a = \sqrt{\gamma\left(\frac{P}{\rho}\right)} = \sqrt{\gamma RT}$; $a = \sqrt{\left(\frac{P}{\rho}\right)} = \sqrt{RT}$	
Mach angle α	$\sin\alpha = \frac{a}{V} = \frac{1}{M}$	

Contd.

Property	Formulae	Remarks, if any
Stagnation point variables P_S, T_S, ρ_S	$P_S = P_1\left[1+\left(\frac{\gamma-1}{2}\right)M_1\right]^{\gamma/\gamma-1}$ or $\Delta P = P_1\left[\left(\frac{\gamma-1}{2}\right)M_1^2\right]^{\gamma/\gamma-1}$ $T_S = T_1\left[1+\left(\frac{\gamma-1}{2}\right)M_1\right] = T_1 + \frac{V_1^2}{2*C_p}$ $\rho_S = \rho_1\left[1+\left(\frac{\gamma-1}{2}\right)M_1\right]^{1/\gamma-1}$ $P_S = P_1 + \frac{1}{2}\rho V_1^2$; $\frac{P_2-P_1}{\frac{1}{2}\rho V_1^2} = 1-\frac{A_1^2}{A_2^2}$ if compressibility is not considered (i.e. ρ = constant) if M < 0.2, then compressibility factor $CF = \left(\frac{2}{\gamma M_1^2}\right)\left\{\left[\left(\frac{T_s}{T_1}\right)\right]^{\gamma/\gamma-1} - 1\right\}$ and $P_S = P_1\left[1+\left(\frac{\gamma-1}{2}\right)M_1^2 * CF\right]^{\gamma/\gamma-1}$ If M > 0.2, then $CF = \left(1+\left(\frac{M_1^4}{4}\right)+\left(\frac{2-\gamma}{24}\right)M_1^4+\ldots\right)$ and $\Rightarrow P_s = P_1 + \left(\frac{\rho V_1^2}{2}\right) * CF$	$P_s, T_s, \rho_s, P_1, T_1, \rho_1$, are pressure, temperature and density at stagnation point and at point 1 on same stream-line, respectively. γ, M_1 are specific weight and Mach numbers at point 1. Actual velocity $V_a = V_1 / \sqrt{CF}$
Area, pressure and velocity relation of a compressible flow	$\frac{dA}{A} = \left[\left(\frac{dV}{V}\right)(M^2-1)\right]$; $dP = \frac{dA}{A}\left[\rho V^2\left(\frac{1}{1-M^2}\right)\right]$	
Velocity and mass flow rate through a nozzle	$V_2 = \sqrt{\left(\frac{2\gamma}{\gamma-1}\right)\left(\frac{P_1}{\rho_1}\right)\left[1-\left(\frac{P_2}{P_1}\right)^{\gamma-1/\gamma}\right]}$; $\overline{m} = C_d A_2 \sqrt{\left(\frac{2\gamma}{\gamma-1}\right)P_1\rho_1\left[\left(\frac{P_2}{P_1}\right)^{2/\gamma} - \left(\frac{P_2}{P_1}\right)^{\gamma+1/\gamma}\right]}$ $\frac{A_{throat}}{A_{exit}} = M_{exit}\left(\frac{P_{exit}}{P_{throat}}\right)^{1+\gamma/2\gamma}$	

Contd.

Property	**Formulae**	**Remarks, if any**
Condition for maximum velocity and mass flow rate through a nozzle.	$\left(\frac{P_2}{P_1}\right) = \left(\frac{2}{\gamma+1}\right)^{\gamma/\gamma-1}$; $V_{2_{max}} = \sqrt{\left(\frac{2\gamma}{\gamma+1}\right)\left(\frac{P_1}{\rho_1}\right)}$ $\bar{m}_{max} = A_2\sqrt{\left(\frac{2\gamma}{\gamma+1}\right)P_1\rho_1\left[\left(\frac{2}{\gamma+1}\right)^{2/\gamma} - \left(\frac{\gamma-1}{\gamma+1}\right)^{\gamma+1/\gamma-1}\right]}$	$V_{2_{max}}$, $\bar{m}_{max}$ are maximum velocity and mass flow rate
Velocity and mass flow rate at the outlet, through a venturimeter.	$V_2 = \sqrt{\frac{\left(\frac{2\gamma}{\gamma-1}\right)\left(\frac{P_1}{\rho_1}\right)\left[1-\left(\frac{P_2}{P_1}\right)^{\gamma-1/\gamma}\right]}{\left[1-\left(\frac{P_2}{P_1}\right)^2\left(\frac{A_2}{A_1}\right)^2\right]}}$; $\bar{m} = C_d\rho_2 A_2\sqrt{\frac{\left(\frac{2\gamma}{\gamma-1}\right)\left(\frac{P_1}{\rho_1}\right)\left[1-\left(\frac{P_2}{P_1}\right)^{\gamma-1/\gamma}\right]}{\left[1-\left(\frac{P_2}{P_1}\right)^2\left(\frac{A_2}{A_1}\right)^2\right]}}$	
Variable properties with Mach number	$\frac{T_1}{T_2} = \left[\frac{1+\left(\frac{\gamma-1}{2}\right)M_1^2}{1+\left(\frac{\gamma-1}{2}\right)M_2^2}\right]$ $\frac{P_1}{P_2} = \left[\frac{1+\left(\frac{\gamma-1}{2}\right)M_1^2}{1+\left(\frac{\gamma-1}{2}\right)M_2^2}\right]^{\gamma/\gamma-1}$	
	$\frac{\rho_1}{\rho_2} = \left[\frac{1+\left(\frac{\gamma-1}{2}\right)M_1^2}{1+\left(\frac{\gamma-1}{2}\right)M_2^2}\right]^{1/\gamma-1}$ $\frac{A_2}{A_1} = \left(\frac{M_1}{M_2}\right)\left(\left[\frac{1+\left(\frac{\gamma-1}{2}\right)M_1^2}{1+\left(\frac{\gamma-1}{2}\right)M_2^2}\right]\right)^{\gamma+1/2(\gamma-1)}$	

CHAPTER 10

Introduction and Impact of Jets

INTRODUCTION TO HYDRAULICS

Hydraulics, the other name of Hydraulic Mechanics, is an engineering application of Fluid Mechanics. The **earth** comprises compositions like air, water, land, life—including humans and an additional external source such as solar. Human beings do construct dams, reservoirs and other storing devices for storing water and power generation. Hydraulic power systems are not only developed for the power generation but also for hydraulic operations of machines like hydraulic lifts, jacks, rams, pumps, presses, turbines, etc. and the operation of these are based on the fluid flow theories.

Almost every hydraulic device work basically on the concept of Pascal law, as the pressure exerts equally in all the directions, with equal magnitude like hydraulic pump used for power steering, hydraulic elevators, presses, hydraulic brake system, etc.

Major part of the hydraulics usage is in hydraulic power generation, which is based on the hydrostatic head, which arises due to constant motion of water through a vast global evaporation cycle from oceans, seas and other water reservoirs and forming as clouds and precipitates as rain

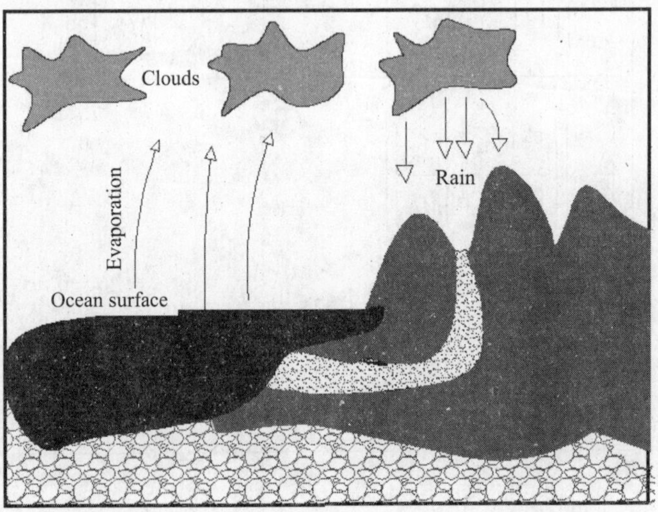

Evaporation cycle

and falls on high altitudes and so flows back to the ocean, due to the influence of gravity. While flowing back, the water does possess high potential energy of this water cycle, which is tapped most efficiently with hydropower, by converting the potential energy to kinetic energy of the water flow, i.e. water falling from high elevation to a lower elevation called hydrostatic head, do possess potential energy depending upon the elevation difference and is used to operate hydraulic machines like the water wheel and the turbine. The amount of work developed in the hydraulic machines is dependent on two things: the distance the water has to fall (head) and how much water is flowing (discharge). Hydroelectric power stations are, therefore, situated where they can take advantage of the greatest fall of a large quantity of water—at the bottom of deep and steep sided valleys or near the base of dams.

Water power can be produced in this way from many natural sources, such as waterfalls and dammed rivers and where no natural sources are available; an artificial reservoir can be made. When energy is plentiful, it is used to pump water into the reservoir; the water is then available as a power source to drive turbines when energy becomes scarce.

Comparison between different types of sources for power generation

Factor	Solar	Wind	Hydraulics
Reliability	Good in day times	Average	Plenty
Maintenance	High	High	Low
Installation	Difficult	Simple	Simple
Operation	Quiet	Noisy	Quiet
Controllability	Good	Difficult	Good.

HYDRAULIC POWER MACHINES

Hydraulic power machines generate power by converting the hydraulic energy (possessed by the water due to its kinetic energy) into mechanical energy, in form of rotation of shafts of the power generators.

This conversion of hydraulic energy to mechanical energy is attained due to the sudden impact of fluids (called **Jet**) on to a stationary/ moving objects (called **Plates/Vanes,** which may be flat or curved in shape), making them (vanes) to rotate with high speeds and thereby rotating the shaft of the generator containing these vanes.

CLASSIFICATION OF IMPACT OF JETS

Fluid streams emerging out from devices like nozzles or falling from a high altitude, in form of jets, do possess high kinetic energy which can give a high thrust force on the objects (plates/vanes) that come in between the stream flow. The magnitude and direction of the thrust force raised due to the impact of the stream jet on to the vanes can be classified based on the relativity of the flow direction, shape, positioning and movement of the vanes and the classification can be derived as given in the chart below, shown in Figure 10.1.

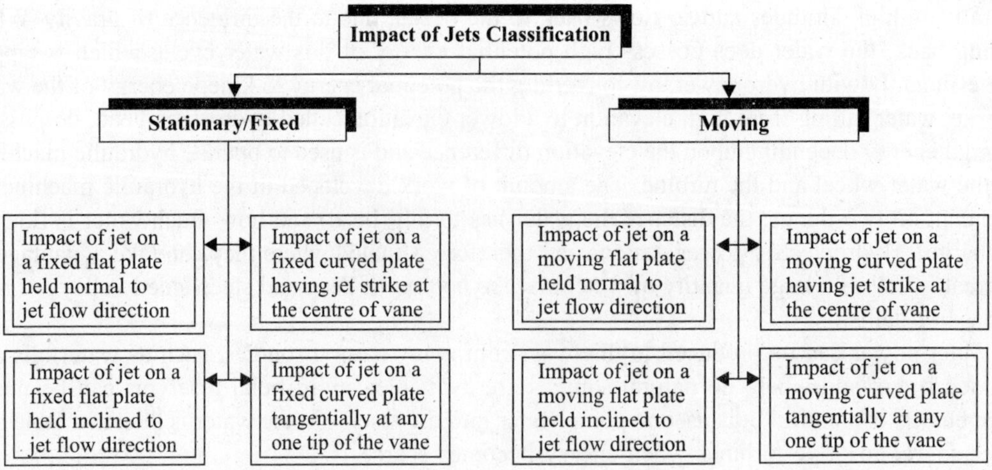

Figure 10.1: Classification of impact of jets

Basically, this chapter is based on **Impulse-Momentum equation,** i.e., sudden impact of a fluid jet on to a vane decreases the jet velocity and thus increases/generates motion to the vane or in other words, the **impulse** (*F*dt* = Impulse) due to impinging of a fluid jet on to a vane (stationary or moving) transforms the hydraulic energy of the jet into the **change in momentum** [*d*((mV)] of the vane.

So mathematically, impulse-momentum equation can be written as $Fdt = (mdV)$ on in other words, it is modification of Newton's second law

$$\Rightarrow F = \frac{d}{dt}(mV)$$

As the mass of the fluid jet impinging on the vane is constant, i.e. m = constant, then we have $Fdt = (mdV)$ where "dV" is change in velocity between the initial and final jet, in the direction of the change in movement and the product of force and time is impulse force (normal thrust force) acting in the direction of the normal, to the surface.

So $\quad F = \bar{m}(V_1 - V_2)$ where $\begin{matrix} V_1 \textit{ is velocity of jet initially} \\ V_2 \textit{ is velocity of jet finally} \end{matrix}$

$$\boxed{\begin{matrix} \Rightarrow F = \rho Q(V_1 - V_2) \\ or \\ F = \rho a V_r (V_1 - V_2) \end{matrix}} \qquad \text{since } \bar{m} = \rho A V = \rho Q \qquad (10.1)$$

V_r is relative velocity of jet w.r.t. plate

Equation 10.1 is the **impulse-momentum equation of impact of jet on plates/vanes.**

IMPACT OF JET ON A FIXED PLATE

Case a: Impact of jet on a fixed flat plate held inclined to the jet flow direction: Consider an amount of mass flow rate ($\bar{m} = \rho AV = \rho Q$) of fluid jet leaving a nozzle exit with a velocity of "V" and strikes an inclined stationary/fixed plate, which is at an angle "ϕ" to the horizontal.

As soon as the fluid jets hits the stationary/fixed plate, the jets split into two parts, one going upstream and other towards downstream, with velocity "μV", where "μ" is the friction coefficient if the plate is not smooth. And hence the actual discharge from the nozzle exit splits into discharges of "Q_1","Q_2" along the upstream and downstream respectively, with an inclination of "ϕ"and"$180-\phi$" with respect to actual jet discharge "Q", as shown in Figure 10.2 and so

$$Q = Q_1 + Q_2 \tag{10.2}$$

Let the thrust force (due to impact of jet on the plate) normal to inclined plate be "F_n" then based on **Impulse-Momentum equation**, refer Eq. (10.1), we have $F_n = \rho a V_r (V_1 - V_2)$.

If the plate is assumed to be smooth, then the velocity of the jet split, will be same as that of the actual discharged jet and so $V_r = V - 0$ since vane is fixed..

And hence the Eq. (10.1) becomes

$$F_n = \rho a V (V_1 - V_2) \tag{10.3}$$

The **normal thrust force** will be at an inclination "θ" to the horizontal, as the plate is inclined and so we can resolve the normal thrust force into horizontal (along the exit of jet flow) and vertical components, so the velocity of the normal force will be,

$$V_1 - V_2 = V\cos\theta = V\sin\phi \text{ since } \theta + \phi = 90^0$$

as shown in Figure 10.2a and hence Eq. (10.3) reduces to

$$\boxed{F_n = \rho a V (V\sin\phi)} \tag{10.4}$$

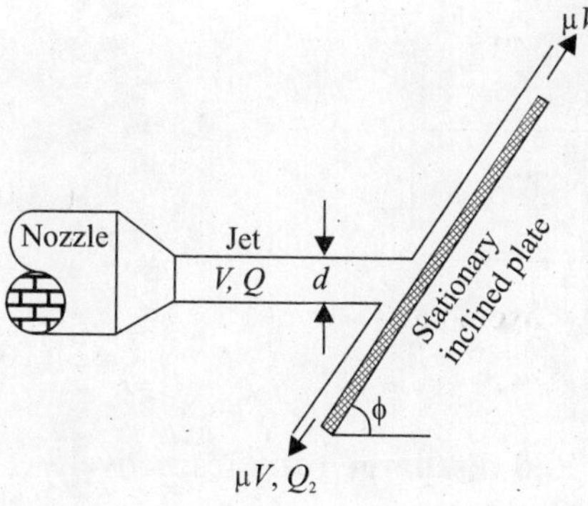

Figure 10.2: Impact of jet on an inclined stationary or fixed plate

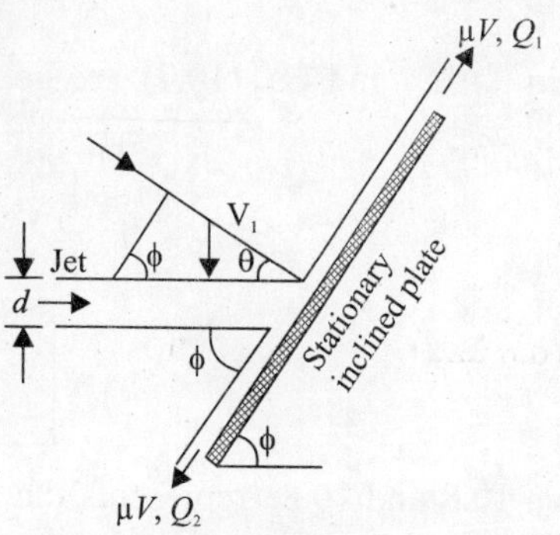

Figure 10.2a: Velocity components on an inclined stationary plate

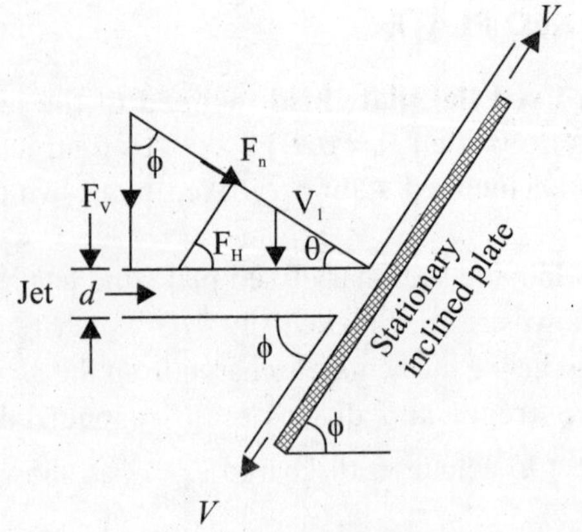

Figure 10.2b: Force components on an inclined stationary or fixed plate

And the **thrust force components,** as shown in Figure 10.2b will be

$$F_H = F_n \sin\phi = \rho a V (V \sin\phi)^2$$
$$F_V = F_n \cos\phi = \rho a V (V \cos\phi \sin\phi) \qquad (10.5)$$

The magnitude and direction of the **Resultant thrust force** becomes

$$F = \sqrt{F_H^2 + F_V^2} \ \& \ \theta = \tan^{-1}\left(F_V / F_H \right) \qquad (10.6)$$

As assumed the plate to be smooth, the frictional resistance will be zero and hence sum of the mass flow rate along the direction parallel to the plate is zero and so $\overline{m}_1 - \overline{m}_2 - \overline{m} = 0$, where $\overline{m}_1$, $\overline{m}_2$, $\overline{m}$ are mass flow rates along the upstream, downstream and from the exit of the nozzle respectively.

So $$\rho Q_1 V - \rho Q_2 V - \rho Q V \cos\phi = 0$$

$$\Rightarrow Q_1 - Q_2 - Q\cos\phi = 0 \qquad (10.7)$$

From Eq. (10.2) and Eq. (10.7), we get

$$Q_1 = \frac{Q}{2}(1 + \cos\phi), Q_2 = \frac{Q}{2}(1 - \cos\phi) \qquad (10.8)$$

$$\Rightarrow \frac{Q_1}{Q_2} = \frac{(1+\cos\phi)}{(1-\cos\phi)} \qquad (10.8a)$$

Equation 10.8 and 10.8a represents the **split discharges and discharge ratio** respectively.

Example 10.1: Determine the discharge of the jet, normal force, its components and resultant force inclination, exerted by a jet of diameter 10 cm flowing with a velocity 12 m/s on a stationary flat plate, inclined at 30° to horizontal. And also find the splitted discharges.

Solution: Given data: $d = 0.1\,\text{m}$, $V = 12$ m/s, $\phi = 30^0$

Area of the water jet is $a = \dfrac{\pi}{4}d^2 = \dfrac{3.14}{4} * 0.1^2 = 0.00785\ \text{m}^2$

Discharge of the jet $Q = aV = 0.00785 * 12 = 0.0942\ \text{m}^3/\text{s}$

We know, the normal force on an inclined flat plate is $F_n = \rho aV(V\sin\phi)$ refer Eq. (10.4)

So $$F_n = \rho aV(V\sin\phi) = 1000 * 0.00785 * 12(12 * \sin 30°) = 565.2\ \text{N}$$

Refer Eq. (10.5) for normal force component

$$F_H = F_n \sin\phi = 1000 * 0.00785 * 12(12 * \sin 30°) = 282.6\,\text{N}$$

$$F_V = F_n \cos\phi = 1000 * 0.00785 * 12(12 * \cos 30° * \sin 30°) = 489.5\,\text{N}$$

Refer Eq. (10.6) for resultant force inclination $\theta = \tan^{-1}\left(F_V / F_H\right) \Rightarrow \theta = \tan^{-1}\left(489.5/282.6\right) = 60°$

Refer Eq. (10.8) for splitted discharges

$$Q_1 = \frac{Q}{2}(1 + \cos\phi) = \frac{0.0942 * (1 + \cos 30°)}{2} = 0.0879\ \text{m}^3/\text{s}$$

$$Q_2 = \frac{Q}{2}(1 - \cos\phi) = \frac{0.0942 * (1 - \cos 30°)}{2} = 0.0063\ \text{m}^3/\text{s}$$

The **discharge, normal force, components, its inclination and splitted discharges are** 0.0942 m³/s, 565.2 N, 282.6 N, 489.5 N, 60°, 0.0879 m³/s, 0.0063 m³/s respectively.

Case b: Impact of jet on a fixed flat plate held normal to the jet flow direction:
In this case, the flat plate is held perpendicular to the jet flow direction and so the inclination of the plate with the horizontal becomes $\phi = 90°$, as shown in Figure 10.3 and so substituting this value of inclination in the Eq. (10.4), Eq. (10.5), Eq. (10.6) and Eq. (10.8), we get normal thrust force, Normal thrust force components, resultant thrust force and its inclination and discharge of the splitted jet; i.e. from Eq. (10.4), we have

$$F_n = \rho aV^2 \tag{10.9}$$

From Eq. (10.5), we have

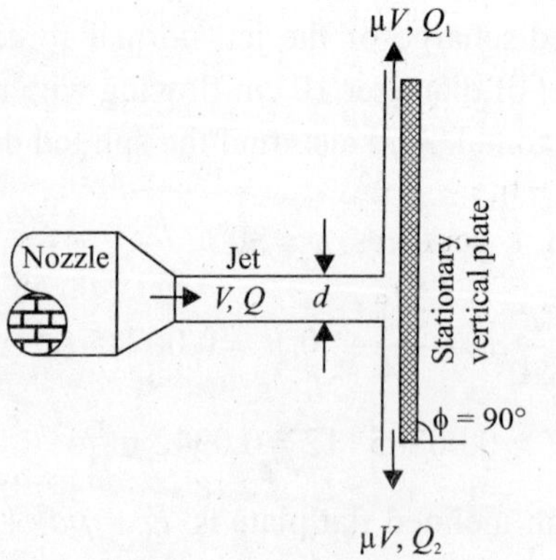

Figure 10.3: Impact of jet on a statuary vertical plate

From Eq. (10.6), we have

$$F_H = F_n \sin\phi = \rho a V^2 \quad F_V = F_n \cos\phi = 0 \qquad (10.10)$$

$$F = F_H \;\&\; \theta = \tan^{-1}\left(0/F_H \right) = 0^0 \qquad (10.11)$$

From Eq. (10.8), we have

$$Q_1 = \frac{Q}{2} = Q_2 \qquad (10.12)$$

And so from the Eq. (10.12) it is clear that the "**discharge ("*Q*") striking a fixed flat plate held normal to the direction of fluid jet flow, deflects perpendicularly with an amount of half the actual discharge, in both upstream and downstream directions of the plate**".

Example 10.2: Determine the discharge of the jet, normal force, its components exerted by a jet of diameter 10 cm flowing with a velocity 12 m/s on a stationary flat plate held normal to jet flow. And also find the split discharges.

Solution: Given data: $d = 0.1$ m, $V = 12$ m/s, $\phi = 90°$

Area of the water jet is $a = \frac{\pi}{4} d^2 = \frac{3.14}{4} * 0.1^2 = 0.00785 \text{ m}^2$

Discharge of the jet $Q = aV = 0.00785 * 12 = 0.0942 \text{ m}^3/\text{s}$

We know, the normal force on an inclined flat plate is $F_n = \rho a V (V \sin\phi)$ refer Eq. (10.4)

So, $F_n = \rho a V (V \sin\phi) = 1000 * 0.00785 * 12(12 * \sin 90°) = 1130.4 \text{ N}$

Refer Eq. (10.5) for normal force component

$$F_H = F_n \sin\phi = 1000 * 0.00785 * 12(12 * \sin 90°) = 1130.4 \text{ N}$$

$$F_V = F_n \cos\phi = 1000 * 0.00785 * 12(12 * \cos 90° * \sin 90°) = 0 \text{ N}$$

Refer Eq. (10.8) for split discharges

$$Q_1 = \frac{Q}{2}(1+\cos\phi) = \frac{0.0942 * (1+\cos 90°)}{2} = 0.0471 \text{ m}^3/\text{s}$$

$$Q_2 = \frac{Q}{2}(1-\cos\phi) = \frac{0.0942 * (1-\cos 90°)}{2} = 0.0471 \text{ m}^3/\text{s}$$

The **discharge, normal force, components, its inclination and split discharges are** 0.0942m^3/s, 1130.4 N, 1130.4 N, 0.0 N, 6000 m^3/s, 6000 m^3/s respectively.

Case c: Impact of jet on a fixed curved plate at the center of symmetry:
Consider an amount of mass flow rate ($\dot{m} = \rho AV = \rho Q$) of fluid jet leaving a nozzle exit with a velocity of "V" and strikes a curved stationary/fixed vane. As soon as the fluid jets hit the stationary/fixed vane, the jets split into two parts, making an angle "ϕ" to the horizontal, one part going upstream and other towards downstream, with velocity "μV", where , "μ" is the friction if the plate is not smooth, leaving from the tips of the curved vane. But as the deflected jets passing on the surface of the curved vane, changes its direction from one point to the other point due to curved profile of the vane and hence the deflected jet leaving through the tips do have constant velocity depending on the angle of deflection of the jet "ϕ", shown in Figure 10.4 and hence the exit deflected jet can be resolved into two horizontal and vertical components of the exit velocity "V", but with change in direction and so

$$V_1 - V_2 = V - (-V\cos\phi) \text{ and } V_r = V - 0 \text{ since vane fixed}$$

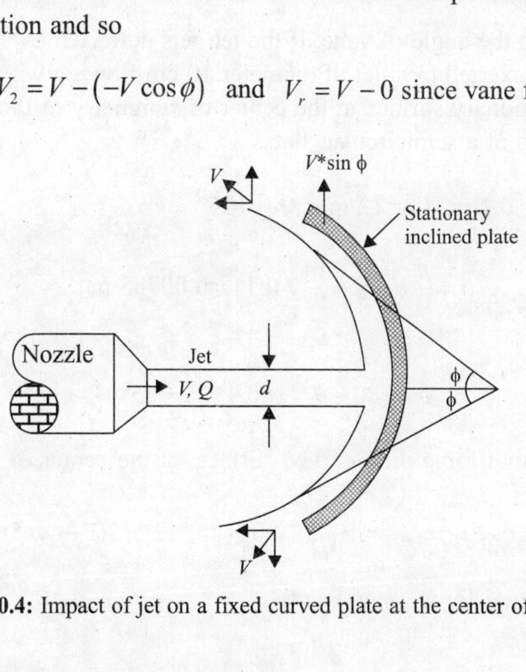

Figure 10.4: Impact of jet on a fixed curved plate at the center of symmetry

Let the thrust force (due to impact of jet on the curved vane) normal to curved vane be "F_n" then based on **Impulse-Momentum equation**, refer Eq. (10.1), we have $F_n = \rho a V_r (V_1 - V_2)$. And so we get normal thrust force as $F_n = \rho a V[V - (-V\cos\phi)]$

$$\Rightarrow F_n = \rho a V^2 [1 + \cos\phi] \tag{10.13}$$

or

If the surface is not smooth $$F_n = \rho a V^2 [1 + \mu \cos\phi] \tag{10.14}$$

To note:

a) If the inclination of the deflected jet is at 0° (i.e. $\phi = 0°$), then the vane will be a semi-circular flat fixed plate and the Eq. 10.13 becomes

$$F_n = 2\rho a V^2 \tag{10.15}$$

and observing Eq. (10.13) and Eq. (10.15), we can note that "**normal thrust force on a fixed semicircular curved vane is always twice than the normal thrust force of a fixed curved vane**".

b) If the inclination of the deflected jet is at 90° (i.e. $\phi = 90°$), then the vane will be a flat fixed plate and the Eq. (10.13) becomes

$$F_n = \rho a V^2 \tag{10.16}$$

and observing Eqs. (10.13) and (10.16), we can note that "**normal thrust force on a fixed curved vane is always more than the normal thrust force of a fixed flat plate**".

Example 10.3: Determine the angle of vane, if the jet gets deflected by 125° by the vane, normal force and its components exerted by a jet of diameter 10 cm flowing with a velocity 12 m/s when impinged on a curved stationary surface at the center of symmetry of the vane. And also find the normal force if the vane is of a semicircular, flat.

Solution: Given data: $d = 0.1$ m, $V = 12$ m/s, $\theta = 125°$

Area of the water jet is $$a = \frac{\pi}{4} d^2 = \frac{3.14}{4} * 0.1^2 = 0.00785 \text{ m}^2$$

Angle of vane $$\phi = 180 - \theta = 180 - 125 = 55^0$$

Refer Eq. (10.13) for normal force on a curved surface, at the center of symmetry is

$$F_n = \rho a V^2 (1 + \cos\phi) \Rightarrow F_n = 1000 * 0.00785 * 12 * 12(1 + \cos 55°) = 1.78 \text{ kN}$$

Refer Eq. (10.15) for normal force components

$$F_H = F_n \sin\phi = 1000 * 0.00785 * 12 * 12(1 + \cos 55°) * \sin 55° = 1.458 \text{ kN}$$

$$F_V = F_n \cos\phi = 1000 * 0.00785 * 12 * 12(1 + \cos 55°) * \cos 55° = 1.021 \text{ kN}$$

If the surface is semicircular:

Then angle of vane $\phi = 180 - 180 = 0°$

Then normal force on a semicircular curved surface, at the centre of symmetry is

$$F_n = 2\rho a V^2 = 2 * 1000 * 0.00785 * 12 * 12 = 2.26 \text{ kN}$$

Normal force components

$$F_H = F_n \sin\phi = 2260 * \sin 0° = 0.0 \text{ kN}$$
$$F_V = F_n \cos\phi = 2260 * \cos 0° = 2.26 \text{ kN}$$

If the surface is flat:

Then angle of vane $\phi = 180 - 90 = 90°$

Then normal force on a flat surface, at the centre of symmetry is

$$F_n = \rho a V^2 = 1000 * 0.00785 * 12 * 12 = 1130.4 \text{ N}$$

Normal force components

$$F_H = F_n \sin\phi = 1130.4 * \sin 90° = 1130.4 \text{ N}$$

$$F_v = F_n \cos\phi = 1130.4 * \cos 90° = 0.0 \text{ N}$$

The vane angle, normal force and its components are 55°, 1.78 kN, 1.458 kN, 1.021 kN respectively.

For a semicircular vane angle, normal force and its components are 0°, 2.26 kN, 0.0 kN, 2.26 kN respectively.

For a flat surface vane angle, normal force and its components are 90°, 1130.4 N, 1130.4 kN, 0.0 N respectively.

Example 10.4: Determine the ratio between outgoing velocity and incoming velocity, if the surface of vane is not smooth and the jet gets deflected by 165° by the vane and normal force exerted by the jet of diameter 2.7 cm flowing with a velocity 12 m/s when impinged on a curved stationary surface at the center of symmetry of the vane is 250 N.

Solution: Given data: $d = 0.027$ m, $V = 12$ m/s, $\theta = 165°$, $F_n = 250$ N

Area of the water jet is $a = \frac{\pi}{4} d^2 = \frac{3.14}{4} * 0.027^2 = 0.000572 \text{ m}^2$

Angle of vane $\phi = 180 - \theta = 180 - 165 = 15°$

Refer Eq. (10.14) for normal force on a curved surface, at the centre of symmetry, having rough surface is

$$F_n = \rho a V^2 (1 + \mu \cos\phi) \Rightarrow 250 \text{ N} = 1000 * 0.000572 * 12 * 12(1 + \mu \cos 15°)$$

$$\Rightarrow (1 + \mu \cos 15°) = 3.035 \Rightarrow \mu = 2.1$$

The ratio of outgoing and incoming velocities is 2.1.

Case d: Impact of jet on a fixed curved plate at any one tip of the plate:

Consider an amount of mass flow rate ($\bar{m} = \rho AV = \rho Q$) of fluid jet leaving a nozzle exit with a velocity of "V" and strikes a curved stationary/fixed vane at any one of the tip, tangentially. As soon as the fluid jets hits the stationary/fixed curved vane, the jet slides over the surface of the vane, by the amount of inclination "ϕ & θ", which the vane tips have been deflected and jet leaves out through the other tip of the vane, as shown in Figure 10.5.

As the deflected jet passing on the surface of the curved vane, changes its direction from one point to the other point due to curved profile of the vane, the deflected jet leaving through the tips do exert a normal thrust force along the surface of the vane, which can be resolved into two horizontal and vertical components in terms of the velocity "V".

Let the thrust force (due to impact of jet on the curved vane) normal to curved vane be , "F" then based on **impulse-momentum equation**, refer Eq. (10.1), we have $F_n = \rho a V_r (V_1 - V_2)$. And so we get normal thrust force as $F_n = \rho a V[V - 0]$

$$\Rightarrow F_n = \rho a V^2 \tag{10.17}$$

We have the **horizontal thrust force components and vertical thrust force components** as difference (as there is change of direction) of two velocity components (inlet and outlet velocity) respectively

$$\begin{aligned} F_H &= \rho a V[V\cos\phi - (-V\cos\theta)] = \rho a V^2(\cos\phi + \cos\theta) \\ F_V &= \rho a V[V\sin\phi - (V\sin\theta)] = \rho a V^2(\sin\phi - \sin\theta) \end{aligned} \tag{10.18}$$

And the **resultant thrust force** as

$$F = \sqrt{F_H^2 + F_V^2} \tag{10.18a}$$

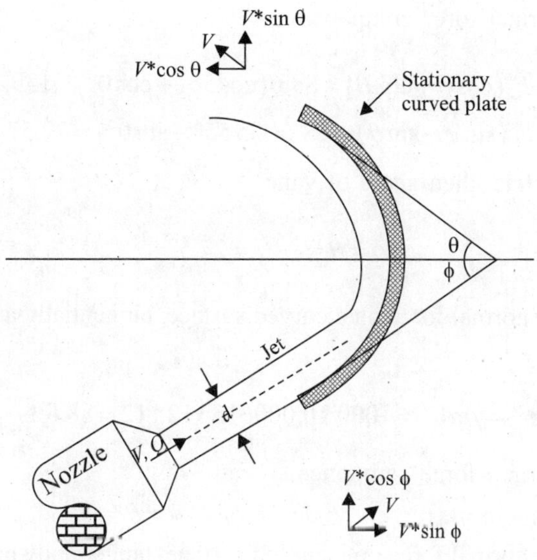

Figure 10.5: Jet strike tangentially, at one of the tip of a statuary curved plate

Note: If the inclination of the deflected jet be at 0° (i.e. $\theta = \phi = 0^\circ$), then the vane will be a semi-circular fixed plate and the Eq. 10.18 becomes

$$F_H = 2\rho a V^2 = F_n = F \tag{10.18b}$$

Example 10.5: Determine the total angle of vane, if the jet gets deflected by 125° on an unsymmetric curved vane having a jet strike tangentially at bottom tip. The flow of jet at the outlet tip is tangent to the horizontal. Find also the normal force and its components exerted by a jet of diameter 2.8 cm flowing with a velocity 12 m/s. And also find the normal force if the vane is of a symmetric, semi-circular.

Solution: Given data: $d = 0.028$ m, $V = 12$ m/s, $\theta = 0°$

Area of the water jet is $\quad a = \frac{\pi}{4} d^2 = \frac{3.14}{4} * 0.028^2 = 0.000615 \text{ m}^2$

Angle of vane $\quad \phi + \theta = 180 - 125 = 55^0 \;\; \Rightarrow \phi = 55^0$

Refer Eq. (10.17) for normal force on an curved surface, tangentially at bottom tip of un-symmetric vane is $F_n = \rho a V^2 = 1000 * 0.000615 * 12 * 12 = 88.6$ N

Refer Eq. (10.18) for normal force components

$$F_H = F_n(\cos\phi + \cos\theta) = 88.6(\cos 55° + \cos 0°) = 139.4 \text{ N}$$
$$F_V = F_n(\sin\phi - \sin\theta) = 88.6(\sin 55° - \sin 0°) = 72.57 \text{ N}$$

If the surface is symmetric, then angle of vane

$$\phi = \theta = {}^{55}\!/_{2} = 27.5°$$

Refer Eq. (10.17) for normal force on a curved surface, tangentially at bottom tip of symmetric vane is

$$F_n = \rho a V^2 = 1000 * 0.000615 * 12 * 12 = 88.6 \text{ N}$$

Refer Eq. (10.18) for normal force components

If the surface is semicircular, then angle of vane $\phi = \theta = 0°$

Refer Eq. (10.17) for normal force on a curved surface, tangentially at bottom tip of symmetric vane is $F_n = \rho a V^2 = 1000 * 0.000615 * 12 * 12 = 88.6$ N.

Refer Eq. (10.18b) for normal force components

$$F_H = F_n(\cos\phi + \cos\theta) = 88.6(\cos 0° + \cos 0°) = 88.6 \text{ N}$$
$$F_V = F_n(\sin\phi - \sin\theta) = 88.6(\sin 0° - \sin 0°) = 0.0 \text{ N}$$

IMPACT OF JET ON A MOVING PLATE

Case a: Impact of jet on a moving flat plate held inclined to the jet flow direction:
Consider a fluid jet with velocity "V", discharging from a nozzle exit and strikes a moving flat plate. The jet strikes the moving plate and deflects into two parts, one part flowing in the upstream direction and the other flowing downstream direction of the moving plate. But as there is a relative motion between the jet velocity ("V") and moving plate velocity ("u"), the effective velocity of the fluid jet striking the moving plate decreases to (V-u) and so the deflected jet portions has a velocity of (V-u), after strike, as shown in Figure 10.6.

Let the thrust force (due to impact of jet on the moving plate) normal to moving inclined plate be "F_n", then based on **impulse-momentum equation**, refer Eq. (10.1), we have

$$F_n = \rho a V_r (V_1 - V_2)$$

If the plate is assumed to be smooth, then the velocity of the jet split, will be same as that of the relative velocity of jet;

so $V_r = (V - u)$ since vane is moving with velocity "u" & $V_1 = (V - u)\sin\phi$, $V_2 = 0$

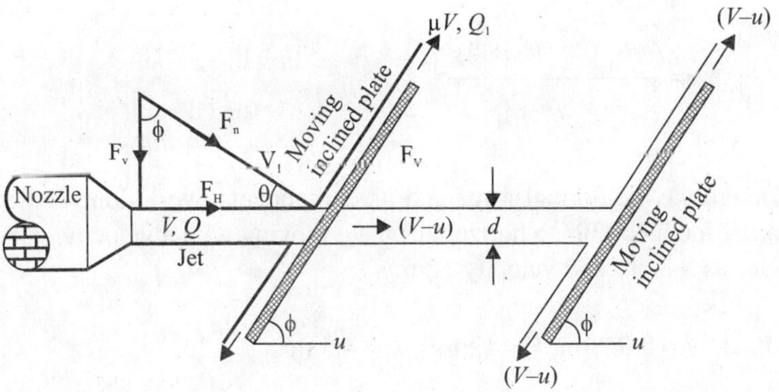

Figure 10.6: Impact of jet on an inclined moving flat plate

The **normal thrust force** will be at an inclination "θ" to the horizontal, as the plate is inclined and so we can resolve the normal thrust force into horizontal (along the exit of jet flow) and vertical components, so the velocity of the normal force will be, and hence

Eq. (10.1) reduces to

$$F_n = \rho a\left[(V-u)^2 \sin\phi\right] \tag{10.19}$$

And the **thrust force components** will be

$$\begin{aligned} F_H &= F_n \sin\phi = \rho a\left[(V-u)\sin\phi\right]^2 \\ F_V &= F_n \cos\phi = \rho a\left[(V-u)^2 \sin\phi\cos\phi\right] \end{aligned} \tag{10.20}$$

The magnitude and direction of the **resultant thrust force** becomes

$$F = \sqrt{F_H^2 + F_V^2} \;\&\; \theta = \tan^{-1}\left(F_V / F_H\right) \tag{10.21}$$

The fluid jet striking a moving plate causes the plate to travel some distance along the horizontal direction and hence some amount of work is done by the jet on the plate and it can be given as

$$Work\ done = F_n * u = \rho a u\left[(V-u)^2 \sin\phi\right] \tag{10.22}$$

Efficiency of the impact of jet on the plate will be

$$\eta = \frac{Work\ done\ on\ the\ plate}{Kinetic\ energy\ of\ the\ jet}$$

So $$\eta = \frac{\rho a u\left[(V-u)^2 \sin\phi\right]}{\frac{1}{2}\rho a V^3} \qquad \Rightarrow \eta = \frac{2u\left[(V-u)^2 \sin\phi\right]}{V^3} \tag{10.22a}$$

Example 10.6: Determine the normal force and its components, work done and efficiency of the jet impinged on a flat inclined (30° to horizontal) plate moving with a velocity of 4.5 m/s. Take the flow of jet diameter as 2.8 cm and velocity 12 m/s.

Solution: Given data: $d = 0.028$ m, $V = 12$ m/s, $u = 4.5$ m/s, $\phi = 30°$

Area of the water jet is $a = \frac{\pi}{4}d^2 = \frac{3.14}{4} * 0.028^2 = 0.000615 \text{ m}^2$

Refer Eq. (10.19a) for normal force is

$$F_n = \rho a\left[(V-u)^2 \sin\phi\right] = 1000 * 0.000615 * \left[(12-4.5)^2 \sin 30°\right] = 17.29 \text{ N}$$

Refer Eq. (10.18) for normal force components

$$F_H = F_n \sin\phi = 17.29 * (\sin 30°) = 8.65 \text{ N}$$
$$F_V = F_n \cos\phi = 17.29 * (\cos 30°) = 14.97 \text{ N}$$

Refer Eq. (10.22) for work done

$$\textit{Work done} = \rho a u\left[(V-u)^2 \sin\phi\right]$$

$$\Rightarrow \textit{Work done} = 1000 * 0.000615 * 4.5\left[(7.5)^2 \sin 30^0\right] = 77.83 \text{ Nm/s}$$

Refer Eq. (10.22a) for efficiency $\eta = \frac{2u\left[(V-u)^2 \sin\phi\right]}{V^2}$

$$\Rightarrow \eta = \frac{2 * 4.5\left[(7.5)^2 \sin 30\right]}{12^3} = 14.65\%$$

So, the normal thrust force and its components, work done and efficiency are 17.29 N, 8.65 N, 14.97 N, 77.8 Nm/s, 14.65% respectively.

Case b: Impact of jet on a moving flat plate held normal to the jet flow direction:
Consider the case of impact of jet on moving inclined plate and if made the plate inclination with horizontal ϕ as 90°, i.e. "$\phi = 90°$", then the equation for impact of jet on a moving plate held normal to the jet flow direction can be deduced and is shown in Figure 10.7.

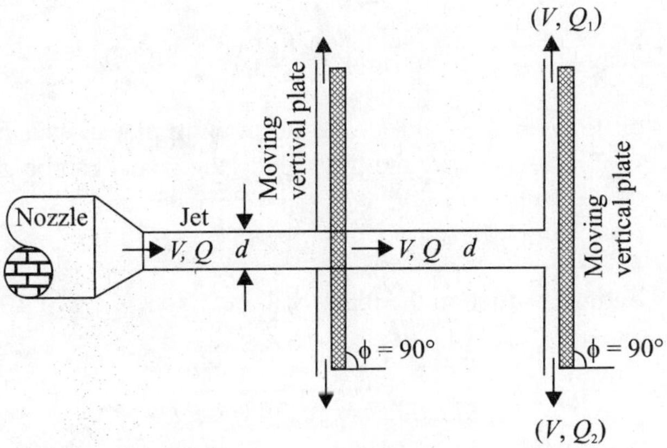

Figure 10.7: Impact of jet on a normal held moving plate

So substituting value of "$\phi = 90°$" in Eq. (10.19a), Eq. (10.20), Eq. (10.22), we get normal thrust force, normal thrust force components and work done;
i.e. from Eq. (10.19a), we have

$$F_n = \rho a\left[(V-u)^2\right] \tag{10.23}$$

From Eq. (10.20), we have

$$\begin{aligned} F_H &= F_n \sin\phi = \rho a\left[(V-u)\right]^2 \\ F_V &= F_n \cos\phi = 0 \end{aligned} \tag{10.24}$$

From Eq. (10.22), we have

$$Work\ done = \rho a u\left[(V-u)^2\right] \tag{10.25}$$

Consider **series of moving flat plates on a circular wheel** and as the fluid jet strikes with a velocity "V" on to a single plate, it displaces to new position and thereby making next consecutive flat plate to come normal to the jet flow direction and thereby no part of the fluid jet is left free without striking the plates. And hence there is a continuous rotation of plates, which make the wheel to turn continuously and so the relative velocity will not alter, i.e. $V_r = (V)$.

For the series of flat plate on a circular wheel, the impulse-momentum equation can be written as, referring Eq. (10.1) we have, the relative velocity between the jet and plate (i.e.(V)). The difference in jet velocities entering and leaving are taken in only one direction, but jet is deflected perpendicular to the actual impinging jet, and so

$$(V_1 - V_2) = V - u$$

And hence the Eq. (10.1) becomes

$$F_n = \rho a(V)(V - u)$$

As there is continuous striking of the fluid jet on the moving plates, causing the plate to rotate continuously, the amount of work done by the fluid jet on the wheel can be given as

$$\boxed{Work\ done = F_n * u = \rho a V\left[(V - u)\right]u} \tag{10.26}$$

And the efficiency of the impact of jet on the plates will be

$$\eta = \frac{Work\ done\ on\ the\ wheel}{Kinetic\ energy\ of\ the\ jet}$$

So
$$\eta = \frac{\rho a V u\left[(V - u)\right]}{\frac{1}{2}\rho a V^3}$$

$$\boxed{\Rightarrow \eta = \frac{2u\left[(V - u)\right]}{V^2}} \tag{10.27}$$

For maximum efficiency, $\frac{d\eta}{du} = 0 \Rightarrow \frac{d}{du}\left(\frac{\rho a V u\left[(V - u)\right]}{\frac{1}{2}\rho a V^3}\right) = 0$

$$\boxed{\Rightarrow V = 2u} \tag{10.28}$$

Substituting Eq. (10.28) in Eq. (10.27), we get maximum efficiency as $\eta_{max} = 50\%$

So, we can make a note that the "**maximum efficiency of a rotating wheel having normal flat plate is 50%, when the plates rotate with half the velocity of the jet coming out from a nozzle exit**".

Example 10.7: Determine the normal force and its components and work done of the jet impinged on a flat vertical plate moving with a velocity of half the velocity of the jet. Take the flow of jet diameter as 2.8 cm and velocity 12 m/s. And also determine the work done and efficiency if series of vanes are mounted.

Solution: Given data: $d = 0.028$ m, $V = 12$ m/s, $u = V/2 = 6$ m/s

Area of the water jet is

$$a=\frac{\pi}{4}d^2=\frac{3.14}{4}*0.028^2=0.000615\text{ m}^2$$

Refer Eq. (10.23) for normal force is

$$F_n=\rho a\left[(V-u)^2\right]=1000*0.000615*\left[(6)^2\right]=22.14\text{ N}$$

Refer Eq. (10.24) for normal force components

$$F_H=F_n\sin\phi=22.14*(\sin 90°)=22.14\text{ N}$$
$$F_V=F_n\cos\phi=22.14*(\cos 90°)=0.0\text{ N}$$

Refer Eq:(10.25) for work done

$$Work\ done=\rho au\left[(V-u)^2\right]$$

$$\Rightarrow Work\ done=1000*0.000615*6\left[(6)^2\right]=132.84\text{ Nm/s}$$

If series of vanes are mounted:

Refer Eq. (10.26) for work done

$$Work\ done=\rho aVu\left[(V-u)\right]$$

$$\Rightarrow Work\ done=1000*0.000615*12*6\left[(6)\right]=265.68\text{ Nm/s}$$

Refer Eq. (10.27) for efficiency $\eta=\dfrac{2u\left[(V-u)\right]}{V^2}\quad\Rightarrow\eta=\dfrac{2*6\left[(6)\right]}{12^2}=50\%$

So, work done on a single vertical flat vane is half the work done on the series of flat vanes.

Example 10.8: A jet of water 50 mm in diameter having a velocity of 20 m/s, strikes normally a flat smooth plate. Determine discharge of the jet and the thrust force on the plate if a) Plate is at rest. b) Plate is moving in same direction with a velocity of 8 m/s. Also find the work done and/second and efficiency of jet, when plate is moving.

Solution: Given data: $d=0.05$ m, $V=20$ m/s, $u=8$ m/s

Area of the water jet is $a=\dfrac{\pi}{4}d^2=\dfrac{3.14}{4}*0.005^2=0.00196\text{ m}^2$

Discharge of the jet $Q=aV=0.0000196*20=0.03925\text{ m}^3\text{/s}$

We know, the normal force on a normal flat plate at rest is

$$F_n = \rho a V^2 = 1000 * 0.00196 * 20 * 20 = 784 \text{ N} \quad \text{refer Eq. (10.9)}$$

We know, the normal force on an normal flat plate at rest is

$$F_n = \rho a\left[(V-u)^2\right] = 1000 * 0.00196 * (20-8)^2 = 282.24N \quad \text{refer Eq:(10.23)}$$

Work done/second if the plate is moving,

$$Work\ \ done = \rho a u\left[(V-u)^2\right] = 282.24 * 8 = 2257 \text{ Nm}$$

Efficiency, if the plate is moving,

$$Efficiency = \frac{Work\ \ done}{Kinetic\ \ energy} = \frac{\rho a u\left[(V-u)^2\right]}{1/2\rho a V^3} = 2257 * 2/1000 * 0.00196 * 20^3 = 28.8\%$$

Case c: Impact of jet on a moving curved plate at the center of symmetry:
Consider a fluid jet with velocity "*V*", discharging from a nozzle exit and strikes a moving curved vane. The jet strikes the moving vane and deflects into two parts, one part flowing in the upstream direction and the other flowing downstream direction of the moving plate and leaves from the tips of the vane. But as there is a relative motion between the jet velocity ("*V*") and moving curved vane velocity ("*u*"), the effective velocity of the fluid jet striking the moving curved vane decreases to (*V-u*) and so the deflected jet portions has a velocity of (*V-u*), after strike, as shown in Figure 10.8.

Let the thrust force (due to impact of jet on the moving curved vane) normal to moving curved vane be "F_n", then based on **impulse-Momentum equation**, refer Eq:(10.1), we have $F_n = \rho a V_r (V_1 - V_2)$.

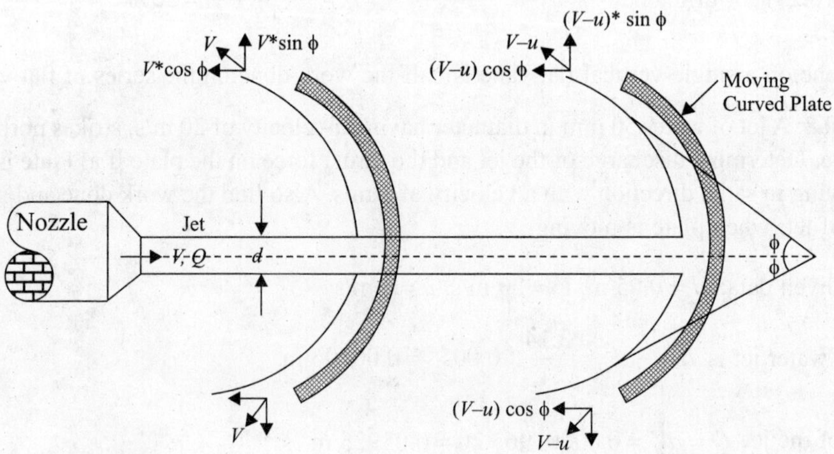

Figure 10.8: Impact of jet on a moving curved plate at center of symmetry

If the plate is assumed to be smooth, then the velocity of the jet split, will be same as that of the relative velocity of jet with its change in direction;

So $$V_r = (V-u),(V_1 - V_2) = \left[(V-u) - ((V-u)\cos\phi)\right]$$

And hence **normal thrust force** becomes

$$F_n = \rho a(V-u)\{(V-u) - [-(V-u)\cos\phi]\}$$

$$\Rightarrow F_n = \rho a(V-u)^2\{1+\cos\phi\} \tag{10.29}$$

The fluid jet striking a moving curved vane causes the vane to travel some distance along the horizontal direction and hence some amount of work is done by the jet on the curved vane and it can be given as

$$Work\ done = F_n * u = \rho au(V-u)^2\{1+\cos\phi\} \tag{10.30}$$

And the **efficiency of the impact of jet on the curved vane** will be

$$\eta = \frac{Work\ done\ on\ the\,vane}{Kinetic\ energy\ of\ the\ jet}$$

So $$\eta = \frac{\rho au(V-u)^2\{1+\cos\phi\}}{\frac{1}{2}\rho aV^3}$$

$$\Rightarrow \eta = \frac{2u\left[(V-u)\right]^2(1+\cos\phi)}{V^3} \tag{10.31}$$

For **maximum efficiency,** $\dfrac{d\eta}{du} = 0 \Rightarrow \dfrac{d}{du}\left(\dfrac{\rho au(V-u)^2\{1+\cos\phi\}}{\frac{1}{2}\rho aV^3}\right) = 0$

$$\Rightarrow \begin{matrix} V = u \\ or \\ V = 3u \end{matrix} \tag{10.31a}$$

Substituting Eq. (10.31a) in Eq. (10.31), we get **maximum efficiency at** "$V = 3u$" as

$$\eta_{\max} = \frac{8}{27}(1+\cos\phi) \tag{10.31b}$$

and we can note, **for a semicircular vane** (i.e. $\phi = 0°$), we get $\eta_{\max} = 0.592$.

Consider **series of moving curved vanes on a circular wheel** and as the fluid jet strikes with a velocity "V" on to a single curved vane, it displaces to new position and thereby making next consecutive curved vane to come normal to the jet flow direction and thereby no part of the fluid jet is left free without striking the vanes. And hence there is a continuous rotation of vanes, which make the wheel to turn continuously and so the relative velocity will be the velocity of jet (i.e. $V_r = (V)$).

For the series of curved vanes on a circular wheel, the impulse-momentum equation can be written as, referring Eq. (10.1), we have $F_n = \rho a V_r (V_1 - V_2)$. If plate assumed to be smooth, then the velocity of the jet split, will be same as that of the relative velocity of jet with its change in direction; so

$$(V_1 - V_2) = \left[(V - u) - ((V - u)\cos\phi)\right].$$

And hence **normal thrust force** becomes

$$F_n = \rho a(V)\{(V - u) - \left[-(V - u)\cos\phi\right]\}$$

$$\Rightarrow F_n = \rho a V(V - u)\{1 + \cos\phi\} \tag{10.32}$$

The fluid jet striking a moving curved vane causes the vane to travel some distance along the horizontal direction and hence some amount of work is done by the jet on the curved vane and it can be given as

$$Work\ done = F_n * u = \rho a u V(V - u)\{1 + \cos\phi\} \tag{10.33}$$

And the **efficiency of the impact of jet on the wheel** will be

$$\eta = \frac{Work\ done\ on\ the\ wheel}{Kinetic\ energy\ of\ the\ jet}$$

So

$$\eta = \frac{\rho a u V(V - u)\{1 + \cos\phi\}}{\frac{1}{2}\rho a V^3}$$

$$\Rightarrow \eta = \frac{2u\left[(V - u)\right](1 + \cos\phi)}{V^2} \tag{10.34}$$

For **maximum efficiency,**

$$\frac{d\eta}{du} = 0 \Rightarrow \frac{d}{du}\left(\frac{2u\left[(V - u)\right](1 + \cos\phi)}{V^2}\right) = 0$$

$$\Rightarrow V = 2u \tag{10.34a}$$

Substituting Eq. (10.34a) in Eq. (10.34), we get "**Maximum efficiency of the system** at

"$V = 2u$" as $$\eta_{\max} = \frac{1}{2}(1+\cos\phi) \tag{10.34b}$$

We can note, **for a semicircular vane** (i.e. $\phi = 0^\circ$), we get $\eta_{\max} = 100\%$. Similarly for flat vane (i.e. $\phi = 90°$), we get $\eta_{\max} = 50\%$.

Example 10.9: Determine the normal force and its components, work done and efficiency of the jet impinged at the centre of a curved symmetric moving plate, with a velocity of twice the velocity of the vane and deflected by 120° by the vane. Take the flow of jet diameter as 2.8 cm and velocity 12 m/s. And also determine the work done and efficiency if series of vanes are mounted.

Solution: Given data: $d = 0.028$ m, $V = 12$ m/s, $u = V/2 = 6$ m/s

Area of the water jet is $$a = \frac{\pi}{4}d^2 = \frac{3.14}{4} * 0.028^2 = 0.000615 \text{ m}^2$$

Vane angle $$\phi = 180 - 120 = 60°$$

Refer Eq. (10.29) for normal force is

$$F_n = \rho a\left[(V-u)^2\right](1+\cos\phi) = 1000 * 0.000615 * \left[(6)^2\right](1+\cos 60) = 33.21 \text{ N}$$

Refer Eq. (10.30) for work done

$$Work\ done = \rho a u\left[(V-u)^2\right](1+\cos\phi)$$

$$\Rightarrow Work\ done = 1000 * 0.000615 * 6\left[(6)^2\right](1+\cos 60) = 199.3 \text{ Nm/s}$$

Refer Eq. (10.31) for efficiency $$\eta = \frac{2u\left[(V-u)\right]^2(1+\cos\phi)}{V^2}$$

$$\Rightarrow \eta = \frac{2*6\left[(6)^2\right]*1.5}{12^2} = 37.5\%$$

If series of vanes are mounted

Refer Eq. (10.32) for normal force is

$$F_n = \rho a V\left[(V-u)\right](1+\cos\phi) = 1000 * 0.000615 * 12\left[(6)\right](1+\cos 60) = 66.42 \text{ N}$$

Refer Eq. (10.33) for work done

$$Work\ done = \rho aVu\left[(V-u)\right](1+\cos\phi) = 66.42*6 = 398.52 \text{ Nm/s}$$

Refer Eq. (10.34) for efficiency

$$\eta = \frac{2u\left[(V-u)\right](1+\cos\phi)}{V^2}$$

$$\Rightarrow \eta = \frac{2*6\left[(6)\right]*1.5}{12^2} = 75\%$$

Case d: Impact of jet on a moving curved plate, striking tangentially at one of the tip

Consider an amount of mass flow rate ($\bar{m} = \rho AV = \rho Q$) of fluid jet leaving a nozzle exit with a velocity of "V" and strikes a curved moving vane at any one of the tip, tangentially. As soon as the fluid jets hits the moving curved vane, the jet slides over the surface of the vane, by the amount of vane inclinations "β_i & β_o", which the vane tips (inlet and outlet) have been deflected from direction of jet flow and then jet leaves out through the other tip of the vane, as shown in Figure 10.9.

As there is a relative motion between the jet velocity ("*V*") and moving curved vane velocity ("*u*"), the effective relative velocity of the fluid jet striking the moving curved vane deduce to vectorial of (*V*-*u*), in the direction parallel to the tangent of the tip of vane and also the deflected jet has a velocity of (*V*-*u*) at the outlet. If frictional losses are not negligible, after strike then relative velocity at outlet may be less than the relative velocity at inlet (i.e. $V_{ro} \le V_{ri}$).

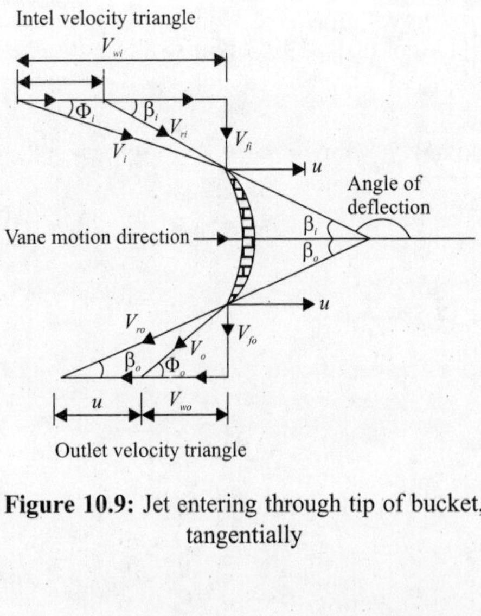

Figure 10.9: Jet entering through tip of bucket, tangentially

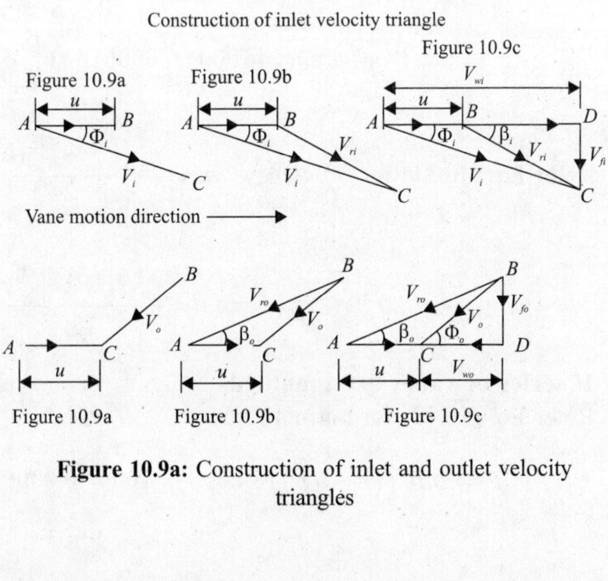

Figure 10.9a: Construction of inlet and outlet velocity triangles

The relative velocities, as they are vectorial form of absolute jet velocity "V" (at inlet and outlet) and absolute blade velocity "u" (at inlet and outlet) can be shown in a graphical representation known as **velocity triangles**.

a) Construction of velocity triangles

To draw **velocity triangles**, the known data is absolute velocity of jet and absolute velocity of blade and direction of both these velocity; for **inlet velocity triangle** direction of both these velocity are in direction of vane motion. So draw a line representing absolute velocity of blade with magnitude and direction and name it as "$\vec{AB}$", then draw another line representing the absolute velocity of jet with magnitude and direction starting from the initial point of the absolute blade velocity line and name it as "$\vec{AC}$", which forms an angle $\angle BAC$, as shown in Figures 10.9a and b. Then close the angle with another line "BC" to form a triangle and this closing side has direction in such a way in the formed triangle there should be one line having direction against the closed loop of the triangle (i.e. $\vec{BC}$).

Absolute velocity can be resolved into two components called **Whirl velocity** V_W and **flow velocity** V_f. Where **whirl velocity** is the component of absolute velocity of the jet along the direction of the vane motion and **flow velocity** is the component of absolute velocity of jet along the perpendicular direction of the vane motion. And hence these velocity components are to be included in the velocity triangle and so extend the absolute blade velocity line $\vec{AB}$, conveniently along the vane motion direction and also extend a perpendicular direction to the vane motion, which completes the inlet velocity triangle, as shown in Figure 10.9c.

For **outlet velocity triangle** direction of both these velocities are in opposite direction to each other, i.e. jet velocity is in opposite direction of vane motion. So draw a line representing absolute velocity of blade with magnitude and direction and name it as "$\vec{AC}$", then draw another line representing the absolute velocity of jet with magnitude and direction starting in negative direction from the end point of the absolute blade velocity line and name it as "$\vec{BC}$", which forms an angle $\angle BCA$, as shown in Figures 10.9a and b. Then close the angle with another line "AB" to form a triangle and this closing side has direction in such a way in the formed triangle, there should be one line direction against the closed loop of the triangle (i.e. $\vec{BA}$) and then extend the absolute velocity of jet line $\vec{AC}$, conveniently along the vane motion direction and also extend a perpendicular direction to the vane motion, which completes the outlet velocity triangle.

b) Force, work done and vane efficiency, in terms of velocities

Consider a jet impinged on one tip (sharp hump splitter) and leave through another tip of a bucket and consider the inlet and outlet velocity triangles of jet impinged on a moving curved vane at one of the tip and leaving through the other tip, as shown in Figure 10.9.

Let β_i, β_o be angle of bucket at the inlet and outlet tips respectively.

ϕ_i, ϕ_o be angle between absolute velocities and whirl velocity components respectively.

V_i, V_o be absolute velocities at the inlet and outlet tips respectively.

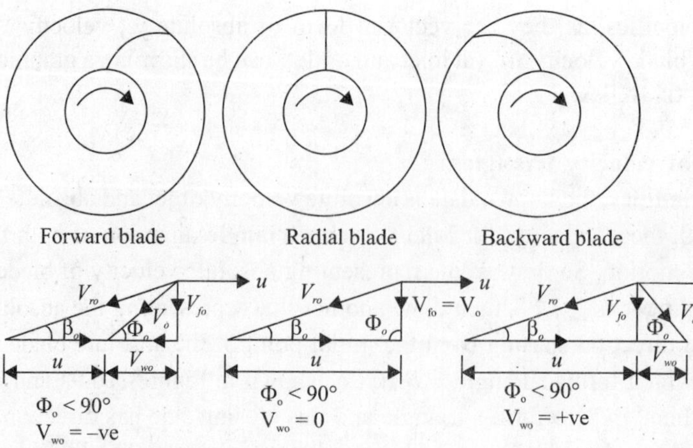

Figure 10.10: Different blade profiles and their outlet velocity triangles of a curved vane having jet strike at one tip and leaving at another

V, u be absolute velocities of the jet before impact and velocity of bucket respectively, where blade velocity, $u = \dfrac{\pi DN}{60}$, where "*D*" and "*N*" are mean diameter of the wheel from the center to the pitch circle and speed of the runner shaft.

If the plate is assumed to be smooth, then the velocity of the jet split will be same as that of the relative velocity of jet with its change in direction and hence

$$V_r = (V - u) \;\&\; (V_1 - V_2) = \left[(V_{ri} \cos\beta_i) - (-V_{ro} \cos\beta_o)\right] \tag{10.35}$$

But $(V_{ri} \cos\beta_i) = (V_{wi} - u) \;\&\; (V_{ro} \cos\beta_o) = (V_{wo} - u)$ and substituting this in Eq. (10.35), we have

$$V_r = (V - u) \;\&\; (V_1 - V_2) = \left[(V_{wi} - u) - (-(V_{wo} - u))\right] \tag{10.35a}$$

Applying **impulse momentum equation**, we have $F = \rho a V_r (V_1 - V_2)$; refer Eq. (10.1)

So **normal thrust force** will be $F_n = \rho a (V - u)\{(V_{wi} - u) - [-(V_{wo} - u)]\}$

$$\begin{aligned} &\Rightarrow F_n = \rho a (V - u)\{V_{wi} + V_{wo}\} \;\; \textit{for forward blade} \\ &\textit{or} \\ &\Rightarrow F_n = \rho a (V - u)\{V_{wi}\} \;\; \textit{for radial blade} \\ &\textit{or} \\ &\Rightarrow F_n = \rho a (V - u)\{V_{wi} - V_{wo}\} \;\; \textit{for backward blade} \end{aligned} \tag{10.36}$$

The $\pm$ symbol depends on the blade profile outlet triangle, as shown in Figure 10.10 for various blade profiles.

If considered a bucket with acute angle (i.e. $\phi_o < 90°$), then **work done by the jet on the bucket per second** will be $WD = Force * Displacement$

$$\Rightarrow WD = F * u$$

If considered a bucket with acute angle (i.e. $f_o < 90°$), then **Work done by the jet on the bucket per second** will be $WD = Force * Displacement$

$$\Rightarrow WD = F * u$$

If there are series of blades, then the whole quantity of fluid strike, without any loss and so, as per the Impulse—Momentum equation, Eq. (13.36) becomes as

$$F = \rho aV\left(V_{wi} - V_{wo}\right) \tag{10.36a}$$

Substituting Eq. (10.36), we get

$$WD = m\left(V_{wi} - V_{wo}\right) * u \quad \text{since } m = \rho aV \tag{10.37}$$

Kinetic energy of the jet is $$KE = \frac{m}{2} * \left(\Delta V^2\right) \tag{10.37a}$$

So **hydraulic efficiency** is ratio of "**work done by the jet on the bucket to the kinetic energy of the jet**" and so

$$\eta_H = \frac{WD}{KE} = \frac{m\left(V_{wi} - V_{wo}\right) * u}{\frac{m}{2}\left(\Delta V^2\right)}$$

$$\Rightarrow \eta_H = \frac{2u\left(V_{wi} - V_{wo}\right)}{\left(\Delta V^2\right)} \tag{10.38}$$

If the **vane assumed not to be smooth**, then the work done per second by the jet will be equal to the change in kinetic energy, i.e.

$$WD = \frac{1}{2} m\left(\Delta V\right)^2$$

$$\Rightarrow m\left(V_{wi} - V_{wo}\right) * u = \frac{1}{2} m\left(v_i - v_o\right)^2$$

$$\Rightarrow 2(V_{wi} - V_{wo}) * u = (V_i - V_o)^2$$

So substituting this in the Eq. (10.8), we have

$$\Rightarrow \eta_H = 1 - \left(\frac{V_o}{V_i}\right)^2 \tag{10.38a}$$

Example 10.10: Determine the normal force, work done and efficiency of the jet impinged tangentially at one tip of a curved symmetric moving plate mounted on a wheel of diameter 3 m, with a velocity of jet twice the velocity of the vane and deflected by 125° by the vane. Take the flow of jet diameter as 2.8 cm. Assume the speed of the frictionless wheel as 250 rpm.

Also find the efficiency for same data, if surface of vane is of rough and so the absolute outlet velocity reduces absolute inlet velocity by 5 m/s.

Solution: Given data:

$D = 3$ m, $N = 250$ rpm, $d = 0.028$ m, $V = 2$ u, $\beta_i\ \beta_o = 180 - 125 = 55°$, $V_{ri} = V_{ro}$

Area of the water jet is $a = \frac{\pi}{4}d^2 = \frac{3.14}{4} * 0.028^2 = 0.000615\ \text{m}^2$

Blade velocity $u = \frac{\pi DN}{60} = \frac{3.14 * 3 * 250}{60} = 39.25$ m/s

So, the velocity of jet $V = 2u = 2 * 39.25 = 78.5$ m/s

Consider the Figure 10.9: From the inlet velocity triangle, we have similar triangle, thereby giving

$$\frac{V_i}{\sin(\beta_i)} = \frac{u_i}{\sin(\beta_i - \phi_i)}$$

$$\Rightarrow \frac{78.5}{\sin(55)} = \frac{39.25}{\sin(55 - \phi_i)}$$

$$\Rightarrow \sin(55 - \phi_i) = 0.4095$$

$$\Rightarrow \phi_i = 55 - 24 = 31$$

$V_{wi} = V_i \cos\phi_i = 78.5 * \cos 31 = 67.287$ m/s and $V_{fi} = V_i \sin\phi_i = 78.5 * \sin 31 = 40.4$ m/s

So, inlet relative velocity $V_{ri} = V_{ro} = \frac{V_{fi}}{\sin\beta_i} = 49.36$ m/s

And outlet whirl velocity is $V_{wo} = V_{ro} \cos \beta_o - u_o = 49.36 * \cos 55 - 39.25 = -10.938$ m/s, negative sign indicates opposite direction and so the vane is forward profile.

Refer Eq. (10.36) for **normal force** is $F_n = \rho a (V - u)\{V_{wi} + V_{wo}\}$

$$\Rightarrow F_n = 1000 * 0.000615(78.5 - 39.25)\{67.287 + 10.938\} = 87.36 \text{ N}$$

Refer Eq. (10.37) for **work done**

$$Work\ \ done = \rho a V u [V_{wi} + V_{wo}]$$

$$\Rightarrow Work\ \ done = 1000 * 0.000615 * 78.5[(67.287 + 10.938)] = 3428.8 \text{ Nm/s}$$

Refer Eq. (10.38) for **hydraulic efficiency**

$$\eta_H = \frac{2u[(V_{wi} + V_{wo})]}{V^2} = \frac{2 * 39.25[(37.287 + 10.938)]}{78.5^2} = 61.4\%$$

Refer Eq:(10.38a) for **hydraulic efficiency** with rough surface of vane

$$\eta_H = 1 - \left(\frac{V_o}{V_i}\right)^2 = 1 - \left(\frac{78.5 - 5}{78.5}\right)^2 = 1.192\%$$

c) Torque exerted by the jet on the wheel having radial curved vanes

Consider series of vanes mounted on a circular wheel, as shown in the Figure 10.11. Then due to continuous impact of a mass jet ("*m*") on the every consecutive vane, there is raise of torque, which rotates the wheel with some speed.

As the velocity of the vane will be tangentially moving around the wheel, the whirl velocity need to be considered, this gives a torque to the rotating wheel. And we know the **torque** is rate of change of angular momentum, i.e. $Torque = \frac{d}{dt}(m\Delta V)R$ and the outlet whirl velocity do have a negative direction and so

$$T = \rho Q(V_{wi} - (-V_{wo}))R$$
$$\Rightarrow T = \rho Q(V_{wi} r_i + V_{wo} r_o) \tag{10.39}$$

And **work done due to series of vanes per second** will be $WD = Torque * angular\ \ velocity$

$$\Rightarrow WD = T * \omega \quad \Rightarrow WD = \rho Q(V_{wi} r_i + V_{wo} r_o) * \omega$$

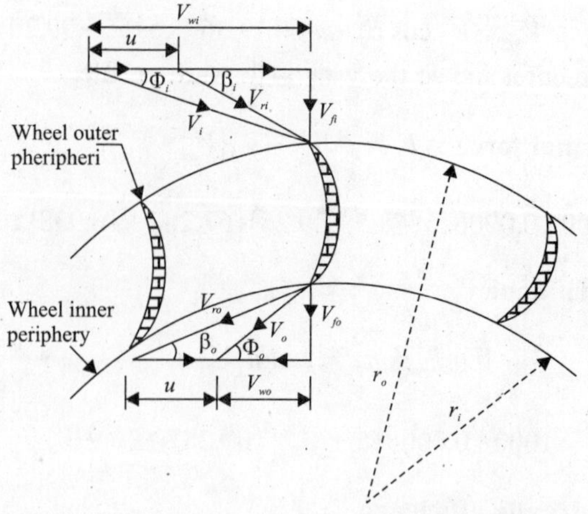

Figure 10.11: Wheel having curved radial vanes mounted on it

But $u_i = r_i * \omega$ & $u_o = r_o * \omega$ and so work done will be

$$WD = \rho Q(V_{wi}u_i \pm V_{wo}u_o) \tag{10.40}$$

and the $\pm$ symbol depends upon the vane profile.

Example 10.11: Determine the velocity of vanes mounted on a circular disc of 200 m, to have maximum efficiency, when a water mass flow rate of 0.0035 m^3/s of the jet impinged tangentially at one tip of a curved symmetric moving vane mounted on the disc and is deflected by 150° by the vane. Take the jet diameter as 2.8 m and the vane has 90% of friction and the outlet relative velocity as 50 m/s. Also find the torque required, if the external diameter is 300 m.

Solution: Given data:

$$Q = 3.5 \text{ m}^3/\text{s},\ D_i = 2 \text{ m},\ D_o = 3 \text{ m},\ N = 250 \text{ rpm}$$
$$d = 0.028 \text{ m},\ \beta_i = \beta_o, V_{ri} = 0.9\ V_{ro}, V_{ro} = 50 \text{ m/s}$$

Area of the water jet is $\quad a = \dfrac{\pi}{4}d^2 = \dfrac{3.14}{4} * 0.028^2 = 0.000615 \text{ m}^2$

Velocity of the jet $\quad V = Q / a = 0.0035 / 0.000615 = 5.691 \text{ m/s}$

Blade angle $\beta_i = 180 - 150 = 30°$ and inlet relative velocity $V_{ri} = 0.9V_{ro} = 0.9 * 50 = 45$ m/s

Consider the figure.10.9: From the inlet velocity triangle and outlet velocity triangle,

$$V_{wi} = u + V_{ri}\cos\beta_i = u + 45 * \cos 30 = (u + 39) \text{ m/s}$$

and $\quad V_{wo} = V_{ro}\cos\beta_o - u_o = 50 * \cos 30 - u = (43 - u) \text{ m/s}$

Refer Eq. (10.38), we have

$$\eta_H = \frac{2u\left[\left(V_{wi} + V_{wo}\right)\right]}{V^2} = \frac{2 * u\left[\left((u+39)-(43-u)\right)\right]}{5.69^2} = \frac{8u^2 - 8u}{5.69^2}$$

For maximum efficiency

$$\frac{d}{du}\eta_H = 0 = \frac{8u - 8}{5.69^2}$$
$$\Rightarrow u = 8/8 = 1.0 \text{ m/s}$$

Hence, **blade velocity** for maximum efficiency is 1.0 m/s.

Refer Eq. (10.39) for **torque** required $T = \rho Q\left(V_{wi} r_i + V_{wo} r_o\right)$

$$T = 1000 * 0.000615 * 5.69\left(40 * 250/2 + 42 * 300/2\right) = 39.542 \text{ kNm}$$

EXERCISE

Objective

1. ________________ is an engineering branch of application of fluid mechanics
   a. Hydraulic machinery.
   b. Fluid mechanics.
   c. Engineering mechanics.
2. The input of a jet on a normal flat vane is maximum, when
   a. the vane moves in the direction of jet
   b. the vane moves opposite to the direction of jet
   c. the vane is stationary
   d. series of such vanes are mounted on a wheel
3. A jet of fluid of area "*a*" and velocity "*V*", impinges on a normal flat plate which itself moves in the direction of jet at velocity *u*. The mass rate of flow striking the plate is
   a) $\rho a V$
   b) $\rho a(V+u)$
   c) $\rho a(V-u)$
   d) $\rho a(V-u)^2$
4. The force of impact of a liquid of a liquid jet of area "*a*" and velocity "*V*" on a moving normal flat vane moving in the direction of jet at velocity *u*
   a) $\rho a(V-u)^2$ b) $\rho a V(V-u)$ c) $\rho a u(V-u)$ d) $\rho a V u$

5. A liquid jet of area "*a*" and velocity "*V*" strikes a series of normal flat vanes mounted on a wheel the vane velocity being *u*. The force of impact is

a) $\rho a V(V-u)$

b) $\rho a(V-u)^2$

c) $\rho a u(V-u)$

d) $\rho a V u$

6. A series of normal flat vanes mounted on periphery of a wheel, the vane speed being *v*. For maximum efficiency, the speed of the liquid jet striking the vanes should be

a) $v/3$
b) $v/2$
c) v
d) $2v$

7. A jet fluid of area a and velocity *V* strikes a flat vane inclined to the jet at θ and moving in the direction of jet at velocity u. The force of impact normal to the plate is

a) $\rho a(V-u)\sin\phi$

b) $\rho a(V-u)^2\sin^2\phi$

c) $\rho a(V-u)^2\sin\phi$

d) $\rho a(V-u)^2\sin\phi\cos\phi$

8. The ratio of discharges of deflected jets, if a jet strikes an inclined fixed flat plate is

a) $\frac{Q_1}{Q_2}=\frac{(1-\cos\phi)}{(1+\cos\phi)}$

b) $\frac{Q_1}{Q_2}=\frac{(1+\cos\phi)}{(1-\cos\phi)}$

c) $\frac{Q_1}{Q_2}=\frac{(1-\cos\phi)}{(1-\cos\phi)}$

d) $\frac{Q_1}{Q_2}=\frac{(1+\cos\phi)}{(1+\cos\phi)}$

9. Curved vanes are preferred to flat vanes in a hydraulic turbine for the following reasons

a) Low cost of maintenance
b) Higher hydraulic efficiency can be obtained
c) Convenience of moulding and machining
d) It is possible to use high velocity jets

10. Water flows over a series of curved vanes mounted on a wheel. The component of velocity of water the radial direction of the wheel is called
    a) Relative velocity
    b) Velocity of flow
    c) Velocity of whirl
    d) Wheel velocity
11. The velocity of whirl in a hydraulic turbo- machine is defined as the component of
    a. Absolute velocity in the tangential direction
    b. Absolute velocity in the radial direction
    c. Relative velocity in the tangential direction
    d. Relative velocity in the radial direction
12. Euler's equation of energy transfer gives
    a) The optimum rotational speed
    b) The theoretical vane angels
    c) The theoretical force of impact on a turbine blade
    d) The theoretical power developed by a hydraulic turbine
13. Power is the product of
    a) Momentum and angular velocity
    b) Torque and tangential velocity
    c) Torque and angular velocity
    d) Hydrodynamic force and angular velocity

Answers

1. a	2. d	3. c	4. a	5. a	6. b
7. c	8. b	9. b	10. b	11. c	12. d
13. c					

Theory:

1. Derive an expression for normal thrust force components and discharge ratio acting on an inclined flat plate, having a horizontal jet strike over it.
2. Show that the resultant thrust force for a semicircular fixed vane having jet strike at one of its tip is $F = \rho a V^2$.
3. Show that the normal thrust force for a semicircular fixed vane having jet strike at the centre of symmetry of a curved symmetric plate is $F = \rho a V^2$.
4. Show that efficiency of a free jet striking normally as series of flat plates mounted on the periphery of a wheel never exceeds 50%.
5. Prove that the maximum hydraulic efficiency of water jet impinged on a symmetric curved moving vane at the centre of symmetry is $\eta_{max} = \frac{8}{27}(1+\cos\phi)$

Problems:

1. A jet of 50 mm diameter impinges on a stationary flat vertical plate with a velocity of 10 m/s. Determine the normal thrust force and work done of the plate jet system. **[196.25 N, 0.0 NM]**
2. A 50 mm diameter jet having a velocity of 18 m/s impinges without shock on a flat plate inclined at an angle of 30° to the horizontal. If the plate is moving at 5 m/s in the direction of jet, determine the normal thrust force, work done and efficiency of the plate jet system. **[165.8 N, 414.58 Nm, 7.24%]**
3. A 75 mm diameter jet moving horizontally with a velocity of 30 m/s impinges on a flat plate inclined, whose normal is at an angle of 45° to the horizontal. Find the normal thrust force, when the plate is stationary, when the plate is moving at 15 m/s in the direction of jet and also power and efficiency of the jet when the plate is moving. **[2812.8 N, 703.2 N, 7458.7 W, 12.5%]**.
4. A jet of water *f* diameter 70 mm moving with a velocity of 40 m/s, strike a curved fixed vane tangentially at one of the end at an angle of 30° to the horizontal. The jet leaves the plate at an angle 20^0 to the horizontal. Determine the normal thrust force components. **[11.117 kN, 0.972 kN]**.
5. A jet of water moving at a 15 m/s impinges on asymmetrical concave vane shaped to deflect the jet through 140°. If the vane is moving at 6 m/s, find the angle of the jet so that there is no shock at the inlet. Also determine the absolute velocity of the exit in magnitude and direction and also work done per unit weight of water. **[12.1°, 4.13 m/s, 49.8°, 104 Nm]**
6. A stationary curved vane deflects a 50 mm diameter jet of water through 150°. Due to friction over the surface, the water leaving the vane has only 80% of its original velocity. Calculate the discharge necessary to produce a force of 2 kN on the vane. **[0.0475 m^3/s]**
7. Water flows over series of curved vanes of a hydraulic turbine wheel, the diameter of which between inlet tips of the vanes is 2 m and that of outlet tips is 1m. The wheel rotates at 240 rpm. Jet of water enters at an angle of 0° to the tangent to the wheel at the inlet with a velocity of 40 m/s and leaves with a velocity of 10m/s at an angle of 30° to the tangent to wheel at outlet tip. Determine vane angles at entry and exit, work done per Newton on the wheel, hydraulic efficiency and power developed by the wheel, when discharge is 0.3 m^3/s. **[64°& 52°, 77.65 Nm/N, 95.3%, 228.5 kW]**

QUICK REFERENCE OF FORMULAE:

Property	Formulae	Remarks, if any
Impulse-Momentum equation of impact of jet	$F = \rho Q(V_1 - V_2)$ *or* $F = \rho a V_r (V_1 - V_2)$	
Normal thrust force, thrust force components, resultant thrust force and discharge ratios, due to impact of jet on a fixed inclined place F_n, F_H & F_V, F, θ, Q_1/Q_2 respectively	$F_n = \rho a V(V\sin\phi)$, $F_H = \rho a V(V\sin\phi)^2$, $F_V = \rho a V(V\cos\phi\sin\phi)$	
	$F = \sqrt{F_H^2 + F_V^2}$ & $\theta = \tan^{-1}\left(F_V/F_H\right)$	
	$Q_1 = \frac{Q}{2}(1+\cos\phi)$, $Q_2 = \frac{Q}{2}(1-\cos\phi)$ and $\frac{Q_1}{Q_2} = \frac{(1+\cos\phi)}{(1-\cos\phi)}$	
Normal thrust force, thrust force components, thrust force and discharge ratios, due to impact of jet on a fixed normal plate F_n, F_H, F_V, F, θ, Q_1/Q_2 respectively	$F_n = \rho a V^2$, $F_H = F_n \sin\phi = \rho a V^2$, $F_V = F_n \cos\phi = 0$	
	$F = F_H$ & $\theta = \tan^{-1}\left(0/F_H\right) = 0^0$	
	$Q_1 = \frac{Q}{2} = Q_2$	
Nromal thrust force, thrust force components, resultant thrust force due to impact of jet on a fixed curved plate at the centre of symmetry F_n, F_H, F_V, F, θ respectively	$F_n = \rho a V^2[1+\cos\phi]$ $F_H = F_n \sin\phi = \rho a V^2[1+\cos\phi]\sin\phi$ $F_V = F_n\cos\phi = \rho a V^2[1+\cos\phi]\cos\phi$	If, friction considered, then $F_n = \rho a V^2$ [1 + μ cos ϕ]
Thurst force components, resultant thrust force due to impact of jet on a fixed curved plate at one of the tip tangentially F_n, F_H & F_V, F, θ, respectively	$F_n = \rho a V^2$; $F = \sqrt{F_H^2 + F_V^2}$ $F_H = \rho a V^2(\cos\phi + \cos\theta)$ $F_V = \rho a V^2(\sin\phi + \sin\theta)$	
Normal thrust force, thrust force components, resultant thrust force, work done and efficiency due to impact of jet on a	$F_n = \rho a\left[(V-u)^2 \sin\phi\right]$	

Contd.

Property	Formulae	Remarks, if any
Moving inclined flat plate F_n, F_H and F_V, F, θ, *WD*, η respectively	$F_H = \rho a\left[(V-u)\sin\phi\right]^2$ $F_V = \rho a\left[(V-u)^2 \sin\phi\cos\phi\right]$ $F = \sqrt{F_H^2 + F_V^2}$ & $\theta = \tan^{-1}\left(F_V / F_H\right)$ $Work\ done = F_n * u = \rho au\left[(V-u)^2 \sin\phi\right]$	
Normal thrust force, thrust force components, resultant thrust force, work done and efficiency, due to impact of jet on a moving Normal plate(s) F_n, F_H, F_V, F, θ, *WD*, η respectively	$\eta = \dfrac{2u\left[(V-u)^2 \sin\phi\right]}{V^3}$ $F_n = \rho a\left[(V-u)^2\right]$, $F_H = \rho a\left[(V-u)\right]^2$, $F_V = 0$ $Work\ done = \rho au\left[(V-u)^2\right]$	
Thrust force, work done and efficiency, due to impact of jet on a moving series of normal plates, on a wheel (F_n, *WD*, η) respectively	$F_n = \rho a(V)(V-u)$ $Work\ done = F_n * u = \rho aVu\left[(V-u)\right]$ $\eta = \dfrac{2u\left[(V-u)\right]}{V^2}$	
Thrust force, work done and efficiency, due to impact of jet on a moving curved plate, at the center of symmetry	$F_n = \rho a(V-u)^2\{1+\cos\phi\}$ $Work\ done = \rho au(V-u)^2\{1+\cos\phi\}$ $\eta = \dfrac{2u\left[(V-u)\right]^2(1+\cos\phi)}{V^3}$	
Thrust force, Work done and efficiency, due to impact of jet on a moving series of curved plates, at the centre of symmetry	$F_n = \rho aV(V-u)\{1+\cos\phi\}$ $Work\ done = \rho auV(V-u)\{1+\cos\phi\}$ $\eta = \dfrac{2u\left[(V-u)\right](1+\cos\phi)}{V^2}$	
Thrust force due to impact of jet on a moving curved plate, striking tangentially at one of the tip	$F_n = \rho a(V-u)\{V_{wi} + V_{wo}\}$ *for forward blade* *or* $F_n = \rho a(V-u)\{V_{wi}\}$ *for radial blade* *or* $F_n = \rho a(V-u)\{V_{wi} - V_{wo}\}$ *for backward blade*	
Thrust force, work done, torque and efficiency, due to impact of jet on a series of moving curved plate, striking tangentially at one of the tip.	$F = \rho aV(V_{wi} - V_{wo}) * u$; $WD = m(V_{wi} - V_{wo}) * u$ $T = \rho Q(V_{wi}r_i + V_{wo}r_o)$; $\eta_H = \dfrac{2u(V_{wi} - V_{wo})}{(V^2)}$	if the surface is not smooth, $\eta_H = 1 - \left(\dfrac{V_o}{V_i}\right)^2$

Table 10.1: Points to remember

1. A semicircular (i.e. $\phi = 0°$) flat fixed plate has normal thrust force as $F_n = 2\,\rho\,aV^2$ *and so we can note that* **"Normal thrust force on a fixed semicircular curved vane is always twice than the normal thrust force of a fixed curved vane."**
2. A flat (i.e. $\phi = 90°$) fixed plate has normal trust force as $F'_n = \phi\,aV^2$ and so, we can note that **"Normal thrust force on a fixed curved vane is always more than the normal thrust force of a fixed flat plate"**.
3. For a tangentially fixed semicircular fixed plate having tangential flow at one of the tip has horizontal force equal to normal thrust force and is given by $F_H = \rho aV^2 = F_n = F$
4. Maximum efficiency of a rotting wheel having normal flat plate is 50%, when the plates rotate with half the velocity of the jet coming out from a nozzle exit (i.e. $V = 2\,u$)."
5. For a jet impining on a symmetric curved moving vane, at the center of symmetry has **Maximum efficiency** at "V = 3 u" and vlaue is $\eta_{max} = 8/27\ (1 + \cos\phi)$.

   We can note, **for a semicircular vane** (i.e. $\phi = 0°$), we get $\eta_{max} = 59.2\%$
6. For a jet impinging on a symmetric series of curved moving vane, at the center of symmetry has **Maximum efficiency** at "V = 2 u" and value $\eta_{max} = 1/2\ (1 + \cos\phi)$.

   We can note, **for a semicircular vane** (i.e. $\phi = 0°$), we get $\eta_{max} = 100\%$

CHAPTER 11

Hydraulic Turbines

INTRODUCTION

Any rotating machine shaft containing series of mechanical elements, which are of rotating or stationary (called **blades/vanes**) mounted on the shaft and the rotating mechanical elements are rotated steadily by means of work addition or extraction from impact of fluid jets on to the blades or vice versa is called a **turbo-machine.** Any rotating machine shaft, which rotates due to the extraction of work from a fluid jet flow on to it, which is impinged by an external means, like a nozzle or due to fall of water from high levels on to the blades is called a **turbine** and rotating machine shaft, which rotates due to the addition of work from a shaft to the fluid jet flow is known as a **compressor or pump**. Or in other words, a turbine is a turbo-machine quite opposite in action to Compressors and pumps.

Turbine acts as a prime mover and the efficiency of power generation of a generator depends on the effective rotation of the turbine shaft and in turn this effectiveness of rotation depends on the amount of thrust force due to the impact of jet onto the blade. And so in order to increase the efficiency of turbines, the blades are given different shape profile and with different inclinations of jet impact on them. When correlated the turbine with any other prime mover, turbines are simple in design and construction, with high load capacity and low maintenance cost, less floor area, lighter foundation.

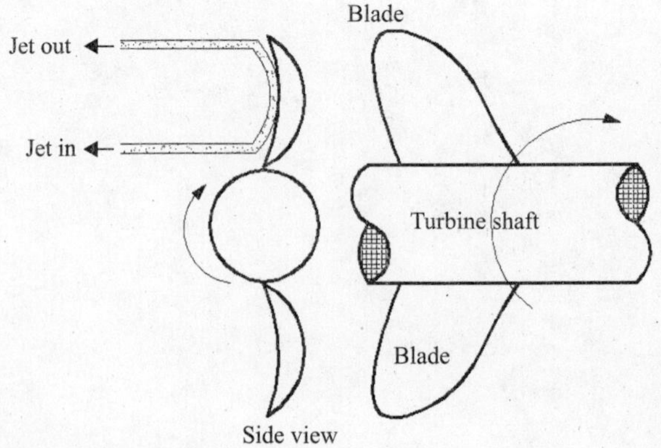

Figure 11.1: Basic working phenomenon of a hydraulic turbine

The impact of fluid jet on to the turbine vane may be of a water fluid jet, steam jet, gas jet or other type of fluid jets, based on the type of turbine and the turbine using water as fluid jet is termed as **hydraulic turbine**.

In hydraulic turbines, the water jet impinges on to **blades/vanes/buckets** on the turbine **shaft/runner/rotor**, thereby rotation of the shaft takes place, as shown in Figure 10.1 and then the jet leaves the blades and falls into the tail race.

CLASSIFICATION OF HYDRAULIC TURBINES

Hydraulic turbines can be classified on the principle of amount of hydraulic energy that impacts the vane (i.e., full available amount of hydraulic energy is utilized in impact or only some amount of available hydraulic energy has been used for impact) and is classified as **impulse turbine** and **reaction turbine**.

An **impulse** turbine has an impact of jet with a high velocity onto its hemispherical vanes, to cause rotation, whereas **reaction** turbines use change of pressure across its blades to rotate the shaft. Further, these two types of classifications can be divided based on the directions of flow of the fluid jet on to the turbine blades, as mentioned in Figure 11.2.

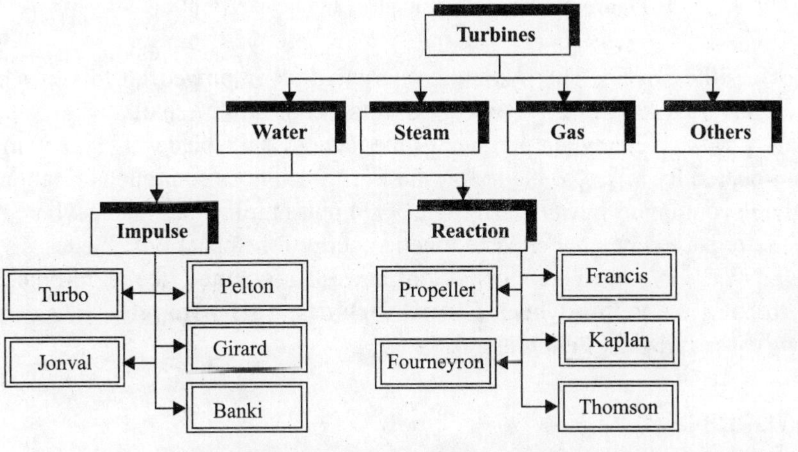

Figure 11.2: Turbines classification

Water/hydraulic turbines and steam turbines are of very much similar in working and classification but with difference of the type of fluid jet; in the first type, water is used as fluid and in the second type, steam is used as fluid and moreover hydraulic turbines can have the flow direction of jet in any direction like tangential, radial or axial but steam turbines do have only axial flow direction of steam jets.

IMPULSE TURBINE

Impulse turbine invented by **Lester A. Pelton** in the 1870s and was named after him. The basic operation of this type of turbine is **energy transfer**, i.e. all the available hydraulic energy of the jet is transferred on to rotor vanes of turbine shaft, in the form of kinetic energy of the jet, by means

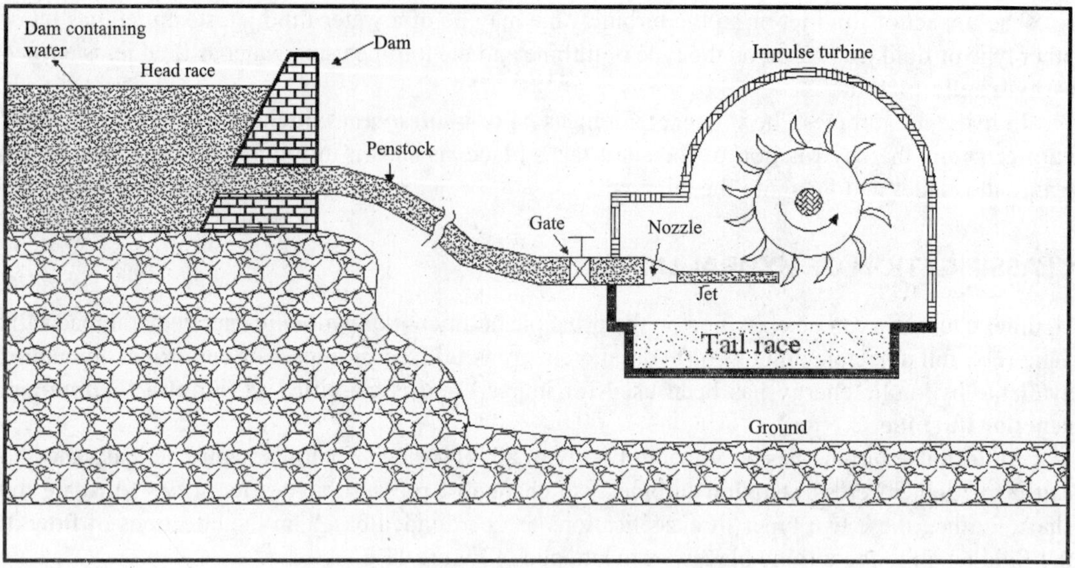

Figure 11.3: Layout of impulse turbine power plant

of external device like nozzle. The hydraulic energy of jet impinged on to vanes/blades of the turbine, is thus converted to mechanical energy of rotation of turbine shaft, due to impulse between jet and the blades, without change in pressure of the jet across the blades, as shown in Figure 11.3.

When surmounted by series of blades on the shaft, results in continuous rotation of the shaft and thereby giving continuous power generation. So impulse turbine is suitable where full hydraulic energy of the water/jet can be converted to kinetic energy for impact onto blades.

This turbine works in atmospheric conditions and is also termed free jet turbine. Examples of impulse type turbines are **Pelton wheel, Girard turbines, Turbo-Impulse, Banki turbines and Jonval**. Among these turbines, Pelton is mostly used.

REACTION TURBINE

Reaction turbine consists of two types of blades, namely, stator blades (called **guide vanes**) and rotor blades (called **runner vanes**). The basic operation of this type of turbine is to add up thrust **force** due to the movement of the jet on the surface of the runner vanes and the **Reaction force** (Newton's third law) produced due to energy transformation from one form to other (like pressure energy to kinetic energy) in the guide vanes.

Consider an example of a Firefighting person holding a fire hose nozzle to shut down a fire accident. When the firefighting person switch on the hose nozzle, the nozzle with hose pulls the fire person back due to the force against the energy transform (pressure energy to kinetic energy) in the nozzle and when the firefighting person holds intact, this reactive force given by the firefighting person also adds up to the jet flow velocity. Similarly, a person shooting a gun need to hold it intact, to give a reactive force against the velocity of the bullet leaving the gun, so that he adds up some more reactive force to the velocity of the bullet.

In reaction turbine, stator blades act as guiding the direction of jet towards the rotor and rotor blades rotate the runner. A part of the available hydraulic energy is transformed to pressure drop change from stator blades to rotor blades and remaining part of available energy is converted to kinetic energy while moving through the runner. Hence, there is a subsequent change in velocity and pressure during the flow of water from stator to rotor. And as there is change in pressure of the jet across the blades, the turbine needs to be enclosed in a casing to prevent interference of atmospheric pressure conditions, as shown in Figure 11.4.

The stator blades guide the jet flow from the penstock to the rotor blades and while doing so they are tilted in such a way that the gap in between two stator blades are of reducing cross-section, which acts like a nozzle and so the jet has drop in pressure and increase in velocity and hence the jet with high velocity do gives a **thrust force** on the rotor blades on the runner and in addition to this thrust force, a reactive force is also added.

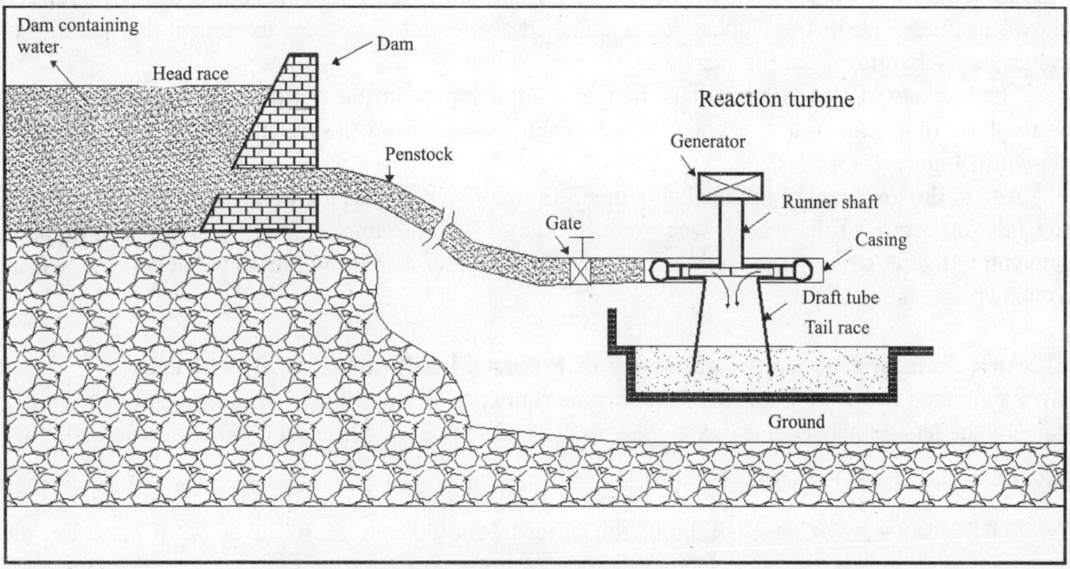

Figure 11.4: Layout of reaction turbine power plant

Examples of reaction type turbines are **Francis, Kaplan, Propeller, Thomson and Fourneyron**. Among these turbines, Francis and Kaplan are widely used.

Table 11.1: Comparison between impulse turbine and reaction turbines

Aspect	**Impulse**	**Reaction**
Conversion of hydraulic energy to kinetic energy	All available hydraulic energy is converted to kinetic energy by an external means like nozzle.	Only a part of available hydraulic energy is converted to kinetic energy by the direction change of vanes on the shaft.
Change of pressure and velocity	Pressure remains constant throughout the travel of fluid but there may or may not be a change in the relative velocity.	Pressure drops from stator blades to rotor blades but the velocity increase.
Action of blades/vanes	Vanes act only when they are in front of the fluid jet impinged.	Vanes are always surrounded by the fluid jet, and hence vanes are always in action.
Casing function	Casing is not required, as this turbine works in atmospheric. But only acts as a preventive device for flow of splitted water waste to the tail race.	Casing is required, as there is pressure drop from inlet to outlet, and is needed to isolate the turbine from effect of atmospheric conditions.

Pelton wheel: Pelton wheel turbine is a type of impulse based turbine and is mostly used for high heads (>250 m). It consists of a runner shaft surmounted by a series of vanes/buckets, evenly placed on its circumference.

Pelton wheel may use one or more number of nozzles to impinge water jet onto the buckets, for striking tangentially to the runner, which cause rotation of runner buckets about a pitch circle, as shown in Figure 11.5. And a spear rod is attached at the end of nozzles to control the quantity of water jet discharging from the nozzle exit.

The buckets of Pelton wheel has a sharp splitter hump, at the symmetry of the blade, which splits the striking jet into two equal halves, each leaving from the outer edge of the bucket, as shown in Figure 11.5a.

And at the lower tip of the bucket, a notch is cut, which avoids the deflection of the leaving jet towards the centre of the wheel, when met by the jet onto the bucket initially and also it helps in preventing the jet striking the consecutive preceding bucket, while the previous bucket is in action, as shown in Figure 11.5b.

a) Work done and hydraulic efficiency of Pelton wheel, in terms of velocities

Work extracted by vanes/buckets of the turbine runner from the fluid jet striking. And hence work done by the jet on the bucket is due to the impulse and so applying impulse momentum equation, we have $F = \rho a V_r (V_1 - V_2)$; refer Eq. (10.1).

Let $\theta = 180 - \beta_o$ be angle of deflection of jet by the bucket. β_i, β_o be angle of bucket at the inlet and outlet tips respectively. ϕ_i, ϕ_o be angle between absolute velocities and whirl velocity

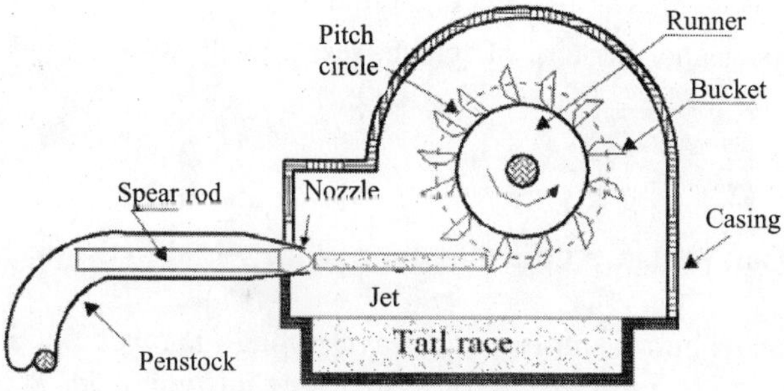

Figure 11.5: Pelton wheel

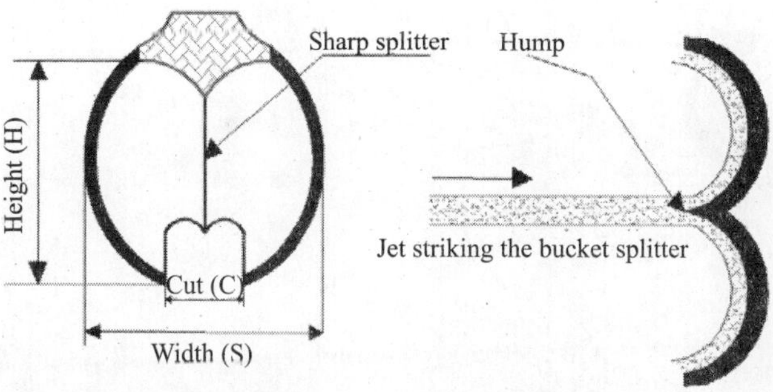

Front view of bucket in direction of jet strike

Top view of bucket in direction of jet strike

Figure 11.5a: Pelton wheel bucket nomenclature

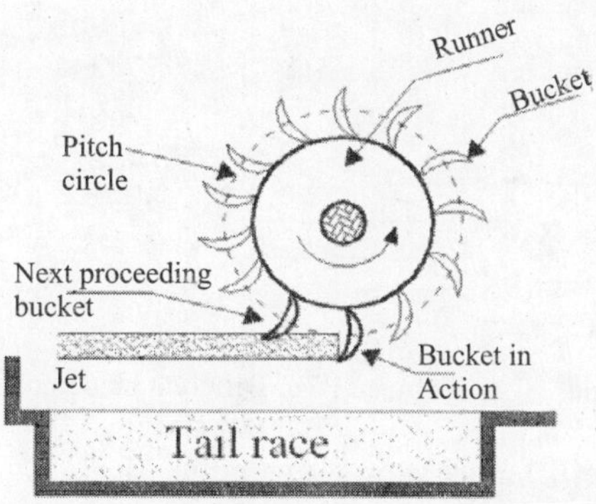

Figure 11.5b

components respectively. V_i, V_o be absolute velocities at the inlet and outlet tips respectively. V, u be absolute velocities of the jet before impact and velocity of bucket respectively, where blade velocity

$$u = \frac{\pi DN}{60}$$

where "D" and "N" are mean diameter of shaft (distance from the centre of the wheel to the pitch circle) and speed of the runner shaft.

Consider a jet impinged on one tip (sharp hump splitter) and leave through another tip, as shown in Figure 11.6. And consider the inlet and outlet velocity triangles of jet impinged on a moving curved vane (impact of jets chapter) at one of the tips and leaving through the other tip, as shown in Figure 10.11. But as the angles $\phi_i, \beta_i = 0°$ and hence the inlet velocity diagram reduces to Figure 11.7a, as shown.

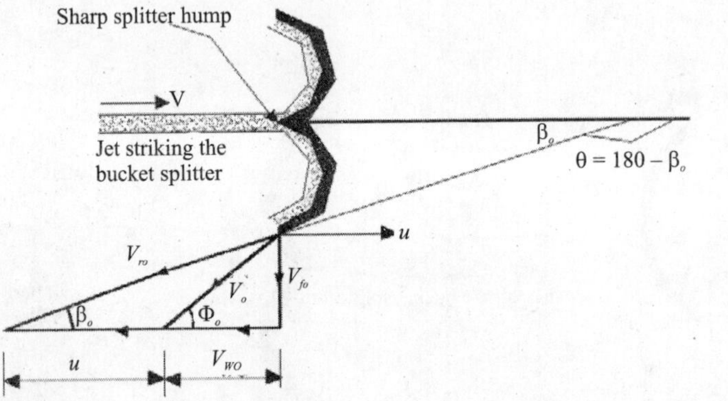

Figure 11.6: Top view of bucket in direction of the jet strike

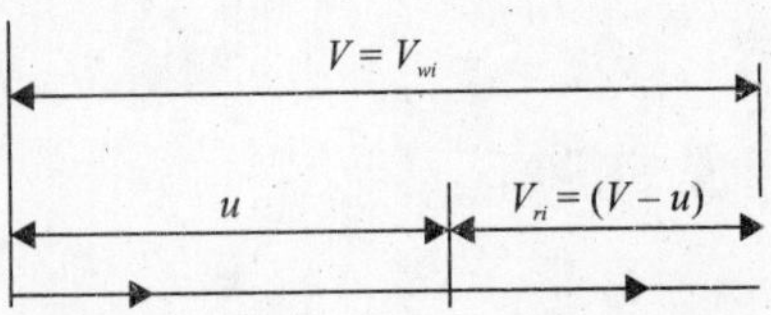

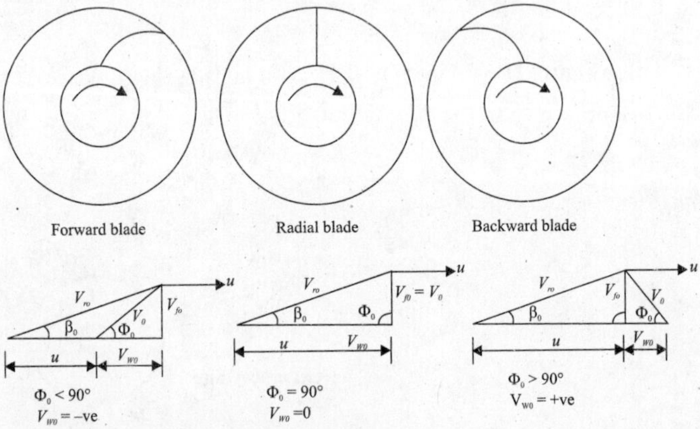

Figure 11.7a: Inlet velocity triangle

Figure 11.7b: Different blade profiles and their outlet velocity triangles

And the outlet triangle will be, as shown in Figure 11.7b, for various blade profiles. As the surface of the bucket may not be smooth and so there may be some loss in relative velocity from inlet to outlet velocity and so

$$V_{ro} = \mu(V - u) \tag{11.1}$$

where "μ" is coefficient of friction and will be less than one, and so

$$\boxed{\mu = \frac{V_{ro}}{V_{ri}}} \tag{11.1a}$$

Work done by the jet on the bucket will be $WD = Force * Displacement \Rightarrow WD = F * u$

Substituting Eq. (11.1), we get

$$\boxed{WD = \frac{\gamma}{g} * Q * (V_{wi} - V_{wo}) * u} \tag{11.2}$$

Kinetic energy of the jet is

$$KE = \frac{\gamma}{2g} * Q * (V^2) \tag{11.3}$$

So **hydraulic efficiency** is ratio of "**work done by the jet on the bucket to the kinetic energy of the jet**" and so

$$\eta_H = \frac{WD}{KE} = \frac{\frac{\gamma}{g} * Q * (V_{wi} - V_{wo}) * u}{\frac{\gamma}{2g} * Q * (V^2)} \tag{11.4}$$

$$\boxed{\Rightarrow \eta_H = \frac{2u(V_{wi} - V_{wo})}{(V^2)}} \tag{11.4a}$$

But from inlet velocity triangle, we have

$$V_{wi} = V \tag{11.5a}$$

and from outlet triangle, we have

$$V_{wo} = u - V_{ro}\cos\beta_o = u - KV_{ri}\cos\beta_o = u - K(V - u)\cos\beta_o \tag{11.5b}$$

Replacing Eq. (11.5a and b) in Eq. (11.4), we get **work done** as

$$\boxed{WD = \frac{\gamma}{g} * Q * [(V - u)(1 + \mu\cos\beta_o)] * u} \tag{11.6a}$$

And hydraulic efficiency

$$\boxed{\eta_H = \frac{2u(V - u)[1 + K\cos\beta_o]}{(V^2)}} \tag{11.6b}$$

Maximum hydraulic efficiency can be obtained by differentiating Eq. (11.6) with blade velocity "*u*" and equate to zero, so

$$\frac{d}{du}\eta_H = \frac{d}{du}\left\{\frac{2u(V-u)[1+K\cos\beta_o]}{(V^2)}\right\} = 0 \quad \Rightarrow u = V/2 \tag{11.7}$$

Hence, the **maximum hydraulic efficiency is possible if the blade velocity is half of the jet velocity.**

And so maximum **hydraulic efficiency** becomes

$$(\eta_H)_{\max} = \frac{1}{2}(1+K\cos\beta_o) \tag{11.8}$$

If $\beta_o = 0°$ or $\beta_o = 180°$ for smooth surface, then $(\eta_H)_{\max} = 100\%$ *or* 0%

Example 11.1: Find the blade profile, work done, hydraulic efficiency of a Pelton wheel having diameter of 2 m and of speed 100 rpm. If the velocity of the jet is 10 m/s having diameter of 1.5 m, been deflected by 120° by the vane. Assume that the surface is smooth.

Also calculate efficiency for rough surface having friction of 0.8.

Solution: Given data: $D = 7$ m, $N = 250$ rpm, $V = 10$ m/s, $\mu = 0.8$, $d = 1.5$ m

Blade outlet angle is $\beta_o = 180 - 130 = 60°$

Area of the jet $\quad a = \frac{\pi}{4}d^2 = \frac{3.14}{4} * 1.5^2 = 1.77 \text{ m}^2$

Discharge of the jet $\quad Q = aV = 1.77 * 10 = 17.7 \text{ m}^3/\text{s}$

Blade velocity $\quad u = \frac{\pi DN}{60} = \frac{\pi * 0.5 * 100}{60} = 2.62 \text{ m/s}$

Refer Eq. (11.5a) and, we have $V_{wi} = V = 10$ m/s and

$$V_{wo} = u - (V-u)\cos\beta_o = 2.62 - (10 - 2.62)\cos 60 = -1.072 \text{ m/s}$$

Negative sign indicates that the **blade profile is forward type**.

Work done

$$WD = \frac{\gamma}{g} * Q * (V_{wi} - V_{wo}) * u = \frac{9810}{9.81} * 1.77 * (10 - 1.072) * 2.62 = 41.402 \text{ kNm/s}$$

Efficiency for smooth surface

$$\eta_H = \frac{2u(V-u)(1+\cos\beta_o)}{(V^2)} = \frac{2*2.62(10-2.62)(1+\cos 60)}{(10^2)} = 58\%$$

Hence, the work done and efficiency are 41.4 kNm/s, 58% respectively.

Efficiency for rough surface

$$\eta_H = \frac{2u(V-u)(1+\mu\cos\beta_o)}{(V^2)} = \frac{2*2.62(10-2.62)(1+0.8*\cos 60)}{(10^2)} = 54.13\%$$

Example 11.2: Find the work done, hydraulic efficiency of a Pelton wheel having head of 1.73 m and of speed 100 rpm. If the flow rate of jet is 12.5 m^3/s having diameter of 1.5 m, has been deflected by 130° by the vane. Assume that the surface is smooth and speed ratio as 0.46.

Also calculate efficiency for rough surface having friction of 0.8.

Solution: Given data: $H_n = 1.73$ m, $N = 250$ rpm, $Q = 12.5$ m^3/s, $\mu = 0.8$, $d = 1.5$ m, $K_u = 0.45$

Blade outlet angle is $\beta_o = 180 - 130 = 60^0$

Area of the jet $a = \frac{\pi}{4}d^2 = \frac{3.14}{4}*1.5^2 = 1.77\,\text{m}^2$

Velocity of the jet $V = Q/a = 12.5/1.77 = 7.08\,\text{m/s}$

Blade velocity $u = K_u\sqrt{2gH_n} = 0.45\sqrt{2*9.81*1.73} = 2.62$ m/s

Work done $WD = \frac{\gamma}{g}*Q*[(V-u)(1+\cos\beta_o)]*u$

Refer Eq. (11.6a)

$$\Rightarrow WD = \frac{9810}{9.81}*0.25*[(7.08-2.62)(1+\cos 60)]*2.62 = 42.99\,\text{Nm/s}$$

Efficiency for smooth surface

$$\eta_H = \frac{2u(V-u)(1+\cos\beta_o)}{(V^2)} = \frac{2*2.62(7.08-2.62)(1+\cos 60)}{(7.08^2)} = 69.9\%$$

Hence, the work done and efficiency are 42.99 Nm/s, 69.9% respectively.

Efficiency for rough surface

$$\eta_H = \frac{2u(V-u)(1+\cos\beta_o)}{(V^2)} = \frac{2*2.62(7.08-2.62)(1+0.8*\cos 60)}{(7.08^2)} = 65.24\%$$

b) Available power and efficiencies, in terms of head

Observe the Pelton wheel set up, with heads mentioned, as shown in Figure 11.8. When water flows from a dam to a turbine, through a penstock and while in motion it converts the available static head of the water in the dam to different types of heads, which act due to gravity, friction and pressure from one point to another point of motion of water, like head loss due to friction H_f, velocity head KE, pressure head P/γ, datum heads Z, etc.

Net head available at the entrance of the turbine

$$H_n = H_g - H_f$$

Let actual discharge from the penstock be Q_a and q be amount of discharge lost in motion of water from one point to other then **hydraulic efficiency** is defined as **"ratio of Power developed by the runner to the net power supplied by the water at the entrance of the turbine"**. i.e.

$$\eta_H = \frac{Power\ developed\ by\ runner}{Net\ power\ supplied\ by\ water\ at\ entrance\ of\ turbine}$$

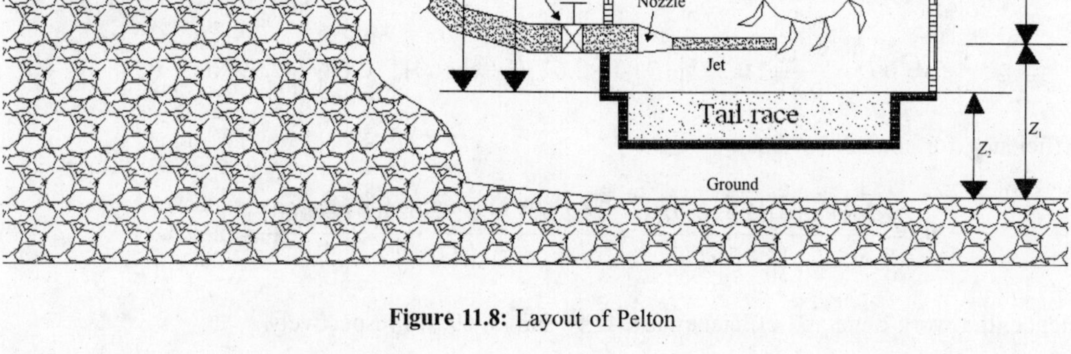

Figure 11.8: Layout of Pelton

$$\Rightarrow \eta_H = \frac{Power\ developed\ by\ runner}{\gamma(Q_a - q)H_n} \tag{11.9a}$$

Mechanical efficiency is defined as **"ratio of power available at the turbine shaft to the power developed by the runner (buckets)"**.

i.e. $$\eta_m = \frac{Power\ available\ at\ turbine\ shaft}{Power\ developed\ by\ runner}$$

$$\Rightarrow \eta_m = \frac{P\ or\ Brake\ power\ or\ turbine\ output\ power}{Power\ developed\ by\ runner} \tag{11.9b}$$

Volumetric efficiency is defined as **"ratio of discharge striking the runner (buckets) to actual water supplied to the turbine"**.

i.e. $$\eta_{vol} = \frac{Discharge\ striking\ runner(buckets)}{Actual\ water\ supplied\ to\,turbine} \Rightarrow \eta_{vol} = \frac{(Q_a - q)}{Q_a} \tag{11.9c}$$

Overall efficiency is defined as **"ratio of power available at the turbine shaft to net power supplied by the water at the entrance of the turbine"**.

i.e. $$\eta_o = \frac{Power\ available\ at\ turbine\ shaft}{Net\ power\ supplied\ by\ water\ at\ entrance\ of\ trubine} = \frac{Brake\ power}{water\ power}$$

$$\Rightarrow \eta_o = \frac{P}{\gamma Q_a H_n} \tag{11.9d}$$

Or in other words, **overall efficiency is the product of hydraulic, mechanical and volumetric efficiencies** and

$$\eta_o = \eta_{hyd} * \eta_{mech} * \eta_{vol} = \frac{P}{\gamma Q_a H_n} \tag{11.9e}$$

Or in other words, **overall efficiency is also called the Pelton wheel efficiency**

i.e. $$\eta_o = \eta_{pw} = \frac{Turbine\ output\ power}{Available\ power\ of\ turbine} \tag{11.9f}$$

where turbine output power will be

$$Turbine\ output\ power\ P_t = \frac{Output\ of\ generator}{Generator\ efficiency} \tag{11.9g}$$

Example 11.3: Find the power developed by (in hp) a Pelton wheel having gross head of 150 m and has loss of head due to friction of thrice the discharge cube and is of speed 100 rpm. If the flow rate of jet is 2.5 m³/s having diameter of 1.5 m, been deflected by 130° by the vane. Assume that the surface is smooth, speed ratio 0.45 and overall efficiency of 80%.

Solution: Given data:

$$H_g = 150 \text{ m}, H_f = 3Q^3, N = 100 \text{ rpm}, Q = 2.5 \text{ m}^3/\text{s}, \eta_o = 60\%, d = 1.5 \text{ m}, K_u = 0.45$$

Velocity of the jet

$$V = Q/a = 2.5/0.196 = 12.7 \text{ m/s}$$

Net head

$$H_n = H_g - H_f = 150 - 3*Q^3 = 150 - 3*2.5^3 = 103 \text{ m}$$

Power developed $\Rightarrow P = \gamma Q H_n \eta_o = 9810*2.5*103*0.6 = 1515.6 \text{ kW}$

Power developed (in hp =746 watt) $\Rightarrow P = 1515.6*1000/746 = 2031 \text{ hp}$

Example 11.4: Find the power, resultant force and overall efficiency of a Pelton wheel having 400 m head and running at a speed of 250 rpm. The jet has been deflected by an angle of 150° by the vane and has relative velocity reduced by 20% due to friction. Assume mechanical efficiency as 10%, co-efficient of velocity as 0.91 and speed coefficient as 0.41, Diameter of the jet as 250mm.

Solution: Given data: $H_n = 400 \text{ m}, N = 250 \text{ rpm}, \eta_m = 0.1, C_V = 0.91, K_u = 0.41, \mu = 0.2, d = 0.25 \text{ m}$

Velocity of jet $V = C_V\sqrt{2gH_n} = 0.91\sqrt{2*9.81*400} = 80.61 \text{ m/s}$

Discharge of jet $Q = aV = \dfrac{\pi}{4}d^2V = \dfrac{\pi}{4}*0.25^2*21.6 = 3.95 \text{ m}^3/\text{s}$

We know **power developed** by a Pelton wheel as $P = \gamma Q H_n$

$$\Rightarrow P = 1000*3.95*400*9.81 = 15512.08 \text{ kW}$$

Velocity of blade $u = K_u\sqrt{2gH_n} = 0.41\sqrt{2*9.81*400} = 36.3 \text{ m/s}$ refer Figure 11.7a. For a Pelton wheel, we have

$$V_{wi} = V = 80.61 \text{ m/s} \;\&\; V_{ri} = (V - u) = 80.61 - 36.3 = 44.28 \text{ m/s}$$

So exit relative velocity will be $V_{ro} = 0.2*V_{ri} = 0.2*44.28 = 8.85 \text{ m/s}$

Vane angle at exit is $\beta_o = 180 - 150 = 30°$

Whirl velocity at exit, see Figure 11.6, we have

$$V_{wo} = u_o - V_{ro} \cos \beta_o \Rightarrow V_{wo} = 36.3 - 8.85 * \cos 30 = 28.63 \text{ m/s}$$

Resultant force $F = \gamma Q / g (V_{wi} - V_{wo})$

$$F = 1000 * 3.95 (80.61 - 28.63) = 205.93 \text{ kN}$$

Example 11.5: Determine the power developed by turbine, diameter of jet and diameter of pipeline of a Pelton wheel setup having flow rate of 3 m³/s and 150 m head from nozzle to head race of the reservoir. The two turbines have two jets per runner and all the four jets have same diameter. Take the pipeline as 2500 m long, efficiency of nozzle as 91%, efficiency of each runner as 90%, the velocity of coefficient of each nozzle as 0.975 and friction factor as 0.0045.

Solution: Given data:

$$Q = 3 \text{ m}^3/\text{s}, H_n = 150 \text{ m}, \eta = 0.9, C_V = 0.975, 4f = 0.0045, \eta_h = 0.9, L = 2500 \text{ m}$$

Efficiency $\eta = \dfrac{H_n}{H_g} = \dfrac{H_g - h_f}{H_g} \Rightarrow 0.9 = \dfrac{150 - h_f}{150} \Rightarrow h_f = 150 - 150 * 0.9 = 15 \text{ m}$

So head loss due to friction is 15 m.

Hydraulic efficiency $\eta_H = \dfrac{\textit{Power developed by Runner}}{\gamma (Q_a - q) H_n}$

refer Eq. (11.9a)

$$\Rightarrow 0.9 = \frac{\textit{Power developed by Runner}}{1000 * 3 * (150 - 15)}$$

$$\Rightarrow \textit{Power developed by Runner} = 0.9 * 1000 * 3 * (150 - 15) = 364.5 \text{ kW}$$

So **power developed by a Pelton wheel** is **364.5 kilowatts.**

Velocity of jet

$$V = C_V \sqrt{2gH_n} = 0.975\sqrt{2 * 9.81 * 135} = 50.18 \text{ m/s}$$

As we have, four jets, discharge per jet is

$$q = \frac{\textit{Total discharge}}{\textit{Number of jets}} = \frac{3}{4} = 0.75 \text{ m}^3/\text{s}$$

Diameter of jet $d = \sqrt{4q/\pi V} = \sqrt{4*0.75/3.14*50.18} = 0.137$ m

When water is flowing through a pipe, there arise head loss due to friction, and so according to Darcy Weisbach equation, we have

$$h_f = \frac{fLV^2}{2gD} = \frac{fL\left(4Q/\pi D^2\right)}{2gD}$$

$$15 = \frac{0.0045*2500\left(4*3/3.14*D^2\right)}{2*9.81*D} \Rightarrow D = 0.53 \text{ m}$$

So **diameter of the pipeline** is 0.53 m.

Example 11.6: A single jet Pelton wheel is required to drive a generator to develop 10000 kW. The available head at the nozzle is 760 m. Assume generator efficiency as 95%, efficiency of runner as 87%, the velocity of coefficient of nozzle as 0.97, mean bucket velocity 0.46 of jet velocity, outlet angle of bucket 15° and relative velocity of the water leaving the buckets 0.85 of the inlet. Determine the flow rate, diameter of jet and force exerted on the buckets. **[AMIE]**

Solution: Given data:

$$P = 10000 \text{ kW}, H_n = 760 \text{ m}, \eta_g = 0.95, C_V = 0.97, \eta_h = 0.87, u = 0.49 \text{ V}, V_{ro} = 0.85\, V_{ri}, \beta_o = 15°$$

Refer Eq. (11.9g), we have turbine output power

$$P_t = \frac{Output\ \ of\ \ generator}{Generator\ \ efficiency} \Rightarrow P_t = \frac{10000*1000}{0.95} = 10526.3 \text{ kW}$$

For **Discharge**, we have

$$\eta_{pw} = \frac{P_t}{\gamma Q H_n} \Rightarrow Q = \frac{P_t}{\gamma \eta_{pw} H_n} = \frac{10526.3*1000}{1000/9.81*0.87*760} = 1.62 \text{ m}^3\text{/s}$$

Velocity of jet $\quad V = C_V\sqrt{2gH_n} = 0.97\sqrt{2*9.81*760} = 118.40$ m/s

Diameter of the jet $\quad d = \sqrt{4Q/\pi V} = \sqrt{4*1.62/\pi*118.40} = 132$ mm

Refer Figure 11.7a, For a Pelton wheel, we have

$$V_{wi} = V = 118.40 \text{ m/s} \ \& \ V_{ri} = (V - u) = (V - 0.46V) = 118.40 - 54.5 = 64 \text{ m/s}$$

So exit relative velocity will be $V_{ro} = 0.85 * V_{ri} = 0.85 * 64 = 54.4$ m/s

Vane angle at exit is $\beta_o = 15°$

Whirl velocity at exit, see Figure 11.6, we have

$$V_{wo} = u_o - V_{ro} \cos\beta_o \Rightarrow V_{wo} = 54.5 - 54.4 * \cos 5 = 1.95 \text{ m/s}$$

Resultant force

$$F = \gamma Q / g\left(V_{wi} - V_{wo}\right)$$

$$F = 1000 * 1.62\left(118.40 - 1.95\right) = 188.7 \text{ kN}$$

c) Pelton wheel design parameters and their proportions

While designing a Pelton wheel, the designer need to take considerations of some parameters mentioned below, which effect the working and efficiency of Pelton wheel, like:

1. Jet Velocity: The velocity of a jet leaving from a nozzle onto a Pelton vane vary from theoretical magnitude to actual, due to the reason that there may be some losses like frictional existing in the nozzle surface and hence the theoretical velocity V_{th} is more than the actual velocity, i.e. $V_{th} \geq V_{actual}$. The **theoretical velocity** is directly proportionate with the **net head** H_n of head race of water in the dam from the turbine and so $V_{th} = \sqrt{2gH_n}$ and **actual velocity** will be

$$V_{act} = C_V \sqrt{2gH_n} \tag{11.10}$$

and $0.97 - C_V - 0.99$

2. Speed ratio K_u speed ratio is defined as the ratio of blade velocity to theoretical velocity of the jet and so

$$K_u = \frac{u}{\sqrt{2gH_n}}$$

$$\text{and } 0.44 - K_u - 0.47 \tag{11.11}$$

where $u = \dfrac{\pi DN}{60}$; *D is mean diameter of Pelton wheel & N speed*

3. Jet ratio (m): It is defined as the ratio of mean diameter of the wheel (from the center to the pitch circle) to the jet diameter, i.e.

$$m = D/d \tag{11.11a}$$

and for maximum efficiency, jet ratio must lie in between 11 and 13.

4. Bucket proportions: Proportions related with bucket/vane nomenclature are listed in Figure 11.8a.

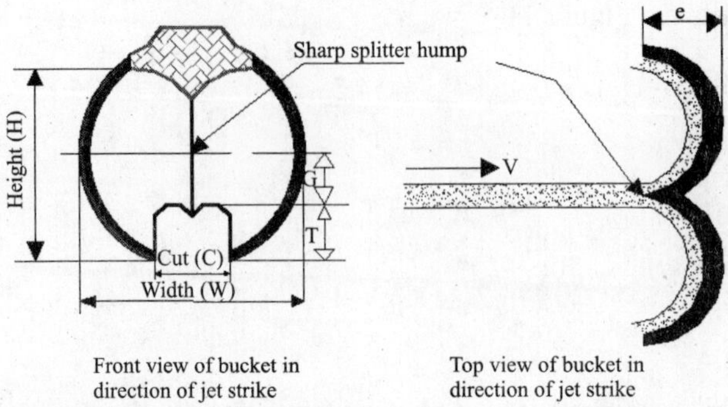

Legend:
W = 2.8 to 3.4 d; H = 2.5 to 3.2 d
e = 0.8 to 1.0 d; C = 1.1 to 1.3 d
G = 0.9 to 1.2 d; T = 0.7 to 0.9 d
Φ = 5 to 15°

Figure 11.8a: Pelton wheel bucket nomenclature and proportions

5. Number of buckets (*Z*): Choosing number of buckets is a critical task. A designer should be able to identify correct number of buckets that are to be mounted on a Pelton wheel, so that there is fully continuous strike of water jet and minimum number of buckets are maintained to minimize losses in friction and also to minimize the weight. The approximate number of buckets to be taken, which holds the above conditions good is given by "Tygun" as

$$Z = \left(15 + \frac{D}{2d}\right) = 15 + m/2 \qquad (11.12)$$

Example 11.7: Find the flow rate, diameter of jet, diameter of wheel, jet ratio and number of jets of a Pelton wheel, working at head of 50 m and develops a power of 650kW, with a speed of 150rpm and has 80% of overall efficiency and assume coefficient of velocity as 0.69 and speed coefficient as 0.41.

Solution: Given data: $H_n = 50$ m, $P = 650$ kW, $N = 150$ rpm, $\eta_o = 0.8$, $C_V = 0.69$, $K_u = 0.41$

We know power developed by a Pelton wheel as $P = \gamma Q H_n \eta_o$

$$\Rightarrow Q = P/\gamma H_n \eta_o = 650*1000/9810*50*0.8 = 1.625 \text{ m}^3/\text{s}$$

Velocity of jet $\quad V = C_V\sqrt{2gH_n} = 0.69\sqrt{2*9.81*50} = 21.6$ m/s

Velocity of blade $\quad u = K_u\sqrt{2gH_n} = 0.41\sqrt{2*9.81*50} = 12.48$ m/s

Diameter of jet $\quad d = \sqrt{4Q/\pi V} = \sqrt{4*1.625/3.14*21.6} = 0.3$ m

Diameter of wheel $D = 60u / \pi N = 60 * 12.48 / .14 * 150 = 1.64$ m

Jet ratio $m = D / d = 1.64 / 0.3 = 5.45$

Number of jets $Z = 15 + m / 2 = 15 + 5.45 / 2 = 18$ numbers

Hence, the **flow rate, diameter of jet and wheel, jet ratio and number of jets are** 1.625 m^3/s, 0.3 m, 1.64 m, 5.45,18 respectively.

FRANCIS TURBINE

Francis turbine named after James B. Francis is a mixed (radial and axial) flow type reaction turbine. In this type of turbine, the jet flows radially inwards from the outer periphery and leaves out axially from the center of the turbine.

It basically consists of penstock, spiral/scroll casing, stay ring and stay vanes, guide vanes, runner with vanes and a draft tube, as shown in Figure 11.9a & b.

Penstock and spiral/scroll casing: water enters the turbine from penstock and penstock is in attachment with a spiral casing, through which the water flows towards the turbine runner. And while having the flow of water towards the runner, a constant velocity of water flow is to be maintained, which is done by the spiral casing, as it has reduced uniform cross-section, throughout the spiral casing.

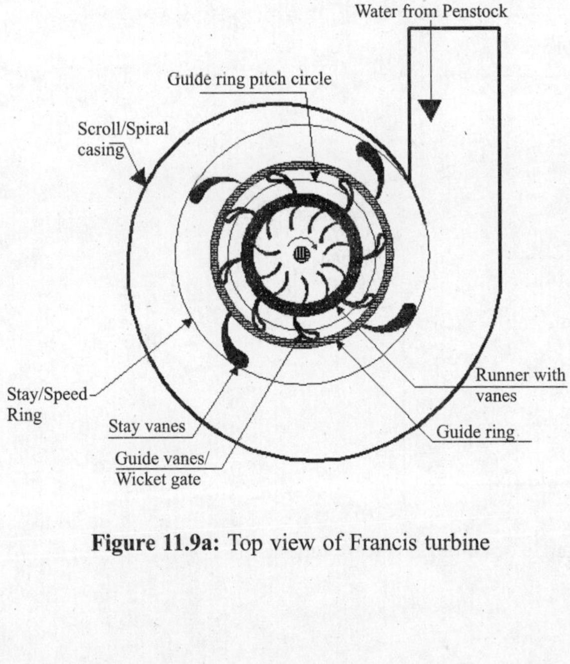

Figure 11.9a: Top view of Francis turbine

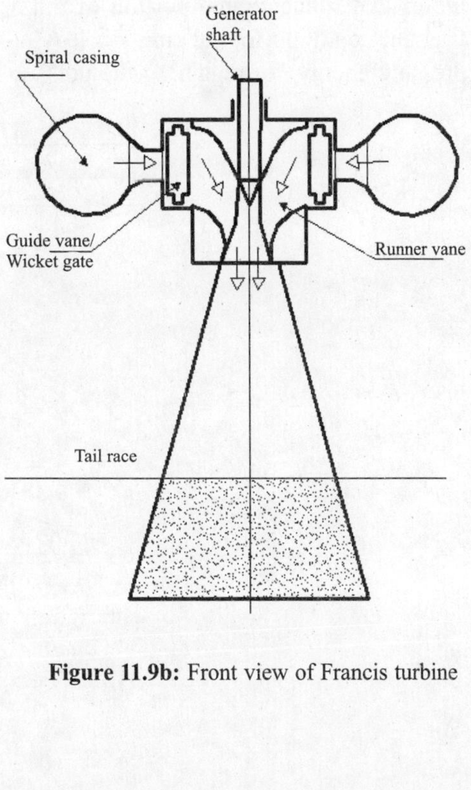

Figure 11.9b: Front view of Francis turbine

Stay/Speed ring and Stay vanes: The water flowing in the casing with constant velocity need to be diverted towards the runner and so a number of fixed vanes called stay vanes are used, which are fixed to a circular ring called stay ring (also called speed ring), as shown in Figure 11.9. And in addition to the direction changing, the stay ring also transmits any unbalanced loads arising due to internal pressure of water motion and due to unbalanced weight of turbine, to the foundation.

Guide ring and guide vanes/wicket gates: Water diverted from stay vanes towards the runner to be pressurized while impinging the water as a jet on to the runner vanes, so that the water jet strikes the vanes with energy and thereby cause rotation of runner. Guide vanes are usually taken in double the number of stay vanes and are pivoted about a guide ring pitch circle and they are free to rotate about their pivots, so that the gap in between to consecutive vanes at their tail ends be reduced, which acts like nozzle, to regulate the quantity of jet strike and also these guide vanes are in aerofoil shaped, to direct the jet strike with an appropriate deflection angle.

Runner with vanes: Runner, the shaft of a turbine generator consists of vanes/buckets (curved), evenly arranged on the circumference of the runner, to perform the rotation of the runner shaft due to the jet strike which may be tangentially, at the vane tips and are so designed that the jet entering the runner vanes are sent out axially to the runner shaft, after the strike completion, towards the draft tube. The runner vanes may be of with different blade profiles, as shown in Figure.11.7b.

Draft tube: Draft tube is a tube-like structure, having a passage for flow of water through it from the exit of the turbine to the tail race and so it can be defined as a transporting device of fluid from runner exit to the tail race, having a circular cross-section at the entrance (at exit of turbine runner) and can be of any cross-section at its exit (at tail race). For transporting the leaving water jet from the exit of runner, some portion of the draft tube is immersed in the tail race, so as to create a vacuum condition for sucking the leaving jet and convert the kinetic energy of the leaving jet to pressure energy, for regaining the pressure energy in tail race.

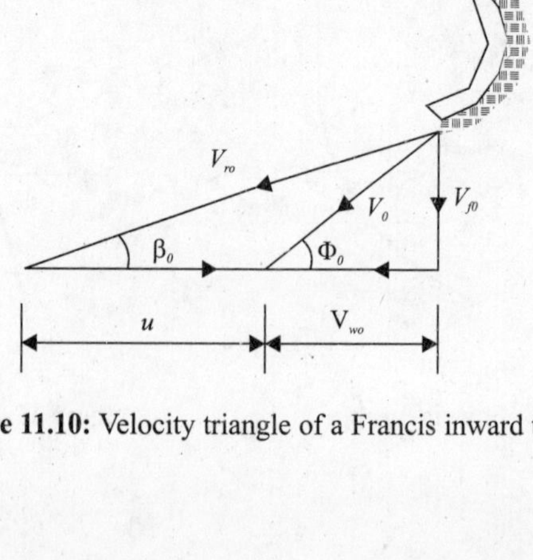

Figure 11.10: Velocity triangle of a Francis inward turbine

a) Work done, efficiencies of Francis turbine and degree of reaction

Work extracted by vanes/buckets of the turbine runner from the fluid jet striking. And hence work done by the jet on the bucket is due to the impulse and so applying impulse momentum equation, we have $F = \rho a V_r (V_1 - V_2)$; refer Eq. (10.1).

Let β_i, β_o be angle of bucket at the inlet and outlet tips respectively.

ϕ_i, ϕ_o be angle between absolute velocities and whirl velocity components respectively.

V_i, V_o be absolute velocities at the inlet and outlet tips respectively.

V, u be absolute velocities of the jet before impact and velocity of bucket respectively, where blade velocity $u = \frac{\pi DN}{60}$, where "*D*" and "*N*" are mean diameter of shaft (distance from the centre of the wheel to the pitch circle) and speed of the runner shaft.

Consider a jet impinged on one tip tangentially inwards and leave through another tip tangentially outwards, as shown in Figure 11.10. And consider the inlet and outlet velocity triangles of jet impinged on a moving curved vane (chapter Impact of Jets) at one of the tip tangentially and leaving through the other tip.

As the surface of the bucket may not be smooth and so there may be some loss in relative velocity from inlet to outlet and so $\mu = \frac{V_{ro}}{V_{ri}}$ where "μ" is coefficient of friction and will be < 1.

Work done per second by the jet on the bucket will be $WD = Force * Displacement$

$\Rightarrow WD = F * u$

So we get

$$WD = \frac{\gamma}{g} * Q * (V_{wi} - V_{wo}) * u \tag{11.13}$$

So **hydraulic efficiency** is ratio of **"output in form of work done by the jet on the bucket to the hydraulic input of the jet"** and so

$$\eta_H = \frac{WD}{input} = \frac{\gamma * Q * (V_{wi} - V_{wo}) * u}{\gamma Q H_n} = \frac{(V_{wi} - V_{wo}) * u}{g H_n} \tag{11.14}$$

and the hydraulic efficiency ranges from 80% to 95%, where H_n is net head.

But if the vane is so designed that the absolute velocity at exit V_o is equal to the flow velocity V_f, as shown in Figure 11.7b (radial blade profile), i.e. V_{wo} = 0, then Eq. (11.13) becomes the maximum and

$$WD = \frac{W}{g} * Q * (V_{wi}) * u \tag{11.14a}$$

thereby Eq. (11.14) reduces to

$$\eta_H = \frac{WD}{input} = \frac{\gamma * Q * (V_{wi}) * u}{\gamma Q H_n} = \frac{(V_{wi}) * u}{g H_n} \qquad (11.14b)$$

Mechanical Efficiency is defined as **"ratio of power available at the turbine shaft to the power developed by the runner (buckets)"**. i.e.

$$\eta_m = \frac{Power \ available \ at \ Turbine \ shaft}{Power \ developed \ by \ Runner}$$

$$\Rightarrow \eta_m = \frac{P}{\gamma * Q * \frac{(V_{wi} - V_{wo}) * u}{g}} \qquad (11.15)$$

Overall efficiency is the product of hydraulic and mechanical efficiencies and so

$$\eta_o = \eta_{hyd} * \eta_{mech} = \frac{P}{\gamma Q_a H_n} \qquad (11.16)$$

Degree of reaction *R*: Degree of reaction is defined as **"ratio of pressure drop in the runner shaft to the work done on the runner"**. If the pressure at inlet and outlet of the runner be P_i & P_o respectively, then

$$R = \frac{Pressure \ drop \ in \ runner \ shaft}{Work \ done \ on \ runner} = \frac{\left(\frac{P_i}{\gamma} - \frac{P_o}{\gamma}\right)}{\frac{(V_{wi} - V_{wo})}{g} * u} \qquad (11.17)$$

Note: If the **Degree of reaction is 50%**, i.e. the pressure drop in the guide vane and runner vane is equal. The value of degree of reaction is useful in designing the characteristics of a turbo-machine.

For an impulse turbine, there is no change in static pressure and hence the degree of reaction is zero (i.e. R = 0). For Francis runner it ranges from 0 to one (i.e. 0 < R < 1) and for Kaplan runner it ranges from 0.5 to 1 (i.e. 0.5 < R < 1).

Example 11.8: Assuming whirl velocity at the outlet to be zero for a reaction turbine working under a head of 150 m and speed of 420 rpm, Determine the flow rate, power and hydraulic efficiency, if the diameter at inlet is 0.12 m and flow area is 0.4 m^2. The angles made by absolute and relative velocities at inlet are 30° and 60° respectively.

Solution: Given data:

$$H_n = 150 \text{ m}, N = 420 \text{ rpm}, D_i = 0.12 \text{ m}, A_f = 0.4 \text{ m}^2, \phi_i = 30°, \beta_i = 60°, V_{wo} = 0$$

Absolute blade velocity

$$u_i = \frac{\pi D_i N}{60} = \frac{3.14 * 0.12 * 420}{60} = 2.64 \text{ m/s}$$

From inlet velocity triangle, see Figure.11.10, we have $\frac{V_{fi}}{V_{wi}} = \tan\phi_i - 0.577$ and also

$$\frac{V_{fi}}{V_{wi} - u_i} = \tan\beta_i = \tan 60 = 1.732$$

So from the two above equations, we can get $0.577V_{wi} = 1.732(V_{wi} - u_i) = 1.732(V_{wi} - 2.64)$

$\Rightarrow V_{wi} = 10.8$ m/s

Discharge $Q = A_{fi} * V_{fi} = 0.4 * 6.23 = 2.49 \text{ m}^3/\text{s}$ where A_{fi} is flow area

Power $$P = \gamma Q\left(\frac{V_{wi}u_i - V_{wo}u_o}{g}\right) \text{ but } V_{wo} = 0$$

So **power** $$P = \gamma Q\left(\frac{V_{wi}u_i}{g}\right) = 9810 * 2.49 * \left(\frac{10.8 * 2.64}{9.81}\right) = 70.99 \text{ kW}$$

Hydraulic efficiency $\eta_h = \left(\frac{V_{wi}u_i}{gH_n}\right) = \left(\frac{10.8 * 2.64}{9.81 * 150}\right) = 1.9\%$ refer Eq. (11.24b)

Example 11.9: Determine the absolute blade velocity, water velocity, whirl velocity and relative velocity of water at inlet, runner blade angles, width of the runner at outlet, weight of water flowing per second, head at inlet, power developed, hydraulic efficiency of a reaction turbine working with external and internal diameters as 1.5 m, 0.95 m respectively. The speed is 310 rpm with constant flow velocity of 3.2 m/s and has inlet width of 300 mm. The guide angle makes 10° with the tangent of the wheel and it discharges out radially.

Solution: Given data:

$D_o = 1.5 \text{ m}, D_i = 0.95 \text{ m}, N = 310 \text{ rpm}, V_{fi} = V_{fo} = 3.2 \text{ m/s}, V_{wo} = 0, \phi_0 = 90°, \phi_i = 10°, B_i = 0.3 \text{ m}$

Consider Figure 11.10a:

Absolute blade velocity at inlet and outlet are $u_i = \frac{\pi D_i N}{60}$ & $u_o = \frac{\pi D_o N}{60}$

$$\Rightarrow u_i = \frac{3.14*0.95*310}{60} = 15.4 \text{ m/s} \;\&\; u_o = \frac{\pi D_o N}{60} = \frac{3.14*1.5*310}{60} = 24.3 \text{ m/s}$$

Absolute water velocity at inlet is $V_i = \dfrac{V_{fi}}{\sin\phi_i} = \dfrac{3.2}{\sin 10} = 18.43$ m/s

Whirl inlet velocity is $V_{wi} = V_i \cos\phi_i = 18.43 * \cos 10 = 18.15$ m/s

Relative inlet velocity is $V_{ri} = \sqrt{(V_{wi} - u_i)^2 + V_{fi}^2} = \sqrt{(18.15 - 15.4)^2 + 3.2^2} = 4.22$ m/s

Inlet blade angle $\beta_i = \tan^{-1}\left(\dfrac{V_{fi}}{V_{wi} - u_i}\right) = \tan^{-1}\left(\dfrac{3.2}{18.15 - 15.4}\right) = 49°$

Outlet blade angle $\beta_o = \tan^{-1}\left(\dfrac{V_{fo}}{u_o}\right) = \tan^{-1}\left(\dfrac{3.2}{24.3}\right) = 7.5°$

Refer Eq. (11.20a) for **width of runner at outlet** $(\pi D_o B_o) = (\pi D_i B_i)$ since $V_{fi} = V_{fo}, K_i = K_o$

$$\Rightarrow B_o = \frac{D_i B_i}{D_o} = \frac{0.95*0.3}{1.5} = 0.19 \text{ m}$$

Discharge $Q = (\pi D_i B_i) V_{fi} = 3.14*1.5*0.19*3.2 = 2.86 \text{ m}^3/\text{s}$

Weight of water per second $W = \gamma Q = 1000 * 2.86 = 2.86$ kN/s

Power $P = \gamma Q\left(\dfrac{V_{wi} u_i}{g}\right) = 9810 * 2.86 * \left(\dfrac{18.15*15.4}{9.81}\right) = 799.4$ kW,

Net head

$$H_n = \left(\frac{V_{wi}u_i \pm V_{wo}u_o}{g}\right) + \frac{V_o^2}{2g} = \left(\frac{V_{wi}u_i}{g}\right) + \frac{V_{fo}^2}{2g} \quad \text{since } V_{fo} = V_o \;\&\; V_{wo} = 0$$

$$\Rightarrow H_n = \left(\frac{18.15*15.4 \pm 0}{9.81}\right) + \frac{3.2^2}{2*9.81} = 29.01 \text{ m}$$

Hydraulic efficiency $\eta_h = \left(\frac{V_{wi}u_i}{gH_n}\right) = \left(\frac{18.15*15.4}{9.81*29.01}\right) = 98.2\%$ refer Eq. (11.14b)

b) Francis turbine design parameters and their proportions:

While designing a Francis turbine, the designer need to take considerations of Power "*P*", Speed "*N*" and head "*H*" and also assuming probable values of some parameters like efficiency $\eta_h, \eta_o, Flow\ ratio(\zeta), n$, where "*n*" is width to diameter ratio:

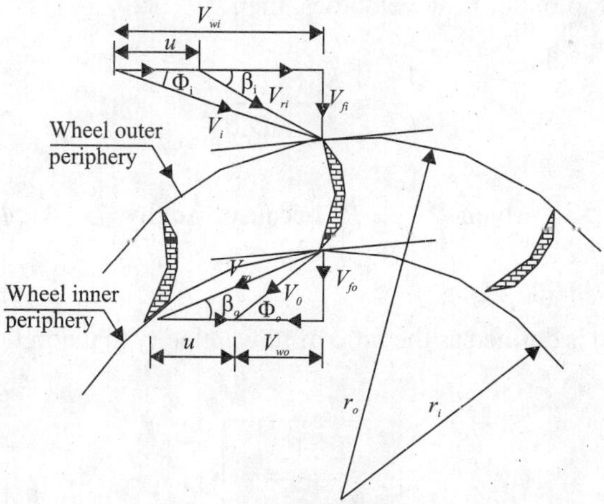

Figure 11.10a: Inward Francis turbine curved radial vanes mounted

1. Determination of required discharge: We know

$$Power = \eta_o(\gamma Q H_n) \tag{11.18}$$

using this relation determine the discharge required.

2. Width to diameter of vane ratio (*n*) and vane thickness coefficient (*K*):

Width to diameter ratio

$$n = \frac{B}{D} \tag{11.19a}$$

is usually taken in the range of 0.1 to 0.45.

And we know the **discharge** is equal to the product of **flow area** and **flow velocity** and so we have $Q = A_f * V_f$ where flow area

$$A_f = \begin{pmatrix} Complete\ area \\ of\ Turbine \end{pmatrix}_{upto\ outer\ periphery} - \begin{pmatrix} Complete\ area \\ of\ Vanes \end{pmatrix} \tag{11.19b}$$

$$\Rightarrow A_f = (\pi D_o B) - (Z * t * B)$$

where Z is number of vanes, t is thickness of vane.

So $Q = \left[(\pi D_o) - (Z * t)\right] * B * V_f$

$$\boxed{\Rightarrow Q = \pi D_o \left[K * B * V_f\right]} \quad (11.20)$$

$K < 1$ vane thickness coefficient.

If taken ratio of inlet to outlet flow velocities, then

$$\frac{V_{fi}}{V_{fo}} = \frac{K_o(\pi D_o B)}{K_i(\pi D_i B)} \quad (11.20a)$$

and if $K_i = K_o$ and $V_{fi} = V_{fo}$, then $2B_o = B_i$ because usually $D_0/3 < D_i < D_o/4$ and taken as $2D_i = D_o$.

3. Flow ratios: Flow ratio is defined as the ratio of flow velocity to theoretical velocity of the jet and so

$$\boxed{\zeta = \frac{V_f}{\sqrt{2gH_n}}} \quad (11.21)$$

and $0.15 - \varsigma - 0.30$

Equating Eq. (11.20) and Eq. (11.1), we get

$$D_o = \left[\frac{Q}{\left(\zeta\sqrt{2gH_n}\right)K_o n\pi}\right]^{0.5} \quad (11.21a)$$

4. Speed ratio K_u **:** Flow ratio is defined as the ratio of flow velocity to theoretical velocity of the jet and so

$$\boxed{K_u = \frac{u}{\sqrt{2gH_n}}} \quad (11.22)$$

and $0.6 - K_u - 0.9$

where $u = \dfrac{\pi DN}{60}$; *D is mean diameter of Francis turbine & N speed*

5. Determine the guide vane angle and runner vane angle using inlet velocity triangle, $\tan\phi_i = \dfrac{V_{fi}}{V_{wi}}$ and

$$\tan \beta_i = \frac{V_{fi}}{V_{wi} - u}$$

6. Determine the blade speed and runner diameters as $u_o = \frac{\pi D_o N}{60}$ & $u_o = 2u_i$ & $D_i = D_o/2$

Example 11.10: Calculate the absolute blade velocities at inlet and outlet, discharge and breadth of the inward flow reaction turbine wheel at the outlet, if it has external and internal diameters of 2.3 m and 1.1 m respectively. The width of the turbine at inlet is 300 mm, running with a speed of 200 rpm, with a flow velocity of 3.5 m/s. Assume that 6% of the circumferential area has been blocked by the thickness of the blades, both at inlet and outlet, flow velocity as 2.0 m/s, Vane thickness coefficient as 0.92.

Solution: Given data:

$$D_o = 2.3 \text{ m}, D_i = 1.1 \text{ m}, B_i = 0.3 \text{ m}, N = 200 \text{ rpm}, V_{fi} = 2.0 \text{ m/s}, K_o = K_i = 0.92$$

Absolute blade velocity at inlet and outlet are

$$u_i = \frac{\pi D_i N}{60} \text{ \& } u_o = \frac{\pi D_o N}{60}$$

$$\Rightarrow u_i = \frac{3.14 * 2.3 * 200}{60} = 24.07 \text{ m/s \& } u_o = \frac{\pi D_o N}{60} = \frac{3.14 * 1.1 * 200}{60} = 11.5 \text{ m/s}$$

Discharge $Q = A_f * V_f = K_o (\pi D_o B_o) V_f$

$$\Rightarrow Q = 0.92(3.14 * 2.3 * 0.3) * 2.0 = 3.98 \text{ m}^3/\text{s}$$

Consider Eq. 11.20a, we have $K_i (\pi D_i B_i) V_{f_i} = K_o (\pi D_o B_o) V_{fo}$

When vane thickness coefficient is constant, the flow velocity will be also constant, so $V_{fi} = V_{fo}$

Vane breadth at the outlet is

$$B_o = \frac{K_i (\pi D_i B_i) V_{f_i}}{K_o (\pi D_o) V_{fo}} = \frac{(D_i B_i)}{(D_o)} = \frac{(1.1 * 0.3)}{(2.3)} = 0.14 \text{ m}$$

Example 11.11: Design a Francis turbine having radial blades with width to diameter ratio at inlet and outlet as 0.5 and 0.7 respectively. Head of 70 m, speed 500 rpm, brake power 300 kW, flow ratio 0.2, speed ratio 0.7, hydraulic efficiency 95% and overall efficiency 85%.

Solution: Given data:

$$n_o = 0.5 \text{ m}, n_i = 0.7 \text{ m}, H_n = 70 \text{ m}, N = 500 \text{ rpm}$$

$$BP = 300 \text{ kW}, \varsigma = 0.2, K_u = 0.7, \eta_h = 0.95, \eta_o = 0.85$$

Flow velocity at inlet and outlet are $V_{f_i} = V_{fo} = \varsigma\sqrt{2gH_n} = 0.2\sqrt{2*9.81*70} = 7.41$ m/s

Refer Eq. (11.9d), we have $\eta_o = \dfrac{BP}{\gamma Q H_n}$, so **Discharge**

$$Q = \frac{BP}{\gamma \eta_o H_n} = \frac{300*1000}{9810*0.85*70} = 0.514 \text{ m}^3/\text{s}$$

$$Q = A_f * V_f = (\pi D_i B_i)V_{fi}$$
$$\Rightarrow 3.98 = (3.14 * D_i * 0.5D_i)7.41$$
$$\Rightarrow D_i = 0.58 \text{m}$$

Breadth at inlet $B_i = 0.5\ D_i = 0.5*0.58 = 0.292$ m

Similarly, diameters of the wheel at outlet

$$Q = (\pi D_o B_o)V_{fo}$$
$$\Rightarrow 3.98 = (3.14 * D_o * 0.7D_o)7.41$$
$$\Rightarrow D_o = 0.494 \text{ m}$$

Breadth at outlet $B_o = 0.7D_o = 0.7*0.58 = 0.406\text{m}$

Absolute blade velocity at inlet and outlet are

$$u_i = \frac{\pi D_i N}{60} \ \& \ u_o = \frac{\pi D_o N}{60}$$

$$u_i = \frac{3.14*2.3*200}{60} = 24.07 \text{ m/s} \ \& \ u_0 = \frac{\pi D_o N}{60} = \frac{3.14*1.1*200}{60} = 11.5 \text{ m/s}$$

Hydraulic efficiency

$$\eta_H = \frac{WD}{input} = \frac{(V_{wi} - V_{wo})*u}{gH_n} \quad \text{refer Eq. (11.24)}$$

$$\Rightarrow 0.95 = \frac{(V_{wi}u_i - 0)}{gH_n} = \frac{(V_{wi}*24.07)}{9.81*70} \quad \Rightarrow V_{wi} = 27.1 \text{ m/s}$$

since $V_{wo} = 0$.

Guide vane angle and runner vane angle, using inlet velocity triangle,

$$\tan\phi_i = \frac{V_{fi}}{V_{wi}} \text{ and } \tan\beta_i = \frac{V_{fi}}{V_{wi} - u}$$

$$\Rightarrow \tan\phi_i = \frac{7.41}{27.1} \quad \text{and} \quad \tan\beta_i = \frac{7.41}{27.1 - 24.07}$$

$$\Rightarrow \phi_i = 15°.29 \qquad \Rightarrow \beta_i = 67°.78$$

Table 11.2: Difference between inward and outward flow reaction turbines

Aspect	Inward flow	Outward flow
Water flow direction	Water enters from outer periphery radially and leaves out from center of the turbine axially	Water enters from inner periphery axially and leaves out from outer periphery radially.
Discharge variation	Discharge does not vary throughout	Discharge increases.
Easiness	Easy and effective speed control	Difficult to control
Usage	Commonly used	Obsolete
Heads	Medium and high heads	Low and medium heads

Table 11.3: Advantages and disadvantages of Francis over Pelton wheel

Aspect	Francis	Pelton wheel
Control of variation in operating head	Easily controlled	Not so easy
Any water usage	Clean water must be used to avoid wear.	Any water can be used.
Cavitation	Ever present.	Minimum.
Decrease of mechanical efficiency with wear.	Slow	Rapid.
Space occupation for same power generation	Small space required.	Large space required.

Kaplan turbine: Kaplan turbine named after Austrian engineer Victor Kaplan, which is purely an axial flow type reaction turbine, suitable for low heads up to 30 m. It has quite similar features in construction like Francis turbine, except that Francis turbine is designed only for certain maximum power output at medium heads with medium flow rate, but to increase in power output with increase in flow rate, one has to increase the widths of runner vane and diameter of the runner of the Francis turbine, which further extends to increase in outlet area (shown in Figure 11.11) and also increase in flow velocity which increase the losses and hence it is not acceptable.

To accommodate this change, Kaplan turbine is so designed that, the runner vanes are not fixed as in case of Francis turbine, but can be guided (rotated) on the runner shaft hub, with the help of a servomotor installed inside the runner shaft, which makes the Kaplan turbine to erect in vertical position only. So not only the guide vanes allow the flow rate but also the runner vanes do accommodate to change in flow rate, thereby changing area for inlet for flow rate to the guide vanes.

It basically consists of penstock, spiral/scroll casing, guide vanes (rotating), runner with hub having rotating vanes on it and a draft tube.

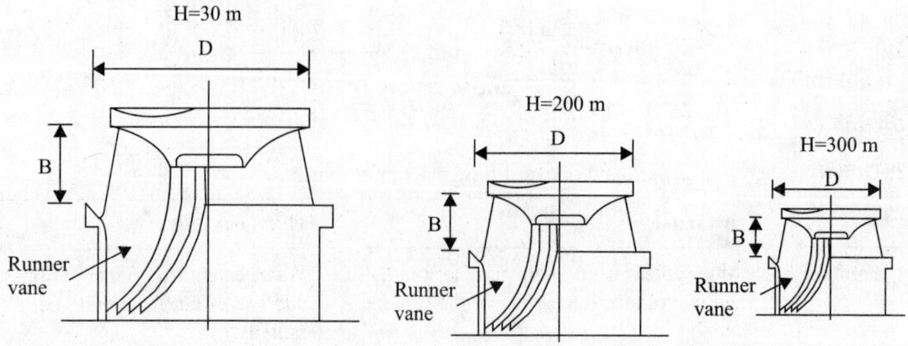

Figure 11.11: Francis turbine relative dimensions of width and diameter w.r.t head

Runner hub with vanes: Runner, the shaft of a turbine generator consists of a hub on which vanes/buckets (curved), evenly arranged on the circumference of the runner hub and is connected with a servomotor mechanism for their rotation, as per the requirement to increase or decrease the flow rate inlet area, as shown in Figure 11.12a.

Construction and Operation: Kaplan turbine has a spiral casing through which the water enters radially towards the runner, with uniform velocity. But the guide vanes (rotating) having a servomotor mechanism connected, do divert the direction of the water flow from the spiral casing to the runner vanes axially, as shown in Figure 11.12.

And the runner vanes are also connected to another servomotor mechanism, which guide the rotation of the runner vanes and thereby adjusting the blade angles for high efficiency working at all loads, let the water out axially towards the runner shaft hence as the water jet enter the blades (runner) axially and leave axially, we do call this turbine an axial flow type reaction turbine.

Compare the velocity triangles of a Francis turbine and Kaplan turbine, as shown in Figure 11.13. Even though there will be no change in velocity triangles at any loads, but as there are two servomotors in a Kaplan turbine, which automatically adjust the blade angles of the guide vane and runner vane, which reduces the eddy losses, which exists in case of a Francis turbine.

a) Kaplan turbine design parameters and their proportions:
Kaplan is just similar to Francis turbine, except that the hub diameter is involved and so while designing a Kaplan turbine, the designer need to take considerations of Hub diameter "*d*" also:

1. Diameter of hub to outside diameter of runner ratio (*n*):

$$n = \frac{d}{D} \tag{11.23}$$

And is usually taken in the range of 0.35 to 0.60.

2. Discharge is equal to the product of **flow area** and **flow velocity** and so we have

$$Q = A_f * V_f$$

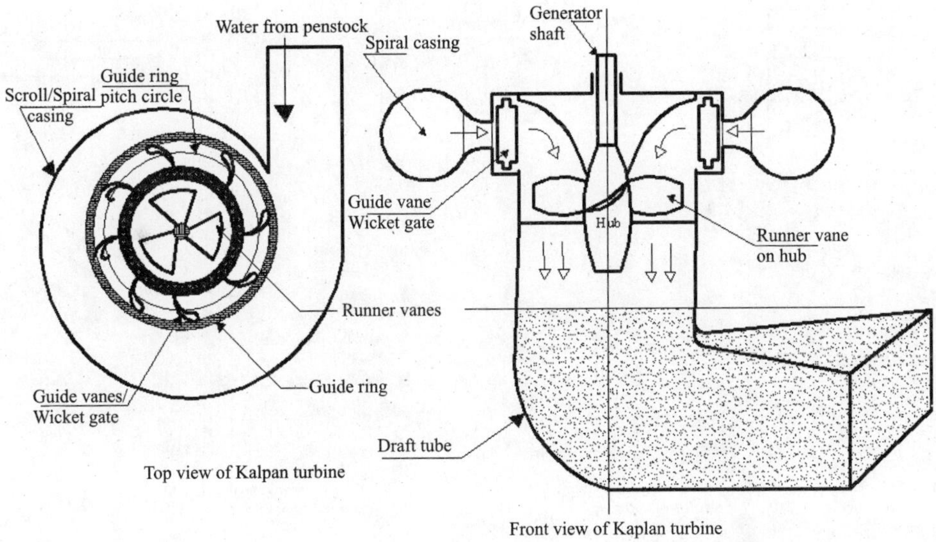

Figure 11.12:

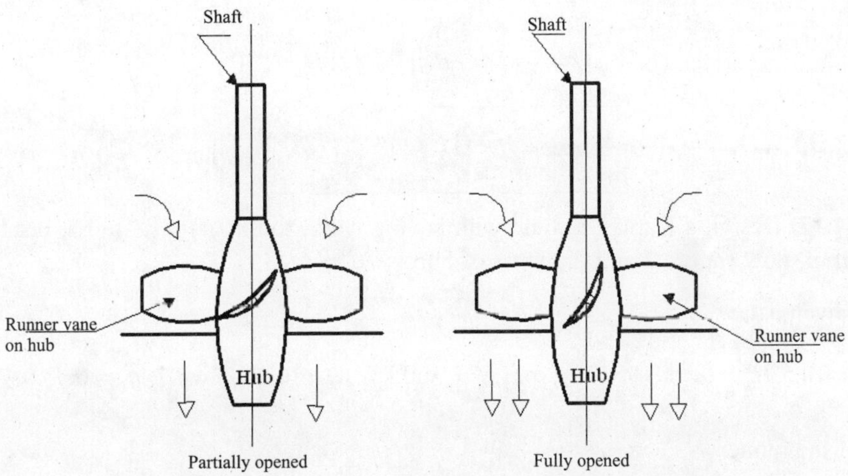

Figure 11.12a: Runner vanes partially and fully opened

where flow area $A_f = \left(\begin{matrix}Complete\ area\\ of\ Turbine\end{matrix}\right) - \left(\begin{matrix}Complete\ area\\ of\ hub\end{matrix}\right)$

$$\Rightarrow A_f = \frac{\pi}{4}\left(D^2 - d^2\right) \tag{11.24}$$

But flow velocity is

$$V_f = \zeta\sqrt{2gH_n}$$

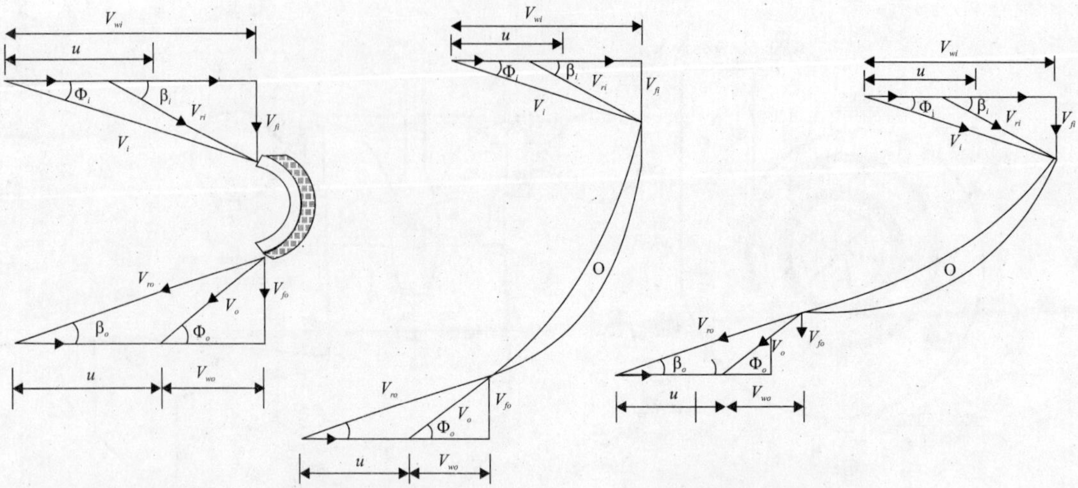

Velocity triangles of a Francis inward turbine at all loads

Velocity triangles of a Kaplan turbine at Full load

Velocity triangles of a Kaplan turbine at partial loads

Figure 11.13: Comparison of Francis and Kaplan velocity triangles.

and so

$$Q = \frac{\pi}{4}\left(D^2 - d^2\right) * \zeta\sqrt{2gH_n}$$

$$\Rightarrow Q = \frac{\pi}{4} D^2\left(1 - n^2\right) * \zeta\sqrt{2gH_n} \quad \text{where } \zeta = 0.70 \qquad (11.25)$$

Example 11.12: Design a Kaplan turbine with having width to diameter ratio 0.7, head of 70 m, speed 500 rpm, flow ratio 0.2 and diameter of hub is 10 m.

Solution: Given data:

$$d = 10\,\text{m},\ n = 0.5, H_n = 70\,\text{m},\ N = 500 \text{ rpm},\ P = 300 \text{ kW},\ \varsigma = 0.2,\ K_u = 0.7,\ \eta_h = 0.95,\ \eta_o = 0.85$$

Diameter of the runner $D_o = \frac{d}{n} = \frac{10}{0.5} = 20\,\text{m}$

Absolute blade velocity is $u_o = \frac{\pi D_o N}{60} = \frac{3.14 * 20 * 500}{60} = 523.3$ m/s

Flow velocity at inlet and outlet are $V_{f_i} = V_{fo} = \varsigma\sqrt{2gH_n} = 0.2\sqrt{2 * 9.81 * 70} = 7.41$ m/s

Refer Eq. (11.25), we have **discharge** $Q = \frac{\pi}{4} D_o^2\left(1 - n^2\right)\varsigma\sqrt{2gH_n}$

$$\Rightarrow Q = \frac{3.14}{4} * 20^2 \left(1 - 0.5^2\right) * 0.2\sqrt{2 * 9.81 * 70} = 1745 \text{ m}^3/\text{s}$$

Example 11.13: Design diameters of runner, absolute blade velocities, discharge and runner vane angles at the hub and outer periphery, for a Kaplan turbine with having width to diameter ratio 0.5, head of 40 m, speed 300 rpm, power 2500 kW, overall efficiency 90%, hydraulic efficiency 85% and diameter of hub is 10 m. Assume no whirl at exit of the runner.

Solution: Given data:

$$n = 0.5, d = 10\,\text{m}, H_n = 40\,\text{m}, N = 300 \text{ rpm}, P = 25000 \text{ kW}$$
$$\eta_o = 0.90, V_{wo} = 0, V_{fi} = V_{fo}, \eta_h = 0.85$$

Outside and inner diameters of the runner $D_o = \frac{d}{n} = \frac{10}{0.5} = 20 \text{ m} \;\&\; D_i = d = 10 \text{ m}$

Absolute runner blade velocities at outer and inner periphery are

$$u_o = \frac{\pi D_o N}{60} \;\&\; u_i = \frac{\pi d N}{60} = \frac{\pi D_i N}{60}$$
$$\Rightarrow u_o = \frac{3.14 * 20 * 300}{60} = 314 \text{ m/s} \;\&\; u_i = \frac{3.14 * 10 * 300}{60} = 157 \text{ m/s}$$

We have **discharge**

$$Q = \frac{P}{\gamma \eta_o H_n} = \frac{25000 * 1000}{1000 * 0.9 * 40} = 694 \text{ m}^3/\text{s}$$

Refer Eq. (11.25), we have **flow velocity at inlet**

$$V_{fi} = Q \Big/ \left(\frac{\pi}{4}\left(D_o^2 - d_b^2\right)\right) = \frac{694 * 4}{3.14\left(20^2 - 10^2\right)} = 2.94 \text{ m/s}$$

Runner vane angle at the outer periphery

Whirl velocity at inlet (outer periphery) of runner blade is

$$\eta_H = \frac{(V_{wi}) * u_0}{gH_n} \Rightarrow V_{wi} = \frac{(gH_n \eta_H)}{u_o} = \frac{(9.81 * 40 * 0.85)}{314} = 1.06 \text{ m/s}$$

From inlet velocity triangle, we have

$$180 - \beta_i = \tan^{-1}\left(\frac{V_{fi}}{u - V_{wi}}\right) = \tan^{-1}\left(\frac{2.94}{314 - 1.06}\right) = 0.5°$$

So runner vane angle at outer periphery is $\beta_i = 179.5°$

Runner vane angle at the inner periphery

From outlet velocity triangle, we have $\beta_o = \tan^{-1}\left(\frac{V_{fo}}{u_i}\right) = \tan^{-1}\left(\frac{2.94}{157}\right) = 1.1^0 \Rightarrow \beta_o = 1.1^0$

Table 11.4: Difference between Francis and Kaplan reaction turbines

Aspect	Francis turbine	Kaplan turbine
Flow direction	Radially inward or outward flow	Only axial flow
Adjustable runner vanes	Runner vanes are fixed	Runner vanes are movable
Number of vanes	16 to 24	3 to 8
Shaft positioning	Horizontal or vertical	Vertical only
Governing Servomotor	Only one servomotor is sufficient for the guide vanes.	Two servomotors required, one for guide vanes and other for runner vanes.
Placing of Servomotors	Outside the turbine	Inside the turbine.
Load and efficiency	Maximum efficiency obtained at full load only	Maximum efficiency obtained even at partial load, due to simultaneous control of guide and runner vanes.
Heads and flow rate	Medium head and flow rate	Low head and high flow rate
Size of governor	Ordinary governor is sufficient as the servomotor is large.	As servomotors are small in size, heavy governors are required.

Table 11.5: Comparison of different types of turbines

Turbine	Inventor	Type	Flow type	Head	Year
Poncelet	Jean V. Poncelet	Impulse	Inward flow		1920
Fourneyron	Benoit Fourneyron	Reaction	Outward radial		1826
Jonval	Feu Jonval	Impulse	Axial		1843
Girard radial		Impulse	Inward radial		1850
Francis	James B. Francis	Reaction	Inward radial		1865
Pelton wheel	Lester A. Pelton	Impulse	Tangential	250 - 450 m, High head (i.e. >250 m)	1870
Banki		Impulse			1900
Kaplan	Victor Kaplan	Reaction	Parallel/Axial	1.5m – 60 m, Low head (i.e. <60 m)	1913
Girard axial		Impulse	Axial		
Modern Francis	J. B. Francis	Reaction	Mixed Axial)	(Radial and Medium head (i.e. 60m – 250m)	
Thomson		Reaction	Inward radial		
Propeller		Reaction	Axial	1.5m – 60 m, Low head (i.e. < 60 m)	
Turgo	Gilkes	Impulse		Medium head (i.e. 60 m – 250 m)	1919

DRAFT TUBE THEORY

For transporting the leaving water jet from the exit of runner, a tube like structure having a passage for flow of water through it from the exit of the turbine to the tail race called draft tube is being used. It has a circular cross-section at the entrance (at exit of turbine runner) and can be of any cross-section at its exit (at tail race) and some portion of the draft tube is immersed in the tail race to create a vacuum condition for sucking the leaving jet and thereby converting the kinetic energy of the leaving jet (which is left as loss) to pressure energy, for regaining the pressure energy in tail race. Different types of draft tubes can be seen in Figure 11.14.

Consider a draft tube attached at the exit of the runner outlet and take three sections 1, 2 and 3 at the inlet of the runner, at outlet of the runner (inlet to the draft tube) and at the exit of immersed draft tube portion respectively, as shown in Figure 11.15.

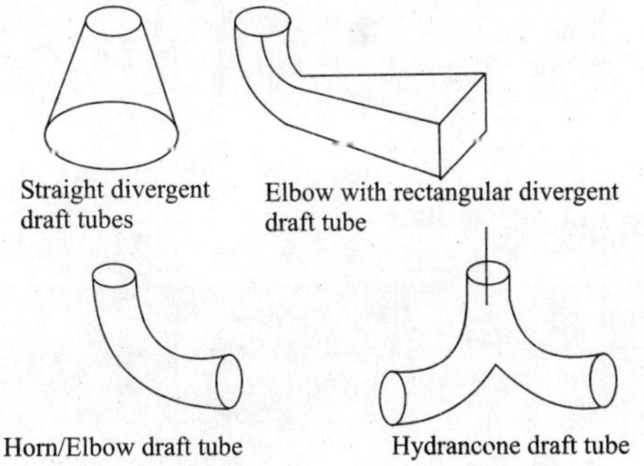

Figure 11.14: Different types of draft tubes

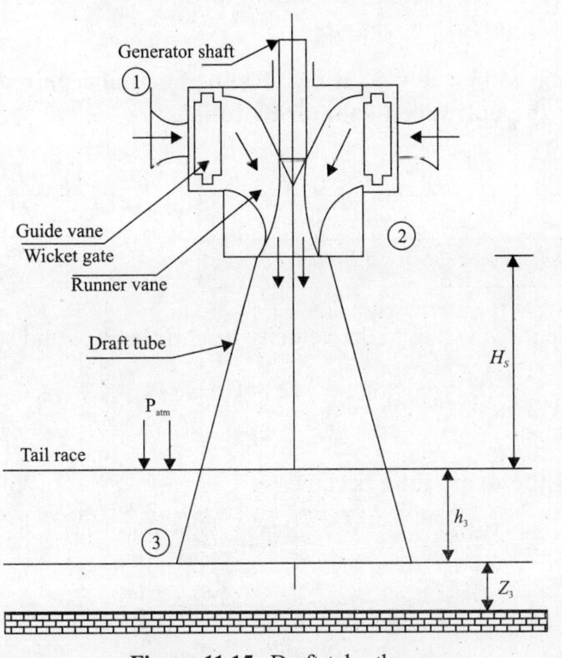

Figure 11.15: Draft tube theory

Let a turbine be with no draft tube attached, then the water jet leaving the runner must be of more or equal to atmospheric pressure (i.e. $P_2 \geq P_{atm}$), to come out of the runner, as there exists atmospheric pressure surrounded due to which the kinetic energy leaving the runner is of no use.

But if the same turbine is attached with a draft tube, as shown in Figure 11.15, then the exit pressure at the runner is of suction (i.e. $P_2 \leq P_{atm}$), so the reduction in kinetic energy is converted to pressure head, which thereby there will be increase in available head and efficiency of the turbine.

Apply Bernoulli's equation between sections 2-2 and 3-3, including the hydraulic frictional losses (h_f) in Figure 11.15, we have

$$\frac{P_2}{\gamma} + \frac{V_2^2}{2g} + (h_3 + H_s) = \frac{P_3}{\gamma} + \frac{V_3^2}{2g} + (0) + h_f$$

$$\Rightarrow \frac{P_2}{\gamma} = \frac{P_3}{\gamma} - \left(\frac{V_2^2 - V_3^2}{2g} - h_f\right) - (H_S + h_3) \tag{11.26}$$

But $\frac{P_3}{\gamma} = \frac{P_{atm}}{\gamma} + h_3$ and so Eq. (11.26) reduces to

$$\Rightarrow \frac{P_2}{\gamma} = \frac{P_{atm}}{\gamma} - \left(\frac{V_2^2 - V_3^2}{2g} - h_f\right) - (H_S) \tag{11.26a}$$

Hence, Eq. (11.26a) proves that the pressure at the runner exit is less than the atmospheric pressure and is said to be suction pressure if used with a draft tube, thereby increase in available head and efficiency.

Efficiency of the draft tube will be defined as the **"Ratio of actual regain of pressure head to the actual velocity head at the entrance of the draft tube"**,

i.e.

$$\eta_D = \frac{\textit{Actual regain of pressure head}}{\textit{Velocity head at draft tube entrance}} \Rightarrow \eta_D = \frac{\left(\frac{V_2^2 - V_3^2}{2g} - h_f\right)}{V_2^2/2g} \tag{11.27}$$

But the hydraulic frictional is due to the velocity head difference and so loss

$$h_f = K\left(\frac{V_2^2 - V_3^2}{2g}\right)$$

And hence **efficiency of the draft tube** becomes

$$\Rightarrow \eta_D = (1 - K)\frac{\left(\frac{V_2^2 - V_3^2}{2g}\right)}{V_2^2/2g} \tag{11.28}$$

Note:

1. If the draft tube is made up of cylindrical cross-section ($V_2 = V_3$), then Eq. (11.26) shows that the exit pressure at the runner is less than the atmospheric pressure by an amount of H_S and so there is no reduction in pressure head, which can be converted in kinetic energy at the exit.
2. If the draft tube is made up of gradual increasing cross-section ($V_2 > V_3$), then Eq. (11.26) shows that the exits pressure at the runner is more than the atmospheric pressure by an amount of $\left(\dfrac{V_2^2 - V_3^2}{2g} - h_f\right)$ and so there is reduction in pressure head, which can be converted in kinetic energy at the exit, thereby increasing the efficiency of turbine. So a gradual increasing draft tube is preferred rather than a cylindrical cross-section tube.

Example 11.14: Determine the pressure head at inlet and efficiency of the conical draft tube having inlet and out diameters as 1.5, 3.3 m respectively. The draft tube is 9 m long and has 2.3 m of it immersed in tail race and it discharges water with a velocity of 3.2 m/s at the outlet. Assume atmospheric pressure as 10.3 m of water and head loss due to friction as 0.3 times of outlet velocity.

Solution: Given data:

$$D_2 = 1.5\,\text{m},\ D_3 = 3.3\,\text{m},\ L = 9\ \text{m},\ h_3 = 2.3\ \text{m},\ V_3 = 3.2\ \text{m/s},\ h_f = 0.3V_3, P_{atm}/\gamma = 10.3\,\text{m}$$

Discharge through the draft tube

$$Q = A_2V_2 = A_3V_3 \Rightarrow Q = \frac{\pi}{4}D_3^2V_3 = \frac{3.14}{4} * 3.3^2 * 3.2 = 27.35\ \text{m}^3/\text{s}$$

Velocity at the inlet of the draft tube is

$$V_2 = A_3V_3 / A_2 = 27.35 * 4 / 3.14 * 1.5 * 1.5 = 15.49\ \text{m/s}$$

Suction head $H_S = L - h_3 = 9 - 2.3 = 6.7$ m

Refer Eq. (11.26a) for **pressure head at the inlet**

$$\frac{P_2}{\gamma} = \frac{P_{atm}}{\gamma} - \left(\frac{V_2^2 - V_3^2}{2g} - h_f\right) - (H_S)$$

$$\Rightarrow \frac{P_2}{\gamma} = 10.3 - \left(\frac{15.49^2 - 3.2^2}{2 * 9.81} - 2.3\right) - (6.7) = 3.0\ \text{m (absolute)}$$

Refer Eq. (11.27) for **efficiency of the draft tube** is

$$\eta_D = \frac{\left(\frac{V_2^2 - V_3^2}{2g} - h_f\right)}{V_2^2 / 2g} \Rightarrow \eta_D = \frac{\left(\frac{15.49^2 - 3.2^2}{2*9.81} - 2.3\right)}{15.49^2 / 2*9.81} = 76.9\%$$

Example 11.15: Determine the total head across a reaction turbine, shaft power, head lost due to friction in turbine and draft tube and power lost in mechanical friction if the reaction turbine is attached with a conical draft tube of diameters as 1.5, 3.3 m at inlet and outlet respectively and discharges 2.9 m^3/s of water through it. The draft tube is 4.5 m long. The turbine has a pressure head of 42 m above atmospheric pressure. The velocity of water at the inlet of draft tube is 3.2 m/s. Assume the draft tube end diameter touches the tail race. Assume hydraulic efficiency as 75% and overall efficiency as 75%.

Solution: Given data:

$$D_2 = 1.5\,\text{m},\ D_3 = 3.3\,\text{m},\ L = 4.5\,\text{m},\ Q = 2.9\,\text{m}^3/\text{s}$$
$$h_3 = 0.0\,\text{m},\ V_2 = 3.2\,\text{m/s},\ \eta_h = 0.75,\ \eta_o = 0.75,\ h_t = 42\,\text{m}$$

Total head across the turbine

$$H_g = h_t + \frac{V_2^2}{2g} + L + h_3 = 42 + 0.52 + 4.5 + 0 = 47.02\ \text{m}$$

Shaft power $\quad P = \gamma Q H_n \eta_o = 9.81 * 2.9 * 46.5 * 0.75 = 992\ \text{kW}$

Net head $\quad H_n = H_g - \left(h_{ft} + h_{fd}\right) = 47.02 - \left(h_{ft} + h_{fd}\right)$

where h_{ft}, h_{fd} are head losts due to friction in turbine and draft tube respectively.

Head utilized by turbine is $H = H_n - \frac{V_3^2}{2g} = 47.02 - \left(h_{ft} + h_{fd}\right) - \frac{8 * Q^2}{\pi^2 D_3^4 g} = 47.01 - \left(h_{ft} + h_{fd}\right)$

Hydraulic efficiency

$$\eta_h = \frac{Head\ \ utilized}{Head\ \ supplied} \Rightarrow 0.75 = \frac{47.01 - \left(h_{ft} + h_{fd}\right)}{H_g} \Rightarrow 0.75 = \frac{47.01 - \left(h_{ft} + h_{fd}\right)}{47.02}$$

$$\Rightarrow \left(h_{ft} + h_{fd}\right) = 47.01 - 0.75 * 47.02 = 11.75\ \text{m}$$

So **head loss due to friction in turbine and draft tube** is 11.75 m.

Power developed by runner = shaft power + power lost in mechanical friction (P_m)

Mechanical efficiency

$$\eta_m = \frac{Shaft\ \ power}{Power\ \ developed\ \ by\ \ runner} \Rightarrow \eta_m = \frac{P}{P+P_m} \Rightarrow 0.75 = \frac{992}{992+P_m}$$

$$\Rightarrow P_m = (992 * 0.75) - 992 = 248 \text{ kW}$$

So **power lost in mechanical friction** is 248 kW.

SPECIFIC PERFORMANCE OF TURBINES

Introduction

Every turbine has varying loads, when in working with a generator but it is to maintain a constant speed at all varying loads. A turbine works efficiently at a particular point like at a particular speed, discharge and output power, to which it is designed for. And hence to analyze or predict the performance of a turbine one must calculate the **specific parameters** like specific speed, discharge, output power and head, to correlate with same type of turbines and to analyze the performance of a turbine with respect to another type of turbine, it must be analyzed for a model quantities called **unit quantities** at that particular point like unit speed, unit discharge, unit power, etc.

Determination and Significance of Unit Quantities

Unit quantities like unit speed, unit discharge and unit power are estimated for a turbine operating at a unit head, for comparing the performance of a turbine with another turbine of similar kind at any working conditions.

Assume that the velocity triangles of a turbine at any working head be of same similar triangles and so consider the inlet velocity triangle of a turbine working at any load having velocity, speed, discharge, flow velocity, power, etc. as u, N, Q, V_f, P respectively and also inlet velocity triangle of a turbine working at a unit head having velocity, speed, discharge, flow velocity, power, etc. as $u_u, N_u, Q_u, V_{fu}, P_u$ respectively, as shown in Figure 11.6.

From similar triangles of Figure 11.16, we have

$$\frac{u}{u_u} = \frac{V_{wi}}{V_{wiu}} = \frac{V_{fi}}{V_{fiu}} = \frac{V_i}{V_{iu}} = \frac{V_{ri}}{V_{riu}} \tag{11.29}$$

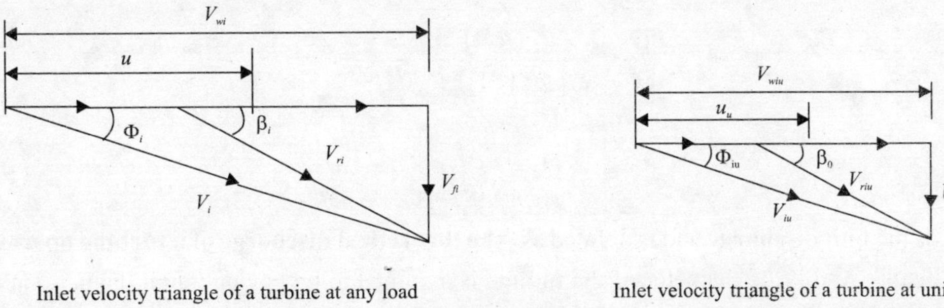

Inlet velocity triangle of a turbine at any load

Inlet velocity triangle of a turbine at unit load

Figure 11.16

Equating the overall efficiency of turbine at any working condition and at unit head condition, we have $\eta_o = \eta_{ou}$

$$\Rightarrow \frac{V_{wi}u}{gH} = \frac{V_{wiu}u_u}{gH_u} = \frac{V_{wiu}u_u}{g*1} \Rightarrow \left(\frac{V_{wi}}{V_{wiu}}\right)\frac{u}{H} = \frac{u_u}{1}$$

$$\boxed{\Rightarrow u_u = \frac{u}{\sqrt{H}}} \quad \text{since } \frac{u}{u_u} = \frac{V_{wi}}{V_{wiu}} \tag{11.30}$$

where u_u is the **unit velocity** and is defined as "**the theoretical velocity of a turbine operating at a unit head**".

We know, as the velocity is product of radius and angular velocity and so $u = \frac{\pi DN}{60}$ and hence substituting this in the Eq. (11.30), we get

$$\frac{\pi D_u N_u}{60} = \frac{\pi * 1 * N_u}{60} = \frac{\pi DN}{60\sqrt{H}}$$

$$\boxed{\Rightarrow N_u = \frac{ND}{\sqrt{H}}} \tag{11.31}$$

where N_u is the **unit speed** and is defined as "**the theoretical speed of a turbine operating at a unit head**". And if the diameter of the turbine is assumed to be having a unit diameter at any head, then Eq. (11.31) reduces to

$$\boxed{\Rightarrow N_u = \frac{N}{\sqrt{H}}} \tag{11.31a}$$

Similarly, **discharge** $Q = A * V_f = K(\pi D) * V_f$

(refer Eq. (11.20)) for a turbine and so $\frac{Q}{Q_u} = \frac{K(\pi D) * V_f}{K(\pi * 1 * 1) * V_{fu}}$. Hence, substituting

$$\frac{u}{u_u} = \sqrt{H} = \frac{V_{fi}}{V_{fiu}} \textit{ from } \text{Eq. (11.40)}$$

we get

$$\frac{Q}{Q_u} = \frac{K(\pi D) * V_{fi}}{K(\pi * 1 * 1) * V_{fi}\sqrt{H}}$$

$$\boxed{\Rightarrow Q_u = \frac{Q}{D\sqrt{H}}} \tag{11.32}$$

where Q_u is the **unit discharge** and is defined as "**the theoretical discharge of a turbine operating at a unit head**". And if the diameter of the turbine is assumed to be having a unit diameter at any head, then Eq. (11.32) reduces to

$$Q_u = \frac{Q}{\sqrt{H}} \tag{11.32a}$$

Similarly power available $P = (\gamma Q H)\eta_o$

(refer Eq. (11.9d)) for a turbine and so $\frac{P}{P_u} = \frac{(\gamma Q H) * \eta_o}{(\gamma Q_u H_u) * \eta_o} = \frac{(QH)}{(Q_u * 1)}$

Hence, substituting $\frac{Q}{Q_u} = D\sqrt{H}$ *from Eq* : (11.32), we get $\frac{P}{P_u} = \frac{(\gamma Q H) * \eta_o}{(\gamma Q_u H_u) * \eta_o} = D\sqrt{H} * H$

$$\Rightarrow P_u = \frac{P}{DH^{3/2}} \tag{11.33}$$

where P_u is the **unit power** and is defined as **"the theoretical power of a turbine operating at a unit head"**. And if the diameter of the turbine is assumed to be having a unit diameter at any head, then Eq. (11.33) reduces to

$$P_u = \frac{P}{H^{3/2}} \tag{11.33a}$$

SPECIFIC PARAMETERS OF TURBINES

Specific paramcters/quantities are required for comparing different type of turbines, by equating them with an imaginary turbine (called specific turbine), which is reduced in dimensions and is capable of developing 1 kilowatt power under unit head.

Specific speed:

Equate the velocity of a turbine at unit speed and at specific speed, i.e. $u_u = u_s$

$$\Rightarrow \frac{\pi D N_u}{60} = \frac{\pi * D_s * N_s}{60} \qquad \Rightarrow N_s = N_u \frac{D}{D_s} \tag{11.34}$$

But by equating the discharges, we do get

$$Q_s = Q_u \Rightarrow \pi D(nD)V_{fu} = \pi D_s (nD_s) V_{fs} \qquad \Rightarrow \frac{Q_u}{Q_s} = \left(\frac{D}{D_s}\right)^2 \tag{11.35}$$

Similarly, equating the power, we have

$$P_s = P_u \Rightarrow \gamma(Q_u * H_u)\eta_o = \gamma(Q_s * H_s)\eta_o \Rightarrow \gamma(Q_u * 1)\eta_o = \gamma(Q_s * 1)\eta_o$$

$$\Rightarrow \frac{P_u}{P_s} = \frac{Q_u}{Q_s} = \left(\frac{D}{D_s}\right)^2 \Rightarrow D_s = \left(\frac{D}{\sqrt{P_u}}\right) \text{ since } P_s = 1$$

But from Eq. (11.33a), we have $P_u = \dfrac{P}{H^{3/2}}$, so $D_s = \dfrac{DH^{3/4}}{\sqrt{P}}$ (11.36)

And substituting Eq. (11.33a) in Eq:(11.34), we get $N_s = N_u \dfrac{\sqrt{P}}{H^{5/4}}$ (11.37)

And so from Eq. (11.37), we can define **specific speed** as **"speed of a turbine geometrically similar to the actual turbine, develops 1 kilo-watt power when working under unit head"**.

HYDRAULIC TURBINES CHARACTERISITC CURVES

Usually, turbines are designed only for a particular design parameters like head "H", discharge "Q", speed "N", power "P" and efficiency "η". But in actual practice there will be a variation of load acting on the turbine and hence there is often a tendency of turbine to deviate from its design conditions.

If the deviation of the turbine parameters, due to variation in load is recorded by testing the actual turbine at the site or on the scale model of that turbine, which will be very much useful in analyzing and predicting the performance of the turbine and hence the recorded variation of turbine parameters are plotted in a graphical representation for clear understanding of the turbine. And the graphical representations of the recorded deviation values of parameters of the turbine are called "characteristic curves". And characteristic curves are of three types mentioned below:

a) Constant head characteristic curves (main characteristic curves).
b) Constant speed characteristic curves (operating characteristic curves)
c) Constant efficiency characteristic curves (iso-efficiency curves).

Constant head characteristic curves: Also known as main characteristic curves, in which the head of the turbine is maintained constant by operating the gate opening but the speed of the turbine is varied by varying the load acting on the turbine and the values are recorded between the speeds and discharge, power, efficiency for impulse and reaction turbine and are shown in Figure 11.7 and 11.17a.

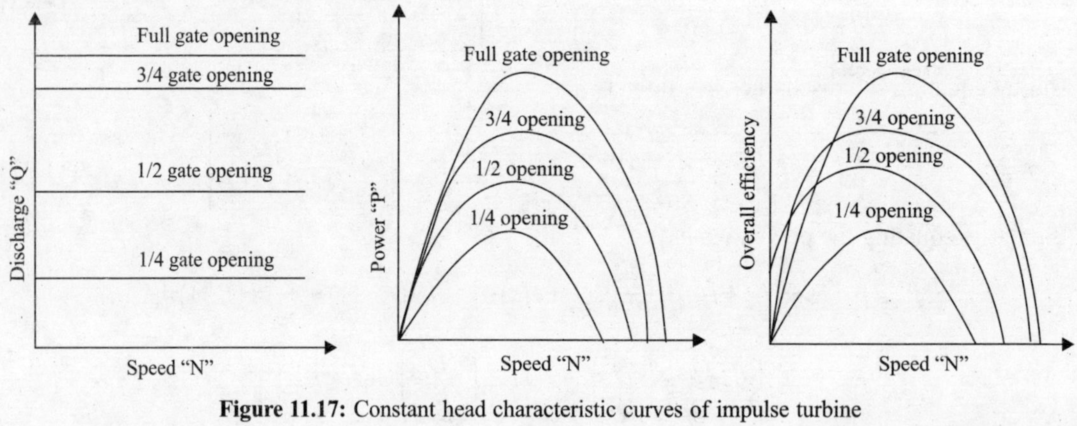

Figure 11.17: Constant head characteristic curves of impulse turbine

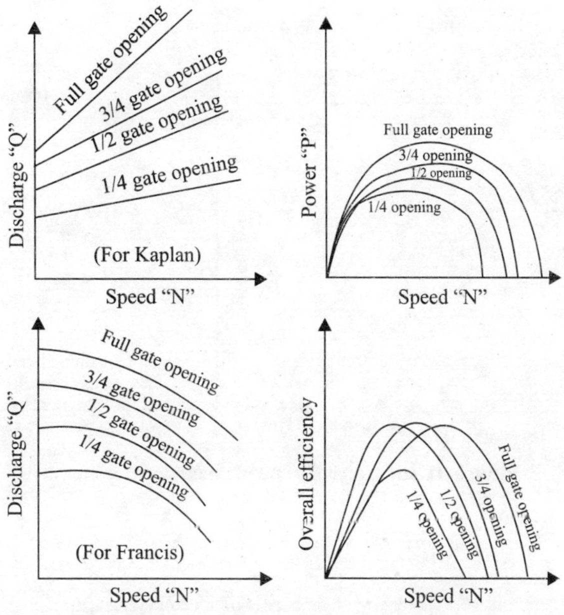

Figure 11.17a: Constant head characteristic curves turbine

For an impulse Pelton, the hydraulic energy is completely converted to kinetic energy and so the discharge regulated by the opening do not affect the speed of the turbine (i.e., irrespective of gate, opening, the speed does not influence because due to continuous strike of the discharge onto the vane) and so the plot have horizontal lines, as shown in Figure 11.7.

For Francis turbine only a part of the hydraulic energy is converted to kinetic energy due to centrifugal rotation of water action, and so the increase in discharge increases the retarding of the runner and thereby speed is inversely proportional to discharge and hence the curve tends to drop down, as shown figure.

For Kaplan turbine only a part of the hydraulic energy is converted to kinetic energy due to centrifugal rotation of water action, but the runner vanes and guide vane so adjust that full centrifugal action of water has been utilized and so the increase in discharge increases and thereby speed increases proportionally and hence the curve tends to shoot up, as shown.

Constant speed characteristic curves: Also termed as operating characteristic curves, in which the speed of the turbine is maintained constant by operating the gate and allowing the discharge to accommodate the change in load variation. And the values are recorded between discharge and efficiency, power keeping the constant speeds for impulse and reaction turbine and are shown in Figure 11.18.

Iso-efficiency curves: Also called constant efficiency (Muschel) characteristic curves, plotted keeping constant efficiency and in between efficiency and speed and in between power and speed, as shown in Figure 11.19.

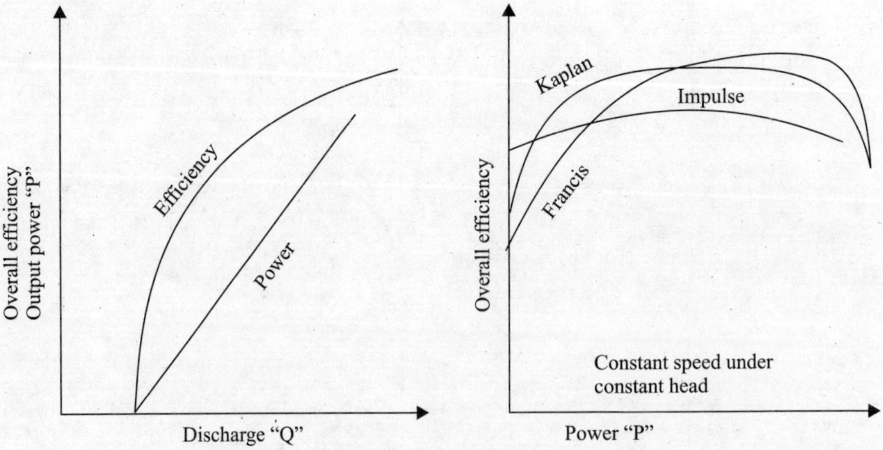

Figure 11.18: Constant speed characteristic curves

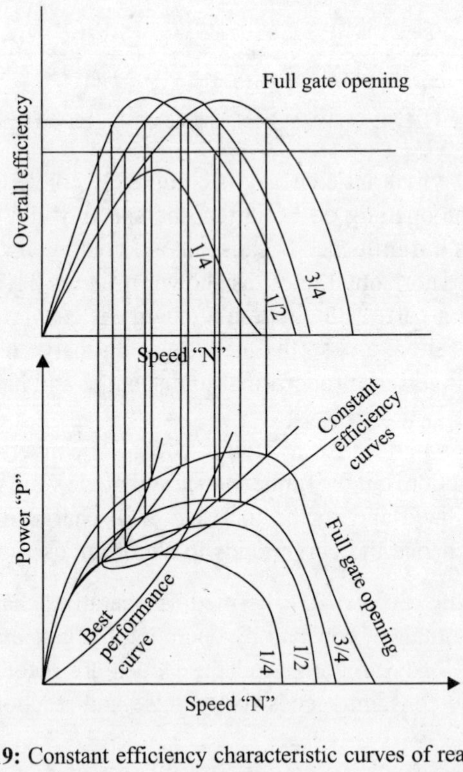

Figure 11.19: Constant efficiency characteristic curves of reaction turbine

Draw a constant efficiency line in the efficiency and speed graph and locate the points of intersections with the gate opening curves and project them on to the respective gate opening curve in the power and speed graph. Join all the points of intersection in the power vs. speed graph, which gives the constant efficiency characteristic curves.

Similarly, by drawing several constant efficiency lines in efficiency and speed graph and projecting them to corresponding power and speed graph, we get series of constant efficiency curves. By doing so, at a particular instant, we get the constant efficiency characteristic curve which passes through the origin and is known as the best performance curve, as shown in Figure 11.19.

GOVERNING MECHANISMS OF TURBINES

Every generator coupled to a turbine need to run at constant speed, in spite of variation in loads acting on the generator and the speed is given by

$$N = {}^{60f}\!/_{p} \tag{11.38}$$

where N = RPM, f = Frequency of power generated in cycles/sec and "p" is the number of pair of poles in generator.

So to maintain the constant speed "N" which is known as **"synchronous speed"**, is based on the turbine that is coupled and is dependent on the design of the turbine and hence governing mechanism of this synchronous speed of generator varies from impulse turbine to reaction turbines.

Governing mechanism of impulse turbine: The governing mechanism of an impulse turbine is shown in Figure 11.20. The governing system takes care of maintaining constant speed of the turbine, irrespective of load variation and also cuts off when there is electrical circuit trip.

It consists of a governor with fly balls, which is sensitive and able to react even for small change in load and moreover it does not hunt (vibrate) up and down after stable condition achieved.

When the load on turbine increases, the speed of the governor increases, tending the governor sleeve to moves up. The upward motion of the sleeve moves the main lever down, about the fulcrum. And as the main lever is connected to the relay valve, the plunger in it is pushed down. The oil supplied by the sump into the relay valve chamber, is compressed by the downward motion of

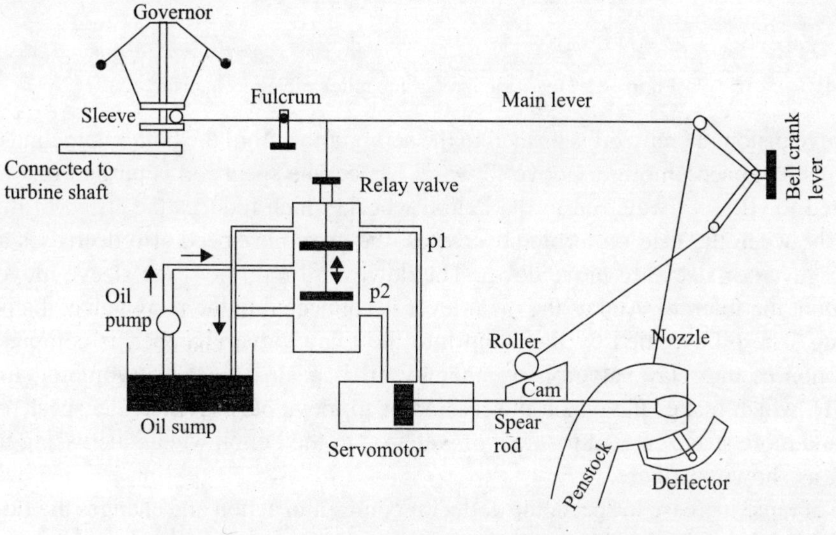

Figure 11.20: Impulse turbine governing mechanism

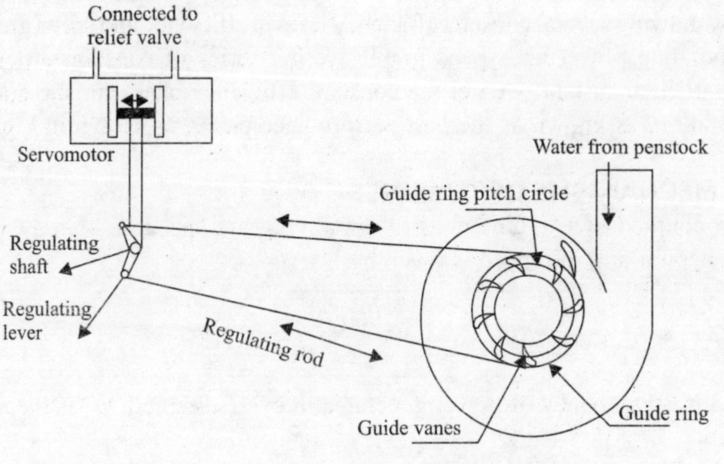

Figure 11.21: Reaction turbine governing mechanism

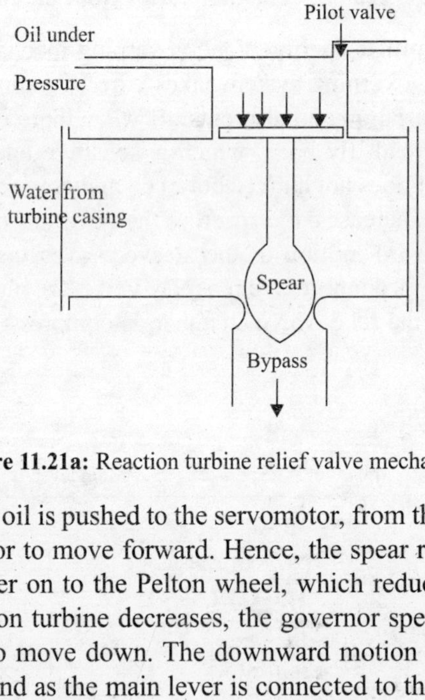

Figure 11.21a: Reaction turbine relief valve mechanism

the relay valve piston, thereby oil is pushed to the servomotor, from the bottom pipeline (p2), which makes the piston in servomotor to move forward. Hence, the spear rod is pushed forward and less space created for flow of water on to the Pelton wheel, which reduces the speed of the turbine.

Similarly, when the load on turbine decreases, the governor speed also decreases and thereby tending the governor sleeve to move down. The downward motion of the sleeve moves the main lever up, about the fulcrum. And as the main lever is connected to the relay valve, the plunger in it is pushed up. The oil supplied by the sump into the relay valve chamber, is compressed by the upward motion of the relay valve piston, thereby oil is pushed to the servomotor, from the top pipeline (p1), which makes the piston in servomotor to move back. Hence, the spear rod is pulled backward and more space created for flow of water on to the Pelton wheel, increasing the speed of the turbine, as shown in figure.

In case of rapid increase in speed, the deflector comes into action and changes the flow direction and thereby cutting off the jet impact on the turbine vanes.

Governing mechanism of reaction turbine: Governing mechanism of reaction turbine is similar to the impulse governing. In impulse, the servomotor piston is connected to the spear rod whereas in reaction turbine, it is connected to a regulating lever. And this regulating lever has regulating rods, which deviates the guide vane angles, to increase or decrease the flow rate, as shown in Figure 11.21a.

In case of rapid increase in speed, the deflector comes into action and changes the flow direction and thereby cutting off the jet impact on the turbine vanes for impulse turbine.

But for reaction turbine, the water from the casing is directly deviated by a relief valve to the tail race from the bypass, without hitting the vanes. This is done by pulling/opening the spear back of the bypass, by reducing the oil pressure acting on the spear, with opening the pilot valve. This pilot valve opens when the bell crank lever moves total downwards.

EXERCISE

Objective

1. Curved vanes are preferred to flat vanes in a hydraulic turbine for the following reason
   a) Low cost of maintenance
   b) Higher hydraulic efficiency can be obtained
   c) Convenience of moulding and machining
   d) It is possible to use high velocity jets
2. The velocity of whirl in a hydraulic turbo-machine is defined as the component of
   a) Absolute velocity in the tangential direction
   b) Absolute velocity in the radial direction
   c) Relative velocity in the tangential direction
   d) Relative velocity in the radial direction
3. A hydraulic turbine converts
   a) Mechanical energy into hydraulic energy
   b) Hydraulic energy into electrical energy
   c) Hydraulic energy to mechanical energy and then to electrical energy
   d) Electrical energy into hydraulic energy
4. Hydraulic efficiency of a water turbine is the ratio of
   a) Power from runner to water power at turbine entrance
   b) Power of shaft to water power at turbine entrance
   c) Power of shaft to power from runner
   d) Power from runner of shaft power
5. The volumetric efficiency of a hydraulic turbine is a ratio of
   a) volumes b) effective c) heads d) powers
6. The difference in elevation between the reservoir water level and the turbine level or tail race level is classified as the following head on turbine
   a) net
   b) effective
   c) operational
   d) gross

7. An under shot water wheel may be classified as the following type turbine
   a) impulse
   b) reaction
   c) radial flow
   d) axial flow
8. An impulse turbine
   a) Completely converts the available hydraulic energy initially into kinetic energy
   b) Should not be run with the wheel exposed to atmospheric pressure
   c) Has runner vanes that convert pressure into mechanical work
   d) Operates under low heads and heavy discharges
9. The casing of an impulse turbine
   a) Increase the effective head on the turbine
   b) Control splashing and spraying of water
   c) Reduce mechanical and windage losses
   d) Reduces hydraulic losses
10. Multiple jets are used in a Pelton wheel for
    a) Increasing the power output with the same specific speed
    b) Increasing the speed of ratio while keeping same specific speed
    c) Increasing the specific speed with the same power output
    d) Increasing the power output with the same head
11. The total deflection angle of a Pelton bucket is made slightly less than 180
    a) Reduce the lateral component of thrust on the shaft
    b) Avoid the deflected jets from striking the back of the oncoming bucket
    c) Prevent racing of the wheel under *n* load
    d) Minimize splashing of water
12. The deflector of a Pelton wheel
    a) Ensures that the buckets receive the full jet impact
    b) Avoids wastage of water during changes of load
    c) Reduces water hammer heads in the penstock
    d) Diffuses the jets into a whirling mass
13. The following is an important component of an impulse turbine
    a) Needle valve
    b) Spiral casing
    c) Draft tube
    d) Relief valve
14. A Pelton wheel is classified as the following type of turbine
    a) Radial flow type
    b) Axial flow type
    c) Tangential flow type
    d) Mixed flow type
15. In an impulse turbine installation, excessive water hammer heads are avoided by using a
    a) Guide wheel
    b) Brake nozzle

c) Needle valve
d) Diffuser

16. Water is admitted to the runner of an impulse turbine through the following device
a) Needle valve
b) Spiral casing
c) Wicket gate assembly
d) Relief valve

17. A typical Pelton wheel bucket deflects the water jets through an angle of
a) 150°
b) 159°
c) 170°
d) 179°

18. The jet ratio of a Pelton wheel is the ratio of
a) Jet diameter to wheel diameter
b) Wheel diameter to jet diameter
c) Wheel diameter to jet velocity
d) Jet velocity to 2gH

19. The approximate jet ratio of Pelton wheel is
a) 20
b) 12
c) 0.985
d) 0.45

20. A brake nozzle serves to quickly stop the runner of the following turbine when the water supply is shut-off
a) Francis
b) Propeller
c) Kaplan
d) Pelton

21. In the following type of turbine, water enters the runner radially and leaves axially
a) Pelton
b) Propeller
c) Kaplan
d) Francis

22. Francis turbine is a reaction turbine of the following type
a) Outward flow type
b) Inward flow type
c) Tangential flow type
d) Axial flow type

23. Kaplan turbine is a reaction turbine of the following type
a) Axial flow type
b) Radial flow type
c) Mixed flow type
d) Tangential flow type

24. The most probable value of the speed ratio of a Francis turbine is
    a) 0.45
    b) 0.8
    c) 1.6
    d) 2.2
25. Most probable value of the speed ratio of a Kaplan turbine is
    a) 0.45
    b) 0.75
    c) 1.15
    d) 2.0
26. The function of the scroll case of a reaction turbine is to
    a) Guide the water to the runner at appropriate angle
    b) Guide the water smoothly to the tail race
    c) Distribute the water evenly around the wheel
    d) Reduce the eddy and shock losses
27. The primary function of wicket gates in reaction turbine installation is to
    a) Avoid water hammer by diverting the water from runner
    b) Control the direction and quantity of water entering the runner
    c) Maintain negative pressure at turbine exit
    d) Reduce whirling of water the runner entrance
28. The draft tube of reaction turbine serves the purpose of
    a) Increasing the effective head on the turbine
    b) Reducing cavitations damage to the turbine
    c) Increasing the kinetic energy of water leaving the runner
    d) Uniformly supplying water to the runner vanes
29. The runner blades of a Kaplan turbine are
    a) Curved than propeller blades
    b) More curved than Pelton blades
    c) More curved than Francis blades
    d) Less curved than Francis blades
30. The runner blades of a Kaplan turbine are adjustable. This facility
    a) Improves the part load efficiency
    b) Improves the full load efficiency
    c) Gives a wide range of operating speeds
    d) Simplifies the governing mechanism
31. The speed ratio of a hydraulic turbine is the ratio of
    a) Velocity head of jet to net head
    b) Tangential velocity of wheel to theoretical jet velocity
    c) Tangential velocity of wheel to net head
    d) Velocity of whirl at exit to net head
32. The speed ratio of a hydraulic turbine is proportional to
    a) 1/D
    b) N^2

c) 1/H
d) H

33. The flow ratio of a hydraulic turbine is the ratio of
a) Velocity of flow to theoretical jet velocity
b) Velocity of flow to net head
c) Velocity of wheel at inlet to velocity of flow
d) Velocity of flow to theoretical jet velocity
34. When there is a sudden full of load on a reaction turbine the immediate action of the governor would be to
a) Move the deflector to intercept the jet
b) Move the needle valve rapidly
c) Open the pressure inclination suitably
d) Adjust the wicket gate inclination suitably
35. Cavitation in a hydraulic turbo-machine is caused by
a) Low velocity
b) Low pressure
c) Local pockets of excessive pressure
d) Regions of low barometric pressure
36. The critical Thomas cavitation number of a turbine is function of
a) Specific speed
b) Head
c) Vapor pressure of water
d) Turbine exit pressure
37. The critical Thomas index of a water turbine
a) Increases with height of turbine above TWL
b) Decreases with saturation vapor pressure of water
c) Increases with specific speed
d) Decreases with net head on turbine
38. Part load operation cover long periods of time, the following type of turbine is the worst choice
a) Pelton
b) Francis
c) Kaplan
d) Propeller
39. If N is the normal running speed of a Kaplan turbine, the probable value of its runaway speed is
a) 1.2 N
b) 1.5 N
c) 1.8 N
d) 2.5 N
40. The ratio of boss diameter to runner diameter of Kaplan turbine is
a) 0.15
b) 0.25
c) 0.35

d) 0.45

41. Two hydraulic turbines are said to be dynamically similar when they are geometrically similar and have
    a) Similar streamlines
    b) Same cavitation index
    c) Same efficiency
    d) Same Reynolds number
42. The diameter of a turbine runner varies as
    a) Square of power for the given head
    b) Square of discharge for the given speed
    c) Square root of speed for the given head
    d) Square root of head for the given speed
43. Match the following

P: Curtis	1: Reaction Turbine
Q: Reteau	2: Gas Turbine
R: Kaplan	3: Velocity compounding
S: Francis	4: Pressure compounding
	5: Impulse water turbine
	6: Axial Turbine
	7: Mixed flow turbine

a) P-2, Q-1, R-7, S-6
b) P-3,Q-1,R-5,S-7
c) P-1,Q-3,R-1,S-5
d) P-3,Q-4,R-7,S-6

Answers

1. b	2. a	3. c	4. a	5. a	6. a
7. a	8. a	9. b	10. d	11. b	12. b
13. a	14. c	15. c	16. a	17. d	18. b
19. b	20. d	21. d	22. b	23. a	24. b
25. a	26. c	27. b	28. c	29. d	30. a
31. b	32. c	33. a	34. d	35. c	36. b
37. d	38. b	39. a	40. b	41. c	42. d
43. d					

Theory

1. What is the primary objective of using more than one jet in a Pelton wheel installation?
2. Why do Pelton wheel buckets have a notch at the centre of their outer periphery?
3. Why cannot reaction turbines be operated by impact of jet on vanes like impulse turbines?
4. What is the difference between the pressures of water at inlet and outlet of a reaction turbine wheel called?
5. What is the chief function of a deflector in a Pelton wheel installation?
6. What component part of a Pelton wheel installation brings the wheel quickly to rest when it is not required to work?

7. The cross-sectional area of a spiral casing of a reaction turbine reduces in the direction of flow. What is the reason?
8. Why is draft tubes not used in impulse wheel installations?
9. Why do draft tubes have enlarging passage area in the direction of flow?
10. What is the principal drawback of a hydraulic turbine?
11. What is the difference between the pressures of water at inlet and outlet of a reaction turbine wheel called?
12. What is the chief function of a deflector in a Pelton wheel installation?

Problems

1. A Pelton wheel is supplied with water under a head of 45 m and at a rate of 48 m^3/min. The buckets deflect the jet through 165° and the mean speed of the bucket is 14 m/s. Determine the power developed by the shaft and overall efficiency of the machine. Assume coefficient of velocity as 0.985 and mechanical efficiency as 0.95. **[18.8 kW, 0.905]**
2. The shaft power of a Pelton wheel, the buckets of which are struck by two jets is 15445 kW. The diameter of each jet is 200 mm. If the net head on the turbine is 400 m, find the overall efficiency of the turbine. Assume coefficient of velocity as 1.0. **[70.85%]**
3. A double jet Pelton wheel is required to generate 7500 k, when the available head at the base of the nozzle is 400 m. The jet is deflected through 165° and the relative velocity of the jet is reduced by 15% in passing over the buckets. Determine the diameter of jet, total flow, force exerted by the jet on buckets tangentially. Assume generator efficiency as 0.95, overall efficiency as 0.8, coefficient of velocity 0.97 and coefficient of speed as 0.46. **[136.5 mm, 2.515 m^3/s, 103.4 kN]**

   Why cannot reaction turbines be operated by impact of jet on vanes like impulse turbines?
4. A reaction turbine has inlet diameter as 3.5 m, outlet diameter as 2.1 m, width 0.5 both at inlet and outlet and the vane angles with the direction of motion as 60° and 150° at inlet and outlet respectively. If the guide blade angle is 15° and discharge is 15 m^3/s, calculate the required speed of the runner at the inlet for smooth flow, torque exerted on runner, power developed, energy extracted/weight and pressure drop through runner. [47 rpm, 309.3 kNm, 1525.3 kW, 10.37 Nm/N, 60.15 kNm^2]
5. An inward flow reaction turbine of inlet diameter 1.2 m operates under a head of 150 m and requires a discharge of 6 m^3/s at a rotational speed of 400 rpm. The guide vane angle is 20° and water leaves the runner blade axially. If the runner is 0.1 m wide at the inlet. Determine the torque, power and efficiency of the turbine shaft. **[156 kNm, 65 kW, 74%]**
6. Find the leading dimensions of the runner of an inward flow reaction turbine to develop 625 k of power with speed of 1000 rpm and working at a head of 100 m. The guide vanes are at angle of 16°, internal diameter is 0.6 times the external diameter and has hydraulic efficiency of 88%, overall efficiency of 86%. Assume blade thickness 5%, Radial velocity of flow constant and axial blade length at inlet is 0.1 times the outer diameter. **[D_i=0.527 m, D_o=.16 m, B_i = 52.7 mm, θ=67.5°, ϕ=31.37°]**

QUICK REFERANCE OF FORMULAE

Property	Formulae	Remarks, if any
Work done and hydraulic efficiency of Pelton wheel, in terms of velocities	$WD = \frac{\gamma}{g} * Q * (V_{wi} - V_{wo}) * u$ $\eta_H = \frac{WD}{KE} = \frac{2u(V_{wi} - V_{wo})}{(V^2)} = \frac{2u(V-u)[1+\mu\cos\beta_o]}{(V^2)}$	$\mu = \frac{V_{ro}}{V_{ri}}; u = \frac{\pi DN}{60}$
Available power and efficiencies, in terms of head of a Pelton wheel.	$Net\ head\ H_n = H_g - H_f$	H_g, H_f are gross, friction heads respectively
	$\eta_H = \frac{Power\ developed\ by\ Runner}{Net\ power\ supplied\ by\ water\ at\ entrance\ of\ turbine}$ $\Rightarrow \eta_H = \frac{Power\ developed\ by\ Runner}{\gamma(Q_a - q)H_n}$ $\eta_m = \frac{Power\ available\ at\ Turbine\ shaft}{Power\ developed\ by\ Runner}$ $\eta_{vol} = \frac{Discharge\ striking\ runner(buckets)}{Actual\ water\ supplied\ to\ turbine} = \frac{(Q_a - q)}{Q_a}$ $\eta_o = \frac{Power\ available\ at\ turbine\ shaft}{Net\ power\ supplied\ by\ water\ at\ entrance\ of\ turbine}$	$\eta_m = \frac{P}{Power\ developed\ by\ runner}$ $\eta_o = \frac{P}{\gamma Q_a H_n}$ $\eta_o = \eta_{hyd} * \eta_{mech} * \eta_{vol}$ $= \frac{P}{\gamma Q_a H_n}$
Pelton wheel proportions	$V_{actual} = C_V\sqrt{2gH_n};\ 0.97 - C_V - 0.99$	Jet velocity
	$K_u = \frac{u}{\sqrt{2gH_n}} = \frac{Blade\ velocity}{Actual\ velocity\ of\ jet}$	Speed ratio; 0.44 – K_u – 0.47
	$m = D/d = \frac{Diameter\ of\ Pelton}{Diameter\ of\ jet};\ 11 - m - 13$	Jet ratio
	$Z = \left(15 + \frac{D}{2d}\right) = 15 + 2\text{ m}$	Number of buckets
Work done, efficiencies of Francis turbine and degree of reaction	$WD = \frac{\gamma}{g} * Q * (V_{wi} - V_{wo}) * u;\ \eta_H = \frac{WD}{input} = \frac{(V_{wi} - V_{wo}) * u}{gH_n}$	If $v_0 = V_f$ i.e. $V_{w0} = 0$, then $\eta_H = \frac{WD}{input} = \frac{(V_{wi}) * u}{gH_n}$

Contd.

Property	Formulae	Remarks, if any
	$\eta_m = \dfrac{Power\ available\ at\ Turbine\ shaft}{Power\ developed\ by\ Runner}$	$\Rightarrow \eta_m = \dfrac{P}{\gamma * Q * \dfrac{(V_{wi} - V_{wo}) * u}{g}}$
	$\eta_o = \eta_{hyd} * \eta_{mech} = {P}/{\gamma Q_a H_n}$	Q_a actual discharge
	$R = \dfrac{Pressure\ drop\ in\ runner\ shaft}{Work\ done\ on\ runner} = \dfrac{\left(\dfrac{P_i}{\gamma} - \dfrac{P_o}{\gamma}\right)}{\dfrac{(V_{wi} - V_{wo})}{g} * u}$	$R = \dfrac{\left(\dfrac{P_i}{\gamma} - \dfrac{P_o}{\gamma}\right)}{\dfrac{(V_{wi} - V_{wo})}{g} * u}$
Francis turbine design proportions	$Power = \eta_o(\gamma Q H_n)$ $n = {B}/{D}$; 0.1-n-0.45 $Q = \pi D_o\left[K * B * V_f\right],\ K = \left[(\pi D_o) - (Z * t)\right]$	*B*, *D* are width of vane and diameter of shaft. **K < 1** vane thickness coefficient.
	$\zeta = \dfrac{Flow\ velocity}{Theoretical\ velocity} = \dfrac{V_f}{\sqrt{2gH_n}}$; $0.15 - \psi - 0.30$ $D_o = \left[\dfrac{Q}{\left(\zeta\sqrt{2gH_n}\right)K_o n\pi}\right]^{0.5}$	Flow ratio
	$K_u = \dfrac{Blade\ velocity}{Theoretical\ velocity} = \dfrac{u}{\sqrt{2gH_n}}$; $0.6 - K_u - 0.9$	Speed ratio usually $2D_1 = D_o$
Kaplan turbine design proportions	$n = {d}/{D}$; 0.35-n-0.60 Flow velocity is $V_f = \zeta\sqrt{2gH_n}$ So $Q = \dfrac{\pi}{4}D^2\left(1 - n^2\right) * \zeta\sqrt{2gH_n}$	d, D are diameters of hub and runner respectively. $A_f = \dfrac{\pi}{4}\left(D^2 - d^2\right)$ & $\zeta = 0.70$
Draft tube efficiency	$\eta_D = \dfrac{Actual\ regain\ of\ pressure\ head}{Velocity\ head\ at\ draft\ tube\ entrance}$ $\Rightarrow \eta_D = \dfrac{\left(\dfrac{V_2^2 - V_3^2}{2g} - H_f\right)}{{V_2^2}/{2g}}$; $\eta_D = (1 - K)\dfrac{\left(\dfrac{V_2^2 - V_3^2}{2g}\right)}{{V_2^2}/{2g}}$	$H_f = K\left(\dfrac{V_2^2 - V_3^2}{2g}\right)$
Unit quantities like unit velocity, speed discharge, power	$u_u = \dfrac{u}{\sqrt{H}}$; $N_u = \dfrac{ND}{\sqrt{H}}$; $Q_u = \dfrac{Q}{D\sqrt{H}}$; $P_u = \dfrac{P}{DH^{3/2}}$	

Contd.

Property	Formulae	Remarks, if any
Specific quantitites like unit velocity, speed, discharge, power	$N_s = N_u \dfrac{D}{D_s}$; $\dfrac{Q_u}{Q_s} = \left(\dfrac{D}{D_s}\right)^2$; $D_s = \dfrac{DH^{3/4}}{\sqrt{P}}$; $N_s = N_u \dfrac{\sqrt{P}}{H^{5/4}}$	
Synchronous speed of a governing mechanism	$N = {60f}/{p}$	f, p are frequency and no. of pair of poles

Table 11.6: Points to remember

Maximum hydraulic efficiency is possible if the blade velocity is half of the jet velocity (i.e. $u = v/2$).
So maximum hydraulic efficiency reduces to $(\eta_H)_{\max} = \dfrac{1}{2}(1 + \mu\cos\beta_o)$.
If $\beta_o = 0°$ or $\beta_o = 180°$ for smooth surface, then $(\eta_H)_{\max} = 100\%$ *or* 0%

CHAPTER 12

Hydraulic Pumps

INTRODUCTION

Pumps are Turbo-machinery rotated steadily by the means of an external agency and adds up energy for a fluid, flowing through it. In turbines high pressure fluid flows with acceleration towards, low pressure. Whereas in pumps, the fluid in low pressure is pressurized to high pressure flow with assistance of an external source, converting mechanical energy to pressure energy or kinetic energy.

CLASSIFICATION OF HYDRAULIC PUMPS

Hydraulic pumps can be classified based on their working nature to lift water from low level to high level, as mentioned in the Figure 12.1.

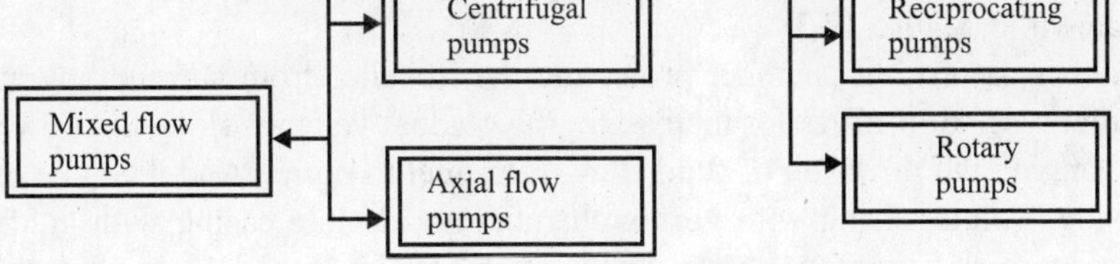

Figure 12.1: Hydraulic pumps—classification

CENTRIFUGAL PUMP

Centrifugal pump is basically classified as roto-dynamic pumps, which lifts a fluid from low pressure to high pressure, due to the combined dynamic forces of centrifugal energy, pressure energy and kinetic energy.

Centrifugal pump acts on the principle that a fluid mass can be lifted up to a certain height by rotating and creating a centrifugal force, in a casing, which throws the fluid mass to a height. A centrifugal pump consists of series of blades called **impeller vanes** surmounted on a shaft (called impeller shaft), which is enclosed in a casing and rotates continuously by means of an external source (called motor). Due to the continuous rotation of impeller vanes, a vacuum is created in between the vanes which suck the fluid at low pressure and thus increases the pressure of the fluid at the center of the impeller shaft due to rotation and throws out the fluid under pressure, radially outwards, due to the centrifugal action created by the vanes, as shown in Figure 12.2.

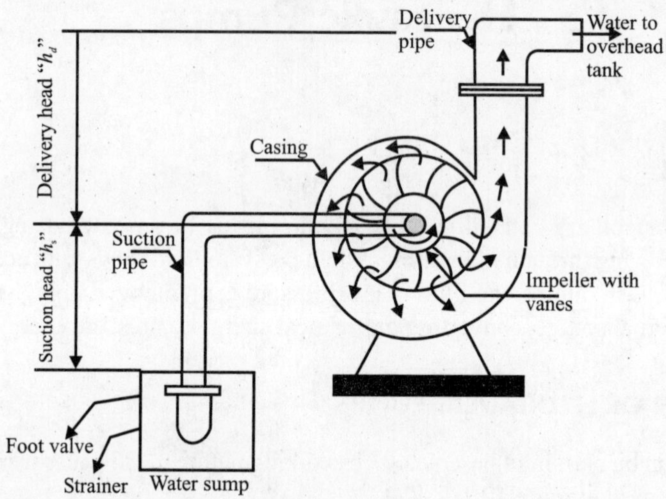

Figure 12.2: Centrifugal pump

COMPONENTS OF CENTRIFUGAL PUMP

1. Impeller: Impeller is a vaned rotor shaft coupled to an external driving source like motor shaft, I.C. Engines, etc. Impellers do change the direction of the flow in the casing and also converts the mechanical energy given by the external source to flow energy of the fluid.

It can be classified into three types like open type, semi-open type and closed/shrouded type impellers, as shown in Figure 12.3.

2. Casing: It is an airtight chamber protecting the impeller from surrounding atmosphere. It can have different types of profiles for increase in the gradual velocity along the flow direction and also helps in changing the direction of fluid flow from inlet to outlet. And it can be of three types, like Volute casing, Volute casing with vortex chamber and Volute casing with guide vanes (also called as diffuser/turbine pump), as shown in Figure 12.4a, 12.4b and 12.4c respectively.

3. Suction and delivery pipes: Suction pipe is a passage connecting liquid in sump at the bottom level to the impeller of the centrifugal pump, for the flow of liquid into the impeller casing for converting the mechanical energy given by the external source to pressure energy. It is provided with a strainer and a foot valve at the bottom and is immersed in the sump, to filter the liquid from foreign particle presence and to stop liquid drain out from the suction pipe respectively.

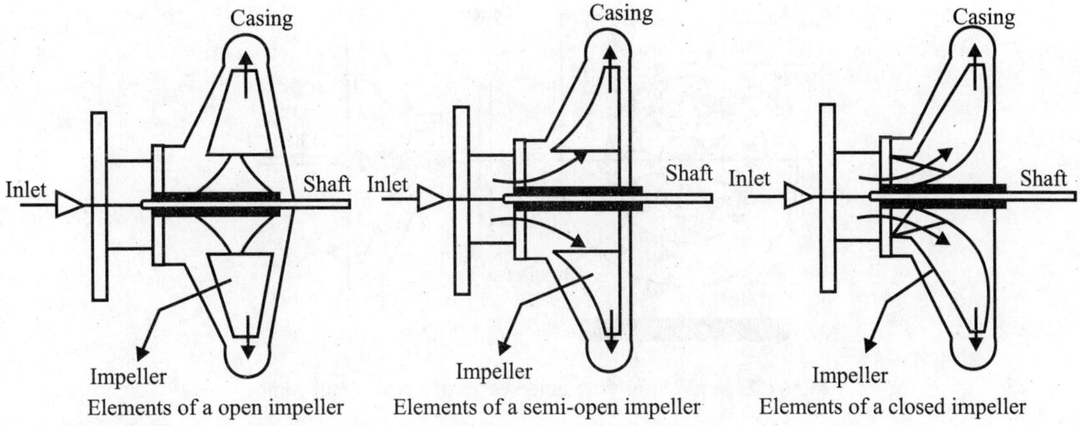

Figure 12.3: Elements of centrifugal pump

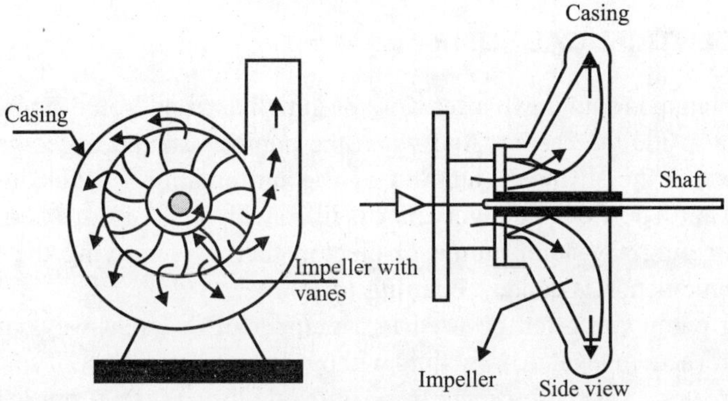

Figure 12.4a: Volute casing centrifugal pump

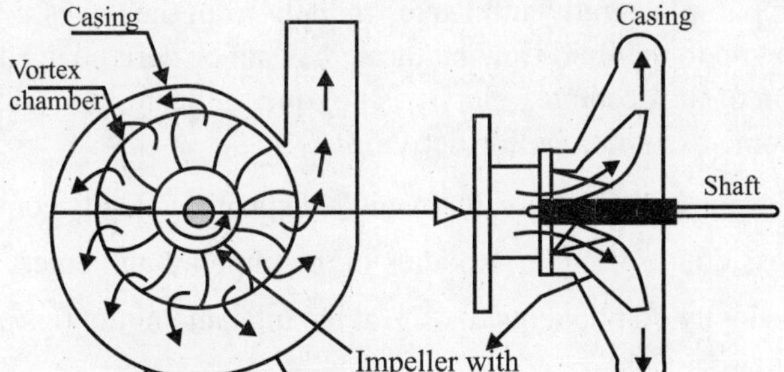

Figure 12.4b: Volume with vortex chamber in a centrifugal pump

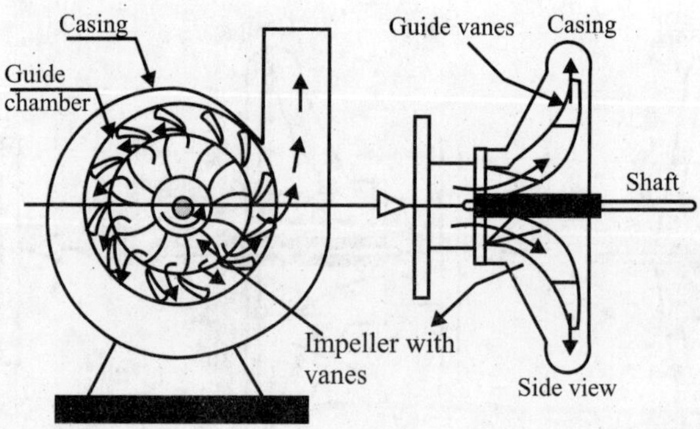

Figure 12.4c: Volume with guide vanes in a centrifugal pump

Delivery pipe is a passage from the outlet of the pump to the overhead tank. It is provided with a regulator valve to regulate the supply of liquid from the pump to the tank.

PRIMING OF A CENTRIFUGAL PUMP

When a centrifugal pump is shut down after work or initially started, some amount of air, vapor and gas may be trapped inside the casing. And when the pump is started once again, the trapped air packets may do reduce the sufficient amount of vaccum creation to suck in the liquid. So the suction pipe, casing and a portion of delivery pipe is filled with liquid, from external opening like air-vent valve or connecting a vaccum pump or ejector, thereby removing the air, vapor and gas trapped and this phenomenon is called **"Priming of pump"**.

A self priming pump can often be used in a number of different ways and they create and maintain a sufficient vacuum level to draw fluid without external assistance. Typically, a self priming centrifugal pump creates a vacuum that can be operated with suction if water is located in the body of the pump.

WORK DONE, HEADS AND EFFICIENCIES OF CENTRIFUGAL PUMP

As the fluid enters the pump centrally and leaves radially from the vanes, work added by impeller vanes/buckets of the pump to the fluid flowing through it can be derived in similar fashion as that of pump. As the direction of the absolute velocity is not due to the absence of guide as in the case of a pump, we include some assumptions for deriving, as such

1. Fluid enters the pump radially, i.e. there will be no whirl velocity component and so $V_i = V_{fi}$
2. There is no loss due to friction or eddies in the impeller and enters without any shock.
3. The relative velocity components V_{ri}, V_{ro} at the inlet and outlet tips are tangent to the vane, as shown in Figure 12.5.

Let β_i, β_o be angle of bucket at the inlet and outlet tips respectively.

ϕ_i, ϕ_o be angle between absolute velocities and whirl velocity components respectively.

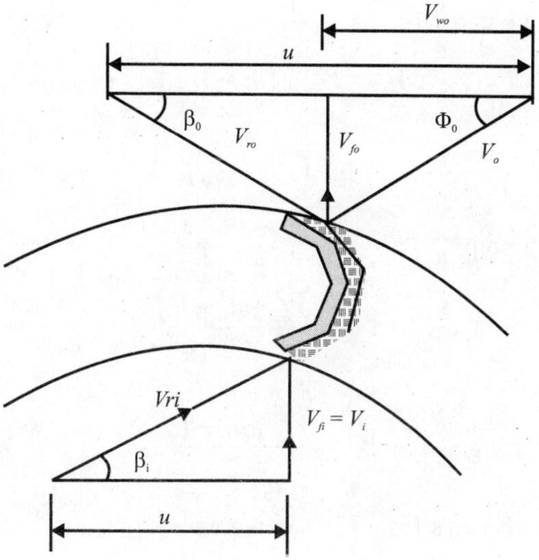

Figure 12.5: Velocity triangles of a centrifugal pump

V_i, V_o be absolute velocities at the inlet and outlet tips respectively.

V, u be absolute velocities of the jet and velocity of bucket respectively, where blade velocity

$u = \frac{\pi DN}{60}$, where "*D*" and "*N*" are mean diameter of shaft (distance from the centre of the wheel to the pitch circle) and speed of the impeller shaft.

Consider a fluid flowing radially inside the pump from the inner tip and leave through another tip. Consider the inlet and outlet velocity triangles of the fluid flow through the pump on a moving curved vane, as shown in figure.

a. Work done per second on the fluid by the impeller bucket *will be*

$$WD = Torque * Angular\ velocity$$

and torque is equal to rate of change of momentum, so we have

$$WD = \frac{\gamma}{g} * Q * (V_{wo} - V_{wi}) * u$$

But $V_{wi} = 0$ and hence

$$WD = \frac{\gamma}{g} * Q * (V_{wo}) * u_o \quad (12.1)$$

And work done per unit weight is termed "**Euler's momentum equation of centrifugal pump**",

i.e.

$$WD = \frac{1}{g} * (V_{wo} - V_{wi}) * u \quad (12.1a)$$

since *weight* = γQ

Observing outlet triangle, we can write as

$$V_{ro}^2 = V_{fo}^2 + (u_o - V_{wo})^2 \quad (12.1b)$$

and

$$V_{fo}^2 = V_o^2 - (V_{wo})^2 \quad (12.1c)$$

So from Eq. (12.1b and 12.1c), we get

$$u_o V_{wo} = \frac{V_o^2 + u_o^2 - V_{ro}^2}{2}$$

Similarly, for inlet triangle, we get $u_i V_{wi} = \dfrac{V_i^2 + u_i^2 - V_{ri}^2}{2}$

Substituting these values in Euler's Eq. (12.1a), we get

$$WD = \left\{ \left(\frac{V_o^2 - V_i^2}{2g} \right) + \left(\frac{u_o^2 - u_i^2}{2g} \right) + \left(\frac{V_{ri}^2 - V_{ro}^2}{2g} \right) \right\} \quad (12.2)$$

$$\Rightarrow WD = \left\{ \begin{pmatrix} Kinetic\ energy \\ or \\ Dynamic\ head \end{pmatrix} + (Centrifugal\ head) + (Static\ pressure) \right\}$$

And Eq. (12.2) is called "**fundamental equation of centrifugal pump**".

All these three heads altogether give rise to pressure head and thereby increase in work done

$\left(\dfrac{V_o^2 - V_i^2}{2g} \right)$. Dynamic head is the pressure head raise by external means and so termed external effect and is due to the energy transfer between kinetic energy of the outlet and inlet absolute velocities, at the impeller.

The purpose of pump impeller is to increase the absolute velocity from inlet to outlet, but it does not necessarily increase in pressure. So, the impeller vanes are so placed that there is a reduction in passage like a diffuser, to reduce the inlet velocity thereby transforming this reduced velocity into pressure raise. And it is also called **Impulse effect**.

$\left(\dfrac{u_o^2 - u_i^2}{2g} \right)$ centrifugal head is an internal effect in the impeller itself. As observed in forced vortex of fluid mechanics, Eq. (5.39) (i.e. $Z = \dfrac{P}{\gamma} = \dfrac{r_o^2 \omega^2}{2g}$), the pressure raise is directly proportional to the product squares of the radius and angular velocity, which is nothing but absolute velocity of the blade. And so there is an increase in pressure from the inlet to the outlet of the blade due to

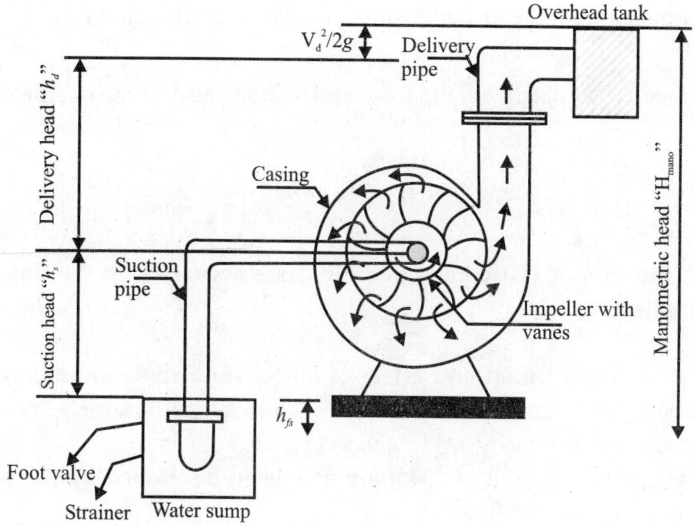

Figure 12.6: Centrifugal pump heads

centrifugal head, which add up to the work done by the pump. $\left(\frac{V_{ri}^2 - V_{ro}^2}{2g}\right)$ is the static pressure raise due to the reaction effect between the relative kinetic energies of the fluid and is also an internal effect.

b. Heads losses of a centrifugal pump:

Losses in a centrifugal pump can be of three types like static head H_S, manometric head H_{mano} and effective head H_e.

Static head (H_S) is **"the sum of suction head of pump from the pump to sump and head of pump from pump to the overhead tank (delivery head)"**.

i.e.
$$H_S = h_s + h_d \tag{12.3a}$$

Manometric head (H_{mano}) is "the total head required by the pump to produce delivery including the losses in the pump".

i.e.
$$H_{mano} = \frac{V_{wo}u_o}{g} - h_L \tag{12.3b}$$

And if no losses present, then manometric head reduces to

$$H_{mano} = \frac{V_{wo}u_o}{g}$$

Or in other words manometric head is the sum of suction head, delivery head, head loss due to friction in suction pipe h_{fs} and delivery pipe h_{fd} and velocity head at the overhead tank $\frac{V_d^2}{2g}$ and so

$$H_{mano} = H_s + h_{fs} + h_{fd} + \frac{V_d^2}{2g} \tag{12.3c}$$

Effective head is **"the sum of static head and all losses occurring in the flow before, through and after the impeller"**.

c. Efficiencies of centrifugal pump and effect of outlet vane angle on manometric efficiency: Efficiencies in a centrifugal pump will be of four, like manometric efficiency η_{mano}, volumetric efficiency η_{val}, mechanical efficiency η_{mech} and overall η_o

Manometric efficiency: is ratio of **"Manometric head developed by the pump to the head imparted by the impeller to the fluid"** and so

$$\eta_{mano} = \frac{H_m}{\left(V_{wo}u_o/g\right)} = \frac{gH_m}{V_{wo}u_o} = \frac{\gamma QH_m}{\gamma Q\left(V_{wo}u_o/g\right)} \tag{12.4}$$

where Q is actual volume of liquid delivered per second.

or
$$\eta_{mano} = \frac{Output\ of\ pump}{Power\ imparted\ by\ impeller} \tag{12.4a}$$

Mechanical efficiency is defined as **"ratio of power actually delivered by impeller to the power supplied to the impeller shaft by the external source"**.

i.e.
$$\eta_m = \frac{\left(V_{wo}u_o/g\right)}{Energy\ head\ given\ to\ shaft} = \frac{\gamma(Q+q)\left(V_{wo}u_o/g\right)}{Power\ given\ to\ impeller} \tag{12.5}$$

When water is passing through the casing there may be leakage in between the casing and impeller, which adds up to the eye of the impeller without passing the vanes, due to which the actual discharge passing through the impeller will be increased by an amount of q, so **volumetric efficiency** is defined as the **"ratio of quantity of discharge per second by the pump to the quantity passing through the impeller"**.

i.e.
$$\eta_{vol} = \frac{Discharged\ quantity\ per\ second}{Actual\ quantity\ passing\ through\ impeller} \Rightarrow \eta_{vol} = \frac{Q_a}{(Q_a+q)} \tag{12.6}$$

Overall efficiency is defined as the **"ratio of power output from the pump to net power supplied by the external source to the pump"**.

i.e. $$\eta_o = \frac{\gamma Q H_m}{Power\ given\ to\ impeller\ shaft} \tag{12.7}$$

Or in other words, overall efficiency is the product of hydraulic, mechanical and volumetric efficiencies and

$$\eta_o = \eta_{mano} * \eta_{mech} * \eta_{vol} = \frac{\gamma Q H_m}{Power\ given\ to\ impeller\ shaft} \tag{12.7a}$$

Effect of outlet vane angle (ϕ_o) on manometric efficiency:

The energy supplied by the impeller to the external source is $V_{wo}u_o/g$ and this energy is the summation of pressure energy H_{mano} and kinetic energy $V_0^2/2g$ and hence

$$\frac{V_{wo}u_o}{g} = H_{mano} + V_o^2/2g \tag{12.8}$$

But from the velocity triangles of Figure 12.5, we have

$$\begin{gathered} V_{fo}^2 = V_o^2 - (V_{wo})^2 \\ and \\ (V_{wo}) = u_o - V_{fo}\cot\phi_o \end{gathered} \tag{12.8a}$$

Substituting Eq. (12.8a) in Eq. (12.8), we get

$$H_{mano} = \frac{u_o - V_{fo}\cot\phi_o V_{wo}u_o}{g} - V_o^2/2g = \frac{(u_o - V_{fo}\cot\phi_o)u_o}{g} - \frac{(u_o - V_{fo}\cot\phi_o)^2 + V_{fo}^2}{2g}$$

$$\Rightarrow H_{mano} = \frac{u_o^2 - V_{fo}^2\operatorname{cosec}^2\phi_o}{2g}$$

Hence, manometric efficiency (refer Eq. (12.4)) $\eta_{mano} = \dfrac{gH_{mano}}{V_{wo}u_o}$

$$\Rightarrow \eta_{mano} = \frac{gH_{mano}}{V_{wo}u_o} = \frac{u_o^2 - V_{fo}^2\operatorname{cosec}^2\phi_o}{2u_o(u_o - V_{fo}\cot\phi_o)} \tag{12.9}$$

Assuming constant flow ratio $\varsigma\left(= V_{fo}/\sqrt{2gH_{mano}}\right) = 0.25$, we can observe that $90° - \phi_o - 20°$ and value of manometric efficiency increases from 0.47 to 0.73 and so the minimum value of ϕ_0 is 20°.

If the value of ϕ_o is reduced less than 20°, then we may have higher manometric efficiency but it creates long and narrow passages, which increases the frictional losses and hence not practicable.

Example 12.1: Determine the work done, relative velocity at the outlet impeller tip, manometric, mechanical efficiency and overall efficiency of a centrifugal pump having outer diameter of the impeller as 720 mm, width as 100 mm, vane angle at outlet of the impeller as 50°, and runs at a speed of 600 rpm and delivers a flow rate of 0.6 m³/s with an effective head of 40 m. Assume water enters radially at inlet and a motor of 500 kW is used to drive the pump.

Solution: Given data:

$$Q = 0.6 \text{ m}^3/\text{s}, H_m = 40 \text{ m}, N = 600 \text{ rpm}, \phi_0 = 50°, B_0 = 0.1\text{m}, D_0 = 0.72\text{m}, P = 500\text{kW}$$

Impeller blade velocity

$$u_o = \frac{\pi D_o N}{60} = \frac{3.14 * 0.72 * 600}{60} = 22.6 \text{ m/s}$$

Since there will be radial flow at the entrance, we have

$$V_{fo} = Q / A = Q / \pi D_o B_0 = 0.6 / 3.14 * 0.72 * 0.1 = 2.65 \text{ m/s}$$

Refer Eq. (12.8a), we have $(V_{wo}) = u_o - V_{fo} \cot\phi_o \Rightarrow (V_{wo}) = 22.6 - 2.65 * \cot 50 = 20.38$ m/s

Refer Eq. (12.1), we have **work done** as

$$WD = \frac{\gamma}{g} * Q * (V_{wo}) * u_o$$

$$\Rightarrow WD = \frac{9810}{9.81} * 0.6 * (20.38) * 22.6 = 276.3 \text{ kNm/s}$$

Refer Eq. (12.1b), we have **relative velocity at outlet as** $V_{ro}^2 = V_{fo}^2 + (u_2 - V_{wo})^2$

$$V_{ro}^2 = 2.65^2 + (22.6 - 20.38)^2$$
$$\Rightarrow V_{ro} = 3.46 \text{ m/s}$$

Refer Eq. (12.4), we have **manometric efficiency** as $\eta_{mano} = \dfrac{H_m}{\left(V_{wo} u_o / g\right)}$

$$\Rightarrow \eta_{mano} = \frac{40}{\left(20.8 * 22.6 / 9.81\right)} = 0.834 = 83.4\%$$

We have **overall efficiency** as $\eta_o = \frac{\gamma Q H_m}{P} = \frac{9810 * 0.6 * 40}{500 * 1000} = 0.47 = 47\%$

We have **mechanical efficiency** as $\eta_m = \frac{\eta_o}{\eta_{mano}} = \frac{0.47}{0.834} = 0.565 = 56.5\%$

Example 12.2: Determine the impeller diameter, work done and vane angle at the outlet of a centrifugal pump having a discharge of 0.3 m³/s against a head of 30 m and with a speed of 1200 rpm. Assume the manometer efficiency as 85% and the losses in the pump to be 0.03 times of square of absolute velocity of water leaving the impeller and impeller area at the outlet as twice of square of the impeller diameter and water enter without whirl.

Solution: Given data:

$$Q = 0.3 \text{ m}^3/\text{s}, H_{mano} = 30 \text{ m}, \eta_{mano} = 0.85, h_L = 0.03 V_1^2, A = 2D_1^2, N = 1200 \text{ rpm}$$

Refer Eq. (12.3b), we have

$$H_{mano} = \frac{V_{wo} u_o}{g} - h_L \text{ but } \eta_{mano} = \frac{H_m}{\left(V_{wo} u_o / g\right)}$$

$$\Rightarrow H_{mano} = \frac{30}{0.85} - h_L = 30$$

$$\Rightarrow h_L = 35.29 - 30 = 5.29 \text{ m}$$

But given, $h_L = 5.29 = 0.03 V_o^2$ $\qquad \Rightarrow V_o = 13.29$ m/s

Since there will be radial flow at the entrance, we have

$$V_{fo} = Q / A = 0.3 / 2D_o^2 = 0.15 / D_o^2$$

And impeller blade velocity $u_o = \frac{\pi D_o N}{60} = \frac{3.14 * D_o * 1200}{60} = 62.8 D_o$

So $\qquad V_{wo} = \frac{g H_m}{\eta_{mano} u_o} = \frac{9.81 * 30}{0.85 * 62.8 D_o} = 5.51 / D_o$

Refer Eq. (12.8a), we have $V_{fo}^2 = V_o^2 - \left(V_{wo}\right)^2 \Rightarrow \left(0.15 / D_o^2\right)^2 = 13.29^2 - \left(5.51 / D_o\right)^2$

$$\Rightarrow 176.6D_o^4 - 30.3D_o^2 - 0.0225 = 0 \Rightarrow D_o = 5.5 \text{ m}$$

So the **diameter of impeller** is 5.5 m.

So, $V_{fo} = 0.15/D_o^2 = 0.15/5.5^2 = 0.0049$ $u_o = 62.8 * 5.5 = 345.4$ $V_{wo} = 5.51/5.5 = 1.01$

Refer Eq. (12.1), we have **work done** as $WD = \frac{\gamma}{g} * Q * (V_{wo}) * u_o$

$$\Rightarrow WD = \frac{9810}{9.81} * 0.3 * (1.01) * 345.4 = 104.65 \text{ kNm/s}$$

Refer Eq. (12.8a), we have $\tan\phi_o = V_{fo}/\left(u_o - (V_{wo})\right) \Rightarrow \tan\phi_o = \frac{0.0049}{(345.4 - 1.01)} = 1.43$

$$\Rightarrow \phi_o = 8.15°$$

So the **vane angle at outlet of impeller** is 8.15 degrees.

MINIMUM STARTING SPEED

When a centrifugal pump is started by an external source, the pump needs to build sufficient amount of pressure energy in the impeller, which is equal or more than the manometric head. And till the pressure is raised, the fluid in the casing will be rotating in forced vortex motion thereby creating a centrifugal head $\left(u_o^2 - u_i^2\right)/2g$ and hence for commencement of the flow of fluid from the pump

$$\frac{\left(u_o^2 - u_i^2\right)}{2g} \geq H_{mano} \Rightarrow \frac{(\pi D_o N/60)^2}{2g} - \frac{(\pi D_i N/60)^2}{2g} \geq H_{mano}$$

$$\Rightarrow \left(\frac{\pi N}{60}\right)^2 \left[\left(D_o^2 - D_i^2\right)\right] = 2gH_{mano} \qquad (12.10)$$

So Eq. (12.10), gives the minimum required speed to start a pump.

DESIGN PARAMETERS AND THEIR PROPORTIONS OF CENTRIFUGAL PUMP

While designing a centrifugal pump, the designer need to take considerations of diameters of inlet, outlet and least diameter of impeller, suction pipe and delivery pipe diameters and also parameters like *speed ratio*(K_u), *flow ratio*(ζ)

1. Determination of speed ratio (K_u): Speed ratio is defined as the **"ratio of peripheral speed at exit (u_0) and theoretical jet velocity corresponding to manometric head"**

So $$K_u = {u_o}/{\sqrt{2gH_{mano}}} \text{ \& varies from 0.95 to 1.25} \qquad (12.11)$$

where $$u_o = \frac{\pi D_o N}{60}$$

2. Flow ratio ζ : Flow ratio is defined as the ratio of flow velocity at the exit to theoretical velocity of the jet and so

$$\zeta = \frac{V_{fo}}{\sqrt{2gH_{mano}}} \qquad (12.12)$$

and $0.1 - \psi - 0.25$

3. Inlet diameter of impeller D_i :

It is kept as 2/3 to 1/3 of D_0, depending upon the specific speed or total manometric head. But in most cases $D_i = {D_o}/{2}$

4. Outlet diameter of impeller D_o : This is determined using Eq. (12.11).

5. Least diameter of impeller D_{min} : Refer Eq. (12.10), i.e. $\left(\frac{\pi N}{60}\right)^2 \left[\left(D_o^2 - D_i^2\right)\right] = 2gH_{mano}$ and substituting $D_i = {D_o}/{2}$, we have

$$D_{min} = \frac{97.68\sqrt{H_{mano}}}{N} \qquad (12.13)$$

6. Suction pipe diameter D_s : Using continuity equation, we have

$$Q = AV \Rightarrow Q = \frac{\pi D_s^2}{4} * V_s \text{ so}$$

$$\Rightarrow D_s = \sqrt{\frac{4Q}{3.14 V_s}} \qquad (12.14)$$

7. Delivery pipe diameter D_d : Using continuity equation, we have

$$Q = AV \Rightarrow Q = \frac{\pi D_d^2}{4} * V_d \text{ so}$$

$$\Rightarrow D_d = \sqrt{\frac{4Q}{3.14V_d}} \tag{12.15}$$

And usually 1.5 m/s $- V_d -$ 3.5 m/s and $V_d \geq V_s$

SPECIFIC SPEED OF CENTRIFUGAL PUMP

As seen in the specific performances of pumps chapter, the specific parameters/quantities are used to correlate two different pumps and hence required to have a centrifugal pump having common terms to both of these, with reduced sizes, called model pump. And specific speed can be defined as "**the speed of a geometrically similar pump, which would deliver 1 litre of liquid per second against 1 metre of head**.

We know, the discharge through a pump can be expressed as $Q = A_o * V_{fo} = K(\pi B_o D_o) * V_{fo}$

$$\Rightarrow Q \; \alpha \; \left(D_o^{\;2}\right) * V_{fo} \;\; \sin ce \;\; B_o \alpha D_o$$

And we know

$$V_{fo} = \varsigma\sqrt{2gH_{mano}} \Rightarrow V_{fo} \; \alpha \sqrt{H_{mano}}$$

and so
$$\frac{Q}{D_o^2 \sqrt{H_{mano}}} = \text{constant} \tag{12.16a}$$

Also we have blade velocity, i.e. $u_o = K_u\sqrt{2gH_{mano}} \Rightarrow u_o = \frac{\pi D_o N}{60} \; \alpha \sqrt{H_{mano}}$

So
$$\frac{\sqrt{H_{mano}}}{D_o N} = \text{constant} \tag{12.16b}$$

Substituting Eq. (12.16b) in Eq. (12.16a), we get

$$N_s = \text{constant} = N\frac{\sqrt{Q}}{H_{mano}^{\;3/4}} \tag{12.17}$$

And power $P = \left(\gamma Q H_{mano} / \eta_o\right)$ so Eq. (12.17) can be reduced to

$$N_s = N\frac{\sqrt{P}}{H_{mano}^{\;5/4}} \tag{12.18}$$

Note: To have model testing of pumps with respect to prototypes of the pump, we require constant values, for model testing of the pump, like **flow coefficient**

$$\frac{Q}{D^3 N} = \text{constant} \tag{12.19}$$

and **power coefficient**

$$\frac{P}{D^5 N^3} = \text{constant} \tag{12.20}$$

Example 12.3: A centrifugal pump running at 750 rpm discharges water at 0.1 m³/s against a head of 10 m at its best efficiency. A second pump of the same homologous series, when working at 500 rpm is to deliver water at 0.05 m³/s at its best efficiency. What will be the design head of the second pump and what is the scale ratio between first and the second pump. **[GATE]**

Solution: Given data:

$Q_1 = 0.1$ m³/s, $Q_2 = 0.05$ m³/s, $N_1 = 750$ rpm, $N_2 = 500$ rpm, $H_{mano1} = 10$ m

As the two pumps are in homologous to each other, we can correlate them as a model and prototype and so referring Eq. (12.17), we have

$$N_s = constant = N\frac{\sqrt{Q}}{H_{mano}^{3/4}}$$

$$\Rightarrow N_1\frac{\sqrt{Q_1}}{H_{mano1}^{3/4}} = N_2\frac{\sqrt{Q_2}}{H_{mano2}^{3/4}}$$

$$\Rightarrow H_{mano2}^{3/4} = N_2\frac{\sqrt{Q_2} * H_{mano1}^{3/4}}{N_1\sqrt{Q_1}} = 500 * \frac{\sqrt{0.05} * 10^{3/4}}{750 * \sqrt{0.1}} = 3.67 \text{ m}$$

Design head of second pump is 3.67 m

Scale ratio $\frac{Q_1}{Q_2} = (L_r)^{2.5} = 2 \Rightarrow (L_r) = 1.32$.

Example 12.4: A centrifugal pump rotating at 1500 rpm delivers 0.2 m³/s at a head of 15 m. Calculate the specific speed of the pump and the power input. Assume overall efficiency of the pump as 0.68.

If this pump were to operate at 900 rpm, what would be the head, discharge and power required for homologous conditions? Assume overall efficiency remains unchanged at new rpm. **[UPSC]**

Solution: Given data: $Q = 0.2\ \text{m}^3/\text{s},\ N = 1500\ \text{rpm},\ H_m = 15\ \text{m},\ \eta_o = 0.68$

Refer Eq. (12.17), we have **specific speed** as

$$N_s = N\frac{\sqrt{Q}}{H_{mano}^{3/4}} \quad N_S = 1500\frac{\sqrt{0.2}}{15^{3/4}} = 88$$

Power input will be

$$P = \frac{\gamma Q H_{mano}}{\eta_o} = \frac{9.81 * 0.2 * 15}{0.68} = 43.28\ \text{kW}$$

At 900 rpm. As this pump is in homologous condition to itself, we can correlate the two speeds as a model and prototype and so referring unit speed, we have $N_u = \frac{N}{H_{mano}^{1/2}}$ **(we use unit quantities for same pump correlation)** $\Rightarrow \frac{N_1}{\sqrt{H_{mano1}}} = \frac{N_2}{\sqrt{H_{mano2}}}$

$$\Rightarrow \frac{900}{\sqrt{H_{mano1}}} = \frac{1500}{\sqrt{15}} \Rightarrow H_{mano1} = 5.4\ \text{m}$$

So the head is 5.4 m

Referring unit **discharge**, we have $\qquad Q_u = \frac{Q}{H_{mano}^{1/2}}$

$$\Rightarrow \frac{Q}{\sqrt{5.4}} = \frac{0.2}{\sqrt{15}} \Rightarrow Q = 0.12\ \text{m}^3/\text{s}$$

Referring unit **power**, we have $\qquad P_u = \frac{P}{H_{mano}^{3/2}}$

$$\Rightarrow \frac{P}{(5.4)^{1.5}} = \frac{43.28}{(15)^{1.5}} \Rightarrow P = 9.35\ \text{kW}$$

Example 12.5: Two geometrically similar pumps are running at the same speed of 1000 rpm. One has an impeller diameter of 0.4 m and discharges 30 lit/s against ahead of 20 m. If the other pump gives half of this discharge rate, determine the head and diameter of the second pump. **[GATE]**

Solution: Given data:

$$D_1 = 0.4 \text{ m}, N_1 = N_2 = 1000 \text{ rpm}, Q_1 = 0.03 \text{ m}^3/\text{s}, H_{m1} = 20 \text{ m}, Q_2 = Q_1$$

As the two pumps are in geometrical similarity to each other, we can correlate them as a model and prototype and so referring Eq. (12.17), we have

$$N_s = \text{constant} = N\frac{\sqrt{Q}}{H_{mano}^{3/4}}$$

$$\Rightarrow N_1\frac{\sqrt{Q_1}}{H_{mano1}^{3/4}} = N_2\frac{\sqrt{Q_2}}{H_{mano2}^{3/4}}$$

$$\Rightarrow H_{mano2}^{3/4} = N_2\frac{\sqrt{Q_2} * H_{mano1}^{3/4}}{N_1\sqrt{Q_1}} = 1000 * \frac{\sqrt{0.015} * 20^{3/4}}{1000 * \sqrt{0.03}} = 12.6m$$

Design head of second pump is 12.6 m.

Also $$D\alpha\frac{\sqrt{H}}{N} \text{ so } \Rightarrow \frac{D_1}{D_2} = \sqrt{\frac{H_1}{H_2}}$$

$$\Rightarrow D_2 = 0.4/1.256 = 0.317 \text{ m}$$

Diameter of the second pump is 0.317 m.

Example 12.6: If a pump is required to discharge 0.3 m^3/s with a speed of 900 rpm against a maximum of 800 specific speed, then determine the manometric head of the pump.

Also determine the number of impellers required for a multistage pump against a head of 10 m.

Solution: Given data: $Q = 0.3 \text{ m}^3/\text{s}, N = 900 \text{ rpm}, N_S = 800$

Refer Eq. (12.17), we have $$N_s = N\frac{\sqrt{Q}}{H_{mano}^{3/4}} \quad 800 = 900\frac{\sqrt{0.3}}{H_{mano}^{3/4}}$$

$$\Rightarrow H_{mano} = 0.524 \text{ m}$$

So, **manometric head for single stage pump** is 0.524 m.

Number of multistages is 10/0.524 = 19.08 ≈ 19.

CENTRIFUGAL PUMP CHARACTERISTIC CURVES

Usually, pumps are designed only for a particular design parameters like head "H", discharge "Q", speed "N", power "P" and efficiency "η". But in actual practice there will be variation of parameters like suction heads, temperature, etc. acting on the pump and hence there is often a tendency of pump to deviate form its design conditions.

If the deviation of the pump parameters, due to variation in design parameters are recorded by testing the actual pump at the site or on the scale model of that pump, which will be very much useful in analyzing and predicting the performance of the pump and hence the recorded variation of pump parameters are plotted in a graphical representation for clear understanding of the pump. And the graphical representations of the recorded deviation values of parameters of the pump are called as "characteristic curves". And characteristic curves are of three types mentioned below:

a) Main and operating speed characteristic curves.
b) Constant efficiencies characteristic curves.
c) Constant head, constant discharge characteristic curves.

Main and operating characteristic curves: These are the curves mainly intended to analyze the performance of power, efficiency and head at constant speed with varying discharge and the values are recorded between the discharge and head, power, efficiency for centrifugal pump and are shown in Figure 12.7a and 12.7b.

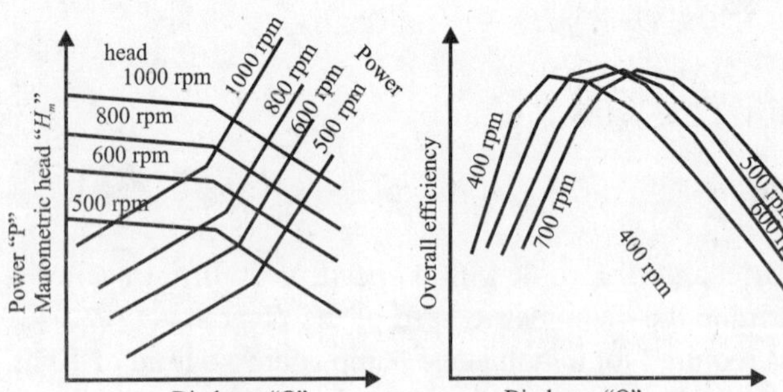

Figure 12.7a: Main characteristic curves

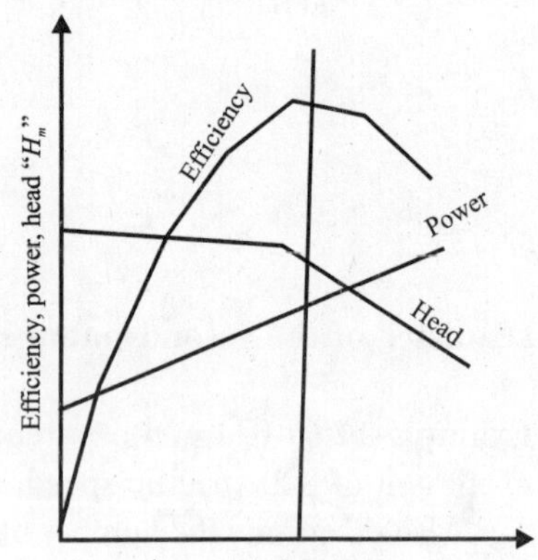

Figure 12.7b: Operating characteristic curves

Operating characteristic curves represents the optimum performance at a particular design conditions and to analyze the max efficiency at a particular point from the main characteristic curves, the operating characteristic curves are be used.

Constant efficiency characteristic curves: These are used in predicting the performance of a pump over wide range of operations like overlapping of constant value of efficiency onto the main characteristic curves between discharge and speed and discharge is obtained and thereby this discharge value is projected onto the head in the main characteristic curves, for that speed.

And this process is repeated for several constant efficiency values. And joining all these points, we get constant efficiency line and thereby best speed line, as shown in Figure 12.7c.

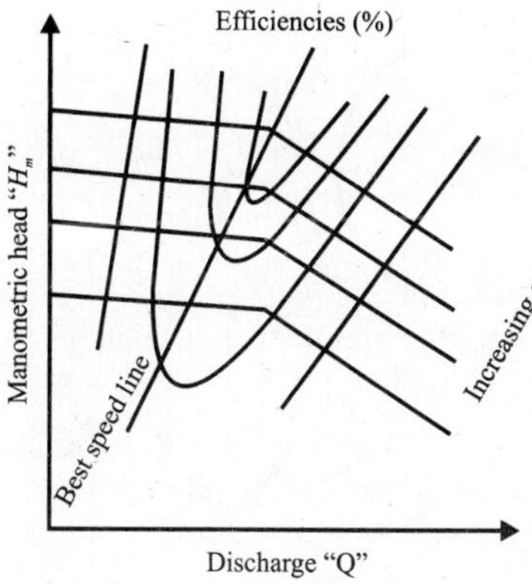

Figure 12.7c: Constant efficiency characteristic curves

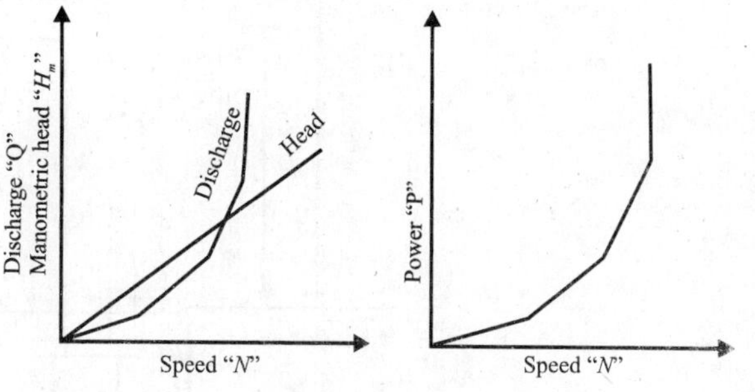

Figure 12.7d: Constant head and constant discharge characteristic curves

Constant head, constant discharge characteristic curves: These curves are very much useful in case of variable speed pumps, with varying speeds. Head is maintained constant to predict the speed vs. discharge and similarly discharge is maintained constant to predict head and speed and also power, by drawing characteristic curves against speed and power, which is parabolic, as shown in Figure 12.7d.

CAVITATION AND ITS EFFECTS

Cavitation is the phenomenon of sudden drop in efficiency, power and head requirement of a pump. It begins to appear when the pressure at suction side of the pump drops below vapour pressure of liquid and is indicated by **net positive suction head (NPSH), Thoma's cavitation factor and suction specific speed.**

When the pressure in the flow passage reaches the vapor pressure of the liquid flowing, then the liquid starts to vaporize and thereby forming small packets of cavity bubbles. When these bubbles reach high pressure zones, they collapse and burst out, there by affecting the surfaces that are near by to them and causing pitting, scoring and eroding of surfaces. This phenomenon is termed "**Cavitation**". This phenomenon is observed in all the turbo-machines and their related components that are subjected to continuous contact of fluid flowing through them, like turbines and pumps.

Net positive suction head (NPSH): It is defined as "the net head (in meters of liquid) required for making the liquid flow through the suction head from the sump to the impeller".

That is, to start lifting up fluid by a pump, it requires a minimum suction head h_s in its casing and suction pipe and delivery pipe, which should be greater or equal to the vapour pressure head (H_v), to avoid cavitation and so observing Figure 12.8,

we have
$$\frac{P_{gage}}{\gamma} = \frac{P_v}{\gamma} + H_{SV} \tag{12.21a}$$

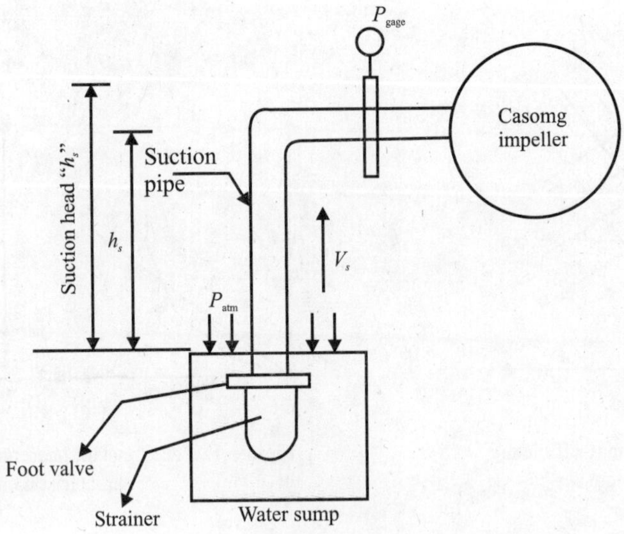

Figure 12.8

Similarly, we have

$$\frac{P_{gage}}{\gamma} = \frac{P_{atm}}{\gamma} - \left[h_s + h_{fs}\right] \quad (12.21b)$$

From Eq. (12.21a) and Eq. (12.21b), we get

$$H_{SV} = H_{atm} - H_S - H_v \quad (12.22)$$

And when the gage pressure $P_{gage} = P_v$, then $H_{SV} = NSPH$ and hence the suction lift limiting value (h_S) should be maximum to have suction with no cavitation,

Therefore,

$$h_s = \left(\frac{P_{atm}}{\gamma} - \frac{P_v}{\gamma}\right) - \left[h_{fs} + \frac{V_s^2}{2g}\right] \quad (12.23)$$

h_{fs} head friction loss in suction pipe.

Thoma's cavitation factor σ **:** Defined as the ratio of NSPH (H_{SV}) to the manometric head (H_{mano}),

So

$$\sigma = \frac{NSPH}{H_{mano}} = \frac{H_{SV}}{H_{mano}} \quad (12.24)$$

Cavitation begins when the Thoma's factor is less than the critical Thoma's factor, i.e. $\sigma < \sigma_c$ and σ_c depends on the design conditions of a pump and is given by $\sigma_c = f\left(\eta, N_s,\ no.\ of\ vanes\right)$

$$\Rightarrow \sigma_c = 0.103\left(N_s / 1000\right)^{4/3} \quad (12.24a)$$

For **Francis turbine**; $\sigma_c\ 0.625(N_s/80.7)^2$ (12.24b)

For **Propeller turbine**; $\sigma_c\ 0.28 + [0.133(N_s/380.7)^3]$ (12.24c)

Suction specific speed N_{ss} **:** It is also a cavitation parameter and is defined as "speed of a geometrically similar pump, which works under condition of developing 1 m of total suction head (in absolute units) and delivers 1 litre of liquid per second".

Refer Chapter 11, we have

$$N_s = \frac{N\sqrt{Q}}{H_{mano}^{3/4}} \tag{12.25a}$$

but the manometric head can be replaced by NSPH and so

$$N_{ss} = \frac{N\sqrt{Q}}{H_{SV}^{3/4}} \tag{12.25b}$$

For **Turbine**;

$$N_{ss} = \frac{N\sqrt{P}}{H_{SV}^{5/4}} \tag{12.25c}$$

Combining Eq. (12.24), Eq. (12.25a) and Eq. (12.25b), we get

$$\sigma = \left(\frac{N_s}{N_{ss}}\right) \tag{12.26}$$

For **Turbine**;

$$\sigma = \left(\frac{N_s}{N_{SS}}\right)^{4/5} \tag{12.26a}$$

Harmful effects of cavitation:

1. Due to continuous of air bubbles, pitting and erosion of surfaces arise.
2. Cavitation cause sudden drop in efficiency, power and head.
3. Noise, vibration and high temperature evolution arise due to suction flow restrictions.
4. Cavitation cause too high specific speed for optimum design.

Example 12.7: Determine the maximum permissible suction head and suction specific speed, if a pump has cavitation parameter as 0.1 to be installed at a location where atmospheric pressure is 89 kPa and vapour pressure is 3.5 kPa absolute and this pump need to work against 30 m head with speed of 500 rpm and need to deliver 0.3 m^3/s of water. Neglect head due to friction.

Solution: Given data:

$$\sigma_c = 0.1, P_{atm} = 89000 \text{ Pa}, P_v = 3500 \text{ Pa}, H_{mano} = 30 \text{ m}, N = 500 \text{ rpm}, Q = 0.3 \text{ m}^3/\text{s}$$

Atmospheric pressure head $H_{atm} = \dfrac{P_{atm}}{\gamma} = \dfrac{89000}{9810} = 9.07$ m of water

Vapour pressure head $H_V = \frac{P_v}{\gamma} = \frac{3500}{9810} = 0.356$ m of water

Refer Eq. (12.24), we have $\sigma = \frac{H_{sv}}{H_{mano}} = \frac{H_{atm} - H_S - H_V}{H_{mano}}$

$$\Rightarrow 0.1 = \frac{9.07 - H_S - 0.356}{30}$$

$$\Rightarrow H_S = 9.07 - 0.356 - 30 * 0.1 = 5.714 \text{ m}$$

So, maximum permissible suction head is 5.714 m.

Suction specific speed, refer Eq. (12.25b), we have $N_{ss} = \frac{N\sqrt{Q}}{H_{SV}^{3/4}}$

NSPH $H_{SV} = H_{atm} - H_S - H_V = 9.07 - 5.714 - 0.356 = 3.0$ m

$$N_{ss} = \frac{500\sqrt{0.3}}{3.0^{3/4}} = 120 \text{ rpm}$$

So, **suction specific speed is 120 rpm.**

Table 12.1: Troubleshooting of a centrifugal pump

Aspect	Reason
Pump fails to start pumping.	Not properly primed. Suction and delivery heads too high. Wrong direction of pump rotation. Impeller may be clogged by foreign particle. Too low speed.
Pump not working up to the capacity.	Air leak into the pump. Too low speed. Discharge head higher than anticipated. Small impeller diameter. Insufficient suction head in case of volatile/hot liquids pumping. Suction pipe not sufficiently submerged.
Pump starts and then stops and also get overheated.	Improper priming. Air packets in the suction line. Lack of lubrication, high speeds. Worn out rings, loose shaft, impeller damage.
Noisy operation and vibrations	Cavitation Misalignment of shafts Worn out parts and bearings Improper foundation Due to rubbing action between moving and stationary elements.

RECIPROCATING PUMP

Reciprocating pump is classified as positive displacement pump. It works on the principle that a liquid can be raised from a low level to high level by sucking the liquid into a chamber and exerting thrust force on the liquid, by a moving member, to push it to the required height.

Basically, it is incorporated in between a pipeline with an external source like motor attached to it. It converts the mechanical energy given by the motor in form of reciprocation of the piston into hydraulic energy, by giving a thrust force on the liquid and thereby the liquid is pushed to a certain height, as shown in Figure 12.9.

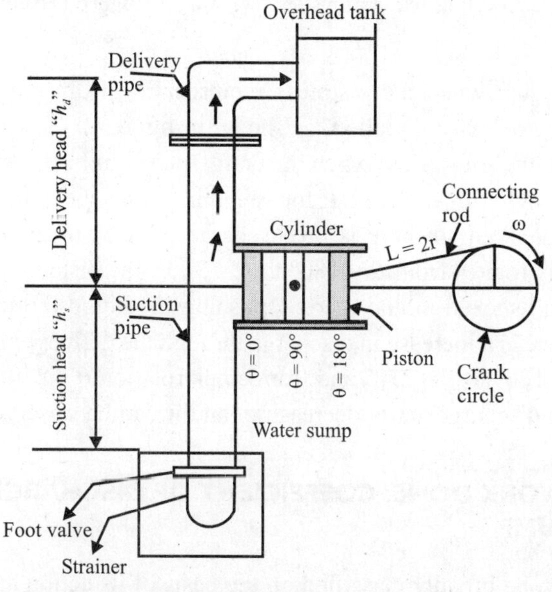

Figure 12.9: Reciprocating pump

Main components and working of reciprocating pump: Reciprocating pump consists of a cylinder chamber, piston, crank, connecting rod, suction pipe and delivery pipe.

a) Cylinder: Cylinder is a circular cross-section fixed chamber having a piston reciprocating in it. It is connected to a suction pipe through which the liquid enters the chamber from one side and on the other side connected to a delivery pipe through which the liquid is pushed out to a height.

b) Piston: Piston is a rigid circular disc sliding to and fro in a cylindrical chamber, with help of an external source. It exerts a thrust force on the liquid in the cylinder and also creates a vacuum conditions in the cylinder chamber for sucking the liquid into the cylinder chamber.

c) Connecting rod: Connecting rod is a rigid bar connecting the piston and crank with help of pins and has a length twice that of the crank, i.e. L=2*r.

d) Crank: Crank is a short bar connecting external source like motor shaft with the connecting rod. It transforms the rotational motion of external source shaft to reciprocating movement of the piston with help of the connecting rod.

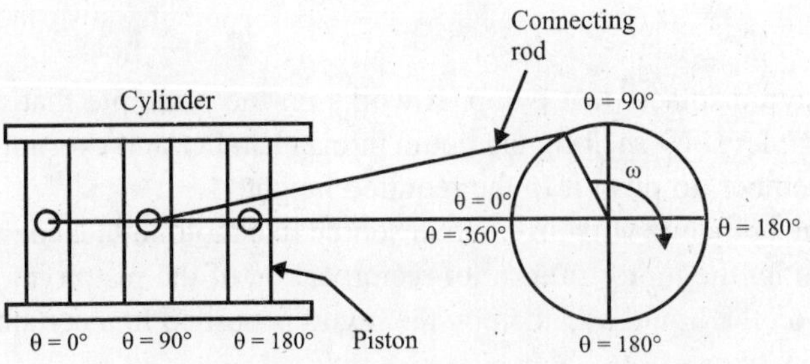

Figure 12.10a: Reciprocating action of piston

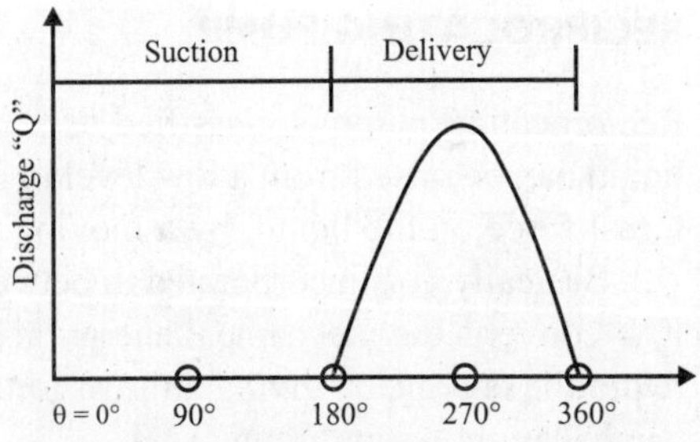

Figure 12.10b: Discharge action of piston w.r.t crank

Observe Figure.12.10a, when the piston is reciprocating from crank angle 0° to 180° (i.e. $\theta = 0°$ to $\theta = 180°$), in clockwise direction, as shown in figure.

The piston moves from left to right when the crank rotates from $\theta = 0°$ to $\theta = 180°$ and thereby it creates a space in the cylindrical chamber for vacuum, which sucks in the liquid from the sump through the suction pipe. And so there is no discharge of fluid in the delivery pipe during this interval. When the crank rotates from $\theta = 180°$ to $\theta = 360°$, the piston surface exerts a continuous thrust force on the liquid sucked in and thereby pushing the liquid through the delivery pipe to a certain height and so there is a increase in discharge through the delivery pipe. But this happens only from crank rotation $\theta = 180°$ to $\theta = 270°$ and from crank rotation $\theta = 270°$ to $\theta = 360°$ as the liquid sucked in exhausts, the discharge starts decreasing, and it can be observed in Figure 12.10b.

CLASSIFICATION, WORK DONE, COEFFICIENT OF DISCHARGE AND SLIP OF RECIPROCATING PUMP

Reciprocating pump can be broadly classified on the basis of its action and number of cylinders it has, as shown in Figure 12.11.

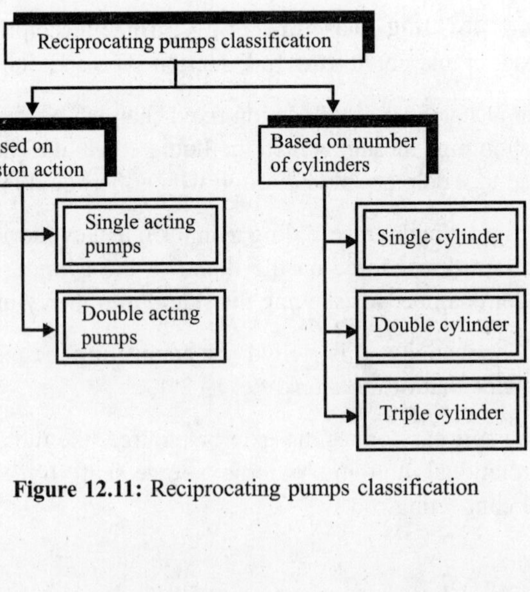

Figure 12.11: Reciprocating pumps classification

It can give the thrust on to the liquid in the cylinder chamber on one side of the piston surface area or on both sides of the piston surface, basing on the piston action.

Similarly, to increase the discharge quantity, one can use multi-cylinder suitably arranged, like single cylinder, double cylinder or triple cylinder, etc.

Single acting and double acting reciprocating pump: Single acting reciprocating pump has action of suction and delivery of liquid on one side of the piston surface only and thus it is ideal in the suction stage $\theta = 0°$ to $\theta = 180°$ and discharges only in the deliver stage $\theta = 180°$ to $\theta = 360°$, as shown in Figure 12.10a.

In case of Double acting Reciprocating pump, there exist two suction pipes and two delivery pipes attached to the cylinder chamber wall, on either side of the piston surfaces. Piston discharges liquid from the second delivery pipe (d_2) when the crank angle rotation is $\theta = 0°$ to $\theta = 180°$ and at the same time it creates a vacuum condition on the other piston surface side, to suck the liquid from the sump through the suction pipe (s_1). While motion of the piston from $\theta = 180°$ to $\theta = 360°$ of the crank rotation, liquid discharges from first delivery pipe (d_1) and at the same time it creates a vacuum condition on the other piston surface side, to suck the liquid from the sump through suction pipe (s_2).

So, as the action of discharging of liquid happens from both the surfaces of the piston, as shown in Figure 12.12 and we term this type of reciprocating pump as double acting reciprocating pump. And as it gives a continuous delivery of liquid, we get the plot against discharge and crank rotation as shown in Figure 12.13b.

Work done by reciprocating pump: As observed the working of a reciprocating pump is just similar to an I.C. Engine except that only a suction stroke and delivery stroke exists for one revolution of crank.

Let the area of the cylinder bore be "*A*" and the stroke length (displacement of piston) of the piston "*L*". "*N*" be the speed with which the piston reciprocates in the cylinder per second.

We know, the discharge is product of area and velocity or in other words volume per second,

So $$Q_{th} = A * V = A * L * N / 60 \qquad (12.27)$$

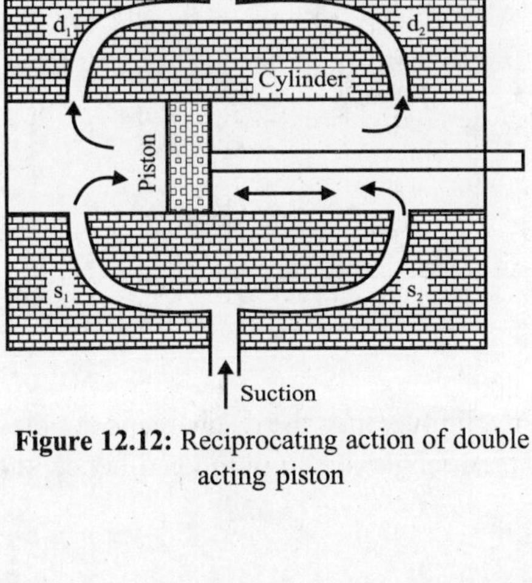

Figure 12.12: Reciprocating action of double acting piston

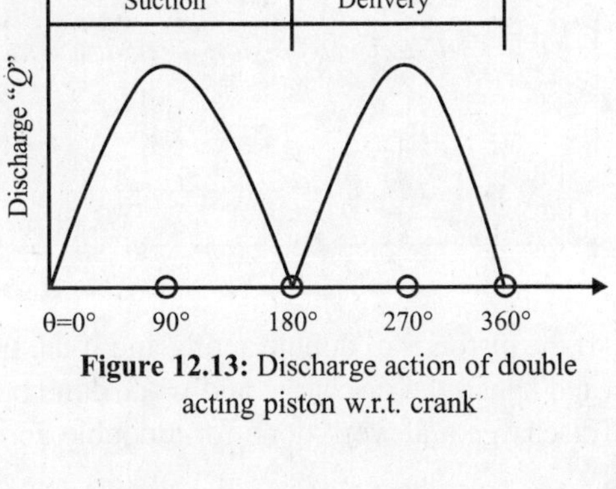

Figure 13.13: Discharge action of double acting piston w.r.t. crank

or

$$Q_{th} = {2ArN}/{60} = {ALN}/{60} \tag{12.27a}$$

Since L = 2*crank radius(r).

The thrust force exerted by the piston surface on the liquid will depend on the specific weight of the liquid, area of the surface contact and total head to which the liquid has to raise, and so $F_{thrust} = \gamma AH_{total}$

So the thrust force during suction stroke is $F_{st} = \gamma Ah_S$ and similarly during delivery stroke $F_{dt} = \gamma Ah_D$ So the total thrust force acting is

$$F = (\gamma Ah_S) + (\gamma Ah_D) = \gamma A(h_S + h_D) \tag{12.28a}$$

We know work done per revolution is the product of Force and displacement,

i.e. $$WD_{th} = Force * displacement$$

Substituting Eq. (12.28a), we have $WD_{th} = \gamma A(h_S + h_D) * L$

So work done per second is $$\Rightarrow WD_{th} = \frac{\gamma ALN(h_S + h_D)}{60} \tag{12.29}$$

Similarly, power $$P = \gamma Q_{th} H_{total} = \frac{\gamma ALN(h_S + h_D)}{60} \tag{12.30}$$

Note:

If the piston rod area ("a") is not negligible in case of double acting reciprocating pump, then the effective area of the piston, discharge and work done theoretical will become

$$A_{eff} = \left[\begin{matrix} Area\ of \\ piston \end{matrix} + \left(\begin{matrix} Area\ of \\ piston \end{matrix} - \begin{matrix} Area\ of \\ piston\ rod \end{matrix}\right)\right] = \left[\frac{\pi}{4}D^2 + \left(\frac{\pi}{4}D^2 - \frac{\pi}{4}d^2\right)\right] = \left[2\left(\frac{\pi}{4}D^2\right) - \left(\frac{\pi}{4}d^2\right)\right]$$

$$Q_{th} = A_{eff} LN/60 = \frac{\left[2\left(\frac{\pi}{4}D^2\right) - \left(\frac{\pi}{4}d^2\right)\right]LN}{60} \ \& \ WD_{th} = \frac{\gamma\left[2\left(\frac{\pi}{4}D^2\right) - \left(\frac{\pi}{4}d^2\right)\right]LN(h_S + h_D)}{60} \tag{12.31}$$

If the piston is of double acting and if the piston rod area is negligible, then the discharge is doubled and hence the discharge and work done becomes twice of their single acting pump and hence, the discharge and work done for a double acting pump are

$$Q_{th} = 2ALN/_{60} \ \& \ WD_{th} = \frac{2\gamma ALN(h_S + h_D)}{60} \quad (12.31a)$$

respectively.

Coefficient of discharge of reciprocating pump: In actual practice, the reciprocating pump has reduction in discharge than that of the theoretical, due to losses emerging from the imperfect action of sucking and delivery of liquid in the pipes. And so the actual discharge is less than the theoretical discharge, i.e. $Q_a < Q_{th}$

The coefficient of discharge is defined as the ratio of actual discharge to the theoretical discharge,

i.e.
$$C_d = \frac{Q_a}{Q_{th}} \quad (12.32)$$

The percentage of coefficient of discharge is termed volumetric efficiency,

i.e.
$$\eta_{vol} = \%C_d \quad (12.32a)$$

Slip of a reciprocating pump: Slip is defined as the amount of reduction in actual discharge from the theoretical discharge,

i.e.
$$Slip = Q_{th} - Q_a \quad (12.33)$$

The percentage of slip is
$$\%Slip = \frac{Q_{th} - Q_a}{Q_{th}} \quad (12.33a)$$

Note: In some cases, the actual discharge of a reciprocating pump may be greater than the theoretical ($Q_a > Q_{th}$), i.e. when the suction pipe length is too high than the delivery pipe length.

When the length of the suction pipe is too high than that of the delivery pipe, then there is a possibility of direct pushing of liquid from the suction pipe to the delivery pipe, without entering the cylinder chamber due to high pressure raise in the suction pipe, which increase the actual discharge. This type of reciprocating pump is termed to have a negative slip ($Q_a > Q_{th}$).

Example 12.8: Determine the theoretical discharge, coefficient of discharge, power developed, slip and slip percentage, if a single acting reciprocating pump delivers 0.003 m^3/s of water while running with a speed of 75 rpm. The diameter of the piston and stroke length of the piston is 175 mm and 210 mm respectively. Assume the suction and delivery heads as 4m and 15 m respectively.

Solution: Given data: $Q = 0.003 \text{ m}^3/\text{s}, N = 75 \text{ rpm}, D = 0.175 \text{ m}, L = 0.21 \text{ m}, h_S = 4 \text{ m}, h_d = 15 \text{ m}$

Refer Eq. (12.27) for **theoretical discharge**

$$Q_{th} = A*L*N/60 \Rightarrow Q_{th} = \frac{\pi}{4*60}D^2*L*N = \frac{3.14}{4*60}*0.175^2*0.21*75 = 0.0063 \text{ m}^3/\text{s}$$

Refer Eq. (12.32) for **coefficient of discharge** $C_d = \frac{Q_a}{Q_{th}} = \frac{0.003}{0.0063} = 0.47$

Refer Eq. (12.30) for **power developed**

$$P = \gamma Q(H_S + H_D)L \Rightarrow P = \frac{9810*3.14*0.175^2*0.21*75(4+15)*0.21}{4*60} = 247.56 \text{ W}$$

Refer Eq. (12.33) for **slip**

$$Slip = Q_{th} - Q_a = 0.0063 - 0.003 = 0.0033 \text{ m}^3/\text{s}$$

Refer Eq. (12.33a) for **percentage of slip** is $\%Slip = \frac{Q_{th} - Q_a}{Q_{th}} = \frac{0.003}{0.0063} = 47.6\%$

Example 12.9: Determine the effective area of the piston, theoretical discharge, coefficient of discharge, work done theoretical, if a double acting reciprocating pump having 25 mm of piston rod diameter delivers 0.003 m³/s of water while running with a speed of 75 rpm. The diameter of the piston and stroke length of the piston is 175 mm and 210 mm respectively. Assume the suction and delivery heads as 4 m and 15 m respectively.

Solution: Given data:

$$d = 0.025 \text{ m}, Q = 0.003 \text{ m}^3/\text{s}, N = 75 \text{ rpm}, D = 0.175 \text{ m}, L = 0.21 \text{ m}, h_S = 4 \text{ m}, h_d = 15 \text{ m}$$

Refer Eq. (13.31) for **effective area of the piston**

$$A_{eff} = \left[2\left(\frac{\pi}{4}D^2\right) - \left(\frac{\pi}{4}d^2\right)\right] \Rightarrow A_{eff} = \left[2\left(\frac{3.14}{4}*0.175^2\right) - \left(\frac{3.14}{4}*0.025^2\right)\right] = 0.0476 \text{ m}^2$$

Refer Eq. (12.31) for **theoretical discharge**

$$\Rightarrow Q_{th} = A_{eff}LN/60 = \frac{0.0476*0.21*75}{60} = 0.0125 \text{ m}^3/\text{s}$$

Coefficient of discharge $C_d = \frac{Q_a}{Q_{th}} = \frac{0.003}{0.0125} = 0.24$

Refer Eq. (12.31) for **work done theoretical**

$$WD = \frac{\gamma A_{eff}LN(H_S + H_D)}{60} \Rightarrow WD = \frac{9810*0.0476*0.21*75(4+15)}{60} = 2.328 \text{ kW}$$

Slip $Slip = Q_{th} - Q_a = 0.0125 - 0.003 = 0.0095\ \text{m}^3/\text{s}$

Percentage of slip is $$\%Slip = \frac{Q_{th} - Q_a}{Q_{th}} = \frac{0.0095}{0.0125} = 76.0\%$$

PISTON ACCELERATION AND EFFECTS ON VARIATION OF VELOCITY ON FRICTION IN PIPES AND PRESSURE IN PIPES

When a crank is rotated by an external motor shaft from $\theta = 0°$ to $\theta = 90°$, the piston gets accelerated but when the crank rotates from $\theta = 90°$ to $\theta = 180°$, the piston gets retarded due to inertia of the crank. So there is increase in suction in the first half of crank rotation and decrease in second half of the suction crank rotation. And this phenomenon also prevails in delivery stroke of crank rotation (i.e. $\theta = 180°$ to $\theta = 360°$), as shown in Figure 12.10a.

As the piston surface is in contact with the liquid, the acceleration variation of the piston exerts a variation in velocity and acceleration of the liquid flowing in the suction and delivery pipes.

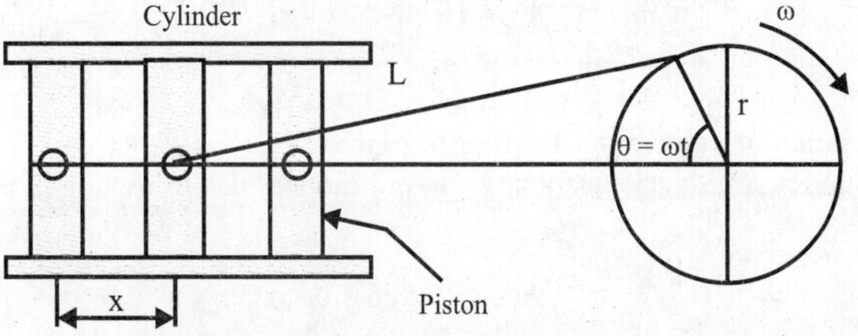

Figure 12.14: Velocity & acceleration of piston

Assuming the piston $(L >>>> r)$ to be reciprocating with simple harmonic motion (i.e. with friction resistance), we have

$$\theta = \omega t = \frac{2\pi N * t}{60} \tag{12.34}$$

where ω is angular speed in rad/s.

Let the piston be displaced by an amount of "x", as shown in Figure.12.14, then

$$x = r - r\cos\theta = r(1 - \cos\omega t) \tag{12.35}$$

Velocity of the piston will be

$$V_p = \frac{dx}{dt} = \frac{d}{dt} r(1 - \cos\omega t) \Rightarrow V_p = r\omega(\sin\omega t) \tag{12.36a}$$

Acceleration of the piston will be

$$Acc_p = \frac{dV_p}{dt} = \frac{d}{dt} r\omega(\sin \omega t) \boxed{\Rightarrow Acc_p = r\omega^2 (\cos \omega t)} \quad (12.36b)$$

The discharge of liquid sucked from the suction pipe must be equal to the liquid flowing into the cylinder chamber,

i.e. $$Q_{suction} = Q_{cylinder}$$

$$\Rightarrow (Area * Velocity)_{suction\ pipe} = (Area * Velocity)_{cylider} \Rightarrow (a_{pipe} * v_{pipe}) = (A * V_p)$$

Substituting Eq. (12.36a), we get the velocity and acceleration of flow in the pipe as

$$\boxed{\begin{gathered} v_{pipe} = \frac{A}{a_{pipe}}(V_p) = \frac{A}{a_{pipe}}(r\omega(\sin\theta)) \\ and \\ acc_{pipe} = \frac{d}{dt}(v_{pipe}) = \frac{A}{a_{pipe}} r\omega^2 (\cos\theta) \end{gathered}} \quad (12.37)$$

a) Effect of variation of velocity on friction in pipes

According to Darcy-Weisbach equation, we have head loss due to friction in pipes as

$$h_f = \frac{flv_{pipe}^2}{2gd}$$, so head loss becomes

$$\boxed{h_f = \frac{flv_{pipe}^2}{2gd} = \frac{fl}{2gd} * \left[\frac{A}{a_{pipe}}(r\omega(\sin\theta))\right]^2} \quad (12.38).$$

At $\theta = 0^\circ, 180^0$; $h_f = 0$ and similarly At

$$\theta = 90^\circ; \ h_f = \frac{fl}{2gd} * \left[\frac{A}{a_{pipe}}(r\omega)\right]^2$$

and so it is clear that the frictional losses will be minimum at starting and end point of a stroke, but it is maximum at the centre of the stroke and hence the maximum head loss due to friction is

$$h_{f_{\max}} = \frac{fl}{2gd} * \left[\frac{A}{a_{pipe}}(r\omega)\right]^2 \quad (12.38a).$$

b) Effect of variation of velocity on pressure head in pipes

Let the pressure intensity acting on the liquid flow in the pipes be p_i, then according to Newton's second law, we have $F = ma$. So substituting Eq. (12.37), we get Newton's second law as

$$F = \left(\frac{\gamma a_{pipe} * l_{pipe}}{g}\right) * \frac{A}{a_{pipe}} r\omega^2 (\cos\theta) \ where\ m = (\rho * volume) = \left(\frac{\gamma a_{pipe} * l_{pipe}}{g}\right)$$

But force = pressure * area

$$\Rightarrow F = p_i * a_{pipe} = \left(\frac{\gamma a_{pipe} * l_{pipe}}{g}\right) * \frac{A}{a_{pipe}} r\omega^2 (\cos\theta)$$

$$\Rightarrow p_i = \left(\frac{\gamma * l_{pipe}}{g}\right) * \frac{A}{a_{pipe}} r\omega^2 (\cos\theta) \tag{12.39}$$

But we know, we know the pressure head in pipes as

$$h = \frac{p_i}{\gamma} \Rightarrow h = \left(\frac{A * l_{pipe}}{g a_{pipe}}\right) * r\omega^2 (\cos\theta) \tag{12.40}$$

$$\theta = 0° \ ; h = \left(\frac{A * l_{pipe}}{g a_{pipe}}\right) * r\omega^2$$

$$\theta = 90°; \ h = 0 \tag{12.41}$$

at

$$\theta = 180°; \ h = -\left(\frac{A * l_{pipe}}{g a_{pipe}}\right) * r\omega^2$$

From Eq. (12.41), we can prove the acceleration do exists in liauid flow in suction. and delivery strokes. And as the pressure head is due to aceleration/retardation, we can term the Eq. (12.41) as the aceleration head in the pipes.

If the connecting rod length is not too high as compared with the crank length, then simple harmonic motion of the piston does not exist and hence the pressure head alters as

$$h = \left(\frac{A * l_{pipe}}{g a_{pipe}}\right) * r\omega^2 \left(\cos\theta + \frac{\cos 2\theta}{n}\right) \tag{12.42}$$

$$\theta = 0° \, h = \left(\frac{A * l_{pipe}}{g a_{pipe}}\right) * r\omega^2 \left(1 + \frac{1}{n}\right)$$

$$\theta = 90°; h = 0 \tag{12.42a}$$

At

$$\theta = 180°; h = -\left(\frac{A * l_{pipe}}{g a_{pipe}}\right) * r\omega^2 \left(1 - \frac{1}{n}\right)$$

Effect of Acceleration on Suction Pipe and Delivery Pipe Indicator Diagrams and Maximum Speed of Reciprocating Pump

Indicator diagram: Indicator diagram is the theoretical diagram indicated by an indicator placed on the cylinder of a reciprocating pump. The indicator diagram is drawn between the crank rotation along the abscissa and pressure heads along vertical ordinate, in one complete revolution.

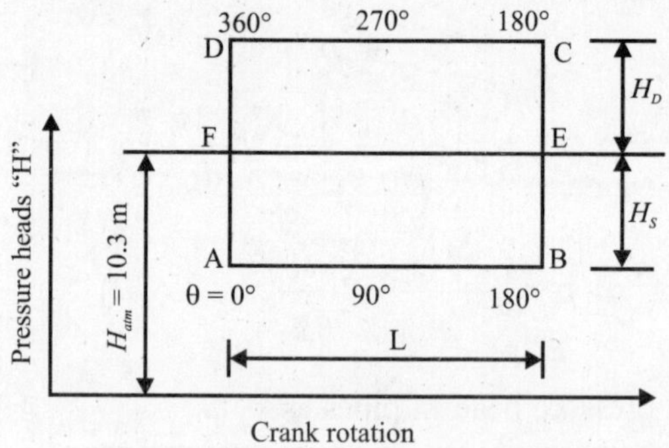

Figure 12.15: Crank rotation

The suction stroke and delivery strokes act below and above the atmospheric conditions by an amount of (H_S, H_D) respectively in a reciprocating pump and so draw a line "EF" indicating the atmospheric pressure head parallel to a reference line "OO". Draw the suction stroke line $\theta = 0°$ to $\theta = 180°$ of length "L" below the atmospheric pressure head, to a scale and name it as line "AB". After suction, delivery stroke starts and so draw the delivery stroke line $\theta = 180°$ to $\theta = 360°$ of length "L" above the atmospheric pressure head, to the same scale with opposite direction and name it as line "CD". Join these two drawn suction stroke line and delivery stroke line to form a rectangle "ABCD", which represents the total work done by a reciprocating pump, as shown in Figure 12.15.

The rectangle "ABCD" is divided by the atmospheric pressure head line "EF" into two parts, namely, rectangle "ABEF" and rectangle "ECDF". Rectangle "ABEF" represents the work done by the pump in suction stroke and similarly rectangle "ECDF" represents work done by the pump in delivery stroke. So the total work done

$$WD_{total} = (H_S * L + H_D * L) \qquad (12.43)$$

Effects of Acceleration on Suction Pipe and Delivery Pipe Indicator Diagrams

When the crank rotates, there arise acceleration/retardation heads in suction and delivery strokes, as seen in Eq. (12.40) and these acceleration heads do modify the theoretical indicator diagram.

Consider Eq. (12.41), which gives acceleration heads in the pipes at different orientations of the crank. It has been observed from this equation that there is a positive acceleration head at the starting of the strokes $(\theta = 0^0 \ \& \ \theta = 180^0)$, zero acceleration head at the intermediate position of the stroke $(\theta = 90^0 \ \& \ \theta = 270^0)$ and negative acceleration head at the end of strokes $(\theta = 180^0 \ \& \ \theta = 360^0)$ Redrawing the indicator diagram, as shown in Figure 12.15, we have the

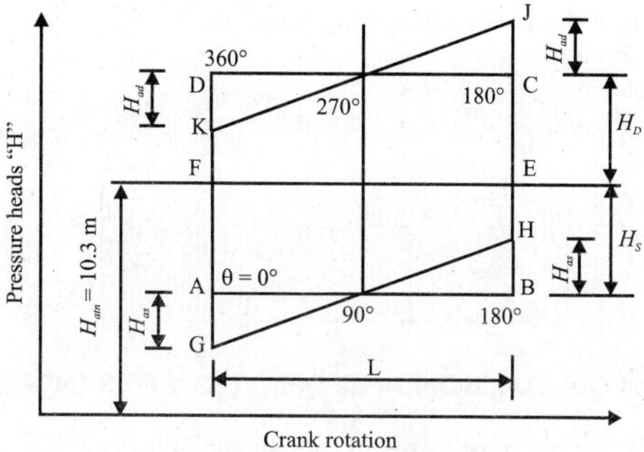

Figure 12.15a: Crank rotation

suction line "AB" modified to new position "GH". And similarly delivery line "CD" modifies to new line "JK", as shown in Figure 12.15a.

We can observe that the work done in suction and delivery strokes are been modified due to presence of the acceleration heads, in the strokes. That is work done in suction stroke is modified to area "GHEF" instead of rectangle "ABCD". Similarly, in case of delivery stroke, work done in delivery stroke is modified to area "JKEF" instead of rectangle "CDEF". But the area of the new work done and the theoretical work done are same in suction and delivery strokes, we can say that the variation of acceleration head do not vary the total work done by the pump in a revolution but still this variation of acceleration head do vary the pressure from point to point in the cylinder of the reciprocating pump.

MAXIMUM SPEED OF RECIPROCATING PUMP

Reciprocating pump is so designed that it can be operated up to a particular maximum speed without the presence of cavitation. And to achieve maximum speed without cavitation, the suction pressure should not drop below certain level, which gives chance to the dissolved gases in the water to escape, thereby causing cavitation.

The pressure heads at the starting of suction stroke and end of the delivery stroke do affect the design conditions of a pump working. The intermediate pressure heads are dependent on these two extreme pressure heads only. So, the starting pressure head at the suction stroke and at the end of delivery stroke must not fall below a limiting value or else the possibility of cavitation may arise, causing pump failure. The limiting pressure head value at starting and end of the suction stroke and delivery stroke respectively is 2.6 m of water absolute.

i.e.
$$H_{sep} = H_S + H_{as} = 10.3 - 2.6 = 7.7 \text{ m } \textit{of water}$$
$$H_{sep} = H_{atm} + (H_D - H_{ad}) = 10.3 + 2.6 = 12.9 \text{ m } \textit{of water} \quad (12.44)$$

Hence, H_{sep} represents the vapour pressure head below atmosphere pressure head at which cavitation does not exist.

So **maximum speed** of the reciprocating pump will be

$$H_{sep} = H_S + H_{as} = H_S + \left(\frac{A * l_s}{g a_s}\right) * r\omega^2$$

$$\Rightarrow H_{sep} = H_S + \left(\frac{A * l_s}{g a_s}\right) * r\left(\frac{2\pi N}{60}\right)^2 \qquad (12.45)$$

EFFECT OF FRICTION ON SUCTION AND DELIVERY PIPES INDICATOR DIAGRAMS

As seen in Eq. (12.38), frictional head losses in suction and delivery pipes gets add up to the total head, if considered and so the indicator diagram gets modified further.

Eq. (12.38) $h_f = \frac{f l v_{pipe}^2}{2gd} = \frac{fl}{2gd} * \left[\frac{A}{a_{pipe}}(r\omega(\sin\theta))\right]^2$ shows a parabolic shape and has zero values at crank rotation $\theta = 0^0, \theta = 180^0$ & $\theta = 360^0$, i.e. at the starting and ending stage of the suction and delivery strokes. But has maximum value at $\theta = 90^0$ & $\theta = 270^0$ i.e. at the middle of the suction and delivery strokes. So the indicator diagram does get modified as shown in Figure.12.15b.

As the height of a parabolic shape has two-thirds the of the its maximum height, we have total work done per one revolution = equal to the area of the indicator diagram and so work done per revolution will be

$$WD_{rev} = L\left(H_S + H_D + \frac{2}{3}h_{fs} + \frac{2}{3}h_{fd}\right).$$

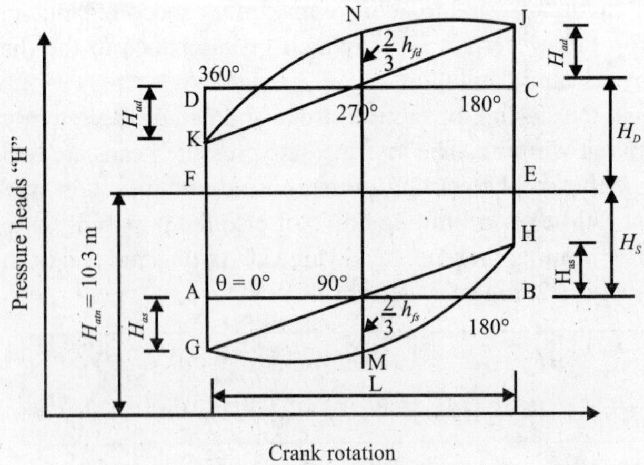

Figure 12.15b: Crank rotation

So total work done per second will be

$$WD_{\sec} = \frac{\gamma ALN}{60}\left(H_S + H_D + \frac{2}{3}h_{fs} + \frac{2}{3}h_{fd}\right) \tag{12.46}$$

Total absolute pressure head in suction and delivery strokes are

$$\begin{aligned} H_{ts} &= H_{atm} - H_s - H_{as} - h_{fs} \\ H_{td} &= H_{atm} + H_s + H_{as} + h_{fs} \end{aligned} \tag{12.47}$$

Hence, from Eq. (12.47), we can observe that H_{ts} has minimum in value at $\theta = 0°$, i.e. at the beginning of the suction stroke and so the cavitation may arise at the junction of suction pipe and the cylinder of the reciprocating pump.

Example 12.10: Determine the maximum speed at which a single acting reciprocating pump having 95 mm diameter of piston, crank radius 150 mm and stroke length of 130 mm. It water delivery to a tank 15 m above the pump, through a pipe diameter of 35 mm, and 17 m long. It takes water from a sump 4 m below the pump through a pipe of 45 mm diameter and 5 m long. If the separation occurs at 80 kPa below atmospheric pressure and assumes the piston has simple harmonic motion.

Solution: Given data:

$$d_s = 0.045 \text{ m},\ l_s = 5 \text{ m},\ d_d = 0.035 \text{ m},\ l_d = 17 \text{ m},\ D = 0.095 \text{ m},\ r = 0.075 \text{ m}$$
$$L = 0.13 \text{ m},\ h_s = 4 \text{ m},\ h_d = 15 \text{ m},\ h_{sep} = 80 \text{ kPa} \ \ \textit{below atmosphere}$$

Area of the piston is $A = \frac{\pi}{4}D^2 = \frac{3.14}{4} * 0.095^2 = 0.007 \text{ m}^2$

Area of the suction pipe is $a_s = \frac{\pi}{4}d_s^{\,2} = \frac{3.14}{4} * 0.045^2 = 0.0015 \text{ m}^2$

Area of the delivery pipe is $a_d = \frac{\pi}{4}d_d^{\,2} = \frac{3.14}{4} * 0.035^2 = 0.00096 \text{ m}^2$

Separation head $h_{sep} = -80 * 1000 / 9810 = 8.15 \text{ m} \ \ \textit{of water}$

For suction: Refer Eq. (12.40), we have $h = \left(\frac{A * l_{pipe}}{g a_{pipe}}\right) * r\omega^2 (\cos\theta)$ but the separation occurs only during the beginning of the stroke (i.e. at 0°), so

$$h_{ssep} = \left(\frac{A*l_{pipe}}{ga_{pipe}}\right)*r\omega^2 = \left(\frac{0.007*5}{9.81*0.0015}\right)*0.075\omega^2 = 0.178\omega^2$$

Pressure head in cylinder at beginning of the stroke will be $= h_{atm} - (h_s + h_{ssep})absolute$

Limiting condition for no separation is

$$h_{sep} = h_{atm} - (h_s + h_{ssep})absolute$$

$$(h_{atm} - h_{sep})absolute = h_{atm} - (h_s + h_{ssep})$$

$$\Rightarrow (h_{sep}) = (h_s + h_{ssep}) \Rightarrow 8.15 = (4 + 0.0178\omega^2) \Rightarrow \omega = 15.27 \text{ rad/s}$$

So, the speed will be $$N = \frac{60\omega}{2\pi} = \frac{60*15.27}{2*3.14} = 145.9 \text{ rpm}$$

Hence, the **speed of the pump without separation during the suction stroke is 145.9 rpm.**

Similarly, for **delivery stroke**

Refer Eq. (12.40), we have $h = \left(\frac{A*l_{pipe}}{ga_{pipe}}\right)*r\omega^2(\cos\theta)$ but the separation occurs only at the end of the stroke (i.e. at 180^0), so $h_{dsep} = -\left(\frac{A*l_{pipe}}{ga_{pipe}}\right)*r\omega^2 = -\left(\frac{0.007*5}{9.81*0.0015}\right)*0.075\omega^2 = -0.178\omega^2$

Pressure head in the cylinder at the beginning of the stroke will be $= h_{atm} + (h_d + h_{ssep})absolute$

Limiting condition for no separation is

$$h_{sep} = h_{atm} - (h_d + h_{ssep})absolute \quad (h_{atm} - h_{sep})absolute = h_{atm} + (h_d + h_{dsep})$$

$$\Rightarrow (h_{sep}) = -(h_d + h_{dsep}) \Rightarrow 8.15 = -(15 - 0.0178\omega^2) \Rightarrow \omega = 36 \text{ rad/s}$$

So, the speed will be $$N = \frac{60\omega}{2\pi} = \frac{60*36}{2*3.14} = 343.9 \text{ rpm}$$

Hence, **the speed of the pump without separation during the suction stroke is 343.9 rpm.**

So **maximum permissible speed to avoid separation in a reciprocating pump is 145.9 rpm.**

AIR VESSELS

As observed in discharge vs. crank rotation (refer Figure 12.10a) of a reciprocating pump, the pump remains ideal in suction stroke and gives useful work only in delivery stroke. Moreover, the pump decreases its efficiency in giving useful work in the second half of the delivery stroke.

So due to acceleration and retardation of the piston, there arises only one-quarter of a complete revolution of crank gives max efficiency of useful work in pumping the water. Hence, to make use of the other three-quarters of the crank revolution effective, air vessels are being used.

Air vessel is a closed tight container, comprising two halves, one half on the top portion filled with compressed air and other bottom half with water due to pumping of water by the pump. Air vessel has air and water filled in it and is separated by a flexible diaphragm, which can move up and down due to the difference in pressure between air and water filled. And also these Air vessels are located just near to the inlet and outlet of the reciprocating pump, as shown in Figure 12.16.

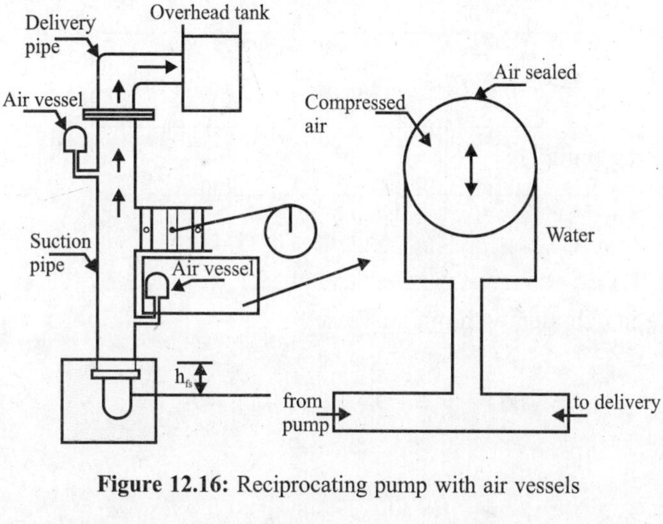

Figure 12.16: Reciprocating pump with air vessels

Working of an air vessel: In suction stroke the piston accelerates (i.e. $\theta = 0°$ to $\theta = 90°$) and hence the flow velocity need to be accelerated. But while having acceleration of flow velocity, the constant discharge from the suction pipe sucked fast and the discharge gets reduced. So an additional amount of water is required to compensate the decrease in discharge quantity, which is done by an air vessel due to the difference between the fluids in it.

Similarly, while retarding of piston in the suction stroke (i.e. $\theta = 90°$ to $\theta = 180°$), the flow velocity is to be decreased and so some amount of discharge needed to be sent out, which is done by air vessel, due to the pressure difference in between the compressed air and the water coming in.

The same action takes place in air vessel at the delivery side also. So to maintain a constant flow velocity within the suction and delivery stroke, the fluid discharge to be monitored to have a uniform flow rate and is done by air vessel. If air vessels are not used, then the flow rate will be intermittent throughout a cycle, of the reciprocating pump. This type of cycle repeats for every revolution of crankshaft.

Air vessel has advantage of maintaining uniform flow rate, reduced friction work, acceleration head reduction, enabling of higher speeds.

Flow into/from air vessel and work done saved due to air vessel
Discharge into/from the cylinder

$$Q_{cylinder} = A_{cylinder} * V_{piston} = A(r\omega \sin\theta) \tag{12.48}$$

refer Eq. (12.36a).

Mean discharge to be maintained by the Air vessel

$$Q_{mean} = A_{cylinder} V_{mean} = A_{cylinder}(r\omega / revolution) = A(r\omega / \pi) \tag{12.49}$$

Discharge into/from the air vessel is

$$Q_{air\ vessel} = Q_{cylinder} - Q_{mean} = Ar\omega\left(\sin\theta - \frac{1}{\pi}\right) \tag{12.50}$$

Note: For double acting pump

$$Q_{air\ vessel} = Q_{cylinder} - Q_{mean} = Ar\omega\left(\sin\theta - \frac{2}{\pi}\right) \tag{12.50a}$$

Work lost in a stroke in cylinder, with out air vessel

$$WD_{without} = \gamma Q H_f = \gamma(AL)\left(\frac{2}{3}h_f\right)$$

$$WD_{without} = \gamma(Ar)\left(\frac{4flV_{piston}^2}{6gd}\right) = \gamma(Ar)\left(\frac{4fl}{6gd}\right)\left(\frac{A}{a}r\omega\right)^2 \tag{12.51}$$

Refer Eq. (12.36a)

Work lost in a stroke with air vessel $WD_{with} = \gamma Q H_f = \gamma(AL)(h_f)_{pipe}$

$$WD_{with} = \gamma(Ar)\left(\frac{flv_{pipe}^2}{gd}\right) = \gamma(Ar)\left(\frac{fl}{gd}\right)\left(\frac{A}{a\pi}r\omega\right)^2 \tag{12.52}$$

Refer Eq. (12.37)

So **work saved** will be

$$WD_{saved} = WD_{without} - WD_{with} = 2\gamma(Ar)\left(\frac{fl}{2gd}\right)\left(\frac{A}{a}r\omega\right)^2\left(\frac{2}{3}-\frac{1}{\pi}\right) \quad (12.53)$$

% **work saved** will be
$$\%WD_{saved} = \frac{WD_{without} - WD_{with}}{WD_{without}} \times 100 = \frac{\left(\frac{2}{3}-\frac{1}{\pi}\right)}{2/3} \times 100 = 84.8\% \quad (12.54)$$

Hence, an amount of 84.8% work done can be saved from friction effect by using Air vessels and thereby increase in usefulness of reciprocating pump.

Example 12.11: A single acting reciprocating pump is to raise a liquid of density 1200 kg/m^3 through a vertical height of 11.5 m, from 2.5 m below pump axis to 9 m above it. The plunger moves with a simple harmonic motion, has diameter 125 mm and stroke of 225 mm. The suction and delivery pipes are 75 mm diameter and 3.5 diameter and 13.5 m long respectively. There is along vessel placed on the delivery pipe near the pump axis but there is no air vessel on the suction pipe. If separation takes place 0.88 bar below atmospheric pressure find:

a. Maximum speed with which the pump can run without separation taking place.

b. Power required to drive the pump, if f = 0.02. **[AMIE]**

Neglect slip for the pump.

Solution: Given data:

$$\rho = 1200 \text{ kg/m}^3, h_s = 11.5 \text{ m}, h_d = 9 \text{ m}, D = 0.125 \text{ m}, f = 0.02$$
$$d_s = d_d = 0.075 \text{ m}, l_s = 3.5 \text{ m}, l_d = 13.5 \text{ m}, r = 0.1125 \text{ m}, L = 0.25 \text{ m}$$

As there is air vessel placed on delivery side, the velocity will be uniform (no acceleration head) and hence separation pressure will be equal to 0.88 bar below the atmospheric pressure,

i.e.
$$h_{sep.} = \frac{P}{\rho g} = 88*1000/1200*9.81 = 7.47 \text{ m} \quad of \quad water$$

Area of the piston is
$$A = \frac{\pi}{4}D^2 = \frac{3.14}{4}*0.125^2 = 0.0122 \text{ m}^2$$

Area of the suction pipe and delivery pipes is $a_s = a_d = \frac{\pi}{4}d_s^{\,2} = \frac{3.14}{4}*0.075^2 = 0.0044 \text{ m}^2$

For suction:

Refer Eq. (12.40), we have

$$h = \left(\frac{A*l_{pipe}}{ga_{pipe}}\right)*r\omega^2(\cos\theta)$$

but the separation occurs only during the beginning of the stroke (i.e. at 0°), so

$$h_{ssep} = \left(\frac{A * l_{pipe}}{g a_{pipe}}\right) * r\omega^2 = \left(\frac{0.0122 * 3.5}{9.81 * 0.0044}\right) * 0.1125\omega^2 = 0.1112\omega^2$$

Pressure head in the cylinder at the beginning of the stroke will be $= h_{atm} - \left(h_s + h_{ssep}\right) absolute$

Limiting condition for no separation is

$$h_{sep} = h_{atm} - \left(h_s + h_{ssep}\right) absolute \quad \left(h_{atm} - h_{sep}\right) absolute = h_{atm} - \left(h_s + h_{ssep}\right) \quad \Rightarrow \left(h_{sep}\right) = \left(h_s + h_{ssep}\right)$$

$$\Rightarrow 7.47 = \left(11.5 + 0.1112\omega^2\right) \Rightarrow \omega = 6.67 \text{ rad/s}$$

So, the speed will be $N = \dfrac{60\omega}{2\pi} = \dfrac{60 * 6.67}{2 * 3.14} = 63.69$ rpm

Hence, the **maximum speed of the pump without separation is 63.69 rpm.**

Discharge $Q = \dfrac{ALN}{60} = 0.00294$ m^3/s

Power developed is $p = \gamma Q H_{total} = \dfrac{\gamma ALN}{60}\left(h_s + h_d + h_{fd} + \dfrac{2}{3}h_{fs}\right)$

$$\Rightarrow P = \frac{\gamma ALN}{60}\left(h_s + h_d + \frac{f l_d v_d^2}{2 g d_d} + \frac{2}{3}\left(\frac{4 f l_s}{2 g d_s}\right)\left(\frac{A}{a_s} r\omega\right)^2\right) = 428.4 \text{ watts}$$

Hence, the **power developed by the reciprocating pump is 428.4 watts.**

EXERCISE

Objective

1. The following is a roto-dynamic pump
   a) Gear pump
   b) Plunger pump
   c) Two throw pump
   d) Volute pump
2. Identify the positive displacement reciprocating pump from the following:
   a) multistage pump
   b) screw pump

c) turbine pump
d) volute pump

3. The following is a rotary pump
a) double acting pump
b) axial flow pump
c) radial piston
d) triple ram pump

4. A centrifugal pump
a) is a positive displacement pump
b) utilizes the hydraulic energy of the fluid
c) increases the energy of the fluid
d) is a turbo-machine whose action is the reverse of an impulse turbine

5. The volute casing of a centrifugal pump
a) Prevents clogging of impeller with suspended matter
b) Converts kinetic energy of fluid leaving the impeller into pressure
c) Improves the performance by creating a free vortex
d) Minimizes the build-up of dangerously high pressure

6. In a centrifugal pump, the following is classified as a mechanical loss
a) Shock loss at vane exit
b) Friction loss in diffusers
c) Friction loss in impeller passages
d) Friction loss in glands and bearings

7. The type of centrifugal impeller best suited for pumping small quantities against high heads is
a) axial flow type
b) radial flow type
c) mixed flow type
d) tangential flow type

8. The centrifugal pump impeller best suited for pumping large quantities against small heads is
a) Multistage type
b) Semi-open type
c) Vane type
d) Axial flow type

9. The manometric head of a centrifugal pump is
a) $(U_1^2-U_2^2)$ 2g
b) $(U_1-U_2)^2$ /2g
c) $V_{w1}U_1/2g$
d) $V_{w1}U_1/g$

10. The manometric head of a centrifugal pump is
a) Sum centrifugal head and pipe losses
b) Additional head imparted to the fluid by the impeller
c) Actual lift of pumpless pump losses
d) Level differences through which the pump lifts the liquid

11. The numerical sum of suction and delivery pressure gage readings in a centrifugal pump setup is equal to the
    a) Actual head
    b) Static head
    c) Manometric head
    d) Centrifugal head
12. The losses that occur in a centrifugal pump are high when the discharge is
    a) High
    b) Low
    c) Both high and low
    d) Of any value
13. The ratio of power delivered to the fluid the staff input power is called
    a) Manometric efficiency
    b) Hydraulic efficiency
    c) Mechanical efficiency
    d) Overall efficiency
14. The manometric efficiency of a centrifugal pump is the ratio of
    a) Fluid power of pump to impeller power
    b) Power of impeller to shaft power
    c) Fluid power of pump to shaft power
    d) Power of impeller to fluid power of pump
15. The ratio of power given to the fluid by the pump to shaft power is called
    a) Manometric efficiency
    b) Hydraulic efficiency
    c) Overall efficiency
    d) Mechanical efficiency
16. For a centrifugal pump
    a) H_m increases as Q, N being constant
    b) H_m decreases as N_2, Q being constant
    c) Q is independent of shaft power, N being constant
    d) Q increase as N, H_m being constant
17. A centrifugal pump fails to lift the liquid if
    a) centrifugal head is less than the liquid
    b) Excessive shock losses occur at exit of impeller
    c) Delivery pipe is leaky
    d) Shaft alignment is defective
18. When a working centrifugal pump suddenly stops delivering the liquid, the conclusion is
    a) Speed is less than design speed
    b) Head is more than design head
    c) Air has leaked into suction side of pump
    d) Impeller is touching the casing
19. When the speed of a centrifugal pump is constant
    a) Shaft power decrease with increase of Q

b) Hm decrease with increase of Q
c) Q increase with increases of H_m
d) Q in independent of H_m

20. A centrifugal pump is connected to a variable speed motor. Its Q is maintained constant. Its Hm is
a) Proportional to N^2 b) Proportional to 1/N
c) Proportional to N d) Independent of N
21. The rotational speed of a centrifugal pump varies as
a) Square of manometric head
b) Square root of manometric head
c) Square of impeller diameter
d) Inverse square root of impeller diameter
22. The specific speed of a centrifugal pump is defined as the speed of a pump of such a size that it
a) Delivers unit discharge under unit head
b) Requires unit power to develop unit head
c) Requires unit power to deliver unit discharge
d) Develops unit head whatever be the discharge
23. The height at which a centrifugal pump can be set above the lower liquid surface
a) Can be any height less than the manometric head
b) Depends on the efficiency of foot valve
c) Is a function of exit vane angle.
d) Is a function of saturation vapour pressure of liquid.
24. The operating characteristics of a centrifugal pump are plotted on a base of
a) H_m, N being constant
b) Q, N being constant
c) Shaft input, Q being constant
d) N, H_m being constant
25. A two-stage centrifugal pump compared to a single stage centrifugal pump of the same size
a) Develops twice the head, Q being the same
b) Develops 2 times the head Q being constant
c) Gives twice the discharge, head being the same
d) Helps the discharge increasing the head
26. The NSPH of a centrifugal pump
a) Depends on h_s
b) Independent of h_s
c) Equal to h_s
d) $h_s - (P_{vap}/W)$
27. The NSPH of a centrifugal pump is equal to
a) σ hs
b) P_{vap}/W
c) σ/H_m
d) σH_m

28. A reciprocating pump
    a) Is suitable for pumping muddy water from river and sea bottoms
    b) Can develop pressure as high as 700 bars
    c) Belongs to the class of radial piston pumps
    d) Is a roto-dynamic pump
29. In a reciprocating pump setup, the crank radius is *r* and the length of connecting rod is L1. Simple harmonic motion may be assumed if
    a) $L_1 >> r$ b) $r >> L_1$ c) $L_1 = r$ d) $L_1/r = 0$
30. The slip of a reciprocating pump is
    a) Qa/Qth b) (Qth-Qa) Qth c) Qth/(Qth –Qa) d) Qth -Qa
31. Negative slip is likely to occur in a reciprocating pump is
    a) Suction pipe is long and delivery pipe is short
    b) Delivery pipe is long and suction pipe is short
    c) Suction valve lifts up before the imperfect valve seating
    d) Leaking occurs due to imperfect valve seating
32. A double acting reciprocating pump
    a) Gives two working strokes as the crank angle varies from 0 to θ
    b) Has a plunger in place of piston and piston rod
    c) Must have two suction valves and two delivery valves
    d) Must have two delivery valves but may have one suction valve
33. Separation is likely to occur in a reciprocating pump is
    a) The speed of pump is too low b) The speed of pump is too high
    c) The delivery pipe is too short d) Valve seating is leaky
34. Separation is likely to occur in a single acting reciprocating pump without air vessel during the suction stroke at
    a) A crank angle of θ = 18.56°
    b) The middle of stroke
    c) The end of stroke
    d) The beginning of stroke
35. The acceleration head of single acting reciprocating pump with no air vessels is
    a) Directly proportional to pipe diameter
    b) Inversely proportional to piston diameter
    c) Directly proportional to crank radius
    d) Inversely proportional to crank speed in rpm
36. A given pump with a given impeller diameter and speed will raise a fluid to a certain height regardless of the fluid density.
    a) No
    b) Yes
    c) Cannot be said
37. The potential energy developed is approximately equal to the ____ at the periphery of the impeller
    a) Velocity head
    b) Potential head
    c) None

38. Higher heads can be achieved by increasing the _______ or the______ of the impeller of a centrifugal pump.
    a) Diameter or rotational speed
    b) Potential head or speed
    c) None
39. Centrifugal pump is also called Radial pump
    a) No
    b) Not relevant
    c) Yes

Answers

1. a	2. c	3. d	4. c	5. b	6. c
7. b	8. a	9. d	10. a	11. c	12. b
13. d	14. a	15. c	16. d	17. a	18. c
19. b	20. c	21. d	22. a	23. d	24. d
25. c	26. d	27. d	28. c	29. a	30. d
31. a	32. c	33. a	34. c	35. c	36. b
37. a	38. a	39.c			

Theory

1. Classify types of hydraulic pump.
2. Explain the working mechanism of centrifugal pump.
3. Explain the components of a centrifugal pump.
4. Derive fundamental equation of centrifugal pump and explain it in detail.
5. Derive manometric head and efficiency of a centrifugal pump.
6. Explain the necessity of having a minimum starting speed for a centrifugal pump.
7. Explain the reasons for priming centrifugal pump.
8. Write a note on cavitation and its effects.
9. Classify types of reciprocating pump.
10. Explain the working mechanism of Reciprocating pump.
11. Explain the term negative slip.
12. Write a note on indicator diagram.
13. Write a note on air vessels and their requirement for reciprocating pump.

Problems

1. The axis of a centrifugal pump is 2.5 m above the water level in the sump and static lift from the pump is 3.5 m. The friction losses in the suction and delivery pipes are 1 m and 7.5 m respectively. The suction and delivery pipes are each of 75 mm diameters. The impeller is 300 mm diameter and 18 mm wide at the outlet and its speed is 1700 rpm. The water at inlet has radial flow and vane angle at outlet is 32° to the tangent to the periphery. Compute the discharge of the pump and power required to drive the pump. Assume manometric efficiency and overall efficiency as 77%, 72% respectively.

[0.038 m^3/s, 8.81 kW]

2. A centrifugal pump of the radial type delivers 5000 lit/min against a head of 8 m, when running at a speed of 1450 rpm. If the outer diameter of the impeller is 300 mm and its width at the outer periphery is 13 mm, find the vane angle at exit, assuming manometric efficiency as 80%. [**71°**]
3. Determine the head imparted to a fluid as it passes through an impeller of 250 mm outlet diameter and 100 mm inlet diameter rotated at 1440 rpm. The outlet vane angle is set back at an angle of 20 degrees to the tangent. Assume radial entrance and velocity of flow as 3m/s. [**20.4 m**]
4. A single stage centrifugal pump runs at 500 rpm and delivers 300 m^3/min of water against a head of 120 m. The pump impeller is 2 m in diameter and it has a positive suction lift (including the velocity and friction heads) of 3 m. Laboratory tests are to be conducted on a model with a 450 mm diameter impeller and on a reduced head of 95 m. Assuming atmospheric pressure head as 10.15m of water and vapor head as 0.4 m of water calculate the speed, discharge and suction lift for the laboratory tests.
   [1977 rpm, 1.51 m^3/min, 4.42 m]
5. A single acting reciprocating pump has piston of diameter 0.3 m and stroke of 0.4m. If the speed of the pump is 60 rpm, and coefficient of discharge is 0.97, determine the percentage slip and actual discharge. **[3%, 27.46 lit/s]**
6. The piston diameter and stroke length of a single acting reciprocating pump are 150 mm and 300 mm respectively. The center of the pump is 5.0 m above the water level in the sump and 33 m below the delivery water level. Both the suction and delivery pipes have the same diameter of 75 mm and are 65 m and 39 m long respectively. If the pump is working at 30 rpm, determine the pressure head on the piston at the beginning, middle and end of the both strokes, the power required to drive the pump. Take the friction coefficient as 0.01 for both pipes and atmospheric pressure head as 10.3m of water. [**1.38 m (abs), 4.672 m(abs), 9.22 m (abs), 66.85 m(abs), 47.07(abs), 19.76 m(abs), 1.06 kW**]
7. A single acting reciprocating pump runs at 60 rpm. The diameter of the piston is 0.15 m and crank radius is 0.15 m. The suction pipe is 0.1 m in diameter and 5 m long. Calculate the maximum permissible value of suction lift H_S, if the separation takes place at 2.6 m of water absolute. [**0.91 m**]
8. The diameter and stroke of a single acting reciprocating pump are 125 mm and 300 mm respectively. The pump is fed by a suction pipe 75 mm in diameter and 7 m long, the suction lift being 4 m. The separation occurs if the absolute pressure head in the cylinder during suction stroke falls below 2.5 m of water. What is the maximum speed at which the pump can be run without separation in suction pipe? Take atmospheric pressure head as 10.m of water. [**34.1 rpm**]
9. A single acting reciprocating pump running at 60 rpm has a piston diameter of 0.25 m and stroke of 0.5 m. The delivery pipe is 0.1 m diameter and 50 m long. If the motion of the pipe is simple harmonic, find the power required to overcome friction of the delivery pipe when no air vessel is fitted, a large air vessel is fitted at the centreline of the pump. Take f = 0.04. **[15.77kW, 2.4kW]**

QUICK REFERENCE OF FORMULAE

Property	Formulae	Remarks, if any
Work Done, Heads and Efficiencies of Centrifugal Pump	$WD = \frac{\gamma}{g} * Q * (V_{wo}) * u_o$; $V_{ro}^2 = V_{fo}^2 + (u_2 - V_{wo})^2$ $WD = \frac{1}{g} * Q * (V_{wo} - V_{wi}) * u$ $WD = \left\{ \left(\frac{V_o^2 - V_i^2}{2g} \right) + \left(\frac{u_o^2 - u_i^2}{2g} \right) + \left(\frac{V_{ri}^2 - V_{ro}^2}{2g} \right) \right\}$	Euler's momentum equation of Centrifugal pump
Heads losses of a Centrifugal pump	$H_S = h_s + h_d$; $H_{mano} = H_s + h_{fs} + h_{fd} + \frac{V_d^2}{2g}$ $H_{mano} = \frac{V_{wo}u_o}{g} - h_L = \frac{V_{wo}u_o}{g} - \frac{V_o^2}{2g}$	H_S, h_s, h_d **are** static head, suction head and delivery head. H_{mano}, h_L are manometric head and head loss due to friction. h_{fs}, h_{fd} are frictional losses in suction and delivery pipes.
Efficiencies of Centrifugal pump and Outlet vane angle	$\eta_{mano} = \frac{Output\ of\ pump}{Power\ imparted\ by\ impeller}$; $\eta_{mano} = \frac{gH_m}{V_{wo}u_o}$ $\eta_m = \frac{\left(V_{wo}u_o / g \right)}{Energy\ head\ given\ to\ head}$ $= \frac{\gamma(Q+q)\left(V_{wo}u_o / g \right)}{Power\ given\ to\ mpeller}$ $\eta_{vol} = \frac{discharge\ quantity/sec}{Actual\ quantity\ passing\ through\ impeller}$ $h_{vol} = \frac{discharged\ quantity/sec}{Actual\ quantity\ passing\ through\ impeller}$ $h_{vol} = \frac{discharged\ quantity/sec}{Actual\ quantity\ passing\ through\ impeller}$ $\eta_{vol} = \frac{Q_a}{(Q_a + q)}$ $\eta_o = \frac{\gamma Q H_m}{Power\ given\ to\ impeller\ shaft}$ $H_{mano} = \frac{u_o^2 - V_{fo}^2 \operatorname{cosec}^2 \phi_o}{2g}$	$\eta_{mano}, \eta_m, \eta_{vol}, \eta_o$ Are monometric efficiency, mechanical efficiency, volumetric efficiency and overall efficiency respectively.

Contd.

Property	Formulae	Remarks, if any
Minimum starting speed	$\left(\frac{\pi N}{60}\right)^2\left[\left(D_o^2 - D_i^2\right)\right] = 2gH_{mano}$	$0.9 - K_u - 1.25$ $0.1 - \psi - 0.25$
Design parameters and their proportion of centrifugal pump	$K_u = {u_o}/{\sqrt{2gH_{mano}}}$ $\zeta = \frac{V_{fo}}{\sqrt{2gH_{mano}}}$ $D_{\min} = \frac{97.68\sqrt{H_{mano}}}{N}$ $D_s = \sqrt{\frac{4Q}{3.14V_s}}$	
Specific speed of centrifugal pump	$N_s = N\frac{\sqrt{Q}}{H_{mano}^{3/4}}$; $N_s = N\frac{\sqrt{P}}{H_{mano}^{5/4}}$	Flow co-efficient
	$\frac{Q}{D^3N}$ = constant $\frac{P}{D^5N^3}$ = constant	Power co-efficient
Cavitation of pumps	$H_{SV} = H_{atm} - H_S - H_V$ $\sigma = \frac{NSPH}{H_{mano}} = \frac{H_{SV}}{H_{mano}}$; $\sigma_c = 0.103(N_s/1000)^{4/3}$ $N_{ss} = \frac{N\sqrt{Q}}{H_{SV}^{3/4}}$; $\sigma = \left(\frac{N_s}{N_{ss}}\right)$	$H_{SV}, H_{atm}, H_S, H_V$ are NSPH, atmospheric head, suction head, vapour pressure head respectively, Thoma's caviation factor, Suction specific speed
Discharge, Force, work done of reciprocating pump	$Q_{th} = A*L*N/60$; $F = \gamma A(h_S + h_D)$ $WD_{th} = \frac{\gamma ALN(h_S + h_D)}{60}$ $P = \gamma Q_{th}H_{total} = \frac{\gamma ALN(h_S + h_D)}{60}$	
	$A_{eff} = \left[\frac{\pi}{4}D^2 + \left(\frac{\pi}{4}D^2 - \frac{\pi}{4}d^2\right)\right] = \left[2\left(\frac{\pi}{4}D^2\right) - \left(\frac{\pi}{4}d^2\right)\right]$ $Q_{th} = \frac{\left[2\left(\frac{\pi}{4}D^2\right) - \left(\frac{\pi}{4}d^2\right)\right]LN}{60}$ & WD_{th} $= \frac{\gamma\left[2\left(\frac{\pi}{4}D^2\right) - \left(\frac{\pi}{4}d^2\right)\right]LN(h_S + h_D)}{60}$	Piston rod area ("a") is not negligible in case of double acting reciprocating pump
Coefficient of discharge and slip of reciprocating pump	$C_d = \frac{Q_a}{Q_{th}}$; $\eta_{vol} = \%C_d$; $Slip = Q_{th} - Q_a$ $\%Slip = \frac{Q_{th} - Q_a}{Q_{th}}$	

Contd.

Property	**Formulae**	**Remarks, if any**
Piston acceleration and velocity, acceleration and friction in pipes	$V_p = r\omega(\sin\omega t)$ $Acc_p = r\omega^2(\cos\omega t)$ $v_{pipe} = \frac{A}{a_{pipe}}(V_p) = \frac{A}{a_{pipe}}(r\omega(\sin\theta))$ *and* $acc_{pipe} = \frac{d}{dt}(v_{pipe}) = \frac{A}{a_{pipe}} r\omega^2(\cos\theta)$ $h_f = \frac{flv^2_{pipe}}{2gd} = \frac{fl}{2gd} * \left[\frac{A}{a_{pipe}}(r\omega(\sin\theta))\right]^2$	
Acceleration pressure head	$h = \frac{p_i}{\gamma} \Rightarrow h = \left(\frac{A * l_{pipe}}{ga_{pipe}}\right) * r\omega^2(\cos\theta)$	
Separation pressure head	$H_{sep} = H_S + H_{as} = H_S + \left(\frac{A * l_s}{ga_s}\right) * r\omega^2$ $H_{sep} = H_{atm} + H_D + H_{ad} = H_{atm} + H_D + \left(\frac{A * l_d}{ga_d}\right) * r\omega^2$	
Maximum speed of reciprocating pump	$H_{sep} = H_S + \left(\frac{A * l_s}{ga_s}\right) * r\left(\frac{2\pi N}{60}\right)^2$	
Work done per sec., total frictional head in suction and delivery pipes	$WD_{sec} = \frac{\gamma ALN}{60}\left(H_S + H_D + \frac{2}{3}h_{fs} + \frac{2}{3}h_{fd}\right)$ $H_{ts} = H_{atm} - H_s - H_{as} - h_{fs}$ $H_{td} = H_{atm} + H_s + H_{as} + h_{fs}$	
Flow into/from air vessel and work saved due to air vessel	$Q_{airvessel} = Q_{cylinder} - Q_{mean} = Ar\omega\left(\sin\theta - \frac{1}{\pi}\right)$ $WD_{saved} = WD_{without} - WD_{with}$ $= 2\gamma(Ar)\left(\frac{fl}{2gd}\right)\left(\frac{A}{a}r\omega\right)^2\left(\frac{2}{3} - \frac{1}{\pi}\right)$	$Q_{airvessel} = Q_{cylinder} - Q_{mean}$ $= Ar\omega\left(\sin\theta - \frac{2}{\pi}\right)$ for double acting

Table 12.2: Points to remember

1. Minimum value of f_o is 20° and ranges from $90° - \phi_o - 20°$ and the corresponding value of manometric efficiency increases from 0.47 to 0.73. If the value of ϕ_o is reduced less than 20°, then we may have higher manometric efficiency but it creates long and narrow passages, which increases the frictional losses and hence not practicable.
2. Reciprocating pump is termed to have a negative slip if $(Q_a > Q_{th})$.
3. At $\theta = 0°$; $h_f = 0$ and similarly at $\theta = 90°$; $h_f = \frac{fl}{2gd} * \left[\frac{A}{a_{pipe}}(r\omega)\right]^2$ and so it is clear that the frictional losses will be minimum at starting and end point of a stroke, but it is maximum at the centre of the stroke.
4. An amount of 84.4% work can be saved from friction effect by using air vessels and thereby increase in usefulness of reciprocating pump.

Laboratory Experiments

EXPERIMENT 1: BERNOULLI'S THEOREM

AIM: To verify Bernoulli's theorem in a continuous liquid flow.

THEORY: Bernoulli equation is concerned with the conservation energies of a fluid stream and their conversion to each other in regions of fluid flow system where net viscous forces are negligible and where other restrictive conditions apply.

Bernoulli stated that in an ideal and incompressible fluid, the total energy of a fluid is constant throughout a streamline, if the flow is steady and continuous or in other words in continues liquid flow, the total head along the flow is the same that is pressure head $\left(p/_{\gamma}\right)$ and velocity head $\left(V^2/_{2g}\right)$ is constant at different points, i.e.

$$\frac{p}{\gamma}+\frac{1}{2g}V^2+Z=\text{Constant}$$

$$\frac{p_1}{\gamma_1}+\frac{1}{2g}V_1^2+Z_1=\frac{p_2}{\gamma_2}+\frac{1}{2g}V_2^2+Z_2$$

And if the fluid is real, then there will be some losses occurring due to friction, etc., then Bernoulli's equation for real fluids becomes

$$\frac{p_1}{\gamma_1}+\frac{1}{2g}V_1^2+Z_1=\frac{p_2}{\gamma_2}+\frac{1}{2g}V_2^2+Z_2+h_L$$

where h_L is a head loss.

Experimental setup: Bernoulli's apparatus consists of inlet supply chamber having a piezometric scale for indicating the flow head and the chamber is interlinked with a horizontal converging duct, with transparent sides made out of perspex sheets for purpose of the flow at its bottom.

The interlinking duct is smoothly varying in cross section, so that the velocity of flow, changes gradually for the purpose of absorbing the flow. Piezometers are provided at suitable regular intervals along the duct for indicating the piezometer heads $\left(\frac{p}{\gamma}+Z\right)$.

A collecting tank having a piezometric scale for indicating the fluid rise for a time period "t" is provided, below the duct outlet for collecting the fluid outflow form the duct and to discharge it to the recycleable tank for pumping up the fluid again to the inlet supply chamber with help of a pump.

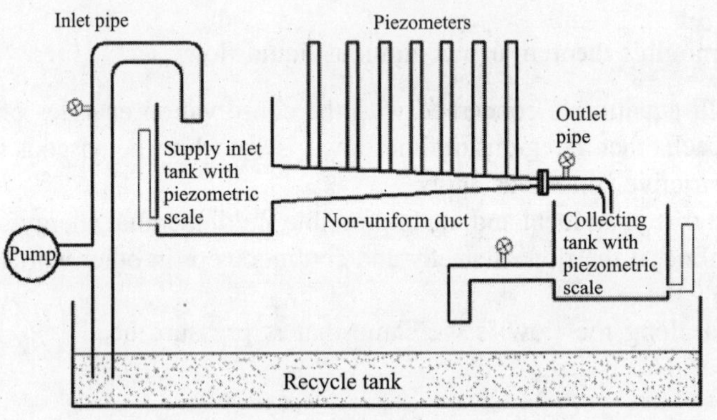

Bernoulli's experiment

Experimental Procedure:

Step 1: Note the distance of the piezometers on the duct from the supply chamber outlet end and also the cross-sectional area at the respective piezometer.

Step 2: Switch on the pump and regulate the valves of the inlet supply chamber so that the there is a steady constant supply head (H_o) in the inlet tank, with respect to some datum.

Step 3: At the steady constant supply head, note down the piezometric heads $\left(P/_\gamma + Z\right)$ in the piezometers on the duct also.

Step 4: Close the outlet valve of the collecting tank and note down the time taken "t" for a raise of fluid level "*h*" in the collecting tank, in seconds.

Step 5: Calculate the discharge "*Q*", velocity "*V*", Total head "*TH*" and Head loss "H_L" using the formulae and also velocities at different section, where the piezometers are located on the duct.

Step 6: Repeat step 2 to step 5, for different constant inlet heads.

Formulae used:

1. Discharge $Q = {}^{Ah}\!/_{t}$ ...m³/sec, where "*A*" is the area of the collecting tank.

2. Velocity $V = Q/a$ ......m/s, where "*a*" is the area of the duct at a particular location.

3. Velocity head $V^2/2g$ ....m

4. Pressure head $\left({}^{P}\!/_{\gamma} + Z\right)$ .....m

5. Total head (TH) = Pressure head $\left({}^{P}\!/_{\gamma} + Z\right)$ + Velocity head $V^2/2g$

6. Head loss (H_L) = Constant head loss (H_o) – Total head (TH)

Basic data:
Area of measuring tank '*A*' in m²
Rise in height of water level '*h*' in m
Area of cross-sections in the duct '*a*' in m²
Acceleration due to gravity '*g*' = 9.81 m/sec²

Tabulation:

S. No.	Area of the duct section "a" in m²	Pressure head at the sections $(P/\gamma + Z)$	Discharge measurement			Velocity '*V*' (m/sec)	Velocity head $V^2/2g$ (m)	Total head (*TH*) (m)	Head loss (H_L) (m)
			Height raised "*h*" in (m)	Time taken for raised of fluid level "*t*" in (sec)	Discharge "*Q*" in (m³/sec)				

Observations and Results:

Precautions:

1. Readings must be taken only after the flow from the supply inlet tank is steady, so that there are no fluctuations in the flow.
2. There should be no air bubbles in the piezometer tubes located on the duct and also the meniscus should be minimized.

Model calculation: (Exp no: ____)

Basic data:

Constant head loss (H_o) = ______________________ m

Area of measuring tank **A** = L * B= ______________________ m^2

Rise in height of water level **h** = ______________________ m

Time taken for "*h*" raise of water level "***t***"= ______________ sec

Area of cross-section of the duct at the piezometer location:
a = l * b = ______________________ m^2

Pressure head $\left(P/_\gamma + Z \right)$ = ______________ m

Calculations:

Discharge $Q = Ah/_t$ = ______________________ ...m^3/sec

Velocity of flow $V = Q / a$ = ______________________ ......m/s

Velocity head $V^2 / 2g$ = ______________________ ....m

Total head (TH) = Pressure head $\left(P/_\gamma + Z \right)$ + Velocity head $V^2 / 2g$

=

= ______________________ m

Head loss (H_L) = Constant head loss (H_o) – Total head (TH)

=

= ______________________ m

EXPERIMENT 2: NOTCH-WEIRS

AIM: To determine the coefficient of discharge of a notch-weir.

THEORY: A **weir** also known as a **low head dam** and is a small overflow-type barrier constructed across a waterway, commonly used to raise the level or to control the flow of water of a river or stream and can be of a Broad-crested weir, Sharp crested weir, Combination weir, V-notch-weir, rectangular weir, Trapezoidal weir, etc.

A notch is an opening in the weirs, through which water flows and they may be of triangular, rectangular or trapezoidal in shapes. The bottom edge of a notch is called **sill** of the notch and the height of water above the sill is called the **head over of the notch**. Notches are used to measure the

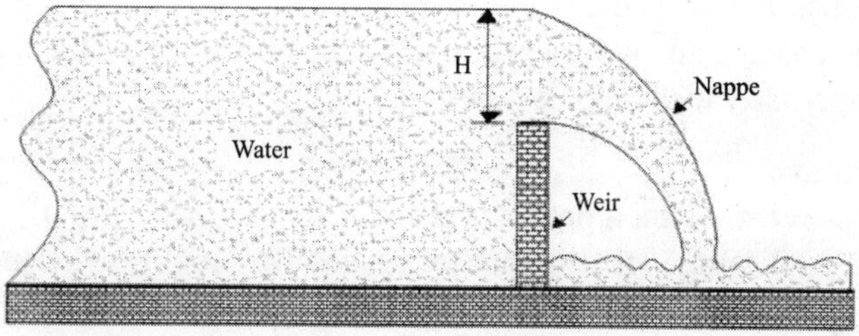

discharge in streams and channels and are generally used in practice to measure small free-surface flows.

For theoretical discharge of different weirs will be varying and is accordingly followed:

For **Rectangular weir**: $$Q_{th} = \frac{2}{3} bH\sqrt{2gH}$$

For **Triangular weir**: $$Q_{th} = \frac{8}{15} H^2 \tan\left(\theta/2\right)\sqrt{2gH}$$

Experimental setup: Notch-weir apparatus consists of a channel having a supply tank at one end and discharge chamber at other end. It has a buffer zone after the supply tank in the channel, for making the flow to smooth and has smooth surface, while passing through the channel. It has a weir plate with a notch (changeable) fixed or fitted across the channel. The flow over the notch falls in a collecting tank through the discharge chamber, for discharge measurements. A hook gauge with scale (sliding mist) with a pointer fitted over the channel to measure the head over the sill.

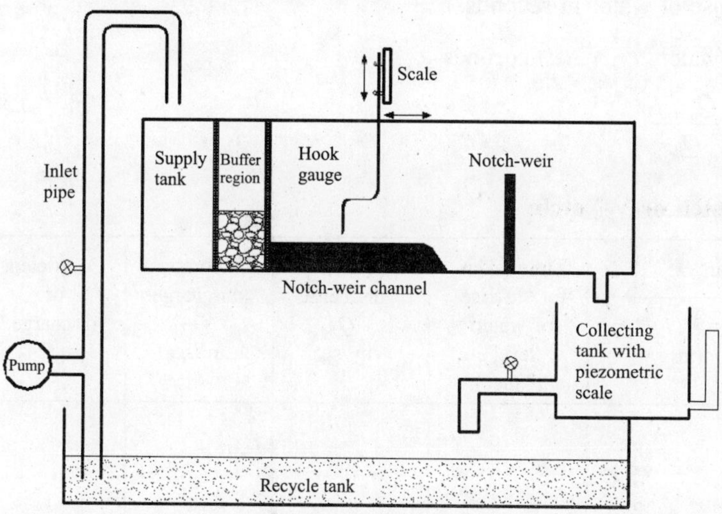

Notch-weir experiment

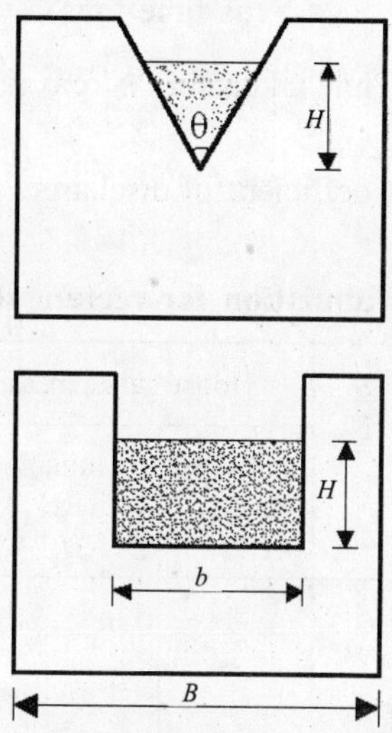

Side view of notch-weirs

A collecting tank has a piezometric scale for indicating the fluid rise for a time period "*t*", and is below the outlet of the channel and it discharge the water into the recyclabletank for pumping up the fluid again to the supply tank with help of a pump.

Experimental Procedure:

Step 1: Fix the notch-weir (V-Notch or rectangular or trapezoidal).

Step 2: Measure the apex angle 'θ' (in case of V-Notch), length of the notch "*b*" (in case of rectangular notch).

Step 3: Open the inlet valve to fill the supply tank and thereby water flows through the baffle region and to then notch, for discharge into the collecting tank.

Step 4: Allow the water level rises up above the sill level and becomes steady.

Step 5: Take the hook gauge scale reading, which is the height of the flow "*H*".

Step 6: Note the time taken "*t*" for known rise of water "*h*" in the collecting tank.

Step 7: Calculate the actual discharge Q_a.

Step 8: Repeat the steps for various water level height "*H*".

Basic data and formulae used:

If **V-Notch-weir**: Apex angle of V-Notch "θ" .... degrees

If **Rectangular notch**: Length of the notch "*b*" .... m

Area of the collecting tank A .....m²

Actual discharge

$$Q = A * h/t \quad units \ m^3/s$$

where *h* is manometric column raise in meters
t is time for "*h*" cm raise of water in seconds.

Time taken for "*h*" cm rise of water "*t*"........ seconds

Coefficient of discharge, $C_d = \dfrac{Q_{actual}}{Q_{th}}$

Tabulation for rectangular notch or V-Notch:

S No.	Hooke gauge measurement "*H*"			Time taken for "h" rise of water level "*t*" (sec)	Actual discharge 'Q_a' m³/sec	Theoretical discharge 'Q_{th}' m³/sec	Coefficient of discharge $C_d = Q_a/Q_t$
	Initial level "H_1"	Initial level "H_2"	$H = H_2 - H_1$ (m)				

GRAPHS:

1. Q_a vs. Q_{th}
2. H vs. Q_a

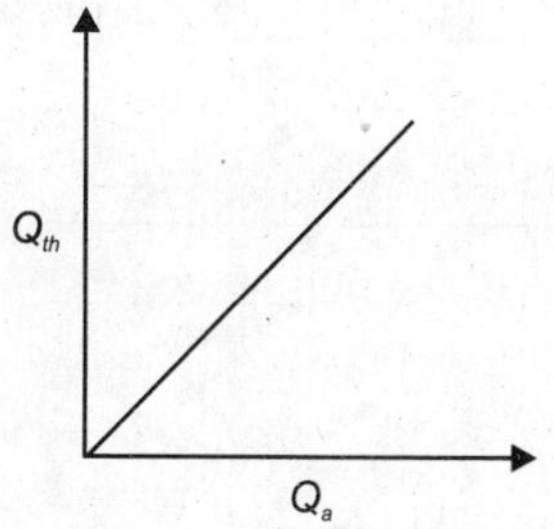

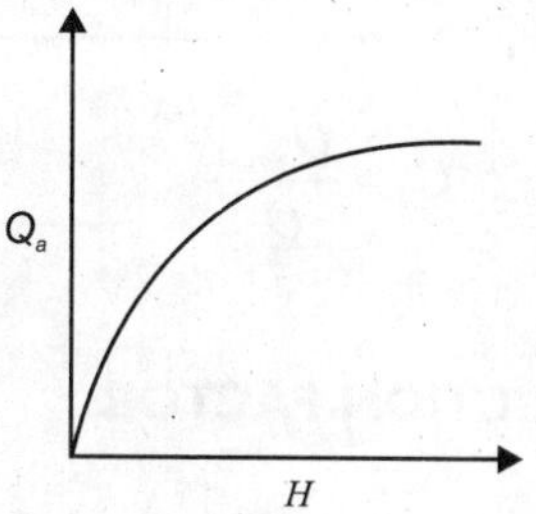

Precautions:

1. Hook gauge readings must be taken when the flow is steady and calm.
2. Make sure the Nappe is well ventilated by providing larger discharge chamber.

Observations and Results:

Model calculation: (Exp no: ____)

Basic data:

If **V-Notch-weir**: Apex angle of V-Notch θ = ________________ degrees

If **Rectangular notch**: Length of the notch b = ________________ m

Area of measuring tank **A** = L * B= ________________ m^2

Rise in height of water level **h** = ________________ m

Time taken for "h" raise of water level "t" = ________________ sec

Calculations:

Hook gauge reading: $H = H_2 - H_1$ = ________________ m

Actual discharge $Q = {Ah}/{t}$ = ________________ ...m^3/sec

Theoretical discharge:

For Rectangular weir: $Q_{th} = \frac{2}{3} bH\sqrt{2gH}$ = ________________

= ________________ m^3/sec

For V-Notch-weir: $Q_{th} = \frac{8}{15} H^2 \tan\left(\theta/_2\right)\sqrt{2gH}$ = ________________________

= ________________________...... m³/sec

Coefficient of discharge: $C_d = \frac{Q_{actual}}{Q_{th}}$ = ________________________

EXPERIMENT 3: FRICTION FACTOR

AIM: To determine the friction factor "f" for different pipes.

EQUIPMENT:

1. Pipe friction experimental set up.
2. Collecting tank with piezometer scale arrangement.
3. Stopwatch.

THEORY: Pipe flow is an important study in domestic water supply, flow between reservoirs, flow of oils through pipelines, etc., considering the losses for effective flow potential. Closed conduits are employed whenever fluids are to be carried out from one point to another under pressures different from atmospheric pressure. The commonly employed conduit for this purpose is circular pipe.

The static pressure drops along the flow due to loss in energy in over coming friction as flow occurs and so frictional head loss is termed as major loss. However, it is also depended on roughness

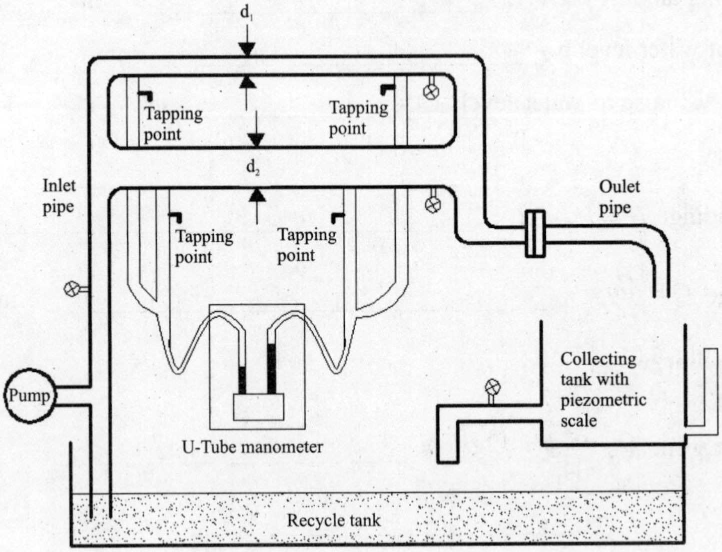

Friction factor experiment.

of the pipe, geometrical factors like varying size and shape of projections and velocity with which the fluid flows.

The value of friction factor "f" is determined experimentally for a pipeline of given material by measuring the head loss due to friction using **Darcy Weisbach** equation

$$h_f = \frac{fLu_{AV}^2}{2gD}$$

where h_f = head loss due to friction, "f" = Darcy's friction factor
L = length of the pipe, u_{AV} = average velocity of flow through pipe
d = diameter of the pipe, g = acceleration due to gravity

In practice, Moody's charts are used to find the value of the friction factor, thereby getting head loss due to friction. But a turbulent flow depends mostly on the roughness projections, which in turn depending on the usage and period of usage, which will be varying and is difficult to predict the friction factor.

Experimental setup: Friction factor consists of a pipe line with a motor pump to pump fluid from the recycle tank and is connected to two different sizes of parallel pipes.

Both these pipes have regulating valves at their ends for opening or closing the respective pipe and also they have an extension of their outlet into the collecting tank. Both these pipes have two tapping point on their lengths and are connected to a common U-Tube manometer for measuring their pressure head, by opening and closing of the respective tapping points, as shown in the figure.

A collecting tank having a piezometric scale for indicating the fluid rise for a time period "t" is provided, below the outlet of the two pipes for collecting the fluid outflow and to discharge it to the recyclable tank for pumping up the fluid again to the pipeline with help of a pump.

Experimental Procedure:

Step 1: Select the required pipe (d_1 or d_2) through which the experiment to be conducted. Open its valve to allow the fluid flow through it and close the valve of the other pipe to stop flow in it.

Step 2: Note down the reading of mercury level in the two limbs of manometer to calculate the head loss due to friction.

Step 3: Note the time taken "t" for known rise of water "h" in the collecting tank.

Step 4: Repeat the same procedure for a pipe selected with different manometric level differences by regulating the valve of the pipeline.

Step 5: Repeat step 1 to 4 for the other pipe also.

Basic data and formulae used:

Density and absolute viscosity of water $\rho_{water} = 1000 \text{ kg/m}^3$ & $\mu_{water} = 1000 \text{ kg/m}-\text{s} = 1 \text{ cp}$

Diameter of the pipes 1 and 2 d_1 & d_2 ... m

Length between the tapping points of pipe 1 and 2 L_1 & L_2 ... m

Area of the collecting tank A ... m^2

Discharge $Q = A * h/t$ units m^3/s

where h is manometric column raise in meters
t is time for "h" cm raise of water in seconds

Velocity $u_{AV} = Q/a$ units m/s where a is area of pipe 1 or 3 in meters

Difference in manometric levels $H = H_2 - H_1$ ….m

If the manometric liquid is heavier than the liquid flowing, then head loss due to friction is

$$h_f = H\left(\frac{S_m}{S_l} - 1\right) \text{ or else } h_f = H\left(1 - \frac{S_l}{S_m}\right)$$

where S_m is manometric liquid specific gravity
S_l is specific gravity of liquid flowing through pipe

Friction factor $f = \dfrac{2gd}{Lu_{AV}^2} h_f$

Reynolds number $R_e = \dfrac{\rho u_{AV} d}{\mu}$

where μ is absolute viscosity of the folowing fluid in Ns/m^2

PRECAUTIONS:

1. Collecting tank reading should not be taken when the water level does not fluctuate in the piezometer.
2. When one pipe is been experimented, the other pipe must be closed.

Tabulation:

S. No.	Manometer reading			Discharge measurement			Velocity V (m/sec)	Friction factor f	Reynolds number R_e
	Height in left column "H_1" in (m)	Height in right column "H_2" in (m)	Difference in manometer levels "H" (m)	Height raised "h" in (m)	Time taken for "h" cm raise of fluid level "t" in (sec)	Discharge "Q" (m^3/sec)			

GRAPHS:

1. V vs. h_f
2. $V^2/2g$ vs. h_f
3. f vs. R_e

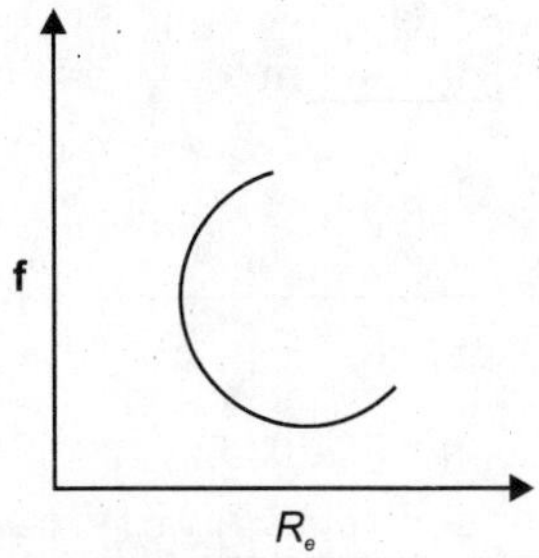

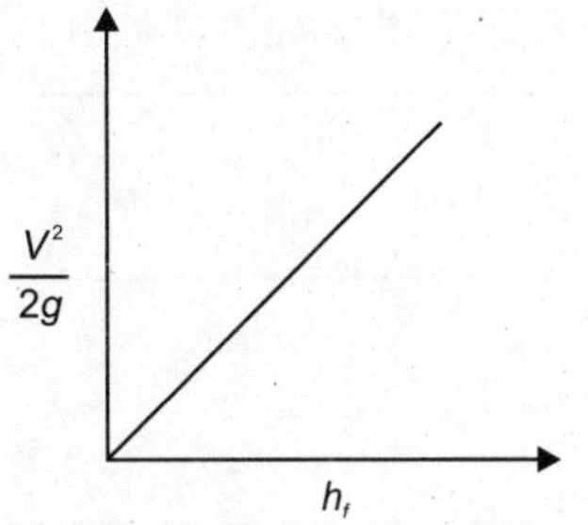

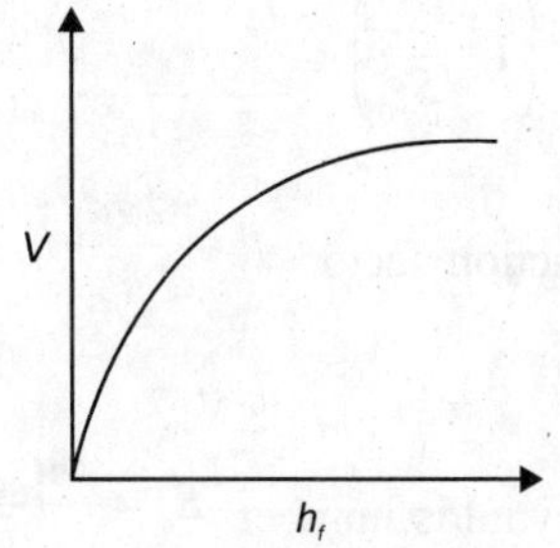

Observations and Result:

Model calculation: (Exp no: ____)

Basic data:

Density and absolute viscosity of water $\rho_{water} = 1000 \text{ kg/m}^3$ & $\mu_{water} = 1000 \text{ kg/m} - \text{s} = 1 \text{ cp}$

Diameter of the pipe 1 d_1 = ______________________________m

Diameter of the pipe 2 d_2 = ______________________________m

Length between the tapping points of pipe 1 L_1 = ______________________m

Length between the tapping points of pipe 2 L_2 = ______________________m

Area of measuring tank **A** = L * B= ______________________________ m^2

Rise in height of water level **h** = ___________________________________ m

Time taken for "*h*" raise of water level "**t**"= ____________________sec

Calculations:

Actual discharge $Q = Ah/t$ = ______________________________________…m^3/sec

Velocity $u_{AV} = Q/a$ = ____m/s

Difference in manometric levels $H = H_2 - H_1$ = ____________________________m

If the manometric liquid is heavier than the liquid flowing, then head loss due to friction is

$$h_f = H\left(\frac{S_m}{S_l} - 1\right) = \_\_\_\_\_\_\_\_\_\_\_\_\_\_\_\_\_\_\_\_ \text{m}$$

If the manometric liquid is lighter than the liquid flowing, then head loss due to friction is

$$h_f = H\left(1 - \frac{S_l}{S_m}\right) = \_\_\_\_\_\_\_\_\_\_\_\_\_\_\_\_\_\_\_\_ \text{m}$$

Friction factor $f = \frac{2gd}{Lu_{AV}^2} h_f =$ ____________________

Reynolds number $R_e = \frac{\rho u_{AV} d}{\mu} =$ ____________________

EXPERIMENT 4: MINOR LOSSES

AIM: To determine the minor losses in a pipe.

EQUIPMENT:

1. Minor losses experimental set up.
2. Collecting tank with piezometer scale arrangement.
3. Stopwatch.

THEORY: When a liquid flows from one point to another point, it comes across different types of losses depending on the shape and geometry of the container through which it is traveling in. Hence, a real fluid flow is to take care of all the head losses arising while in motion like losses due to **friction, curved bends (elbows), sudden expansion and sudden contraction** of the pipeline. The major head loss is due to friction which is determined by the Darcy Weisbach equation

$h_f = \frac{fLu_{AV}^2}{2gD}$ and other remaining losses can be treated as minor.

The minor losses are expressed as

$$h_L = K\frac{(V)^2}{2g},$$

where K is loss factor depend on obstruction type

V is mean velocity of flow

If loss is due to **sudden expansion**, then head loss

$$h_E = K_E\frac{(V_1)^2}{2g}$$

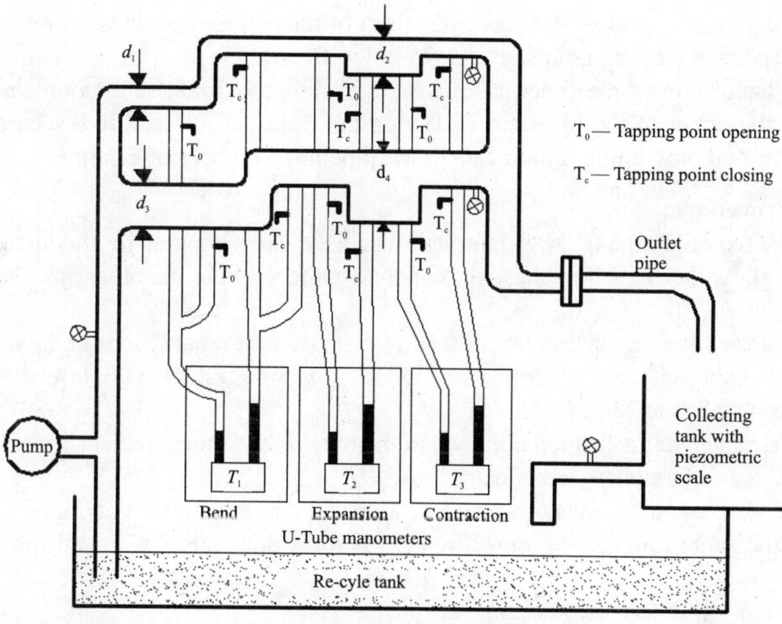

Minor losses experiment.

where V_1 smaller pipe velocity
V_2 bigger pipe velocity

If the loss is due to gradual expansion (diffuser), then the head loss $h_E = K_E \dfrac{(V_1)^2}{2g}$ where K_E depends on ratio of d_1/d_2 and angle of divergence.

If loss is duc to **sudden contraction**, then head loss $h_C = K_C \dfrac{(V_1)^2}{2g}$ where $K_C = \left(\dfrac{1}{C_C} - 1\right)^2$ and termed as loss coefficient.

If the loss is due to **bend**, then the head loss $h_B = K_B \dfrac{(V)^2}{2g}$ where $K_B = 0.6 to 0.9$ depends on ratio of radius of curvature of the bend and diameter of the pipe $\left(r/D\right)$ and angle of bend (90^0).

Experimental setup: Minor losses apparatus consists of a pipeline with a motor pump to pump fluid from the recycle tank and is connected to different sizes of parallel pipes having bends, sudden expansions and sudden contractions.

All these pipes have regulating valves at their ends for opening or closing the respective pipe and also they have an extension of their outlet into the collecting tank. Each pipe have two tapping points at every form of minor loss occurrences and are connected to a common U-Tube manometer for measuring the pressure head at that particular form of minor loss occurrences, by opening and closing of the respective tapping points, as shown in the figure.

A collecting tank having a piezometric scale for indicating the fluid rise for a time period "*t*" is provided, below the outlet of the pipes for collecting the fluid outflow and to discharge it to the recyclable tank for pumping up the fluid again to the pipeline with help of a pump.

Experimental Procedure:

Step 1: Select the required pipe (d_1 or d_2) through which the experiment to be conducted. Open its valve to allow the fluid flow through it and close the valve of the other pipes to stop flow in it.

Step 2: Select the required minor loss form (bend or expansion or contraction) in the pipe through which the experiment to be conducted. Open its tapping points and close the remaining other form of minor losses.

Step 3: Note down the reading of mercury level in the two limbs of manometer (T_1 or T_2 or T_3) to calculate the difference in manometric levels H

Step 4: Note the time taken "*t*" for known rise of water "*h*" in the collecting tank.

Step 5: Repeat the same procedure (step 2 to step 4) for a pipe selected for all the minor loss forms.

Step 6: Repeat step 1 to 5 for the other pipe also.

Basic data and formulae used:

Diameter of the pipes 1, 2, 3 and 4 d_1, d_2, d_3 and d_4 … m

Area of the collecting tank A … m^2

Discharge $$Q = A * h/t \text{ units m}^3\text{/s}$$

where h is manometric column raise in meters
t is time for "*h*" cm raise of water in seconds

Velocity $$V_{1 \, or \, 2 \, or \, 3 \, or \, 4} = Q / a \text{ units } m/s$$

where a is area of pipe 1 or 2 or 3 or 4 in meters

Difference in manometric levels $H = H_2 - H_1$ ….m

If the manometric liquid is heavier than the liquid flowing, then lead loss due to minor loss forms is

$$h_L = H\left(\frac{S_m}{S_l} - 1\right) \text{ or else } h_L = H\left(1 - \frac{S_l}{S_m}\right)$$

where S_m is manometric liquid specific gravity
S_l is specific gravity liquid flowing through pipe

Bend loss coefficient $K_B = \dfrac{2gh_B}{(V)^2}$, Expansion coefficient $K_E = \dfrac{2gh_E}{(V_1)^2}$

Contraction loss coefficient $K_C = \dfrac{2gh_C}{(V_1)^2}$ and $K_C = \left(\dfrac{1}{C_C} - 1\right)^2$

Tabulation for curved bend losses:

S. No.	Manometer reading			Discharge measurement			Velocity 'V_1' (m/sec)	Velocity 'V_1' (m/sec)	Loss Co-efficent K_B
	Height in left column "H_1" in (m)	Height in right column "H_2" in (m)	Difference in manometer levels "H" (m)	Height raised "h" in (m)	Time taken for "h" cm raise of fluid level "t" in (sec)	Discharge "Q" (m³/sec)			

Tabulation for sudden expansion losses:

S. No.	Manometer reading			Discharge measurement			Velocity 'V_1' (m/sec)	Velocity 'V_1' (m/sec)	Loss Co-efficent K_E
	Height in left column "H_1" in (m)	Height in right column "H_2" in (m)	Difference in manometer levels "H" (m)	Height raised "h" in (m)	Time taken for "h" cm raise of fluid level "t" in (sec)	Discharge "Q" (m³/sec)			

Tabulation for sudden contraction losses:

S. No.	Manometer reading			Discharge Measurement			Velocity 'V_1' (m/sec)	Velocity 'V_2' (m/sec)	Loss Co-efficient K_B
	Height in left column "H_1" in (m)	Height in right column "H_2" in (m)	Difference in manometer levels "H" (m)	Height raised "h" in (m)	Time taken for "h" cm raise of fluid level "t" in (sec)	Discharge "Q" (m³/sec)			

PRECAUTIONS:

1. Collecting tank reading should not be taken when the water level does fluctuate in the piezometer.
2. When one pipe is been experimented, the other pipe must be closed.
3. All other tapping points, except the minor loss form are to be closed.

Observations and Result:

Model calculation: (Exp no: ____)

Basic data:

*Diameter of the pipe 1 d_1 = ______________________ m

*Diameter of the pipe 2 d_2 = ______________________ m

*Diameter of the pipe 3 d_3 = ______________________ m

*Diameter of the pipe 4 d_4 = ______________________ m

Area of measuring tank **A** = L * B= ______________________ m²

Rise in height of water level **h** = ______________________ m

Time taken for "h" raise of water level "**t**"= ______________ sec

Calculations:

Actual discharge $Q = Ah/t$ = ______________________________ ...m³/sec

Velocity $u_{AV} = Q/a$ =____m/s

Difference in manometric levels $H = H_2 - H_1$ =____________________m

If the manometric liquid is heavier than the liquid flowing, then head loss due to friction is

$$h_f = H\left(\frac{S_m}{S_l} - 1\right) = \_\_\_\_\_\_\_\_\_\_\_\_\_\_\_\_\_\_\_\_\_\_\_\_\_\_\_\_\_\_\_\_\_\ \text{m}$$

If the manometric liquid is lighter than the liquid flowing, then head loss due to friction is

$$h_f = H\left(1 - \frac{S_l}{S_m}\right) = \_\_\_\_\_\_\_\_\_\_\_\_\_\_\_\_\_\_\_\_\_\_\_\_\_\_\_\_\_\_\_\_\_\ \text{m}$$

*Bend loss coefficient $K_B = \dfrac{2gh_B}{(V)^2}$ =______________________________m

*Expansion coefficient $K_E = \dfrac{2gh_E}{(V_1)^2}$ =______________________________m

*Contraction loss coefficient $K_C = \dfrac{2gh_C}{(V_1)^2}$ =_________________________m

*Coefficient of contraction $K_C = \left(\dfrac{1}{C_C} - 1\right)^2$ =_________________________

*** Wherever applicable.**

EXPERIMENT 5: MOUTHPIECE

AIM: To determine the coefficient of discharge of a mouthpiece.

EQUIPMENT:

1. A supply tank with piezometric scale and attached with an external mouthpiece.
2. Collecting tank with piezometric scale.
3. Stopwatch.

THEORY: Mouthpiece is a short pipe protruding from the side of a tank, whose length is at least three times the diameter of the pipe and should be short enough to be able to neglect friction losses.

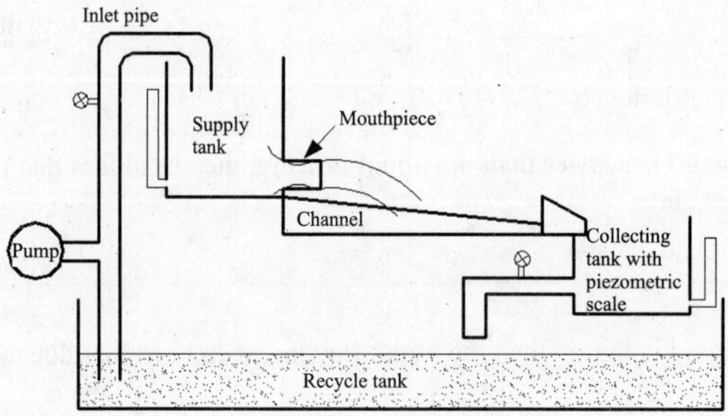

Mouthpiece experiment

When water flows through the mouthpiece, at the inlet of the mouthpiece, there is a sudden contraction of flow up to some length and then flow expands to occupy full diameter of the mouthpiece. Due to the sudden contraction and sudden expansion of the flow, there arise some loss of energy and hence the actual discharge becomes lesser than the theoretical, which give rise to the coefficient of discharge.

As the flow separates only in the mouthpiece for a short distance and while coming out it attains the diameter of the mouthpiece, the contraction coefficient is minimized. Hence, coefficient of contraction C_c in the case of mouthpiece is equal to unity. In other words, there is no contraction at the point where the jet emerges out form the mouth piece.

Due to the separation of flow in the initial stages of mouthpiece, the expansion process is associated with eddies having separation and reverse flow, thereby decreasing the theoretical velocity of the flow by an amount of coefficient of velocity C_v.

Theoretical discharge consists of the coefficient due to contraction and also coefficient of velocity and hence $Q_{th} = C_C * C_V * a * V = C_C * C_V * a\sqrt{2gH}$, where "*a*" area of cross-section of the mouthpiece. And substituting $C_c = 1$ and knowing the coefficient of discharge $C_d = C_C * C_V$.

So the theoretical discharge becomes $Q_{th} = C_d * a\sqrt{2gH}$.

Note: In case an **orifice** C_c exists in small and C_v is close to unity but in case of mouthpiece $C_c = 1$, C_v exists in small and is less than unity.

Experimental setup: Mouthpiece apparatus consists of a supply tank with piezometric scale and has an opening at one of its side. An external mouthpiece of some diameter and length, fitted at this opening to let the water from the supply tank to discharge out. And the discharged water is passed through a sloping down channel into a collecting tank, as shown in the figure.

The collecting tank has a piezometric scale for indicating the fluid rise for a time period "*t*", and is below the outlet of the sloping down channel and it discharge the water into the recyclabletank for pumping up the fluid again to the supply tank with help of a pump.

Experimental Procedure:

Step 1: Open the inlet pipe and let water enter the supply tank and adjust the head of water to a constant value in the piezometric scale attached to the tank and note the reading as "H".

Step 2: Collect "h" m of water in the collecting tank and note the time of collection "t" seconds and drain of the water in the collecting tank.

Step 3: Increase the supply tank head "H" and repeat steps 1 and 2, at least for 5 more times.

Basic data and formulae used:

Diameter of mouthpiece "d"........ m

Area of the collecting tank A .....m^2

Discharge $$Q = A * h/t \text{ units m}^3/\text{s}$$

where h is manometric column raise in meters

t is time for "h" cm raise of water in seconds

Velocity $V = Q/a$ Units m/s

where a is area of mouthpiece in meter square

Time taken for "h" cm rise of water 't' in seconds

Actual discharge $Q_a = Ah/t$ ....m^3/sec

Theoretical discharge $Q_{th} = a\sqrt{2gH}$ ... m^3/sec

Coefficient of discharge $$C_d = \frac{Q_a}{Q_{th}} = \frac{Ah/t}{a\sqrt{2gH}}$$

Tabulation for mouthpiece:

S. No.	Liquid head in supply tank "H" (m)	Time for "h" cm rise of water level "t" (sec)	Actual discharge Q_a (m^3/sec)	Theoretical discharge Q_t (m^3/sec)	Coefficient of discharge C_d
1.					
2.					
3.					
4.					
5.					

GRAPHS:

1. H vs Q_a
2. Q_{th} vs. Q_a

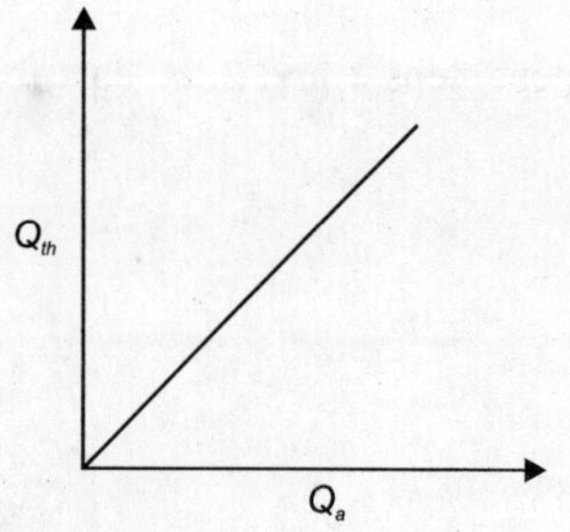

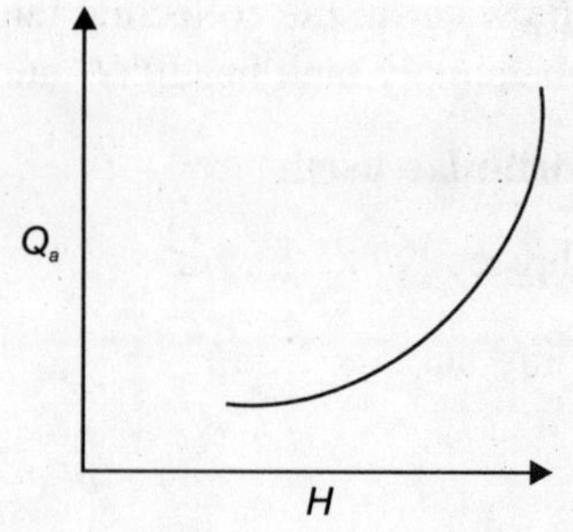

Observations and Result:

Model calculation: (Exp no: ____)

Basic data:

Diameter of mouthpiece d = ______________________ m

Liquid head in supply tank H = ______________________ m

Area of measuring tank **A** = L * B= ______________________ m^2

Rise in height of water level **h** = ______________________ m

Time taken for "*h*" raise of water level "**t**" = ______________ sec

Calculations:

Actual discharge $Q = {Ah}/{t}$ = ______________________ …m^3/sec

Velocity $u_{AV} = Q/a$ = ______________________ m/s

Theoretical discharge $Q_{th} = a\sqrt{2gH}$ = ______________________ m^3/sec

Coefficient of discharge $C_d = \dfrac{Q_a}{Q_{th}}$ = ______________________

EXPERIMENT 6: DISCHARGE THROUGH ORIFICE

AIM: To determine the coefficient of discharge of an orifice.

EQUIPMENT:

1. A supply tank with piezometric scale and attached with an orifice.
2. Collecting tank with piezometric scale.
3. Stopwatch.

THEORY: Orifice is an opening in the wall or at bottom of a tank/reservoir to discharge the fluid from a tank/reservoir. It may also be an opening in a plate normal to the axis of a pipe the plate being at the end of the pipe or in intermediate location. The thickness of the wall or plate is very small relative to the size of the opening. Orifices are generally round, though there are square, triangular and rectangular ones they are unusual.

As the flow discharges from an orifice, the flow gets contracted to smaller dimension and the maximum contraction takes place slightly downstream side of the orifice, where the flow is more or less horizontal. This contracted section is termed as "**Vena-contracta**". The ratio of the diameter of the orifice jet d_J and diameter of the vena-contracta region d_V is termed as "**Coefficient of contraction**", i.e. $C_C = {}^{d_V}/_{d_J}$.

Application of Bernoulli's equation between the free surface of the supply tank and the center of the orifice yields: $V = a\sqrt{2gH}$. Substituting $C_V = 1$ and knowing the coefficient of discharge $C_d = C_C * C_V$ theorctical dischargc becomes $Q_{th} = C_d * a\sqrt{2gH}$.

Note:

1. In case an **orifice** C_c exists in small and C_v is close to unity but in case of mouthpiece C_c = 1, C_v exists in small and is less than unity.
2. The value of coefficient of discharge can be found by finding the C_C & C_V values also.

Experimental setup: Orifice apparatus consists of a supply tank with piezometric scale and has an opening at one of its side. An orifice of some diameter, fitted at this opening to let the water from the supply tank to discharge out. And the discharged water is passed through a sloping down channel into a collecting tank, as shown in the figure.

The collecting tank has a piezometric scale for indicating the fluid rise for a time period "*t*", and is below the outlet of the sloping down channel and it discharge the water into the recyclable tank for pumping up the fluid again to the supply tank with help of a pump.

Experimental Procedure:

Step 1: Open the inlet pipe and let water enter the supply tank and adjust the head of water to a constant value in the piezometric scale attached to the tank and note the reading as "*H*".

Step 2: Collect "*h*" m of water in the collecting tank and note the time of collection "*t*" seconds and drain of the water in the collecting tank.

Step 3: Increase the supply tank head "*H*" and repeat steps 1 and 2, at least for 5 more times

Basic data and formulae used:

Diameter of orifice "*d*"........ m

Area of the collecting tank *A* .....m^2

Actual discharge $Q = A * h/t$ units m^3/s

where h is manometric column raise in meters
t is time for "h" cm raise of water in seconds

Velocity $V = Q / a$ units m/s

where a is area of mouthpiece in meter square

Time taken for "h" cm rise of water 't' in seconds

Actual discharge $Q_a = Ah/t$ ....…m^3/sec

Theoretical discharge $Q_{th} = a\sqrt{2gH}$ … m^3/sec

Coefficient of discharge $C_d = \frac{Q_a}{Q_{th}} = \frac{Ah/t}{a\sqrt{2gH}}$

Tabulation:

S.No.	Liquid head of supply tank "H"	Time for "h" rise of water level "t" (sec)	Actual discharge Q_a (m^3/sec)	Theoretical discharge Q_t (m^3/sec)	Coefficient of discharge C_d
1.					
2.					
3.					
4.					
5.					

Graphs:

1. H vs Q_a
2. Q_{th} vs Q_a

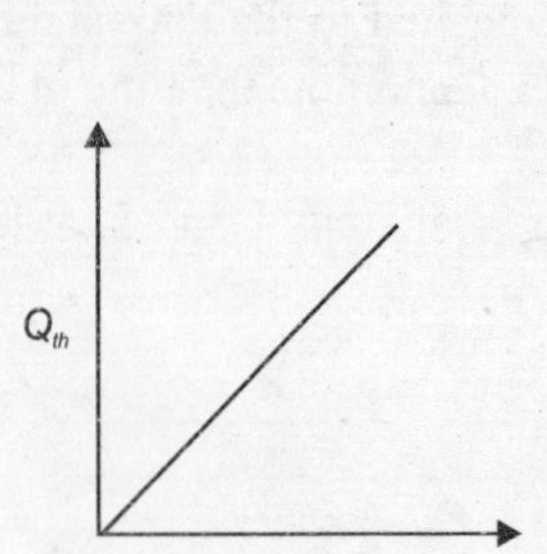

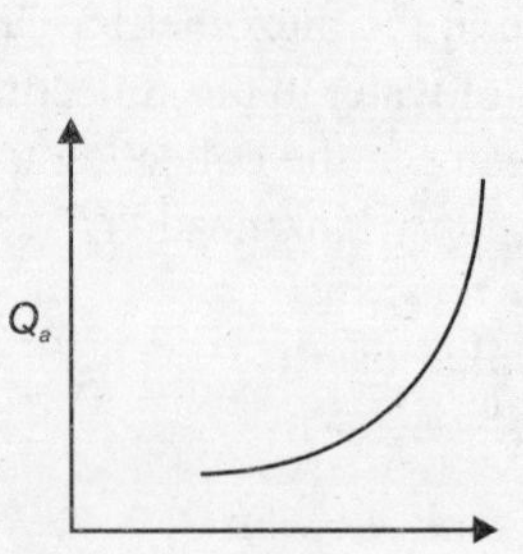

PRECAUTIONS:

1. The head should be kept constant till one complete set of readings is taken.
2. No readings should be taken while the water level in balancing tank fluctuates.

Observations and Results:

Model calculation: (Exp no: ____)

Basic data:

Diameter of orifice d = ____________________ m

Supply head H = ____________________ m

Area of measuring tank **A** = L * B = ____________________ m²

Rise in height of water level **h** = ____________________ m

Time taken for "*h*" raise of water "**t**" = ______________ sec

Calculations:

Area of orifice $a = \pi d_o^2 / 4$ = ____________________ m^2

Actual discharge $Q = Ah/t$ = ____________________ m^3/sec

Theoretical discharge $Q_{th} = a\sqrt{2gH}$ = ____________________ m^3/sec

Coefficient of discharge $C_d = \dfrac{Q_a}{Q_{th}}$ = ____________________

=

EXPERIMENT 7: VENTURIMETER

AIM: To determine the coefficient of discharge of a venturimeter.

EQUIPMENT:

1. Venturimeter experimental set up.
2. Collecting tank with piezometer and gate valve.
3. Stopwatch.

THEORY: Venturimeter is a closed conduit flow-measuring device used to measure the rate of discharge in a pipeline and has been named after Italian Engineer **Venturi**. It consists of three parts, namely, a short converging part, a throat followed by a diverging part (as shown in figure). The diameter of the converging and diverging portions at their inlet and outlet respectively has the same diameter as that of the pipeline, to which venturimeter is fitted.

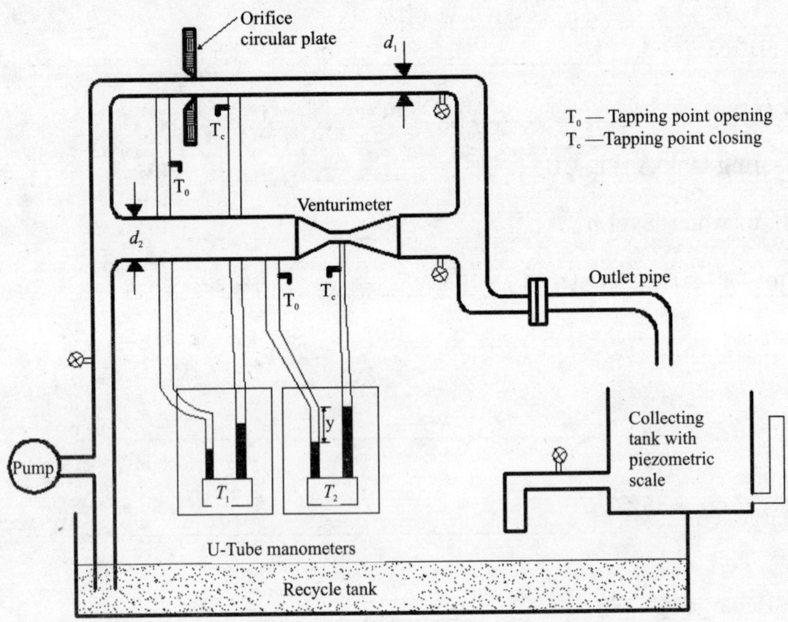

Venturi and orificemeter experiment

Generally, the length of the diverging section is 2 to 3 times the diameter of the converging cone. The throat to pipe diameter ratio (d_2/d_1) should be such that the pressure of the following fluid avoids cavitation. The throat ratio generally will be 1/2 to 1/3.

Applying the Bernoulli's equation between the inlet of the convergent and middle of the throat and introducing the continuity equation, we get the discharge from the venturimeter as: $Q = C * \sqrt{H}$

where
$$C = \frac{A_1 A_2}{\sqrt{A_1^2 - A_2^2}} * \sqrt{2g}$$

and is called as constant of venturimeter and has "A_1" and "A_2" are the cross-sectional areas of the pipe at inlet and throat respectively. "*H*" is the difference in pressure head between the inlet and throat in terms of head of fluid flowing through the meter.

Note:

1. If Venturimeter is containing a **heavier manometric liquid** when compared to the liquid passing in the venturimeter, then $H = y\left[\frac{S_{ML}}{S_{LP}} - 1\right]$; where y, S_{ML}, S_{LP} are differential gauge reading, specific gravities of manometric liquid and liquid in the pipe respectively.
2. If Venturimeter is containing a **lighter manometric liquid** when compared to the liquid passing in the venturimeter then $H = y\left[1 - \frac{S_{LP}}{S_{ML}}\right]$; where y, S_{ML}, S_{LP} are differential gauge reading, specific gravities of manometric liquid and liquid in the pipe respectively.

Experimental Procedure:

Step 1: Select the required pipe (d_1 or d_2) through which the experiment to be conducted. Open its valve to allow the fluid flow through it and close the valve of the other pipes to stop flow in it.

Step 2: Switch on the pump and let the water flow through the venturimeter in the pipe. Open its tapping point at the inlet and outlet.

Step 3: Note down the reading of mercury level in the two limbs of manometer (T_1 or T_2) to calculate the difference in manometric levels "y" .

Step 4: Note the time taken "t" for known rise of water "h" in the collecting tank.

Step 5: Calculate the actual discharge Q_a.

Step 6: Repeat step 1 to 5 for various manometric differences.

Basic data and formulae used:

Diameter of the pipe at inlet section "d_1" .... m

Diameter of the pipe at throat section "d_2" .... m

Area of the collecting tank A .....m^2

Actual discharge $Q = A * h/t$ units m^3/sec

where h is manometric column raise in meters
t is time for "h" cm raise of water in seconds

Time taken for "h" cm rise of water 't' in seconds

Actual discharge $Q_a = Ah/t$ ....m^3/sec

Theoretical discharge $Q_{th} = C * \sqrt{H}$ ... m^3/sec

Constant of venturimeter, $C = \frac{A_1 A_2}{\sqrt{A_1^2 - A_2^2}} * \sqrt{2g}$

Coefficient of discharge, $C_d = \frac{Q_{actual}}{Q_{th}}$

Tabulation:

S. No.	Manometer reading			Time for "h" cm rise of water level "t" (sec)	Actual discharge Q_a (m^3/sec)	Theoretical discharge Q_t (m^3/sec)	Coefficient of discharge C_d
	Initial level y_1	Final level y_2	Difference in level "y" in (m)				

GRAPHS:

1. H vs Q_a
2. Q_{th} vs. Q_a

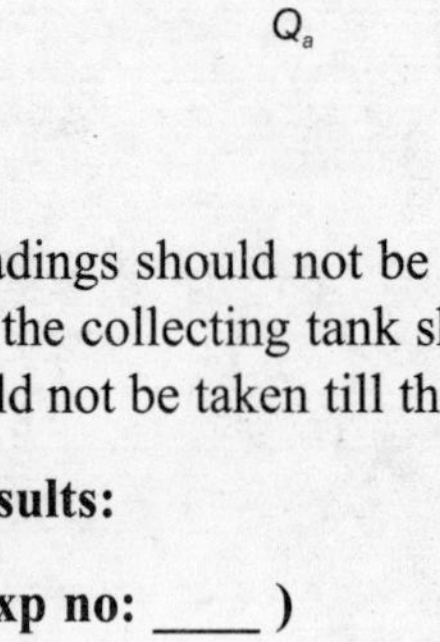

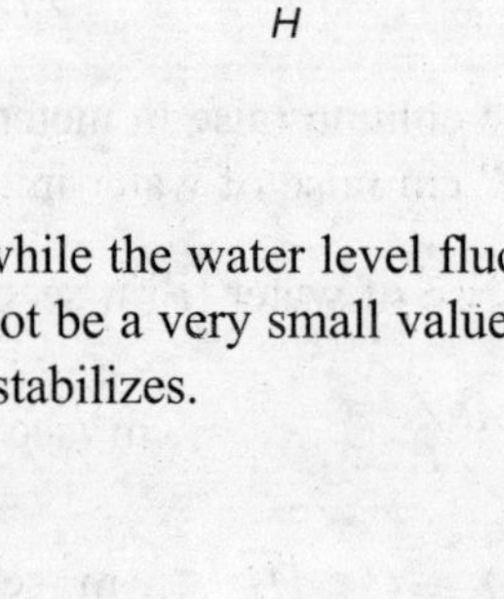

PRECAUTIONS:

1. Manometer readings should not be taken while the water level fluctuates in the limbs.
2. The rise '*h*' in the collecting tank should not be a very small value.
3. Readings should not be taken till the flow stabilizes.

Observations and Results:

Model calculation: (Exp no: ____)

Basic data:

Diameter of the pipe at inlet section d_1 = ____________________m

Diameter of the pipe at throat section d_2 = ____________________m

Area of measuring tank **A** = L * B = ____________________ m^2

Rise in water level h = _______ m

Time taken for "h" raise of water "t"= _______sec

Calculations:

Difference in pressure head $y = y_2 - y_1$ = ____________________m

*Pressure head $H = y\left[\frac{S_{ML}}{S_{LP}} - 1\right]$ = ____________________m

*Pressure head $H = y\left[1 - \frac{S_{LP}}{S_{ML}}\right]$ = ____________________m

Actual discharge $Q = Ah/t$ = ____________________ ...m^3/sec

Area at the inlet section $A_1 = \pi d_1^2/4$ = ____________________ m^2

Area at the throat section $A_2 = \pi d_2^2/4$ = ____________________ m^2

Constant of venturimeter, $C = \frac{A_1 A_2}{\sqrt{A_1^2 - A_2^2}} * \sqrt{2g}$ = ____________________

Theoretical discharge $Q_{th} = C * \sqrt{H}$ = ____________________ m^3/sec

Coefficient of discharge, $C_d = \frac{Q_{actual}}{Q_{th}}$ = ____________________m

EXPERIMENT 8: ORIFICE METER

AIM: To determine the coefficient of discharge of an orifice meter.

EQUIPMENT:

1. Orifice meter experimental set up.
2. Collecting tank with piezometer and gate valve.
3. Stopwatch.

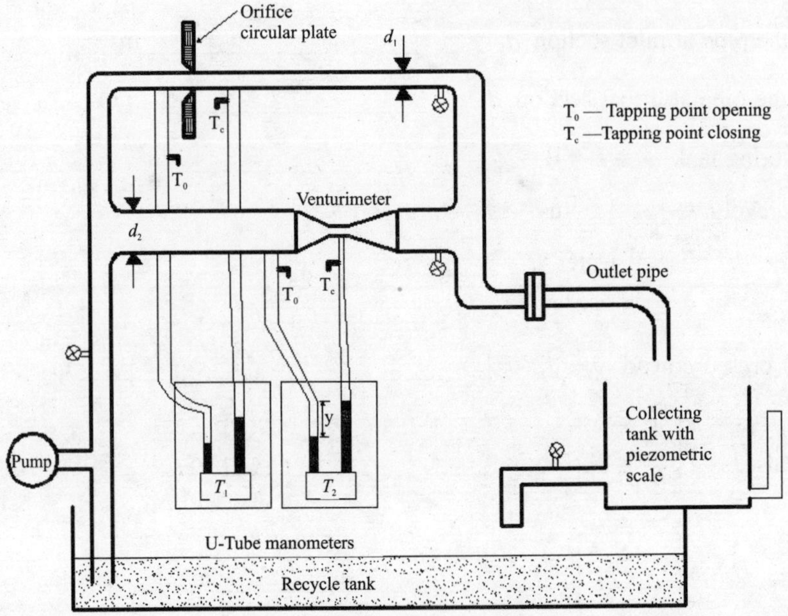

Venturi and orifice meter experiment

THEORY: Orifice meter is a device used to measure the discharge of fluid flow through a **pipeline**.

Orifice meter is a flat circular plate having a hole with diameter D_0 at the throat and having sharp edges to minimize frictional losses by minimizing the contact between the hole and the liquid, through which the liquid flows and works on the same principal of venturimeter. As the orifice has sharp edges at the throat, the fluid coming out of the throat will tend to reduce its diameter of flow, till some length and then it expands to the diameter of the pipeline (as shown in figure). The region where the fluid flow reduces its diameter is called **Vena-contraction**.

If Bernoulli's equation is applied between sections at the Inlet of the pipeline and Vena-contraction region, as shown in figure, and introducing the continuity equation, we get the discharge from the

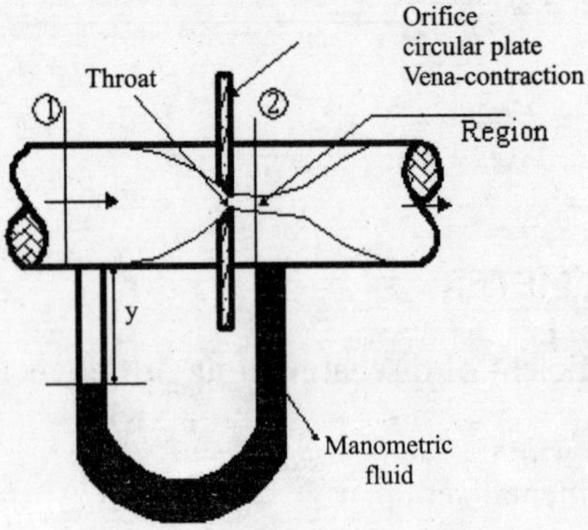

Orifice meter fitted to a pipe line

orifice meter as: $(Q_{th})_{Orifice} = C\left[\frac{A_1 A_0\sqrt{2gH}}{\sqrt{A_1^2 - A_0^2}}\right]$ and $C = C_C\left[\frac{\sqrt{A_1^2 - A_0^2}}{\sqrt{A_1^2 - A_V^2}}\right]$ where $C_C = \frac{A_V}{A_0}$ be coefficient of contraction and is the ratio of areas at Vena-contracta A_V and throat A_o of orificemeter respectively.

Note:

1. If orifice-meter is containing a **heavier manometric liquid** when compared to the liquid passing in the orificemeter, then $H = y\left[\frac{S_{ML}}{S_{LP}} - 1\right]$; where y, S_{ML}, S_{LP} are differential gauge reading, specific gravities of manometric liquid and liquid in the pipe respectively.
2. If orificemeter is containing a **lighter manometric liquid** when compared to the liquid passing in the orifice meter, then $H = y\left[1 - \frac{S_{LP}}{S_{ML}}\right]$; where y, S_{ML}, S_{LP} are differential gauge reading, specific gravities of manometric liquid and liquid in the pipe respectively.

Experimental Procedure:

Step 1: Select the required pipe (d_1 or d_2) through which the experiment to be conducted. Open its valve to allow the fluid flow through it and close the valve of the other pipes to stop flow in it.

Step 2: Switch on the pump and let the water flow through the orificemeter in the pipe. Open its tapping point at the inlet and outlet.

Step 3: Note down the reading of mercury level in the two limbs of manometer (T_1 or T_2) to calculate the difference in manometric levels "y"

Step 4: Note the time taken "t" for known rise of water "h" in the collecting tank.

Step 5: Calculate the actual discharge Q_a

Step 6: Repeat step 1 to 5 for various manometric differences.

Basic data and formulae used:

Diameter of the pipe at inlet section "d_1" .... m

Diameter of the pipe at orifice section "d_0" .... m

Diameter of the pipe at Vena-contracta section "d_V" .... m

Area of the collecting tank A .....m^2

Actual discharge $Q = A * h/t$ units m^3/s

where h is manometric column raise in meters
t is time for "h" cm raise of water in seconds

Time taken for "*h*" cm rise of water '*t*' in seconds

Actual discharge $Q_a = {Ah}/{t}$ ....m³/sec

Theoretical discharge $(Q_{th})_{Orifice} = C\left[\dfrac{A_1 A_0 \sqrt{2gH}}{\sqrt{A_1^2 - A_0^2}}\right]$ and $C = C_C\left[\dfrac{\sqrt{A_1^2 - A_0^2}}{\sqrt{A_1^2 - A_V^2}}\right]$ .. ... m³/sec

Coefficient of contraction $C_C = \dfrac{A_V}{A_0}$

Coefficient of discharge, $C_d = \dfrac{Q_{actual}}{Q_{th}}$

Tabulation:

S. No.	Manometer reading			Time for "*h*" cm rise of water level "*t*" (sec)	Actual discharge Q_a (m³/sec)	Theoretical discharge Q_t (m³/sec)	Coefficient of discharge C_d
	Initial level y_1	Final level y_2	Difference in level "*y*" in (m)				

GRAPHS:

1. H vs Q_a
2. Q_{th} vs. Q_a

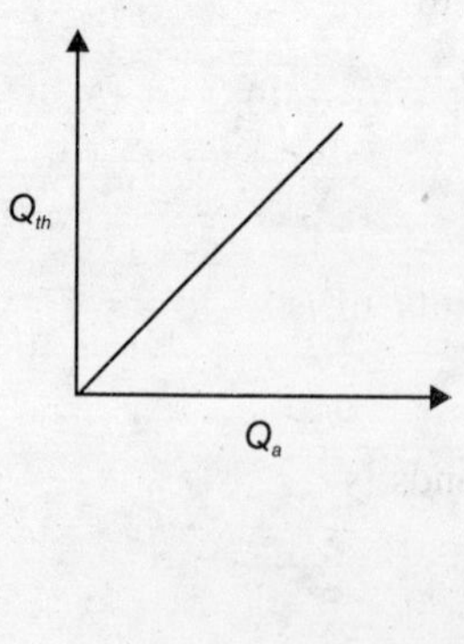

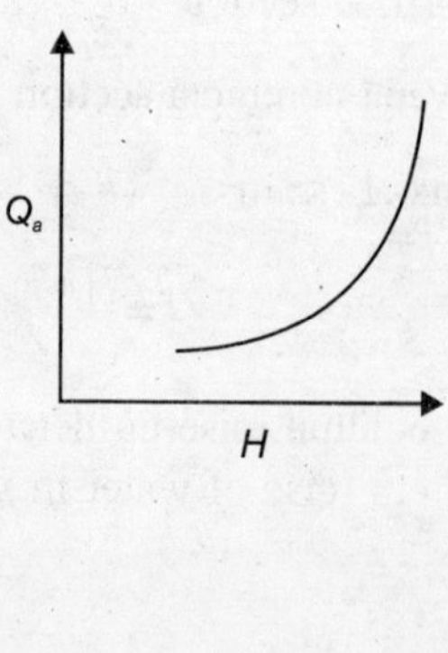

PRECAUTIONS:

1. Manometer readings should not be taken while the water level fluctuates in the limbs.
2. The rise '*h*' in the collecting tank should not be a very small value.
3. Readings should not be taken till the flow stabilizes.

Observations and Results:

Model calculation: (Exp no: ____)

Basic data:

Diameter of the pipe at inlet and throat sections d_1 = ______ m; d_2 = ______ m, respectively.

Diameter of the pipe at Vena-contracta section d_V = ______________________ m

Area of measuring tank **A** = L * B = ______________________ m^2

Rise in water level *h* =_______m; Time taken for "*h*" raise of water "*t*"= ______sec

Calculations:

Difference in pressure head $y = y_2 - y_1$ =___________________________m

*Pressure head $H = y\left[\dfrac{S_{ML}}{S_{LP}} - 1\right]$ = ______________________ m

*Pressure head $H = y\left[1 - \dfrac{S_{LP}}{S_{ML}}\right]$ = ______________________ m

Actual discharge $Q = Ah/t$ = ____________________________________ ...m^3/sec

Area at the inlet section $A_1 = \pi d_1^2/4$ = ______________________m^2

Area at the throat section $A_o = \pi d_o^2/4$ = ______________________m^2

Area at the Vena-contracta section $A_V = \pi d_V^2/4$ = ______________________m^2

Coefficient of contraction $C_C = \dfrac{A_V}{A_0}$ = ____________________________

Constant of orifice meter $C = C_C\left[\frac{\sqrt{A_1^2 - A_0^2}}{\sqrt{A_1^2 - A_V^2}}\right]$ = __________________________

Theoretical discharge $(Q_{th})_{Orifice} = C\left[\frac{A_1 A_0\sqrt{2gH}}{\sqrt{A_1^2 - A_0^2}}\right]$ = __________________________ m³/sec

Coefficient of discharge, $C_d = \frac{Q_{actual}}{Q_{th}}$ = ____________________________________ m

EXPERIMENT 9: IMPACT OF JET

AIM: To investigate the effect of a jet flow on fixed surfaces and to determine the coefficient of impact of water on them.

EQUIPMENT:

1. Impact of jet experimental set up.
2. Weights.
3. Stopwatch.

THEORY: Sudden impact of a fluid jet on to a surface, decrease the jet velocity and thus imparts motion to the surface or in other words the **Impulse** due to impinging of a fluid jet on to a vane (stationary or moving) transforms the hydraulic energy of the jet into the **change in momentum** of the surface which is termed as **Impulse-Momentum equation.**

We know mathematically, momentum equation can be written based on Newton's second law as **Force = rate of change of momentum.** i.e.

$$F = \frac{d}{dt}(mV)$$

where m is mass of jet
V is velocity of jet

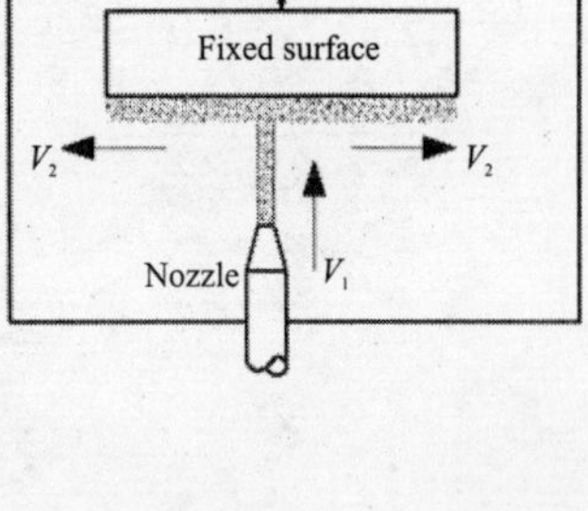

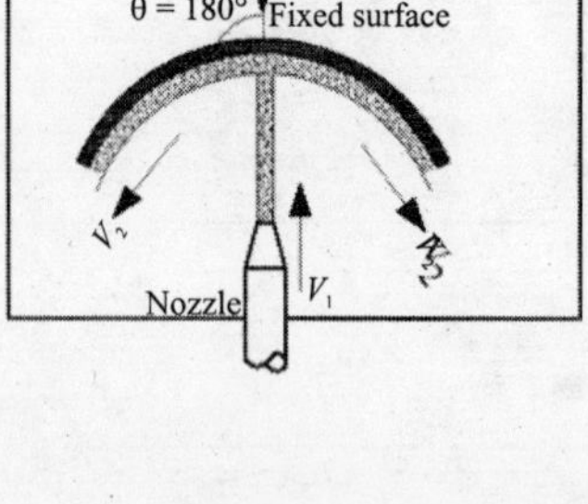

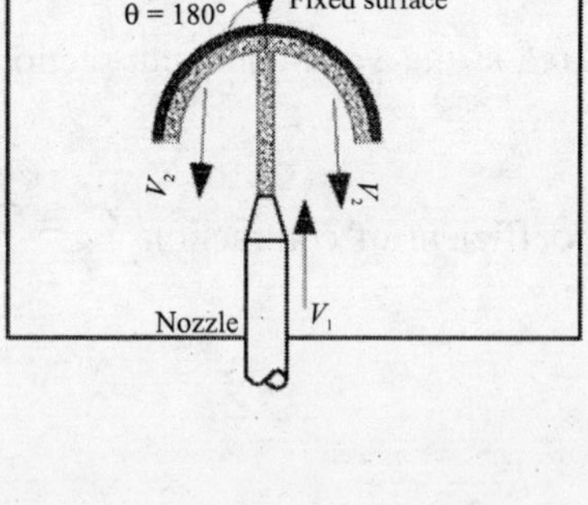

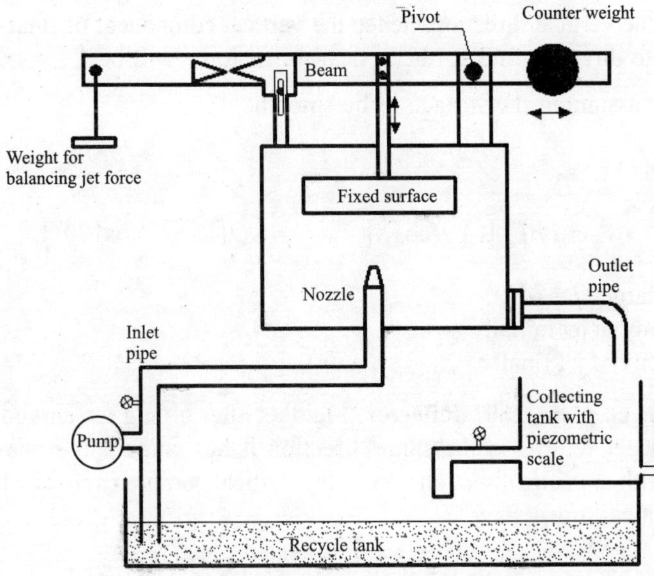

Impact of jet experiment

As the mass of the fluid jet impinging on the surface is constant, we have $Fdt = (mdV)$ where "dV" is change in velocity between the initial and final Jet. The product of force and time is termed as **Impulse** ($F * dt$ = Impulse) and this impulse force (thrust) acts in the direction of the fluid jet flow.

$$\Rightarrow F = \overline{m}(V_1 - V_2)$$

where $\overline{m}$ is mass flow rate
V_1 is velocity of jet initial
V_2 is velocity of jet finally

$$\Rightarrow F = \rho Q(V_1 - V_2) \text{ since } \overline{m} = \rho AV = \rho Q$$

Flat circular plate, 90° deflector: Fluid jet after hitting the flat circular plate gets deflected at 90^0 and has no longer any momentum in the vertical direction, hence the vertical component of final jet velocity is zero, i.e. $V_2 = 0$. So theoretical thrust force

$$F_{th} = \rho a V^2$$

where ρ is density of fluid
a is area of jet stream
V is velocity of jet

Flat curved plate, 120° deflector: Fluid jet after hitting the curved plate gets deflected at 120° and has momentum in the vertical direction, hence the vertical component of final jet velocity is V cos 120° in the opposite direction and so theoretical thrust force will be $F_{th} = \rho Q[V - (V \cos 120°)]$ $\Rightarrow F_{th} = 3\rho a V^2/2$ assuming the surface to be smooth.

If surface not smooth

$$F_n = \rho a V^2[1 + \mu \cos\phi] \Rightarrow F = \rho Q\left(V_1 - V_2 \cos 120^0\right)$$

where Q is Discharge $Q = aV$
V_1 is velocity of jet initially
V_2 is velocity of jet finally

Hemispherical curved plate, 180° deflector: Fluid jet after hitting the curved plate gets deflected at 180^0, i.e., the water returns in the same direction it has come and so has momentum in the vertical direction with opposite direction, hence the vertical component of final jet velocity is V cos 180° and so theoretical thrust force

$$F_{th} = \rho Q(V - V\cos 180°) = 2\rho QV$$

where ρ is density of fluid
a is area of jet stream
V is velocity of jet

Experimental setup: Impact of jet apparatus consists of a transparent chamber having a pipe attached with a nozzle from the bottom and a surface hanging into the chamber from the top. The surface (changeable) is fixed to a beam with help of a rod and this beam is pivoted to the transparent chamber at the top, as shown in the figure. The surface has the flexibility to reciprocate up and down.

One end of the beam has a counterweight which can slide on the beam and acts like a spring to bring back the surface exactly facing the jet at its center, after the impingement of jet is stopped. And the other end a small pan attached to a small rod, on which weights can be hanged to measure the force of strike by the jet impinged.

A collecting tank having a piezometric scale for indicating the fluid rise for a time period "*t*" is provided, below the transparent chamber outlet for collecting the fluid outflow and to discharge it to the recyclable tank for pumping up the fluid again to the transparent chamber with help of a pump.

Experimental Procedure:

Step 1: Adjust (slide) the counter weight such that the beam is in horizontal when no load acts on the fixed surface and fix it tight with the screw on it.

Step 2: Put load (W) on the pan, so that scale losses its horizontal position and note the weight.

Step 3: Release the water jet from the nozzle opening and see that the beam is brought back to its original horizontal position by jet force.

Step 4: Note the time taken "*t*" for known rise of water "*h*" in the collecting tank.

Step 5: Repeat steps from 2 to 4 for different loads added on to the pan.

Basic data and formulae used:

Density of water $\rho_{water} = 1000$ kg/m^3

Angle of the vane $\theta = 90^0, 120^0, 180^0$

Diameter of the nozzle d_{nozzle} ……m

Discharge $Q = A * h / t$ units m^3/s

where h is manometric column raise in meters
t is time for "h" cm raise of water in seconds

Velocity $V = Q / a$ *units* m/s

Actual Force, $F_{actual} = Weights\ \ added\ * gravity = W * 9.81$ ….N

Coefficient of impact $C_{impact} = F_{actual} / F_{th}$ *units dimensionless*

Area of the collecting tank A … .m^2

Tabulation for circular flat plate, 90^0 deflector:

S. No.	Weight 'W' in (kg)	Discharge measurement			Velocity 'V' (m/sec)	F_{actual} (N)	$F_{th} = \rho a V^2$ (N)	Coefficient of impact C_{impact}
		Height raised "h" in (m)	Time taken for "h" cm raise of fluid level "t" in (sec)	Discharge "Q" (m^3/sec)				

Tabulation for circular flat plate, 120^0 deflector:

S. No.	Weight 'W' in (kg)	Discharge measurement			Velocity 'V' (m/sec)	F_{actual} (N)	$F_{th} = \rho a V^2$ (N)	Coefficient of impact C_{impact}
		Height raised "*h*" in (m)	Time taken for "*h*" cm raise of fluid level "*t*" in (sec)	Discharge "*Q*" (m^3/sec)				

Tabulation for circular flat plate, 180^0 deflector:

S. No.	Weight 'W' in (kg)	Discharge measurement			Velocity 'V' (m/sec0	F_{actual} (N)	$F_{th} = raV^2$ (N)	Coefficient of impact C_{impact}
		Height raised "*h*" in (m)	Time taken for "*h*" cm raise of fluid level "*t*" in (sec)	Discharge "*Q*" (m^3/sec)				

GRAPHS:

1. F_{act} vs. F_{th}
2. F_{act} vs. V
3. F_{act} vs. V^2

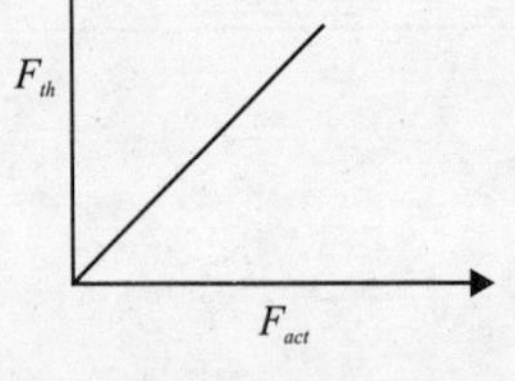

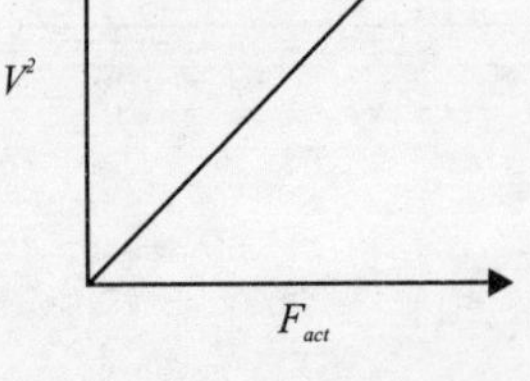

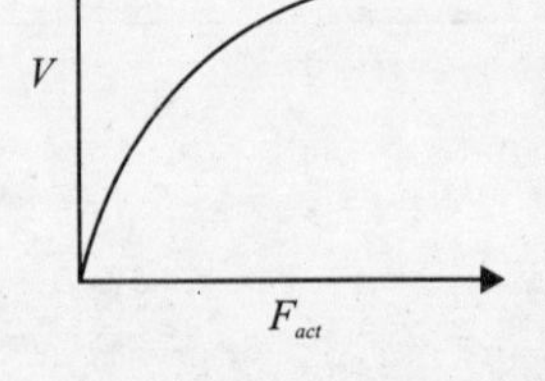

Observations and Result:

Model calculation: (Exp no: ____)

Basic data:

Density of water $\rho_{water} = 1000$ kg/m^3; Angle of the vane $\theta =$ ____________ deg.

Diameter of the nozzle d_{nozzle} = ____________________ m

Area of measuring tank A = L * B= ____________________ m^2

Rise in water level h =_______m; Time taken for "h" raise of water "t"= ______sec

Weight added W = ____________________ kg.

Calculations:

Actual discharge $Q = Ah/t$ = ______________________________ ...m^3/sec

Area at the exit of the nozzle $a = \pi d_{nozzle}^2/4$ =____________________ m^2

Velocity $V = Q/a$ =______________________________ m/s

Actual force, $F_{actual} = W * 9.81$ = ____________________ N

*Theoretical force $F_{th} = \rho a V^2$ = ____________________ m

*Theoretical force $F_{th} = 3\rho a V^2/2$ – ____________________ m

*Theoretical force $F_{th} = 2\rho a V^2$ = ____________________ m

Coefficient of impact $C_{impact} = F_{actual} / F_{th}$ = ______________________________ m

* Means whichever is applicable.

EXPERIMENT 10: CENTRIFUGAL PUMP

AIM: To draw the characteristic curves of centrifugal pump.

EQUIPMENT:

1. Centrifugal pump.
2. Pressure and vacuum gauges.
3. Collecting tank with piezometer.

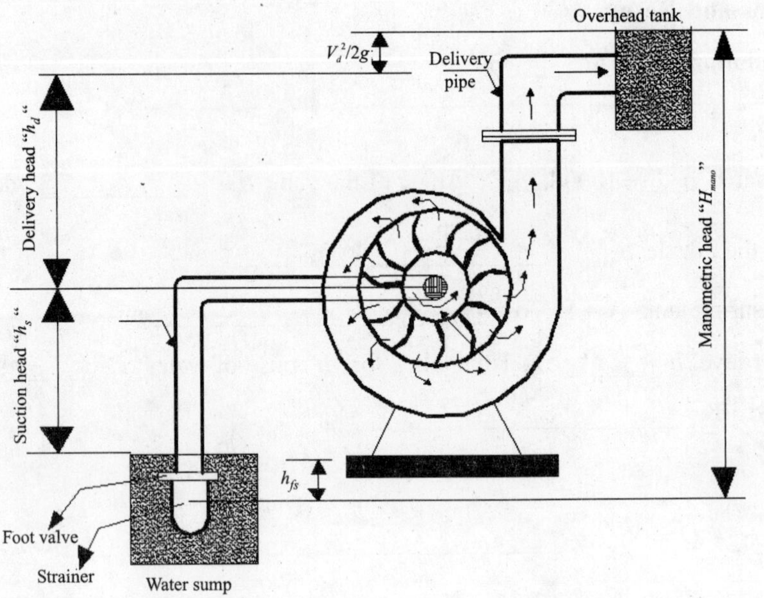

4. Energy meter.
5. Stopwatch.

THEORY: Pumps are turbo-machinery rotated steadily by the means of an external agency and adds up energy for a fluid, flowing through it. Centrifugal pumps is basically classified as Roto-dynamic pumps, which lifts a fluid from low pressure to high pressure, due to the combined dynamic forces of centrifugal energy, pressure energy and kinetic energy.

Centrifugal pump acts on the principle that a fluid mass can be lifted up to a certain height by rotating and creating a centrifugal force, in a casing, which throws the fluid mass to a height. A centrifugal pump consists of series of blades called **Impeller vanes** surmounted on a shaft (called impeller shaft), which enclosed in a casing and rotates continuously by means of an external source (called motor).

Due to the continuous rotation of impeller vanes, a vaccum is created in between the vanes which suck the fluid at low pressure and thus increases the pressure of the fluid at the center of the impeller shaft due to rotation and throws out the fluid under pressure, radially outwards, due to the centrifugal action created by the vanes.

Experimental Procedure:

Step 1: Note down the difference in height of two gauges "*h*" and energy meter constant "*c*".

Step 2: Switch on the pump and open the discharge valve slightly, note down the corresponding pressure and vacuum gauge readings.

Step 3: Note the time taken "*t*" for known rise of water "*h*" in the collecting tank.

Step 4: Calculate the actual discharge *Q*.

Step 5: Note down the time taken 't_e' for "*x*" revolutions of energy meter.

Step 6: Repeat the steps 2 to 5 by increasing the discharge for 5 observations.

Basic data and formulae used:

Area of the collecting tank A ....m^2

Specific weight of water $\gamma = 9810$ N/m^3

Difference in height of two gauges "z".........m

Energy meter constant "c"............. rev/kwh

Suction head $h_s = \dfrac{P_{vaccum}}{\gamma} = \dfrac{13.6 * P_{vaccum}}{1000}$ ....m; P_{vaccum} in mm of H_g

Delivery head $h_d = \dfrac{P_{gauge}}{\gamma} = \dfrac{9.81 * P_{gauge}}{\gamma} * 10^4$ .... meters of water; P_{gauge} in kgf/cm^2

Discharge $Q = A * h/t$ .... m^3/sec

Total manometric head $H_m = h_s + h_d + z = \left(\dfrac{P_{vacuum}}{\gamma}\right) + \left(\dfrac{P_{gauge}}{\gamma}\right) + (z_d - z_s)$ ....m

Output $= \left(\gamma Q H_m / 1000\right)$ ... kW

Input = Power given to impeller shaft $= \left(\dfrac{x}{c * t_e}\right) * 3600$ ...kw

No of revolutions of energy meter "$x = 5$".

Efficiency $\eta = \dfrac{output}{input} = \dfrac{\left(\gamma Q H_m / 1000\right)}{Power\ given\ to\ impeller\ shaft} * 100$

Tabulation for centrifugal pump:

S No.	Total "$H_m = h_s + h_d + z$"					Time for "h" rise of water level "t" (sec)	Discharge 'Q' m^3/sec	Time taken for "x" revolutions of energy meter "t_e" (sec)	I/p	O/p	Efficiency η (%)
	Vaccum Pressure P_{vacuum} (mm of H_h)	Suction head h_s (m)	Gauge Pressure P_{gauge} (kgf/cm^2)	Delivery head h_d (m)	Total Head "H_m" (m)						

Graphs:

Q vs. H_m; Q vs. η Q vs. O/p; Q vs. I/p

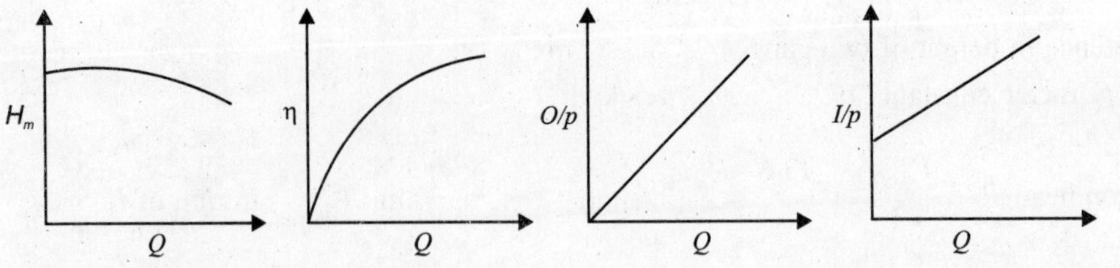

Observations and Results:

Model calculation: (Exp no: ____)

Basic data:

No of rev~~olut~~ions of energy meter x =____________________

Time taken for "x" revolutions of energy meter "t_e"= ________sec

Difference in height between suction and delivery "z"=____________m

Specific weight of water $\gamma_{water} = 9810$ N/m^3

Area of measuring tank **A** = L * B = ________________________ m^2

Rise in height of measuring tank "h" =_____m; Time taken for "h" raise of water "t"=______sec

Energy meter constant c = ______________________rev/kwh.

Vacuum gauge reading P_{vacuum} =______________ mm of H_g

Delivery gauge reading P_{gauge} =______________ kgf/cm^2

Calculations:

Suction head $h_s = \frac{P_{vacuum}}{\gamma} = \frac{13.6 * P_{vacuum}}{1000}$ = ________________________________m

Delivery head $h_d = \frac{P_{gauge}}{\gamma} = \frac{9.81 * P_{gauge}}{\gamma} * 10^4$ = ____________________ m of H_2O.

Actual discharge $Q = Ah/t$ = ____________________________________ ...m^3/sec

Total manometric head $H_m = h_s + h_d + z$ = ______________________________ m

Output $= \left(\gamma Q H_m / 1000\right)$ = ______________________________ kW

Input $= \left(\dfrac{x}{c * t_e}\right) * 3600$ = ______________________________ kW

Efficiency $\eta = \dfrac{output}{input} * 100$ = ______________________________ %

EXPERIMENT 11: RECIPROCATING PUMP

AIM: To draw the characteristic curves of reciprocating pump.

EQUIPMENT:

1. Reciprocating pump.
2. Pressure and vacuum gauges.
3. Collecting tank with piezometer.
4. Energy meter.
5. Stopwatch.

THEORY: Reciprocating pump is classified as positive displacement pump. It works on the principle that a liquid can be raised from a low level to high level by sucking the liquid into a chamber and exerting thrust force on the liquid, by a moving member, to push it to the required height.

Basically, it is incorporated in between a pipeline with an external source like motor attached to it. It converts the mechanical energy given by the motor in form of reciprocation of the piston into hydraulic energy, by giving a thrust force on the liquid and thereby the liquid is pushed to a certain height.

The relation between the efficiency, speed or head will help in the selection of the speed or head at which the pump is to be run for maximum efficiency.

Reciprocating pump is just similar to an I.C. engine except that only a suction stroke and delivery stroke exists for one revolution of crank.

Let the area of the cylinder bore be "*A*" and the stroke length (displacement of piston) of the piston "*L*". "*N*" be the speed with which the piston reciprocates in the cylinder per second.

We know, the discharge is product of area and velocity or in other words volume per second,

So $Q_{th} = A * V = A * L * N / 60$ and for double acting cylinders $Q_{th} = \dfrac{ALN}{60} + (A - a)\dfrac{LN}{60}$,

where "*a*" is the area of the piston rod.

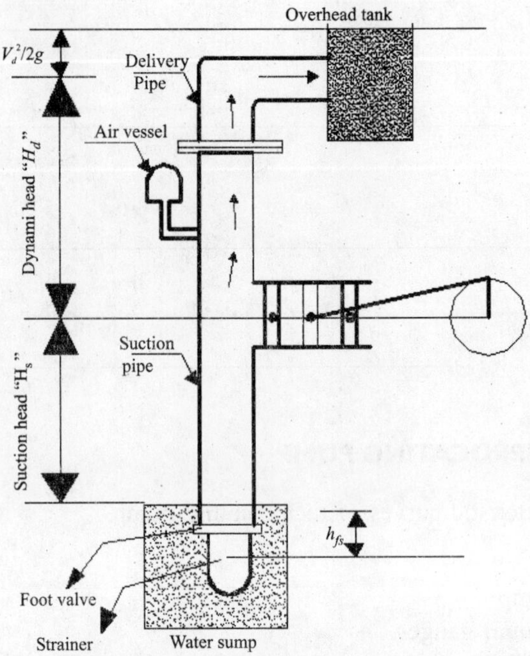

Reciprocating pump heads

As the actual discharge will be always less than the theoretical discharge due to losses, Slip is defined as the amount of reduction in actual discharge from the theoretical discharge, i.e.

$Slip = Q_{th} - Q_a$ and the percentage of slip is $\dfrac{Q_{th} - Q_a}{Q_{th}}$.

Experimental Procedure:

Step 1: Note down the difference in height of two gauges "*h*" and energy meter constant "*c*".

Step 2: Switch on the pump and open the discharge valve slightly, note down the corresponding pressure and vacuum gauge readings.

Step 3: Note the time taken "*t*" for known rise of water "*h*" in the collecting tank.

Step 4: Calculate the actual discharge *Q*

Step 5: Note down the time taken 't_e' for "*x*" revolutions of energy meter.

Step 6: Repeat the steps 2 to 5 by increasing the discharge for 5 observations.

Basic data and formulae:

Area of the collecting tank A ....m^2

Specific weight of watr g = 9810 N/m^3

Difference in height of two gauges "*z*" = ... m

Energy meter constant '*c*' .... rev/kWh

Suction head $h_s = \frac{P_{vacuum}}{\gamma} = \frac{13.6 * P_{vacuum}}{1000}$ …… ……m; P_{vacuum} in mm of H_g

Delivery head $h_d = \frac{P_{gauge}}{\gamma} = \frac{9.81 * P_{gauge}}{\gamma} * 10^4$ .... meters of water; P_{gauge} in kgf/cm²

Discharge $Q = \frac{A * h}{t}$ .... … m³/sec

Total manometric head $H_m = h_s + h_d + z = \left(\frac{P_{vaccum}}{\gamma}\right) + \left(\frac{P_{gauge}}{\gamma}\right) + (z_d - z_s)$ …….m

Output = $\left(\frac{\gamma Q H_m}{1000}\right)$ ....kw

Input = Power given impeller shaft = $\left(\frac{x}{c * t_e}\right) * 3600$ ………kW

No of revolutions of energy meter “x=5”.

Efficiency $\eta = \frac{output}{input} = \frac{\left(\frac{\gamma Q H_m}{1000}\right)}{Power\ given\ to\ impeller\ shaft} * 100$

Tabulation for reciprocating pump:

S No.	Total “$H_m = h_s + h_d + z$”					Time for “h” rise of water level “t” (sec)	Discharge ‘Q’ m³/sec	Time taken for “x” revolutions of energy meter “t_c” (sec)	I/p	O/p	Efficiency η (%)
	Vaccum Pressure p_{vacuum} (mm of H_h)	Suction head h_s (m)	Gauge Pressure P_{gauge} (kgf/cm²)	Delivery head h_d (m)	Total Head “H_m” (m)						

Graphs:

H_m vs. Q;
H_m vs. η;
H_m vs. O/p;
H_m vs. I/p

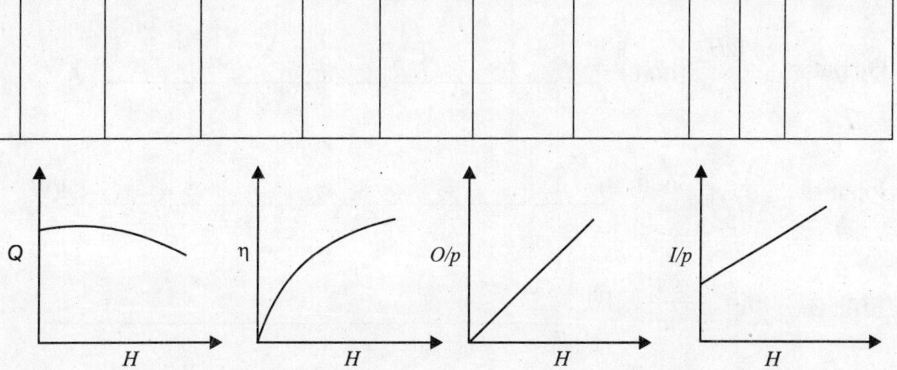

Observations and Results:

Model calculation: (Exp no: ____)

Basic data:

No. of revolutions of energy meter x = ____________________

Time taken for "x" revolutions of energy meter "t_e"= ________sec

Difference in height between suction and delivery "z"=____________m

Specific weight of water $\gamma_{water} = 9810 \text{ N/m}^3$

Area of measuring tank **A** = L * B= ____________________ m^2

Rise in height of measuring tank "h" =_____m; Time taken for "h" raise of water "t"=______sec

Energy meter constant c = ____________________rev/kwh.

Vacuum gauge reading P_{vacuum} =____________ mm of H_g

Delivery gauge reading P_{gauge} =____________ kgf/cm^2

Calculations:

Suction head $h_s = \dfrac{P_{vacuum}}{\gamma} = \dfrac{13.6 * P_{vacuum}}{1000}$ =____________________m

Delivery head $h_d = \dfrac{P_{gauge}}{\gamma} = \dfrac{9.81 * P_{gauge}}{\gamma} * 10^4$ =____________________ m of H_2O.

Actual discharge $Q = Ah/t$ = ____________________...m^3/sec

Total manometric head $H_m = h_s + h_d + z$ =____________________m

Output = $\left(\gamma Q H_m / 1000\right)$ =____________________kW

Input = $\left(\dfrac{x}{c * t_e}\right) * 3600$ =____________________kW

Efficiency $\eta = \dfrac{output}{input} * 100$ =____________________%

EXPERIMENT 12: PELTON WHEEL TURBINE

AIM: To draw the characteristic curves of Pelton wheel turbine.

EQUIPMENT:

1. Pelton wheel turbine with brake drum arrangement.
2. A pump set.
3. Pipeline with venturimeter and pressure gauge.
4. Spring balance and weights.
5. Tachomet.

THEORY: Any rotating machine shaft, which rotates due to the extraction of work from a fluid jet flow on to it, which is impinged by an external means, like a nozzle or due to fall of water from high levels on to the blades is called as a **Turbine.** Pelton wheel comes under the category of impulse turbines, which has been widely used. **Lester A. Pelton** first demonstrated the principle of this turbine in 1870. It can also be classified as high head turbine, as it works under high heads (> 250 m) and requires low quantities of water.

It consists of a runner shaft surmounted by a series of Vanes/Buckets having the shape of double hemispherical cup, evenly placed on its circumference and may use one or more number of nozzles to impinge water jet onto the buckets, for striking tangentially on to the runner to cause rotation of buckets about a pitch circle.

The flow of water through the nozzle can be regulated by means of sphere by adjusting the wheel sphere. The casing is usually provided even though it has no hydraulic function to perform but to prevent the splashing of water.

Very often hydraulic turbines work under varying conditions of head, speed, output and gate opening, which change the performance of the turbine. Different tests are performed on the turbine to predict its behavior under varying conditions. The loading on the turbine can be accomplished by rope break drum. The results obtained from the test are expressed as unit and specific quantities. Characteristic curves are usually drawn to see the effect of varying conditions on performance. The test results are graphically represented and the resulting curves are termed as characteristic curves. These are plotted in terms of unit quantities for the sake of convenience. These curves are three types.

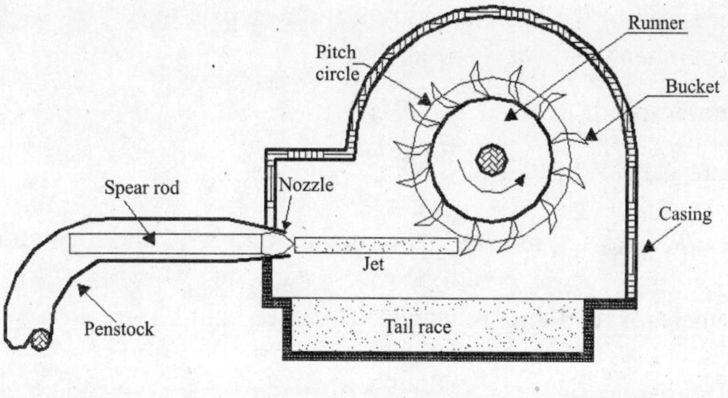

Pelton wheel

1. Constant head (main characteristic curves).
2. Constant speed (operating characteristic curves).
3. Iso-efficiency curves (universal characteristic curves).

Experimental Procedure:

I. CONSTANT HEAD METHOD:

Step 1: Start the pump after priming and allow water to flow through turbine.

Step 2: Adjust the sphere wheel for selected gate opening (e.g., 1/4 gate opening) and select a pressure and adjust the gate valve such that the required pressure is maintained.

Step 3: Convert this pressure into pressure head '*H*'.

Step 4: Note down the readings in the pressure gauges of the venturimeter and calculate the differential pressure head '*h*'

Step 5: Place the initial load on load hanger and note down the load placed and spring balance readings.

Step 6: Note down the speed of the runner with help of tachometer.

Step 7: Increase the load and note down speed of the runner.

Step 8: Repeat the steps 5 and 6 for uniform increments of load.

Step 9: Change the gate opening from ¼ to ½ and repeat the steps 4 to 8.

Step 10: Repeat the experiment for ¾ and full gate openings also.

II. CONSTANT SPEED METHOD:

Step 1: Start the pump after priming and allow water to flow through the turbine.

Step 2: Select the speed of the turbine at which the test is to be conducted.

Step 3: Place the initial load on the load hanger and adjust the gate valve such that required constant speed of the runner is maintained.

Step 4: Note down the readings in the pressure gauges of the venturimeter and calculate the differential pressure head '*h*'.

Step 5: Increase the load and again the adjust the gate till the constant is reached and again note down the readings of the venturimeter pressure gauges and calculate the differential pressure head '*h*'.

Step 6: Repeat step '5' for uniform increments of load.

Step 7: Select another speed of the turbine and repeat the steps 3 to 6.

Step 8: Repeat the experiment for 2 or 3 speeds.

Basic data and formulae used:

Specific weight of water "γ" = 9810"....... N/m^3

Brake drum radius $R = R_1 + R_2$ .... m; Radius of brake drum "R_1" and rope radius "R_2"

Venturimeter inlet diameter d_i and throat diameter d_t ....m

Differential pressure head of venturimeter "h" = $\dfrac{\Delta P}{\gamma} * g$ ....m

Rate of discharge through turbine "$Q = \frac{A_i A_t}{\sqrt{A_i^2 - A_t^2}} \sqrt{2gh}$ .... m^3/s

Output $= \frac{2\pi N}{60 \times 1000}(W - S)g$ .... KWatts; "N" in rpm; (W-S) in kgf.

Pressure head "H" $= \frac{P_1}{\gamma} * g$ ....m

Input $= \frac{\gamma Q H_1}{1000}$ ....KWatts

Overall efficiency $\eta = \frac{output}{input} * 100$

Unit speed $N_u = \frac{N}{\sqrt{H}}$; Unit power $P_u = \frac{P}{H^{3/2}}$; Unit discharge $Q_u = \frac{Q}{\sqrt{H}}$

Graphs:

1. CONSTANT HEAD CHARACTERISTIC CURVES:
   1. Unit speed vs Unit discharge
   2. Unit speed vs Overall efficiency
   3. Unit speed vs Unit power
2. CONSTANT SPEED CHARACTERISTIC CURVES:
   1. Head vs Out put
   2. Head vs Overall efficiency

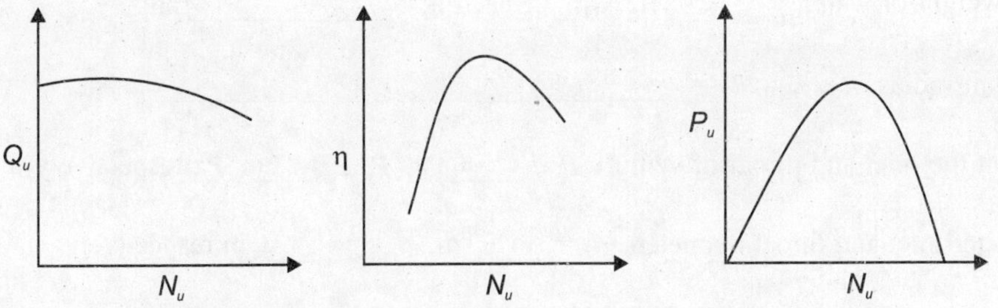

Observations and Results:

Tabulation for constant head method:

S No.	Gate opening	Pressure P_1 (kgf/cm^2)	Pressure head 'H' (m)	Load 'W' (kgs)	Spring balance reading 'S' (kg)	Speed 'N' (rpm)	Venturimeter reading "Q"				I/p (kw)	O/p (kw)	η
							P_2	P_3	h	Q (m^3/s)			

Tabulation for constant speed method:

S No.	Speed selected (rpm)	Pressure P_1 (kgf/cm^2)	Pressure head 'H' (m)	Load 'W' (kgs)	Spring balance reading 'S' (kg)	Speed 'N' (rpm)	Venturi meter reading "Q"				I/p (kW)	O/p (kW)	η	N_u	P_u	Q_u
							P_2	P_3	h	Q (m^3/s)						

Model calculation for constant head method: (Exp no: ____)

Basic data:

Specific weight of water $\gamma_{water} = 9810$ N/m^3; Speed N =_____________rpm

Brake drum radius $R = R_1 + R_2$ =_________________________________ .... m,

Pressure at the inlet and throat of venturi P_2 =_____Pa, P_3 =_______Pa respectively

Venturimeter inlet and throat diameters d_i =______m, d_t =_______m respectively.

Gauge pressure P_1 =____________kgf/cm^2.

Calculations:

Area at inlet of venturimeter $A_i = \pi d_i^2 / 4$ =__________________________m^2

Area at throat of venturimeter $A_t = \pi d_t^2/4$ =______________________________m^2

Differential pressure head of venturimeter $h = \frac{(P_3 - P_2)}{\gamma} * g$ =______________________________m

Discharge through turbine $Q = \frac{A_i A_t}{\sqrt{A_i^2 - A_t^2}}\sqrt{2gh}$ =______________________________m^3/s

Input = $\frac{\gamma QH}{1000}$ =______________________________KWatts

Output = $\frac{2\pi N}{60 \times 1000}(W - S) * g$ =______________________________KWatts.

Pressure head $H = \frac{P_1}{\gamma} * g * 10^4$ =______________________________m

Overall efficiency $\eta = \frac{output}{input} * 100$ =______________________________%

Unit speed $N_u = \frac{N}{\sqrt{H}}$ =______________________________

Unit power $P_u = \frac{P}{H^{3/2}}$ =______________________________

Unit discharge $Q_u = \frac{Q}{\sqrt{H}}$ =______________________________

EXPERIMENT 13: FRANCIS TURBINE

AIM: To draw the characteristic curves of Francis turbine.

EQUIPMENT:

1. Francis turbine with brake drum arrangement.
2. A pump set.
3. Pipeline with venturimeter and pressure gauges.
4. Spring balance and weights.
5. Tachometer.

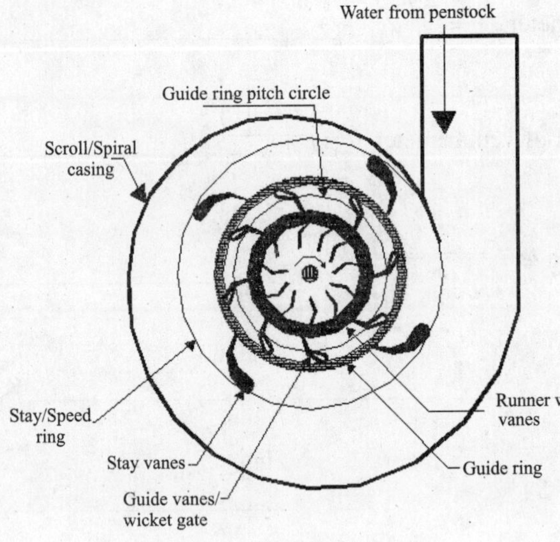

Top view of Francis turbine

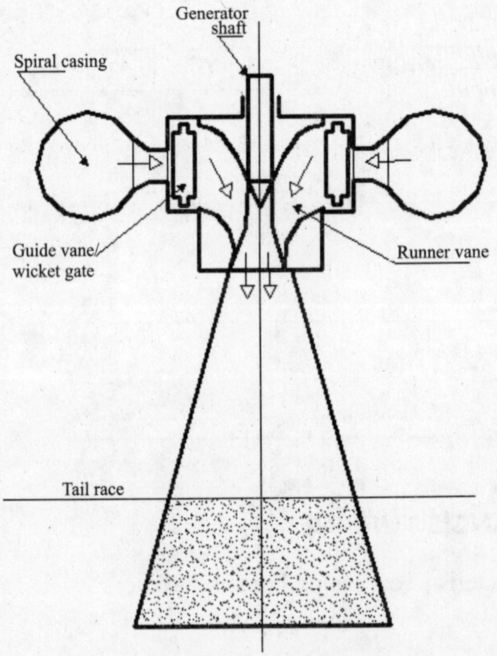

Front view of Francis turbine

THEORY: Francis turbine named after James B. Francis is a mixed (radial and axial) flow type reaction turbine. It is classified as a Reaction turbine consisting of two types of blades, namely, **Stator** blades (called Guide vanes) and **Rotor** (called Runner vanes) having the basic operation to add up **Thrust force** due to the movement of the jet on the surface of the runner vanes and the **Reaction force** (Newton's third law) produced due to energy transformation from one form to other (like pressure energy to kinetic energy) in the guide vanes.

In this type of turbine, the jet flows radially inwards from the outer periphery and leaves out axially from the center of the turbine. It basically consists of penstock, spiral/scroll casing, stay ring and stay vanes, guide vanes, runner with vanes and a draft tube, as shown in figures.

Penstock and spiral/scroll casing: Water enters the turbine from penstock and penstock is in attachment with a spiral casing, through which the water flows towards the turbine runner. And while having the flow of water towards the runner, a constant velocity of water flow is to be maintained, which is done by the spiral casing, as it has reduced uniform cross-section, throughout the spiral casing.

Stay/Speed ring and Stay vanes: The water flowing in the casing with constant velocity need to be diverted towards the runner and so a number of fixed vanes called stay vanes are used, which are fixed to a circular ring called stay ring (also called speed ring), as shown in Figure 11.9. And in addition to the direction changing, the stay ring also transmits any unbalanced loads arising due to internal pressure of water motion and due to unbalanced weight of turbine, to the foundation.

Guide ring and guide vanes/wicket gates: Water diverted from stay vanes towards the runner to be pressurized while impinging the water as a jet on to the runner vanes, so that the water jet strikes the vanes with energy and thereby cause rotation of runner. Guide vanes are usually taken in double the number of stay vanes and are pivoted about a guide ring pitch circle and they are free to rotate about their pivots, so that the gap in between to consecutive vanes at their tail ends be reduced, which acts like nozzle, to regulate the quantity of jet strike and also these guide vanes are in aerofoil shaped, to direct the jet strike with an appropriate deflection angle.

Runner with vanes: Runner, the shaft of a turbine generator consists of vanes/buckets (curved), evenly arranged on the circumference of the runner, to perform the rotation of the runner shaft due to the jet strike which may be tangentially, at the vane tips and are so designed that the jet entering the runner vanes are sent out axially to the runner shaft, after the strike completion, towards the draft tube.

The runner vanes may be with different blade profiles, as shown in Figure 11.7b.

Draft tube: Draft tube is a tube like structure, having a passage for flow of water through it from the exit of the turbine to the tail race and so it can be defined as a transporting device of fluid from runner exit to the tail race, having a circular cross-section at the entrance (at exit of turbine runner) and can be of any cross-section at its exit (at tail race). For transporting the leaving water jet from the exit of runner, some portion of the draft tube is immersed in the tail race, so as to create a vacuum condition for sucking the leaving jet and convert the kinetic energy of the leaving jet to pressure energy, for regaining the pressure energy in tail race.

The draft tube has two purposes:

1. It permits suction head to be established at the runner exit, thus making it possible to install the turbine above the tail race level with out loss of head.
2. It converts a large proportion of velocity energy rejected from the runner into useful pressure energy, i.e., it acts as a recuperates of pressure energy.

Characteristic curves are usually drawn to see the effect of varying conditions on performance. The tests are graphically represented and the resulting curves are termed as characteristic curves. These are plotted in terms of unit quantities for the sake of convenience. These curves are three types.

1. Constant head (main characteristic curves).
2. Constant speed (operating characteristic curves).
3. Iso-efficiency curves (universal characteristic curves).

Experimental Procedure:

I. CONSTANT HEAD METHOD:

Step 1: Start the pump after priming and allow water to flow through turbine.
Step 2: Select a pressure and adjust the gate valve such that the required pressure is maintained.
Step 3: Convert this pressure into pressure head '*H*'.
Step 4: Note down the readings in the pressure gauges of the venturimeter and calculate the differential pressure head '*h*'.
Step 5: Place the initial load on load hanger and note down the load placed and spring balance readings.
Step 6: Note down the speed of the runner with help of tachometer.
Step 7: The load and note down speed of the runner.
Step 8: Repeat the steps 5 and 6 for uniform increments of load.
Step 9: Change the gate opening for another head and repeat the steps 4 to 8.

II. CONSTANT SPEED METHOD:

Step 1: Start the pump after priming and allow water to flow through the turbine.
Step 2: Select the speed of the turbine at which the test is to be conducted.
Step 3: Place the initial load on the load hanger and adjust the gate valve such that required constant speed of the runner is maintained.
Step 4: Note down the readings in the pressure gauges of the venturimeter and calculate the differential pressure head '*h*'.
Step 5: Increase the load and again the adjust the gate till the constant is reached and again note down the readings of the venturimeter pressure gauges and calculate the differential pressure head '*h*'.
Step 6: Repeat step '5' for uniform increments of load.
Step 7: Select another speed of the turbine and repeat the steps 3 to 6.
Step 8: Repeat the experiment for 2 or 3 speeds.

Basic data and formulae used:

Specific weight of water "γ = 9810"……. N/m^3

Brake drum radius $R = R_1 + R_2$ .... m; Radius of brake drum "R_1" and rope radius "R_2"

Venturimeter inlet diameter d_i and throat diameter d_t ....m

Differential pressure head of venturimeter "h" = $\frac{\Delta P}{\gamma} * g$ ....m

Rate of discharge through turbine "$Q = \frac{A_i A_t}{\sqrt{A_i^2 - A_t^2}} \sqrt{2gh}$ .... m^3/s

Output $= \frac{2\pi N}{60 \times 1000}(W - S)g$ .... KWatts; "N" in rpm; (W-S) in kgf.

Head "H" = $\frac{P_1}{\gamma} * g$ ....m

Input $= \frac{\gamma QH}{1000}$ ....KWatts

Overall efficiency $\eta = \frac{output}{input} * 100$

Unit speed $N_u = \frac{N}{\sqrt{H}}$

Unit power $P_u = \frac{P}{H^{3/2}}$

Unit discharge $Q_u = \frac{Q}{\sqrt{H}}$

GRAPHS:

1. CONSTANT HEAD CHARACTERISTIC CURVES:
   1. Unit speed vs Unit discharge
   2. Unit speed vs Unit power
   3. Unit speed vs Overall efficiency

2. CONSTANT SPEED CHARACTERISTIC CURVES:
   1. Head vs Output
   2. Head vs Overall efficiency

Observations and Results:

Tabulation for constant head method:

S No.	Gate opening	Pressure P_1 (kgf/cm^2)	Pressure head 'H' (m)	Load 'W' (kgs)	Spring balance reading 'S' (kg)	Speed 'N' (rpm)	Venturimeter reading "Q"				I/p (kW)	O/p (kW)	η	N_u	P_u	Q_u
							P_2	P_3	h	Q (m^3/s)						

Tabulation for constant speed method:

S No.	Speed Selected (rpm)	Pressure P_1 (kgf/cm^2)	Pressure head 'H' (m)	Load 'W' (kg)	Spring balance reading 'S' (kg)	Speed 'N' (rpm)	Venturimeter reading "Q"				I/p (kw)	O/p (kw)	η	N_u	P_u	Q_u
							P_2	P_3	h	Q (m^3/s)						

Model calculation for constant head method: (Exp no: ____)

Basic data:

Specific weight of water $\gamma_{water} = 9810$ N/m^3 ; Speed N =_____________rpm

Brake drum radius $R = R_1 + R_2$ =___________________________________ .... m,

Pressure at the inlet and throat of venturi P_2 =_____Pa, P_3 =_______Pa respectively

venturimeter inlet and throat diameters d_i =______m, d_t =_______m respectively.

Gauge pressure P_1 =_____________kgf/cm^2.

Calculations:

Area at inlet of venturimeter $A_i = \pi d_i^2 / 4$ = ______________________ m²

Area at throat of venturimeter $A_t = \pi d_t^2 / 4$ = ______________________ m²

Differential pressure head of venturimeter $h = \frac{(P_3 - P_2)}{\gamma} * g$ = ______________________ m

Discharge through turbine $Q = \frac{A_i A_t}{\sqrt{A_i^2 - A_t^2}} \sqrt{2gh}$ = ______________________ m³/s

Input = $\frac{\gamma Q H}{1000}$ = ______________________ KWatts

Output = $\frac{2\pi N}{60 \times 1000}(W - S) * g$ = ______________________ KWatts

Pressure head $H = \frac{P_1}{\gamma} * g * 10^4$ = ______________________ m

Overall efficiency $\eta = \frac{output}{input} * 100$ = ______________________ %

Unit speed $N_u = \frac{N}{\sqrt{H}}$ = ______________________

Unit power $P_u = \frac{P}{H^{3/2}}$ = ______________________

Unit discharge $Q_u = \frac{Q}{\sqrt{H}}$ = ______________________

Chapter 9

1. b	2.a				

Chapter 10

1. a	2. d	3. c	4. a	5. @$e(-\rho aV(V-u))$
6. b	7. c	8. b	9. b	10. b
11. a	12. c	13. d		

@ means additional option is being added.

CHAPTER 11

1. **&**	2. **&**	3. **&**	4. **&**	5.**&**	6. c
7. a	8. a	9. a	10. a	11. a	12. b
13. d	14. b	15. b	16. a	17. c	18. c
19. a	20. d	21. b	22. b	23. d	24. d
25. b	26. a	27. b	28. a	29. c	30. b
31. c	32. d	33. a	34. b	35. c	36. a
37. d	38. c	39. b	40. d	41. b	42. a
43. b	44. c	45. d			

& means delete, as these questions are added in the previous chapter.

Chapter 12

1. a	2. c	3. d	4. c	5. b	6. c
7. b	8. a	9. d	10. a	11. c	12. b
13. d	14. a	15. c	16. d	17. a	18. c
19. b	20. c	21. d	22. a	23. d	24. d
25. c	26. d	27. d	28. c	29. a	30. d
31. a	32. c	33. a	34. c	35. c	

Index

PRIET